STANDARD GRAPHS OF FUNCTIONS

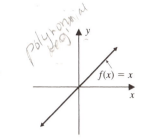

Polynomial deg 1

$f(x) = x$

1. Identity function

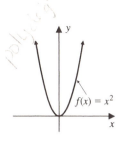

poly

$f(x) = c$

2. Constant function
(*c* is a constant)

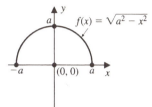

polydeg

$f(x) = x^2$

3. Squaring function

$f(x) = \sqrt{x}$

4. Square root function

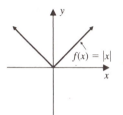

$f(x) = |x|$

5. Absolute value function

$f(x) = x^3$

6. Cubing function

$f(x) = \sqrt[3]{x}$

7. Cube root function

a $f(x) = \sqrt{a^2 - x^2}$

$-a$ $(0, 0)$ a

8. Upper half-circle function
(center: $(0, 0)$; radius a)

$f(x) = b^x$
$b > 1$

$f(x) = b^x$
$0 < b < 1$

(a)   (b)

9. Exponential functions

$f(x) = \log_b x$
$b > 1$

$f(x) = \log_b x$
$0 < b < 1$

(a)   (b)

10. Logarithmic functions

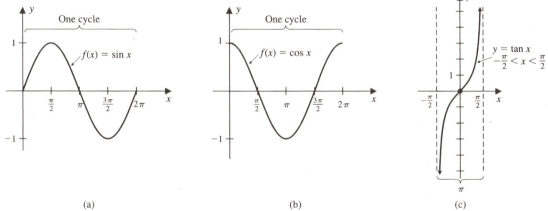

One cycle

$f(x) = \sin x$

One cycle

$f(x) = \cos x$

$y = \tan x$
$-\frac{\pi}{2} < x < \frac{\pi}{2}$

(a)   (b)   (c)

11. Trigonometric functions

Precalculus

EIGHTH EDITION

M.A. Munem

J. P. Yizze

Macomb College

KENDALL/HUNT PUBLISHING COMPANY
4050 Westmark Drive   Dubuque, Iowa 52002

Book Team

Chairman and Chief Executive Officer   Mark C. Falb
Senior Vice President, College Division   Thomas W. Gantz
Director of National Book Program   Paul B. Carty
Editorial Development Manager   Georgia Botsford
Developmental Editor   Angela Willenbring
Vice President, Production and Manufacturing   Alfred C. Grisanti
Assistant Vice President, Production Services   Christine E. O'Brien
Prepress Editor   Carrie Maro
Senior Managing Editor, College Field   Paul Gormley
Acquisitions Editor, College Field   David Mattaliano

Cover photo courtesy of NASA.

Photos on pp. 1, 8, 76, 91, 124, 243, 278, 301, 403, 427, 501, 508, 611, and 634 © 2006 JupiterImages Corporation.
Photo on p. 163 © Dave G. Houser/CORBIS.
Photo on p. 571 © Bettmann/CORBIS.

ISBN 0-7575-2123-1

Printed in the United States of America
10 9 8 7 6 5 4 3 2 1

Contents

Preface

As in previous editions of *Precalculus,* the goal of the eighth edition is to provide the essential concepts and skills expected in a text that covers this level of college mathematics. In preparing this edition, several improvements—some obvious, others subtle—have been made. Our treatment is still accessible, attractive, and interesting to all students who have appropriate pre-requisite skill levels, regardless of their purposes for studying topics in algebra and trigonometry at the precalculus level. Improvements evolved from student feedback, thoughtful responses of users of previous editions, as well as market survey results. Important to this book is that we have tested our ideas and suggestions from other professionals in the classroom. The content and exposition represent our vision of how the subject should be taught and learned. The book contains elements of reform but within the context of a traditional curriculum. This text serves both students preparing to take calculus and students requiring college mathematics at this level to fulfill their degree requirements. In response to requests of users and reviewers, and recognizing recent developments in teaching mathematics at this level, we have included the following.

Features

A glance at the precalculus textbooks published over the past years shows that the topics, and even their order, are somewhat standard. Where these textbooks differ is more in matters of approach, emphasis, and pedagogy.

In this edition, we present topics in a variety of ways—*geometrically, graphically, numerically, symbolically,* and *analytically.* In addition, we have included the *verbal* or *descriptive* point of view. Special attention has been paid to pointing out the *connection* between *algebraic* and *geometric* interpretations of important concepts. This approach is effective in promoting *visual thinking.* We attempt to appeal to the various cognitive learning styles encountered in typical classroom situations. Several features contribute to the text's effectiveness. For instance, each section begins with a clear delineation of the learning objectives for the material being presented.

Areas that receive particular emphasis in this textbook are the following:

1. **Graphs:** Sketching graphs by hand-drawing and equation recognition is a fundamental part of learning mathematics at this level. The text provides the option to use graphing tools to supplement and extend this process. The ability to generate accurate graphs efficiently increases the use of visualization and thus helps convey ideas along with word *explanations. Interpreting graphs, exploring new ideas* from graphs, and *extracting information* from graphs continue to be of primary importance in this book, as in all reform mathematics.

2. **Revised Examples:** This edition helps students improve their mathematical skills by providing a variety of illustrative examples that are accompanied by full-color annotations to assist the students in tracking each step in each solution. These examples are used to introduce concepts and to demonstrate problem-solving techniques. Several examples, whether drill or applied, are revised, and many new ones have been added. Each example has a descriptive title so that students can see at a glance the type of problem being illustrated.

3. **Revised Problem Sets:** Problem sets have been reorganized and rewritten. More than 20% of the problems are new. They have been checked for accuracy and appropriateness by instructors who class-tested this edition of the book. The problem sets have been divided into three categories:

   a. The first category, *Mastering the Concepts*, is designed to provide the practice needed to master the *basic concepts in this book*. The problems are generally modeled after the examples and follow the order of presentation in the book.

   b. The second category, *Applying the Concepts*, provides a broad range of applications and models that require students to *apply the concepts*. Special emphasis has been put on the design of these problems.

   c. The third category, *Developing and Extending the Concepts*, challenges students to *develop and advance the concepts*. These problems encourage *critical thinking* and provide an excellent opportunity to stimulate class discussions that foster *collaborative learning* among students.

   In addition, the option to expand an exploration of the material through the use of technology is included in the problem sets.

   The problem sets also offer the opportunity for writing assignments, such as a report from an individual, for solving an applied problem or a class project.

4. **Chapter Review and Chapter Test:** The popular feature of ending each chapter with a review problem set and a comprehensive chapter test has continued.

5. **Margin Notes:** In this edition we continue to include notes to students in the margins. These notes provide *additional insights,* help students *avoid common errors,* and give students an opportunity to be involved in the *development of the material.*

6. **Four-Color Design:** New to this edition is a four-color design for easy reference of important definitions. Properties and procedures are highlighted in color. Artwork is clarified with the use of multiple colors.

7. **Real-World Applications and Models:** As in prior editions, we have placed emphasis on real-world applied problems drawn from a variety of disciplines consistent with the current goals of reform in mathematics. These problems are integrated into the material throughout the book. We believe we offer an *approach to mathematical modeling that will capture the students' attention* consistent with the current goals of contemporary mathematics. Based on our experiences in classroom teaching, we have selected certain types of problems and organized their solutions through a sequence of questions that help students through the difficult thought processes of developing mathematical models and using them to interpret real-world situations. Students have many opportunities to put function concepts and graphs to use in a variety of life-like situations. Solving these problems enables students to practice their skills and acquire insights needed in calculus and other fields.

8. **Curve Fitting:** In this edition, we include data modeling problems that help students discover how to curve-fit data from a wide variety of real-world situations. We spent a considerable amount of time in *libraries, contacting companies in business and medical fields*, and *searching the Internet* for interesting real-world data to motivate the concepts of applied mathematics at this level. Students learn how to use equations that best fit paired data for given models in order to make estimations and predictions.

9. **Use of Technology:** The generic term *grapher* is used to refer to any of the various graphing calculators or computer software mathematics packages available. When properly used, calculators and computer software are useful tools for discovering and understanding concepts and solving problems. We use the icon G to indicate clearly when technology can be used. However, technology doesn't make pencil and paper obsolete; hand calculations and sketches

are often preferable to technology for illustrating and reinforcing some concepts because they force the students to engage their thought processes in mathematical procedures. Although the use of graphers is optional, it is likely that many students and instructors will want to make use of them as instructional tools to integrate the exposition of graphical and numerical viewpoints. There continue to be three options for using technology:

a. Omit its use throughout the text.

b. Use it whenever it appears as an integrating learning tool.

c. Use it on a selective, limited basis according to individual judgment.

10. Other useful features from earlier editions have been retained and others have been added, including Appendix I, which contains a basic review of number systems and synthetic division. Also, frequently used facts from plane geometry are summarized with examples in Appendix II.

Organizational Changes

A variety of organizational changes have been made. The extensive revision and updating of the exposition, examples, and problems sets throughout the text allow the use of technology to be optional, without compromising full coverage of the mathematics. Noteworthy are the following revisions and highlights.

1. **Analytic Geometry:** The material on standard forms of conic sections with center or vertices at the origin, including applications, has been moved from Chapter 9 and is now incorporated into Chapter 1 as part of graphing equations.

2. **Functions:** Chapter 2 stresses multiple ways of representing functions verbally, numerically, visually, and algebraically.

3. **Polynomial and Rational Functions:** Chapter 3 has been extensively rewritten and reorganized to conform to what we learned from testing our ideas in the classroom. A review of complex number arithmetic is included in Appendix I for those who need it.

4. **Trigonometric Functions of Real Numbers:** In the previous edition, Chapter 5 began with the introduction of right-triangle trigonometry. Here, the unit circle approach precedes the definition of trigonometric functions of angles. With the unit circle definitions completed, the identities need not carry the restriction in which they are limited to be acute angles. Also, the new material on graphing trigonometric functions is presented earlier in a more natural way.

5. **Combination, Permutation, and Probability:** Chapter 10 has been expanded to cover combinations, permutations, and an introduction to probability.

Supplements

The eighth edition is accompanied by the following supplement packages:

For the Instructor

Instructor's Solution Manual
This manual provides solutions to all problems in the text.

Instructor's Resource Manual
a. Printed Test Manual: This manual includes six different test forms for each chapter. Three of these tests have multiple-choice questions, and the other three have standard problem solving equations. Answers to all tests are provided.

b. Computerized Testbank: A computerized testbank that provides a variety of formats to enable the teacher to create tests by choosing questions either manually or randomly by section, question types, level of difficulty, and other criteria is available.

For the Student

New to this edition is a free Student Solutions Manual on CD, which accompanies every copy of the textbook. The CD includes worked-out solutions to every odd-numbered problem in the book.

Acknowledgments

In preparing this edition, we have drawn from our own experiences in teaching precalculus, as well as the feedback provided by our colleagues and students. The suggestions obtained from instructors using earlier editions were of great help. We wish to thank all of these people and, in particular, to express our gratitude to the following:

> Robert Boner, Western Maryland College
> Carol Flakus, Lower Columbia College
> Gilbert Steiner, Fairleigh Dickinson University

We are grateful to the following individuals who assisted in planning the revisions of our text:

> Margaret Dolgas, University of Delaware
> Richard Jetton, Arkansas State University
> Jane Pinnow, University of Wisconsin, Parkside
> Elise Price, Tarrant County Junior College

And we wish to thank the following instructors who reviewed material that influenced the eighth edition.

> Georgianna Aviola, College of William and Mary
> Elizabeth Baton, University of North Texas at Denton
> Arlene Blasius, SUNY College at Old Westbury
> Harold Bowman, University of Nevada, Las Vegas
> Harvey Greenwald, California Polytechnic State University
> Richard Marshall, Eastern Michigan University
> Elmo Moore, Humboldt State University
> Stephen Murdock, Tulsa Junior College

Special thanks are due to our many colleagues at Macomb College, especially Professor Bill Knott, who taught from previous editions of the book and made several suggestions for organizing the material in this edition. We also thank Robert Eckert and Christine Guastella, who provided expert advice, checked problems, and assisted in proofreading. Special recognition is given for the outstanding work of Lisa Durak, who helped prepare the Instructors Solution Manual and Student Solution Manual and checked the accuracy of every answer in the book.

Finally, we are very grateful to the staff of Kendall/Hunt Publishing for their cooperation and assistance throughout the project. We would also like to express our sincere gratitude to the staff of Carlisle Publishing Services who carried out the details of production in a dedicated, professional, and caring way.

M. A. Munem
J. P. Yizze

Mathematical Models and Coordinate Geometry

In aviation, the jet stream is a narrow current of air encountered between altitudes of 25,000 and 35,000 feet. One day, a flight from Los Angeles to New York (a distance of 2460 miles) flying with the jet stream (with a tail wind) took the same amount of time as a flight from New York to Denver (a distance of 1626 miles) flying against the jet stream (with a head wind). Assume that both planes fly 550 miles per hour in still air. How fast was the jet stream and how long was the flight for both planes? An equation involving fractions is used to solve this problem. See Example 8 on page 48.

In this chapter, we focus on topics that are important to the study of functions and their graphs. We also introduce mathematical modeling.

Objectives

1. Solve Linear Equations
2. Solve Equations Involving Rational Expressions
3. Solve Equations Involving Absolute Value
4. Solve Literal Equations or Formulas
5. Solve Applied Problems Using Mathematical Modeling

1.1 Linear Equations and Mathematical Modeling

In this section, we review techniques learned in algebra for solving **linear equations** *in one variable.* Methods for solving equations involving rational expressions and absolute values are also covered. In addition, we introduce mathematical modeling.

To **solve** an equation in one variable, we need to find all values that, when substituted for the variable, make the statement of equality a true statement. These values are called **solutions,** or **roots,** of the equation. For example, -3 is the solution for

$$-2x + 7 = 13 \qquad \text{because} \qquad -2(-3) + 7 = 13$$

is a true statement and no other value for x satisfies this equation.

As we shall see, solving an equation involves the use of equivalent equations. Two or more equations that have the same solutions are called **equivalent equations.**

For example,

Equivalent equations

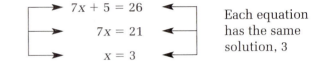

$$7x + 5 = 26$$
$$7x = 21$$
$$x = 3$$

Each equation has the same solution, 3

Solving Linear Equations

First-degree, or **linear, equations in one variable,** such as

$$4x - 7 = 1, \qquad 2y - 7 = y + 1, \qquad \text{and} \qquad 2x - 5x = 4 - 3x$$

involve only the first power of the variable and constants.

To solve this type of equation, we transform the equation into an equivalent one in which the variable is isolated on one side of the equal sign by applying the following properties:

Properties that Generate Equivalent Equations

> Assume a, b, and c represent numbers.
>
> **(i) Addition and Subtraction Properties:**
>
> If $a = b$, then $a + c = b + c$ and $a - c = b - c$.
>
> **(ii) Multiplication and Division Properties:**
>
> If $a = b$ and $c \neq 0$, then $ac = bc$ and $\dfrac{a}{c} = \dfrac{b}{c}$.

EXAMPLE 1   **Solving a Linear Equation**

Solve the linear equation

$$4(x - 3) = 2(3x + 1) + 5x.$$

Solution   To solve this equation, we first simplify each side and then apply the above properties as follows:

$$4(x - 3) = 2(3x + 1) + 5x$$
$$4x - 12 = 6x + 2 + 5x$$
$$4x - 12 = 11x + 2$$
Simplify each side

$$4x = 11x + 14$$
Add 12 to each side

$$-7x = 14$$
Subtract 11x on each side

$$x = -2$$
Divide each side by -7

So the solution of the equation is -2.

Solving Equations Involving Rational Expressions

If an equation involves rational expressions, we can convert it to a simpler form by multiplying both sides of the equation by a common denominator (we normally use the least common denominator, LCD) of all the fractions. Then we solve the resulting equation. The proposed solution should always be checked if variables are in the denominator of the original equation, since it may be an **extraneous root**—that is, a misleading value that does not satisfy the original equation.

EXAMPLE 2   **Solving a Rational Equation**

Solve the equation $\dfrac{3}{x} - \dfrac{1}{2x} = \dfrac{2}{x - 1}$.

Solution   The solution of the equation proceeds as follows:

$$\frac{3}{x} - \frac{1}{2x} = \frac{2}{x - 1}$$
Given

$$2x(x - 1)\left(\frac{3}{x} - \frac{1}{2x}\right) = 2x(x - 1)\left(\frac{2}{x - 1}\right)$$
Multiply both sides by the LCD

$$2(x - 1)(3) - (x - 1) = 2x(2)$$
Distribute

$$6(x - 1) - (x - 1) = 4x$$
$$6x - 6 - x + 1 = 4x$$
$$5x - 5 = 4x$$
Simplify

$$x = 5$$
Add 5 and subtract 4x on each side

Next, we check the proposed solution by substituting $x = 5$ into the given equation:

$$\frac{3}{5} - \frac{1}{10} = \frac{2}{4} \quad \text{or} \quad \frac{6 - 1}{10} = \frac{2}{4} \quad \text{or} \quad \frac{5}{10} = \frac{2}{4}$$

which is true. Therefore, the solution is 5.

Solving Equations Involving Absolute Value

The number line provides us with a way of introducing the concept of *absolute value* from a geometric perspective. If a is a real number, then $|a|$, called the *absolute value* of a, represents the distance on a number line between point a and the 0 point. If $a \neq 0$, then $-a$ and a are the two numbers whose distance to 0 is $|a|$ units.

For instance, -4 and 4 are the two numbers whose distance to the 0 point is 4 units (Figure 1), and we write

$$|-4| = |4| = 4.$$

Figure 1

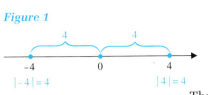

Thus we are led to the following definition:

Definition
Absolute Value

If a is a real number, the **absolute value** of a, denoted by $|a|$, is defined as follows:

$$|a| = \begin{cases} a & \text{if } a \text{ is positive or zero, that is, if } a \geq 0 \\ -a & \text{if } a \text{ is negative, that is, if } a < 0 \end{cases}$$

For instance,

$$|7| = 7 \qquad \text{because 7 is positive}$$
$$|-7| = -(-7) = 7 \qquad \text{because } -7 \text{ is negative}$$
$$|\sqrt{3} - 3| = -(\sqrt{3} - 3) = 3 - \sqrt{3} \quad \text{because } \sqrt{3} - 3 \text{ is negative.}$$

The following properties of absolute value should be noted:

(i) $|a| \geq 0$ for any real number a.

(ii) $|0| = 0$.

(iii) $|a| = |-a|$ for any real number a.

Sometimes it is necessary to rewrite an absolute value expression in a form that contains no absolute value.

EXAMPLE 3   **Rewriting an Absolute Value Expression**

Rewrite $|w - 2|$ in a form without absolute value.

Solution   According to the definition of absolute value,

$$|w - 2| = \begin{cases} w - 2 & \text{if } w - 2 \geq 0; \text{ that is, if } w \geq 2 \\ -(w - 2) = 2 - w & \text{if } w - 2 < 0; \text{ that is, if } w < 2 \end{cases}$$

In other words,

$$|w - 2| \text{ can be replaced with } w - 2 \text{ if } w \geq 2$$

Figure 2

and

$$|w - 2| \text{ can be replaced with } 2 - w \text{ if } w < 2 \text{ (Figure 2).}$$

We can use absolute value to find the distance between numbers on a number line. Consider the number line in Figure 3, which displays the locations of -5 and 4.

Figure 3

The distance between -5 and 4 can be found in either of the following ways:

$$|4 - (-5)| = |9| = 9$$

or

$$|(-5) - 4| = |-9| = 9.$$

This observation leads to the following general property:

Distance between Two Points

> The **distance** between two points on a number line whose coordinates are a and b is given by
>
> $$|a - b| \quad \text{or} \quad \text{by} \ |b - a|$$

EXAMPLE 4

Finding the Distance on a Number Line

Find the distance on a number line between each pair of points with the given coordinates. Display the result on a number line.

(a) -5 and 9

(b) -8 and -1

(c) a and 3, where $a < 3$

Solution

The solution of each part is given in the following table:

Given Numbers	Distance between Points	Graphical Display
(a) -5 and 9	$\|9 - (-5)\| = \|14\| = 14$ or $\|(-5) - 9\| = \|-14\| = 14$	(number line: 14 between -5 and 9)
(b) -8 and -1	$\|-1 - (-8)\| = \|7\| = 7$ or $\|-8 - (-1)\| = \|-7\| = 7$	(number line: 7 between -8 and -1)
(c) a and 3, where $a < 3$	$\|3 - a\| = 3 - a$ or $\|a - 3\| = -(a - 3) = 3 - a$	(number line: $3 - a$ or $a - 3$ between a and 3)

To solve equations involving absolute value, we use the following properties, which follow from the definition of absolute value.

Absolute Value Properties

> Let p be a real number.
>
> (i) If $p > 0$, then $|u| = p$ if and only if $u = -p$ or $u = p$.
>
> (ii) If $p < 0$, then $|u| = p$ has no solution.
>
> (iii) If $p = 0$, then $|u| = p$ if and only if $u = 0$.

EXAMPLE 5   **Solving an Absolute Value Equation**

Solve the equation $|2x - 3| = 5$.

Solution   We proceed by replacing u by $2x - 3$ and p by 5 in Property (i) as follows:

$$|u| = p \text{ if and only if } u = -p \text{ or } u = p$$

so that

$$2x - 3 = -5 \quad \text{or} \quad 2x - 3 = 5$$

By solving these equations, we have

$$
\begin{array}{ccc}
2x - 3 = -5 & \text{or} & 2x - 3 = 5 \\
2x = -2 & & 2x = 8 \\
x = -1 & & x = 4
\end{array}
$$

Thus, the solution includes -1 and 4.

Solving Literal Equations or Formulas

Many applied problems involve equations or formulas with more than one variable. For instance,

$$d = rt, \quad A = p(1 + rt) \quad \text{and} \quad S = 2\pi rh + 2\pi r^2$$

are formulas from physics, business, and geometry, respectively. To solve a **literal** *equation* or a *formula* such as these for a specific variable or letter, we follow the same strategy used for solving linear equations, by isolating the specified variable on one side and the remaining letters on the other side.

EXAMPLE 6   **Solving a Formula**

Solve the formula $S = 2\pi rh + 2\pi r^2$ for h.

Solution   We solve this equation for h as follows:

This formula expresses the surface area S of a right circular cylinder in terms of its height h and its radius r.

$$2\pi rh + 2\pi r^2 = S$$
$$2\pi rh = S - 2\pi r^2$$
$$h = \frac{S - 2\pi r^2}{2\pi r}$$

Solving Applied Problems Using Mathematical Modeling

The process of describing a real-world situation in mathematical terms is called *mathematical modeling.* This process may involve extracting and interpreting numerical data, geometric figures, formulas, or equations developed from specific information. Applied problems often are presented in written or verbal form. Although there are no set rules for constructing mathematical models for such situations, the following general guidelines may help:

---

1. **Understand the Problem:** Study the problem carefully to determine what information is given and what information is being sought.
2. **Organize the Information:**
   (a) First, represent each quantity with an appropriate mathematical expression, taking into account restrictions or limitations. If appropriate, sketch a geometric figure to represent the situation.
   (b) Express each relationship in the problem by a table, an equation, or a formula.
3. **Analyze the Information:** Perform the necessary mathematical procedure for the model such as solving an equation or inequality or using a graph.
4. **Interpret the Information:** Draw conclusions from the mathematical model about the real-world situation and determine whether the model provides a good tool for understanding and interpreting the situation.

EXAMPLE 7   **Modeling an Investment Problem**

A person invests part of $5000 in a certificate of deposit (CD) that yields 5.5% simple annual interest and the rest in a mutual fund (MF) that yields 6% simple annual interest. At the end of the year, the combined interest from the two investments is $282.50. How much was invested in the CD and how much in the MF?

Solution   Here we let x represent the money invested in the CD, so $(5000 - x)$ represents the remaining amount invested in the MF. The quantities involved in the problem are listed with their mathematical representations in the following table:

Investment Type	Principal *Amount invested*	Interest Rate	Simple Interest Earned in 1 Year
CD	x dollars	0.055	$0.055x$
MF	$(5000 - x)$ dollars	0.06	$0.06(5000 - x)$

Since the combined interest is $282.50, we have

$$\begin{pmatrix} \text{Interest} \\ \text{from CD} \end{pmatrix} + \begin{pmatrix} \text{Interest} \\ \text{from MF} \end{pmatrix} = \begin{pmatrix} \text{Total} \\ \text{interest} \end{pmatrix}$$

$$0.055x + 0.06(5000 - x) = 282.50$$

$$0.055x + 300 - 0.06x = 282.50$$

$$-0.005x = -17.5$$

$$x = 3500$$

Therefore, $3500 was invested in the CD. Since $5000 - x = 5000 - 3500 = 1500$, it follows that $1500 was invested in the MF.

EXAMPLE 8   **Modeling a Jet Stream Problem**

In aviation, the jet stream is a narrow current of air encountered between altitudes of 25,000 and 35,000 feet. One day, a flight from Los Angeles to New York (a distance of 2460 miles) flying with the jet stream (with a tail wind) took the same amount of time as a flight from New York to Denver (a distance of 1626 miles) flying against the jet stream (with a head wind). Assume that both planes fly 550 miles per hour in still air. How fast was the jet stream and how long was the flight for both planes? Round off the answers to two decimal places.

Solution   The tail wind increases the plane's speed by applying force in the direction of travel, whereas the head wind decreases the plane's speed by applying force against the direction of travel.

Let r represent the speed (in miles per hour) of the jet stream, then

$550 + r$ represents the speed (in miles per hour) of the plane with the jet stream,

and

$550 - r$ represents the speed (in miles per hour) of the plane against the jet stream. The table below summarizes the information.

Flight	Rate or Speed (miles per hour)	Distance (in miles)	Time $= \dfrac{\text{Distance}}{\text{Speed}}$
With the jet stream from Los Angeles to New York	$550 + r$	2460	$\dfrac{2460}{550 + r}$
Against the jet stream from New York to Denver	$550 - r$	1626	$\dfrac{1626}{550 - r}$

Since both flights took the same amount of time, it follows that

$$\left(\begin{array}{c}\text{Time traveling with jet stream}\\ \text{from Los Angeles to New York}\end{array}\right) = \left(\begin{array}{c}\text{Time traveling against jet stream}\\ \text{from New York to Denver}\end{array}\right).$$

Thus, the equation that describes the model is

$$\frac{2460}{550 + r} = \frac{1626}{550 - r}.$$

The equation is solved as follows:

$2460(550 - r) = 1626(550 + r)$ Multiply both sides by $(550 + r)(550 - r)$

$1{,}353{,}000 - 2460r = 894{,}300 + 1626r$ Distribute

$-4086r = -458{,}700$

$r = 112.26$ (approx.) $\Bigg\}$ Isolate r

Therefore, the jet stream was about 112.26 miles per hour. Since the flights took the same amount of time for both planes, we can substitute 112.26 for r in either

$$\frac{2460}{550 \; + \; r} \quad \text{or} \quad \frac{1626}{550 \; - \; r}$$

to get

$$\frac{2460}{550 \; + \; 112.26} = 3.71 \text{ (approx.)}$$

or

$$\frac{1626}{550 \; - \; 112.26} = 3.71 \text{ (approx.)}$$

So the flights took about 3.71 hours.

PROBLEM SET 1.1

Mastering the Concepts

In problems 1–4, solve each equation.

1. (a) $34 - 3x = 4(3 - 2x) + 27$
    (b) $8(5x - 1) + 36 = -3(x + 5)$
2. (a) $3y - 2(y + 1) = 2(y - 1)$
    (b) $7(t - 3) = 4(t + 5) - 47$
3. (a) $5 + 8(u + 2) = 23 - 2(2u - 5)$
    (b) $11 - 7(1 - 2p) = 9(p + 1)$
4. (a) $11 = 3v - 8(7 - v) + 23$
    (b) $6(y - 10) + 3(2y - 7) = -45$

In problems 5–10, solve each equation. Be sure to check for extraneous roots.

5. $\dfrac{3x - 2}{3} + \dfrac{x - 3}{2} = \dfrac{5}{6}$

6. $\dfrac{x - 14}{5} + 4 = \dfrac{x + 16}{10}$

7. (a) $\dfrac{1}{y} + \dfrac{2}{y} = 3 - \dfrac{3}{y}$

   (b) $\dfrac{x}{x + 1} = 5 - \dfrac{1}{x + 1}$

8. (a) $\dfrac{2t}{t - 2} = \dfrac{4}{t - 2} + 1$

   (b) $\dfrac{9}{5y - 3} = \dfrac{5}{3y + 7}$

9. $\dfrac{2}{y - 3} - \dfrac{1}{y + 5} = \dfrac{1}{y^2 + 2y - 15}$

10. $\dfrac{1}{x - 2} - \dfrac{3}{x + 2} = \dfrac{2}{x^2 - 4}$

In problems 11–16, rewrite each expression without absolute value notation.

11. (a) $|\pi - 5|$
    (b) $|\sqrt{7} - 2|$
12. (a) $|\sqrt{5} - 3|$
    (b) $|2 - \sqrt{11}|$
13. (a) $|x - 5|$
    (b) $|x + 3|$
14. (a) $|y + 0.1|$
    (b) $\left| t - \dfrac{5}{2} \right|$
15. $|x + 5|$
16. $|3x - 2|$

In problems 17–22, use absolute value to find the distance between each pair of numbers.

17. 2 and -6
18. -7 and -3
19. -4 and 4
20. -3 and 3
21. $-4t$ and $6t$
22. $-8t$ and $-6t$

In problems 23–28, solve each equation.

23. (a) $|x - 2| = 4$
(b) $|2t + 6| = 18$

24. (a) $|3x + 9| = 12$
(b) $|14 - 7y| = |-21|$

25. (a) $\left|1 - \dfrac{2x}{3}\right| = 5$
(b) $\left|3 - \dfrac{2x}{5}\right| - 2 = 5$

26. (a) $|6 - 3y| - 5 = 7$
(b) $|7 - 2t| + 4 = |-9|$

27. (a) $|x - 2(x - 2)| = |-3|$
(b) $|y - 9| = |3y + 1|$

28. (a) $|t - 2(t + 5)| = |-2|$
(b) $|y - 9| = |2y - 3|$

In problems 29–34, solve each literal equation for the specified variable.

29. (a) $b - c = a; b$
(b) $A = \ell w; w$

30. (a) $b - c = a; c$
(b) $d = rt; r$

31. (a) $A = \dfrac{1}{2}bh; b$
(b) $V = 32t + 4; t$

32. (a) $C = 2\pi r; r$
(b) $4w = 2t - 8; t$

33. (a) $A = \dfrac{1}{2}(a + b)h; h$
(b) $z = \dfrac{5}{3} + \dfrac{zy}{4}; y$

34. (a) $P = 2\ell + 2w; \ell$
(b) $\dfrac{4w}{x} - 1 = \dfrac{y}{x}; x$

Applying the Concepts

In problems 35–44, set up an appropriate equation to model the situation and then solve each problem.

35. Investment: An investor wishes to borrow $10,000. There is an annual simple interest rate charge of 7% on part of the loan, and an annual simple interest rate charge of 6.25% on the other part. If the annual interest on both parts totals $656.50, how much money is borrowed at each rate?

36. Investment: A family invests a total of $85,000 in two tax-free municipal bonds to reduce its income tax. One bond pays 5% tax-free simple annual interest, and the other bond pays 5.5% tax-free simple annual interest. The total nontaxable income from both investments at the end of 1 year is $4475. How much did the family invest in each bond?

37. Investment: A retiree deposited $6200 in two different savings accounts. The annual interest rate on one account was 5.5% and the annual interest rate on the other account was 5.8%. At the end of the first year, the interest earned on the first account was $92.40 more than the interest earned on the second. Determine the amount deposited in each account.

38. Budgeting: A small business allocated four times as much for utility expenses as for interest expenses. The annual total amount budgeted for these expenses was $82,000. How much was budgeted for utilities?

39. Mixture: A medicine contains 25% alcohol. How much water should be added to 120 cubic centimeters of the medicine if the final mixture is to contain 20% alcohol?

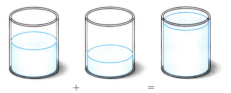

40. Mixture: A bottle of mineral water contains 16 ounces of liquid with a 4% calcium content. How many ounces of this liquid should be poured out and replaced with a 6% calcium solution to obtain a 4.5% calcium solution?

41. Cycling and Jogging Rates: A jogger and a bicycle rider head for the same destination at the same time and from the same point (Figure 4). The average speed of the bicycle rider is 3 times the speed of the jogger. At the end of 2 hours, the bicycle rider is 12 miles ahead of the jogger. How fast is the bicycle rider traveling, assuming that the speeds are constant?

Figure 4

12 miles

Starting point

42. Price of a Graphing Calculator: The model $P = R - C$ is used to express the relationship between the *profit* P, the *revenue* R, and the *cost* C. When the price of a graphing calculator is $75 per unit, an electronics store sells 95 units per day. If the daily profit is $665, what is the cost of each calculator?

43. Time Required to Do a Job: A computer can be used to perform a large stock portfolio analysis in 6 minutes. With the help of a new computer, the analysis can be run in 2 minutes. How long would it take the new computer to run the analysis alone?

44. Filling a Swimming Pool: Two hoses are being used to fill a swimming pool. The larger hose supplies twice as much water as the smaller hose. If both hoses can fill the swimming pool in 8 hours, how long would it take each hose alone to fill the pool?

Developing and Extending the Concepts

In problems 45–50, solve each equation.

45. (a) $3x + 3 = 7(x - 1) - 4x$
  (b) $3x + 3 = 4(x - 1) - (x - 7)$

46. (a) $x - 10 = 2\left(x + \dfrac{1}{2}\right) - (x + 1)$
  (b) $x = \dfrac{1}{5}(x - 1) + \dfrac{4}{5}(x + 5)$

47. (a) $\dfrac{1}{x - 1} = \dfrac{1}{x}$
  (b) $\dfrac{x}{x + 1} = \dfrac{x + 1}{x}$

48. (a) $\dfrac{2x + 1}{4x} = \dfrac{x - 3}{2x + 1}$
  (b) $\dfrac{x}{x + a} = \dfrac{x}{x + b}$; for x

49. (a) $|x| = x + 1$
  (b) $|x| = x$

50. (a) $\dfrac{|x|}{x} = -1$
  (b) $|x| = x - 1$

51. Try different values of x and y to illustrate the validity of each equation. Then explain why these equations are true for all values of x and y, except $y \neq 0$ in part (b).
  (a) $|xy| = |x||y|$
  (b) $\left|\dfrac{x}{y}\right| = \dfrac{|x|}{|y|}$
  (c) $|x - y| = |y - x|$

52. Under what conditions are the following equations true?
  (a) $\dfrac{|x|}{y} = \dfrac{x}{|y|}$
  (b) $|xy| = x|y|$
  (c) $|x - y| = |x + y|$

Objectives

1. Solve Quadratic Equations
2. Solve Polynomial Equations by Factoring
3. Solve Equations Involving Radicals
4. Solve Applied Problems

1.2 Nonlinear Equations

Quite often mathematical models lead to situations described by *nonlinear equations*. In this section, we review methods for solving such types of equations.

Solving Quadratic Equations

Any equation that can be written in the form

$$ax^2 + bx + c = 0$$

where a, b, and c are constants and $a \neq 0$, is called a **second-degree,** or **quadratic, equation in one variable.**
Examples of quadratic equations are

$$x^2 = 9, \quad 3x^2 - 2x = 1, \quad \text{and} \quad x^2 + 2x - 1 = 0$$

The form

$$ax^2 + bx + c = 0$$

is referred to as the **standard form** of a quadratic equation.
For instance the equation

$$2x^2 - 3x = -1 \quad \text{has standard form} \quad 2x^2 - 3x + 1 = 0.$$

We'll review four methods for solving quadratic equations with real number coefficients:

1. *The factor method*
2. *The square root extraction method*
3. *The completing-the-square method*
4. *The quadratic formula*

1. The Factor Method

At times, we solve quadratic equations in standard form by first factoring the quadratic expression as a product of two first-degree polynomials and then applying the following property.

Zero Factor Property

$a \cdot b = 0$ if and only if $a = 0$ or $b = 0$, or both a and b equal 0.

Table 1 gives the step-by-step procedure for using the factor method to solve quadratic equations along with an example of the procedure.

TABLE 1   Factor Method for Solving a Quadratic Equation

	Step-by-Step Procedure	Example: Solve $x^2 + 2x = 3$
Step 1	Write the equation in standard form (zero is on one side of the equal sign)	$x^2 + 2x - 3 = 0$ (subtract 3 from each side)
Step 2	Factor the quadratic expression	$(x + 3)(x - 1) = 0$ (factor)
Step 3	Use the zero factor property by setting each first-degree factor equal to zero	$x + 3 = 0$ or $x - 1 = 0$ (set each factor equal to zero)
Step 4	Solve the resulting linear equation(s) from Step 3 to obtain the solution of the given equation	$x = -3$ or $x = 1$ (solve each linear equation for x) So the solution of the equation $x^2 + 2x = 3$ includes -3 and 1

Notice that in Table 1 we can check the two answers: first, by substituting 3 for x in the given equation to get $(-3)^2 + 2(-3) = 9 - 6 = 3$, which is true; then, by substituting 1 for x in the given equation to obtain $1^2 + 2(1) = 3$, which is also true.

2. The Square Root Extraction Method

To understand the technique of using square roots to solve quadratic equations, let us consider the equation

$$x^2 = 9.$$

This equation can be solved by **extracting the square roots** of each side to get

$$x = -\sqrt{9} \quad \text{or} \quad x = \sqrt{9}$$
$$x = -3 \quad \quad \quad x = 3$$

So the solutions of $x^2 = 9$ are -3 and 3.

Similarly we can solve the equation

$$x^2 = -16$$

by the root extraction method to get

$$x = \sqrt{-16} \quad \text{or} \quad x = -\sqrt{-16}$$
$$x = 4i \qquad\qquad\quad x = -4i$$

so that $4i$ and $-4i$ are the roots of the equation.

In general, we have the following property

Property
Square Root Extraction

> The solutions of $x^2 = k$ are $x = -\sqrt{k}$ and $x = \sqrt{k}$, $k \neq 0$

EXAMPLE

Solving an Equation by Root Extraction

Solve the following equation by root extraction.

$$(3x - 2)^2 = 100$$

Solution

By extracting the square roots of each side of the given equation, we get

$$3x - 2 = \sqrt{100} \quad \text{or} \quad 3x - 2 = -\sqrt{100}$$
$$3x - 2 = 10 \qquad\qquad\quad 3x - 2 = -10$$
$$3x = 12 \qquad\qquad\qquad\quad 3x = -8$$
$$x = 4 \qquad\qquad\qquad\quad x = -8/3$$

So the solution includes 4 and $-8/3$.

3. The Completing-the-Square Method

Some quadratic equations are very difficult to solve by the factor method. One method for solving such equations is to first rewrite them in the form

$$(x + a)^2 = k$$

and then solve the latter equation by using root extraction. This method, called the **completing-the-square method**, depends on adding an appropriate constant term to an expression of the form $x^2 + bx$ to *produce a perfect square trinomial*, that is, a trinomial that is factorable as the square of a binomial.

If two terms of a quadratic expression are $x^2 + bx$, *completing the square entails adding

$$\left(\frac{1}{2} \cdot b\right)^2 = \left(\frac{b}{2}\right)^2 \text{ to } x^2 + bx \text{ to get}$$

a perfect square trinomial

$$x^2 + bx + \left(\frac{b}{2}\right)^2 = \left(x + \frac{b}{2}\right)^2.$$

Table 2 displays two examples of this procedure.

TABLE 2   Examples of Producing a Perfect Square Trinomial

Given	Add Term	Result
$x^2 + 4x$	$\left(\dfrac{1}{2} \cdot 4\right)^2 = 4$	$x^2 + 4x + 4 = (x + 2)^2$
$x^2 - 3x$	$\left[\dfrac{1}{2}(-3)\right]^2 = \dfrac{9}{4}$	$x^2 - 3x + \dfrac{9}{4} = \left(x - \dfrac{3}{2}\right)^2$

The completing-the-square method for solving a quadratic equation depends on this procedure and the square root extraction method. The method is given in Table 3, along with an example.

TABLE 3   The Completing-the-Square Method for Solving a Quadratic Equation

Step-by-Step Procedure	Example: Solve $2x^2 - 2x - 1 = 0$
Step 1 Write the equation in a form in which only the variable terms are on one side of the equal sign	$2x^2 - 2x = 1$ (add 1 to each side)
Step 2 Proceed to the next step if the coefficient of the x^2 term is 1; otherwise, divide each side of the equation by the coefficient of the x^2 term	$x^2 - x = \dfrac{1}{2}$ (divide each side by 2 to make the x^2 coefficient equal to 1)
Step 3 Complete the square by adding the square of one-half the coefficient of the x term to each side	$x^2 - x + \dfrac{1}{4} = \dfrac{1}{2} + \dfrac{1}{4}$ add $\left(\dfrac{1}{2}(-1)\right)^2 = \dfrac{1}{4}$ to each side to complete the square)
Step 4 Factor the resulting perfect square trinomial to obtain the square of a binomial	$\left(x - \dfrac{1}{2}\right)^2 = \dfrac{3}{4}$ (factor and simplify)
Step 5 Use the technique of extracting square roots to solve the equation in Step 4	$x - \dfrac{1}{2} = -\dfrac{\sqrt{3}}{2}$ or $x - \dfrac{1}{2} = \dfrac{\sqrt{3}}{2}$ (extract the square roots of each side) $x = \dfrac{1}{2} - \dfrac{\sqrt{3}}{2}$ or $x = \dfrac{1}{2} + \dfrac{\sqrt{3}}{2}$ (solve for x) So the solution of the original equation includes $\dfrac{1 - \sqrt{3}}{2}$ and $\dfrac{1 + \sqrt{3}}{2}$ (approx. -0.37) and (approx. 1.37).

4. The Quadratic Formula

We can generalize the method of completing the square to arrive at a formula that enables us to solve any quadratic equation (problem 50, on page 62).

Quadratic Formula

> If $ax^2 + bx + c = 0$, with a, b, and c real numbers and $a \neq 0$, then the solutions of the equation can be determined by the formula
>
> $$x = \frac{-b \pm \sqrt{b^2 - 4ac}}{2a}$$
>
> where $b^2 - 4ac$ is called the **discriminant** of the quadratic equation.

EXAMPLE 2

Using the Quadratic Formula

Solve each equation by using the quadratic formula.

(a) $2x^2 - 8x + 3 = 0$ (b) $5x^2 + 2x + 1 = 0$

Solution

(a) Here, $a = 2$, $b = -8$, and $c = 3$; so by using the quadratic formula we have

$$x = \frac{-b \pm \sqrt{b^2 - 4ac}}{2a}$$

$$= \frac{-(-8) \pm \sqrt{64 - 4(2)(3)}}{2(2)}$$

$$= \frac{8 \pm \sqrt{64 - 24}}{4}$$

$$= \frac{8 \pm \sqrt{40}}{4} = \frac{8 \pm 2\sqrt{10}}{4} = \frac{2(4 \pm \sqrt{10})}{4} = \frac{4 \pm \sqrt{10}}{2}.$$

Therefore, the solutions are

$$2 + \frac{\sqrt{10}}{2} \text{ (approx. 3.58)}   \text{and}   2 - \frac{\sqrt{10}}{2} \text{ (approx. 0.42)}.$$

(b) Using the quadratic formula with $a = 5$, $b = 2$, and $c = 1$, we have

$$x = \frac{-2 \pm \sqrt{4 - 4(5)(1)}}{10}$$

$$= \frac{-2 \pm \sqrt{-16}}{10} = \frac{-2 \pm 4i}{10}$$

$$= \frac{2(-1 \pm 2i)}{10} = \frac{-1 \pm 2i}{5}.$$

Note that the solutions are complex conjugates of each other. In Chapter 3 we will discuss complex roots of polynomial equations in more detail.

So the solutions are the complex numbers

$$\frac{-1}{5} + \frac{2}{5}i   \text{and}   \frac{-1}{5} - \frac{2}{5}i.$$

Note that the quadratic formula can be used to solve *all* quadratic equations, including those for which the factor method can be employed.

Solving Polynomial Equations by Factoring

The factor method for quadratic equations can be extended to solve a factored polynomial equation with a degree higher than two. This idea is illustrated in the next example.

EXAMPLE 3   **Solving a Polynomial Equation by Factoring**

Solve the given equation by factoring.

$$x^3 - 4x = 0$$

Solution   We begin by factoring the left side of the equation as follows:

$$x^3 - 4x = 0 \qquad \text{Given}$$
$$x(x^2 - 4) = 0 \qquad \text{Remove common factor } x$$
$$x(x + 2)(x - 2) = 0 \qquad \text{Factor } x^2 - 4$$

Now we apply the zero property by setting each factor equal to zero and solving the resulting equation.

$$x = 0 \qquad \text{or} \qquad x + 2 = 0 \qquad \text{or} \qquad x - 2 = 0$$
$$x = 0 \qquad\qquad\qquad x = -2 \qquad\qquad\qquad x = 2$$

So the solutions are 0, −2, and 2.

Notice that the third-degree equation in Example 3 above has three factors and three solutions. This is *not* an accident. As we shall see later in Chapter 3, a polynomial equation can have at most as many solutions as its degree.

Recall from Section 1.1 that one way to solve equations involving rational expressions is to first multiply each side of the equation by the least common denominator (LCD) thus eliminating the fractions from the equation. In the next example, we solve an equation involving rational expressions that simplifies to a quadratic equation.

EXAMPLE 4   **Solving an Equation Containing Rational Expressions**

Solve the equation

$$\frac{6}{x - 1} + \frac{6}{x} = 5$$

Solution   The solution of the equation proceeds as follows:

$$\frac{6}{x - 1} + \frac{6}{x} = 5 \qquad \text{Given}$$

$$x(x - 1)\left[\frac{6}{x - 1} + \frac{6}{x}\right] = x(x - 1)5 \qquad \text{Multiply both sides by the LCD}$$

$$x(x - 1)\left[\frac{6}{x - 1}\right] + x(x - 1)\frac{6}{x} = 5x(x - 1) \qquad \text{Distribute}$$

$$\left.\begin{array}{r} 6x + 6(x - 1) = 5x^2 - 5x \\ 6x + 6x - 6 = 5x^2 - 5x \\ 12x - 6 = 5x^2 - 5x \\ 5x^2 - 17x + 6 = 0 \end{array}\right\} \qquad \text{Simplify}$$

We can solve this latter equation by the factor method as follows:

$$(5x - 2)(x - 3) = 0 \qquad \text{Factor}$$

$$5x - 2 = 0 \qquad \text{or} \qquad x - 3 = 0$$

$$5x = 2 \qquad\qquad\qquad x = 3$$

$$x = 2/5$$

Set each factor equal to zero and solve the resulting linear equation

The solutions appear to be

$$\frac{2}{5} \quad \text{and} \quad 3.$$

As we pointed out in Section 1.1, whenever an equation has variables in the denominator, the process of multiplying each side by the LCD may produce an extraneous root. However, on checking the two values, 2/5 and 3, in the original equation we find that neither is extraneous. So the solution of the equation includes both 2/5 and 3.

Solving Equations Involving Radicals

Next, we consider real number solutions of algebraic equations that contain *radicals* or *rational exponents*. Examples of these equations are

$$\sqrt{x} = 5, \quad \sqrt{5y + 1} = 4, \quad \sqrt{x} = \sqrt{x + 16} - 4$$

and

$$(3u + 1)^{1/3} = u + 1$$

A method of solving equations of this type is to isolate one of the radicals and then raise each side of the equation to the same positive integer power that *eliminates* the isolated radical. It may be necessary to repeat this step until all the radicals are eliminated. Then the resulting equation is simplified and solved.

It is important to realize that this process of eliminating radicals may introduce extraneous roots when dealing with the even-numbered roots. Therefore, it is necessary to *check all the proposed solutions in the original equation whenever a radical with an even index is involved*, to determine whether each solution should be accepted or rejected.

EXAMPLE 5

Solving a Radical Equation Involving Cube Root

Solve the equation.

$$\sqrt[3]{y^2 - 1} = 2$$

Solution

Proceeding as follows

$$(\sqrt[3]{y^2 - 1})^3 = 2^3 \qquad \text{Cube each side of the equation}$$

$$y^2 - 1 = 8$$

$$y^2 = 8 + 1 \qquad \text{Add 1 to each side}$$

$$y^2 = 9$$

$$y = \pm\sqrt{9} \qquad \text{Take the square root}$$

$$y = \pm 3$$

Therefore, the solutions are

$$y = -3 \quad \text{and} \quad y = 3.$$

EXAMPLE 6   **Solving an Equation Involving Square Root**

Solve the equation.

$$\sqrt{x + 4} - 2 = \sqrt{x}$$

Solution   We proceed as follows:

$$(\sqrt{x + 4} - 2)^2 = (\sqrt{x})^2 \qquad \text{Square each side of the given equation}$$

$$x + 4 - 4\sqrt{x + 4} + 4 = x \qquad \text{Multiply}$$

$$x - 4\sqrt{x + 4} + 8 = x \qquad \text{Collect terms on the left side}$$

$$-4\sqrt{x + 4} = -8 \qquad \text{Isolate the radical expression on one side}$$

$$\sqrt{x + 4} = 2 \qquad \text{Divide each side by } -4$$

We repeat the process by squaring each side of the latter equation to obtain

$$(\sqrt{x + 4})^2 = 2^2$$

$$x + 4 = 4 \qquad \text{Multiply}$$

$$x = 0 \qquad \text{Solve for } x$$

Check: Since even roots are involved, we check $x = 0$ to determine whether it is extraneous. Substituting 0 into the given equation, we get

$$\sqrt{0 + 4} - 2 = \sqrt{0}$$

$$2 - 2 = 0$$

which is true. So the solution is $x = 0$.

Equations such as

$$x^{\frac{2}{3}} + x^{\frac{1}{3}} - 6 = 0$$

reduce to a quadratic equation.

If we substitute u for $x^{\frac{1}{3}}$ in this equation, we get the quadratic equation

$$u^2 + u - 6 = 0.$$

This latter equation is solved as follows:

$$u^2 + u - 6 = 0$$

$$(u + 3)(u - 2) = 0 \qquad \text{Factor}$$

$$u + 3 = 0 \quad \text{or} \quad u - 2 = 0 \quad \left.\right\} \quad \begin{array}{l}\text{Set each factor equal to 0,}\\\text{then solve the results.}\end{array}$$

$$u = -3 \qquad\qquad u = 2$$

Next, by replacing u with $x^{\frac{1}{3}}$, we get

$$x^{\frac{1}{3}} = -3 \ \text{ or } \ x^{\frac{1}{3}} = 2$$

$$\left(x^{\frac{1}{3}}\right)^3 = (-3)^3 \qquad \left| \qquad \left(x^{\frac{1}{3}}\right)^3 = 2^3 \right.$$

$$x = -27 \qquad\qquad\quad x = 8$$

so that the solutions of $x^{\frac{2}{3}} + x^{\frac{1}{3}} + 6 = 0$ are -27 and 8.

Solving Applied Problems

In the next two examples, quadratic equations are used to model real-world situations.

EXAMPLE 7   **Solving a Physics Model**

A ball is thrown vertically upward with an initial speed of 48 feet per second. Disregarding air resistance, its height (in feet) above the ground after t seconds is given by the equation

$$h = 48t - 16t^2.$$

(a) How long will it take the ball to reach a height of 32 feet on the way up?

(b) How long will it take the ball to reach the ground?

Solution   (a) We want to find the time t for the ball to reach a height of 32 feet on the way up. So we substitute 32 for h to get

$$32 = 48t - 16t^2.$$

To solve this equation we rewrite it in standard form and then solve it by the factor method as follows:

$$16t^2 - 48t + 32 = 0$$

$$t^2 - 3t + 2 = 0 \qquad \text{Divide each side by 16}$$

$$(t - 1)(t - 2) = 0 \qquad \text{Factor}$$

$$\left.\begin{array}{lcl} t - 1 = 0 & \text{or} & t - 2 = 0 \\ t = 1 & & t = 2 \end{array}\right\} \quad \text{Solve}$$

Since we want the time t for the ball to reach a height of 32 feet on the way up, the desired solution is 1 second. The ball begins to fall and returns to a height of 32 feet after 2 seconds.

(b) Since the height $h = 0$ when the ball hits the ground, we substitute 0 for h and solve for t by using the factor method:

$$0 = 48t - 16t^2$$

$$0 = 16t(3 - t) \qquad \text{Factor}$$

$$\left.\begin{array}{lcl} 16t = 0 & \text{or} & 3 - t = 0 \\ t = 0 & & t = 3 \end{array}\right\} \quad \text{Solve}$$

The solution $t = 0$ represents the starting point on the ground, so it takes the ball 3 seconds to return to the ground.

EXAMPLE 8   **Modeling a Volume Problem**

A rectangular sheet of aluminum whose length is 2 inches more than its width is to be made into a tray by cutting 2-inch squares from each corner and bending up the flaps.

(a) Express the volume V in terms of x, where x (in inches) is the original width of the sheet of aluminum.

(b) Find the width x when the volume is 336 cubic inches.

Solution

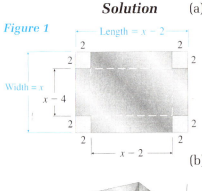

Figure 1

(a) A diagram of the situation is shown in Figure 1. From the figure we can see that the width of the tray is $x - 4$ and the length is $x - 2$. Since V represents the volume, it follows that

$$\text{Volume} = (\text{Length}) \times (\text{Width}) \times (\text{Height})$$
$$= (x - 2)(x - 4)(2)$$
$$= 2x^2 - 12x + 16$$

(b) To find x when $V = 336$, we replace V by 336 in the equation to get

$$336 = 2x^2 - 12x + 16.$$

$2x^2 - 12x - 320 = 0$	Write in standard form
$x^2 - 6x - 160 = 0$	Divide each side by 2
$(x - 16)(x + 10) = 0$	Factor
$x - 16 = 0 \quad \text{or} \quad x + 10 = 0$ $\left. \begin{array}{c} \\ \\ \end{array} \right\}$ Solve	
$x = 16 \qquad\qquad x = -10$	

Here we use only the positive solution since the dimensions of the tray must be positive numbers.

So a width of 16 inches yields a tray with volume 336 cubic inches.

PROBLEM SET 1.2

Mastering the Concepts

In problems 1–4, solve each quadratic equation by factoring.

1. (a) $x^2 - x - 12 = 0$
   (b) $x^2 - x - 20 = 0$
2. (a) $6x^2 + 6x - 36 = 0$
   (b) $2y^2 + 10y - 28 = 0$
3. (a) $3t^2 + 6t - 24 = 0$
   (b) $42 - 7x - 7x^2 = 0$
4. (a) $40 - 2x - 2x^2 = 0$
   (b) $66 - 27x - 3x^2 = 0$

In problems 5–8, solve each quadratic equation by extracting square roots.

5. (a) $x^2 = 36$
   (b) $x^2 = 8$
   (c) $x^2 = \dfrac{16}{25}$
   (d) $x^2 = -9$
6. (a) $x^2 = 49$
   (b) $x^2 = 6$
   (c) $x^2 = \dfrac{1}{12}$
   (d) $x^2 = \dfrac{-3}{25}$
7. (a) $(x - 7)^2 = 49$
   (b) $\left(x + \dfrac{1}{2} \right)^2 = 12$
8. (a) $(x + 5)^2 = -25$
   (b) $\left(x - \dfrac{2}{3} \right)^2 = \dfrac{16}{9}$

In problems 9–12, fill in the blank with the number that makes the given expression a perfect square trinomial. Also, find the factored form of the result.

9. (a) $x^2 + 12x +$ _____
   (b) $x^2 - 5x +$ _____
10. (a) $x^2 - 10x +$ _____
    (b) $x^2 + 3x +$ _____
11. (a) $x^2 - \dfrac{x}{3} +$ _____
    (b) $x^2 + \dfrac{3}{4}x +$ _____
12. (a) $x^2 + 0.2x +$ _____
    (b) $x^2 - 1.6x +$ _____

In problems 13–16, use the method of completing the square to solve each equation.

13. (a) $x^2 - 2x - 2 = 0$ (b) $x^2 + 10x + 3 = 0$
14. (a) $7y^2 - 4y - 1 = 0$ (b) $2p^2 + 6p - 7 = 0$
15. (a) $9t^2 - 30t = -21$ (b) $5m^2 - 8m = 17$
16. (a) $2r^2 + 3r = 2$ (b) $4y^2 + 7y = 5$

In problems 17–20, solve each equation by using the quadratic formula.

17. (a) $2x^2 + x - 1 = 0$
(b) $3x^2 - 5x + 1 = 0$

18. (a) $2y^2 - 6y + 3 = 0$
(b) $3u^2 + 4u - 4 = 0$

19. $1.5x^2 - 7.5x - 4.3 = 0$

20. $5.13x^2 - 7.14x - 3.57 = 0$

In problems 21–24, solve each equation by factoring.

21. (a) $x^3 - 4x^2 - 5x = 0$
(b) $x^5 - 2x^4 - 3x^3 = 0$

22. (a) $6x^3 + x^2 - 2x = 0$
(b) $x^4 - 4x^2 = 0$

23. (a) $x^4 - 16 = 0$
(b) $x^3 + 5x^2 - 16x - 80 = 0$

24. (a) $x^4 - 81 = 0$
(b) $x^4 - 4x^2 - x^3 + 4x = 0$

In problems 25–28, solve each equation.

25. $x = \dfrac{4}{x - 3}$

26. $\dfrac{x^2}{x + 1} = 2$

27. $\dfrac{1}{x - 2} - \dfrac{1}{x} = \dfrac{1}{60}$

28. $\dfrac{5}{x + 5} + \dfrac{5}{x - 5} = \dfrac{3}{4}$

In problems 29–42, solve each equation.

29. (a) $\sqrt{x - 5} = 2$
(b) $\sqrt{3x + 1} = 4$

30. (a) $\sqrt[3]{6x - 4} = 2$
(b) $\sqrt[3]{3x + 4} = 3$

31. (a) $4 + \sqrt{y - 5} = 7$
(b) $6 + \sqrt{x - 1} = 7$

32. (a) $x - 5 = \sqrt{x + 7}$
(b) $\sqrt{5x + 9} = x - 1$

33. $\sqrt{x - 2} = \sqrt{2x - 2}$

34. $\sqrt{y} = \sqrt{y + 16} - 4$

35. $\sqrt{m + 3} = \sqrt{m - 2} + 1$

36. $\sqrt{r^2 + 5r - 10} = r$

37. $\sqrt{u^2 + 3u} = u + 1$

38. $\sqrt{\sqrt{c + 4} + c} = 4$

39. $4 - (2t + 1)^{3/2} = 5$

40. $(4x + 5)^{(-1/4)} = 2$

41. $(t^2 + 6t)^{1/3} = 3$

42. $(5x^2 + 7x - 3)^{3/2} = 1$

Applying the Concepts

43. Designing an Advertising Flier: A 20 by 30 centimeter brochure is to be used for an advertising flier. The margins at the sides, top, and bottom all have the same width. Find the width of the margins if the printed area is 416 square centimeters.

44. Enclosing a Region: Suppose a livestock rancher wants to enclose a rectangular feedlot. One side of the lot is centered against the back of the barn, which is 50 feet in length, and no fencing is needed along the back of the barn (Figure 2). The rancher needs an area of 2800 square feet to accommodate the cattle. Determine the dimensions necessary to enclose the required area of the lot if 170 feet of fencing are used.

Figure 2

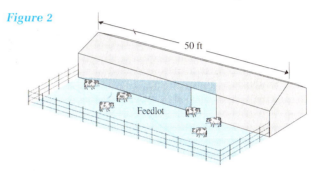

45. Height of a Toy Rocket: The height h (in feet) reached by a model rocket t seconds after it is launched is given by

$$h = -16t^2 + 96t.$$

(a) How long will it take the rocket to initially reach a height of 10 feet above ground? Round off the answer to two decimal places.

(b) How long will it take the rocket to hit the ground?

46. Physics: A ball is thrown upward from the top of a building 80 feet high. After t seconds, the height h (in feet) is given by

$$h = 80 + 32t - 16t^2.$$

(a) How long will it take the ball to reach a height of 70 feet above the ground?

(b) How long will it take the ball to reach the ground?

47. Geometry: The length of a rectangle is 1 inch more than its width.

(a) Express the area A of the rectangle in terms of its width x.

(b) Find the dimensions of the original rectangle if the area will increase by 30 square inches when the length is doubled (the width does not change).

48. Geometry: An open box is formed from a square piece of cardboard with side length of x inches by cutting 2-inch squares from each corner and folding up the sides (Figure 3).

Figure 3

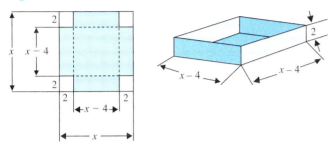

(a) Express the volume V of the box in terms of x.

(b) If the volume of the box is 72 cubic inches, what are the dimensions of the original cardboard?

Developing and Extending the Concepts

49. Use the discriminant to determine the number and kind of roots for each quadratic equation.

(a) $2x^2 - 4x + 1 = 0$

(b) $x^2 - 6x + 9 = 0$

(c) $2x^2 + 3x + 2 = 0$

50. (a) Show that if $a \neq 0$, then $ax^2 + bx + c = 0$ can be written in the form

$$x^2 + \frac{b}{a}x = -\frac{c}{a}.$$

(b) Complete the square to obtain

$$\left(x + \frac{b}{2a}\right)^2 = \frac{b^2 - 4ac}{4a^2}.$$

(c) Solve the equation in part (b) for x.

51. Given a quadratic equation with real coefficients

$$ax^2 + c = 0$$

determine the type of solutions of the equation if:

(a) $ac = 0$ (b) $ac < 0$

52. The number of diagonals d of a polygon with n sides is given by the formula

$$d = \frac{n(n-3)}{2}.$$

Determine the number of sides of a polygon with 27 diagonals.

In problems 53–56, use the given substitution to transform each equation to an equivalent equation that is quadratic in form. Solve the resulting equation and then use the result to solve the original equation.

53. (a) $y^{-2} + y^{-1} = 2$; use $u = y^{-1}$

(b) $x^{\frac{1}{2}} - 3x^{\frac{1}{4}} + 2 = 0$; use $u = x^{\frac{1}{4}}$

54. (a) $2t^{-2} - 5t^{-1} - 3 = 0$; use $u = t^{-1}$

(b) $x^{\frac{2}{3}} - 3x^{\frac{1}{3}} - 10 = 0$; use $u = x^{\frac{1}{3}}$

55. $(t^2 - t)^2 - 4(t^2 - t) = 12$; use $u = t^2 - t$

56. $\left(3x - \dfrac{2}{x}\right)^2 + 6\left(3x - \dfrac{2}{x}\right) + 5 = 0$;

use $u = 3x - \dfrac{2}{x}$

In problems 57–60, solve each equation for y.

57. (a) $\dfrac{x^2}{4} + \dfrac{y^2}{9} = 1$ (b) $(x - 1)^2 + (y + 3)^2 = 9$

58. (a) $x^2 - \dfrac{y^2}{25} = 1$ (b) $(y - 7)^2 = 6x - 12$

59. (a) $\sqrt{y - 2} = w$ (b) $\sqrt[3]{2y + a} = b$

60. (a) $\sqrt{5y + 1} = a$ (b) $\sqrt{\dfrac{y}{2} + 1} = t$

Objectives

1. Order Real Numbers and Use Interval Notation
2. Solve Linear Inequalities
3. Solve Absolute Value Inequalities
4. Solve Nonlinear Polynomial Inequalities
5. Solve Rational Inequalities
6. Solve Applied Problems

1.3 Inequalities

In this section, we review methods for solving different types of inequalities. We begin by examining the notion of intervals.

Ordering Real Numbers and Using Interval Notation

A number line enables us to establish *order* relationships between real numbers. If point a lies to the left of point b on a number line, we say that a is *less than b* (or equivalently, b is *greater than a*) and we write

$$a < b \quad \text{(or } b > a) \quad \text{(Figure 1)}$$

For example, $5 > 3$ and $-8 < -5$.

Figure 1

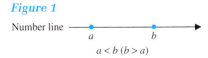

The order relationship is defined more formally as follows:

<table>
<tr><td>**Definition**
Order</td><td>Assume that a and b are real numbers. We say that a **is less than** b, or equivalently, b **is greater than** a, if $b - a$ is a positive number. This **order** is denoted by the **inequality** $a < b$ (or $b > a$).</td></tr>
</table>

For instance, -5 is less than 2, since $2 - (-5) = 7$ (which is positive); so we write

$$-5 < 2 \text{ or } 2 > -5.$$

The inequality symbol $\leq$ means **less than or equal to,** whereas the symbol $\geq$ means **greater than or equal to.**

Also, $a > 0$ means that a is **positive,** and $a < 0$ means that a is **negative.**

Certain sets of numbers, defined in terms of the order relation, can be expressed in a special notation called **interval notation.** Interval notation falls into two categories: (1) *bounded intervals* and (2) *unbounded intervals.*

Bounded Intervals

If a and b are real numbers, with $a < b$, there are four types of **bounded intervals** consisting of real numbers between a and b. Table 1 lists the four types and shows three equivalent ways of denoting each type—one by using regular inequality notation, another by using a graphical representation, and a third method using **interval notation.** In each case, $a < b$.

TABLE 1   Bounded Interval Notation

Terminology for Bounded Intervals	Inequality Notation	Number Line Representation	Interval Notation
Open interval	$a < x < b$		(a, b)
Closed interval	$a \leq x \leq b$		$[a, b]$
Half-open interval	$a \leq x < b$		$[a, b)$
Half-open interval	$a < x \leq b$		$(a, b]$

The numbers a and b are called the **end points** of the closed interval $[a, b]$ from a to b. The brackets indicate the *inclusion* of a and b in the interval. The parentheses for the open interval (a, b) from a to b indicate the *exclusion* of the end points a and b.

Illustrations of bounded intervals, including the three ways of denoting each interval, are given in Figure 2 (a, b and c).

Figure 2

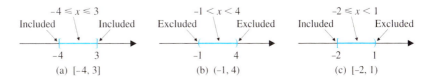

(a) $[-4, 3]$ (b) $(-1, 4)$ (c) $[-2, 1)$

Figure 3

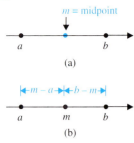

(a)

(b)

At times, we need to locate the midpoint m of a bounded interval with end points a and b where $a < b$ (Figure 3a). Since the distance between a and m is equal to the distance between m and b (Figure 3b), it follows that

$$m - a = b - m$$

$$2m - a = b \qquad \text{Add } m \text{ to each side}$$

$$m = \frac{b + a}{2} \quad \text{or} \quad \frac{a + b}{2} \qquad \text{Solve for } m$$

Thus, we have the following property.

Property

Midpoint of a Bounded Interval

Think of the midpoint of the interval as the "average" of the endpoints.

> If a and b are real numbers, where $a < b$, then the **midpoint** of the interval bounded by a and b is given by
>
> $$m = \frac{a + b}{2}.$$

For instance, the midpoint of interval $[-4, 3]$ is given by

$$\frac{-4 + 3}{2} = -\frac{1}{2};$$

the midpoint of interval $[-7, -1]$ is given by

$$\frac{(-7) + (-1)}{2} = -4 \text{ ; and}$$

the midpoint of interval $[w, 2]$ is given by

$$\frac{w + 2}{2}.$$

Unbounded Intervals

If a is a real number, there are five types of **infinite**, or **unbounded**, intervals consisting of real numbers as listed in Table 2.

Note that ∞ ("infinity") and −∞ are convenient symbols; they are not real numbers.

TABLE 2   Unbounded Interval Notation

Inequality Notation	Number Line Representation	Interval Notation
$a < x$	⟵(⟶ _a_	(a, ∞)
$a \leq x$	⟵[⟶ _a_	$[a, \infty)$
$x < a$	⟵)⟶ _a_	$(-\infty, a)$
$x \leq a$	⟵]⟶ _a_	$(-\infty, a]$
$-\infty < x < \infty$	⟵ ⟶ 0	$(-\infty, \infty)$

The interval $(0, \infty)$ consists of the positive real numbers, and the interval $(-\infty, 0)$ consists of the negative real numbers. Note that the interval $(-\infty, \infty)$ is also represented by $\mathbb{R}$, the set of all real numbers.

EXAMPLE 1 **Illustrating Unbounded Intervals**

Illustrate the given intervals on a number line and describe them with inequality notation.

(a) $[2, \infty)$ (b) $(-\infty, 3)$

(c) all numbers in both interval $[3, \infty)$ and interval $(-1, 5]$

Solution   (a) and (b) Figures 4a and 4b show the intervals for parts (a) and (b) along with their inequality descriptions.

(c) By examining Figure 4c we see that the values common to both of the given intervals lie in the interval where they overlap, that is, they are in interval $[3, 5]$. These are the values of x that satisfy

$$3 \leq x \leq 5$$

Figure 4

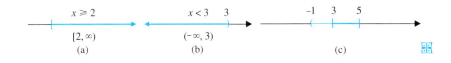

$$x \geq 2 \qquad\qquad x < 3 \quad 3 \qquad\qquad -1 \quad 3 \quad 5$$
$$[2, \infty) \qquad\qquad (-\infty, 3) \qquad\qquad$$
$$(a) \qquad\qquad\qquad (b) \qquad\qquad\qquad (c)$$

Solving Linear Inequalities

Inequalities such as

$$3x + 2 \leq 5 \quad \text{and} \quad 5x - 2 \geq 7x + 11$$

are called **linear inequalities** in one variable. To **solve** a linear inequality, we convert to an equivalent form in which the unknown is isolated on one side of the inequality by employing the same reasoning we used when solving a linear equation.

The process of converting to equivalent inequality forms is based on the following properties.

Properties of Inequalities

Assume a, b, and c are real numbers.

The Addition and Subtraction Properties:

(i) If $a < b$, then $a + c < b + c$ and $a - c < b - c$

The Multiplication and Division Properties:

(ii) If $a < b$ and $c > 0$, then

$$ac < bc \quad \text{and} \quad \frac{a}{c} < \frac{b}{c}$$

(iii) If $a < b$ and $c < 0$, then

$$ac > bc \quad \text{and} \quad \frac{a}{c} > \frac{b}{c}$$

To *graph* the solution of an inequality on a number line is to locate the set of all real numbers that satisfy the inequality.

EXAMPLE 2   **Solving Linear Inequalities**

Solve each inequality, express the solution in interval notation, and graph it on a number line.

(a) $2x - 5 > 3$ (b) $2 - 3x \leq 11$ (c) $-5 \leq 3x + 1 < 7$

Solution

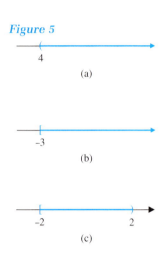

Figure 5

(a)

(b)

(c)

(a) $2x - 5 > 3$ Given

$\quad 2x > 8$ Add 5 to each side; Property (i)

$\quad\ x > 4$ Divide each side by 2; Property (ii)

So the solution includes all real numbers greater than 4.
The interval notation of the solution is $(4, \infty)$, and the graph is shown in Figure 5a.

(b) $2 - 3x \leq 11$ Given

$\quad -3x \leq 9$ Subtract 2 from each side; Property (i)

$\quad\ x \geq -3$ Divide each side by -3 and reverse the inequality sign; Property (iii)

So the solution consists of all real numbers greater than or equal to -3.
The interval notation of this solution is $[-3, \infty)$. Its graph is shown in Figure 5b.

(c) $-5 \leq 3x + 1 < 7$ Given

$\quad -6 \leq 3x < 6$ Subtract 1 from all three parts; Property (i)

$\quad -2 \leq x < 2$ Divide each term by 3; Property (ii)

The solution consists of all real numbers between -2 and 2, including -2 but excluding 2. In interval notation, the solution is $[-2, 2)$, and its graph is shown in Figure 5c.

Solving Absolute Value Inequalities

To solve inequalities involving absolute values we use the following properties:

Properties of Inequalities Involving Absolute Values

If $p > 0$, then:

(i) $|u| < p$ if and only if $-p < u < p$ (Figure 6a)

(ii) $|u| > p$ if and only if $u < -p$ or $u > p$ (Figure 6b)

Figure 6

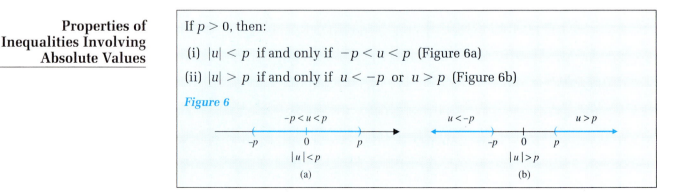

These properties also apply if the signs $<$ and $>$ are replaced by $\leq$ and $\geq$, respectively.

EXAMPLE 3   **Solving Absolute Value Inequalities**

Solve each absolute value inequality, write the solution in interval notation, and graph it on a number line.

(a) $|x - 1| \le 2$ (b) $|2x - 3| > 5$

Solution   (a) To solve $|x - 1| \le 2$, we apply Property (i) with $x - 1$ in place of u, and 2 in place of p to get

$$-p \le u \le p \qquad \text{Property (i)}$$
$$-2 \le x - 1 \le 2 \qquad \text{Substitute}$$
$$-1 \le x \le 3 \qquad \text{Add 1 to each part}$$

Thus, the solution consists of all numbers in the interval $[-1, 3]$ (Figure 7a).

(b) We apply Property (ii) with $2x - 3$ in place of u, and 5 in place of p to get the following inequalities:

$$u < -p \quad \text{or} \quad u > p \qquad \text{Property (ii)}$$
$$2x - 3 < -5 \quad \Big| \quad 2x - 3 > 5 \qquad \text{Substitute}$$
$$2x < -2 \quad \Big| \quad 2x > 8$$
$$x < -1 \quad \Big| \quad x > 4 \qquad \text{Solve the results}$$

Therefore, the solution consists of all numbers in *either* of the intervals $(-\infty, -1)$ or $(4, \infty)$ (Figure 7b).

Figure 7

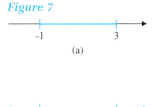

(a)

(b)

Solving Nonlinear Polynomial Inequalities

Inequalities such as

$$x^2 + 2x < 15, \quad x^2 \ge 2x + 3 \quad \text{and} \quad x(x + 4)(x - 1) \le 0$$

are examples of **nonlinear polynomial inequalities.**

    A polynomial inequality is in *standard form* when it is written so that 0 is on one side of the inequality. Thus standard forms for the above inequalities are, respectively,

$$x^2 + 2x - 15 < 0,$$
$$x^2 - 2x - 3 \ge 0,$$

and

$$x(x + 4)(x - 1) \le 0$$

Note that an inequality can always be converted to standard form by adding (or subtracting) an appropriate expression on both sides of the inequality. When solving nonlinear polynomial inequalities, we use standard forms.

    Nonlinear polynomial inequalities can be solved algebraically by a strategy which we shall refer to as the **cut-point method.** The step-by-step procedure for this method is given on the following page.

Later, we'll study another method for solving such inequalities that makes use of graphs of equations.

Cut points are also referred to as critical numbers.

> **Cut-Point Method: Step-by-Step Procedure:**
>
> **Step 1 Rewrite:** Write the given inequality in standard form.
>
> **Step 2 Replace** the inequality symbol with = and solve the related equation.
>
> **Step 3 Separate:** Mark the number solutions found in Step 2 on the number line. These numbers divide the number line into intervals. We call these numbers the *cut-points*.
>
> **Step 4 Test:** Select a *test number* from *within* each interval. Substitute this test number in the original inequality.
>
> (i) If the resulting inequality is true, then all numbers in the interval are to be included in the solution.
>
> (ii) If the resulting inequality is not true, then all numbers in the interval are excluded from the solution.
>
> **Step 5 Complete the Solution:** Consider the numbers found in Step 2. These numbers are included in the solution if the symbol is ≤ or ≥, and excluded if the symbol is < or >.

At this point, we can solve quadratic inequalities because we are able to find all real solutions of any quadratic equation.

EXAMPLE 4   Solving a Quadratic Inequality

Solve the given inequality by the cut-point method.

$$6x^2 + 7x \le 3$$

Express the solution in interval notation and display this solution on a number line.

Solution   We carry out the cut-point procedure as follows:

Step 1. Rewrite: First we rewrite the inequality in standard form by subtracting 3 on each side of the inequality to get the standard form

$$6x^2 + 7x - 3 \le 0.$$

Step 2. Replace: In the standard form, we replace the symbol ≤ with = to obtain the associated equation

$$6x^2 + 7x - 3 = 0.$$

Next we solve this equation by the factor method to get

$$(2x + 3)(3x - 1) = 0$$

so that

$$x = -\frac{3}{2} \quad \text{or} \quad x = \frac{1}{3}.$$

Step 3. **Separate:** Next we mark the numbers $-3/2$ and $1/3$ on a number line. These cut-point numbers divide the number line into three intervals denoted by A, B, and C (Figure 8).

Figure 8

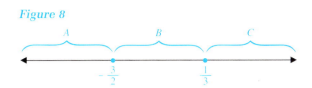

Step 4. **Test:** (See Table 3.) In this step we select a number from within each interval say, -2 in A, 0 in B, and 1 in C. Then we test each number to determine if it satisfies the inequality established in Step 1. Notice that in this situation we are trying to find values of x that make the expression $6x^2 + 7x - 3$ negative. If the test number satisfies the inequality, then all values in the interval satisfy it. On the other hand, if the test number does not satisfy the inequality, then no values in that interval will satisfy it.

TABLE 3

Interval	Test Number	Test Value of $6x^2 + 7x - 3$	Sign of $6x^2 + 7x - 3$	Conclusion
$A: x < -\dfrac{3}{2}$	-2	$6(-2)^2 + 7(-2) - 3 = 7$	$+$	Does not satisfy
$B: -\dfrac{3}{2} < x < \dfrac{1}{3}$	0	$6 \cdot 0^2 + 7 \cdot 0 - 3 = -3$	$-$	Satisfies
$C: \dfrac{1}{3} < x$	1	$6 \cdot 1^2 + 7 \cdot 1 - 3 = 10$	$+$	Does not satisfy

Step 5. **Complete the Solution:** Since the inequality is *non-strict* ($\leq$), then the numbers $-3/2$ and $1/3$ are also included in the solution. So that the solution of the inequality includes all values of x such that

$$-\frac{3}{2} \leq x \leq \frac{1}{3}$$

or in interval notation,

$$\left[-\frac{3}{2}, \frac{1}{3}\right] \quad \text{(Figure 9)}.$$

Figure 9

The cut-point method can also be applied to solve polynomial inequalities of degree higher than two such as

$$x^3 + x^2 - 12x < 0 \quad \text{and} \quad x^4 - 3x^3 - 4x^2 \geq 0$$

The process is facilitated if the polynomial inequality can be expressed in factored form as the next example illustrates. A method for handling nonfactored higher-degree polynomial inequalities is covered in Chapter 3.

EXAMPLE 5   **Solving a Nonlinear Polynomial Inequality**

Solve the given inequality by the cut-point method. Express the answer in interval notation

$$x^3 + 3x^2 > 4x$$

Solution   We carry out the steps in the cut-point method as follows:

Step 1. **Rewrite:** First we rewrite the inequality in standard form by subtracting $4x$ on each side.

$$x^3 + 3x^2 - 4x > 0$$

Step 2. **Replace:** In this step we replace the symbol $>$ with $=$ to obtain the equation $x^3 + 3x^2 - 4x = 0$, which is solved by factoring as follows:

$$x^3 + 3x^2 - 4x = x(x + 4)(x - 1) = 0$$

Setting each factor equal to zero, we obtain

$$x = 0 \quad \text{or} \quad x + 4 = 0 \quad \text{or} \quad x - 1 = 0.$$

So the solutions are 0, -4, and 1. They are the cut-points.

Figure 10

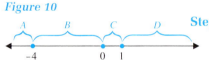

Step 3. **Separate:** Next we mark the numbers 0, -4, and 1 on a number line. These cut-point numbers divide the number line into four intervals denoted by A, B, C, and D (Figure 10).

Step 4. **Test:** (See Table 4.) Next we select a test number from each interval, say -5 in A, -1 in B, $1/2$ in C, and 2 in D. We are seeking all values of x that make the expression

$$x^3 + 3x^2 - 4x = x(x + 4)(x - 1)$$

a positive number.

TABLE 4

Interval	Test Number	Test Value of $x(x + 4)(x - 1)$	Sign of $x(x + 4)(x - 1)$	Conclusion
$A: x < -4$	-5	$-5(-1)(-6) = -30$	$-$	Does not satisfy
$B: -4 < x < 0$	-1	$-1(3)(-2) = 6$	$+$	Satisfies
$C: 0 < x < 1$	$\dfrac{1}{2}$	$\dfrac{1}{2}\left(\dfrac{9}{2}\right)\left(-\dfrac{1}{2}\right) = -\dfrac{9}{8}$	$-$	Does not satisfy
$D: 1 < x$	2	$2(6)(1) = 12$	$+$	Satisfies

Figure 11

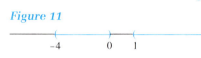

Step 5. **Complete the Solution:** In this situation, since we have a *strict* inequality ($>$), the cut-points are excluded. Thus the solution includes all real numbers x such that $-4 < x < 0$ or $x > 1$. In interval notation, the solution includes all numbers in intervals $(-4, 0)$ or $(1, \infty)$ (Figure 11).

Solving Rational Inequalities

Inequalities involving rational expressions such as

$$\frac{x - 1}{x - 2} \geq 0 \quad \text{and} \quad \frac{3x + 1}{x - 2} < 2$$

can be solved by modifying Step 2 and Step 5 in the cut-point method described above. In this case, the cut-points in Step 2 are the numbers that make *either* the numerator or denominator equal to zero for the rational expression resulting from Step 1 of the process. In Step 5 we only consider the cut-points determined from the numerator. The next example illustrates the procedure.

EXAMPLE 6   **Solving a Rational Inequality**

Solve the given inequality by the cut-point method.

$$\frac{3x + 1}{x - 3} < 2$$

Express the solution in interval notation and show the solution set on a number line.

Solution   We carry out the steps in the cut-point method as follows:

Step 1. **Rewrite:** First we convert to standard form by subtracting 2 from each side to get

$$\frac{3x + 1}{x - 3} - 2 < 0.$$

Next we combine terms on the left side to obtain

$$\frac{3x + 1 - 2(x - 3)}{x - 3} = \frac{x + 7}{x - 3} < 0.$$

Figure 12

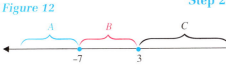

Step 2. **Replace:** Here we set the numerator and denominator equal to zero in order to obtain the cut-points. Thus

$$x + 7 = 0 \qquad \text{or} \qquad x - 3 = 0$$
$$x = -7 \qquad\qquad\quad x = 3$$

Step 3. **Separate:** Next we mark the numbers -7 and 3 on a number line. These cut-point numbers divide the number line into three intervals denoted by A, B, and C (Figure 12).

Step 4. **Test:** (See Table 5.) Next we select a test number from each interval, say -8 in A, 0 in B, and 4 in C. We are seeking all values of x that make the expression $(x + 7)/(x - 3)$ negative.

TABLE 5

Interval	Test Number	Test Value of $\dfrac{x + 7}{x - 3}$	Sign of $\dfrac{x + 7}{x - 3}$	Conclusion
$A: x < -7$	-8	$\dfrac{-8 + 7}{-8 - 3} = \dfrac{1}{11}$	$+$	Does not satisfy
$B: -7 < x < 3$	0	$\dfrac{0 + 7}{0 - 3} = -\dfrac{7}{3}$	$-$	Satisfies
$C: x > 3$	4	$\dfrac{4 + 7}{4 - 3} = 11$	$+$	Does not satisfy

Figure 13

Step 5. **Complete the Solution:** Because this is a strict inequality, the solution excludes -7, the numerator's cut-point. So the solution includes all real numbers x such that $-7 < x < 3$, or in interval notation, we have $(-7, 3)$ (Figure 13).

Other rational inequalities of the form

$$\frac{x^2 - 3x + 2}{x^2 + 6x + 9} \geq 0$$

can also be solved by the cut-point-method.

First we factor the numerator and the denominator as follows:

$$\frac{(x - 1)(x - 2)}{(x + 3)^2} \geq 0$$

Notice that, since $(x + 3)^2 > 0$ for all $x \neq -3$, we need only apply the method to the polynomial in the numerator.

In reference to Step 5 in the process, it should be pointed out that the real solutions for the equation $(x - 1)(x - 2) = 0$, 1 and 2 satisfy the *equality* part of the original inequality and must be included in the final solution of the inequality. However, since division by zero is never allowed, the denominator's cut-point, -3, which is the solution of $(x + 3)^2 = 0$, is not included in the final solution of the given inequality.

Solving Applied Problems

Inequalities are often used in solving applied problems. In the following models, we continue to use the guidelines on page 47 to understand, organize, analyze, and interpret information about models from different fields. Here, we use inequalities to describe some of these models. For instance, wildlife biologists use a model such as

$$50t^2 + 200t + 250 \geq 5050$$

to predict the time t when the population of a certain endangered species will be at least 5050. Other models can also be described by inequalities.

EXAMPLE 7   **Modeling a Travel Problem with a Linear Inequality**

On a recent trip across the country, a jetliner flew from San Francisco to Detroit at an average speed of 500 miles per hour. It then continued to New York at an average speed of 550 miles per hour. If the entire trip covered at most 3075 miles and the total flying time was 6 hours, what was the time spent on the first leg of the trip?

Solution   This trip consists of two legs, first, San Francisco to Detroit and, second, Detroit to New York. Let t represent the time (in hours) traveled on the first leg. Since the total time was 6 hours, it follows that $6 - t$ represents the time (in hours) traveled on the second leg. This information is summarized in the following table.

	Speed r (miles per hour)	Time t (hours)	Distance = rt (miles)
First leg of the trip	500	t	$500t$
Second leg of the trip	550	$6 - t$	$550(6 - t)$

Since the total distance traveled is at most 3075, we have

$$\left(\begin{array}{c}\text{distance from}\\\text{San Francisco to}\\\text{Detroit is } 500t\end{array}\right) + \left(\begin{array}{c}\text{distance from Detroit}\\\text{to New York}\\\text{is } 550(6-t)\end{array}\right) \left(\begin{array}{c}\text{at most}\\\leq\end{array}\right) \left(\begin{array}{c}\text{total distance is}\\3075\end{array}\right)$$

Thus we have

$$500t + 550(6-t) \leq 3075$$

$$500t + 3300 - 550t \leq 3075 \qquad \text{Distribute}$$

$$\left.\begin{array}{r}-50t \leq -225\\ t \geq 4.5\end{array}\right\} \quad \text{Solve for } t$$

So at least 4.5 hours (that is, 4 hours and 30 minutes) was spent on the first leg of the trip.

EXAMPLE 8 Modeling a Fireworks Problem with a Nonlinear Inequality

A fireworks shell is launched from a mortar on a fireboat with an initial speed of 64 feet per second. The height h (in feet) of the shell after t seconds from being launched is given by the model

$$h = 64t - 16t^2.$$

During what time interval will the shell be at least 48 feet above the ground?

Solution When the shell is at least 48 feet above the ground, the time t will satisfy the inequality

$$h = 64t - 16t^2 \geq 48.$$

By using the cut-point method to solve the inequality, we find that the solution includes all values of t such that $1 \leq t \leq 3$ or in interval notation, $[1, 3]$.

Thus the shell will be at least 48 feet above ground level during the time interval from 1 second up to and including 3 seconds.

◈ PROBLEM SET 1.3

Mastering the Concepts

In problems 1–4, express each statement using inequality notation.

1. (a) -2 is less than or equal to 3
   (b) -5 is greater than -7
2. (a) x is at most 4
   (b) t is at least -8
3. (a) x lies to the left of a and to the right of 6
   (b) x is nonnegative
4. (a) x is less than 4, and y is at least 4
   (b) x is greater than -5, and y is at most -5

In problems 5 and 6, illustrate each interval on a number line and describe it using inequality notation.

5. (a) $(0, 6)$
   (b) $[-1, 4)$
   (c) $(1, \infty)$
   (d) $(-\infty, 5]$
6. (a) $[1, 7)$
   (b) $(-2, 3]$
   (c) $(-\infty, 0)$
   (d) $[4, \infty)$

In problems 7–16, solve each inequality and express the solution in interval notation. Also graph the solution on a number line.

7. (a) $2x \leq -4$
   (b) $-2x \leq 4$
8. (a) $2x > -6$
   (b) $-2x > 6$

9. (a) $3x < 9$
   (b) $4x + 3 \geq 12$
10. (a) $5x \geq 15$
    (b) $3x - 2 > 7$
11. (a) $-21w \leq -63$
    (b) $-8t - 4 \leq -16 - 2t$
12. (a) $-4x > -12$
    (b) $5 + x < -x + 3$
13. $2 < 16 + 4x < 10$
14. $-5 \leq 3x + 2 < 4$
15. $\dfrac{x}{3} + 2 \leq \dfrac{x}{4} - 2x$
16. $\dfrac{x + 6}{5} \leq 4 - \dfrac{3x}{7}$

In problems 17–26, solve each absolute value inequality. Express the solution in interval notation and graph it on a number line.

17. (a) $|x| \leq 3$
    (b) $|x| > 2$
18. (a) $|x| < 1$
    (b) $|x| > 4$
19. (a) $|x - 2| < 3$
    (b) $|x - 2| \geq 3$
20. (a) $|x + 1| \leq 2$
    (b) $|x + 1| > 2$
21. (a) $|x + 2| \geq 5$
    (b) $|x + 2| < 5$
22. (a) $|x + 1| > 7$
    (b) $|x + 1| \leq 7$
23. $|2x + 3| < 1$
24. $|3x - 6| \leq 4$
25. $|5x - 3| \geq 2$
26. $|5x - 3| < 2$

In problems 27–34, solve each nonlinear inequality by the cut-point method. Express the solution in interval notation and display it on a number line.

27. (a) $x^2 - x - 2 < 0$
    (b) $x^2 + 2x \geq 3$
28. (a) $x^2 - x - 6 \geq 0$
    (b) $x^2 + 5x < -6$
29. $-x^2 \leq 3x - 40$
30. $31x - 10x^2 \geq -14$
31. $x(x - 1)(x + 2) < 0$
32. $x^3(x - 1)(x + 2) \geq 0$
33. $x^5 - 9x^3 \geq 0$
34. $x^4 - 6x^3 + 5x^2 \leq 0$

In problems 35–38, solve each rational inequality by the cut-point method. Express the solution in interval notation and graph it on a number line.

35. $\dfrac{x - 1}{x - 4} < 0$
36. $\dfrac{x + 3}{x - 3} > 0$
37. $\dfrac{x + 1}{x - 3} \leq 1$
38. $\dfrac{x + 2}{x - 1} \geq 2$

Applying the Concepts

39. **Temperature Range:** The average daily range of temperatures T for a city during a summer month varies from at least 72°F to no more than 98°F.
    (a) Use an inequality to describe the temperature range for one day.
    (b) If the average temperature decreases 1°F per day, use an inequality to describe the temperature at the end of one week.

40. **Electrical Energy Consumption:** The average American home uses at least 90 but no more than 120 kilowatt-hours of electricity per month.
    (a) Use an inequality to express the average range of kilowatt-hours per day, assuming that one month is 30 days.
    (b) Use an inequality to express the average range of kilowatt-hours per hour; per week; per year.

41. **Plumber's Wages:** A plumber charges $18 per hour plus $30 per service call for home repair work. How long must the plumber work to earn at least $84 for a service call?

42. **Investment:** Suppose that $5600 is invested in a mutual fund that pays simple interest for 1 year. If an investor wishes to earn at least $280 in interest for the year, what interest rate is required?

43. **Diet Program:** A clinic advertises that a person can reduce his or her weight by at least 1.5 pounds per week by exercising and special dieting. At the beginning of the diet program, a person weighs 195 pounds. What is the maximum number of weeks before this person's weight will be reduced to 170 pounds?

44. **Business Profit:** A tire manufacturing company shows a profit; that is, the revenue obtained from the sales of its tires is greater than the cost of producing them. If each tire costs $30 in materials, and if it costs $7500 a week in labor and overhead to produce and sell the tires, how many tires must be sold each week at $85 each in order for the company to show a profit?

45. **Stock-Brokerage:** A stockbroker uses a computer program to be alerted on the price changes of stocks. Suppose that a certain stock is selling for $18.25 per share and the price of the stock changes by at least $1.25. The prices x

that would set the alert are given by the inequality

$$|18.25 - x| \geq 1.25.$$

Solve this inequality and interpret the solution.

46. **Body Temperature:** During and after surgery, a patient's temperature T (in degrees Fahrenheit) is given by

$$|T - 98.6| \leq 2.3.$$

Determine the interval over which the patient's temperature varies.

47. **Cellular Phone Sales:** In a certain region, the daily revenue R (in dollars) of cellular phones is given by the model

$$R = 128x - 0.1x^2, x \geq 0.$$

where x is the quantity sold per day. Determine the number of cellular phone units to be sold per day to generate a daily revenue of at least $40,960.

48. **Height of Olympic Arrow:** In the 1992 Summer Olympics in Spain, the Olympic flame was lit by a flaming arrow. As the arrow moved x feet horizontally from the archer, its height h (in feet) was approximated by the model.

$$h = -0.002x^2 + 0.8x + 6.6, x \geq 0$$

Find the horizontal distance of the arrow from the archer so that the height of the arrow is at least 76.6 feet above ground level. Round off the answer to one decimal place.

Developing and Extending the Concepts

In problems 49 and 50, two given sets of numbers are combined to form a third set by using the *set operations* of **union** and **intersection.** If A and B are sets of numbers, then the set consisting of all numbers that belong to A or to B (or to both) is called the **union of A and B** and is written as $A \cup B$. The set of all numbers that are in both A and B is called the **intersection of A and B** and is written as $A \cap B$. For example, if we are given intervals $(-\infty, 2]$ and $(1, 9)$, then

$$(-\infty, 2] \cup (1, 9) = (-\infty, 9) \qquad \text{(Figure 14a)}$$
$$(-\infty, 2] \cap (1, 9) = (1, 2] \qquad \text{(Figure 14b)}$$

Figure 14a

$(-\infty, 2] \cup (1, 9) = (-\infty, 9)$

Figure 14b

$(-\infty, 2] \cap (1, 9) = (1, 2]$

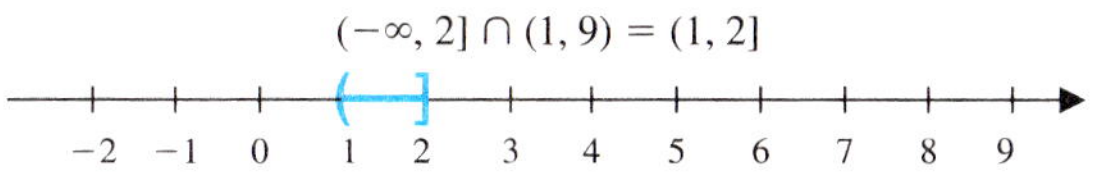

Rewrite each set using inequality notation and graph the set on a number line.

49. (a) $[0, \infty) \cap [-1, 2)$
    (b) $(-2, 4) \cup (3, 8]$
50. (a) $(-\infty, 2) \cup (-\infty, 5)$
    (b) $(-\infty, 2) \cap (-\infty, 5)$
51. Insert $=$ or $<$ between the expressions

$$|x + y| \quad \text{and} \quad |x| + |y|$$

so that the result is true for the given conditions. Give examples to support each assertion.
    (a) $x > 0, y > 0$
    (b) $x < 0, y < 0$
    (c) $x < 0, y > 0$
    (d) $x > 0, y < 0$
52. The **triangle inequality** asserts that

$$|a + b| \leq |a| + |b|$$

for any real numbers a and b. Use the triangle inequality to verify each of the following statements:
    (a) If $|x| < 3$ and $|y| < 1$, then

$$|x + y| < 4.$$

    (b) If $|x - y| < \dfrac{1}{10}$ and $|y - t| < \dfrac{1}{10}$, then

$$|x - t| < \dfrac{1}{5}.$$

53. Explain why each of the following statements is true.
    (a) If $x^3 \leq 0$, then $x \leq 0$.
    (b) The inequality $x^4 + x^2 + 5 < 0$ has no solution.
    (c) The solution of $\dfrac{x^2 + 4}{x^2} > 0$ includes all nonzero numbers.
    (d) The inequality $\dfrac{x}{x + 1} > \dfrac{4x + 3}{x + 1}$ has no solution.

Objectives

1. Plot Points in the Plane
2. Find the Distance and the Midpoint between Two Points
3. Graph Equations by Point-Plotting
4. Determine the Symmetry of a Graph
5. Use Technology to Graph
6. Solve Applied Problems

In applications, the axes may have different labels. For instance, to explore the relation between time t and the distance d traveled by a car, we might label the horizontal axis as the t axis and the vertical axis as the d axis.

1.4 The Cartesian Coordinate System

A number line associates points on the line with real numbers. We now extend this concept by introducing the *Cartesian coordinate system* (also called the *rectangular coordinate system*), which enables us to associate points in the plane with *ordered pairs* of real numbers.

Plotting Points in the Plane

Figure 1 displays the **Cartesian, or rectangular, coordinate system.** It consists of two perpendicular number lines: one horizontal (positive portion to the right and negative portion to the left) and one vertical (positive portion above and negative portion below), called the **coordinate axes.** These axes intersect at a point called the **origin,** which is the zero point on each number line. The horizontal axis is often labeled as the **x axis** and the vertical as the **y axis.** A plane endowed with a Cartesian coordinate system is called a **Cartesian plane** or the **xy plane.**

Any **ordered pair** of real numbers (a, b) can be represented as a point P in the coordinate system. The first number, a, which is called the **x coordinate** (or *abscissa*) of P, indicates the location of P relative to the x axis. The second number, b, which is called the **y coordinate** (or *ordinate*) of P, indicates the location of P relative to the y axis. The point P with coordinates (a, b) is called the **graph** of the ordered pair (a, b). We plot, or locate, the position of (a, b) in the plane by placing a dot at the point P (Figure 1). The association between point P and ordered pair (a, b) seems so natural that in practice we write

$$P = (a, b)$$

The coordinate axes divide the plane into four regions called **quadrants,** which are described and displayed in Figure 2.

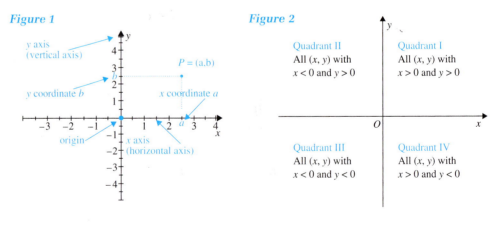

Figure 1

Figure 2

Quadrant II
All (x, y) with $x < 0$ and $y > 0$

Quadrant I
All (x, y) with $x > 0$ and $y > 0$

Quadrant III
All (x, y) with $x < 0$ and $y < 0$

Quadrant IV
All (x, y) with $x > 0$ and $y < 0$

Figure 3

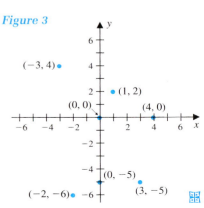

EXAMPLE 1   **Plotting Points**

Plot the points $(1, 2)$, $(-3, 4)$, $(-2, -6)$, $(4, 0)$, $(0, 0)$, $(0, -5)$, and $(3, -5)$, and specify their quadrant locations.

Solution   Figure 3 above shows that the points $(1, 2)$, $(-3, 4)$, $(-2, -6)$, and $(3, -5)$ lie, respectively, in quadrants I, II, III, and IV. The points $(4, 0)$, $(0, 0)$, and $(0, -5)$ lie on the coordinate axes.

Finding the Distance and the Midpoint between Two Points

The distance between two points $P_1 = (x_1, y_1)$ and $P_2 = (x_2, y_2)$ can be found in terms of their coordinates by using the following formula:

The Distance Formula

If $P_1 = (x_1, y_1)$ and $P_2 = (x_2, y_2)$ are two points in a Cartesian plane, then the **distance** d between P_1 and P_2 is given by

$$d = d(P_1, P_2) = \sqrt{(x_2 - x_1)^2 + (y_2 - y_1)^2}$$

Proof   To establish this formula, we construct a right triangle $P_1P_2P_3$ by connecting P_1 and P_2 and drawing line segments from P_1 and P_2 parallel to the x and y axis, respectively, as shown in Figure 4. We see that the horizontal distance between P_1 and P_3 is given by

Figure 4

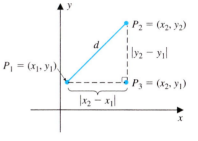

$$|\overline{P_1P_3}| = |x_2 - x_1|$$

and the vertical distance between P_2 and P_3 is given by

$$|\overline{P_2P_3}| = |y_2 - y_1|$$

By applying the Pythagorean theorem, the length d of the hypotenuse is

$$
\begin{aligned}
d = d(P_1, P_2) &= \sqrt{|\overline{P_1P_3}|^2 + |\overline{P_2P_3}|^2} \\
&= \sqrt{|x_2 - x_1|^2 + |y_2 - y_1|^2} \\
&= \sqrt{(x_2 - x_1)^2 + (y_2 - y_1)^2}
\end{aligned}
$$

Note that the formula is true even if the two given points line up vertically or horizontally. Also, the distance between two points is the same regardless of which point is chosen as (x_1, y_1) or (x_2, y_2).

EXAMPLE 2   **Finding the Distance between Two Points**

Find the distance between the two given points. Round off the answer to two decimal places.

(a) $P_1 = (-1, -2)$ and $P_2 = (3, -4)$

(b) $P = (71.37, -27.04)$ and $Q = (14.86, 11.73)$

Solution   (a) The distance between P_1 and P_2 is given by

$$
\begin{aligned}
d(P_1, P_2) &= \sqrt{[3 - (-1)]^2 + [-4 - (-2)]^2} \\
&= \sqrt{4^2 + (-2)^2} = \sqrt{20} = 2\sqrt{5} \text{ or approx. } 4.47.
\end{aligned}
$$

(b) The distance between P and Q is given by

$$d(P, Q) = \sqrt{(71.37 - 14.86)^2 + (-27.04 - 11.73)^2} \text{ or approx. } 68.53.$$

We can determine the coordinates of the midpoint of a line segment in a Cartesian plane by using the coordinates of the end points. Figure 5 shows a line segment with end points $P_1 = (x_1, y_1)$ and $P_2 = (x_2, y_2)$.

Let $M = (a, b)$ be the midpoint of $\overline{P_1P_2}$. By extending the notion of the midpoint for two points on a number line (see Section 1.3), the value of a is found by taking the *average* of the x coordinates to get

Figure 5

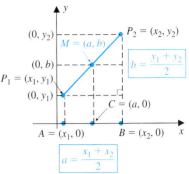

$$a = \frac{x_1 + x_2}{2}.$$

Similarly, the value of b is obtained by using the y coordinates to get

$$b = \frac{y_1 + y_2}{2}.$$

In general, we have the following formula:

The Midpoint Formula

The **midpoint** M of the line segment with end points $P_1 = (x_1, y_1)$ and $P_2 = (x_2, y_2)$ is given by

$$M = \left(\frac{x_1 + x_2}{2}, \frac{y_1 + y_2}{2} \right)$$

EXAMPLE 3   **Calculating a Midpoint**

Find the midpoint of the line segment with end points $(-4, 1)$ and $(3, 5)$.

Solution   Using the midpoint formula, the coordinates of the midpoint M of the line segment are given by

$$M = \left(\frac{x_1 + x_2}{2}, \frac{y_1 + y_2}{2} \right)$$

$$= \left(\frac{-4 + 3}{2}, \frac{1 + 5}{2} \right)$$

$$= \left(-\frac{1}{2}, 3 \right).$$

Graphing Equations by Point-Plotting

Two quantities are sometimes related by means of an equation or formula that involves two variables. Here, we discuss how to represent such an equation geometrically on a Cartesian coordinate system.

Definition
Graph of an Equation

The **graph** of an equation in two variables x and y consists of all points (x, y) in a Cartesian plane whose coordinates x and y satisfy that equation.

At times, the graph of an equation in two variables can be sketched by first plotting some points and then connecting them with a smooth curve. This technique, called the *point-plotting method* of graphing equations, is outlined as follows.

Graphing by
Point-Plotting

Step 1	Determine some ordered pairs of numbers that satisfy the given equation. To do so, it is helpful to express one of the variables in terms of the other, and list the ordered pairs in a table.
Step 2	Plot these ordered pairs as points in a coordinate system.
Step 3	Connect the points to form a curve.

EXAMPLE 4   **Graphing a Linear Equation by Plotting Points**

Sketch the graph of $y = 3x - 4$ by point-plotting.

Solution   To graph $y = 3x - 4$ by point-plotting, we proceed as follows:
First we assign values to x and then substitute them into the equation to determine corresponding y values.
For instance, if we select

$$x = 1, \text{ then } y = 3(1) - 4 = -1.$$

The resulting pair of numbers $(1, -1)$ is listed in Table 1, along with three other pairs of numbers that satisfy the equation. Next we plot the points corresponding to the pairs listed in Table 1 and then we connect them with a smooth curve. The resulting graph turns out to be a straight line (Figure 6).

TABLE 1

x	y
0	−4
1	−1
2	2
3	5

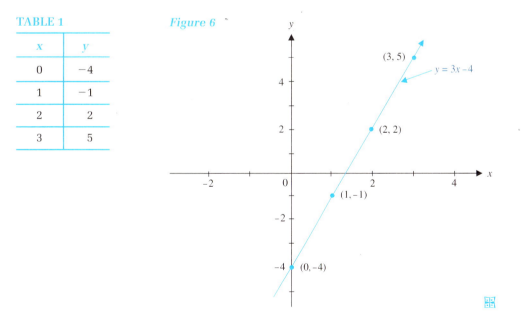

Figure 6

When graphing an equation in two variables, we often need to know the points where the graph intersects the coordinate axes. The x and y coordinates of these points are called **intercepts** as explained in Table 2.

TABLE 2   Intercepts of the Graph of an Equation That Relates x and y

Terminology	Description	Illustration	Determination of Intercepts
x intercepts	The x coordinates of points where the graph intersects the x axis	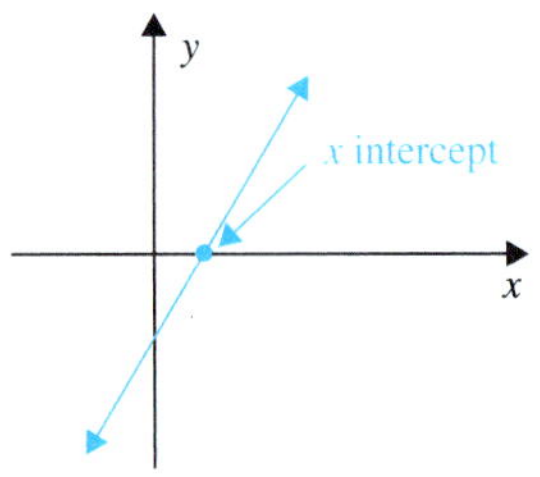	Set $y = 0$ and solve for x
y intercepts	The y coordinates of points where the graph intersects the y axis	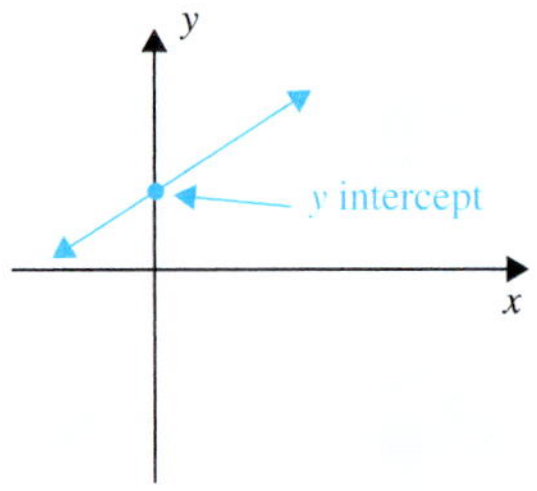	Set $x = 0$ and solve for y

Thus for

$$y = 3x - 4 \text{ (see Example 4),}$$

the x intercept is obtained by setting

$$y = 0$$

to get

$$0 = 3x - 4 \quad \text{or} \quad x = \frac{4}{3}.$$

Similarly, to get the y intercept, we set

$$x = 0$$

to get

$$y = -4,$$

which is shown in Figure 6 on page 79.

EXAMPLE 5   **Graphing a Nonlinear Equation by Point-Plotting**

Use point-plotting to sketch the graph of $y = x^2 - 9$. Also, locate the x and y intercepts.

Solution   Countless ordered pairs of numbers satisfy this equation. A sample of these pairs is determined by substituting arbitrary x values into the equation and calculating the corresponding y values (Table 3). By plotting these ordered pairs as points in a coordinate system and then connecting them with a smooth curve, we obtain the graph (Figure 7).

TABLE 3

x	y
−4	7
−3	0
−2	−5
−1	−8
0	−9
1	−8
2	−5
3	0
4	7

Figure 7

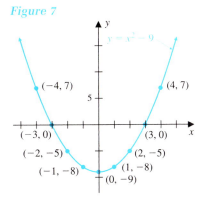

Upon examining Table 3, we see that the x intercepts are −3 and 3, and the y intercept is −9. These intercepts are plotted in Figure 7.

Determining the Symmetry of a Graph

In addition to point-plotting, there are other tools that help us to accurately sketch graphs of equations. For instance, by examining the graph in Figure 7 above, we observe that the number of plotted points to the right of the y axis is equal to the number of the reflecting points on the graph across that axis. Because of this, we say that the graph is symmetric with respect to the y axis.

Graphs that are symmetric with respect to the y axis, the x axis, or the origin are defined as follows:

Definition
Symmetric Graphs

Symmetry with Respect To:	Whenever Point (x, y) Is on the Graph, Then Its Reflection:	Example
y axis	(−x, y) is on the graph	$y = \dfrac{1}{1 + x^2}$ (−x, y) ── (x, y)
x axis	(x, −y) is on the graph	$y^2 = x$ (x, y) (x, −y)
Origin	(−x, −y) is on the graph	$y = x^3 - x$ (x, y) (−x, −y)

To be able to use these types of symmetry as an aid in graphing an equation, we need to know how to recognize when they will occur by examining the equation. Table 4 explains how this is done.

TABLE 4   Equation Tests for Symmetry

Symmetry with Respect To:	Occurs if an Equivalent Equation Is Obtained When:	Example Each Pair of Equations is Equivalent
y axis	x is replaced by $-x$	$y = \dfrac{1}{x^2 + 1}$ and $y = \dfrac{1}{(-x)^2 + 1}$
x axis	y is replaced by $-y$	$y^2 = x$ and $(-y)^2 = x$
Origin	x and y are replaced by $-x$ and $-y$, respectively	$y = x^3 - x$ and $(-y) = (-x)^3 - (-x)$ $-y = -x^3 + x$ $-y = -(x^3 - x)$

EXAMPLE 6   **Using Symmetry as an Aid to Graphing**

Use symmetry to sketch the graph of each equation.

(a) $y = x^2$

(b) $y = x^3$

Solution   (a) Substituting $-x$ for x, we obtain

$$y = (-x)^2 \text{ or } y = x^2$$

which is equivalent to the original equation. So the graph is symmetric with respect to the y axis.

To obtain the graph, we first use point-plotting to sketch the part of the graph in the first quadrant and the origin (Figure 8a). Finally, we use the symmetry with respect to the y axis by reflecting the graph in Figure 8a about the y axis to obtain the complete graph (Figure 8b).

Figure 8

x	y
0	0
1	1
2	4
3	9

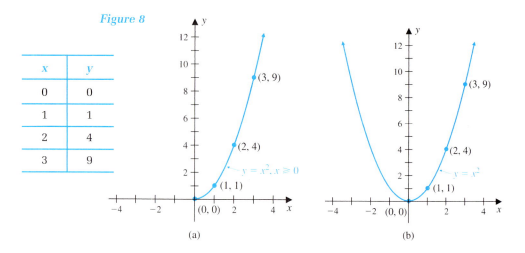

(a)

(b)

(b) The substitution of $-x$ for x and $-y$ for y produces the equation

$$-y = (-x)^3 \text{ or } -y = -x^3.$$

Since this equation is equivalent to the equation $y = x^3$, the graph is symmetric with respect to the origin. Thus, we first sketch the part of the graph in the first quadrant and the origin (Figure 9a). Then we reflect the result about the origin to get the complete graph (Figure 9b).

Figure 9

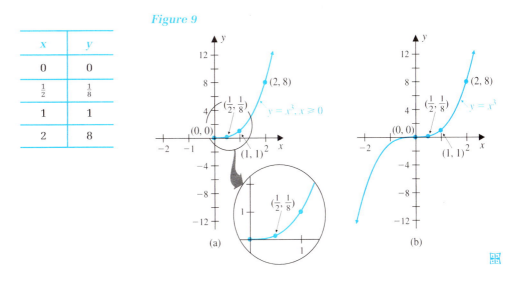

x	y
0	0
$\frac{1}{2}$	$\frac{1}{8}$
1	1
2	8

(a)   (b)

Using Technology to Graph

At times graphing equations by the point-plotting method can present misleading results if not enough points are used. For instance, suppose that we are to graph the equation

$$y = x^4 - 8x^2 - 9$$

and we begin by using the ordered pairs listed in a tabular format. Figure 10a shows a graph obtained by connecting the points from Table 5 with a smooth curve.

However, if we continue to plot several more points we actually get the graph shown in Figure 10b. When Figure 10a is compared with Figure 10b, it is apparent that Figure 10a is not an accurate representation of the graph of $y = x^4 - 8x^2 - 9$.

TABLE 5

Figure 10

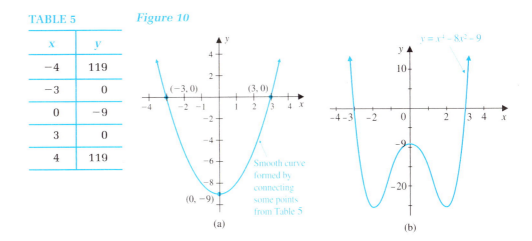

x	y
−4	119
−3	0
0	−9
3	0
4	119

(a)   (b)

Obviously, the more points we plot, the more accurate a graph becomes; however, the task can be very tedious and laborious.

In order to *hand-draw* graphs of complicated equations such as $y = x^4 - 8x^2 - 9$, we rely on more sophisticated methods learned in calculus. At our level of exploration, we will use graphing technology in the form of *graphers* to obtain such graphs.

Graphers make the process of obtaining accurate graphs much easier. The term *grapher* refers to both **graphing calculators** and **computer graphing software**.

These tools generate graphs by plotting several points in close proximity to each other with speed and accuracy.

Throughout the book, we use the symbol whenever a grapher is to be used.

The portion of the graph that is displayed at any one time by a grapher is called a **viewing rectangle** or **viewing window.** Although graphers differ in features (the instruction manual should be consulted), most have a WINDOW menu that allows us to choose the minimum and maximum values of both x and y for the viewing window. Figure 11 illustrates a viewing window selection.

Figure 11

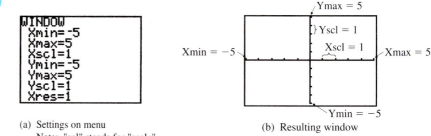

(a) Settings on menu
Note: "scl" stands for "scale"

(b) Resulting window

Figure 12   G

$y = 4.19x^3 - 8.83x^2 + x + 2.43$

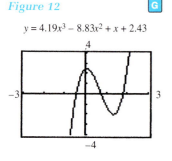

For instance, Figure 12 shows the graph of $y = 4.19x^3 - 8.83x^2 + x + 2.43$ generated by a grapher. The viewing window selections are also displayed.

Throughout this text, all graphs are either hand-drawn or outputs of graphers. Those graphs produced by graphers are enclosed by viewing windows with displayed WINDOW selections, as in Figure 12.

Solving Applied Problems

The next example illustrates the use of graphs in mathematical modeling. Here we plot given data to get a clearer view of the trend in the average annual 30-year fixed mortgage rates from 1995 to 2004 for the nation.

EXAMPLE 7   **Modeling by Using Graphs**

The average percent annual 30-year fixed mortgage rate A (in the United States) for each year n from 1995 through 2004 is listed in Table 6.

TABLE 6

Year n	1995	1996	1997	1998	1999	2000	2001	2002	2003	2004
Average percent 30-year fixed mortgage rate (A)	7.93	7.81	7.60	6.94	7.44	8.05	6.97	6.54	5.83	5.89

(a) Plot the data in the table, using n as the horizontal axis and A as the vertical axis.

Organizing and graphing data are important tools in data analysis.

(b) Connect each consecutive pair of points in part (a) with a line segment. The resulting graph is referred to as a **line graph** of the data. How does the pattern of the line graph relate to the data in Table 6?

Solution   (a) Figure 13a shows the points plotted from the data in Table 6. The plot of the data is called a **scattergram**, or **scatter diagram**.

Figure 13

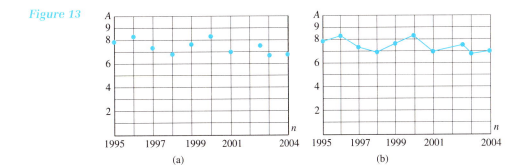

(a)   (b)

The spacing between successive units on the horizontal or vertical axis is called the scale of the axis. Note that the scales on the two coordinate axes in Figure 13 are different from each other.

(b) By connecting the points plotted in Figure 13a consecutively with line segments, we obtain the graph in Figure 13b. The geometric pattern of the line graph shows the trend in the average 30-year fixed mortgage rates from 1995 through 2004.

If two quantities x and y are related by the equation

$$y = \frac{k}{x}$$

where k is a constant, we say that y is **inversely proportional** to x, or y **varies inversely** as x. In the next example, we use a formula called *Boyle's law* to model the inverse relationship between the volume and pressure of a gas.

EXAMPLE 8   **Using an Inverse Variation Model from Physics**

Boyle's law states that under constant temperature, the volume V of a gas is inversely proportional to the pressure P. That is,

$$V = \frac{k}{P} \text{ where } k \text{ is a constant.}$$

(a) Suppose that for a certain gas, $k = 3975$, V is expressed in cubic inches, and P represents pounds per square inch. Express V in terms of P.

(b) What is the restriction on P?

(c) Use point-plotting to graph the equation in part (a) with the restriction in part (b), where P is scaled on the horizontal axis and V is scaled on the vertical axis.

(d) Use the result in part (c) to examine the trend of the volume as the pressure increases.

Solution   (a) Since $V = k/P$ and $k = 3975$, it follows that $V = 3975/P$.

(b) Since P represents pressure, then $P > 0$.

(c) By plotting a few points and then connecting them with a smooth curve, we get the graph displayed in Figure 14.

Figure 14

P	V
10	397.5
20	198.75
40	99.38
100	39.75
500	7.95
1000	3.98

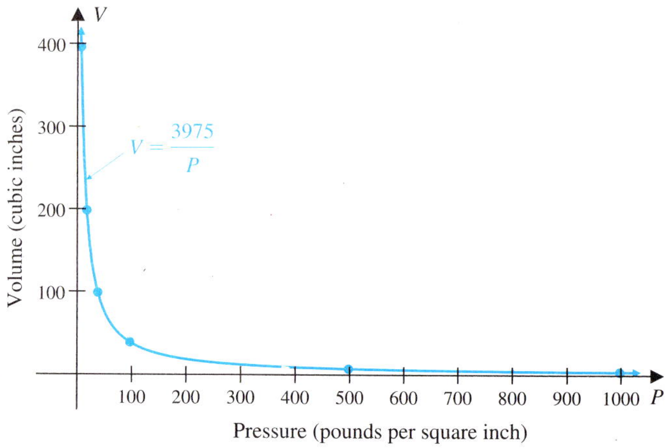

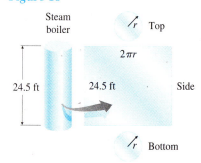

$V = \dfrac{3975}{P}$

Volume (cubic inches)

Pressure (pounds per square inch)

Could the pressure equal 0 in this mode? Could the volume ever equal 0?

(d) Upon examining the graph in Figure 14 we see that as the pressure increases, the volume correspondingly decreases.

EXAMPLE 9

Modeling the Cost of a Steam Boiler

A company manufactures cylindrical-shaped steam boilers in different sizes. The cost of each boiler is \$75 per square foot of its total surface area. Suppose that a boiler is 24.5 feet long and has a radius of r feet (Figure 15).

Figure 15

(a) Construct an equation that expresses the cost C (in dollars) of the boiler in terms of the radius r.

(b) What restriction is there on r?

(c) Adjust the equation in part (a) to express the new cost K in terms of the radius if the price per square foot of the side of the boiler is reduced by 20%.

Solution

(a) We derive the equation that expresses the cost C in terms of the radius r as follows:

$$C = 75 \, [(\text{Area of the side}) + (\text{Area of top}) + (\text{Area of bottom})]$$
$$= 75[(2\pi r \cdot 24.5) + \pi r^2 + \pi r^2]$$
$$= 75 \, (49\pi r + 2\pi r^2)$$
$$= 75\pi(49r + 2r^2)$$
$$= 3675\pi r + 150\pi r^2$$

(b) Since r represents the radius, $r > 0$.

(c) If the side is discounted by 20%, then the new cost K becomes

$$K = C - (20\% \text{ of cost of the side})$$
$$= 75\pi(49r + 2r^2) - 0.20(75 \cdot 49\pi r)$$
$$= 75\pi(49r + 2r^2) - 735\pi r$$
$$= 3675\pi r - 735\pi r + 150\pi r^2$$
$$= 2940\pi r + 150\pi r^2.$$

G   A viewing window of the graph of the equation in part (a) is displayed in Figure 16a. The pattern of the graph visually conveys the fact that the bigger the radius, the more costly the boiler.

Figure 16b displays a viewing window of the graph of this equation in part (c) above. Upon comparing the two graphs, we *see* that as the radius increases, the cost with the discount increases at a slower rate than the cost without the discount.

Figure 16

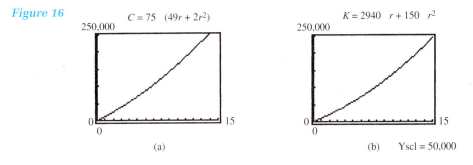

(a)   (b)   Yscl = 50,000

◈ PROBLEM SET 1.4

Mastering the Concepts

In problems 1 and 2, plot the points on the same coordinate system and indicate which quadrant or coordinate axis contains each point.

1. $(3, 4), (-2, 4) (\sqrt{5}, -1), (0, \sqrt{7})$

2. $(3, -\sqrt{2}), (0, -8/3), (-2, -\sqrt{11}), (-\pi, \sqrt[3]{71})$

In problems 3–8:

(i) Use the distance formula to find $d(A, B)$, the distance between the points A and B. Round off the numerical problems to two decimal places.

(ii) Use the midpoint formula to find the midpoint of the line segment $\overline{AB}$.

3. (a) $A = (1, 1), B = (-3, 2)$
     (b) $A = (-2, 5), B = (3, -1)$

4. (a) $A = (1, -2), B = (7, 10)$
     (b) $A = (2, 3), B = (-1/2, 1)$

5. (a) $A = (3, -4), B = (-5, -7)$
     (b) $A = \left(-\dfrac{2}{5}, \dfrac{1}{5}\right), B = \left(\dfrac{1}{5}, \dfrac{3}{5}\right)$

6. (a) $A = (5, -t), B = (7, 5)$
     (b) $A = (t, 8), B = (t, 7)$

7. (a) $A = (t, u + 1), B = (t + 1, u)$
     (b) $A = (-3.65, 8.22), B = (4.73, 5.32)$

8. (a) $A = \left(w, \dfrac{w}{4}\right), B = (3w, w)$
     (b) $A = (-6.13, 1.87), B = (5.25, -4.28)$

In problems 9–16, use the point-plotting method to sketch the graph of each equation. Also, locate the x and y intercepts.

9. $y = -2x + 3$

10. $y = \dfrac{1}{3}x - 2$

11. $y = 3x^2 + 1$

12. $y = -\dfrac{1}{2}x^2 + 4$

13. $y = \sqrt{x} - 3$

14. $y = \sqrt{-x} + 3$

15. $y = -x^3 - 1$

16. $y = 4x^3 - 4$

In problems 17–20, plot the point P in the Cartesian plane and then locate and give the coordinates of the following points:

(a) Q, symmetric to P with respect to the x axis

(b) R, symmetric to P with respect to the y axis

(c) S, symmetric to P with respect to the origin

17. $P = (1, 4)$

18. $P = (-4, -2)$

19. $P = (-5, 2)$

20. $P = (3, -1)$

In problems 21–24, part of the graph of an equation is displayed as shown. Sketch the complete graph assuming the graph is symmetric to the

(a) x axis   **(b)** y axis   **(c)** origin

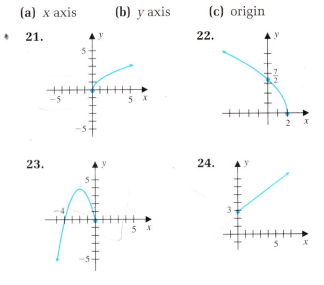

21.

22.

23.

24.

In problems 25–32, use the equation tests in Table 5 to determine symmetry with respect to the x axis, y axis, or origin. Then use the symmetry, if it exists, to sketch the graph of each equation.

25. $y = |x| + 2$ **26.** $y = 3|x| - 1$

27. $x = y^2 + 2$ **28.** $x + 1 = 2y^2$

29. $y = \sqrt{9 - x^2}$ **30.** $y = -\sqrt{9 - x^2}$

31. $y = \sqrt[3]{x}$ **32.** $y = 2x^3 + x$

In problems 33–38, match each equation with its graph.

33. $y = 2x^2 - 1$ **34.** $y = \sqrt{x} - 1$

35. $y = \sqrt{x - 1}$ **36.** $x = -y^2 + 1$

37. $y = -x^3 + x$ **38.** $y = x^3 - 2x$

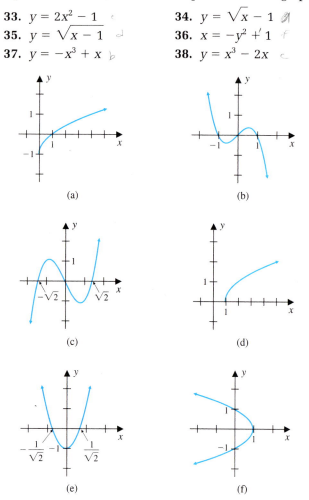

(a)   (b)

(c)   (d)

(e)   (f)

G In problems 39 and 40, use a grapher to sketch the graph of the given equation in each of the specified viewing windows. Describe the differences between the resulting graphs in parts (a) and (b). Which of the two graphs appear to be more accurate? Explain.

39. $y = x^3 - 3x^2 + 1$
  (a) Xmin: -2; Xmax: 2; Xscl: 1;
      Ymin: -2; Ymax: 8; Yscl: 1
  (b) Xmin: -5; Xmax: 5; Xscl: 1;
      Ymin: -5; Ymax: 5; Yscl: 1

40. $y = -x^3 + 7x + 10$
  (a) Xmin: -5; Xmax: 5; Xscl: 1;
      Ymin: -20; Ymax: 10; Yscl: 1
  (b) Xmin: -5; Xmax: 5; Xscl: 1;
      Ymin: -5; Ymax: 20; Yscl: 1

G In problems 41 and 42, use a grapher to sketch the graph of each equation.

41. (a) $y = 2x^2 + 4x - 6$
      (b) $y = x^3 + 4x^2 - 11x - 30$

42. (a) $y = x^2 + 2x - 35$
      (b) $y = 4x^3 + x$

G In problems 43–48, use a grapher to sketch the graph of each equation. Use the graph to decide whether the resulting curve is symmetric with respect to the x axis, y axis, or origin. Verify the answer by using the equation tests for symmetry given in Table 5.

43. $y = 1/x^3$ **44.** $y = x^3 + x$

45. $y = 5x - x^2$ **46.** $y = x^2 + 3x$

47. $y = 3x^2 + 1/x^2$ **48.** $y = x^2 - 1/x^2$

Applying the Concepts

In problems 49–52, data are given for a specific situation.

(a) Construct a scattergram for the data. The variable in the top row of each table represents the horizontal axis, and the variable in the second row represents the vertical axis.

(b) Also draw a line graph for the data (see Example 7). Use the line graph to describe the trend indicated in each table.

49.

Temperature, T Degrees Fahrenheit	28	31	34	35	37	39	41	43
Humidity, H Percent		42	44	48	50	54	57	58	61

50.

Pressure, P Pounds per Square Inch	20	25	30	35	40	45	50	55
Volume, V Cubic Inches	75	60	50	42	37	33	30	25

51.

Miles Driven by a Taxi, M	1	2	3	5	8	9
Fare, F Dollars	1.70	2.90	4.20	6.50	10.10	11.30

52.

Long Distance Time, T Minutes	1	3	6	10	15	30
Charges, C Dollars	0.37	0.60	0.94	1.40	1.95	3.50

In problems 53–56, the mathematical models are based on the notion of **direct variation.** We say that **y is directly proportional to x,** or that **y varies directly as x** if there is a constant k such that

$$y = kx$$

holds for all values of x. The constant k is called the **constant of variation,** or **constant of proportionality.** For instance, the formula for the area A of a circle of radius r, $A = \pi r^2$, indicates that A varies directly as r^2.

53. Engineering: Automobile Skid Marks: The speed V at which an automobile stops when its brakes are applied varies directly as the square root of the length d of the skid marks. Suppose that skid marks from a car involved in an accident measured 88 feet when the car had been going 40 miles per hour.
  (a) Find a mathematical model that expresses V in terms of d.
  (b) Estimate the speed of a car if its skid marks are 173 feet. Was this car exceeding the speed limit of 55 miles per hour?
  (c) Graph the equation in part (a) with appropriate restrictions on d.
  (d) Use the result in part (c) to discuss the relationship between the stopping speed and the length of the skid marks.

54. Astronomy: Planetary Motion: Kepler's third law of planetary motion states that the square of time t (in days) required for a planet to make one revolution around its sun varies directly as the cube of the planet's average distance d (in millions of miles) from its sun.
  (a) Construct a mathematical model that expresses t in terms of d if $t = 365$ when $d = 93$.
  (b) Graph the equation in part (a) with appropriate restrictions on d.
  (c) Use the result in part (b) to discuss the relationship between the time required for one revolution around the sun and the distance from the sun.

55. Pollution: The amount of pollution A (in tons) entering the atmosphere varies directly as the number of people N living in a certain area.

Suppose a population of 50,000 people generates 35,000 tons of pollutants entering the atmosphere.
  (a) Write a mathematical model that expresses A in terms of N.
  (b) Predict how many tons of pollutants enter the atmosphere in a city of 600,000 people.
  (c) Graph the result in part (a) with appropriate restrictions on N.
  (d) Use the result in part (c) to explain the relationship between the amount of pollution and the population.

56. Manufacturing: A manufacturer of graphers determines that the daily cost C (in dollars) varies directly as the number of units n produced, where $3000 \le n \le 6000$.
  (a) Find a mathematical model that expresses C in terms of n if the cost is \$263,500 when 4200 units are produced.
  (b) What will the cost be if the production is increased to 5600 units?
  (c) Graph the equation in part (a) with the given restrictions on n.
  (d) Use the result in part (c) to explain the relationship between the cost and number of graphers produced.

57. Physics: Ohm's law states that the current I (in amperes), voltage E (in volts), and resistance R (in ohms) in a simple electric circuit are related by the formula

$$I = \frac{E}{R}.$$

  (a) What restriction should be imposed on R?
  (b) If the voltage is 110, graph the resulting equation with the restriction from part (a).
  (c) What is the value of the current if the resistance is 8 ohms? 10 ohms? 12 ohms?
  (d) If the trend continues, discuss the relationship between the current and the resistance.

58. Advertising: An advertising agency determines that its revenue R (in thousands of dollars) is given by the model

$$R = 50x^2 - 200x + 800$$

where x represents the number of clients.
  (a) What restriction should be imposed on x?
  (b) Sketch the graph of the equation that represents this model using the restriction from part (a).

(c) How much revenue will be generated if the agency has 150 clients? 200 clients? 500 clients?

(d) If the trend continues, discuss the relationship between the revenue and number of clients.

59. Ice Cream Cone: An ice cream machine dispenses soft ice cream to fill a cone-shaped shell with a height h equal to twice the radius r (in inches) of its base (Figure 17).

Use $V = \dfrac{1}{3}\pi r^2 h$

Figure 17

$h = 2r$

(a) Express the volume V of the ice cream cone in terms of r.

(b) What restriction is there on r?

(c) Determine the volume if the radius is 1 inch; 1.5 inches; 2 inches. Round off each answer to two decimal places.

60. [G] (Refer to problem 59.)

(a) Use a grapher to graph the equation in part (a) with the restriction in part (b).

(b) Use the graph to discuss the trend of the volume as the radius increases.

61. Cost of a Walkway: The length and width of a backyard are 50 feet and 40 feet, respectively. A homeowner wishes to construct a concrete walk of uniform width x feet that borders the yard and to sod the interior with grass. Suppose that the cost of laying concrete is \$5 per square foot and the cost of sodding grass is \$3 per square foot.

(a) Draw a sketch of the backyard and concrete border. Use the sketch to determine the restrictions on x.

(b) Construct an equation that expresses the total cost C (in dollars) of the concrete and sod in terms of x.

(c) Adjust the equation in part (b) to express the cost K in terms of x if the price per square foot of the concrete is reduced by 10%.

62. [G] (Refer to problem 61.)

(a) Use a grapher to graph the equation in part (b) with the restrictions in part (a).

(b) Use a grapher to graph the equation in part (d) and compare it to the graph in part (c). Use the graphs to compare the trends of the cost as the width x increases.

Developing and Extending the Concepts

63. Assume $a > 0$ and $b < 0$. State the quadrant in which each point lies.

(a) (a, b) (b) $(a, -b)$

(c) $(b, -a)$ (d) $(-a, -b)$

64. The center of a rectangle is at the origin. If the four sides are parallel to the coordinate axes and if the coordinates of two vertices are $(-4, 3)$ and $(4, 3)$, find the coordinates of the other two vertices.

65. (a) Which of the three points $P = (1, 5)$, $Q = (2, 4)$, and $R = (3, 3)$ is the farthest from the origin?

(b) Suppose that we interchange the x and y coordinates of each point in part (a). That is, suppose that $P_1 = (5, 1)$, $Q_1 = (4, 2)$, and $R_1 = (3, 3)$. Is the answer the same as the answer in part (a)? Explain.

66. Use the distance formula to determine whether the triangle ABC with vertices $A = (-5, 1)$, $B = (-6, 5)$, and $C = (-2, 4)$ is an isosceles triangle (a triangle with two sides of equal length).

67. Suppose that P_1, P_2, and P_3 are points in the plane such that P_2 lies on the line segment $\overline{P_1 P_3}$. Then P_1, P_2, and P_3 are said to be **collinear.** This condition exists if and only if

$$d(P_1, P_3) = d(P_1, P_2) + d(P_2, P_3)$$

(a) Draw a diagram to illustrate whether the points $P_1 = (-3, -2)$, $P_2 = (1, 2)$, and $P_3 = (3, 4)$ are collinear.

(b) Use the distance formula to confirm the observation of part (a).

68. Use the distance formula and the Pythagorean theorem to determine whether $P = (-4, 1)$, $Q = (6, -1)$, and $R = (-3, 6)$ are the vertices of a right triangle.

69. Graph the following four equations on the same coordinate system.

$$y_1 = \sqrt{x} \quad y_2 = \sqrt{-x} \quad y_3 = -\sqrt{x} \quad y_4 = -\sqrt{-x}$$

Describe the symmetric relationships among the graphs.

70. Given the equation $ax^2 + by^2 = 1$, where a and b are positive constants:

(a) In what quadrants will the graph lie?

(b) Explain why it is sufficient to graph the equation in quadrant I and then use symmetry to get the complete graph.

(c) Select specific values for a and b to illustrate the situation.

71. Given the equation $y^2 = x^2$:

(a) Test the equation for symmetry.

(b) Graph the equation.

72. (a) Explain why the graph of $y^2 = 1 - x$ is different from the graph of $y = \sqrt{1 - x}$.

(b) Explain why the graph of $y^3 = 1 - x$ is the same as the graph of $y = \sqrt[3]{1 - x}$.

(c) Use the results from parts (a) and (b) to compare the graph of $y^n = 1 - x$ to the graph of $y = \sqrt[n]{1 - x}$ according to whether n is a positive even integer or a positive odd integer.

Objectives

1. Graph Linear Equations
2. Find and Interpret the Slope of a Line
3. Find Equations of Lines
4. Graph Equations of Circles
5. Solve Applied Problems
6. Use Curve Fitting–Linear Regression

1.5 Straight Lines and Circles

Proficiency in graphing equations by hand is a major objective in this text. Besides point-plotting, it is possible at times to sketch graphs efficiently by being familiar with certain equation forms. Our goal in this section is to recognize equations whose graphs are straight lines or circles and to determine equations of lines and circles with given characteristics.

Graphing Linear Equations

The graph of the first-degree equation

$$Ax + By = C$$

where A and B are constants, with A and B not both equal to 0, is a straight line. Conversely, any straight line in a Cartesian coordinate system has an equation of this form. For this reason, such equations are called **linear equations,** and x and y are said to be **linearly related.** For instance,

$$y = 2x, \quad y = 3, \quad x = 2, \quad \text{and} \quad 3x - 2y = 6$$

We use the terms straight line and line interchangeably.

are examples of linear equations.

We know from geometry that a straight line is defined by any two of its points. Consequently, we can sketch the graph of a linear equation by plotting any two points and then drawing a line through them.

EXAMPLE 1   **Graphing a Linear Equation**

Graph the linear equation

$$3x - 4y = 12$$

and locate the x and y intercepts.

Solution   First we find two pairs of numbers that satisfy the equation.

For instance, if we select $x = 0$, then $3(0) - 4y = 12$

$$-4y = 12$$
$$y = -3$$

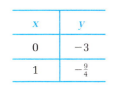

x	y
0	-3
1	$-\frac{9}{4}$

Figure 1

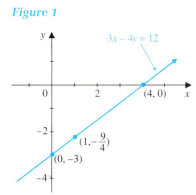

Similarly, if $x = 1$ is selected, then $3(1) - 4y = 12$

$$3 - 4y = 12$$
$$-4y = 9$$
$$y = -\frac{9}{4}$$

Next we display the two pairs in a table, plot the two points, and then draw a line through them to obtain the graph of the equation (Figure 1).

To find the x intercept, we set $y = 0$ to get $3x = 12$ or $x = 4$.

The y intercept was found above when we set $x = 0$ to get $y = -3$. Both intercepts are shown in Figure 1.

Figure 2

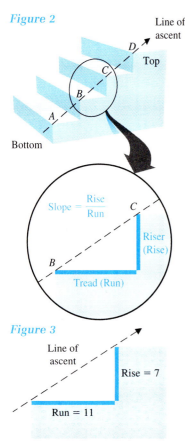

Finding and Interpreting the Slope of a Line

A key geometric feature of a straight line is how steeply it rises or falls as we move horizontally from left to right. To measure this steepness, we use an important characteristic called the *slope* of a line.

The idea of the slope of a line is nicely illustrated by examining a straight flight of stairs. Each step of the stairway consists of a **riser** and a tread (or **run**) (Figure 2).

The **slope** of the line of ascent is given by the ratio

$$\text{Slope} = \frac{\text{Length of riser}}{\text{Length of tread}}$$

or simply

$$\text{Slope} = \frac{\text{Rise}}{\text{Run}}.$$

Figure 3

For instance, if a stairway has risers of length 7 inches and treads of length 11 inches (Figure 3), the slope of the line of ascent is given by

$$\frac{\text{Rise}}{\text{Run}} = \frac{7}{11}.$$

This illustration leads us to the general definition of the slope of a line.

Definition

Slope of a Line

If $P_1 = (x_1, y_1)$ and $P_2 = (x_2, y_2)$ are two different points on a nonvertical line, the **slope** m of the line is given by

$$m = \frac{\text{Change in } y}{\text{Change in } x} = \frac{\Delta y}{\Delta x} = \frac{y_2 - y_1}{x_2 - x_1} \quad \text{(Figure 4)}$$

where the symbol Δy is read as "delta y" and Δx is read as "delta x."

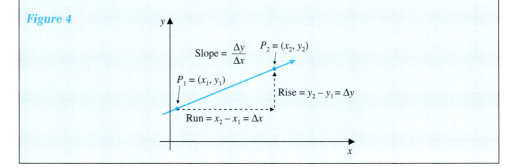

Figure 4

Note that since the slope measures the ratio of the amount of change in y to the change of x from x_1 to x_2, it is interpreted as the *average rate of change* of y with respect to x.

Also the slope of a line is the same no matter which two distinct points are selected to compute its value (problem 65).

EXAMPLE 2

Finding and Interpreting a Slope

Sketch the line containing points P_1 and P_2, find the slope, and interpret the slope in terms of the direction of the line.

(a) $P_1 = (-2, -1)$ and $P_2 = (2, 5)$ 　　　(b) $P_1 = (-1, 3)$ and $P_2 = (4, 1)$

Solution

The graphs and slopes, along with a description of the direction of each line, are given in Table 1.

TABLE 1

Given Points	Sketch of Line	Slope of Line, $m = \Delta y/\Delta x$	Direction of Line
(a) $P_1 = (-2, -1)$; $P_2 = (2, 5)$	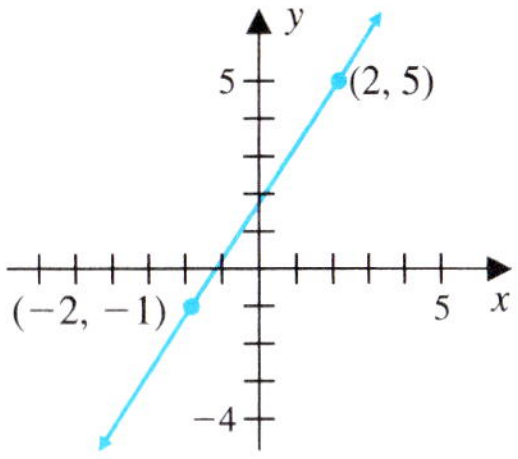	$\dfrac{5 - (-1)}{2 - (-2)} = \dfrac{5 + 1}{2 + 2} = \dfrac{6}{4} = \dfrac{3}{2}$	Rises (to the right)
(b) $P_1 = (-1, 3)$; $P_2 = (4, 1)$	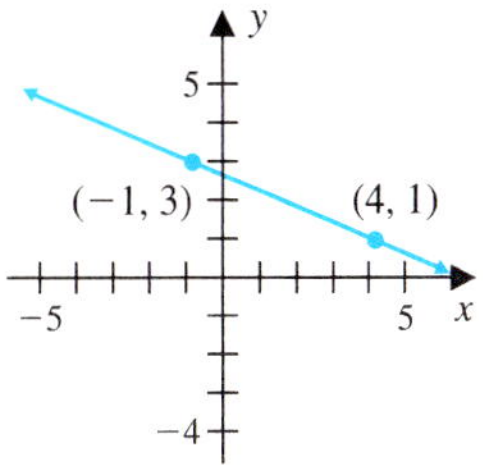	$\dfrac{1 - 3}{4 - (-1)} = \dfrac{-2}{5}$	Falls (to the right)

If either $A = 0$ or $B = 0$ in the first-degree equation $Ax + By = C$, we obtain equations of the form

$$y = b \quad \text{or} \quad x = a,$$

respectively, where b and a are constants. Table 2 shows an example of each situation.

TABLE 2

Equation	Graph	Slope of Line	Direction of Line
(a) $y = 4$	$\begin{array}{c\|c} x & y \\ \hline -2 & 4 \\ 3 & 4 \end{array}$	$\dfrac{4 - 4}{3 - (-2)} = \dfrac{0}{5} = 0$	Horizontal
(b) $x = 4$	$\begin{array}{c\|c} x & y \\ \hline 4 & 3 \\ 4 & -2 \end{array}$	$\dfrac{3 - (-2)}{4 - 4} = \dfrac{5}{0}$ (Undefined)	Vertical

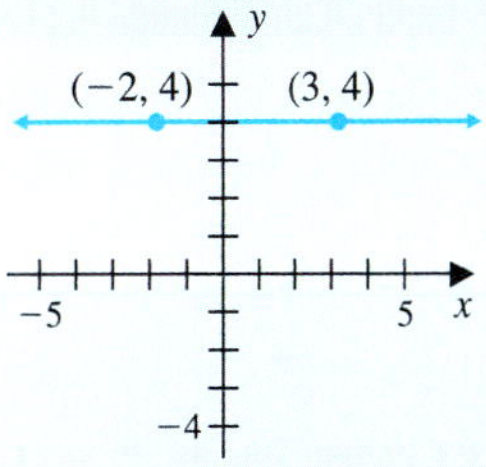

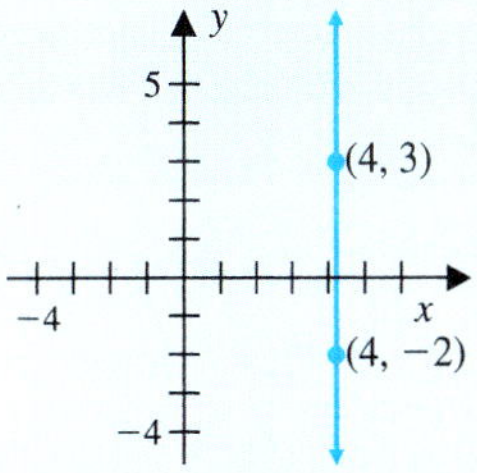

The results shown in Tables 1 and 2 are generalized in Table 3.

TABLE 3 **Slope and Direction of a Line**

Slope Value (m)	Line Direction	Sample Graph
Positive	Rises from left to right	$m > 0$
Negative	Falls from left to right	$m < 0$
Zero	Horizontal	$m = 0$
Not defined	Vertical	m undefined

Finding Equations of Lines

Given sufficient information about a line in the Cartesian system, we can find an equation whose graph is that line. Here we shall investigate two standard forms of equations of lines:

1. Point–slope form,

    and

2. Slope–intercept form.

1. The Point–Slope Equation of a Line

Suppose that we are given the slope m and a point $P_1 = (x_1, y_1)$ on a nonvertical line (Figure 5). If $P = (x, y)$ represents any other point on the line, then the slope of the line containing P_1 and P is given by

Figure 5

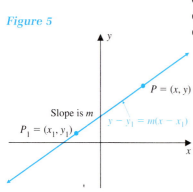

$$\frac{y - y_1}{x - x_1} = m$$

So
$$y - y_1 = m(x - x_1)$$

Any point $P = (x, y)$ whose coordinates satisfy the above equation lies on the line. Conversely, any point that lies on the line has coordinates that satisfy the equation. Therefore, we have the following:

Point–Slope Equation of a Line

An equation for the line that contains the point $P_1 = (x_1, y_1)$ and has the slope m is given by

$$y - y_1 = m (x - x_1)$$

This equation is referred to as a **point–slope equation of the line.**

EXAMPLE 3   **Finding a Point–Slope Equation of a Line**

Graph the line that contains point (2, 3) and has slope 5, and then find a point–slope equation for the line.

Solution

Figure 6

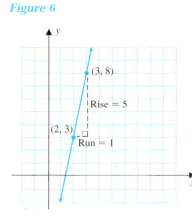

To graph the line, we need to find another point in addition to the given point (2, 3). Since the slope $m = 5 = 5/1$, we can locate a second point by starting at the point (2, 3), and then moving 1 unit to the right (run) and 5 units up (rise) to get point (3, 8).

Then the graph is obtained by drawing a line through points (2, 3) and (3, 8), as shown in Figure 6.

Substituting $x_1 = 2$, $y_1 = 3$, and $m = 5$ into the point–slope equation

$$y - y_1 = m(x - x_1)$$

we get

$$y - 3 = 5(x - 2).$$

2. The Slope–Intercept Equation of a Line

Suppose that a nonvertical line has slope m. Since the line is not parallel to the y axis, it must intersect this axis at some point $(0, b)$ (Figure 7). In other words, b is the y intercept of the line. Applying a point–slope equation with $(x_1, y_1) = (0, b)$, we get

Figure 7

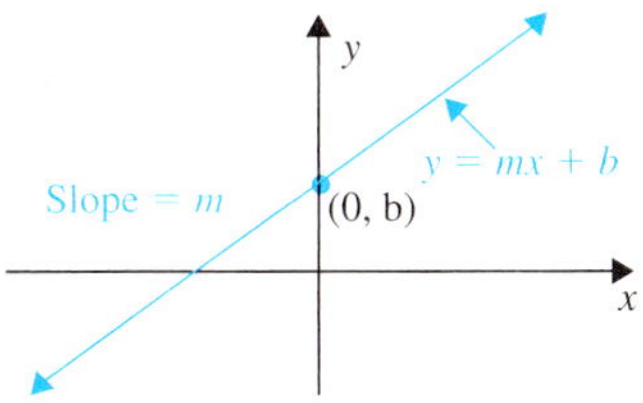

Slope = m $y = mx + b$ (0, b)

$$y - b = m(x - 0)$$

or

$$y = mx + b.$$

This latter equation is called the *slope–intercept equation of the line,* and we have the following result:

Slope–Intercept Equation of a Line

The graph of the equation

$$y = mx + b,$$

which is called the **slope–intercept equation of the line,** is a line with slope m and y intercept b.

A nonvertical line can have only one slope–intercept equation, since a line can have only one slope m and one y intercept b.

When an equation of a line is written in slope–intercept form, the slope and y intercept can be determined by inspection. Consequently, we can tell from the equation the direction of the line it represents. Table 4 shows some examples.

TABLE 4   Examples of the Slope–Intercept Form of Linear Equations

Equation	Slope	y Intercept	Direction
$y = 3x + 5$	$m = 3$	$b = 5$	Rises
$y = -2x + 7$	$m = -2$	$b = 7$	Falls
$y = x - 4 = (1)x + (-4)$	$m = 1$	$b = -4$	Rises
$y = 3 = (0)x + 3$	$m = 0$	$b = 3$	Horizontal

By rewriting a linear equation in slope–intercept form, we can easily find its slope and y intercept.

For instance, suppose we are given the linear equation

$$3x + 5y = 20.$$

We solve for y to get the slope–intercept equation

$$y = \left(-\frac{3}{5}\right)x + 4,$$ so the line has slope $-\frac{3}{5}$, y intercept 4, and the line falls.

The next example shows how we can find an equation of a line in a Cartesian system when we are given the location of two of its points.

EXAMPLE 4   **Finding the Slope–Intercept Equation Given Two Points**

Find the slope–intercept equation of the line that contains the points

$$(-2, 5) \quad \text{and} \quad (3, -4).$$

Solution   The slope of the line is given by

$$m = \frac{5 - (-4)}{-2 - 3} = -\frac{9}{5}.$$

Since the point $(-2, 5)$ belongs to the line, we can use $(x_1, y_1) = (-2, 5)$ in a point–slope equation to get

$$y - 5 = -\frac{9}{5}[x - (-2)] \quad \text{or} \quad y - 5 = -\frac{9}{5}(x + 2).$$

Solving the latter equation for y, we obtain

$$y = -\frac{9}{5}x - \frac{18}{5} + 5 \qquad \text{Distribute } -\frac{9}{5} \text{ and add 5 to each side}$$

$$y = -\frac{9}{5}x + \frac{7}{5} \qquad \text{Combine like terms}$$

Thus, the slope is $-\frac{9}{5}$ and the y intercept is $\frac{7}{5}$.

We use the two equation forms of lines given above to determine equations of parallel and perpendicular lines. Since parallel lines have the same direction, they have the same slope. Also, two lines that intersect at a right angle are said to be *perpendicular,* and their slopes satisfy the condition included in the following properties:

Properties

Slopes of Parallel and Perpendicular Lines

Suppose L_1 and L_2 are two distinct nonvertical lines with slopes m_1 and m_2, respectively.

1. L_1 is **parallel** to L_2 if and only if $m_1 = m_2$.

2. L_1 is **perpendicular** to L_2 if and only if $m_1 = -\dfrac{1}{m_2}$,

   or equivantly,

$$m_1 m_2 = -1 \quad \text{(problem 66).}$$

EXAMPLE 5   **Finding Equations of Parallel or Perpendicular Lines**

Let L be a line whose equation is given by $2x - y + 6 = 0$. Find the slope–intercept equation of the line that contains the point $(3, 2)$ and is:

(a) Parallel to L (b) Perpendicular to L

In both situations, graph the two lines in the same coordinate system.

Solution   The slope–intercept equation for L is obtained by solving for y in terms of x to get

$$y = 2x + 6$$

Figure 8

So the slope is $m_1 = 2$ and the y intercept is $b = 6$.

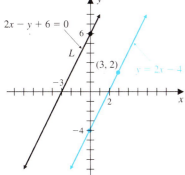

(a) Since parallel lines have the same slope, the required line has slope 2. Thus, a point–slope equation of the line parallel to L and containing the point $(3, 2)$ is given by

$$y - 2 = 2(x - 3)$$

We solve this equation for y to get the slope–intercept equation

$$y = 2x - 4$$

Figure 8a shows the graphs of the two lines.

(b) The slope m_2 of a line perpendicular to L, which has slope $m_1 = 2$, must satisfy the condition

$$m_1 m_2 = -1 \quad \text{or} \quad m_2 = -1/m_1, \quad \text{so} \quad m_2 = -1/2.$$

Hence, a point–slope equation of the line perpendicular to L and containing the point $(3, 2)$ is given by

$$y - 2 = -\frac{1}{2}(x - 3)$$

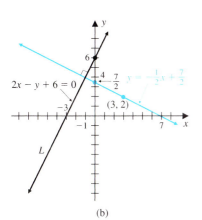

Solving for y, we get the slope–intercept equation

$$y = -\frac{1}{2}x + \frac{7}{2}$$

Figure 8b shows the graphs of the two lines.

Graphing Equations of Circles

It is possible to recognize and graph equations of circles with relative ease. We know from geometry that a circle consists of all points that are at a fixed distance r, its *radius*, from a fixed point C, its *center* (Figure 9a). Suppose that a circle is located in a Cartesian plane so that its center C is (h, k) and its radius is r. If a point $P = (x, y)$ is on the circle, its distance from $C = (h, k)$ has to be r units (Figure 9b).

Figure 9

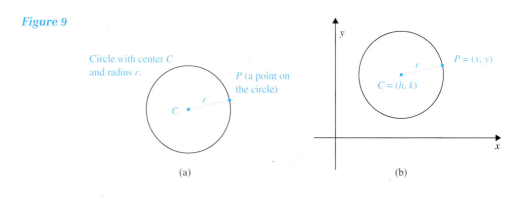

By the distance formula, x and y satisfy the equation

$$\sqrt{(x - h)^2 + (y - k)^2} = r$$

or

$$(x - h)^2 + (y - k)^2 = r^2$$

Conversely, any ordered pair (x, y) that satisfies this last equation defines a point $P = (x, y)$ that lies on the circle with center $C = (h, k)$ and radius r. Thus we have the following general result:

Standard Equation of a Circle

> Any point (x, y) on the circle with center (h, k) and radius r must satisfy the equation
>
> $$(x - h)^2 + (y - k)^2 = r^2$$
>
> The equation is called the **standard equation of the circle.**

A circle with radius r and center at the origin must satisfy the *standard equation*

$$x^2 + y^2 = r^2$$

because $(h, k) = (0, 0)$.

Table 5 lists two examples of standard equations of circles along with their graphs.

TABLE 5   Examples of Graphs of Circles

Equation of a Circle	Center (h, k)	Radius	Graph
$(x - 1)^2 + (y - 2)^2 = 9$	$(1, 2)$	3	
$x^2 + y^2 = 4$ or $(x - 0)^2 + (y - 0)^2 = 4$	$(0, 0)$	2	

A circle whose center is at (0, 0) and has a radius of 1 unit has the standard equation $x^2 + y^2 = 1$ and is called the unit circle.

Given the radius and the center of a circle, we can determine its standard equation as illustrated in the next example.

EXAMPLE 6   **Finding the Standard Equation of a Circle**

Find the standard equation for the circle of radius 4 with center at the point $(4, -1)$ and sketch its graph.

Solution   Substituting 4 for r and $(4, -1)$ for (h, k) in the equation

$$(x - h)^2 + (y - k)^2 = r^2$$

we get

$$(x - 4)^2 + [y - (-1)]^2 = 4^2$$

or

$$(x - 4)^2 + (y + 1)^2 = 16$$

The graph is shown in Figure 10.

Figure 10

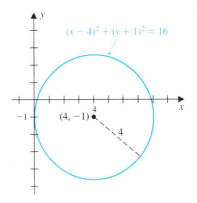

When the equation of a circle is given in an algebraic form that is not recognizable, we rewrite it in standard form by utilizing the technique of completing-the-square, which was presented in Section 1.2. Example 7 illustrates the method.

EXAMPLE 7   **Finding the Center and Radius of a Circle**

Show that the graph of the equation

$$x^2 + y^2 + 6x - 2y - 15 = 0$$

is a circle by first converting it to a standard equation of a circle. Then determine the center and radius and sketch the circle.

Solution   We proceed as follows:

$$(x^2 + 6x) + (y^2 - 2y) = 15$$ Isolate variable terms on one side by grouping the x terms and the y terms

Figure 11

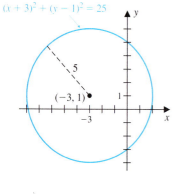

$(x + 3)^2 + (y - 1)^2 = 25$

$$(x^2 + 6x + 9) + (y^2 - 2y + 1) = 15 + 9 + 1$$ Complete the square in each pair of parentheses; add $9 + 1$ to both sides

$\left(\frac{6}{2}\right)^2 \qquad \left(\frac{2}{2}\right)^2$

$$(x + 3)^2 + (y - 1)^2 = 25$$ Factor and simplify

or

$$[x - (-3)]^2 + (y - 1)^2 = 5^2$$ Put in standard form

We recognize the form of this last equation. Its graph is a circle of radius

$$r = 5 \text{ with center } (h, k) = (-3, 1) \text{ (Figure 11)}$$

[G] When graphing an equation in x and y on a grapher, we first "isolate y" by solving the given equation for y in terms of x. Sometimes the process is straightforward. For instance, to graph $2x - y + 3 = 0$, we first solve for y to get $y = 2x + 3$. At times, solving for y may result in two different equations. If this occurs we super-impose two graphs, one for each of the two equations, in the same viewing win-dow in order to get the graph of the original equation.
For instance, to graph

$$x^2 + y^2 = 4$$

Figure 12

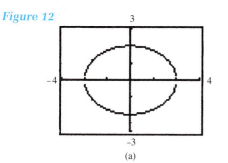

(a)

on a grapher, we first solve for y in terms of x to get

$$y = \pm\sqrt{4 - x^2}$$

Thus we have the two resulting equations

$$y_1 = \sqrt{4 - x^2} \text{ and } y_2 = -\sqrt{4 - x^2}$$

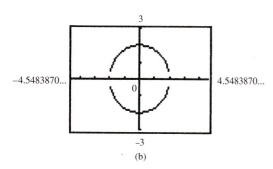

(b)

Figure 12a displays a view window that superimposes both the graph of $y_1 = \sqrt{4 - x^2}$ (above and on the x axis) and also the graph of $y_2 = -\sqrt{4 - x^2}$ (below and on the x axis) on the same coordinate system. These two graphs together constitute the graph of $x^2 + y^2 = 4$, a circle with center $(0, 0)$ and radius 2. Note that the grapher may pro-duce a drawing that is distorted, as is the case in Figure 12a). Here the image is more oval than circular. This type of distortion can usually be corrected by using a feature on the grapher that changes the aspect ra-tio of the screen. The result of such a correction is shown in Figure 12b.

Solving Applied Problems

Many practical situations are modeled by linear equations and their graphs.

EXAMPLE 8   **Modeling the Resale Value of an Automobile**

The resale value R (in dollars) of a certain automobile is modeled by the linear equation

$$R = P(1 - 0.07t)$$

where P is the original price (in dollars) of the automobile and t is the number of years after it was purchased.

(a) Express the resale value R as an equation in terms of t if the original price was $16,000.

(b) Find the resale value of the car after 1 year, 3 years, and 5 years.

(c) Graph the equation from part (a) by representing t on the horizontal axis and R on the vertical axis.

(d) Find and interpret the intercepts.

(e) Find and interpret the slope of the line.

(f) Does this model make sense if the car is sold after 15 years? Explain.

Solution   (a) Since the original price of the automobile was $16,000, we substitute $P = 16,000$ into the given equation to get

$$R = 16,000(1 - 0.07t)$$
$$R = -1120t + 16,000.$$

The latter equation is in slope–intercept form.

(b) After 1 year, $t = 1$, so $R = -1120(1) + 16,000 = 14,880$

After 3 years, $t = 3$, so $R = -1120(3) + 16,000 = 12,640$

After 5 years, $t = 5$, so $R = -1120(5) + 16,000 = 10,400$

Thus, the resale values after 1 year, 3 years, and 5 years are, respectively, $14,880; $12,640; $10,400.

(c) Since t represents elapsed time, it follows that $t \geq 0$. The graph of the linear equation from part (a) with the restriction $t \geq 0$ is displayed in Figure 13.

Figure 13

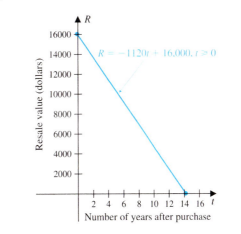

(d) The R intercept is found by setting $t = 0$ to get

$$R = -1120(0) + 16{,}000 = 16{,}000.$$

Because of the way we labeled the vertical and horizontal axes, we refer to the y intercept as the R intercept and the x intercept as the t intercept.

This means that when no time has elapsed, the automobile is new, so its resale value is the same as the original price, $16,000. The t intercept is found by setting $R = 0$ to get

$$0 = -1120t + 16{,}000$$

$$1120t = 16{,}000$$

$$t = \frac{16{,}000}{1120} = 14.29 \text{ (approx.)}.$$

This means that after about 14.29 years the automobile will have no resale value according to this model.

(e) The slope of the line can be read directly from the slope–intercept equation $R = -1120t + 16{,}000$ to be -1120. This means that for each elapsed year (1 unit change in t) there is a reduction of $1120 in value.

(f) If the automobile is more than 15 years old, then according to the model, we have

$$t \geq 15$$

$$-1120t \leq -1120(15) \qquad \text{Multiply each side by } -1120$$

$$-1120t + 16{,}000 \leq -1120(15) + 16{,}000 \qquad \text{Add 16,000 to each side}$$

$$R \leq -800 \qquad \text{Substitute and simplify}$$

If the automobile is still operational after 15 years, it is unlikely that it would have a negative resale value. So the model does not make sense after 14.29 years.

Using Curve Fitting–Linear Regression

Recall from Section 1.4 that scattergrams are valuable tools for envisioning relationships that exist between two variables for given discrete real-world data. At times, it is possible to have linear equations serve as models for such data. If the pattern of the data points in a scattergram visually appears to be linear, then the situation suggests that there is a line that "fits" the data. However, a different visual inspection of the pattern may not produce a unique line to fit the given data. That is, if a line appears to "fit" the data points well, there may be another line that "fits it better." Most graphers are programmed to determine the equation of the line in the form $y = ax + b$, where a and b are constants, that "best fits" the data by utilizing a statistical method called the **least-squares fit.** Such an equation is referred to as a **linear regression model.** Once this equation is determined, we can use it for prediction as illustrated in the next example.

EXAMPLE 9   G **Determining a Linear Regression Model**

A study shows that the percentages of women in state politics throughout the United States is on the rise. The data in the table represent percentages of women in politics from 1994 through 2004.

Year	1994	1995	1996	1997	1998	1999	2000	2001	2002	2003	2004
t	0	1	2	3	4	5	6	7	8	9	10
Percentage P	17.00	17.80	18.3	19.1	20.3	20.7	21.3	21.8	22.4	23.1	23.7

(a) Draw a scatter diagram of the data in which the values of t are located on the horizontal axis and the values of P are on the vertical axis.

(b) Use a grapher to find the linear regression model $P = at + b$ for the line that best fits the data and graph this equation. Round off a and b to three decimal places.

(c) Assuming this trend continues, use the equation from part (b) to predict the percentage of women in politics in the year 2010 to one decimal place.

Solution   (a) To draw a scattergram we plot points from the data given in the table using the percentage P as the vertical axis and the years t as the horizontal axis. From the scattergram in Figure 14a, it appears that a linear relation exists between t and P.

(b) By using a grapher we find the linear regression model to be

$$P = 0.667t + 17.164.$$

It is graphed in Figure 14b along with the scattergram.

Figure 14

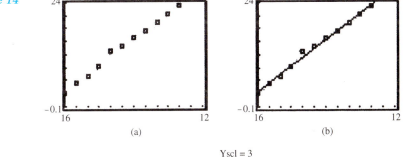

(a)   (b)

Yscl = 3

(c) To predict the percentage of women in politics in the year 2010, we substitute $t = 16$ into the regression equation from part (b) and round off to one decimal place to get $P = 27.8$. So there will be approximately 27.8% women in politics in 2010 according to this model.

◇ PROBLEM SET 1.5

Mastering the Concepts

In problems 1–10, graph each linear equation, and locate the x and y intercepts.

1. $2x + y = 2$

2. $-x + 2y = 4$

3. $y = -2x + 2$

4. $y = \dfrac{1}{2}x + 3$

5. $x = -2$

6. $y = 4$

7. $\dfrac{x}{3} + \dfrac{y}{4} = 1$

8. $\dfrac{y}{2} - \dfrac{3x}{5} = 1$

9. $5x + 8y = 7$

10. $3x - 2y = 1$

In problems 11–16, sketch the line containing points P_1 and P_2, find the slope, and interpret the slope in terms of the direction of the line.

11. $P_1 = (2, 3)$, $P_2 = (-1, -2)$

12. $P_1 = (4, 1)$, $P_2 = (3, -1)$

13. $P_1 = (-1, 7)$, $P_2 = (3, 7)$

14. $P_1 = (0, 1)$, $P_2 = (1, -3)$

15. $P_1 = (-3, -1)$, $P_2 = (-12, 11)$

16. $P_1 = (0, 1)$, $P_2 = (-1, 3)$

In problems 17 and 18, sketch the line that contains the point P and has the slope m and then find a point–slope equation for the line.

17. (a) $P = (4, -3)$, $m = 3$
  (b) $P = (-2, 3)$, $m = 0$

18. (a) $P = (-3, 1)$, $m = -\dfrac{2}{3}$

  (b) $P = (-2, 0)$, $m = -\dfrac{1}{2}$

In problems 19 and 20, find the slope–intercept equation of the line whose graph is given.

19. (a)   (b)

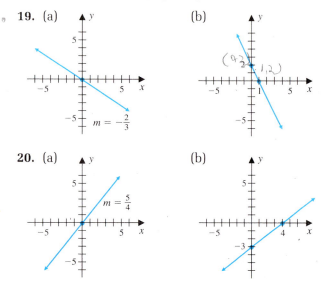

$m = -\dfrac{2}{3}$

20. (a)   (b)

$m = \dfrac{5}{4}$

In problems 21–26, if possible, find the slope–intercept equation of the line that satisfies the given conditions. Then graph the line.

21. Contains points $P_1 = (-2, 5)$ and $P_2 = (2, -3)$
22. Contains points $P_1 = (3, 3)$ and $P_2 = (1, -2)$
23. Contains point $P = (2, 1)$; parallel to the line with equation $x - 3y = 6$
24. Contains point $P = (-1, 4)$; parallel to the line with equation $3x + y = 5$
25. Contains point $P = (-3, 2)$; perpendicular to the line with equation $4x - 2y = 7$
26. Contains point $P = (-3, -5)$; perpendicular to the line with equation $3x + 2y = -2$

In problems 27–30, find the standard equation of the circle with the given characteristics and graph it.

27. (a) Center: $(3, -2)$; radius: 5
  (b) Center: $(-1, 3)$; radius: 2
28. (a) Center: $(3, 4)$; contains the point $(0, 0)$
  (b) Center: $(6, 1)$; contains the point $(-2, 1)$
29. (a) End points of a diameter: $(-3, 7)$, $(7, 1)$
  (b) End points of a diameter: $(2, -3)$, $(-4, 7)$

30. (a) Center: $(4, 7)$; tangent to the y axis
  (b) Center: $(h, 4)$ in quadrant I; radius: 4; tangent to the x axis

In problems 31 and 32, write the standard equation of each of the graphed circles.

31. (a)   (b)

32. (a)   (b)

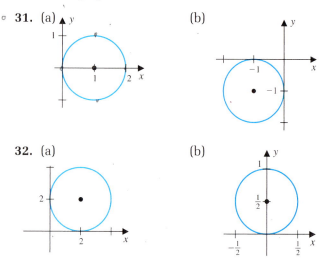

In problems 33–38, find the center and radius of the circle with the given equation. Then sketch its graph.

33. $(x - 1)^2 + (y + 2)^2 = 25$
34. $(x + 2)^2 + (y - 1)^2 = 4$
35. $x^2 + y^2 - 3x + 4y + 4 = 0$
36. $x^2 + y^2 + 8x - 6y = 15$
37. $2x^2 + 2y^2 + 50x + 20y = -24$
38. $3x^2 + 3y^2 + 12x - 18y = -12$

G In problems 39–42, solve for y in terms of x and then use a grapher to graph the given equation. Superimpose two graphs on the same viewing window, if necessary.

39. $4x - 16y^2 = 16$
40. $3x^2 + y^2 = 9$
41. $(y - 1)^2 + (x + 2)^2 = 25$
42. $(y + 3)^2 + x^2 = 16$

Applying the Concepts

43. Grade of a Ramp: The *grade of a ramp* is the slope of the ramp. It is equal to the vertical distance the road rises divided by the length of the horizontal over which the road rises. At an exit ramp off an expressway, cars ascend the ramp to an overpass that is 38 feet above the expressway. The horizontal distance from the beginning of the exit ramp to the point on the expressway below the overpass is 1640 feet (Figure 15). Find the grade of the exit ramp to the nearest hundredth percent.

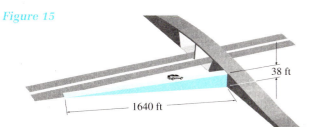

Figure 15

38 ft

1640 ft

44. Pitch of a Roof: The *span of a roof* is the horizontal distance between the outside extensions of the roof, while the *rise* is the vertical distance from the top of the rafters to the center of the span (Figure 16). The *pitch* is the slope of the rafters with respect to the span. Determine the rise for a house whose span is 30 feet with a pitch of 1/4, 1/3, 1/2, or 7/12.

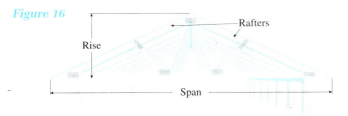

Figure 16

Rafters

Rise

Span

45. Temperature Conversion: The relationship between temperature in degrees Celsius C and degrees Fahrenheit F is given by the equation

$$C = \frac{5}{9}(F - 32)$$

(a) Graph this equation using F as the horizontal axis and C as the vertical axis, and interpret it.
(b) Find and interpret the intercepts.
(c) Find and interpret the slope of the line.
(d) Is there any temperature that is the same on both scales?

46. Weight vs. Age: A study conducted by a weight clinic determines that the average weight W (in pounds) of a man is related to his age A (in years) by the mathematical model

$$W = \frac{9}{5}A + 114, \quad 20 \le A \le 45.$$

(a) Predict the average weight of a 35-year-old man.
(b) Using this model, predict the age of a man whose weight is 186 pounds.
(c) Graph this equation using A as the horizontal axis and W as the vertical axis, and interpret it.
(d) Find the slope of the line and interpret it.
(e) Do you think this model is applicable to men over 45 years old? How about under 20 years old?

47. Manufacturing: In planning to manufacture a laser printer for a computer, a manufacturer determines that the number n of printers produced and the total cost C (in dollars) of producing these printers are related by a linear equation. Projections are that the total cost of producing 50 printers is $9000 and that 125 printers will cost $15,000 to produce.

(a) Find an equation that expresses C in terms of n.
(b) Graph the equation and interpret it.
(c) Give an interpretation of the slope of this equation.
(d) According to this model, what is the total cost of producing 200 laser printers?
(e) In economics, the difference between the cost of producing n items and $n + 1$ items is called the marginal cost. Compute the marginal cost for this situation and interpret it.

48. Baseball Statistics: Suppose that a major league baseball player has hit 7 home runs in the first 29 games of the season, and assume this pace continues throughout the 162-game season.

(a) If the relationship is linear, express the number of home runs, N, in terms of the number t of games played.
(b) Give an interpretation of the slope of the equation.
(c) Use the model from part (a) to predict approximately how many home runs the player will hit for the entire season.

49. Physiology: A jogger's heart rate of N beats per minute is related to the jogger's speed V (in feet per second) by a linear mathematical model. Assume a jogger's heart rate is 75 beats per minute at a speed of 12 feet per second and 80 beats per minute at a speed of 15 feet per second.

(a) Find an equation that expresses N in terms of V.
(b) Graph the equation and interpret it.
(c) Find the slope and interpret it.
(d) Find the speed V if the heart rate N is 85 beats per minute.

50. Physics: Hooke's law for a perfectly elastic spring states that the distance d (in meters) a spring is stretched is directly proportional to the stretching force F (in newtons). Suppose that a force of 3 newtons stretches the spring a distance of 0.2 meter, and a force of

9 newtons stretches the spring a distance of 0.6 meter.

(a) Find a linear equation that expresses F in terms of d.

(b) Give an interpretation of the slope of this equation.

(c) What is the length of the spring if a force of 6 newtons is applied?

(d) How much force is needed to stretch the spring a distance of 0.5 meter?

51. **Real Estate:** The management of an apartment complex charges $475 per month for each of its 90 apartments, and the complex is filled. Because of rising maintenance costs, the management needs to raise the rent to $500 for each apartment. As a result, the number of rented apartments decreases to 65. Assume that this trend continues and the relationship between the rent R and the number n of apartments rented is linear, where $n \geq 20$.

(a) Find an equation that expresses R in terms of n.

(b) Graph the equation and interpret the slope.

(c) What is the monthly rent per unit if 75 apartments are rented?

(d) How many apartments are rented if the monthly rent per apartment is $525?

52. **State Income Tax:** In the year 2004, a state's individual income tax was 4% on all adjusted gross income over $2900.

(a) Assuming the relationship between an individual's income tax T and the adjusted gross income I is linear, find an equation that expresses T in terms of I for someone earning more than $2900.

(b) Graph the equation and interpret the slope.

(c) Use the model to find an individual's adjusted gross income for 1 year if the income tax owed was $1460.

53. **Straight-Line Depreciation:** A straight-line depreciation is one way a business can find the value of an item for tax purposes. An athletic club buys weight-lifting equipment for $6400 and determines that the equipment is expected to last for 6 years. Suppose that the equipment will have a salvage value of $500 at the end of the 6-year period, and assume that the decline in value is the same each year.

(a) Find an equation that expresses the salvage value V in terms of t years of elapsed time.

(b) What are the restrictions on t?

(c) Find and interpret the slope of the line.

(d) Determine by how many dollars the equipment depreciates each year.

54. **Recycling:** Suppose that a recycling center recycled 495 tons of newspapers in the month of January and 525 tons of newspapers in the month of April. Suppose the increase was at a uniform rate. Let T denote the number of tons of newspaper recycled in the nth month (where $n = 1$ represents January). Assume the relationship between T and n is linear, where $1 \leq n \leq 12$.

(a) Find an equation that expresses T in terms of n.

(b) Find and interpret the slope of the line.

(c) In how many months is the amount of paper 565 tons?

(d) Does this model apply for more than 1 year?

55. [G] **Shoe Prices:** Suppose that each year from 1996 to 2004 a shoe manufacturer designs new running shoes. The following table shows the retail price P (in dollars) for each year.

Year	t	Price P per Pair (in dollars)
1996	0	80
1997	1	87
1998	2	94
1999	3	102
2000	4	107
2001	5	113
2002	6	120
2003	7	128
2004	8	135

(a) Draw a scattergram of the data. Locate the t values on the horizontal axis and the P values on the vertical axis.

(b) Use a grapher to find the linear regression equation $P = at + b$ for the line that best fits the data and graph the equation. Round off a and b to three decimal places.

(c) Assuming this trend continues, use the equation from part (b) to predict the retail price of a pair of shoes in the year 2005. Round off to the nearest dollar.

56. [G] **Mortgage Rates:** The following table lists data showing the average annual 30-year fixed mortgage rate R in a certain region for month n of the year 2000.

Month	n	Mortgage Rate Percentage R
Jan	1	6.79
Feb	2	6.81
Mar	3	7.04
Apr	4	6.92
May	5	7.15
Jun	6	7.55
July	7	7.63
Aug	8	7.94
Sept	9	7.82

(a) Draw a scattergram of the data. Locate R on the vertical axis and n on the horizontal axis.

(b) Use a grapher to find the linear regression equation

$$R = an + b$$

for the line that best fits the data and graph the equation. Round off a and b to three decimal places.

(c) Assuming this trend continues, use the equation from part (b) to predict the mortgage rate percentage in the month of November of 2000. Round off the answer to two decimal places.

Developing and Extending the Concepts

57. Suppose that A, B, C, and D are nonzero real numbers. Determine whether the two lines with equations

$$Ax + By + C = 0 \quad \text{and} \quad Ax + By + D = 0$$

are parallel, perpendicular, or neither.

58. Use slopes to determine whether $P = (9, 6)$, $Q = (-1, 2)$, and $R = (1, -3)$ are vertices of a right triangle. If they are, which vertex is the right angle?

59. Describe the graph of $x^2 + y^2 = r^2$ if
(a) $r = 0$ (b) $r < 0$ (c) $r > 0$

60. Which of the following is an equation of a circle? Give the reasons in each case.
(a) $2x^2 + 2y^2 + 4x - 7y + 2 = 0$
(b) $2x^2 + 2y^2 + 4x - 7y + 10 = 0$

61. Find an equation for
(i) the upper half, (ii) the lower half,
(iii) the right half, and (iv) the left half of each circle.
Then graph each equation.
(a) $x^2 + y^2 = 16$
(b) $(x - 2)^2 + y^2 = 9$

62. Given the equation of a circle

$$(x - h)^2 + (y - k)^2 = r^2$$

indicate whether point $P = (x_1, y_1)$ is inside, outside, or on the circle if:
(a) $(x_1 - h)^2 + (y_1 - k)^2 < r^2$
(b) $(x_1 - h)^2 + (y_1 - k)^2 > r^2$
(c) $(x_1 - h)^2 + (y_1 - k)^2 = r^2$
(d) Examine the results in parts (a)–(c) and give specific examples to illustrate your assertions.

63. Suppose that $P = (3, 4)$ is a point on the circle whose equation is $x^2 + y^2 = 25$. From plane geometry, the tangent line to the circle at point P is perpendicular to the radius $\overline{OP}$, where O is the center of the circle. Find the equation of the tangent line at point P.

64. Let L be a line that is neither horizontal nor vertical, and assume the x and y intercepts are the nonzero values a and b, respectively. Then (x, y) is on L provided that

$$\frac{x}{a} + \frac{y}{b} = 1.$$

This equation is called the **intercept equation** of a line. Write this equation in the slope–intercept form, and show that the slope is $-b/a$. Also, find the intercept form of the equation of a line for each of the following situations:
(a) x intercept 4; and y intercept -6
(b) x intercept 1; and y intercept $2/5$
(c) Line contains points $(1, -2)$ and $(3, 5)$

65. Suppose that we choose the two points $P_1 = (x_1, y_1)$ and $P_2 = (x_2, y_2)$ on a nonvertical line L, and then choose two other points $P_1' = (x_1', y_1')$ and $P_2' = (x_2', y_2')$ on L (Figure 17).
(a) Prove that right triangles $P_1P_2P_3$ and $P_1'P_2'P_3'$ are similar.
(b) Use the result in part (a) to fill in the following equal ratios of corresponding sides.

Figure 17

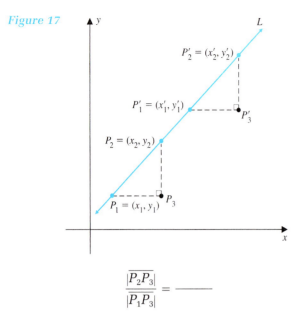

$$\frac{|\overline{P_2P_3}|}{|\overline{P_1P_3}|} = \underline{\qquad}$$

Then rewrite this equation in terms of the coordinates of the points.

(c) What can be concluded if P_1 and P_2 or P'_1 and P'_2 are the two points selected to compute the slope of line L?

66. Figure 18 shows the slopes of perpendicular lines L_1 and L_2 to be m_1 and m_2 respectively. Where $P_1 = (1, m_1)$ is a point on L_1 and $P_2 = (1, m_2)$ is a point on L_2. Use the distance formula, together with the Pythagorean theorem, to show that

$$m_1m_2 = -1.$$

Hint: Since triangle P_1OP_2 is a right triangle, then we have

$$|\overline{OP_1}|^2 + |\overline{OP_2}|^2 = |\overline{P_1P_2}|^2$$

By the distance formula, we can write these terms as

$$|\overline{OP_1}|^2 = (1 - 0)^2 + (m_1 - 0)^2$$
$$= 1 + m_1^2$$
$$|\overline{OP_2}|^2 = (1 - 0)^2 + (m_2 - 0)^2$$
$$= 1 + m_2^2$$
$$|\overline{P_1P_2}|^2 = (1 - 1)^2 + (m_2 - m_1)^2$$
$$= m_2^2 - 2m_1m_2 + m_1^2$$

Figure 18

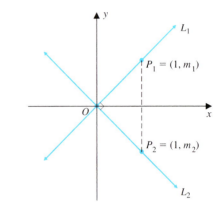

Objectives

1. Graph Parabolas with Vertex at $(0, 0)$
2. Graph Ellipses with Center at $(0, 0)$
3. Graph Hyperbolas with Center at $(0, 0)$
4. Solve Applied Problems

1.6 Introduction to Conics

In this section we introduce other shaped curves called **conic sections** (or **conics,** for short). The Greek mathematician Appolonius is credited with first formally describing conics around 200 B.C. They are formed by intersecting a plane with a right circular cone that has two symmetric parts called *nappes* and an *axis,* as depicted in Figure 1a.

When the plane is perpendicular to the axis of the cone, a *circle* is formed (Figure 1b). By varying the angle at which the plane intersects the axis, its intersection forms either a *parabola* (Figure 1c), an *ellipse* (Figure 1d), or a *hyperbola* (Figure 1e). These plane curves are illustrated in Figure 2.

Our objective here is to examine the geometric features of parabolas, ellipses, and hyperbolas and to recognize certain standard equations for their curves when they are located in a Cartesian plane. Later in Chapters 2 and 3 we'll learn more about equations that have graphs that are parabolic in shape. In Chapter 9, we extend the study of conics by exploring advanced topics that deal with various equation forms and graphs.

Figure 1

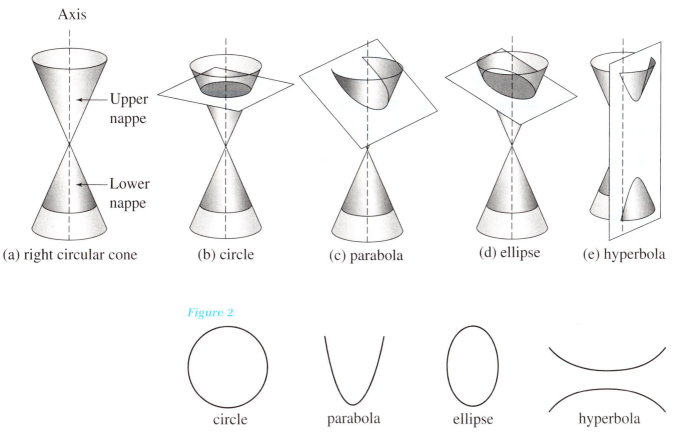

(a) right circular cone   (b) circle   (c) parabola   (d) ellipse   (e) hyperbola

Figure 2

circle   parabola   ellipse   hyperbola

Graphing Parabolas with Vertex at (0, 0)

Our aim here is to describe parabolas from a geometric view and to establish standard equation forms for certain parabolas located in a Cartesian system. The designs of certain suspension bridges, cables, and satellite dishes utilize parabolic shapes.

Definition
Parabola

A **parabola** is the set of all points P in a plane, such that the distance from P to a fixed point, the **focus**, is equal to the distance from P to a fixed line, the **directrix**.

Figure 3

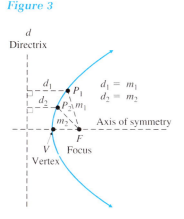

In Figure 3 the points P_1 and P_2 are on the parabola with focus at point F and vertical directrix d to the left of the focus. The distance from P_1 to d, denoted by d_1, is equal to the distance from P_1 to F, denoted by m_1, that is, $d_1 = m_1$. Also, $d_2 = m_2$. The line containing the focus and perpendicular to the directrix is the **axis of symmetry** of the parabola, and the point V where the parabola intersects its axis of symmetry is called the **vertex.**

To derive the standard equation of a parabola with vertex at (0, 0), and vertical directrix to the left, we choose a coordinate system in such a way that the directrix is vertical and the origin is midway between the focus and directrix, as

Figure 4

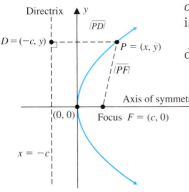

displayed in Figure 4. If the distance between the focus F and the origin is c (where $c > 0$), then the distance from the origin to the directrix is also c. Thus the focus F is located at $(c, 0)$ and the equation of the directrix is $x = -c$.

Next we let $P = (x, y)$ represent any point on the parabola. By definition, the distance from P to F is equal to the distance from P to the directrix:

that is,

$$|\overline{PF}| = |\overline{PD}|$$
$$\sqrt{(x - c)^2 + y^2} = |x + c|$$

Distance formula and absolute value property

$$\left.\begin{array}{c}(x - c)^2 + y^2 = (x + c)^2 \\ x^2 - 2cx + c^2 + y^2 = x^2 + 2cx + c^2\end{array}\right\}$$

Square both sides and multiply

After simplifying the equation, we get

$$y^2 = 4cx \qquad c > 0$$

We refer to this latter equation as the **standard equation** for this parabola. The equation of the directrix is $x = -c$ and the focus is at $(c, 0)$.

Every point (x, y) on the parabola satisfies the equation $y^2 = 4cx$. Conversely, if (x, y) is a point satisfying the equation, then by reversing the steps above, we find that the point (x, y) is on the parabola.

Similar derivations can be given for three other *standard equations* of parabolas with vertices at the origin and horizontal or vertical directrices. Table 1 summarizes the results.

TABLE 1   Standard Equations of Parabolas with Vertex at the Origin and Vertical or Horizontal Directrices ($c > 0$)

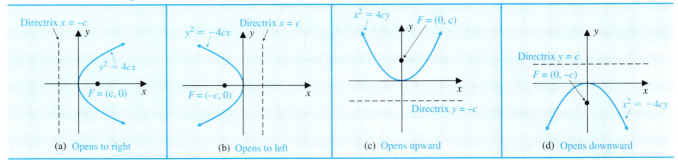

| (a) Opens to right | (b) Opens to left | (c) Opens upward | (d) Opens downward |

EXAMPLE 1   **Graphing a Parabola**

Given the parabola with equation $y^2 = -24x$, find the axis of symmetry, focus, and directrix. Also, sketch the graph.

Figure 5

Solution   The standard equation of this parabola has the form $y^2 = -4cx$, with

$$-4c = -24 \text{ or } c = 6.$$

Hence it opens to the left with focus given by

$$F = (-c, 0) = (-6, 0)$$

and its axis of symmetry is the x axis. The vertex is at $(0, 0)$, and the equation of the directrix, which is vertical, is given by $x = c$ or $x = 6$ (Figure 5).

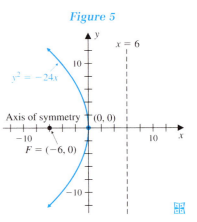

It is possible at times to derive the standard equation of a given parabola, as illustrated in the following example.

EXAMPLE 2   **Finding the Standard Equation of a Parabola**

Find the standard equation of a parabola with vertex at the origin, focus at $(0, 3)$, and axis of symmetry is the y axis. Sketch the graph.

Solution   The focus $(0, 3)$ is above the vertex $(0, 0)$ and the y axis is the axis of symmetry, so the parabola opens upward and has the standard equation, given by

$$x^2 = 4cy$$

Since c is the distance between the focus and vertex, we have $c = 3$. So the equation of the parabola is $x^2 = 12y$ (Figure 6).

Figure 6

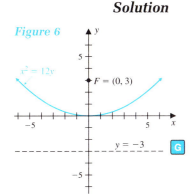

 If a grapher is used to graph the equation of a parabola of the form $x^2 = 4cy$ or $x^2 = -4cy$, then we simply rewrite the equation by isolating y on one side of the equation and then enter the latter equation form in the grapher. For instance, if

$$x^2 = 8y$$

then we rewrite the equation as

$$y = \left(\frac{1}{8}\right)x^2$$

and enter this form in the grapher.

However, if the equation is given in the form $y^2 = 4cx$ or $y^2 = -4cx$, then we proceed as we did with the equation of a circle. For example, if we want to graph

$$y^2 = -2x$$

we first solve for y to get the two resulting equations

$$y_1 = \sqrt{-2x} \quad \text{and} \quad y_2 = -\sqrt{-2x}.$$

These two equations are graphed on the same window. The superimposition of the two graphs gives us the graph of the original equation (Figure 7).

Figure 7

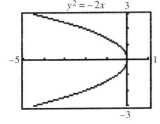

Graphing Ellipses with Center at (0, 0)

Ellipses provide mathematical models for a variety of physical phenomena ranging from art to astronomy. The geometric definition of an ellipse follows.

Definition

Ellipse

> An **ellipse** is the set of all points P in a plane such that the sum of the distances from P to two fixed points is a constant. The fixed points are called the **foci** (plural of **focus**) of the ellipse.

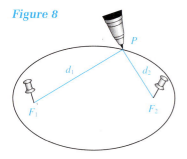

The definition can be used to draw an ellipse (refer to Figure 8). Suppose F_1 and F_2 are the foci of the ellipse. Fasten the ends of a string of length $k > |F_1F_2|$ at F_1 and F_2. By holding the string taut with the point of a pencil, the pencil traces out the ellipse. Note that as P moves about, $|\overline{PF_1}| + |\overline{PF_2}| = k$, that is, $d_1 + d_2 = k$, the constant referred to in the definition.

An ellipse is a curve that has the features described and denoted as follows.

Curve Features
Ellipse

(Refer to Figure 9.)

1. **Foci:** Points F_1 and F_2, two fixed points
2. **Center:** C, the midpoint of line segment $\overline{F_1F_2}$, which has length denoted as $2c$.
3. **Major Axis:** Line segment containing F_1 and F_2 with end points on the ellipse.
   Its length is denoted as $2a$. ($2a$ is the constant used to construct the ellipse.)
4. **Vertices:** Points V_1 and V_2, the endpoints of the major axis.
5. **Minor Axis:** Line segment which is the perpendicular bisector of the major axis and has end points B_1 and B_2 on the ellipse. Its length is denoted as $2b$.
6. The curve is *symmetric* with respect to the major axis and with respect to the minor axis.
7. $0 < b < a$ and $a^2 = b^2 + c^2$.

Figure 9

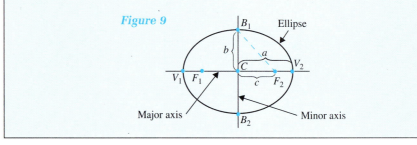

By locating an ellipse in a Cartesian plane so that the center is at $(0, 0)$ and the major axis lies on either the x axis or y axis, we obtain the standard equations given in Table 2.

TABLE 2   Standard Equations of Ellipses with Center at (0, 0), Horizontal or Vertical Major Axis, and Foci F_1 and F_2; $a > b$, $c^2 = a^2 - b^2$

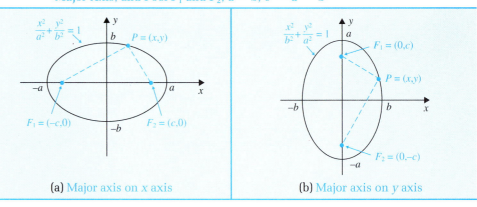

(a) Major axis on x axis   (b) Major axis on y axis

EXAMPLE 3   **Graphing an Ellipse with Center at (0, 0)**

Show that the equation $4x^2 + y^2 = 4$ represents an ellipse by rewriting it in standard form. Then find the vertices and foci and sketch the graph.

Solution   After dividing both sides of the equation by 4, we get the equation

$$\frac{x^2}{1} + \frac{y^2}{4} = 1$$

which is the standard equation of an ellipse with center at the origin and a vertical major axis. Since $b^2 = 1$ and $a^2 = 4$, we have $c^2 = a^2 - b^2 = 4 - 1 = 3$. So $b = 1$, $a = 2$, and $c = \sqrt{3}$. Thus the vertices are

$$V_1 = (0, 2) \quad \text{and} \quad V_2 = (0, -2)$$

and the foci are

$$F_1 = (0, \sqrt{3}) \text{ and } F_2 = (0, -\sqrt{3})$$

The graph is shown in Figure 10.

Figure 10

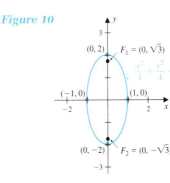

Graphing Hyperbolas with Center at (0, 0)

Hyperbolas are of practical importance in fields such as optics, engineering, and navigation. For instance, a spacecraft moving with more than enough kinetic energy to escape the sun's gravitational pull traces out one branch of a hyperbola.
    The geometric definition of a hyperbola follows:

Definition

Hyperbola

Figure 11

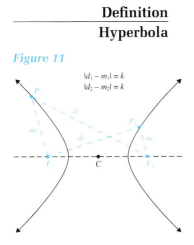

A **hyperbola** is the set of all points P in the plane such that the absolute value of the difference of the distances from P to two fixed points F_1 and F_2 is a constant positive number. Here F_1 and F_2 are called the **focal points,** or **foci,** of the hyperbola. The midpoint C of the line segment $F_1 F_2$ is called the **center** of the hyperbola.

Figure 11 illustrates a hyperbola with foci F_1 and F_2, and with center C. P_1 and P_2 represent two points on the hyperbola that satisfy the condition

$$|d_1 - m_1| = |d_2 - m_2|$$
$$= k \quad \text{(a constant)}.$$

A hyperbola has the following features.

Curve Features

Hyperbola

Figure 12

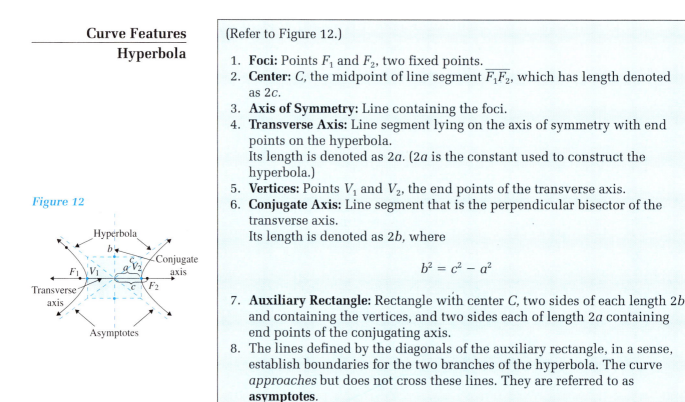

(Refer to Figure 12.)

1. **Foci:** Points F_1 and F_2, two fixed points.
2. **Center:** C, the midpoint of line segment $\overline{F_1F_2}$, which has length denoted as $2c$.
3. **Axis of Symmetry:** Line containing the foci.
4. **Transverse Axis:** Line segment lying on the axis of symmetry with end points on the hyperbola.
   Its length is denoted as $2a$. ($2a$ is the constant used to construct the hyperbola.)
5. **Vertices:** Points V_1 and V_2, the end points of the transverse axis.
6. **Conjugate Axis:** Line segment that is the perpendicular bisector of the transverse axis.
   Its length is denoted as $2b$, where

$$b^2 = c^2 - a^2$$

7. **Auxiliary Rectangle:** Rectangle with center C, two sides of each length $2b$ and containing the vertices, and two sides each of length $2a$ containing end points of the conjugating axis.
8. The lines defined by the diagonals of the auxiliary rectangle, in a sense, establish boundaries for the two branches of the hyperbola. The curve *approaches* but does not cross these lines. They are referred to as **asymptotes**.

If a hyperbola is positioned in a Cartesian plane so that its center is (0, 0) and its transverse axis is either on the x axis or the y axis, then its standard equation takes on, respectively, each of the forms given in Table 3.

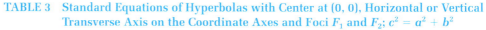

TABLE 3   Standard Equations of Hyperbolas with Center at (0, 0), Horizontal or Vertical Transverse Axis on the Coordinate Axes and Foci F_1 and F_2; $c^2 = a^2 + b^2$

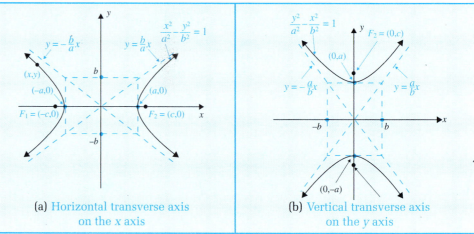

(a) Horizontal transverse axis on the x axis

(b) Vertical transverse axis on the y axis

EXAMPLE 4   **Graphing a Hyperbola**

Show that the equation $25x^2 - 16y^2 = 400$ represents a hyperbola by rewriting it in standard form. Then find the vertices, foci, and equations of the asymptotes. Also, sketch the graph.

Solution   After dividing both sides of the equation by 400, we obtain the equation

$$\frac{x^2}{16} - \frac{y^2}{25} = 1$$

which is the standard equation of a hyperbola with center at the origin, horizontal transverse axis (the hyperbola opens left and right), and vertical conjugate axis.

Since $a = 4$ and $b = 5$, it follows that the transverse axis has length $2a = 8$, and the conjugate axis has length $2b = 10$. The end points of the transverse and conjugate axes determine the auxiliary rectangle whose extended diagonals give us the asymptotes (Figure 13).

The vertices are $V_1 = (-4, 0)$ and $V_2 = (4, 0)$.

To find the foci, we calculate

$$c^2 = a^2 + b^2 = 16 + 25 = 41$$

so $c = \sqrt{41}$, and the foci are $F_1 = (-\sqrt{41}, 0)$ and $F_2 = (\sqrt{41}, 0)$. The equations of the asymptotes are given by

$$y = \frac{b}{a}x = \frac{5}{4}x \quad \text{and} \quad y = -\frac{b}{a}x = -\frac{5}{4}x.$$

The graph of the hyperbola is displayed in Figure 13.

Figure 13

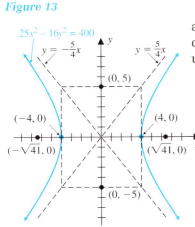

Given certain geometric features of a hyperbola, it is possible to find its equation.

EXAMPLE 5   Finding an Equation of a Hyperbola

Find the standard equation of a hyperbola with vertices at $(0, 3)$ and $(0, -3)$ and foci at $(0, 5)$ and $(0, -5)$. Sketch the graph and find its asymptotes.

Solution   Since the vertices and foci are on the y axis, the transverse axis is on the y axis. We know that $a = 3$ and $c = 5$, because the distance between the center $(0, 0)$ and each vertex is 3, and the distance between the center $(0, 0)$ and each focus is 5. So

$$b^2 = c^2 - a^2$$
$$= 25 - 9$$
$$= 16, \text{ or } b = 4.$$

Because the transverse axis is vertical (the hyperbola opens up and down), we use the standard equation

$$\frac{y^2}{a^2} - \frac{x^2}{b^2} = 1$$

Figure 14

to get

$$\frac{y^2}{9} - \frac{x^2}{16} = 1.$$

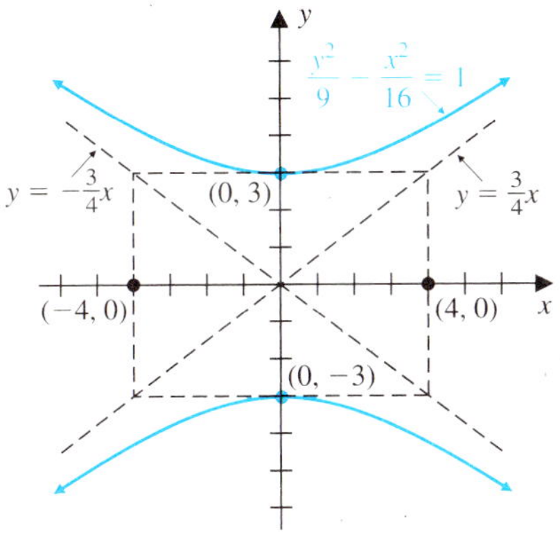

After constructing the auxiliary rectangle using $a = 3$ and $b = 4$, we are able to graph the hyperbola and obtain the equations of the asymptotes, which are

$$y = \frac{a}{b}x = \frac{3}{4}x \quad \text{and} \quad y = -\frac{a}{b}x = -\frac{3}{4}x \quad \text{(Figure 14)} \ \ \text{}$$

Solving Applied Problems

Conics occur in many applications and models. For example, parabolic shapes are used in suspension bridges, reflecting telescopes, satellite dish antennas, and radar equipment. Elliptical shapes provide mathematical models for a variety of situations such as the design of the Colosseum in Rome, the Whispering Gallery of the National Statuary Hall in Washington, D.C., the travel of planets around the sun, the travel of communication satellites around the earth, and the design of water tanks in which ultrasonic devices are used to crush kidney stones. Hyperbolic forms are encountered in the study of the paths of comets, the LORAN system of navigation for ships and aircraft, optics, and some modern hyperbolic cooling towers for electric power stations.

Parabolic dish antenna

The Colosseum in Rome is in the form of an ellipse.

Hyperbolic cooling towers for an electric power station

EXAMPLE 6   **Modeling a Parabolic Dish Antenna**

Figure 15a shows a satellite dish antenna with a parabolic cross section. The open end of the dish has a diameter of 8 feet, and the dish is 1.5 feet deep. Because of the reflecting property of parabolas, the receiver is to be placed at the focus of the parabolic cross section.

(a) Sketch a parabolic cross section of the dish in a Cartesian plane so that its vertex is placed at the origin, it opens upward, and the y axis is its axis of symmetry.

(b) Find the standard equation of the parabola in part (a).

(c) How far from the vertex should the receiver be placed?

Solution   (a) The cross section is the parabola sketched in the Cartesian plane in Figure 15b. Its vertex is at the origin, it opens upward, and the axis of symmetry is the y axis. The location of the receiver is at the focus F.

Figure 15a

Figure 15b

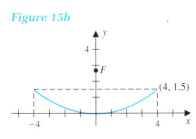

(b) The standard equation of the parabola in part (a) is of the form $x^2 = 4cy$, where c is the required distance from the center of the dish to the receiver at the focus.

From the given data, $(4, 1.5)$ is a point on the parabola, so its coordinates satisfy the equation $x^2 = 4cy$. Thus

$$c = \frac{x^2}{4y} = \frac{16}{4(1.5)} = \frac{8}{3}.$$

Therefore, the standard equation of the parabola is given by

$$x^2 = 4\left(\frac{8}{3}\right)y \quad \text{or} \quad x^2 = \frac{32}{3}y.$$

(c) Since the focus F is at $(0, c) = (0, 8/3)$, it follows that the receiver should be placed 2 feet and 8 inches from the vertex.

Civil engineers use elliptical shapes to design supporting arcs for bridges, as the next example illustrates.

EXAMPLE 7   **Constructing an Elliptical Bridge Support**

Figure 16

An arch in the shape of the upper half of an ellipse is to support a bridge over a roadway 100 feet wide. The center of the arch is 30 feet above the center of the roadway, and the paved portion of the roadway is within a strip that is 80 feet wide (Figure 16a).

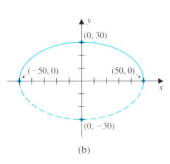

(a)

(a) Sketch the ellipse in an xy coordinate system, with horizontal major axis along the roadway and center at the origin.

(b) Find the standard equation of the ellipse in part (a).

(c) Confirm that the foci of the ellipse are located at the edges of the paved portion of the roadway.

(d) What is the clearance above the roadway to the overpass at the edge of the paved roadway?

Solution

(a) First we set up an xy coordinate system so that the center of the ellipse is at the origin, the semimajor axis is horizontal on the x axis and has length 50, and the semiminor axis, which is vertical, has length 30 (Figure 16b).

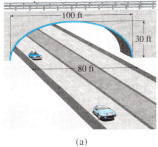

(b)

(b) The standard equation of the ellipse is

$$\frac{x^2}{2500} + \frac{y^2}{900} = 1$$

(c) To find the coordinates of the foci, we first find c:

$$c = \sqrt{a^2 - b^2} = \sqrt{2500 - 900} = 40$$

So the coordinates of the foci are $(-40, 0)$ and $(40, 0)$. This confirms that the foci of the ellipse are located at the edges of the paved portion of the roadway, which is known to be 80 feet wide.

(d) We want to find the value of y when $x = 40$. Using the equation

$$\frac{x^2}{2500} + \frac{y^2}{900} = 1$$

and replacing x by 40, we have

$$\frac{1600}{2500} + \frac{y^2}{900} = 1$$

so that

$$y^2 = 324.$$

Since y is positive in this context, we get

$$y = \sqrt{324}$$
$$= 18.$$

Therefore, the clearance above the edge of the paved roadway to the overpass is 18 feet.

The following example illustrates the LORAN system which makes use of hyperbolic curves in the navigation of ships.

EXAMPLE 8   **Locating a Ship Using LORAN**

Two Coast Guard stations, S_1 and S_2, are located 150 miles apart on an east–west line (Figure 17a). A distress signal from a ship is received at slightly different times by the two stations. It is determined that the difference in the distances from the ship to each station is consistently 100 miles.

(a) Sketch the position of the ship in a Cartesian plane so that the two stations lie on the x axis and the midpoint between them coincides with the origin. Describe the path of the ship.

(b) Find the standard equation of the curve in part (a).

(c) If, at a certain time, the ship is east of the line containing the conjugate axis and 40 miles north of the line between S_1 and S_2, locate the position of the ship relative to the origin of the coordinate system. Round off the answer to the nearest mile.

Figure 17a

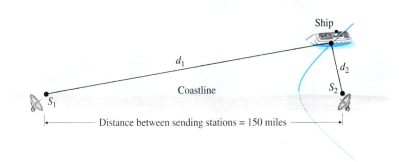

(a)

Solution   (a) First we sketch an xy coordinate system, with the origin midway between two stations S_1 and S_2 on the x axis. The ship is located at some point $P = (x, y)$ in the plane, where

$$\left| |\overline{PS_1}| - |\overline{PS_2}| \right| = |d_1 - d_2|$$
$$= 100 \text{ (Figure 17b)}.$$

(b) By definition, the path of the ship is hyperbolic. Because of the orientation, the standard equation of the hyperbolic path has the form

$$\frac{x^2}{a^2} - \frac{y^2}{b^2} = 1.$$

Because the stations are 150 miles apart, they are located at the foci 75 miles from each side of the origin. Therefore, $c = 75$.
Since $|d_1 - d_2| = 2a$, then

$$2a = 100,$$
$$a = 50.$$

Figure 17b

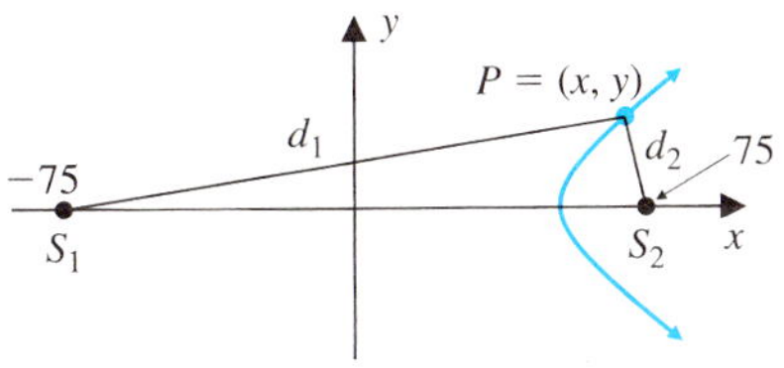

(b)

Also, $c^2 = a^2 + b^2$, so

$$b^2 = c^2 - a^2$$
$$= (75)^2 - (50)^2$$
$$= 3125.$$

Thus the standard equation is

$$\frac{x^2}{2500} - \frac{y^2}{3125} = 1.$$

(c) Because the ship is located 40 miles north of the line connecting S_1 and S_2, the coordinates of P are given by $P = (x, y) = (x, 40)$. Thus we substitute $y = 40$ into the equation to get

$$\frac{x^2}{2500} - \frac{1600}{3125} = 1.$$

Solving for x and rounding off the answer to the nearest mile, we find that $x = -61$ or $x = 61$. However, since the ship is east of the conjugate axis, we select $x = 61$.
Therefore, relative to the origin, the ship is located at $(61, 40)$.

◈ PROBLEM SET 1.6

Mastering the Concepts
In problems 1–6, find the axis of symmetry, vertex, focus, and directrix of each parabola. Also, sketch the graph.

1. $y^2 = 8x$

2. $x^2 = 2y$

3. $x^2 + 4y = 0$

4. $y^2 + 4x = 0$

5. $-y^2 - 3x = 0$

6. $2y - 7x^2 = 0$

In problems 7–10, find the standard equation of the parabola with the given information. Identify the axis of symmetry and sketch the graph.

7. Vertex at the origin and focus at $(0, 4)$

8. Vertex at the origin and focus at $(-4, 0)$

9. Focus at $(0, -2)$ and directrix $y = 2$

10. Focus at $(2, 0)$ and directrix $x = -2$

11. **G** Use a grapher to graph parabolas $x^2 = 4cy$ for $c = 0.25, 0.90, 2,$ and 4 in the same viewing window. What happens to the graph as c gets larger and larger?

12. Use the results in problem 11 above to determine what happens to the graph of $x^2 = 4cy, c > 0$, as c gets smaller and smaller.

In problems 13–18, find the vertices and foci, and sketch the graph of each ellipse.

13. $\dfrac{x^2}{16} + \dfrac{y^2}{9} = 1$

14. $\dfrac{y^2}{25} + \dfrac{x^2}{16} = 1$

15. $4y^2 + 16x^2 = 64$

16. $4x^2 + 9y^2 = 36$

17. $0.1x^2 + y^2 - 1 = 0$

18. $0.25x^2 + 0.5y^2 - 1 = 0$

In problems 19–24, find the standard equation of an ellipse with the given characteristics and sketch the graph.

19. Vertices at $(-5, 0)$ and $(5, 0)$; foci at $(-3, 0)$ and $(3, 0)$

20. Vertices at $(0, -5)$, and $(0, 5)$; foci at $(0, -4)$ and $(0, 4)$

21. Foci at $(0, -13)$ and $(0, 13)$; length of semiminor axis = 5

22. Foci at $(0, -12)$ and $(0, 12)$; length of semimajor axis = 13

23. Vertices at $(0, -6)$ and $(0, 6)$; containing the point $(3, 2)$

24. Vertices at $(-5, 0)$ and $(5, 0)$; containing the point $(4, \frac{12}{5})$

G In problems 25 and 26, use a grapher to graph each given equation.

25. $16x^2 + 25y^2 = 400$

26. $16x^2 + 15y^2 = 240$

In problems 27–34, find the coordinates of the vertices and foci, and find the equations of the asymptotes of each hyperbola. Also, sketch the graph.

27. $\dfrac{x^2}{9} - y^2 = 1$

28. $x^2 - \dfrac{y^2}{9} = 1$

29. $y^2 - x^2 = 9$

30. $16y^2 - 4x^2 = 48$

31. $36x^2 - 9y^2 = 1$

32. $36x^2 - 49y^2 = 1764$

33. $4y^2 + 20 - 5x^2 = 0$

34. $3y^2 = x^2 + 1$

In problems 35–40, find the standard equation for the hyperbola that satisfies the given conditions and sketch the graph.

35. Vertices at $(-16, 0)$ and $(16, 0)$; asymptotes

$$y = \pm \frac{5}{4}x$$

36. Vertices at $(-3, 0)$ and $(3, 0)$; asymptotes $y = \pm 2x$

37. Transverse axis of length 8 units; foci at $(0, 5)$ and $(0, -5)$

38. Center at the origin; a vertex at $(3, 0)$; a focus at $(4, 0)$

39. Center at $(0, 0)$; a horizontal transverse axis; contains the points $(2, 5)$ and $(3, -10)$

40. Center at $(0, 0)$; a horizontal transverse axis; contains the points $(4, 3)$ and $(-7, 6)$

G In problems 41 and 42, use a grapher to graph each hyperbola by solving for y in terms of x.

41. $\dfrac{x^2}{4} - \dfrac{y^2}{9} = 1$

42. $\dfrac{y^2}{1} - \dfrac{x^2}{10} = 1$

Applying the Concepts

43. Satellite Dish Antenna: Figure 18 shows a satellite dish antenna with a parabolic cross section. The opening of the dish has a diameter of 12 feet, and the dish is 3 feet deep. The receiver is to be placed at the focus.

Figure 18

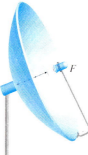

(a) Sketch a parabolic cross section of the dish in a Cartesian plane so that its vertex is placed at the origin, it opens to the right, and the x axis is its axis of symmetry.

(b) Find the standard equation of the parabola in part (a).

(c) At what distance from the vertex should the receiver be placed?

44. Parabolic Mirror: Suppose that the diameter of a parabolic mirror is 10 inches, and assume that the mirror is 5 inches deep at the center.

(a) Sketch a parabolic cross section of the mirror in a Cartesian plane so that its vertex is at the origin, it opens to the right, and the x axis is its axis of symmetry.

(b) Find the standard equation of the parabola in part (a).

(c) How far is the focus from the center of the mirror?

45. Reflecting Telescope: One of the world's largest reflecting telescopes is called the Hale telescope in honor of the American astronomer George E. Hale (1868–1938). It is located at the Palomar

Mountain Observatory northeast of San Diego, California. It uses a mirror 200 inches in diameter. A cross section of the mirror through a diameter is a parabola. The distance from the focus to the vertex is 666 inches (Figure 19).

Figure 19

(a) Sketch the parabola in a Cartesian plane so that its vertex is at the origin and it opens upward.
(b) Find the standard equation of the parabola in part (a).
(c) What is the depth of the parabolic cross section of the mirror?

46. **Verrazano Narrows Bridge:** The Verrazano Narrows suspension bridge is held up by parabolic main cables. It has twin supporting towers 700 feet above the water level. The distance between these towers is 2627 feet, and the low point on a supporting cable is 225 feet above the water level (Figure 20).

Figure 20

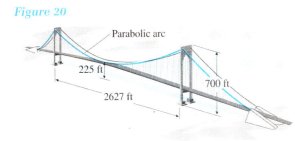

Parabolic arc
225 ft
700 ft
2627 ft

(a) Set up a Cartesian system to model the situation so that the *x* axis coincides with the water level and the vertex of the parabola is on the *y* axis.
(b) Find the standard equation of the parabolic main cable.
(c) How much higher than the low point is a point on the main cable 900 feet from one supporting tower?

47. **Parabolic Stone Bridge:** The surface of a roadway over a stone bridge follows a parabolic curve with the vertex in the middle of the bridge. The span of the bridge is 60 feet and the road surface is 1 foot higher in the middle than at the ends (Figure 21). How much higher than the ends is a point on the roadway 15 feet from an end?

Figure 21

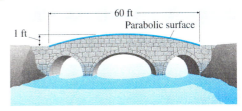

60 ft
Parabolic surface
1 ft

48. **Trajectory of a Ball:** A ball thrown horizontally from the top edge of a building has a parabolic path whose vertex is at the top edge of the building and whose axis of symmetry is along the side of the building. The ball passes through a point 100 feet from the building when it is a vertical distance of 16 feet from the top (Figure 22).

Figure 22

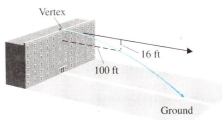

Vertex
16 ft
100 ft
Ground

(a) Assuming that the building is 64 feet high, sketch the parabola in a Cartesian plane so that the intersection of the ground and the edge of the building is at (0, 0) and the edge of the building lies on the *y* axis.
(b) Find the standard equation of the parabola in part (a).
(c) How far from the building will the ball land?

49. **Dimensions of an Arch:** An arch in the shape of the upper half of an ellipse with a horizontal major axis supports a railroad bridge over a river. The base of the arch is 26 meters across, and the highest part of the arch is 5 meters above the water level of the river (Figure 23).

(a) Sketch the ellipse in an *xy* coordinate system with the horizontal major axis on the *x* axis at the level of the river, and center at the origin.
(b) Find the standard equation of the ellipse in part (a).

Figure 23

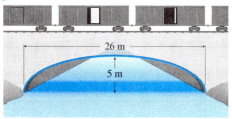

(c) What is the clearance between the river and the arch at a point where one of the foci is located?

(d) Find the height of the arch 3 meters from the center of the base.

50. **Dimensions of a Park:** The Ellipse, a park located near the Washington Monument, is bounded by an elliptical path with a semimajor axis of 744 feet and a semiminor axis of 634 feet.

    (a) Sketch the ellipse in an *xy* coordinate system with the horizontal major axis on the *x* axis and center at the origin.

    (b) Find the standard equation of the ellipse in part (a).

    (c) What is the distance between the two foci of this ellipse?

51. **Earth's Orbit:** The earth moves in an elliptical orbit with the sun at one focus. In early July, the earth is farthest from the sun, at a distance of 94,448,000 miles. In early January, it is closest to the sun, at a distance of 91,341,000 miles.

    (a) Sketch the orbit in an *xy* coordinate system with the horizontal major axis on the *x* axis and center at the origin.

    (b) Find the standard equation of the ellipse in part (a).

    (c) What is the distance between the sun and the other focus on the earth's orbit?

52. **Communications Satellite:** A communications satellite travels in an elliptical orbit around the earth. The radius of the earth is about 6400 kilometers, and its center is located at one focus of the orbit. Suppose that the satellite moves in such a way that when it is closest to the center of the earth, it is 2000 kilometers from the surface, and when it is farthest from the center, it is 10,000 kilometers from the surface.

    (a) Sketch the orbit in an *xy* coordinate system with the major axis lying on the *x* axis and with center at the origin.

    (b) Determine the standard equation of the ellipse in part (a).

    (c) What is the height of the satellite above the surface of the earth at a point corresponding to the focus of the elliptical orbit?

53. **Whispering Gallery:** The Mormon Tabernacle in Salt Lake City, Utah, has a whispering gallery. A vertical cross section forms the upper half of an ellipse with a horizontal major axis 250 feet long and a semiminor axis measured vertically from floor to ceiling of 75 feet.

    (a) Sketch the ellipse in an *xy* coordinate system with the horizontal major axis on the *x* axis and center at the origin.

    (b) Find the standard equation of the ellipse in part (a).

    (c) What is the distance between the foci of the ellipse?

54. **Pool Table:** Suppose that a pool table is constructed so that its shape is elliptical, with a major axis of 6 feet and a minor axis of 4 feet. Assume that a single pocket is located at one of the foci.

    (a) Sketch the ellipse in an *xy* coordinate system with the horizontal major axis on the *x* axis and center at the origin.

    (b) Find the standard equation of the ellipse in part (a).

    (c) Where should one of the balls be located so that when it is hit, it bounces off the cushion directly into the pocket? Explain.

55. **LORAN Navigation:** Two LORAN stations are 180 miles apart on an east–west line. A ship receiving signals from these stations determines that the difference in the distances from the ship to each station is consistently 70 miles.

    (a) Sketch the path of the ship in an *xy* coordinate system. Locate the position of the ship so that the origin is midway between the stations, which lie on the *x* axis.

    (b) Find the standard equation of the curve in part (a).

    (c) If, at a certain time, the ship is east of the conjugate axis and 50 miles north of the line between the two stations, locate the position of the ship relative to the coordinate system. Round off the answer to the nearest mile.

56. **Locating Explosions:** Two microphones are located at two detection stations, *A* and *B,* 1150 meters apart on an east–west line (Figure 24). An explosion occurs at an unknown location *P.* The sound of the explosion is detected by the microphone at station *A* exactly 2 seconds before it is detected by the microphone at station *B.* Assume that the sound travels at a uniform speed of 330 meters per second.

    (a) Sketch the curve that describes the possible location of *P* in an *xy* coordinate system so

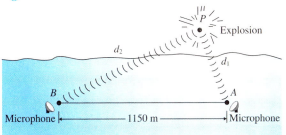

Figure 24

that the origin is at the middle of the horizontal axis between A and B.

(b) Find the standard equation of the curve in part (a).

(c) Can the explosion at P be pinpointed at an exact location? Explain.

57. Comet Orbit: Figure 25 shows a hyperbolic orbit in a Cartesian plane of a comet passing through the solar system with the sun at one focus. Suppose that the comet is moving away from the earth (located at the origin) on a path that is asymptotic to the line $y = 2x$.

Figure 25

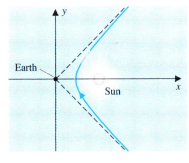

(a) Find an equation of the hyperbola if the closest distance between the comet and the earth is 41.6 million miles.

(b) Approximate the distance between the sun and the earth to the nearest million miles.

58. Airplane Flight: An airplane starts at a point directly north of the origin and flies northeasterly in a hyperbolic path. Transmitters located at the origin and 150 miles directly north of the origin send out synchronized radio signals. Instruments in the airplane measure the difference between the arrival times of the signals. By knowing the speed of the radio signals, it is determined that the distance from the plane to the transmitter at the origin is 50 miles farther than the distance to the other transmitter.

(a) Sketch the path of the airplane in a coordinate system that satisfies the given conditions.

(b) Find the standard equation of the curve in part (a).

(c) Find the location of the airplane when it is 40 miles directly east of the line between the two transmitters.

Developing and Extending the Concepts

59. The **focal chord** of a parabola, also called the **latus rectum**, is the line segment perpendicular to the axis of symmetry at the focus with end points on the parabola (Figure 26).

Figure 26

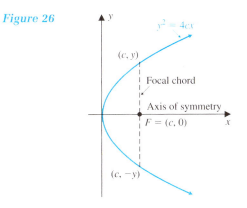

(a) Show that if c is the distance between the vertex and the focus of a parabola, then its focal chord has length $4c$.

(b) Find the length of the focal chord for each of the following parabolas:

    (i) $y^2 = 8x$

    (ii) $x^2 + 4y = 0$

    (iii) $y^2 - 4x = 0$

    (iv) $2y - 7x^2 = 0$

60. (Refer to problem 59.) It can be shown that the slope m of the tangent line to the parabola $x^2 = 4cy$ at the point (a, b) is given by

$$m = \frac{a}{2c}.$$

Show that the tangent lines at the end points of the focal chord of

$$x^2 = 8y$$

are perpendicular to each other.

61. Find the standard equation of an ellipse so that if $P = (x, y)$ is a point on the ellipse, then the sum of the distances from P to $F_1 = (0, -12)$ and to $F_2 = (0, 12)$ is 26 units.

62. Suppose that the foci of an ellipse are $F_1 = (0, -c)$ and $F_2 = (0, c)$, where $c > 0$, and $2a$ is the constant referred to in the definition of an ellipse. If we let $P = (x, y)$ be any point on the ellipse

Figure 27

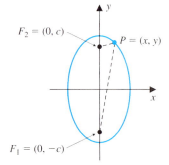

(Figure 27), then in triangle F_1PF_2, we have $|PF_1| + |PF_2| = 2a$.

(a) Use the distance formula to derive an equation that relates x and y.

(b) Simplify the equation found in part (a) and use the substitution $b^2 = a^2 - c^2$ to show that it is equivalent to the equation $(x^2/b^2) + (y^2/a^2) = 1$.

63. Given the equation: $\dfrac{x^2}{a^2} + \dfrac{y^2}{b^2} = 1$

(a) What is the graph of this equation if $a = b$?

(b) Is this the equation of an ellipse? If so, where are the foci? Explain.

64. (a) A line segment that contains a focus, is perpendicular to the major axis, and has end points on the ellipse is called a **focal chord,** or **latus rectum,** of the ellipse (Figure 28). Show that the length of a focal chord of the ellipse with the equation $(x^2/a^2) + (y^2/b^2) = 1$ is $(2b^2)/a$.

Figure 28

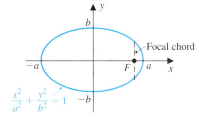

(b) Find the length of a focal chord of an ellipse whose equation is given by $9x^2 + 16y^2 = 144$.

65. Figure 29 shows a rectangle inscribed in the region above the x axis and under the half-ellipse with equation $9x^2 + 16y^2 = 144$, where $y \geq 0$.

Figure 29

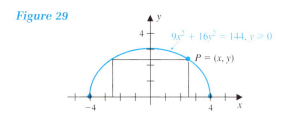

(a) Find the area A of the rectangle in terms of the coordinates of the point $P = (x, y)$ where the upper right corner of the rectangle is on the graph of the ellipse.

(b) Rewrite the equation in part (a) so that the area A is expressed as a function of x.

66. (a) Find the domain of the function

$$y = \sqrt{b^2\left(1 - \frac{x^2}{a^2}\right)}.$$

(b) What is the relationship between the graph of this function and the graph of the ellipse with equation $b^2x^2 + a^2y^2 = a^2b^2$?

67. Find the standard equation of a hyperbola such that the difference of the distances from a point $P = (x, y)$ on the hyperbola to the two points $F_1 = (-5, 0)$ and $F_2 = (5, 0)$ is 8.

68. (a) A **focal chord** of a hyperbola is a line segment with end points on the hyperbola, containing a focus, and perpendicular to the transverse axis (Figure 30). Show that the length of a focal chord of the hyperbola $(x^2/a^2) - (y^2/b^2) = 1$ is $(2b^2)/a$.

Figure 30

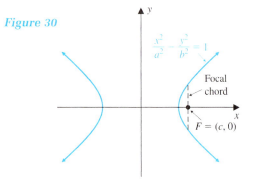

(b) Find the length of the focal chord for each of the following hyperbolas:

(i) $\dfrac{x^2}{9} - \dfrac{y^2}{4} = 1$ (ii) $\dfrac{y^2}{25} - \dfrac{x^2}{10} = 1$

69. (a) Graph the *unit hyperbola* $x^2 - y^2 = 1$ and determine equations for the asymptotes.

(b) Verify that any point of the form $\left(\dfrac{2^u + 2^{-u}}{2}, \dfrac{2^u - 2^{-u}}{2}\right)$ where u represents any real number lies on the graph of $x^2 - y^2 = 1$.

70. The standard equations of hyperbolas centered at the origin can be written in one of the forms

$$\frac{x^2}{a^2} - \frac{y^2}{b^2} = 1 \quad \text{or} \quad \frac{y^2}{a^2} - \frac{x^2}{b^2} = 1$$

(a) Sketch the graphs of both hyperbolas in the same coordinate plane for $a = 4$ and $b = 3$.

(b) Discuss the relationship between the two graphs.

71. Given the hyperbola with standard equation

$$\frac{x^2}{a^2} - \frac{y^2}{b^2} = 1 \text{ where } c^2 = a^2 + b^2$$

(a) Draw the rectangle with center at the origin and side lengths determined by the transverse and conjugate axes.

(b) Graph the circle with equation $x^2 + y^2 = c^2$ in the same coordinate system used in part (a).

(c) Locate the points of intersection of the rectangle and circle drawn in parts (a) and (b).

(d) Explain why the circle in part (b) contains both foci of the hyperbola.

72. Determine the type of curve represented by the equation

$$\frac{x^2}{h^2} + \frac{y^2}{h - 9} = 1$$

in each of the following cases:

(a) $h > 9$　　　(b) $0 < h < 9$　　　(c) $h < 0$

◈ CHAPTER 1 REVIEW PROBLEM SET

In problems 1–4, solve each equation.

1. (a) $2(2x - 1) = 3(x + 1)$

(b) $\dfrac{2x + 3}{5} - \dfrac{2x - 1}{3} = \dfrac{8}{15}$

(c) $|3x - 2| = 4$

(d) $2x^2 - x - 1 = 0$

2. (a) $12(x - 2) + 8 = 5(x - 1) + 2x$

(b) $\dfrac{10 - x}{x} + \dfrac{3x + 3}{3x} = 3$

(c) $\sqrt[3]{2x + 5} = 3$

(d) $|x - 1| = 2x$

3. (a) $|2 - 3x| = 5$

(b) $\sqrt{1 - x} = 2 - \sqrt{1 - 5x}$

(c) $x(x - 2) = 15$

(d) $3x^2 + 6x - 7 = 0$

4. (a) $|7x - 1| = -3$

(b) $(5x - 1)^{3/2} = 8$

(c) $3x^2 - 2x + 12 = 0$

(d) $|x - 2| = |2 - x|$

In problems 5 and 6, solve each literal equation for the specified variable.

5. (a) $ay = 5y + 7; y$

(b) $y^2 + xy + 5 = 0; y$

6. (a) $4y = 8x + 3y; x$

(b) $\dfrac{w}{3} = \dfrac{a}{4} + \dfrac{z}{2}; z$

7. Write each expression without absolute value notation.

(a) $|3 - \sqrt{7}|$　　　(b) $|5x - 3|$　　　(c) $|1 - 7x|$

8. Use absolute value to find the distance between each pair of numbers.

(a) -3 and 7　　　(b) $-7t$ and $-4t$

(c) $-5x$ and $3x$

9. Fill in the blanks with the $<$, $>$, or $=$ symbol.

(a) If $x < 0$, then $-x$ _____ 0.

(b) If $x > 0$, then $-x$ _____ 0.

(c) If $x > 0$, then $|x|$ _____ $-x$.

(d) If $x < 0$, then $|x|$ _____ $-x$.

10. For what values of x is each of the given statements true?

(a) $|x| = |-x|$　　(b) $\dfrac{|-x|}{-x} = -1$　　(c) $|1 - x| = |x|$

In problems 11 and 12, graph each interval on a number line and describe it using inequality notation.

11. (a) $[-4, 5]$

(b) $(-\infty, -2]$

12. (a) $(-\infty, 3]$

(b) $[7, \infty)$

In problems 13 and 14, solve each inequality and express the solution in interval notation. Also, graph the solution on a number line.

13. (a) $5x - 9 > 2x + 3$

(b) $\dfrac{2x}{3} + \dfrac{1}{5} > \dfrac{7}{15} + \dfrac{4x}{5}$

(c) $\dfrac{5}{x} < \dfrac{10}{3x}$

14. (a) $3(x - 5) \le 6 - 3x$

(b) $\dfrac{x}{2} + \dfrac{5}{3} < \dfrac{5x}{6} + 1$

(c) $\dfrac{5x - 1}{-5} > \dfrac{3}{10}$

In problems 15 and 16, solve each absolute value inequality. Express the solution in interval notation and graph it on a number line.

15. (a) $|3t - 7| < 1$ (b) $|3x + 2| \geq 8$
(c) $|x + 4| \leq |x - 1|$

16. (a) $|3 - 2x| \leq 17$ (b) $|1 - 3x| \geq 7$
(c) $|x + 6| < |x - 4|$

In problems 17 and 18, solve each nonlinear inequality and write the answer in interval notation. Also, graph the solution on a number line.

17. (a) $x^2 - 8x - 9 > 0$
(b) $\dfrac{x + 7}{x - 3} - 1 \leq 0$

18. (a) $2x^2 - 7x + 3 \leq 0$
(b) $\dfrac{3x - 2}{x + 1} > 2$

In problems 19 and 20:

(i) Use the distance formula to find the distance between P and Q.

(ii) Use the midpoint formula to find the midpoint of the line segment $\overline{PQ}$.

19. (a) $P = (2, 1)$; $Q = (4, -5)$
(b) $P = (-3, 2)$; $Q = (6, -1)$

20. (a) $P = (-6, -3)$; $Q = (2, 1)$
(b) $P = (5.82, -3.71)$; $Q = (4.73, 6.28)$
(Round off to two decimal places.)

In problems 21 and 22, find the center and radius of the circle with the given equation. Sketch the graph.

21. (a) $(x + 1)^2 + (y + 3)^2 = 16$
(b) $x^2 + y^2 - 6x + 2y - 26 = 0$

22. (a) $\left(x - \dfrac{3}{2}\right)^2 + (y + 4)^2 = 25$
(b) $x^2 + y^2 - 6x + 4y - 12 = 0$

In problems 23 and 24, use the point-plotting method to sketch the graph of each equation. Also, find the x and y intercepts. Discuss the symmetry of each graph.

23. (a) $y = |x + 5|$ **24.** (a) $y = \sqrt{4 - x^2}$
(b) $y = x^2 - |x|$ (b) $y = x^2 - 4x$

In problems 25 and 26, determine the slope of the line containing the given points.

25. (a) $P_1 = (-1, -2)$; $P_2 = (3, 4)$
(b) $P_1 = (-2, 3)$; $P_2 = (3, 1)$

26. (a) $P_1 = (2, -2)$; $P_2 = (-1, 5)$
(b) $P_1 = (-3, 1)$; $P_2 = (2, 5)$

In problems 27 and 28, graph each linear equation and locate the x and y intercepts. Also, find the slope of each line.

27. (a) $x - 2y + 3 = 0$
(b) $2x + 3y - 5 = 0$

28. (a) $4x + y - 9 = 0$
(b) $-3x - 5y + 10 = 0$

In problems 29 and 30, determine the value of k so that the given condition is satisfied for the given line.

29. (a) $kx + 7y = 9$; the slope of the line is -3.
(b) $3x - ky = 2$; the line is parallel to $3x + 2y = 11$.

30. (a) $5kx - 3y = 4$; the line is perpendicular to $2x + y = 0$.
(b) $kx - 3y = 10$; the line is parallel to the x axis.

In problems 31 and 32, find an equation of each line in slope–intercept form and sketch the graph.

31. (a) $m = 3$ and contains $P_1 = (4, 5)$
(b) Contains $P_1 = (3, -2)$ and $P_2 = (5, 6)$

32. (a) $m = 0$ and contains $P_1 = (-2, 3)$
(b) Perpendicular to the line $8x + 7y + 3 = 0$ and contains $P_1 = (2, -4)$

33. Temperature Conversion: A temperature of $10°$ on a Celsius scale corresponds to a temperature of $50°$ on a Fahrenheit scale. Also, $40°C$ corresponds to $104°F$. Assume the relationship between F and C is linear.
(a) Find the equation that relates to F and C.
(b) Find the value of F that corresponds to $52°C$.
(c) Sketch the graph of the equation in part (a).
(d) Interpret the slope of the line.

34. Chemistry: The concentration C of carbon monoxide in the atmosphere is related to the number of years t since 1991 by a linear equation. Assume there were 9.9 parts per million (ppm) of carbon monoxide in the atmosphere in 1991, and there were 6.8 ppm in the atmosphere in 2000.
(a) Find the equation that relates C to t.
(b) Use the equation to predict C in 2010.
(c) Graph the equation in part (a).
(d) Interpret the slope of the line.

35. Investment: A small company had $24,000 to invest. It invested some of the money in a bank that paid 5% annual simple interest. The rest of the money was invested in stocks that paid dividends equivalent to 8% annual simple interest. At the end of 1 year, the combined income from these investments was $1620. How much money was originally invested in stocks?

36. **Investment:** A businesswoman had $18,000 to invest. She invested some of it in a bank certificate that paid 6% annual simple interest, and the rest in another certificate that paid 5% annual simple interest. At the end of 1 year, the combined income from these investments was $1030. How much money did she originally invest at 6% interest?

37. **Parking Fees:** The manager of a parking garage agreed to pay a troop of Boy Scouts a fixed amount for washing some cars—enough to give each boy in the troop $10. When 25 boys failed to show up, those already present agreed to do the work, which meant that each boy who worked got $15. How many boys are in the troop?

38. **Landscaping:** A landscaper wants to put a cement walk of uniform width around a rectangular garden that measures 20 by 60 feet. The landscaper has enough cement to cover 516 square feet. How wide should the walk be in order to use all the cement?

39. **Projectile:** An arrow is shot upward from the ground with an initial speed of 112 feet per second. Its height h (in feet) after t seconds is given by the model

$$h = 112t - 16t^2.$$

  (a) What restriction should be imposed on t?
  (b) Sketch the graph of the equation that represents this model.
  (c) What is the height of the arrow after 4 seconds?
  (d) When does the arrow hit the ground?
  (e) Find the time it takes for the arrow to first reach a height of 144 feet.

40. **Manufacturing:** Suppose that a company's profit P (in dollars) for a given week when producing n items in that week is given by the model

$$P = -n^2 + 100n - 1000$$

  (a) What restriction should be imposed on n?
  (b) Use a grapher to sketch the graph of the equation that represents this model.
  (c) What is the company's profit in 1 week if 40 items were produced and sold?
  (d) How many units should be produced in 1 week if the company's profit for that week is $1500?

41. **G Airfare:** Airlines use the guidelines given in the following table to determine the price p (in dollars) of a ticket based on a round trip air distance d (in miles) between two cities in the United States.

Distance d (in miles)	Price p (in dollars)
200	88
750	155
1145	205
1310	225
1670	267
2100	375
2425	465

  (a) Construct a scattergram for these data, representing the distance d on the horizontal axis and the price of a ticket p on the vertical axis.
  (b) Assume that the set of data can be modeled by a linear equation. Use a grapher to find the linear regression equation of the form

$$p = ad + b$$

  for the line that best fits the data and graph the equation. Round off a and b to three decimal places.
  (c) Use the equation in part (b) to determine the price of a ticket from Detroit to Los Angeles, a distance of 2250 miles. Round off the answer to the nearest dollar.

42. **G Bowling Average:** The data in the table below shows the relationship between bowling averages and six age groups.

Age in Years (A)	Bowling Average (B)
25	195
37	180
42	170
50	165
54	160
60	140

  (a) Construct a scattergram to display the information. Consider the age A to be the horizontal axis and the average B as the vertical axis.
  (b) Use a grapher to find the linear regression equation

$$B = aA + b$$

  for the line that best fits the data and graph the equation. Round off a and b to three decimal places.

(c) Assuming this trend continues, use the equation in part (b) to predict the bowling average of a 65-year-old person and of a 70-year-old person. Round off the answers to a whole number.

In problems 43 and 44, sketch the graph of each parabola, and identify the focus and directrix.

43. (a) $x^2 = -8y$

(b) $y^2 = 36x$

44. (a) $y^2 = -5x$

(b) $x^2 = 7y$

In problems 45 and 46, sketch the graph of each ellipse and identify the vertices and foci.

45. (a) $\dfrac{x^2}{9} + \dfrac{y^2}{36} = 1$

(b) $\dfrac{x^2}{6} + \dfrac{y^2}{4} = 1$

46. (a) $\dfrac{x^2}{16} + \dfrac{y^2}{12} = 1$

(b) $\dfrac{x^2}{2} + \dfrac{y^2}{12} = 1$

In problems 47 and 48, sketch the graph of each hyperbola and identify the vertices and foci. Also, find the equations of the asymptotes.

47. (a) $\dfrac{x^2}{4} - \dfrac{y^2}{9} = 1$

(b) $16x^2 - 4y^2 = 64$

48. (a) $\dfrac{y^2}{9} - \dfrac{x^2}{4} = 1$

(b) $16y^2 - 4x^2 = 64$

Write the standard equation of each conic using the given information.

49. (a) An ellipse with foci at $(-2, 0)$ and $(2, 0)$, and vertices at $(-3, 0)$ and $(3, 0)$

(b) A hyperbola with foci at $(-5, 0)$ and $(5, 0)$, and vertices at $(-3, 0)$ and $(3, 0)$

50. **G** Solve for y in terms of x to obtain two equations. Then use a grapher to superimpose the graphs of the two equations on the same viewing window to get the graph of $x + 4y^2 = 4$.

◈ CHAPTER 1 TEST

1. Solve each equation.

(a) $x + 3(3x - 1) = 4(x + 2) + 4$

(b) $(x + 6)(x - 2) = -7$

(c) $|3x + 2| = 5$

(d) $\sqrt{x - 2} = x - 4$

(e) $3y - 5x = wx$; for y

2. (a) Graph each interval on a number line, and describe the interval using inequality notation.

(i) $(-2, 5]$ (ii) $(-\infty, 4]$

(b) Write each number set in interval notation and graph it on a number line.

(i) $-2 \le x < 4$ (ii) $x \ge 3$ or $2 \le x \le 5$

3. Solve each inequality. Express the solution in interval notation and graph the solution on a number line.

(a) $5x - 2 < -8$

(b) $|x + 5| \le 7$

(c) $|3x - 2| > 4$

(d) $3x^2 + 9x \le 30$

4. Use the point-plotting method to sketch the graph of each equation. Find the intercepts of the graph, and discuss the symmetry of each graph.

(a) $y = |x - 2|$

(b) $y = -3x^2$

(c) $y = x^3 + x$

5. Find the radius and center of the circle with equation $x^2 + y^2 + 6x - 2y - 15 = 0$. Sketch the graph.

6. Let $P_1 = (-3, 4)$ and $P_2 = (-5, 1)$

(a) Find the distance between P_1 and P_2.

(b) Find the coordinates of the midpoint M of the line segment $\overline{P_1P_2}$.

(c) Find the slope of the line containing points P_1 and P_2.

(d) Find an equation of the line containing P_1 and P_2. Express the answer in slope–intercept form.

(e) Sketch the graph of the line and locate its intercepts.

(f) Find an equation of the line that contains P_1 and is:

(i) Parallel to the line $3x + y = 1$

(ii) Perpendicular to the line $3x + y = 1$

7. A manufacturer finds that the cost C (in dollars) of producing a new product line and the number of items, x, produced are related by a linear equation. Suppose that the cost of producing 50 items will be $3750 and the cost of producing 400 items will be $12,500. Use this information to find the cost of producing 470 items.

8. If the temperature of a gas is constant, the volume V occupied by that gas varies inversely as the pressure P to which the gas is subjected. If the volume of a gas is 8 cubic feet when the pressure is 12 pounds per square inch, find the volume of the gas if the pressure is 16 pounds per square inch.

9. An investment of $8000 is split into a savings account paying 4% simple annual interest and a mutual fund paying 7% simple annual interest. If the total annual interest from the two accounts is $485, how much is invested at each rate?

10. Find the vertices (or vertex) and the foci (or focus) of the given conic. If the conic is a hyperbola, find the equations of its asymptotes. Sketch the graph.
    (a) $y^2 = -16x$
    (b) $9x^2 + y^2 = 9$
    (c) $x^2 - 4y^2 = 4$

11. Find the standard equation of the conic that satisfies the given conditions.
    (a) An ellipse with vertices at $(-2, 0)$ and $(2, 0)$; foci at $(-\sqrt{3}, 0)$ and $(\sqrt{3}, 0)$
    (b) A hyperbola with vertices at $(-4, 0)$ and $(4, 0)$; foci at $(-6, 0)$ and $(6, 0)$

12. A parabolic communications antenna has a focus 2 feet from the vertex of the antenna. Find the width of the antenna 3 feet from the vertex in the direction of the focus.

13. Engineering: A bridge has a parabolic arch that is 17 meters high in the center and 50 meters wide at the bottom. Find the height of the arch 10 meters from the center. Round off the answer to two decimal places.

14. Space Science: A spacecraft, in one of its orbits about the earth, had a minimum altitude of 200 miles and a maximum altitude of 1000 miles. The path of the spacecraft is elliptical, with the center of the earth at one focus. Find the standard equation of the path if the radius of the earth is assumed to be about 4000 miles and the center of the earth is positioned at the origin.

15. Navigation: Two Coast Guard stations are located 600 kilometers apart at points $A = (0, 0)$ and $B = (0, 600)$. A distress signal from a ship is received at slightly different times by the two stations. It is determined that the ship is 200 kilometers farther from station A than it is from station B. Determine the standard equation of a hyperbola that contains the location of the ship.

Functions and Graphs

Chapter Contents

Functions are used in modeling real-world situations. For example, a youth soccer club decides to organize a 1-hour fund-raising ice skating party at the local ice rink. The rink charges a nonrefundable rental fee of $150 per hour. The fee includes one attendant for the first 40 skaters. For each additional group of skaters up to and including 40, another attendant is required at a cost of $30 per attendant. The rink also limits the number of skaters to 155 people, and the soccer club charges $5 per ticket. Find the worst possible loss and the maximum possible profit. This problem appears as Example 8 on page 164.

Ice skating rink.

Before embarking on the study of algebra and trigonometry it is essential to have certain basic concepts firmly in mind. Chief among them is the idea of a function. Functions are used to represent relationships between quantities. Such representations are especially important in modeling real-life situations. They help us understand the dependence of one quantity on another. In one way or another, functions underlay many aspects of mathematics and this chapter focuses on this important concept. Graphs of functions are an integral part of this chapter because they give us additional insight into properties and behaviors of functions.

Objectives

1. Define a Function
2. Use Function Notation and Terminology
3. Determine the Domain
4. Determine Difference Quotients
5. Solve Applied Problems

2.1 The Function Concept

A *function* establishes a relationship or correspondence between two sets of elements. For each element in a first set there corresponds one and only one element in a second set.

To help understand the concept of a function, let us consider some real-life illustrations:

1. For each videocassette in a store, there is one corresponding price.
2. For each car in a state, there is one corresponding license number.
3. For each person, there is one corresponding weight at any given time.

There are three characteristics common to all functions: (1) an input set, (2) an output set, and (3) a rule that determines the association between the members of the two sets. The following table lists these characteristics for the above three examples:

Examples of Functions

In the Set of Inputs	In the Set of Outputs	Rule (established by)
1. Videocassettes	Prices	Store pricing
2. Cars	License numbers	Licensing procedure
3. Persons	Weights	Scale reading

Defining a Function

The diagrams in Figures 1a and 1b depict correspondences from one set of numbers, labeled *D*, to another set of numbers, labeled *R*. Figure 1a lists corresponding values for the squares of some numbers, whereas Figure 1b lists corresponding values whose squares equal each given number.

Figure 1

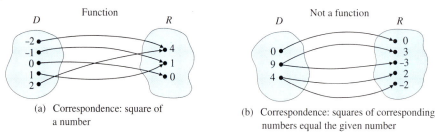

(a) Correspondence: square of a number

(b) Correspondence: squares of corresponding numbers equal the given number

The correspondence in Figure 1a defines a function, since each value in set *D* has one and only one corresponding value in set *R*. However, the correspondence in Figure 1b does *not* define a function, because at least one value in set *D* has more than one corresponding value in set *R*. For instance, 9 has two corresponding values, 3 and -3.

The common features—input, rule, and output—are incorporated in the definition of a function.

Definition
Function

A **function** from a set *D* to a set *R* is a rule of correspondence that assigns to each element in set *D* one and only one element in set *R*. The members of sets *D* and *R*, respectively, are referred to as the *inputs* and *outputs* of the function.

In this definition:

1. The set D of all possible inputs is called the **domain** of the function.
2. The set R of all the corresponding outputs is called the **range** of the function.

Notice that the *rule* of a function must satisfy the condition that exactly one output is produced for each input. However, different inputs may produce the same output, as illustrated in Figure 1a.

EXAMPLE 1

Even though the domain numbers 5 and 1 have the same corresponding range number, 3, this does not violate the definition of a function.

Identifying Functions

Table 1 summarizes the results of a 5-point quiz given to a class of 21 students.

(a) Does the correspondence in Table 1 define a function from set S to set N?

(b) Does the correspondence in Table 1 define a function from set N to set S?

If the correspondence is a function, identify the domain and range.

TABLE 1

Set S Possible Scores	5	4	3	2	1	0
Set N Number of Students Who Earned the Score	3	1	8	6	3	0

Solution

(a) Figure 2a is a diagram that depicts the correspondence from set S to set N. Each number (input) in set S has one and only one corresponding number (output) in set N. Consequently, the correspondence in Table 1 defines a function from set S to set N. The domain is S, the set of possible scores, and the range is N, the set of the numbers of students who earned the scores.

(b) Figure 2b depicts the correspondence from set N to set S. It shows that the number 3 has two corresponding values, 1 and 5, so the correspondence does not define a function from set N to set S.

Figure 2

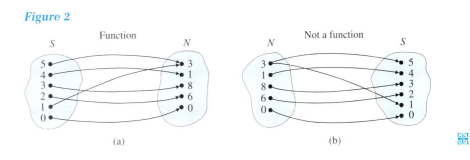

(a)   (b)

Functions may be presented *verbally* (by a description in words), *numerically* (by a table of data), *visually* (by a chart or graph), or *algebraically* (by an equation or formula). For example, Table 2 gives the average annual inflation rate from 1995 through 2003. We can interpret the data in Table 2 as a function if we consider each year as a member of the domain and each inflation rate as a member of the range. For each year, there is one and only one average annual inflation rate. An examination of the data indicates the trend of these rates from 1995 through 2003.

TABLE 2   History of Annual Inflation Rates

Year	1995	1996	1997	1998	1999	2000	2001	2002	2003
Average Annual Inflation Rate (percent)	3.23	2.44	3.63	2.75	1.76	3.38	2.83	1.69	2.27

Functions also arise naturally as charts. For instance, Figure 3 shows the temperature (in degrees Fahrenheit) in a midwestern city over a 24-hour period in May. This chart defines a function in which the times t constitute the domain and the temperatures T make up the range. The location of each point on the chart, such as (1 PM, 65°F), provides us with an ordered pair of numbers that enables us to identify each time (domain member) and its corresponding temperature (range member).

Figure 3

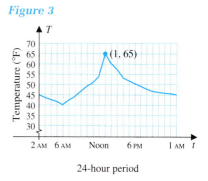

Real-world situations are often modeled by functions defined by tables or charts such as those mentioned above. However, much of the emphasis in mathematics is given to functions defined by equations that relate sets of numbers to each other. For example, if we represent the radius of a circle by r, then its area A is given by the equation

$$A = \pi r^2$$

As we shall see, not every equation represents a function.

For each value of the radius r there corresponds one value for the area A, and we say that "A is a function of r." The domain of this function, which is the set of all possible radii r, includes all positive real numbers, and the range is the set of the associated values of the areas A, which consists of all positive real numbers.

Using Function Notation and Terminology

Function notation provides us with a convenient shorthand for writing functions defined by equations. Suppose such a function is given. We can denote this function by a letter such as f, and let x represent any value in the domain of f.

Then $f(x)$, read "f of x" or "the value of f at x," denotes the value in the range determined by the value x according to the rule of function f.

Thus

The parentheses in f(x) do not denote multiplication. The entire symbol is part of a shorthand language.

$f(x)$ is the range value for the function f associated with the domain value x.

$f(x)$ is the output produced by input x.

For example, if $f(x) = x^2$, then

$f(-1)$ is the output produced by input -1, and we write $f(-1) = (-1)^2 = 1$

$f(0)$ is the output produced by input 0, and we write $f(0) = 0^2 = 0$

$f(2)$ is the output produced by input 2, and we write $f(2) = 2^2 = 4$

Any choice of letters of the alphabet such as f, g, and F can be used to designate functions. Sometimes, instead of writing $f(x) = x^2$, we might write $y = x^2$ where it is understood that the value of y is calculated after choosing a value of x.

That is, the value of y *depends* on the choice of the value of x. Because of this dependence of the values of y on the values of x, we refer to x as the **independent variable** and to y as the **dependent variable.**

In dealing with a function defined by $y = f(x)$, it is important to understand the following notation:

1. The *function f* is a rule of correspondence.
2. The *output* value y or $f(x)$ is a number depending on the *input* value of x.
3. The *equation* $y = f(x)$ relates the dependent variable y to the independent variable x.

When a function is given by an equation such as $f(x) = x^2$, this equation can be thought of as a set of directions as illustrated below:

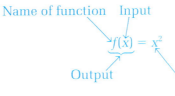

Name of function   Input

$$f(x) = x^2$$

Output   Directions that tell you what to do to get the output

These directions can be applied to any input quantities, whether they are actual numbers or algebraic expressions, as the following examples show.

EXAMPLE 2   **Finding Output Values of a Function**

Let f be the function defined by

$$f(x) = 7x + 2.$$

Find each value.

(a) $f(1)$ (b) $f(-2)$ (c) $[f(4)]^2$ (d) $\sqrt{f(2)}$

Solution   Here, f is defined by the equation $f(x) = 7x + 2$.

(a) We replace x by 1 to get $f(1) = 7(1) + 2 = 7 + 2 = 9$.

$$\text{So } f(1) = 9.$$

(b) Replacing x by -2, we get $f(-2) = 7(-2) + 2 = -14 + 2 = -12$.

$$\text{So } f(-2) = -12.$$

(c) Similarly, $[f(4)]^2 = [7(4) + 2]^2 = (28 + 2)^2 = 30^2 = 900$.

$$\text{Thus } [f(4)]^2 = 900.$$

(d) Finally, $\sqrt{f(2)} = \sqrt{7(2) + 2} = \sqrt{14 + 2} = \sqrt{16} = 4$.

$$\text{Thus } \sqrt{f(2)} = 4 .$$

Algebraic expressions can be substituted for the independent variable of a function. Thus, to compute $f(a + 3)$ for $f(x) = 7x + 2$, the output corresponding to input $a + 3$ is found by replacing x with $a + 3$ to get

$$f(a + 3) = 7(a + 3) + 2$$
$$= 7a + 21 + 2$$
$$= 7a + 23.$$

EXAMPLE 3 **Replacing the Independent Variable with an Algebraic Expression**

Let $g(x) = 3x^2 - 2$. Find and simplify each expression.

(a) $g(2t)$ (b) $g(k + 1)$ (c) $g(k) + 1$

Solution (a) To compute $g(2t)$, the output corresponding to input $2t$, we replace x with $2t$ to get

$$g(2t) = 3(2t)^2 - 2 = 3(4t^2) - 2$$
$$= 12t^2 - 2.$$

(b) $g(k + 1) = 3(k + 1)^2 - 2 = 3(k^2 + 2k + 1) - 2$
$$= 3k^2 + 6k + 3 - 2$$
$$= 3k^2 + 6k + 1$$

Note that
$g(k + 1) \neq g(k) + 1$.

(c) $g(k) + 1 = (3k^2 - 2) + 1 = 3k^2 - 2 + 1$
$$= 3k^2 - 1$$

Sometimes a function is defined by using different equations on different intervals. Such a **piecewise-defined** function is illustrated in the following example.

EXAMPLE 4 **Evaluating a Piecewise-Defined Function**

Let f be defined as

$$f(x) = \begin{cases} -2x + 3 & \text{if } x < 2 \\ 4 & \text{if } x = 2 \\ 3x^2 - 1 & \text{if } x > 2 \end{cases}$$

Find each value:

(a) $f(0)$ (b) $f(-3)$ (c) $f(2)$ (d) $f(4)$

Solution (a) We observe that when $x = 0$, the equation

$$f(x) = -2x + 3$$

is used because $0 < 2$; thus

$$f(0) = -2(0) + 3 = 3$$

(b) Also when $x = -3$, the same equation is used because $-3 < 2$ and so

$$f(-3) = -2(-3) + 3 = 6 + 3 = 9.$$

(c) When $x = 2$, the equation $f(x) = 4$ is used; thus

$$f(2) = 4.$$

(d) When $x = 4$, the equation used for f is $f(x) = 3x^2 - 1$ since $4 > 2$; thus

$$f(4) = 3(4)^2 - 1 = 47.$$

Determining the Domain

Unless otherwise specified, the domain of a function defined by an equation consists of all real number values of the input variable that result in real number output values. *The domain is often determined by examining the equation that defines the function. It must exclude real numbers such as those that result in even roots of negative numbers and those that cause division by zero,* as illustrated in the next example.

EXAMPLE 5   **Determining the Domains of Functions**

Find the domain of each function by examining its defining equation.

(a) $f(x) = x^2 + 3$ (b) $g(x) = \sqrt{x}$ (c) $h(x) = \dfrac{1}{x + 2}$

Solution   The solution for each part is given in Table 3.

TABLE 3

Function	Domain	Reason
(a) $f(x) = x^2 + 3$	$\mathbb{R}$	$x^2 + 3$ is a real number no matter what real value is assigned to x
(b) $g(x) = \sqrt{x}$	All real numbers x such that $x \geq 0$	x must be nonnegative in order for $\sqrt{x}$ to be a real number
(c) $h(x) = \dfrac{1}{x + 2}$	All real numbers such that $x \neq -2$	Division by 0 is not defined

In Section 2.2 we will discuss how graphs are used to find both the domain and range of a function.

Determining Difference Quotients

Suppose we are given a function f defined by

$$y = f(x).$$

*A line formed by two points on a curve is referred to as a **secant line** of the curve.*

Figure 4a displays the use of function notation to denote the line containing the points $P = (x_1, f(x_1))$ and $Q = (x_2, f(x_2))$ on the graph of f. The slope m of this line is given by

$$m = \frac{\Delta y}{\Delta x}$$

$$= \frac{f(x_2) - f(x_1)}{x_2 - x_1}, x_1 \neq x_2.$$

If we substitute t for x_1 and $t + h$ for x_2 as shown in Figure 4b, then the slope m can be rewritten as

$$m = \frac{f(t + h) - f(t)}{h}, h \neq 0.$$

The difference quotient serves as the basis for introducing important concepts in calculus.

This expression is called the **difference quotient** of f.

Figure 4

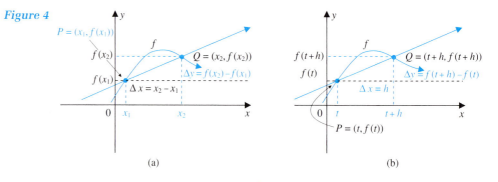

(a)   (b)

EXAMPLE 6   **Finding Difference Quotients**

Find the difference quotient for each function.

(a) $f(x) = 3x - 5$ (b) $f(x) = 2x^2 + 1$

Solution   (a) If $f(x) = 3x - 5$, then

$$\frac{f(t + h) - f(t)}{h} = \frac{[3(t + h) - 5] - [3t - 5]}{h}$$

$$= \frac{3t + 3h - 5 - 3t + 5}{h}$$

$$= \frac{3h}{h} = 3.$$

(b) For $f(x) = 2x^2 + 1$, we have

$$\frac{f(t + h) - f(t)}{h} = \frac{[2(t + h)^2 + 1] - [2t^2 + 1]}{h}$$

$$= \frac{[2(t^2 + 2th + h^2) + 1] - [2t^2 + 1]}{h}$$

$$= \frac{2t^2 + 4th + 2h^2 + 1 - 2t^2 - 1}{h}$$

$$= \frac{4th + 2h^2}{h}$$

$$= \frac{h(4t + 2h)}{h} = 4t + 2h$$

Solving Applied Problems

Our experience with moving objects provides us with an intuitive idea of average speed. For instance, consider a rock falling from a cliff 150 feet high. Assume the distance s (in feet) the rock has fallen after time t (in seconds) is given in Table 4.

TABLE 4

Time t (seconds)	1	1.5	1.9	1.99	2	2.01	2.1	2.5	3.0
Distance $s(t)$ (feet)	16	36	57.76	63.36	64	64.64	70.56	100	144

The average speed for a given time interval is determined by using

Here, displacement refers to distance.

$$\text{Average speed} = \frac{\text{Displacement}}{\text{Elapsed time}}$$

$$= \frac{\Delta s}{\Delta t}$$

$$= \frac{s_2 - s_1}{t_2 - t_1}.$$

For instance, the average speed for the time interval from $t = 2$ to $t = 3$, that is, for the interval [2, 3], is given by

$$\text{Average speed} = \frac{s(3) - s(2)}{3 - 2}$$

$$= \frac{144 - 64}{3 - 2}$$

$$= 80 \text{ feet/second.}$$

Notice that the

$$\text{Average speed} = \frac{s(2 + \text{Elapsed time}) - s(2)}{\text{Elapsed time}}$$

for interval [2, 2 + Elapsed time].

In general, if the distance s is defined by a function

$$s = s(t),$$

where t represents elapsed time, then the average speed during the time interval from t_1 to t_2 is given by

$$\text{Average speed} = \frac{s(t_2) - s(t_1)}{t_2 - t_1}.$$

If we let $t_1 = t$ and $t_2 = t + h$, where h is the elapsed time, then the **average speed** for interval $[t, t + h]$ is given by

$$\text{Average speed} = \frac{s(t + h) - s(t)}{h}$$

which is the difference quotient for s. This expression is the same as the *slope* of the secant line connecting the points $(t_1, s(t_1))$ and $(t_2, s(t_2))$.

Note that, in the difference quotient, h is usually thought of as a small positive number, so that the average speed is found for a very short interval of time.

EXAMPLE 7   **Finding Average Speeds**

A sandbag is dropped from a hot air balloon 500 feet above ground. Ignoring the air resistance, suppose the distance s (in feet) from the sandbag to the ground (see the diagram) after t seconds is given by

$$s(t) = -16t^2 + 500.$$

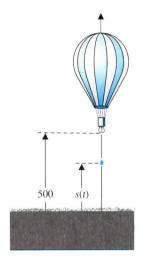

500    $s(t)$

(a) Find the difference quotient

$$\frac{s(t + h) - s(t)}{h}.$$

(b) Find the numerical value of the difference quotient in part (a) for
(i) $t = 3$ and $h = 1$
(ii) $t = 3$ and $h = 0.5$
(iii) $t = 3$ and $h = 0.3$
(iv) $t = 3$ and $h = 0.1$
(c) Explain the results in part (b) in terms of average speeds of the falling sandbag.

Solution   (a) The difference quotient for the function

$$s(t) = -16t^2 + 500$$

is found as follows:

$$\frac{s(t + h) - s(t)}{t} = \frac{[-16(t + h)^2 + 500] - [-16t^2 + 500]}{h}$$

$$= \frac{-16t^2 - 32th - 16h^2 + 500 + 16t^2 - 500}{h}$$

$$= \frac{-32th - 16h^2}{h}$$

$$= \frac{-16\,h\,(2t + h)}{h}$$

$$= -16(2t + h)$$

(b) Using the result from part (a) we have:

(i) $\dfrac{s(3 + 1) - s(3)}{1} = -16[2(3) + 1]$

$$= -112$$

(ii) $\dfrac{s(3 + 0.5) - s(3)}{0.5} = -16[2(3) + 0.5]$

$$= -104$$

(iii) $\dfrac{s(3 + 0.3) - s(3)}{0.3} = -16[2(3) + 0.3]$

$= -100.8$

(iv) $\dfrac{s(3 + 0.1) - s(3)}{0.1} = -16[2(3) + 0.1]$

$= -97.6$

(c) Each of the values of the difference quotients in part (b) turns out to be negative because the object is falling or losing height. By reading Table 5 from the bottom up, we observe that as the elapsed time h increases from 0.1 second to 1 second over time interval $[3, 3 + h]$, the average speed of descent increases since the sandbag is accelerating due to the effect of gravity.

TABLE 5

Time Interval From	Time Elapsed After 3 Seconds (h)	Average Speed of Descent
3 seconds to 4.0 seconds	1 second	112 feet/second
3 seconds to 3.5 seconds	0.5 second	104 feet/second
3 seconds to 3.3 seconds	0.3 second	100.8 feet/second
3 seconds to 3.1 seconds	0.1 second	97.6 feet/second

By generalizing the notion of average speed, we arrive at the concept of *average rate of change.* For any function $y = f(x)$, the **average rate of change of y with respect to** x over the interval $[x_1, x_2]$ is the *slope* of the secant line connecting the points $(x_1, f(x_1))$ and $(x_2, f(x_2))$ on the graph of f (Figure 5). That is,

$$\text{Average rate of change} = \frac{f(x_2) - f(x_1)}{x_2 - x_1} = \frac{f(x_1 + h) - f(x_1)}{h} \text{ , where } x_2 = x_1 + h.$$

Figure 5

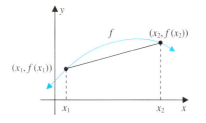

Note that when finding average rate of change, the independent variable x does not necessarily represent elapsed time as it did for average speed.

For instance, a microbiologist may determine the average rate of change at which the number of bacteria in a colony changes over a time interval, which is the same as the average speed. However, an engineer may determine the average rate of change at which the length of a metal bar changes over a temperature interval.

EXAMPLE 8   **Finding Average Rate of Change**

Suppose f is defined by

$$f(x) = x^2.$$

(a) Find and simplify the rate of change function

$$Q(x) = \frac{f(x + h) - f(x)}{h} \text{ for } h = 0.01.$$

(b) Graph f and the rate of change function Q from x to $x + 0.01$ on the same coordinate system.

(c) Determine $Q(3)$ and interpret it.

Solution   (a) $Q(x) = \dfrac{f(x + 0.01) - f(x)}{0.01} = \dfrac{(x + 0.01)^2 - x^2}{0.01}$

$$= \frac{x^2 + 0.02x + (0.01)^2 - x^2}{0.01}$$

$$= \frac{0.02x + (0.01)^2}{0.01} = \frac{0.01[2x + 0.01]}{0.01}$$

$$= 2x + 0.01$$

(b) Figure 6 shows the graphs of f and Q.

Figure 6

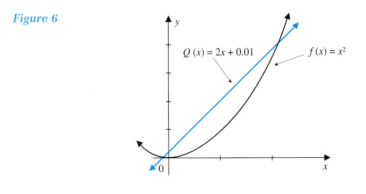

$Q(x) = 2x + 0.01$ $f(x) = x^2$

(c) $Q(3) = 2(3) + 0.01 = 6.01$

This value represents the average rate of change of f over the interval [3, 3.01].

◆ PROBLEM SET 2.1

Mastering the Concepts

In problems 1 and 2, indicate which of the correspondences depicted in the diagrams are functions. If the correspondence is not a function, explain why.

1. (a)    (b)

2. (a)   (b)

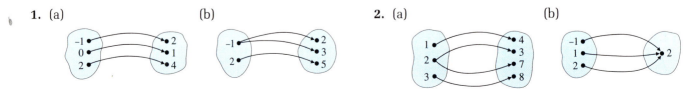

3. Table 6 lists the top five ticket sales for a high school play.

TABLE 6

Set T Number of tickets sold	20	15	14	10	8
Set N Number of students who sold that many tickets	1	3	1	2	5

    (a) Does Table 6 define a function from set T to set N?

    (b) Does Table 6 define a function from set N to set T?

4. Table 7 lists the number of employees of a company between the ages of 55 and 60.

TABLE 7

Set A Age	55	56	57	58	59	60
Set N Number of employees of that age	5	3	1	0	2	4

    (a) Does Table 7 define a function from set A to set N?

    (b) Does Table 7 define a function from set N to set A?

In problems 5–14, find each value.

5. $f(x) = -5x + 3$
    (a) $f(2)$ (b) $f(-2)$
    (c) $f(3/10)$ (d) $f(-3/10)$

6. $g(x) = 3x - 7$
    (a) $g(-5)$ (b) $g(5)$
    (c) $g(-2/3)$ (d) $g(2/3)$

7. $F(x) = x^2 - 4x + 2$
    (a) $F(-3)$ (b) $F(3)$
    (c) $F(2/3)$ (d) $F(-2/3)$

8. $G(x) = -x^2 + 3x - 5$
    (a) $G(-5)$ (b) $G(5)$
    (c) $G(3/5)$ (d) $G(-3/5)$

9. $V(t) = \dfrac{2t + 1}{t - 2}$
    (a) $V(1)$ (b) $V(-1)$
    (c) $V(1/2)$ (d) $V(-1/2)$

10. $W(r) = \dfrac{-r + 3}{5 - 2r}$
    (a) $W(2)$ (b) $W(-2)$
    (c) $W(-1/4)$ (d) $W(1/4)$

11. $h(w) = \sqrt{w - 5} + 3$ [Round off to three decimal places in parts (c) and (d).]
    (a) $h(6)$ (b) $h(9)$
    (c) $h(7.256)$ (d) $h(2\pi)$

12. $f(y) = 6 - \sqrt{4y + 1}$ [Round off to three decimal places in parts (b), (c) and (d).]
    (a) $f(0)$ (b) $f(1/4)$
    (c) $f(2.875)$ (d) $f(\pi^2)$

13. $g(x) = \begin{cases} x + 2 & \text{if } x \le 3 \\ x + 4 & \text{if } x > 3 \end{cases}$
    (a) $g(-1)$ (b) $g(3)$
    (c) $g(0)$ (d) $g(4)$

14. $f(x) = \begin{cases} 2x^2 + 5 & \text{if } x \le -1 \\ 3 - x^2 & \text{if } x > -1 \end{cases}$
    (a) $f(-2)$ (b) $f(0)$
    (c) $f(7)$ (d) $f(-1)$

In problems 15–22, find and simplify each expression.

15. $h(t) = -2t - 5$
    (a) $h(a + 3)$ (b) $h(a) + h(3)$
    (c) $h\left(\dfrac{4}{a + 3}\right)$ (d) $\dfrac{h(4)}{h(a + 3)}$

16. $g(w) = (1/3)w - 2$
    (a) $g(3y + 6)$ (b) $g(3y) + g(6)$
    (c) $3g(y) + 6$ (d) $3g(y) + g(6)$

17. $f(x) = 2x^2 - 5x + 1$
    (a) $f(a + 1)$ (b) $f(a) + f(1)$
    (c) $2f(a) - f(2a)$ (d) $f(a^2)$

18. $G(r) = -r^2 + 3r - 6$
    (a) $G(\sqrt{t})$ (b) $\sqrt{G(t)}$
    (c) $G(t + h) - G(t)$ (d) $G(1/h)$

19. $H(x) = 3 - 2|x|$
    (a) $H(7a)$ (b) $7H(a)$
    (c) $H(a^2)$ (d) $H(-a^2)$

20. $V(t) = |t| - |t + 3|$
    (a) $V(x - 3)$ (b) $V(x) - V(3)$
    (c) $V(5x) + 3$ (d) $V(5x + 3)$

21. $Q(y) = \dfrac{1}{y + 3}$
    (a) $Q\left(\dfrac{1}{y + 3}\right)$ (b) $\dfrac{Q(1)}{Q(y + 3)}$
    (c) $\dfrac{1}{Q(y) + Q(3)}$ (d) $\dfrac{Q(1)}{Q(y) + Q(3)}$

22. $F(t) = \dfrac{4}{t-3}$

  (a) $F\left(\dfrac{3y+4}{y}\right)$
  (b) $\dfrac{F(3y+4)}{F(y)}$

  (c) $\dfrac{3F(y)+F(3y)}{F(y)}$
  (d) $\dfrac{3y+4}{F(y)}$

In problems 23–28, find each function value.

23. $f(x) = \begin{cases} x+3 & \text{if } x \le 1 \\ 4x^2 & \text{if } x > 1 \end{cases}$

  (a) $f(3)$
  (b) $f(-2)$
  (c) $f\left(-\dfrac{1}{2}\right)$
  (d) $f(2.5)$

24. $g(x) = \begin{cases} 2x^2+1 & \text{if } x < 3 \\ 10-x & \text{if } x \ge 3 \end{cases}$

  (a) $g(-3)$
  (b) $g(2)$
  (c) $g(0)$
  (d) $g(10)$

25. $h(x) = \begin{cases} 3x+1 & \text{if } x < -2 \\ 0 & \text{if } -2 \le x \le 2 \\ -2x+3 & \text{if } x > 2 \end{cases}$

  (a) $h(0)$
  (b) $h(3)$
  (c) $h(-2)$
  (d) $h(-5)$

26. $f(x) = \begin{cases} -x & \text{if } x < 0 \\ 4-2x & \text{if } 0 \le x < 3 \\ 2x+1 & \text{if } x \ge 3 \end{cases}$

  (a) $f(0)$
  (b) $f(5)$
  (c) $f(1.5)$
  (d) $f(-1)$

27. $f(x) = \begin{cases} -2 & \text{if } x < 1 \\ 3x & \text{if } 1 \le x < 3 \\ 2-x^2 & \text{if } x \ge 3 \end{cases}$

  (a) $f(3)$
  (b) $f\left(\dfrac{1}{2}\right)$
  (c) $f(-2)$
  (d) $f(11)$

28. $h(x) = \begin{cases} 1-x & \text{if } 0 < x \le 1 \\ 3+x & \text{if } 1 < x \le 2 \\ -1 & \text{if } 2 < x \le 3 \end{cases}$

  (a) $h(1)$
  (b) $h(2)$
  (c) $h(-1)$
  (d) $h(5)$

In problems 29–38, find the domain of each function by examining its equation.

29. $f(x) = 5-3x$
30. $g(x) = 5x-7$

31. $h(x) = \dfrac{7}{x+4}$
32. $H(x) = \dfrac{-3}{-x-2}$

33. $F(x) = \dfrac{5}{x^2-9}$
34. $f(x) = \dfrac{x^2+3}{x^2+2x-8}$

35. $f(x) = \sqrt{3x-2}$
36. $F(x) = \sqrt{3-5x}$

37. $g(x) = \dfrac{x}{x^2-x}$
38. $h(x) = \dfrac{x-2}{x^2-4}$

In problems 39–42, find the difference quotient
$$\dfrac{f(t+h)-f(t)}{h}, h \ne 0.$$

39. (a) $f(x) = 4x+1$
   (b) $f(x) = x^2-3$

40. (a) $f(x) = -2x+3$
   (b) $f(x) = -4x^2+7$

41. (a) $f(x) = \dfrac{1}{x}$
   (b) $f(x) = \dfrac{3}{x-1}$

42. (a) $f(x) = -2x^2+3x-1$
   (b) $f(x) = x^2 - \dfrac{1}{x}$

Applying the Concepts

43. Average Speed: A helicopter drops a crate of supplies from a height of 180 feet. The distance s (in feet) above ground after t seconds is given by the function
$$s(t) = 180 - 16t^2$$
  (a) Find the difference quotient
$$\dfrac{s(t+h)-s(t)}{h}$$
  (b) Find the numerical value of the difference quotient in part (a) for
   (i) $t=1$ and $h=0.5$
   (ii) $t=1$ and $h=0.3$
   (iii) $t=1$ and $h=0.1$
  (c) Explain the results in part (b) in terms of average speeds of the falling crate.

44. Average Speed: A projectile is fired directly upward from the ground with an initial speed of 100 feet per second. Its distance s (in feet) above ground after t seconds is given by the function
$$s(t) = 100t - 16t^2$$
  (a) Find the difference quotient
$$\dfrac{s(t+h)-s(t)}{h}$$
  (b) Find the numerical value of the difference quotient in part (a) for
   (i) $t=2$ and $h=0.7$
   (ii) $t=2$ and $h=0.5$
   (iii) $t=2$ and $h=0.3$
  (c) Explain the results in part (b) in terms of average speeds of the projectile.

45. Rate of Change—Population Growth: The population P of a small town after t weeks is given by the model

$$P(t) = 3t^2 - 12t + 200, \qquad 0 < t < 10$$

(a) Find the difference quotient $\dfrac{P(t + h) - P(t)}{h}$

(b) Find the numerical value of the difference quotient in part (a) for
   (i) $t = 5$ and $h = 2$
   (ii) $t = 5$ and $h = 1$
   (iii) $t = 5$ and $h = 0.7$

(c) Explain the results in part (b) in terms of average rates of change of the population.

46. Rate of Change—Bacteria Treatment: The number N of bacteria (in thousands) in a culture t hours after treatment with an antibiotic is given by the model

$$N(t) = -t^2 + 12t + 100, \qquad 0 < t < 30$$

(a) Find the difference quotient $\dfrac{N(t + h) - N(t)}{h}$

(b) Find the numerical value of the difference quotient in part (a) for
   (i) $t = 10$ and $h = 2$
   (ii) $t = 10$ and $h = 1$
   (iii) $t = 10$ and $h = 0.5$

(c) Explain the results in part (b) in terms of average rates of change of the number of bacteria. What is happening as the elapsed time gets closer to the instant in time when

$$t = 10.$$

Developing and Extending the Concepts

In problems 47 and 48, indicate which of the given situations always defines a function from set A to set B. Explain.

47. (a) For a 50-item multiple choice final examination in Calculus I, the following correspondence is defined:
   Set A consists of the possible scores (number of correct answers) on the examination.
   Set B consists of the numbers of students who had the same score.
   Each possible score is associated with the number of students who had that score.

   (b) Set A consists of all possible gross taxable incomes reported to the Internal Revenue Service in a given year.
   Set B consists of the amounts paid in Federal income tax for the same year.
   Each gross taxable income is associated with the corresponding amount of taxes paid for that income.

48. (a) Set A consists of the heights of a group of people.
   Set B consists of the weights of the same group of people.
   Each height is associated with the weights of people of that height.

   (b) On a 50-item multiple choice examination, the following correspondence is defined:
   Set A consists of the item numbers on the examination.
   Set B consists of the numbers of students who answered each item incorrectly.
   Each test item is associated with the number of students who answered that item incorrectly.

49. Explain why the equation $y^2 = x$ does not define a function in which x represents the domain variable and y represents the range variable.

50. Birthdate: Suppose that more than 366 people are in a room. For each calendar date we associate every person in the room who has that birthdate.

(a) Explain why this association does not define a function with domain all calendar dates and range all the people.

(b) Suppose we "reverse" the association; that is, we consider the people to be the domain and the birthdays to be the range. Would this reverse association define a function? Explain.

51. Determine whether each function satisfies

$$f(a + b) = f(a) + f(b):$$

(a) $f(x) = 3x$
(b) $f(x) = 2x + 5$

52. Determine whether each function satisfies

$$f(ka) = kf(a):$$

(a) $f(x) = 3x$
(b) $f(x) = 2x + 5$

53. In calculus, the notation $g(x)\big|_a^b$ is defined by

$$g(x)\bigg|_a^b = g(b) - g(a)$$

Evaluate each of the following expressions:

(a) $(-3x + 5)\big|_{-3}^1$

(b) $5\left(\dfrac{x^{-2}}{-2} + x^3\right)\bigg|_{-1}^4$

54. Find the difference quotient of the function

$$f(x) = mx + b$$

and compare it to the slope of the line with equation

$$y = mx + b.$$

Objectives

2.2 Graphs of Functions

In Section 1.4, we used the Cartesian coordinate system to construct graphs for relationships between variables usually denoted by x and y. In this section, we apply these ideas to functions.

The **graph of a function** f is the set of points $(x, f(x))$, such that x is in the domain of f. We shall see throughout this text that an accurate graph of a function provides us with a visual representation that displays important characteristics of the function. In a sense, we "read" the graph to learn more about the function.

Graphing Functions

If a function is defined by an equation, then the graph of the function is the same as the graph of the equation. To graph such an equation we rely on point-plotting, equation recognition when known, and symmetry when possible.

EXAMPLE 1   **Graphing a Function Defined by an Equation**

Sketch the graph of $f(x) = 2x - 3$.

Figure 1

Solution   The graph of the function is the same as the graph of the equation,

$$y = 2x - 3.$$

We recognize this to be an equation of the line with

slope $m = 2$ and y intercept -3.

The graph is shown in Figure 1.

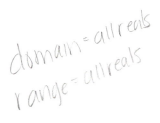
domain = all reals
range = all reals

Generalizing Example 1, the graph of any function f defined by the linear equation

$$f(x) = mx + b$$

is the line with slope m and y intercept b. For this reason, we refer to such a function as a **linear function.**

Identifying Graphs That Represent Functions

Graphs of functions can be easily distinguished from other graphs, as shown by the graphs of

$$y = \sqrt{9 - x^2} \text{ and } x^2 + y^2 = 9$$

in Figures 2a and 2b, respectively.

Figure 2

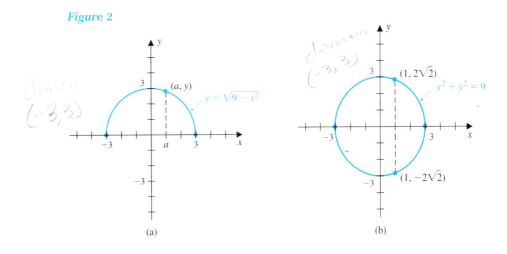

(a)   (b)

In Figure 2a, we see that each input value a, where $-3 \le a \le 3$, corresponds to exactly one output value y, so the equation

$$y = \sqrt{9 - x^2}$$

represents a function of x. Notice that the vertical line through any value of

$$x = a$$

intersects the graph at exactly one point.

In contrast, Figure 2b shows that the input value 1 has two different corresponding output values, $-2\sqrt{2}$ and $2\sqrt{2}$, so that the equation

$$x^2 + y^2 = 9$$

does not represent a function of x. In this situation; there is a vertical line,

$$x = 1$$

that intersects the graph at two points.

These observations lead to the following generalization:

Vertical-Line Test

> A set of points in a Cartesian plane is the graph of a function if and only if no vertical line intersects the graph more than once.

The following diagram illustrates the connection between a relation that does not represent a function and its graph.

Pairs of Numbers Used to Graph	
a	b
a	c

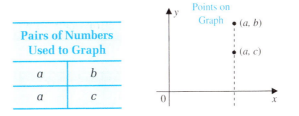

EXAMPLE 2   **Applying the Vertical-Line Test**

Determine which of the graphs in Figure 3 is the graph of a function.

Figure 3

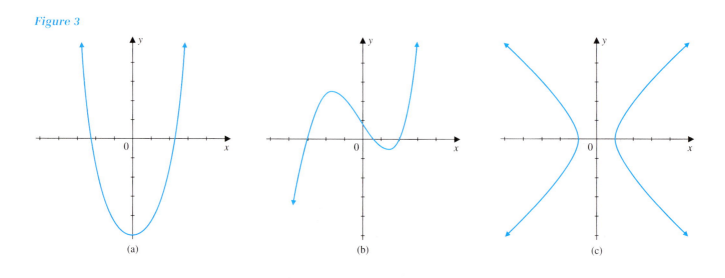

(a)   (b)   (c)

Solution   (a) and (b) The graphs in Figures 3a and 3b are the graphs of functions, because every vertical line crosses the graph at most once.

(c) However, the graph in Figure 3c is not the graph of a function, because we can draw a vertical line that crosses the graph more than once.

Using a Graph to Determine the Domain and Range

The graph of a function helps us to recognize its domain and range. As illustrated in Figure 4, the domain of a function f is the set of all x coordinates of points on its graph, and the range is the set of all y coordinates of points on its graph.

In Figure 4a, a solid dot is used to indicate that a point is on the graph of f, while in Figure 4b, an open dot is used to indicate that a point is not on the graph of f. An open or solid dot at the end of a graph indicates that the graph terminates there.

Figure 4

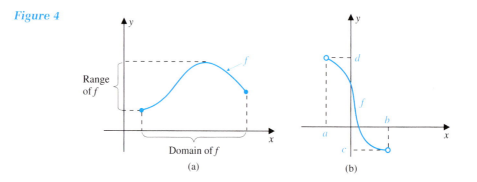

(a)   (b)

At times it is possible to read the graph of a function to determine its domain and range.

EXAMPLE 3

Using Graphs to Determine the Domain and Range of a Function

Use the complete graph of each function in Figure 5 to determine the domain and range. Express the domain and range in interval notation.

Figure 5

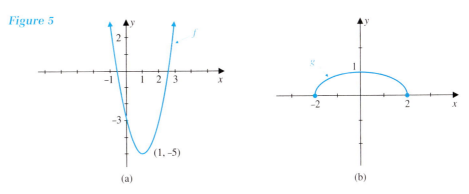

(a)   (b)

Solution

(a) Figure 5a suggests that the domain of f contains all real numbers and the range includes all real numbers y such that $y \geq -5$. In interval notation, the domain is $(-\infty, \infty)$ and the range is interval $[-5, \infty)$.

(b) Figure 5b suggests that the domain of g includes all real numbers x such that $-2 \leq x \leq 2$ and the range consists of all real numbers y such that $0 \leq y \leq 1$. So, in interval notation, the domain is interval $[-2, 2]$ and the range is interval $[0, 1]$.

As we have seen, we can often determine the domain of a function by examining its equation. However, it is easier to find the range when the graph is available.

EXAMPLE 4

Finding the Domain and Range of a Function

Find the domain of

$$f(x) = \sqrt{x - 2},$$

then use the graph of f to find its range.

Solution

Note that $f(x)$ is a nonnegative real number only if $x - 2 \geq 0$, that is, only if $x \geq 2$. So the domain of f consists of all real numbers x such that $x \geq 2$. In interval notation, this means that x can be any number in interval $[2, \infty)$.

The graph of f is the graph of the equation $y = \sqrt{x - 2}$.

By plotting the points in the table and connecting them with a smooth curve, we obtain the graph of f (Figure 6).

By reading the graph, we observe that $y \geq 0$. Thus the range of f consists of all real numbers y such that $y \geq 0$, or in interval notation, $[0, \infty)$.

Figure 6

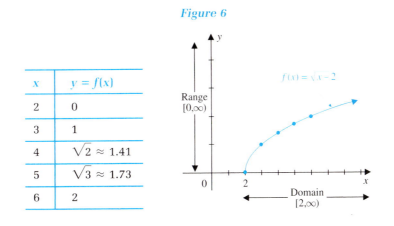

x	$y = f(x)$
2	0
3	1
4	$\sqrt{2} \approx 1.41$
5	$\sqrt{3} \approx 1.73$
6	2

Solving Applied Problems

Functions are used in modeling real-life situations, as illustrated in the next example.

EXAMPLE 5   **Modeling the Cost of Laying Cable**

A cable television company has its master antenna located at point A on the bank of a straight river 0.6 mile wide (Figure 7). It is going to run a cable from A to a point P on the opposite bank of the river x miles from point Q and then straight along the bank to a town T situated 1.8 miles downstream from Q, a point on the shore directly opposite point A so that $\overline{AQ}$ is perpendicular to $\overline{QP}$. It costs \$5 per foot to run the cable underwater and \$3 per foot to run the cable along the bank.

Figure 7

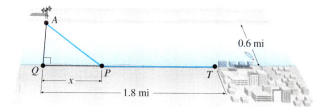

(a) Express the cost C (in dollars) as a function of the distance x (in miles) between points P and Q.

(b) What are the restrictions on x?

(c) What is the cost if the cable is run directly underwater from A to Q and then along the bank to T?

(d) What is the cost if the cable is run totally underwater?

Solution   First, we denote the lengths of line segments $\overline{AQ}$, $\overline{AP}$, $\overline{QP}$, and $\overline{PT}$ as $|\overline{AQ}|$, $|\overline{AP}|$, $|\overline{QP}|$, and $|\overline{PT}|$, respectively. Notice that $|\overline{QP}| = x$, and $|\overline{AQ}| = 0.6$.

(a) The cost C can be found as follows:

$C = $ (Cost underwater from A to P) + (Cost along bank from P to T)

$= [(\$5 \text{ per foot}) \times |\overline{AP}| \text{ (in feet)}] + [\$(3 \text{ per foot}) \times |\overline{PT}| \text{ (in feet)}]$

$= [5\sqrt{(0.6)^2 + x^2} \cdot 5280] + [3(1.8 - x) \cdot 5280]$ Pythagorean theorem; 5280 feet per mile

Thus
$$C = 5280[5\sqrt{0.36 + x^2} + 5.4 - 3x].$$

This equation expresses C as a function of x.

(b) Upon examining Figure 7 we see that x is restricted to values such that $0 \le x \le 1.8$.

(c) If the cable is run from A to Q and then to T, $x = 0$. So
$$C = 5280[5\sqrt{0.36} + 5.4]$$
$$= 44,352.$$

Thus, this option would cost \$44,352.

(d) If the cable is run entirely underwater from A to T, then $x = 1.8$. So
$$C = 5280[5\sqrt{0.36 + (1.8)^2} + 5.4 - 3(1.8)]$$
$$= 50,090.48 \text{ (approx.)}.$$

Thus, it would cost \$50,090.48 if the cable is run totally underwater.

Graphs give us insight regarding mathematical models and graphers provide us with an efficient way of obtaining graphs of functions that are difficult to get by using point-plotting.

G For instance, consider the cost model in Example 5, given by the function

$$C = 5280[5\sqrt{0.36 + x^2} + 5.4 - 3x], 0 \le x \le 1.8.$$

Selections for **Ymin** *and* **Ymax** *were made so that a smooth connected graph is shown for $0 \le x \le 1.8$. Generally, appropriate selections for* **Ymin** *and* **Ymax** *on a grapher require practice and experience.*

Figure 8

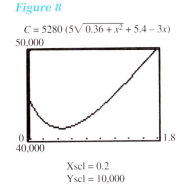

$C = 5280 (5\sqrt{0.36 + x^2} + 5.4 - 3x)$

Xscl = 0.2
Yscl = 10,000

Figure 8 displays a view window of the graph of this function.

The graph suggests that as the distance x increases from 0 to 1.8 miles, the cost appears to decrease at first and then increase to reach its highest value. This highest value occurs when $x = 1.8$, that is, when the cable is run totally underwater. The location of the lowest point would enable us to find the least expensive route for the cable. Later in Chapter 3, we will discuss ways of finding the lowest point on a graph.

PROBLEM SET 2.2

Mastering the Concepts

In problems 1 and 2, indicate whether the graph represents a function.

1. (a) (b)

2. (a) (b)

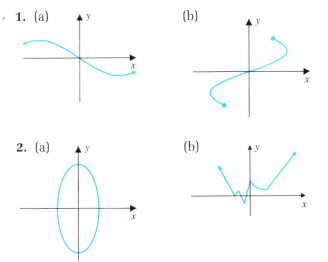

In problems 3–10, sketch the graph of each equation, and use the graph to identify those equations that define y as a function of x.

3. $x^2 + y^2 = 25$
4. $y = 2x^2 - 1$
5. $-2x^2 + y = 4$
6. $(x - 1)^2 + y^2 = 4$
7. $y = 3$
8. $y = \sqrt{4 - x}$
9. $y = \sqrt{x + 5}$
10. $x = 2$

In problems 11 and 12, use the graph to determine the domain and range of each function. Express the domain and range in interval notation.

11. (a) (b)

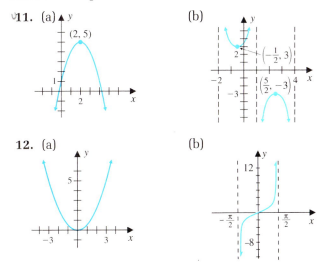

12. (a) (b)

In problems 13–22, graph each function; then find the domain and range.

13. $h(x) = \left(-\dfrac{1}{2}\right)x + 3$
14. $f(x) = \sqrt{49 - x^2}$
15. $f(x) = -\sqrt{49 - x^2}$
16. $G(t) = 4 - 3t$
17. $g(x) = 3x^2 + 2$
18. $h(x) = 5 - x^2$
19. $F(x) = |x| + 4$
20. $G(x) = |x + 4|$

21. $f(x) = (x - 6)^2$ **22.** $g(x) = x^2 - 6$

In problems 23 and 24, graph each function defined by the given table. Find the domain and range. Assume the entries in the top row are from the domain and those in the second row are from the range.

23. (a)

x	-2	-1	0	1	2	3
y	-3	-1	1	3	5	7

(b)

t	-2	-1	0	1	2	3
w	-3	0	1	0	-3	-8

24. (a)

r	1	2	3	4	5
A	1	$\sqrt{2}$	$\sqrt{3}$	2	$\sqrt{5}$

(b)

x	1/2	1	2	3	4
y	2	1	1/2	1/9	1/16

Applying the Concepts

25. Fuel Economy: Figure 9 displays a graph that relates the fuel economy F (in miles per gallon) of a car to different average speeds V (in miles per hour) of the car between 40 and 80 miles per hour.

Figure 9

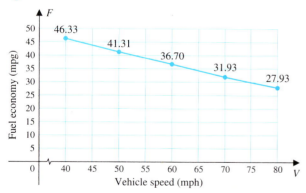

(a) Explain why the graph defines a function of V.
(b) Determine the domain and range.
(c) Estimate the fuel economy when the car has an average speed of 40 miles per hour; 60 miles per hour; 75 miles per hour. Round off to two decimal places.
(d) Estimate the average speed of the car when the fuel economy is 41.31 miles per gallon; 35 miles per gallon; 30 miles per gallon.

26. Biology: Figure 10 shows the relationship between the weight W (in ounces) and the length L (in inches, up to 16 inches) of a certain type of fish.

Figure 10

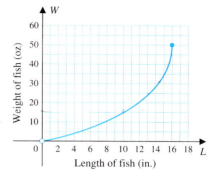

(a) Explain why the graph defines a function of L.
(b) Determine the domain and range.
(c) Estimate the weight (in ounces) if the length (in inches) is 10; 13; 16.
(d) Estimate the length (in inches) if the weight (in ounces) is 15; 30; 40.

27. Global Temperature: Table 1 lists the annual average global temperatures (in degrees Fahrenheit) for 10-year periods from 1890 to 2000.

TABLE 1

Year (t)	Average Temperature (T) Degrees Fahrenheit
1890	58.2
1900	58.5
1910	58.7
1920	58.3
1930	58.8
1940	59.2
1950	59.1
1960	59.0
1970	59.15
1980	58.4
1990	59.8
2000	60.1

(a) Explain why Table 1 defines a function, where t represents members of the domain and T represents members of the range.
(b) Graph the function defined by the data in Table 1.
(c) Discuss the trend in the change of global temperature.

28. Stock Market: Table 2 shows the Dow Jones Industrial average for various times during one day.

TABLE 2

Time of Day (t)	10 am	11 am	noon	1 pm	2 pm	3 pm	4 pm
Dow Jones Average (I)	10562	10564	10562	10561	10563	10552	10547

(a) Explain why Table 2 defines a function, where t represents members of the domain and I represents members of the range.

(b) Graph the function defined by the data in Table 2.

(c) Discuss the trend in the change of the Dow Jones average.

29. Dimensions of a Play Area: A homeowner wants to enclose a rectangular play area next to the house using 40 meters of fencing. The house wall is to form one side of the play area (Figure 11).

Figure 11

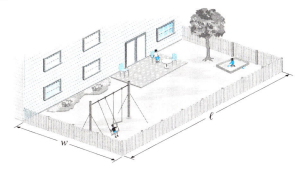

(a) Express the length l of the play area as a function of its width w.

(b) Express the area A of the play area as a function of its width w.

(c) What are the restrictions on w?

(d) Find the length and area of the play area if the width is 5 meters, 10 meters, or 15 meters.

30. Outdoor Track: An outdoor track is to be constructed in the shape shown in Figure 12 and is to have a perimeter of P meters.

Figure 12

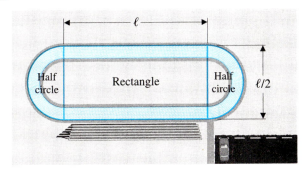

(a) Express the perimeter P as a function of ℓ. That is, write

$$P = f(\ell)$$

(b) Use the result from part (a) to express ℓ as a function of P. That is, write

$$\ell = g(P)$$

(c) Express the total area A enclosed by the track as a function of ℓ. That is, write

$$A = h(\ell)$$

(d) Use the results from parts (b) and (c) to express the area A as a function of P. That is, write

$$A = F(P)$$

(e) Find the perimeter and area if the length is 30 meters, 70 meters, or 100 meters. Round off the answers to two decimal places.

31. Offshore Oil Well: An offshore oil well is located at point A, which is 13 kilometers from the nearest point Q on a straight shoreline. The oil is to be piped from A to a point P on the shoreline x miles from point Q and then straight along the shoreline to a terminal at point T, which is 10 kilometers from point Q (Figure 13). Suppose that it costs \$90,000 per kilometer to lay pipe underwater, and it costs \$60,000 per kilometer to lay the pipe along the shoreline.

Figure 13

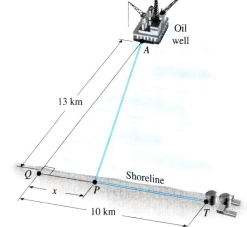

(a) Express the cost C (in dollars) as a function of x, where x is the distance between Q and P.

(b) What restrictions are there on x?

(c) What is the cost if the pipe is laid from A to Q and then from Q to T?

(d) What is the cost if the pipe is laid totally underwater from A to T? Round off to two decimal places.

32. Landscaping: A landscape designer is planning to plant a rectangular tulip bed of dimensions $2x$ and $2y$ on a circular plot of land of radius 5 yards, as displayed in Figure 14. It is determined that the cost of planting the tulip bed is $20 per square yard. Let C (in dollars) be the total cost of planting the tulip bed.

Figure 14

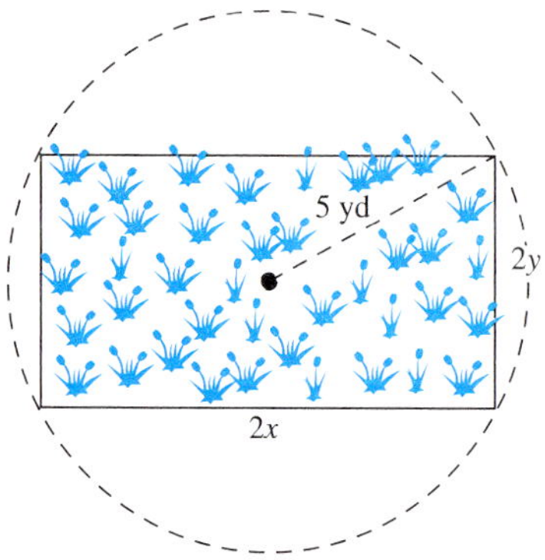

(figure: rectangle of dimensions $2x$ by $2y$ inscribed in a dashed circle, with "5 yd" labeling the radius, and tulips inside the rectangle)

(a) Draw a circle of radius 5 whose center is at the origin of a Cartesian coordinate system, and find its equation.

(b) Use the result from part (a) to express the total cost C of planting the tulip bed as a function of x.

(c) What are the restrictions on x?

(d) Find the cost if x is 2 feet, 3 feet, or 4 feet.

G In problems 33–36, use a grapher to sketch the graphs.

33. (Refer to Problem 29).
Sketch the graph of the function from part (b) using the restrictions in part (c), and interpret the graph.

34. (Refer to Problem 30).
Sketch the graph of the function from part (d) and interpret it.

35. (Refer to Problem 31).
Graph the function in part (a) with the restrictions in part (b). Use the graph to describe what happens to the cost as x increases from 0 to 10 kilometers.

36. (Refer to Problem 32).
Graph the function in part (b) with the restrictions in part (c). Use the graph to describe the cost as x increases.

Developing and Extending the Concepts

37. (a) In a Cartesian system, is the y axis the graph of a function? What is its equation?

(b) In a Cartesian system, is the x axis the graph of a function? What is its equation?

38. The area A of a circle of radius r is given by the formula

$$A = \pi r^2.$$

Explain why the graph in Figure 15 is *not* the graph of this function.

Figure 15

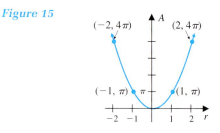

39. (a) Use the quadratic formula to solve the equation

$$y = x^2 + 2x$$

for x in terms of y.

(b) For what values of y does the equation of part (a) have a real solution?

(c) Use part (b) to determine the range of the function

$$f(x) = x^2 + 2x.$$

40. (a) How are the graphs of

$$y = \sqrt{x} \text{ and } x = y^2$$

alike? How are they different?

(b) Do the graphs in part (a) have points in common? What are they?

(c) What restriction on

$$y = \sqrt{x}$$

indicates why its graph is different from that of

$$x = y^2?$$

41. (a) Give an example of a function f whose domain as well as its range is $[0, \infty)$.

(b) Sketch the graph of the function f.

42. (a) Give an example of a function f for which the interval $(-\infty, 3]$ is its domain.

(b) Sketch the graph of the function f.

(c) What is the range of f?

2.3 Graph Features

In this section, we identify certain properties of functions by reading their graphs. In addition, we expand our ability to recognize standard graphs of special functions from their equations as we did earlier for circles and straight lines.

Using Graphs to Solve Inequalities

The location of a graph with respect to the x axis can be used to identify when a function has positive or negative values. For instance, upon examining the graph of the function f in Figure 1, we are led to the conclusions in Table 1.

Figure 1

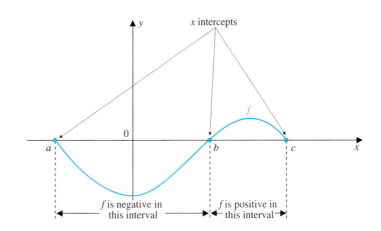

TABLE 1

Value	Graph Feature	Function Value(s)	Terminology
$x = a$	Point on x axis	$f(a) = 0$	a is an x intercept
$a < x < b$, i.e., x is in interval (a, b)	Graph below x axis	Negative, i.e., $f(x) < 0$	f is negative in interval (a, b)
$x = b$	Point on x axis	$f(b) = 0$	b is an x intercept
$b < x < c$, i.e., x is in interval (b, c)	Graph above x axis	Positive, i.e., $f(x) > 0$	f is positive in interval (b, c)
$x = c$	Point on x axis	$f(c) = 0$	c is an x intercept

By reading the graph in Figure 1, we see that the x intercepts are located at a, b, and c. These values are also called *zeros* of the function f.

Definition
Zero of a Function

A real number x is called a real **zero** of a function f if

$$f(x) = 0,$$

Notice that a real zero is also a real *solution* or *root* of the equation

$$f(x) = 0.$$

In a sense, the values of a, b, and c for the function graphed in Figure 1 have three labels:

1. Each is an *x intercept* of the graph of f
2. Each is a *zero* of the function f
3. Each is a *solution* of $f(x) = 0$

Table 2 gives examples of the equivalency between x intercepts, zeros, and solutions of equations.

TABLE 2

Function *F*	Graphic Illustration	*x* Intercept(s)	Zero(s)	Solution(s) of $f(x) = 0$
1. $f(x) = x - 2$		2	2	solution of $x - 2 = 0$ is 2
2. $f(x) = x^2 - 9$		-3 and 3	-3 and 3	solutions of $x^2 - 9 = 0$ are -3 and 3
3. $f(x) = x(x + 1)(x - 2)$		-1, 0, and 2	-1, 0, 2	solutions of $x(x + 1)(x - 2) = 0$ are 0, -1, and 2.

It is possible to use graphs to solve inequalities. To solve an inequality by graphing we need to know the location(s) of the x intercepts of the graph of the associated function. The technique is illustrated in the next example.

EXAMPLE 1   **Solving an Inequality Graphically**

Figure 2 shows the graph of the function *f*.

(a) By reading the graph of f in Figure 2, find the x intercepts and the zeros of f, and then determine the intervals where f is positive or negative.

(b) Use the results in part (a) to solve the given inequalities.

    (i) $x^3 + 3x^2 - 22x - 24 > 0$

    (ii) $x^2 + 3x^2 - 22x - 24 \leq 0$

Figure 2

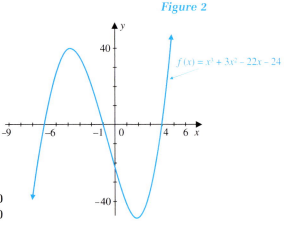

Solution   (a) From the graph, we see that the x intercepts are -6, -1, and 4, so the zeros are also -6, -1, and 4.

As Figure 2 shows, the function f is positive where the graph is above the x axis in each of the intervals $(-6, -1)$ and $(4, \infty)$. It is negative in each of the intervals $(-\infty, -6)$ and $(-1, 4)$ because the graph lies below the x axis in these intervals.

(b) Notice that the expression on the left side of each of the inequalities is the same expression used to define the function f.
(i) The inequality

$$f(x) = x^3 + 3x^2 - 22x - 24 > 0$$

for values of x wherever f is positive. Thus the solution includes all values of x in intervals $(-6, -1)$ or $(4, \infty)$, that is, all values of x such that

$$-6 < x < -1 \text{ or } x > 4.$$

(ii) Similarly the solution of the inequality

$$f(x) = x^3 + 3x^2 - 22x - 24 < 0$$

includes all values of $x < -6$ or $-1 < x < 4$. Also, the solution of the equation

$$f(x) = x^3 + 3x^2 - 22x - 24 = 0$$

corresponds to the x intercepts of the graph of f, namely, -6, -1, and 4. Combining these results, we conclude that the solution of the given inequality

$$x^3 + 3x^2 - 22x - 24 \leq 0$$

includes all values of x such that $x \leq -6$ or $-1 \leq x \leq 4$. In interval notation, the solution includes intervals

$$(-\infty, -6] \text{ and } [-1, 4].$$

G A grapher can help when it is necessary to produce graphs that are difficult to draw by hand. For instance, to solve the inequality

$$\left|2x - 3\right| \leq \left|x + 3\right|$$

Figure 3

graphically, we first rewrite it in the equivalent form

$$\left|2x - 3\right| - \left|x + 3\right| \leq 0.$$

Next, we graph

$$f(x) = \left|2x - 3\right| - \left|x + 3\right| \quad \text{(Figure 3).}$$

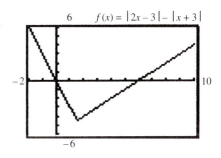

The graph has x intercepts at 0 and 6.

Upon examining the graph, we observe that the values of x that satisfy $f(x) \leq 0$ are those for which the graph is below or on the x axis. These values of x are between and including 0 and 6. So the solution of the inequality includes all values of x such that $0 \leq x \leq 6$, or in interval notation $[0, 6]$.

Determining Increasing and Decreasing Behavior

Consider the graph of the function f shown in Figure 4. The graph of f *rises* as the values of x increase from a to b, *falls* as the values of x increase from b to c, *rises* again as the values of x increase from c to d, and remains *constant* as the values of x increase from d to e.

Figure 4

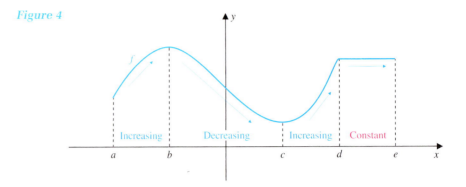

The function is said to be *increasing* on each of the intervals $[a, b]$ and $[c, d]$. It is said to be *decreasing* on the interval $[b, c]$. It is *constant* on the interval $[d, e]$ where the graph of f is a horizontal line.

In general, we have the following concepts:

Properties

Increasing, Decreasing, and Constant Function Behavior

Terminology	Algebraic Condition	Graphic Illustration
f is **increasing** in interval I	$f(a) < f(b)$ whenever $a < b$ and a and b are in I	
f is **decreasing** in interval I	$f(a) > f(b)$ whenever $a < b$ and a and b are in I	
f is **constant** in interval I	$f(a) = f(b)$ whenever a and b are in I	

We describe where increasing, decreasing, or constant behavior occurs in terms of the x values.

EXAMPLE 2

Determining Intervals Where a Function Is Increasing, Decreasing, or Constant

Figure 5 shows the graph of a function f. Find the intervals where f is increasing, decreasing, or constant.

Figure 5

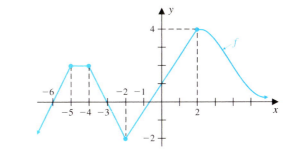

Solution

By reading the graph, we conclude that the function is increasing in each of the intervals $(-\infty, -5]$ and $[-2, 2]$, decreasing in each of the intervals $[-4, -2]$ and $[2, \infty)$, and constant in the interval $[-5, -4]$.

Identifying Even and Odd Functions

In Section 1.4 we introduced the notion of symmetry of the graph of an equation. Now we connect that discussion to function notation. A function whose graph is symmetric with respect to the y axis is called an *even function;* a function whose graph is symmetric with respect to the origin is called an *odd function.* Even and odd functions are described by using function notation as explained in the following property:

Property
Even and Odd Functions

Terminology	Graphic Interpretation	Graphic Illustration	Function Condition
f is an **even function**	Graph of f is symmetric with respect to y axis		$f(-x) = f(x)$ for all x values in the domain
f is an **odd function**	Graph of f is symmetric with respect to origin		$f(-x) = -f(x)$ for all x values in the domain

For example, points $(2, 4)$ and $(-2, 4)$ are on the graph of

$$f(x) = x^2$$

and are equidistant from the y axis. The reason for this symmetry about the y axis is the fact that

$$f(-x) = (-x)^2 = x^2 = f(x)$$

for every value of x that is, because f is an even function.

On the graph of

$$f(x) = x^3$$

we find pairs of points such as $(3, 27)$ and $(-3, -27)$. These points are equidistant from the origin and are located on the opposite sides of the origin.

This occurs because

$$f(-x) = (-x)^3 = -x^3 = -f(x)$$

for all values of x, that is, because f is an odd function.

EXAMPLE 3   **Identifying Even and Odd Functions Graphically**

Consider the graphs of the functions in Figure 6.

Figure 6

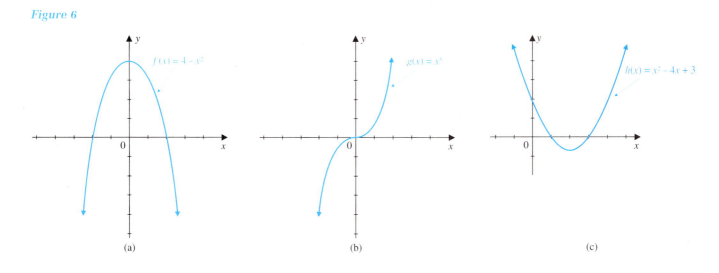

(a)   (b)   (c)

Determine whether each graph is the graph of an even function, an odd function, or a function that is neither even nor odd.

Solution   The graph of f in Figure 6a is that of an even function, because it is symmetric with respect to the y axis. The graph of g in Figure 6b is that of an odd function, because it is symmetric with respect to the origin. The function h whose graph is given in Figure 6c is neither even nor odd, because the graph does not show symmetry with respect to the y axis or the origin.

Graphing functions becomes easier if we can identify symmetry *before* graphing by determining whether a function is even or odd from the equation that defines the function.

EXAMPLE 4   **Identifying Even and Odd Functions Algebraically**

Use the given function equation to determine whether the function is even, odd, or neither. Also discuss the symmetry of each graph.

(a) $f(x) = 3x^4$

(b) $g(x) = x^3 - 2x$

(c) $h(x) = x^2 + x$

Solution   (a) Since

$$f(-x) = 3(-x)^4$$
$$= 3x^4$$
$$= f(x)$$

it follows that f is an even function. So we expect the graph of f to be symmetric with respect to the y axis (see Figure 7a).

(b) Here,

$$g(-x) = (-x)^3 - 2(-x)$$
$$= -x^3 + 2x$$
$$= -(x^3 - 2x)$$
$$= -g(x)$$

so g is an odd function and its graph is symmetric with respect to the origin (see Figure 7b).

(c) Since

$$h(-x) = (-x)^2 + (-x)$$
$$= x^2 - x \neq x^2 + x = h(x)$$

it follows that h is not an even function because $h(-x) \neq h(x)$.
Also

$$-h(x) = -(x^2 + x)$$
$$= -x^2 - x \neq x^2 - x = h(-x).$$

So $h(-x) \neq -h(x)$, thus h is not an odd function. Consequently the graph of h does not exhibit symmetry with respect to the y axis or origin (see Figure 7c).

Figure 7

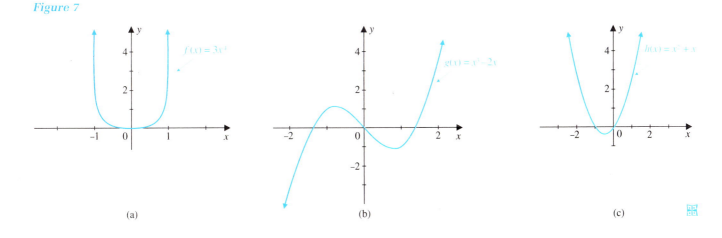

(a)   (b)   (c)

Recognizing Standard Graphs of Special Functions

Certain functions and their graphs play a special role in the study of functions. The recognition of these *special functions* and their graphs is important for two reasons:

1. They provide an easy transition from familiar ideas to new concepts.
2. Knowing their graphs, which we shall refer to as *standard graphs,* can greatly simplify sketching graphs of other functions with equations that are variations of their equations.

These special functions and their graphs are displayed in Figure 8. Their properties are listed in Table 3.

Figure 8   Standard Graphs of Functions

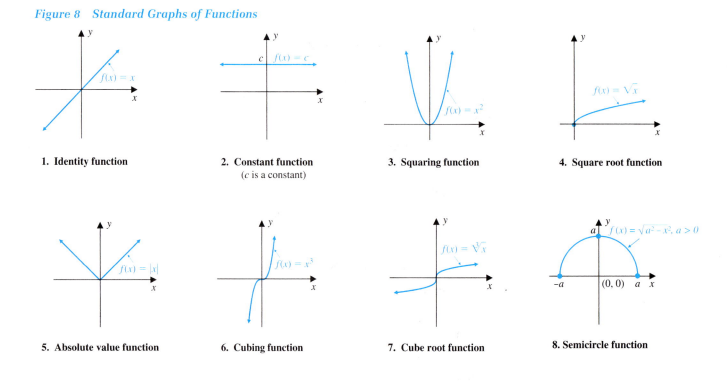

1. Identity function

2. Constant function
(*c* is a constant)

3. Squaring function

4. Square root function

5. Absolute value function

6. Cubing function

7. Cube root function

8. Semicircle function

TABLE 3   Standard Graphs: Properties of Functions

Function	Domain	Range	Increases in Interval	Decreases in Interval	Symmetry with Respect To	Even or Odd
1. $f(x) = x$	$\mathbb{R}$	$\mathbb{R}$	$(-\infty, \infty)$	—	Origin	Odd
2. $f(x) = c$, a constant	$\mathbb{R}$	One number, c	Constant	Constant	y axis	Even
3. $f(x) = x^2$	$\mathbb{R}$	$[0, \infty)$	$[0, \infty)$	$(-\infty, 0]$	y axis	Even
4. $f(x) = \sqrt{x}$	$[0, \infty)$	$[0, \infty)$	$[0, \infty)$	—	—	—
5. $f(x) = |x|$	$\mathbb{R}$	$[0, \infty)$	$[0, \infty)$	$(-\infty, 0]$	y axis	Even
6. $f(x) = x^3$	$\mathbb{R}$	$\mathbb{R}$	$(-\infty, \infty)$	—	Origin	Odd
7. $f(x) = \sqrt[3]{x}$	$\mathbb{R}$	$\mathbb{R}$	$(-\infty, \infty)$	—	Origin	Odd
8. $f(x) = \sqrt{a^2 - x^2}, a > 0$	$[-a, a]$	$[0, a]$	$[-a, 0]$	$[0, a]$	y axis	Even

Graphing Piecewise-Defined Functions

In Section 2.1 we introduced the notion of a piecewise-defined function. The graph of such a function is obtained by sketching the graphs of the individual equations that define the function with their restrictions on the same coordinate system.

EXAMPLE 5   **Using Standard Graphs to Graph a Function Defined by Two Equations**

Use the appropriate standard graphs to sketch the graph of the piecewise-defined function f given by

$$f(x) = \begin{cases} x^2 & \text{if } x < 1 \\ x & \text{if } x \geq 1 \end{cases}$$

Also identify the domain and range.

Solution   If $x < 1$, then $f(x) = x^2$. This portion of the graph coincides with the standard graph of the squaring function $y = x^2$ for x in the interval $(-\infty, 1)$.

For $x \geq 1$, the graph of f coincides with the standard graph of the identity function $y = x$ restricted to the interval $[1, \infty)$. Figure 9 displays the graph of the given function f. The domain of f is the set of real numbers, and the range consists of all numbers in the interval $[0, \infty)$.

Figure 9

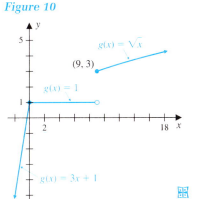

EXAMPLE 6   **Graphing a Function Defined by Three Equations**

Sketch the graph of the piecewise-defined function g given by

$$g(x) = \begin{cases} 3x + 1 & \text{if } x \leq 0 \\ 1 & \text{if } 0 < x < 9. \\ \sqrt{x} & \text{if } x \geq 9 \end{cases}$$

Determine the domain and range.

Figure 10

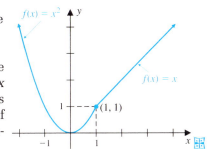

Solution   The graph of g, which consists of parts of three graphs with equations $y = 3x + 1$, $y = 1$, and $y = \sqrt{x}$, is displayed in Figure 10.

From the graph we see that the domain includes all real numbers, and the range includes all values of y such that $y \leq 1$ or $y \geq 3$.

The graph in Figure 9 has no breaks or gaps in it. In this case, we say that the function f is **continuous** for all values of x. On the other hand, Figure 10 shows a graph that is "disconnected" at $x = 9$, so we say that the function is **discontinuous** at $x = 9$.

Another example of a discontinuous function is the **greatest-integer function**, which is defined by the equation

$$f(x) = [\![x]\!]$$

where $[\![x]\!]$ is the greatest integer less than or equal to x.

For instance:

$[\![4]\!] = 4$, since 4 is the greatest integer less than or equal to 4

$\left[\!\left[5\frac{1}{2}\right]\!\right] = 5$, since 5 is the greatest integer less than or equal to $5\frac{1}{2}$

$\left[\!\left[-\frac{5}{4}\right]\!\right] = [\![-1.25]\!] = -2; [\![0.7]\!] = 0;$ and $[\![0]\!] = 0$

The graph of $f(x) = [\![x]\!]$ is displayed in Figure 11. This function is discontinuous wherever x equals an integer.

Figure 11

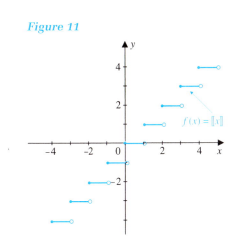

Solving Applied Problems

Functions are used to model real-life situations, as illustrated in the following examples. The first example uses a function defined by a graph.

EXAMPLE 7   **Modeling a Deer Population**

The number n of deer in a forest t years after the beginning of a 10-year conservation study is shown in Figure 12.

(a) Determine the time intervals over which the deer population decreased and increased.

(b) Estimate the lowest and highest populations and when they occurred.

Figure 12

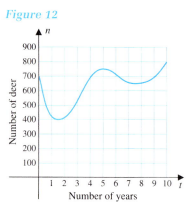

Solution   (a) Upon examining the graph, we see that the function decreased over the intervals [0, 1.5] and [5, 7.5]. So the deer population declined over the first 1.5 years of the study, and then again between 5 and 7.5 years later. The function increased over the intervals [1.5, 5] and [7.5, 10]. This means that the deer population increased between 1.5 and 5 years after the study began and again later, between 7.5 and 10 years after the beginning of the study.

(b) The lowest population (smallest value of n) is 400, and it occurred when $t = 1.5$, that is, after 1.5 years. The highest population (largest value of n) is 800, and it occurred when $t = 10$, that is, after 10 years.

In the next example, we use a piecewise-defined function to model a situation.

EXAMPLE 8   **Modeling a Fund-Raising Event**

A youth soccer club decides to organize a 1-hour fund-raising ice skating party at the local rink. The rink charges a nonrefundable rental fee of $150 per hour. The fee includes one attendant for the first 40 skaters. For each additional group of skaters up to and including 40, another attendant is required at a cost of $30 per attendant. The rink also limits the number of skaters to 155 people, and the soccer club charges $5 per ticket.

(a) Construct a piecewise-defined function that expresses the profit P (in dollars) as a function of the number of sold tickets t.

(b) What are the restrictions on t?

(c) Graph the function with the restrictions from part (b).

(d) Find the range and interpret it. Find the worst possible loss and the maximum possible profit.

Solution   Table 4 summarizes the situation.

TABLE 4

Number of Sold Tickets (t)	Income *Dollars*	Cost *Dollars*	Profit (P) = Income − Cost
$0 \leq t \leq 40$	$5t$	150	$5t - 150$
$40 < t \leq 80$	$5t$	$150 + 30 = 180$	$5t - 180$
$80 < t \leq 120$	$5t$	$150 + 30 + 30 = 210$	$5t - 210$
$120 < t \leq 155$	$5t$	$150 + 30 + 30 + 30 = 240$	$5t - 240$

(a) By using the information in Table 4, we are able to write P as a piecewise-defined function of t as follows:

$$P = \begin{cases} 5t - 150 & \text{if} \quad 0 \leq t \leq 40 \\ 5t - 180 & \text{if} \quad 40 < t \leq 80 \\ 5t - 210 & \text{if} \quad 80 < t \leq 120 \\ 5t - 240 & \text{if} \quad 120 < t \leq 155 \end{cases} \qquad \text{where } t \text{ is an integer}$$

(b) Since the number of skaters is limited to 155, it follows that t is an integer such that $0 \leq t \leq 155$.

(c) Even though t technically represents integers, we'll graph the function in a Cartesian plane so that t represents all real numbers in the interval [0, 155] (Figure 13).

Figure 13

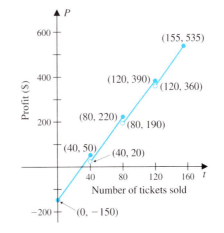

(d) The range includes values of P such that $-150 \leq P \leq 535$. There is a loss if P is negative. If no tickets are sold ($t = 0$), the club would lose \$150, its worst possible loss. The maximum profit, or largest value of P, is \$535, which occurs if 155 tickets are sold.

◆ PROBLEM SET 2.3

Mastering the Concepts

In problems 1–6, read the graph of the given function to determine the x intercepts and zeros of the function, then solve each of the inequalities.

1. $f(x) = x^3 - 8$ (see Figure 14)
    (a) $x^3 - 8 > 0$
    (b) $x^3 - 8 \leq 0$

4. $f(x) = x^3 + 2x^2 - x - 2$ (see Figure 17)
    (a) $x^3 + 2x^2 - x - 2 \geq 0$
    (b) $x^3 + 2x^2 - x - 2 < 0$

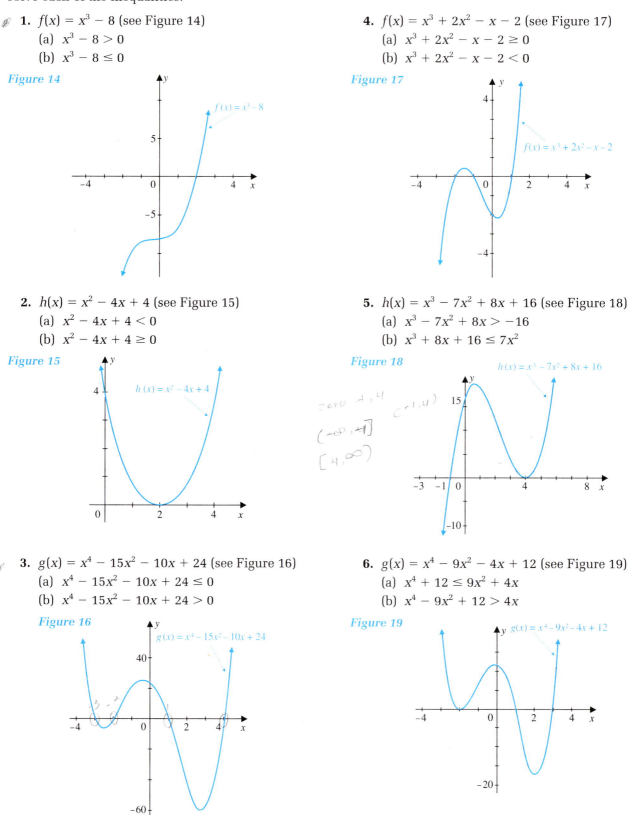

Figure 14

$f(x) = x^3 - 8$

Figure 17

$f(x) = x^3 + 2x^2 - x - 2$

2. $h(x) = x^2 - 4x + 4$ (see Figure 15)
    (a) $x^2 - 4x + 4 < 0$
    (b) $x^2 - 4x + 4 \geq 0$

5. $h(x) = x^3 - 7x^2 + 8x + 16$ (see Figure 18)
    (a) $x^3 - 7x^2 + 8x > -16$
    (b) $x^3 + 8x + 16 \leq 7x^2$

Figure 15

$h(x) = x^2 - 4x + 4$

Figure 18

$h(x) = x^3 - 7x^2 + 8x + 16$

(handwritten) zero -1, 4   (-1, 4)
(-∞, -1]
[4, ∞)

3. $g(x) = x^4 - 15x^2 - 10x + 24$ (see Figure 16)
    (a) $x^4 - 15x^2 - 10x + 24 \leq 0$
    (b) $x^4 - 15x^2 - 10x + 24 > 0$

6. $g(x) = x^4 - 9x^2 - 4x + 12$ (see Figure 19)
    (a) $x^4 + 12 \leq 9x^2 + 4x$
    (b) $x^4 - 9x^2 + 12 > 4x$

Figure 16

$g(x) = x^4 - 15x^2 - 10x + 24$

Figure 19

$g(x) = x^4 - 9x^2 - 4x + 12$

In problems 7–12, use the given graph to find the intervals where the function is increasing, decreasing, or constant.

7.

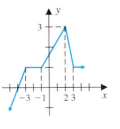

8.

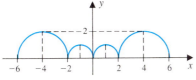

9.

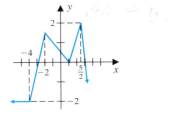

10.

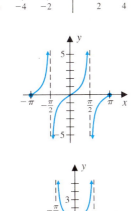

11.

12.

In problems 13–18:

(a) Determine without graphing whether each function is even, odd, or neither by using function notation.

(b)  Demonstrate the validity of the answer in part (a) by graphing the function with a grapher.

13. $f(x) = 5x^3 - 2x$

14. $f(x) = 5x^2 + x^3$

15. $f(x) = -x^2 - |x|$

16. $G(x) = \dfrac{\sqrt{x^2 + 4}}{|x|}$

17. $f(x) = \dfrac{1}{x} + x^2$

18. $h(x) = \dfrac{5}{x^2 + 4}$

19. Determine whether each graph in problems 7–12 is the graph of an even function, an odd function, or neither.

20. Explain why the equation $x^2 + y^2 = 4$ does not represent an even function, even though its graph is symmetric with respect to the y axis.

In problems 21–26:

(a) Determine without graphing whether each function is even, odd, or neither by using function notation.

(b) Demonstrate the validity of the answer in part (a) by graphing the function and examining the graph for symmetry.

21. $h(x) = 4x$

22. $g(x) = -\dfrac{1}{2}x$

23. $f(x) = 8x^2 + 3$

24. $g(x) = -4x^2 + 8$

25. $F(x) = x^2 - x$

26. $G(x) = \dfrac{1}{x}$

In problems 27–36, sketch the graph of each function. Use the graph to determine whether the function is even or odd, and to indicate the intervals where it is increasing, decreasing, or constant.

27. $f(x) = 3x$

28. $g(x) = -2x$

29. $h(x) = -4x + 5$

30. $f(x) = 3x + 2$

31. $F(x) = -5x^2 + 10$

32. $G(x) = -7x^2 + 1$

33. $F(x) = \sqrt{4 - x^2}$

34. $H(x) = -\sqrt{4 - x^2}$

35. $g(x) = -x^2$

36. $H(x) = x^2 - 1, x \geq 0$

In problems 37–40, use standard graphs to sketch the graph of each piecewise-defined function. (Refer to Figure 8.)

37. $g(x) = \begin{cases} 2 & \text{if } x < 4 \\ \sqrt{x} & \text{if } x \geq 4 \end{cases}$

38. $h(x) = \begin{cases} x & \text{if } x \leq 0 \\ \sqrt{x} & \text{if } x > 0 \end{cases}$

39. $F(x) = \begin{cases} x^3 & \text{if } x \leq 0 \\ \sqrt{1 - x^2} & \text{if } 0 < x \leq 1 \end{cases}$

40. $f(x) = \begin{cases} \sqrt[3]{x} & \text{if } x < -1 \\ |x| & \text{if } x \geq -1 \end{cases}$

In problems 41–44, sketch the graph of each piecewise-defined function. Determine the domain and range. Identify each point of discontinuity.

41. $f(x) = \begin{cases} x + 3 & \text{if } x \leq 1 \\ 4x^2 & \text{if } x > 1 \end{cases}$

42. $g(x) = \begin{cases} 2x^2 + 1 & \text{if } x < 3 \\ 10 - x & \text{if } x \geq 3 \end{cases}$

43. $h(x) = \begin{cases} 3x + 1 & \text{if } x < -2 \\ 0 & \text{if } -2 \leq x \leq 2 \\ -2x + 3 & \text{if } x > 2 \end{cases}$

44. $f(x) = \begin{cases} -x & \text{if } x < 0 \\ 4 - 2x & \text{if } 0 \leq x < 3 \\ 2x + 1 & \text{if } x \geq 3 \end{cases}$

In problems 45 and 46, graph the function and determine its range.

45. $f(x) = [\![3x]\!]$

46. $f(x) = \left[\!\left[\dfrac{x}{3} \right]\!\right]$

G In problems 47–50, use a grapher to find the x intercepts of the associated function, then solve each inequality.

47. $|3x - 1| \leq |2x + 1|$

48. $|2x + 3| > |x - 3|$

49. $|x + 2| < 5 - |x - 1|$

50. $|2x + 1| \geq |x - 1|$

Applying the Concepts

51. Cost Trend: The function defined by the graph in Figure 20 indicates the average cost C (in thousands of dollars) for a manufacturer to produce x (in hundreds) items of a product, up to 1000 items.

Figure 20

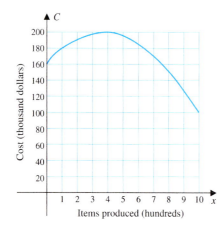

(a) Find the average cost if the number of items produced is 100, 400, 700, 800, 1000.

(b) Find the number of items that would yield an average cost of $140,000.

(c) Find the intervals where the function is increasing and decreasing, and interpret this behavior in terms of average cost and items produced.

(d) Find the highest and lowest average costs and when they occur.

52. Water Level: Figure 21 indicates the height h (in feet) of the tide above mean sea level in a harbor at a time t over a 24-hour period.

Figure 21

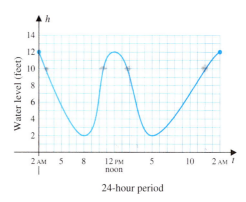

24-hour period

(a) Find the height at 5AM, noon, 5PM, 2AM.

(b) Approximate the times when the water level above mean sea level will be 10 feet.

(c) Determine the time intervals when the water level is rising and falling.

(d) Find the highest and lowest levels and when they occur.

53. Water Charges: The monthly charge C (in dollars) for water in a small town is given by the piecewise-defined function

$$C(x) = \begin{cases} 18 & \text{if } 0 \leq x \leq 20 \\ 18 + 0.1(x - 20) & \text{if } x > 20 \end{cases}$$

where x is in hundreds of gallons.

(a) Sketch the graph of C.

(b) What is the monthly fixed charge?

(c) Suppose a proposed rate change would fix the monthly charge at $12 for the first 1500 gallons and charge an additional 25¢ for each hundred gallons beyond 1500. Write a piecewise-defined function that expresses the monthly charge K (in dollars) in terms of x (in hundreds of gallons of water) for the proposed rate change.

(d) Graph the function from part (c).

(e) Compare the charges under each rate structure.

54. Telephone Charges: Suppose the cost of an international telephone call between two countries during business hours is $0.83 for the first minute and $0.75 for each additional minute (or portion of a minute). So the cost C (in dollars) of a call that lasts t minutes is given by the equation

$$C = 0.83 - 0.75[\![1 - t]\!]$$

(a) Sketch the graph of this function.

(b) What is the fixed charge?

(c) A proposed rate change would charge $0.65 up to 1 minute and $0.85 for each additional minute (or portion of a minute). Under this proposal, use the greatest-integer function to express the cost K (in dollars) as a function of time t, the number of minutes the call lasts.

(d) Graph the function from part (c).

(e) Compare the costs of a call under each rate structure.

55. Car Rental: A car rental agency offers special rates for a 4-day convention. It charges C dollars for renting a car for N days as specified in Table 5.

TABLE 5

Number of Days (N)	Charge (C) Dollars
1 day, or part of a day	20
2 days, or 1 day and part of a day	38
3 days, or 2 days and part of a day	56
4 days, or 3 days and part of a day	74

(a) Determine a piecewise-defined function that expresses C in terms of N.

(b) What are the restrictions on N?

(c) Graph the function in part (a) with the restrictions from part (b).

(d) Suppose that a competing car agency offers a flat rate of $19.50 per day or any part of a day. Express the cost K (in dollars) as a piecewise-defined function of N, where $0 < N \leq 4$, and graph it.

(e) Compare the competing rate structures.

56. Stockbroker's Fee: Suppose that the fee C (in dollars) charged by a discount stockbroker for executing a trade of N shares is given in Table 6.

(a) Find the fee if the broker executes a trade of 50 shares, 100 shares, 275 shares, 923 shares.

(b) Determine a piecewise-defined function that expresses C in terms of N.

(c) What are the restrictions on N?

TABLE 6

Number of Shares (N)	Fee (C) Dollars
0 < N < 100	25 + 0.10 per share traded
100 ≤ N < 500	30 + 0.05 per share traded
N ≥ 500	45 + 0.02 per share traded

(d) Graph the function in part (b) with the restrictions from part (c).

(e) What would the graph look like if the fee structure in Table 5 were changed so that a flat fee of $55 is charged for trades involving at least 100 shares? Compare this graph to the graph from part (d). Compare the two fee structures.

57. Post Office Rates: The 2000 U.S. postal rates R (in dollars) for mailing a first-class item weighing up to and including 6 ounces are summarized in Table 7.

TABLE 7

Weight w Ounces	Rate R Dollars
0 < w ≤ 1	0.33
1 < w ≤ 2	0.55
2 < w ≤ 3	0.77
3 < w ≤ 4	0.99
4 < w ≤ 5	1.21
5 < w ≤ 6	1.43

(a) Use the greatest-integer function to determine a function that expresses the rate R in terms of w.

(b) What are the restrictions on w?

(c) Graph the function in part (a) with the restrictions from part (b).

(d) If this rate structure applies to items that weigh up to 11 ounces, then what would be the rate for a letter weighing 7 ounces? 9.5 ounces? 10 ounces?

58. Parking Charges: A public parking structure with an 8-hour limit charges its customers $2.50 for the first hour (or part of the hour) and $1.50 for each additional hour (or part of the hour) up to 8 hours.

(a) Use the greatest-integer function to determine a function that expresses the cost C (in dollars) of parking as a function of t, the number of hours parked.

(b) Sketch the graph of the function with the appropriate restrictions on t.

(c) A rate change is made that charges the customer $3.25 for the first hour and $1.25 for each additional hour or part of an hour up to 8 hours. Under this change, use the greatest-integer function to determine a function that expresses the cost K (in dollars) in terms of t, the number of hours parked, where $0 < t \leq 8$.

(d) Graph the function in part (c).

(e) Compare the parking charges under each rate structure.

Developing and Extending the Concepts

59. (a) If f is an odd function, does the graph of f necessarily contain the origin? Support your assertion.

(b) Sketch the graph of the function

$$f(x) = \frac{1}{x}.$$

(c) Is your assertion in part (a) consistent with the graph in part (b)?

60. (a) If g is an even function, does the graph of g necessarily intersect the y axis? Support your assertion.

(b) Sketch the graph of the function

$$g(x) = \frac{1}{x^2}.$$

(c) Is your assertion in part (a) consistent with the graph in part (b)?

61. Given the piecewise-defined equation

$$y = \begin{cases} x & \text{if } x \leq 5 \\ x - 1 & \text{if } x > k \end{cases}$$

(a) Does the equation represent a function if $k = 1$? Explain.

(b) Does the equation represent a function if $k = 6$? Explain.

(c) What conditions must be placed on the value of k in order to have a function?

62. (a) Assume the graph of

$$f(x) = x^n + x^2$$

is symmetric with respect to the y axis. What condition must n, a positive integer, satisfy? Explain.

(b) It is possible for the graph of

$$g(x) = x^n + x^3,$$

where n is a positive integer, to be symmetric with respect to the y axis? Explain.

63. (a) Is it possible to have a function with a graph that is symmetric with respect to the x axis? Explain.

(b) Is it possible for a function to be both even and odd? Explain.

64. Rewrite each function as a piecewise-defined function that does not involve absolute value; then sketch the graph. Find any points of discontinuity.

(a) $f(x) = \dfrac{|x|}{x}$

(b) $g(x) = 4|x| - 1$

(c) $h(x) = 2x + \dfrac{|x - 3|}{x - 3}$

65. Assume that the following table is a representation of an even function. Complete the table.

x	$f(x)$
-3	82
-2	___
-1	2
0	0
1	___
2	17
3	___

Plot the resulting points and connect them with a smooth curve. Is this graph symmetric with respect to the y axis? Explain.

66. Assume that the following table is a representation of an odd function. Complete the table.

x	$f(x)$
-3	-22
-2	___
-1	4
0	0
1	___
2	13
3	___

Plot the resulting points and connect them with a smooth curve. Is this graph symmetric with respect to the origin? Explain.

Objectives

1. Graph by Vertical Shifting
2. Graph by Horizontal Shifting
3. Graph by Reflecting
4. Graph by Vertical Scaling
5. Graph by Using More Than One Transformation
6. Solve Applied Problems

2.4 Transformations of Graphs

In this section we present graphing techniques, called *transformations,* where we geometrically *transform* standard graphs to obtain graphs of other functions. Our graphing techniques rely on remembering the standard graphs in Section 2.3.

Graphing by Vertical Shifting

To understand the idea behind vertical shifting, we begin by considering the functions

$$f(x) = x^2, g(x) = x^2 + 2, \text{ and } h(x) = x^2 - 1.$$

Table 1 lists some x values along with the corresponding y values for each of the three functions.

TABLE 1

x	$f(x) = x^2$	$g(x) = x^2 + 2$	$h(x) = x^2 - 1$
-2	4	6	3
-1	1	3	0
0	0	2	-1
1	1	3	0
2	4	6	3

Next we graph all three functions on the same coordinate system by plotting the points from Table 1 for each of the three functions and then connecting them with smooth curves (Figure 1). By examining the numerical patterns in Table 1 and the geometric feature of the three graphs, we make the following observations.

1. The three graphs have the same shape but different locations.

2.

Figure 1

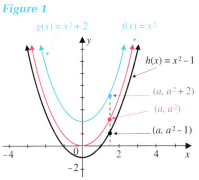

	Input Value	Resulting Outputs	
	$\rightarrow a$	$g(a) = a^2 + 2 \leftarrow$	For each point on the graph of f, there is a point on the graph of g 2 units up
x coordinates are the same	$\rightarrow a$	$f(a) = a^2 \Leftarrow$	
			For each point on the graph of f, there is a point on the graph of h 1 unit down
	$\rightarrow a$	$h(a) = a^2 - 1 \leftarrow$	

Thus we conclude that the graph of $g(x) = x^2 + 2$ can be obtained by *vertically shifting* or *translating* the graph of $f(x) = x^2$ upward 2 units. Similarly, the graph of $h(x) = x^2 - 1$ can be obtained by vertically shifting the graph of $f(x) = x^2$ downward 1 unit.

This example is generalized in Table 2.

TABLE 2 Graphing by Vertical Shifting

Operation on $y = f(x)$	New Function Equation	Geometric Effect on Graph of $y = f(x)$	Illustration
Add a positive constant k to $f(x)$	$y = f(x) + k$	Shifts the graph of f vertically k units upward	
Subtract a positive constant k from $f(x)$	$y = f(x) - k$	Shifts the graph of f vertically k units downward	

EXAMPLE 1   Vertically Shifting a Graph

Use the graph of $f(x) = |x|$ to sketch the graph of each function.

(a) $F(x) = |x| + 2$

(b) $G(x) = |x| - 3$

Solution (a) The graph of F can be obtained by shifting the graph of f upward 2 units (Figure 2a).

(b) The graph of G can be obtained by shifting the graph of f downward 3 units (Figure 2b).

Figure 2

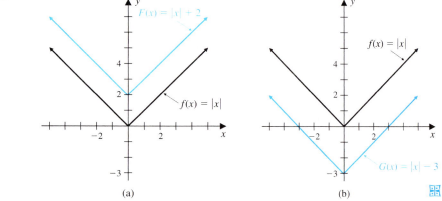

(a)   (b)

Graphing by Horizontal Shifting

We introduce the concept of horizontal shifting by making some comparisons between the functions

$$f(x) = x^2 \text{ and } g(x) = (x - 1)^2.$$

Tables 3a and 3b list some points on the graphs of f and g. Notice the numerical patterns in Table 3. If we add 1 to each x value for f, the resulting x value for g produces the same y value.

TABLE 3

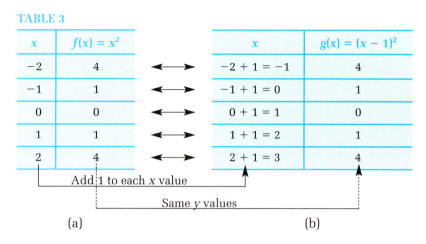

x	$f(x) = x^2$		x	$g(x) = (x - 1)^2$
-2	4	←→	$-2 + 1 = -1$	4
-1	1	←→	$-1 + 1 = 0$	1
0	0	←→	$0 + 1 = 1$	0
1	1	←→	$1 + 1 = 2$	1
2	4	←→	$2 + 1 = 3$	4

Add 1 to each x value

Same y values

         (a)   (b)

The graphs of f and g are obtained by plotting the points in Table 3 and then connecting them with smooth curves (Figure 3).

Figure 3

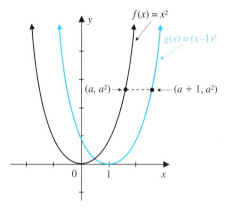

By examining the numerical pattern shown in Table 3, along with the geometric features of the graphs in Figure 3, we are led to the following observations:

1. The shapes of the graphs are the same, but their locations are different.
2.

	Input Value	Resulting Outputs	
For each point on the graph of f, there is a point on the graph of g 1 unit to the right	a	$f(a) = a^2$	Same y coordinates
	$a + 1$	$g(a + 1) = [(a + 1) - 1]^2$ $= a^2$	

In other words, the graph of g can be obtained by shifting the graph of f horizontally 1 unit to the right.

This example leads to the general results given in Table 4.

TABLE 4 Graphing by Horizontal Shifting

Operation on $y = f(x)$	New Function Equation	Geometric Effect on Graph of $y = f(x)$	Illustration
Add a positive constant h to x	$y = f(x + h)$	Shifts the graph of f horizontally h units to the left	
Subtract a positive constant h from x	$y = f(x - h)$	Shifts the graph of f horizontally h units to the right	

EXAMPLE 2 **Horizontally Shifting a Graph**

Use the graph of $y = \sqrt{x}$ to sketch the graph of each function.

(a) $f(x) = \sqrt{x + 1}$

(b) $g(x) = \sqrt{x - 2}$

Solution   (a) The graph of

$$f(x) = \sqrt{x + 1}$$

can be obtained by shifting the graph of

$$y = \sqrt{x}$$

to the left 1 unit (Figure 4a).

(b) The graph of

$$g(x) = \sqrt{x - 2}$$

can be obtained by shifting the graph of

$$y = \sqrt{x}$$

to the right 2 units (Figure 4b).

Figure 4

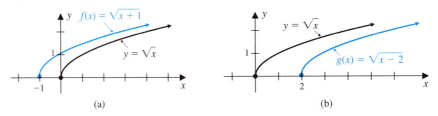

(a)   (b)

Graphing by Reflecting

Another type of transformation is called a *reflection.* Let us consider the functions

$$f(x) = \sqrt{x} \quad \text{and} \quad g(x) = -\sqrt{x}$$

Table 5 lists some points on the graphs of f and g. By plotting these points on the same coordinate system and connecting them with smooth curves, we obtain both graphs in (Figure 5).

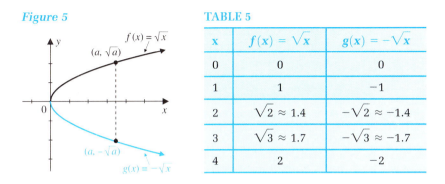

Figure 5

TABLE 5

x	$f(x) = \sqrt{x}$	$g(x) = -\sqrt{x}$
0	0	0
1	1	−1
2	$\sqrt{2} \approx 1.4$	$-\sqrt{2} \approx -1.4$
3	$\sqrt{3} \approx 1.7$	$-\sqrt{3} \approx -1.7$
4	2	−2

By examining the numerical values in Table 5 and the geometric features in Figure 5, we discover that the graphs of f and g are symmetric with respect to the x axis. Our observations are summarized as follows:

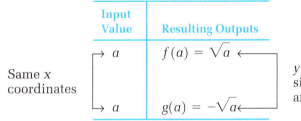

So the graph of $g(x) = -\sqrt{x}$ can be obtained by reflecting the graph of $f(x) = \sqrt{x}$ about the x axis.

Similarly, the graphs of

$$f(x) = \sqrt{x}$$

and

$$h(x) = \sqrt{-x}$$

shown in Figure 6 are symmetric about the y axis. We prove this symmetry as follows by assuming that $a > 0$:

Figure 6

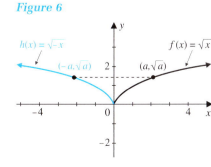

	Input Value	Resulting Outputs
	a	$f(a) = \sqrt{a}$
x coordinates have opposite signs	$-a$	$h(-a) = \sqrt{-(-a)}$ $= \sqrt{a}$

Same *y* coordinates

So the graph of $h(x) = \sqrt{-x}$ can be obtained by reflecting the graph of

$$f(x) = \sqrt{x} \text{ about the } y \text{ axis}$$

These illustrations lead to the general results given in Table 6.

TABLE 6 Graphing by Reflecting

Operation on $y = f(x)$	New Function Equation	Geometric Effect on Graph of $y = f(x)$	Illustration
Multiply $f(x)$ by -1	$y = -f(x)$	Reflects the graph of f about the x axis	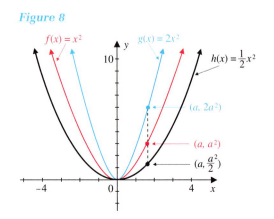
Replace x by $-x$	$y = f(-x)$	Reflects the graph of f about the y axis	

EXAMPLE 3   **Reflecting a Graph About the x Axis**

Use the graph of $y = x^2$ to sketch the graph of $f(x) = -x^2$.

Solution   The graph of $f(x) = -x^2$ is obtained by reflecting the graph of $y = x^2$ about the x axis (Figure 7).

Figure 7

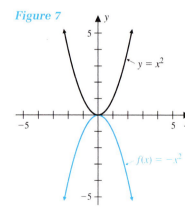

Graphing by Vertical Scaling

Translations and reflections change only the position of the graph, not its shape. Now we consider a transformation, called *vertical scaling,* that actually changes the shape of a graph. We begin by forming a table of values for the three functions

$$f(x) = x^2, \quad g(x) = 2x^2, \quad \text{and} \quad h(x) = \frac{1}{2}x^2$$

Next we use Table 7 and the point-plotting technique to graph all three functions on the same coordinate system (Figure 8).

TABLE 7

x	$f(x) = x^2$	$g(x) = 2x^2$	$h(x) = \frac{1}{2}x^2$
-2	4	8	2
-1	1	2	$\frac{1}{2}$
0	0	0	0
1	1	2	$\frac{1}{2}$
2	4	8	2

Figure 8

By comparing the numerical patterns in Table 7 and the graphs we make the following observations:

1. In comparison to the graph of f, the graph of g is vertically stretched, whereas the graph of h is vertically compressed or flattened.

2.

	Input Value	Resulting Outputs	
	$\rightarrow a$	$g(a) = a^2$	y coordinates of graph of g are *double* those of graph of f
Same x coordinates	$\rightarrow a$	$f(a) = a^2$	
	$\rightarrow a$	$h(a) = \dfrac{1}{2} a^2$	y coordinates of graph of h are *half* those of graph of f

Thus we conclude that the graph of $g(x) = 2x^2$ can be obtained by vertically stretching the graph of $f(x) = x^2$ by a factor of 2. Similarly, the graph of $h(x) = (1/2)x^2$ can be obtained by vertically compressing or flattening the graph of f by a factor of $1/2$.

This example leads us to the general results given in Table 8.

TABLE 8

Operation on $y = f(x)$	New Function Equation	Geometric Effect on Graph of $y = f(x)$	Illustration
Multiply $f(x)$ by a, where $a > 1$	$y = af(x)$	Vertically stretches the graph of f by a factor of a	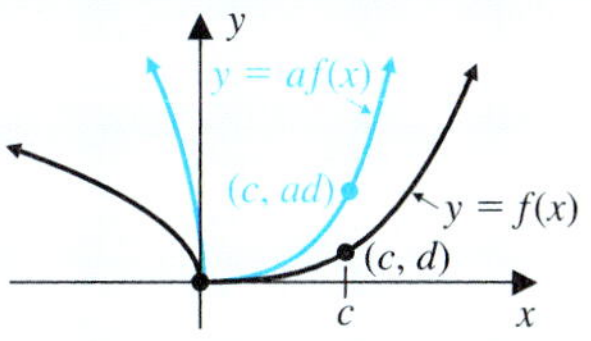
Multiply $f(x)$ by a, where $0 < a < 1$	$y = af(x)$	Vertically compresses the graph of f by a factor of a	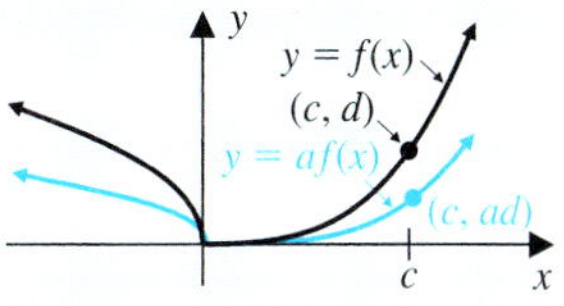

EXAMPLE 4   **Vertically Scaling a Graph**

Use the graph of $y = x^3$ to sketch the graph of each function.

(a) $g(x) = 5x^3$ (b) $h(x) = (1/5)x^3$

Solution   (a) The graph of $g(x) = 5x^3$ can be obtained by vertically stretching the graph of $y = x^3$ by a factor of 5 (Figure 9a).

(b) The graph of

$$h(x) = (1/5)x^3$$

can be obtained by vertically compressing the graph of $y = x^3$ by a factor of 1/5 (Figure 9b).

Figure 9

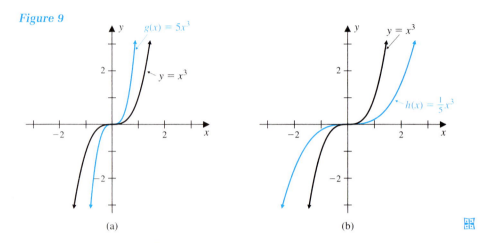

(a)   (b)

Graphing by Using More Than One Transformation

When combining more than one transformation in the same graph, the order in which these transformations are performed is important. Generally, the graph of a function of the form

$$F(x) = af(x - h) + k$$

can be sketched from the graph of $y = f(x)$ by following the sequence of transformations in the order specified in Table 9.

TABLE 9   Sketching the Graph of $F(x) = af(x - h) + k$ from the Graph of $y = f(x)$

Order of Transformations	Equation Change	Geometric Change
Step 1 *Vertically stretch or compress by a factor of $\lvert a \rvert$*	$y = af(x)$	Stretch if $\lvert a \rvert > 1$; compress if $0 < \lvert a \rvert < 1$
Step 2 *Reflect about x axis if $a < 0$*	$y = af(x)$	Reflect about x axis
Step 3 *Horizontally shift*	$y = af(x - h)$	Horizontal shift to the right if $h > 0$, to the left if $h < 0$
Step 4 *Vertically shift*	$y = af(x - h) + k$	Vertical shift up if $k > 0$, down if $k < 0$

EXAMPLE 5   **Using Multiple Transformations to Sketch a Graph**

Use transformations of the graph $y = |x|$ to sketch the graph of the function

$$f(x) = -2|x + 1| + 3.$$

Solution   We begin with the graph of $y = |x|$ and then proceed by following the order of transformations given in Table 9.

Step 1. Vertically stretch the graph of $y = |x|$ by a factor of 2 to obtain the graph of $y_1 = 2|x|$ (Figure 10a).

Step 2. Reflect the graph of $y_1 = 2|x|$ about the x axis to get the graph of $y_2 = -2|x|$ (Figure 10b).

Step 3. Horizontally shift the graph of $y_2 = -2|x|$ to the left 1 unit to get the graph of $y_3 = -2|x + 1|$ (Figure 10c).

Step 4. Vertically shift the graph of $y_3 = -2|x + 1|$ up 3 units to obtain the desired graph of $f(x) = -2|x + 1| + 3$ (Figure 10d).

Figure 10

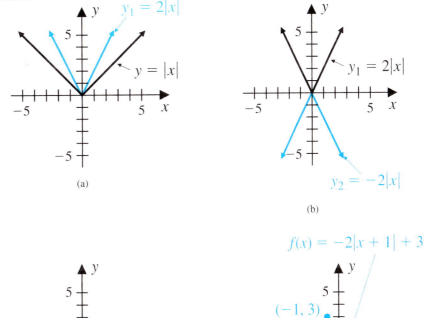

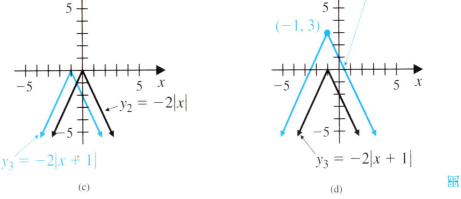

(a)   (b)   (c)   (d)

A reflection about the y axis is carried out in place of or in addition to Step 2 as the next example demonstrates.

EXAMPLE 6   **Reflecting a Graph about the *y* Axis**

Use transformations of the graph of $y = \sqrt{x}$ to sketch a graph of

$$f(x) = \sqrt{2 - x}.$$

Solution   First we rewrite the function as

$$f(x) = \sqrt{-(x - 2)}.$$

Next we follow the order of transformations given in Table 9 as follows:

Step 1. There is no vertical stretching or compressing.

Step 2. Reflect the graph of $y = \sqrt{x}$ about the *y* axis to graph

$$y_1 = \sqrt{-x} \text{ (Figure 11a)}.$$

Step 3. Horizontally shift the graph of $y_1 = \sqrt{-x}$ to the right 2 units to obtain the desired graph of

$$f(x) = \sqrt{-(x - 2)} \text{ (Figure 11b)}.$$

If we follow a different order of transformations, we may get an erroneous result (see problem 53).

Figure 11

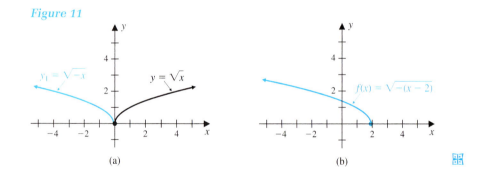

(a)   (b)

G We can use a grapher to study the effects of transformations on functions such as

$$f(x) = \sqrt{x}(x + 3)$$

$$\text{Suppose } g(x) = f(x - 2)$$

$$h(x) = -f(x)$$

$$\text{and } k(x) = -f(x - 2) + 3$$

Figure 12

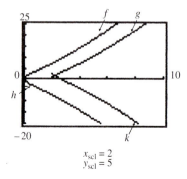

$$x_{\text{scl}} = 2$$
$$y_{\text{scl}} = 5$$

Figure 12 displays the graphs of the functions *f*, *g*, *h*, and *k*

$$\text{since} \qquad f(x) = \sqrt{x}(x + 3),$$

$$g(x) = f(x - 2) = \sqrt{x - 2}(x + 1), \text{ so its graph}$$

has the same shape as the graph of *f*, except it is shifted 2 units to the right.

$$h(x) = -f(x) = -\sqrt{x}(x + 3), \text{ and its}$$

graph is the mirror image of the graph of *f* across the *x* axis.

$$k(x) = -f(x - 2) + 3 = -\sqrt{x - 2}(x + 1) + 3, \text{ and its graph}$$

is the mirror image of the graph of *g* across the *x* axis, but it is also vertically shifted 3 units upward.

The next example illustrates how to start with a standard graph and follow a given sequence of transformations to construct another graph and determine its equation.

EXAMPLE 7   **Constructing the Function Equation from Given Transformations**

Determine the function equation for F if the graph of F is obtained from the graph of $f(x) = x^3$ as follows:

First, the graph of f is reflected about the x axis.
The resulting graph is shifted 4 units to the right.
Finally, the latter graph is shifted 3 units down.

Solution   Construction of the equation for F is developed as follows:

Figure 13

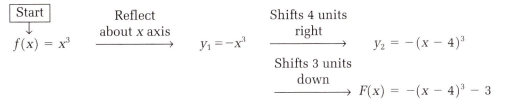

The graphs of f and F are shown in Figure 13.

Solving Applied Problems

Transformations of graphs can be used in mathematical modeling, as the next example illustrates.

EXAMPLE 8   **Graphing a Model from Physics**

A sandbag is dropped from a hot air balloon 5000 feet above ground. Its height h (in feet) above ground level t seconds after it is dropped is given by the function

$$h(t) = -16t^2 + 5000.$$

(a) Use transformations to get the graph of h from the graph of $f(t) = t^2$.

(b) Suppose that another sandbag of the same size is dropped from the balloon 2 seconds after the first sandbag is dropped. Construct a model for its height H (in feet) above ground level as a function of t by using the function h.

(c) Use the graph of h to obtain the graph of H.

Solution   Since t represents elapsed time, and h and H represent heights, we restrict the graphs to points where $t \geq 0$, $h(t) \geq 0$, and $H(t) \geq 0$.

(a) We can get the graph of

$$h(t) = -16t^2 + 5000$$

from the graph of $f(t) = t^2$ by first vertically stretching the graph of f by a scale factor of 16, then reflecting the resulting graph across the t axis, and finally vertically shifting the latter graph up 5000 units to obtain the graph of h (Figure 14).

Figure 14

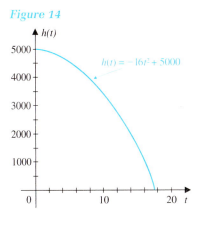

(b) Since the second sandbag is the same size and is dropped 2 seconds later, then its time in seconds is represented by $t - 2$, and its equation is obtained from that of h by replacing t with $t - 2$ to get

$$H(t) = -16(t - 2)^2 + 5000, \ t \geq 2$$

(c) The graph of

$$H(t) = -16(t - 2)^2 + 5000$$

can be obtained by shifting the graph of $h(t) = -16t^2 + 5000$, 2 units to the right (Figure 15).

Figure 15

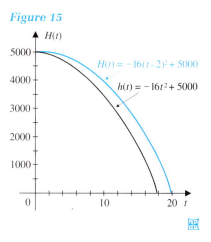

PROBLEM SET 2.4

Mastering the Concepts

In problems 1–6, use transformations to sketch the graphs of f for the given values of each constant on the same coordinate system.

1. $f(x) = \sqrt{x} + k, \ k = -2, 1, 2$
2. $f(x) = \sqrt{1 - x^2} + k, \ k = -2, 1, 2$
3. $f(x) = (x + h)^3, \ h = -2, 1, 2$
4. $f(x) = \sqrt[3]{x + h}, \ h = -2, 1, 2$
5. $f(x) = a|x|, \ a = -3, -1, -\frac{1}{2}, \frac{1}{2}, 3$
6. $f(x) = ax^3, \ a = -3, -1, -\frac{1}{2}, \frac{1}{2}, 3$

In problems 7–12, use transformations to explain how the graph of f is related to the given function, and sketch the graph of f.

7. $y = x^2; \ f(x) = 2x^2 + 3$
8. $y = \sqrt{x}; \ f(x) = 3\sqrt{x} - 2$
9. $y = |x|; \ f(x) = 3|x - 1| - 2$
10. $y = x^3; \ f(x) = -(x + 1)^3 + 2$
11. $y = \sqrt{x}; \ f(x) = -2\sqrt{x - 1} + 3$
12. $y = \sqrt[3]{x}; \ f(x) = 2\sqrt[3]{x - 1} - 3$

In problems 13 and 14, use the given graph of f along with transformations to graph each of the following functions:

(a) $y = 4f(x)$

(b) $y = f(x) - 2$

(c) $y = f(x - 3)$

(d) $y = -\dfrac{1}{2}f(x) + 3$

(e) $y = 2f(x + 1) - 3$

13.

14.

In problems 15 and 16, the graph of a function f is given along with its equation. The graphs of G, T, and S, obtained by transformations of the graph of f, are shown on the same coordinate system. Find equations for G, T, and S.

15.

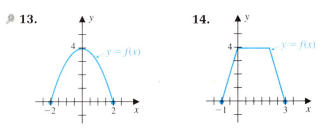

16.

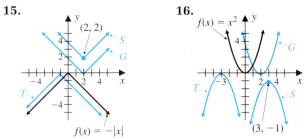

In problems 17–24, the graph of function G is obtained from the graph of the given function f as specified. Determine the function equation for G. Graph f and G on the same coordinate system.

17. Shift the graph of $f(x) = x$ up 3 units.
18. Shift the graph of $f(x) = |x|$ down 3 units.
19. Shift the graph of $f(x) = x^2$ to the right 2 units.
20. Shift the graph of $f(x) = x^3$ to the left 2 units.
21. Reflect the graph of $f(x) = \sqrt{x}$ across the x axis, and then shift it to the right 5 units.

22. Reflect the graph of $f(x) = \sqrt{x}$ across the x axis, and then shift it $\frac{1}{2}$ unit to the right.

23. Shift the graph of $f(x) = x^2$ up 1 unit and to the left 2 units, and then vertically scale it by a multiple of $\frac{1}{3}$.

24. Reflect the graph of $f(x) = x^2$ across the x axis, and then vertically scale it by a multiple of 2.

In problems 25–44, use a standard graph along with transformations to graph each function. Identify the equation of the standard graph. Also, find the domain and range of the function.

25. $f(x) = x + 2$

26. $g(x) = \sqrt{x} - 2$

27. $G(x) = 1 - x^3$

28. $h(x) = 1 + \sqrt[3]{x}$

29. $f(x) = \sqrt{x + 4}$

30. $g(x) = |x + 3|$

31. $F(x) = 2|x - 1|$

32. $h(x) = -2|x + 1|$

33. $H(x) = -5\sqrt{x + 1} + 1$

34. $f(x) = 7 + |x - 2|$

35. $f(x) = -(1/3)(x + 1)^3$

36. $g(x) = \dfrac{(x - 3)^2}{2}$

37. $g(x) = 4x^2 - 1$

38. $f(x) = 2(x - 1)^2 + 1$

39. $h(x) = 2(x - 3)^2 + 4$

40. $h(x) = (x + 1)^3 - 2$

41. $G(x) = 3\left(x - \dfrac{1}{2}\right)^3 - 1$

42. $H(x) = -2\sqrt{x + 1} + 3$

43. $g(x) = (1/2)\sqrt{1 - x^2}$

44. $F(x) = 3\sqrt{4 - x^2} + 1$

In problems 45 and 46, use transformations of an appropriate standard graph to sketch the graph of each given function.

45. $f(x) = (5 - x)^3$

46. $g(x) = \sqrt[3]{5 - x}$

Developing and Extending the Concepts

47. Suppose that the point (a, b) lies on the graph of $y = f(x)$. What are the coordinates of the resulting point location in terms of a and b after (a, b) is transformed according to the graph of each equation?
(a) $y = f(x - 2)$ (b) $y = f(x + 4)$
(c) $y = -f(-x) + 1$

48. Suppose that the function f is decreasing on the interval $(-1, 1)$ and increasing on the intervals $(-\infty, -1)$ and $(1, \infty)$. On what interval(s) is the function defined by $y = -f(-x)$ decreasing?

In problems 49–52, suppose that h, k, and a, $a \neq 0$, are constants. Let:

(a) $g(x) = f(x) + k$
(b) $g(x) = f(x + h)$
(c) $g(x) = af(x)$

49. Suppose that f is increasing. Which of the given g functions is increasing?

50. Suppose that f is decreasing. Which of the given g functions is decreasing?

51. Suppose that f is even. Which of the given g functions is even?

52. Suppose that f is odd. Which of the given g functions is odd?

53. In Example 6, the graph of $f(x) = \sqrt{2 - x}$ was obtained by reflecting the graph of $y = \sqrt{x}$ about the y axis, followed by shifting 2 units to the right.
(a) Reverse the order of transformations; that is, shift the graph of $y = \sqrt{x}$ to the right 2 units and then reflect the result about the y axis.
(b) The graph obtained in part (a) is not the same as the graph in Example 6. Explain the differences and why they occurred.
(c) How can such erroneous results be avoided?

54. Suppose that a function f is described by the following table.

x	-2	-1	0	1	2	3	4	5	6
$f(x)$	-3	-1	0	3	-5	6	7	9	8

Use the numerical representation of f given by the table to make a table of the numerical representation of g if the functions f and g are related by the given equation.

(a) $g(x) = f(x) + 4$
(b) $g(x) = f(x - 3)$
(c) $g(x) = f(x + 1) + 2$
(d) $g(x) = -f(x - 2) - 1$

G In problems 55–58, a function f is given.

(a) Determine the function equation for G.
(b) Use a grapher to sketch graphs of f and G on the same coordinate system.
(c) Describe how the graph of G can be obtained from the graph of f if transformations are used.

55. $f(x) = \sqrt[3]{x}(x + 2)$; $G(x) = f(x - 1)$

56. $f(x) = \sqrt[4]{x}(x - 3)$; $G(x) = f(x + 1)$

57. $f(x) = \dfrac{x^2}{x + 1}$; $G(x) = f(x - 1) + 2$

58. $f(x) = \dfrac{\sqrt[3]{x}}{x^2 + 1}$; $G(x) = f(x - 1) + 2$

Applying the Concepts

59. Physics Problem: A rescue helicopter drops a crate of supplies from a height of 160 feet. After t seconds, the height h (in feet) of the crate is given by the model

$$h(t) = 160 - 16t^2, t \geq 0$$

(a) Use transformations to obtain the graph of h from the graph of $f(t) = t^2$.

(b) Suppose that another crate of the same size is dropped from the helicopter 3 seconds after the first crate is dropped. Construct a model for its height H (in feet) above ground level as a function of t by using the given function h.

(c) Use the graph of h to obtain the graph of H.

60. Projectile Problem: A projectile is fired directly upward from the ground with an initial speed of 96 feet per second. Its distance d (in feet) above ground after t seconds is given by the model

$$d(t) = 96t - 16t^2, t \geq 0$$

(a) Show that d can be rewritten in the form

$$d(t) = -16(t - 3)^2 + 144$$

Then obtain the graph of d from the graph of $f(t) = t^2$.

(b) Suppose that another projectile of the same size is fired directly upward from the ground with the same initial speed, 2 seconds later. Construct a model for its distance D (in feet) above ground as a function of t by using the function d from part (a).

(c) Use the graph of $f(t) = t^2$ to obtain the graph of D.

2.5 Combinations of Functions

Just as two numbers can be added, subtracted, multiplied, and divided to produce other numbers, so two functions can be combined using these operations to produce other functions. In this section, we discuss these operations along with another, called *composition,* which has no analogy in the arithmetic of numbers.

Determining the Sum, Difference, Product, and Quotient Functions

If $f(x) = x^2 - 1$ and $g(x) = 2x + 1$, we form the *sum, difference, product,* and *quotient* of f and g as follows:

Sum:

$$f(x) + g(x) = (x^2 - 1) + (2x + 1) = x^2 + 2x$$

Difference:

$$f(x) - g(x) = (x^2 - 1) - (2x + 1) = x^2 - 2x - 2$$

Product:

$$f(x) \cdot g(x) = (x^2 - 1)(2x + 1) = 2x^3 + x^2 - 2x - 1$$

Quotient:

$$\frac{f(x)}{g(x)} = \frac{x^2 - 1}{2x + 1}, x \neq -\frac{1}{2}$$

The result of each operation is a *new* function, denoted by $f + g$, $f - g$, $f \cdot g$ and f/g, respectively, and is formalized in the following definition.

Definitions
───────────
**Sum, Difference,
Product, and
Quotient Functions**

Let f and g be any two functions. We define the functions $f + g$, $f - g$, $f \cdot g$, and f/g as follows:

(i) **Sum function:** $\qquad$ $(f + g)(x) = f(x) + g(x)$

(ii) **Difference function:** $\qquad$ $(f - g)(x) = f(x) - g(x)$

(iii) **Product function:** $\qquad$ $(f \cdot g)(x) = f(x) \cdot g(x)$

(iv) **Quotient function:** $\qquad$ $\left(\dfrac{f}{g}\right)(x) = \dfrac{f(x)}{g(x)}, \quad$ for $g(x) \neq 0$

The domains of the sum, difference, product, and quotient functions consist of all values of x *common* to the domains of f and g, except for the quotient function, in which case the values of x for which the denominator is 0 are also excluded.

EXAMPLE 1

Finding the Sum, Difference, and Product Functions

Let $f(x) = 3x^3 + 7$ and $g(x) = x^2 - 3x - 4$. Find each of the following functions.

(a) $(f + g)(x)$ $\qquad$ (b) $(f - g)(x)$ $\qquad$ (c) $(f \cdot g)(x)$

Solution

(a) $(f + g)(x) = f(x) + g(x) = (3x^3 + 7) + (x^2 - 3x - 4)$

$\qquad\qquad\qquad = 3x^3 + x^2 - 3x + 3$

(b) $(f - g)(x) = f(x) - g(x) = (3x^3 + 7) - (x^2 - 3x - 4)$

$\qquad\qquad\qquad = 3x^3 - x^2 + 3x + 11$

(c) $(f \cdot g)(x) = f(x) \cdot g(x) = (3x^3 + 7) \cdot (x^2 - 3x - 4)$

$\qquad\qquad\qquad = 3x^5 - 9x^4 - 12x^3 + 7x^2 - 21x - 28$

EXAMPLE 2

Finding the Quotient Function and Its Domain

Find the quotient function f/g and its domain for each pair of functions.

(a) $f(x) = 3x^3 + 7$ and $g(x) = x^2 - 4$

(b) $f(x) = \sqrt{9 - x^2}$ and $g(x) = \sqrt{x - 1}$

Solution

(a) $\qquad\qquad\qquad \left(\dfrac{f}{g}\right)(x) = \dfrac{f(x)}{g(x)} = \dfrac{3x^3 + 7}{x^2 - 4}$

The domain for both f and g consists of all real numbers. However, the expression $(3x^3 + 7) / (x^2 - 4)$ represents a real number for all real values of x except those for which the denominator $x^2 - 4 = (x - 2)(x + 2) = 0$. Thus the domain of f/g includes all real numbers except 2 and -2.

(b) $\qquad\qquad\qquad \left(\dfrac{f}{g}\right)(x) = \dfrac{f(x)}{g(x)} = \dfrac{\sqrt{9 - x^2}}{\sqrt{x - 1}}$

The domain of f is the interval $[-3, 3]$, and the domain of g is the interval $[1, \infty)$. The values of x common to the domains of f and g consist of all numbers in the interval $[1, 3]$. However, we must exclude 1 from the domain of f/g because $g(1) = 0$. Thus the domain of f/g includes all numbers in the interval $(1, 3]$.

Forming Composite Functions

The *composition* of two functions provides us with another way of combining two functions to form a third function. The idea is to apply the functions in a specific order.

Suppose we are given two functions $y = f(x)$ and $y = g(x)$. We define a new function h as follows:

First, put an input number, say x, into the g function.

Second, take the output of the g function, $g(x)$, and use it as input into the f function.

The resulting output, $f[g(x)]$, is the output of the function $h(x) = f[g(x)]$; the function h is referred to as a *composite function*.

For instance, if $f(x) = x^3$ and $g(x) = 2x + 3$, then

$$h(1) = f[g(1)]$$
$$= f[2(1) + 3]$$
$$= f(5)$$
$$= 5^3 = 125$$
$$h(2) = f[g(2)]$$
$$= f[2(2) + 3]$$
$$= f(7)$$
$$= 7^3 = 343$$
$$h(-1) = f[g(-1)]$$
$$= f[2(-1) + 3]$$
$$= f(1)$$
$$= 1^3 = 1.$$

In a sense, the function h is obtained by a "chain reaction" in which we first apply the g function followed by the f function. The composite function h is sometimes written as

$$h = f \circ g \quad \text{Read "} f \text{ circle } g.\text{"}$$

Although the symbol $f \circ g$ may suggest some kind of product, it should not be confused with the actual product $f \cdot g$ of f and g.

For $f(x) = x^3$ and $g(x) = 2x + 3$, the equation for the composite function h can be determined by the following substitutions:

$h(x) = (f \circ g)(x) = f[g(x)]$	Input x
$= f(2x + 3)$	Output of g function is input of f function
$= (2x + 3)^3$	Output of f function
$(f \circ g)(x) = (2x + 3)^3$	Output of composite function h

These illustrations lead to the following definition

Definition

Composition of Functions

Let f and g be two functions. The **composition** of g *followed by* f, which is denoted by $f \circ g$, is the function defined by

$$(f \circ g)(x) = f[g(x)].$$

$f \circ g$ is called the **composite function,** whose domain consists of all values of x in the domain of g for which $g(x)$ is in the domain of f.

Figure 1 illustrates how the composite function $f \circ g$ is formed by applying the functions g and f in sequence: First apply g to x; then apply f to $g(x)$.

Figure 1

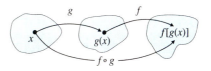

EXAMPLE 3   **Finding the Composition of Two Functions**

Let

$$f(x) = 5x^2 \quad \text{and} \quad g(x) = 2x - 3.$$

Find each of the following results:

(a) $(f \circ g)(4)$ (b) $(g \circ f)(4)$ (c) $(f \circ g)(x)$

(d) $(g \circ f)(x)$ (e) $(f \circ f)(x)$ (f) $(g \circ g)(x)$

Solution

(a) $(f \circ g)(4) = f[g(4)] = f(2 \cdot 4 - 3) = f(5) = 5 \cdot 5^2 = 125$

(b) $(g \circ f)(4) = g[f(4)] = g(5 \cdot 4^2) = g(80) = 2 \cdot 80 - 3 = 157$

(c) $(f \circ g)(x) = f[g(x)] = f(2x - 3) = 5(2x - 3)^2$

$$= 5(4x^2 - 12x + 9) = 20x^2 - 60x + 45$$

(d) $(g \circ f)(x) = g[f(x)] = g(5x^2) = 2 \cdot 5x^2 - 3 = 10x^2 - 3$

(e) $(f \circ f)(x) = f[f(x)] = f(5x^2) = 5(5x^2)^2 = 125x^4$

(f) $(g \circ g)(x) = g[g(x)] = g(2x - 3) = 2(2x - 3) - 3$

$$= 4x - 6 - 3 = 4x - 9$$

If either function f or g in a composition is not defined for some real numbers, then as the next example illustrates, the domain of the composite function $f \circ g$ cannot always be determined by examining the final form $(f \circ g)(x)$. Any numbers that are excluded from the domain of g must also be excluded from the domain of $f \circ g$.

EXAMPLE 4   **Finding the Domain of a Composite Function**

Let $f(x) = 2x + 3$ and $g(x) = \sqrt{x}$.

(a) Find $(f \circ g)(x)$ and the domain of $f \circ g$.

(b) Find $(g \circ f)(x)$ and the domain of $g \circ f$.

Solution

(a)
$$(f \circ g)(x) = f[g(x)] = f(\sqrt{x}) = 2\sqrt{x} + 3$$

The domain and range of g are both $[0, \infty)$. Since the domain of f consists of all real numbers $\mathbb{R}$, all output values from the g function are acceptable input for the f function. Thus the domain of $f \circ g$ consists of all values x in the interval $[0, \infty)$.

(b)
$$(g \circ f)(x) = g[f(x)] = g(2x + 3) = \sqrt{2x + 3}$$

The domain and range of f consist of all real numbers $\mathbb{R}$. Since the domain of g is $[0, \infty)$, the only acceptable input values for the g function must be nonnegative. So the output values from the f function must satisfy $2x + 3 \geq 0$, that is, $x \geq -3/2$. Therefore the domain of $g \circ f$ is the interval $[-3/2, \infty)$.

The order of forming a composite function is important. Notice in Example 4 that $f \circ g$ is not equal to $g \circ f$ For instance,

$$(f \circ g)(1) = 2\sqrt{1} + 3 = 5 \quad \text{whereas} \quad (g \circ f)(1) = \sqrt{2 \cdot 1 + 3} = \sqrt{5}$$

Care must be taken to make sure that the correct domain is used when graphing a composite function.

EXAMPLE 5   **Graphing a Composite Function**

Let $f(x) = \sqrt{x^2 + 2}$ and $g(x) = \sqrt{x - 1}$.

(a) Find the equation $y = (f \circ g)(x) = f[g(x)]$.

(b) Graph the equation determined in part (a).

(c) Find the domain of $f \circ g$

(d) What is the domain suggested by the graph found in part (b), and how does it compare to the answer in part (c)?

(e) What adjustment has to be made to the sketch in part (b) to obtain the correct graph of $f \circ g$?

Solution   For descriptive purposes, let us agree to call g the *inner function* and f the *outer function* because of the positions they occupy in the expression $f[g(x)]$.

(a) For $f(x) = \sqrt{x^2 + 2}$ and $g(x) = \sqrt{x - 1}$, we have

$$y = f[g(x)]$$
$$= f(\sqrt{x - 1})$$
$$= \sqrt{(\sqrt{x - 1})^2 + 2}$$
$$= \sqrt{x - 1 + 2}$$
$$= \sqrt{x + 1}$$

that is, $y = \sqrt{x + 1}$

(b) Figure 2a shows the graph of $y = \sqrt{x + 1}$.

(c) Since the inner function, $g(x) = \sqrt{x - 1}$, is defined for $x \geq 1$ and the outer function f is defined for all real numbers, it follows that the domain of $f \circ g$ includes all real numbers x such that $x \geq 1$ or $[1, \infty)$.

(d) The graph in Figure 2a suggests that the domain includes all x such that $x \geq -1$; however, the domain of $f \circ g$ includes all x such that $x \geq 1$ or $[1, \infty)$.

(e) By restricting x in the graph in Figure 2a so that $x \geq 1$, we get the graph of $f \circ g$, which is displayed in Figure 2b.

Figure 2

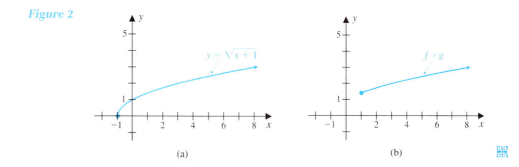

(a)   (b)

🇬 Most graphers have the capability of combining giving functions. For example, suppose that

$$y_1 = \sqrt{x^2 + 8} \text{ and } y_2 = \sqrt{x^2 - 9} \text{ are given.}$$

For each function y_3, defined in terms of y_1 and y_2 by equations such as

$$y_3 = y_1 + y_2, \text{ (Figure 3a) } y_3 = y_1 \cdot y_2 \text{ (Figure 3b) or } y_3 = y_1 \circ y_2 \text{ (Figure 3c)}$$

we merely enter y_1 and y_2, and then let the grapher form y_3.

Figure 3

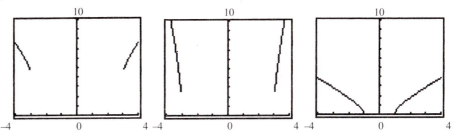

Expressing a Function as a Composition of Other Functions

In calculus we often need to interpret a given function as a composition of two functions. In a sense we "unravel" the equation used to define the given function. Example 6 illustrates the process.

EXAMPLE 6   **Finding the Components of a Composite Function**

Express

$$h(x) = (5x - 1)^2$$

as a composition $h = f \circ g$; that is, find f and g so that $h(x) = f[g(x)]$.

Solution   In order to see that h can be obtained as a composition $h = f \circ g$, we must be able to recognize an inner function g and an outer function f in the equation that defines h. Of course, g is the first function and f is the second function applied in the composition $f \circ g$. The diagram in Figure 4 helps us to define f and g so that $h(x) = f[g(x)]$. We can see that one possibility for $f \circ g$ is given by

$$g(x) = 5x - 1 \text{ and } f(x) = x^2.$$

Figure 4

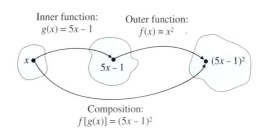

Inner function:
$g(x) = 5x - 1$

Outer function:
$f(x) = x^2$

Composition:
$f[g(x)] = (5x - 1)^2$

It is worthwhile noting that this is not the only solution. Instead, we may let

$$g(x) = 5x \text{ and } f(x) = (x - 1)^2 \text{ then}$$

$$f[g(x)] = f(5x) = (5x - 1)^2 = h(x).$$

Solving Applied Problems

The composition of functions arises in many applications and models involving several functional relationships simultaneously. In such cases one quantity may have to be expressed as a function of another as the next example illustrates.

EXAMPLE 7   **Modeling a Medical Procedure**

A certain medical procedure requires that a spherical balloon be inserted into the patient's stomach. Suppose that after t seconds, the radius r of the balloon measures $52/(2t + 13)$ centimeters; that is,

$$r = \frac{52}{2t + 13}.$$

So the radius is a function of time. The volume V (in cubic centimeters) is a function of the radius r and is given by the formula

$$V = \frac{4}{3}\pi r^3.$$

(a) Express the volume V of the balloon as a function of t.

(b) What is the volume of the balloon after 5 seconds? Round off to one decimal place.

(c) When is the volume 57.6π cubic centimeters? Round off to one decimal place.

Solution   (a) First we write

$$V = \frac{4}{3}\pi r^3 = f(r)$$

and

$$r = \frac{52}{2t + 3} = g(t).$$

Next we form the composite function of V in terms of time t as follows:

$$V = (f \circ g)(t) \quad = f[g(t)]$$

$$= f\left(\frac{52}{2t+3}\right) = \frac{4}{3}\pi\left(\frac{52}{2t+3}\right)^3.$$

(b) After 5 seconds, the volume V of the balloon is given by

$$V = \frac{4}{3}\pi\left(\frac{52}{2(5)+3}\right)^3$$

$$= 268.1 \text{ (approx.)}.$$

Therefore, the volume is about 268.1 cubic centimeters.

(c) To find the time t when $V = 57.6\pi$, we solve the equation

$$57.6\pi = \frac{4}{3}\pi\left(\frac{52}{2t+3}\right)^3 \quad \text{for } t$$

to get $\quad 0.75(57.6) = \left(\frac{52}{2t+3}\right)^3 \quad$ Multiply both sides by $\dfrac{3}{4\pi}$

$$\sqrt[3]{0.75(57.6)} = \frac{52}{2t+3} \qquad \text{Take the cube root of each side}$$

$$\frac{1}{\sqrt[3]{0.75(57.6)}} = \frac{2t+3}{52} \qquad \text{Take the reciprocal of each side}$$

$$\frac{1}{2}\left(\frac{52}{\sqrt[3]{0.75(57.6)}} - 3\right) = t \qquad \text{Isolate } t \text{ on one side}$$

$$t = 5.9 \text{ (approx.)}.$$

Therefore, after about 5.9 seconds the volume is 57.6π cubic centimeters.

EXAMPLE 8   **Modeling a Manufacturer's Profit**

A pharmaceutical company finds that the cost function C for producing a certain antibiotic is $C(x) = 0.003x^2 + 80x + 500{,}000$, where $C(x)$ is the cost (in dollars), x is the number of units of the antibiotic produced, and $0 \le x \le 30{,}000$. The firm sets the selling price at \$200 per unit.

(a) Express the total revenue R as a function of x.

(b) Write an expression for the total profit P as a function of x.

(c) Determine when the profit is 0.

Solution   (a) If x units are sold at \$200 per unit, then the revenue $R(x)$ (in dollars) is given by

$$R(x) = 200x,\ 0 \le x \le 30{,}000$$

(b) Since

$$\text{Profit} = \text{Revenue} - \text{Cost}$$

the profit function P in this situation is given by the difference function,

$$P(x) = R(x) - C(x)$$
$$= 200x - (0.003x^2 + 80x + 500{,}000)$$
$$= -0.003x^2 + 120x - 500{,}000,\ 0 \le x \le 30{,}000$$

(c) To determine when the profit $P(x)$ is zero, we solve the quadratic equation

$$P(x) = -0.003x^2 + 120x - 500{,}000 = 0$$

to get $x = 4725$ or $x = 35{,}275$. Because of the restrictions on x, the value 4725 is the only acceptable solution. Thus, the profit is 0 when 4725 units are sold.

G   A viewing window of the graph of the profit function P from Example 8 on page 191 is shown in Figure 5. We established in part (c) of Example 8 that the x intercept is 4725. By reading the graph we see that there is a loss if $P < 0$ (when $x < 4725$) and a profit if $P > 0$ (when $x > 4725$). In this context we refer to the value $x = 4725$ as the *break-even point*. As more units are sold beyond 4725, the profit increases to a "peak" level and then begins to decrease.

Figure 5

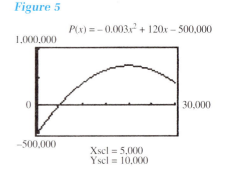

In Chapter 3 we will investigate techniques that enable us to find this highest profit value and the value of x when it occurs. For this particular situation, a maximum profit of \$700,000 occurs when 20,000 units are sold.

◆ PROBLEM SET 2.5

Mastering the Concepts

In problems 1–8, find each of the following functions:

(a) $(f + g)(x)$ (b) $(f - g)(x)$ (c) $(f \cdot g)(x)$

1. $f(x) = 3x + 1$ and $g(x) = -2x - 7$

2. $f(x) = -4x - 5$ and $g(x) = x + 6$

3. $f(x) = x^2 - 3x + 2$ and $g(x) = 4x^2 + 1$

4. $f(x) = 7x^2 - 1$ and $g(x) = -3x^2 + 8x - 5$

5. $f(x) = x^3 + 5x$ and $g(x) = 2x^3 - 3x + 4$

6. $f(x) = x^2 + 5$ and $g(x) = -x^3 + x^2 - 7x + 2$

7. $f(x) = \dfrac{2}{x - 5}$ and $g(x) = \dfrac{x}{3 - 4x}$

8. $f(x) = \dfrac{2x + 3}{x - 5}$ and $g(x) = \dfrac{2 - 7x}{3x + 1}$

In problems 9–14, find each quotient function and its domain.

(a) $\dfrac{f}{g}$ (b) $\dfrac{g}{f}$

9. $f(x) = 3x + 1$ and $g(x) = -2x - 7$

10. $f(x) = -4x - 5$ and $g(x) = 6 - x$

11. $f(x) = x^2 + 4x + 3$ and $g(x) = x^3 + 1$

12. $f(x) = 2x^2 - 8x$ and $g(x) = x^3 + 1$

13. $f(x) = \sqrt{2 - x}$ and $g(x) = \dfrac{x}{x - 1}$

14. $f(x) = \sqrt{x + 3}$ and $g(x) = \dfrac{2x - 3}{x}$

In problems 15 and 16, find each of the following values:

(a) $(f \circ g)(2)$ (b) $(g \circ f)(2)$
(c) $(f \circ f)(2)$ (d) $(g \circ g)(2)$
(e) $g[f(4)]$ (f) $f[g(3)]$
(g) $f[g(5)]$ (h) $g[f(5)]$
(i) $f[f(-1)]$ (j) $g[g(-1)]$

15. $f(x) = 2x^2 + 6$ and $g(x) = 7x + 2$

16. $f(x) = 5 - 3x^2$ and $g(x) = -3x + 1$

In problems 17 and 18, use the given graphs of f and g to compute each of the following values:

(a) $(g \circ f)(2)$ (b) $(g \circ g)(3)$
(c) $(f \circ g)(3)$ (d) $(g \circ f)(7)$
(e) $(f \circ f)(2)$ (f) $(g \circ g)(7)$

17.   18.

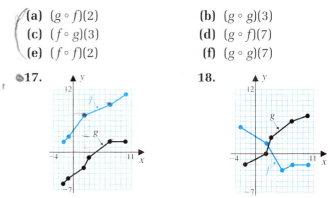

19. Table 1 lists values of two functions f and g. Find each of the following values:

(a) $(g \circ f)(2)$ (b) $(f \circ f)(0)$
(c) $(f \circ g)(3)$ (d) $(g \circ f)(-3)$

TABLE 1

x	$f(x)$	x	$g(x)$
-3	-5	-5	-15
-2	-3	0	0
-1	-1	1	-9
0	1	3	-1
1	3	5	15
2	5		
3	7		

20. Table 2 lists values of two functions f and g. Find each of the following values:

(a) $(g \circ f)(-2)$ (b) $(f \circ g)(1)$

(c) $(f \circ f)(-1)$ (d) $(g \circ g)(-3)$

TABLE 2

x	$f(x)$	x	$g(x)$
-2	1	-3	-2
-1	3	-2	1
0	5	-1	0
3	11	1	10
4	5	2	9
7	3	3	5
10	-4		

In problems 21–30, find each of the following functions and their domains.

(a) $y = f[g(x)]$

(b) $y = g[f(x)]$

21. $f(x) = 2x$ and $g(x) = 5x - 3$

22. $f(x) = -3x$ and $g(x) = 3x$

23. $f(x) = 2x^2 + 5$ and $g(x) = 7\sqrt{x}$

24. $f(x) = \sqrt{2x - 1}$ and $g(x) = x^2 + 9$

25. $f(x) = 11x + 2$ and $g(x) = \dfrac{x}{11} - \dfrac{2}{11}$

26. $f(x) = x^3 + 1$ and $g(x) = \sqrt[3]{x - 1}$

27. $f(x) = \sqrt{x - 1}$ and $g(x) = x + 5$

28. $f(x) = 2x - 2$ and $g(x) = \sqrt{x + 2}$

29. $f(x) = \dfrac{1}{3x + 2}$ and $g(x) = \dfrac{3}{2x - 5}$

30. $f(x) = \dfrac{x}{3x - 5}$ and $g(x) = \dfrac{1 - x}{3 + 2x}$

In problems 31–36, find two functions f and g so that the given function h can be expressed as a composition $h = f \circ g$.

31. $h(x) = (5x - 2)^3$

32. $h(x) = (x^2 + 2x - 1)^4$

33. $h(t) = (t^2 - 2)^{-2}$

34. $h(w) = \left(\dfrac{w + 1}{w - 1}\right)^3$

35. $h(x) = \sqrt[3]{x} + x^{-1}$

36. $h(t) = (t^2 + 3)^{4/5}$

37. Let $D(t) = 3t + 1$ and $R(x) = -5x + 2$

(a) Find $(D \circ R)(x)$

(b) For what value of x does $D[R(x)] = 2$?

38. Let $S(r) = r^3 + 2$ and $D(t) = \sqrt[3]{t + 7}$

(a) Find $(S \circ D)(t)$

(b) For what value of t does $S[D(t)] = 13$?

In problems 39–44:

(a) Find the equation $y = f[g(x)]$.

(b) Sketch the graph of the equation determined in part (a).

(c) Find the domain of $f \circ g$.

(d) What adjustment, if any, has to be made to the sketch in part (b) in order to obtain the correct graph of $f \circ g$?

39. $f(x) = x^2$ and $g(x) = \sqrt{x}$

40. $f(x) = \sqrt{x}$ and $g(x) = x^2$

41. $f(x) = \dfrac{1}{x}$ and $g(x) = \dfrac{1}{x}$

42. $f(x) = \dfrac{1}{x - 1}$ and $g(x) = \dfrac{1}{x}$

43. $f(x) = \sqrt{4 - x^2}$ and $g(x) = \sqrt{2 - x}$

44. $f(x) = \sqrt{25 - x^2}$ and $g(x) = \sqrt{x - 5}$

Applying the Concepts

45. Meteorology: A spherical weather balloon is being inflated at a constant rate. The radius r (in inches) of the balloon is increasing with time t (in minutes) according to the function $r = g(t) = 1.85t$. The volume V of the balloon is given by the formula $V = f(r) = (4/3)\pi r^3$.

(a) Find a composite function that expresses the volume V of the balloon as a function of time t, and interpret its meaning.

(b) The surface area S of the balloon is given by the function $S = h(r) = 4\pi r^2$. Find a composite function that expresses the surface area S as a function of time t, and interpret its meaning.

46. Ecology: An offshore oil well begins to leak, and the oil slick starts to spread on the surface of the water in a circular pattern in such a way that the radius r is increasing at the rate of 0.5 kilometer per hour.

(a) Express the radius r as a function f of the number of elapsed hours t.

(b) The area of a circle of radius r is given by the function $A(r) = \pi r^2$. Form the composite function $A \circ f$, and interpret the meaning of the function.

(c) Find the area of the oil slick 5 hours after the beginning of the leak.

47. Baseball: A baseball diamond is a square 90 feet on each side. Suppose that a ball is hit directly down the third base line at the rate of 55 feet per second. Let x denote the distance (in feet) the ball travels, y denote the distance (in feet) of the ball from first base, and t denote the elapsed time (in seconds) since the ball was hit (Figure 6). Here, y is a function of x, say $y = f(x)$, and x is a function of t, say $x = g(t)$.

Figure 6

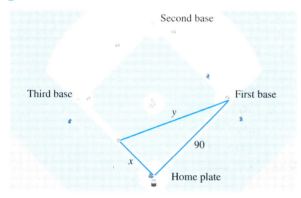

(a) Find equations that define $f(x)$ and $g(t)$.

(b) Find the expression for $(f \circ g)(t)$ and interpret $f \circ g$.

48. Automobile Rebate: A car dealer offers a $1000 rebate and a 15% discount off the sticker price x (in dollars) of a new car at the end of a model year.

(a) Express the cost R of the car as a function of x if only the rebate is given.

(b) Express the cost D of the car as a function of x if only the discount is given.

(c) Find $(D \circ R)(x)$ by calculating the rebate first, then the discount. Graph $D \circ R$.

(d) Find $(R \circ D)(x)$ by calculating the discount first, then the rebate. Graph $R \circ D$.

(e) Which of the options in parts (c) and (d) is a better deal? Interpret the results if the sticker price of the car is $17,500.

49. Converting Water Measurement: In Table 3, $f(x)$ represents the cubic feet of water in an irrigation pond after x days. In Table 4, $g(x)$ represents the gallons of water corresponding to x cubic feet in the same pond.

TABLE 3		TABLE 4	
x (days)	**f(x) (cu. ft.)**	**x (cu. ft.)**	**g(x) (gallons)**
1	50,000	10,000	74,810
2	40,000	20,000	149,620
3	30,000	30,000	224,430
4	20,000	40,000	299,240
5	10,000	50,000	374,050

(a) Evaluate

$$(g \circ f)(3) \quad \text{and} \quad (g \circ f)(5)$$

and interpret the results of these calculations.

(b) Interpret the general meaning of

$$(g \circ f)(x) \;.$$

50. Geometry Problem: A 10-meter ladder is leaning against a vertical wall. Suppose the ladder begins sliding down the wall in such a way that the foot of the ladder is moving away from point 0 at the base of the wall at the rate of 1 meter per second as shown in Figure 7.

Figure 7

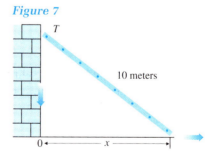

(a) Express the height $h = |\overline{OT}|$ as a function of the horizontal distance x (in meters).

(b) Express x as a function f of the time t (in seconds).

(c) Evaluate $(h \circ f)(4)$ and interpret the result of this calculation.

(d) Interpret the general meaning of $(h \circ f)(t)$.

51. Manufacturing: A company's cost C (in dollars) for producing a certain product is given by

$$C(x) = x^2 + 30x + 500 \qquad 0 \le x \le 45$$

where x is the number of units manufactured and sold. The selling price is set at $90 per unit.

(a) Determine the revenue R as a function of x. Then use a grapher to sketch graphs of the

cost and revenue functions in the same coordinate system. Interpret the result.
(b) **G** Determine the profit P as a function of x and use a grapher to graph it. Interpret the graph in terms of profit and loss.

52. Marketing: The cost C (in dollars) of buying x bushels of corn is given by

$$C(x) = 5x + 1000 \qquad \text{where } 0 \le x \le 2500$$

The revenue R (in dollars) from selling x bushels of corn is given by

$$R(x) = 8x - \frac{x^2}{1000}.$$

(a) **G** Use a grapher to sketch graphs of the cost and revenue functions in the same coordinate system. Interpret the result.
(b) **G** Determine the profit function P and use a grapher to graph it. Interpret the graph in terms of profit and loss.

Developing and Extending the Concepts

53. Assume that f and g are increasing functions for all real numbers.
(a) Is
$$f + g$$
necessarily an increasing function?
(b) Give an example of two increasing functions so that
$$f - g$$
is a decreasing function for all real numbers.

54. Assume that f and g are decreasing functions for all real numbers.
(a) Is
$$f + g$$
necessarily a decreasing function? Explain.
(b) Give an example of two decreasing functions so that
$$f - g$$
is an increasing function for all real numbers.

55. Assume f and g are even functions. Are
$$f + g, f - g, f \cdot g, \text{ and } f/g$$
all necessarily even functions? Explain and give examples to support your assertion.

56. Assume f and g are odd functions. Are
$$f + g, f - g, f \cdot g, \text{ and } f/g$$

all necessarily odd functions? Explain and give examples to support your assertion.

57. (a) If g is an even function, is $f \circ g$ necessarily even?
(b) If f and g are both odd functions, is $f \circ g$ necessarily odd? Explain.

58. Composing a function f with itself is called *function iteration*. The successive functions
$$f \circ f, \quad f \circ f \circ f, \quad f \circ f \circ f \circ f,$$
and so on are called the *iterates* of f. If $n \ge 2$ is an integer, we use the notation
$$f^{(n)} = f \circ f \circ f \circ f \circ f \circ f \circ f \circ \ldots \circ f \ (n \text{ times})$$
for the *nth iterate* of f. For instance,
$$f^{[2]} = f \circ f, \quad f^{[3]} = f \circ f \circ f,$$
and so on.
(a) Find $f^{[2]}(x)$ if $f(x) = 2x + 1$.
(b) Find $f^{[3]}(x)$ if $f(x) = -5x + 2$.
(c) Find $f^{[10]}(x)$ if $f(x) = \dfrac{1}{x}$.

59. The composition $f \circ g \circ h$ of three functions $f, g,$ and h is defined by $(f \circ g \circ h)(x) = f[g[h(x)]]$. Find $(f \circ g \circ h)(x)$ for the following functions:
(a) $f(x) = x - 3,$
$g(x) = x^3,$
$h(x) = \sqrt[5]{x}$
(b) $f(x) = x^2 + 4,$
$g(x) = 3x + 2,$
$h(x) = 2x - 5$

60. (a) Give an example of two functions f and g so that $(f \circ g)(x) = (f \cdot g)(x)$.
(b) Give an example of two functions f and g so that $(f \circ g)(x) = (g \circ f)(x)$.

61. Let $f(x) = x^3 - 6x^2$ and let $h(x) = -\dfrac{1}{2}f\left(\dfrac{1}{3}x\right)$.
(a) Determine a function g so that
$$h(x) = -\frac{1}{2}(f \circ g)(x).$$
(b) **G** Use a grapher to sketch the graph of h.

62. Let $f(x) = x^3 - x^2 - 10x - 8$ and let
$$h(x) = -f(2x).$$
(a) Determine a function g so that
$$h(x) = -(f \circ g)(x).$$
(b) **G** Use a grapher to sketch the graph of h.

◈ CHAPTER 2 REVIEW PROBLEM SET

1. Indicate whether each correspondence depicted in the diagrams is a function. Explain your reasoning.

(a)

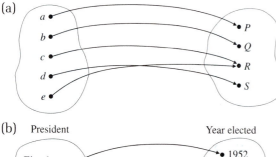

(b)

2. Rewrite each equation in the function form $y = f(x)$.
  (a) $3x - 7y + 3 = 0$
  (b) $17x^2 + 3y - 20 = 0$

3. Determine which of the graphs shown represents a function.

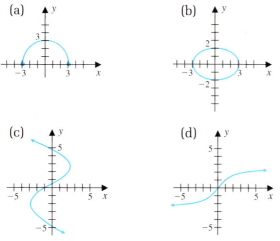

(a)

(b)

(c)

(d)

4. Sketch the graph of each equation and determine whether each graph represents a function.
  (a) $|y| = 4x^2$ (b) $|y| = 2$

5. Let $f(x) = x^3$, $g(x) = 2x + 3$, and $h(x) = x^2 - 5x$. Find each value:
  (a) $f(-1)$ (b) $g(-1)$
  (c) $h(-5)$ (d) $f(\sqrt[3]{2})$
  (e) $h(\sqrt{2})$ (f) $g(a + b) - g(a)$
  (g) $\dfrac{g(a+b) - g(a)}{b}$

6. Let $f(x) = 2x - 5$, $g(x) = 3x^2 + 4x$, and $h(x) = \sqrt{4x - 3}$.

Find each value:
  (a) $f(-2)$ (b) $g(-3)$
  (c) $h(3)$ (d) $f(t + 2)$
  (e) $g(b) - g(1)$ (f) $g(t + k)$
  (g) $[f(a)]^2$

7. Let $f(x) = \begin{cases} x^3 & \text{if } x < 1 \\ \sqrt{x} & \text{if } x \geq 1 \end{cases}$
  (a) Find $f(-2)$, $f(0)$, $f(1)$, and $f(4)$.
  (b) Find the domain and range of f.
  (c) Sketch the graph of f.

8. The figure below shows the graph of a function f.

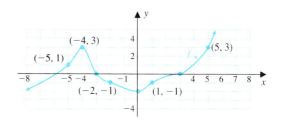

  (a) Find $f(-5)$, $f(-4)$, $f(1)$, and $f(5)$.
  (b) In what intervals is f increasing?
  (c) In what intervals is f decreasing?
  (d) Use the graph to solve the equation $f(x) = 3$.

9. Find the difference quotient
$$\frac{f(t + h) - f(t)}{h}$$
for each function.
  (a) $f(x) = 5x - 7$
  (b) $f(x) = 3x^2 + 2$

10. Temperature Change: A machine shop determines the temperature T (in degrees Fahrenheit) of a machine after t minutes of operation is given by the model
$$T(t) = t^2 - 8t + 10, \; t \geq 0$$
  (a) Find the difference quotient
$$\frac{T(t + h) - T(t)}{h}$$
  (b) Find the numerical value of the difference quotient in part (a) for
    (i) $t = 3.8$ and $h = 0.3$
    (ii) $t = 3.8$ and $h = 0.6$

(c) Interpret the meaning of each answer in part (b) from

    (i) $t = 3.8$ to $t = 3.8 + 0.3$ and

    (ii) $t = 3.8$ to $t = 3.8 + 0.6$

In problems 11 and 12, use the graph to find the x intercept and zeros of the function then solve the associated inequality.

11. [G] $f(x) = x^3 - 4x$; $x^3 - 4x \le 0$

12. $f(x) = |2x - 3| - 5$; $|2x - 3| \ge 5$

In problems 13 and 14, determine whether each function is even, odd, or neither. Describe the corresponding symmetry of the graph, if any.

13. (a) $f(x) = x^3 - 3x$

    (b) $g(x) = -x^4 + 3x^2 + 1$

14. (a) $g(x) = x^4 + 4x^2$

    (b) $f(x) = -\dfrac{1}{x}$

15. Describe a sequence of transformations that will transform the graph of f into the graph of g.

    (a) $f(x) = x^2 + 1$; $g(x) = (x + 2)^2 + 4$

    (b) $f(x) = x^2 - x$; $g(x) = (x + 1)^2 - (x + 1) + 3$

    (c) $f(x) = x^2 - 5$; $g(x) = 5 - x^2$

    (d) $f(x) = \sqrt{x + 4}$; $g(x) = -2\sqrt{x + 4} + 2$

16. The figure below shows the graph of a function f.

Use this graph to sketch each graph of g:

(a) $g(x) = f(x) - 3$

(b) $g(x) = f(x - 3)$

(c) $g(x) = -3f(x)$

17. The graph of g is obtained from the graph of the given function f as specified. Determine the function equation for g. Graph f and g on the same coordinate system.

    (a) Shift the graph of $f(x) = -x^2 + 2$ horizontally 3 units to the right and then vertically downward 4 units.

    (b) Shift the graph of $f(x) = \sqrt[3]{x}$ horizontally 5 units to the left, vertically stretch it by a factor of 2, and then vertically shift the result upward 3 units.

    (c) Shift the graph of $f(x) = \sqrt{-x}$ horizontally 4 units to the left and then reflect it across the x axis.

18. Graph each of the following functions and find the x intercepts and the zeros of each function.

    (a) $f(x) = |x - 6| - 2$

    (b) $f(x) = -2(x - 1)^2 - 4$

19. For the given functions f and g, find an expression for:

    (i) $(f + g)(x)$ (ii) $(f - g)(x)$

    (iii) $(f \cdot g)(x)$ (iv) $(f \circ g)(x)$

    (v) $(g \circ f)(x)$ (vi) $(f/g)(x)$

    (a) $f(x) = x + 2$; $g(x) = x - 1$

    (b) $f(x) = x^3$; $g(x) = -x$

    (c) $f(x) = x^2$; $g(x) = 2x + 1$

    (d) $f(x) = 2x^2 - 1$; $g(x) = 2x^2 + 1$

20. Let $f(x) = x^3$ and $g(x) = \sqrt[3]{x}$.

    (a) Find $(f \circ g)(x)$ and its domain.

    (b) Find $(g \circ f)(x)$ and its domain.

    (c) Are $(f \circ g)(x)$ and $(g \circ f)(x)$ the same? Explain.

21. Express the given function h as the composition of two other functions f and g so that $h = f \circ g$.

    (a) $h(x) = (7x + 2)^5$

    (b) $h(t) = \sqrt{t^2 + 17}$

22. (a) Let $f(x) = 7 - 5x$. Find $g(x)$ so that $(f \circ g)(x) = x$.

    (b) Let $f(x) = 2x + 1$. Find $g(x)$ so that we have $(f \circ g)(x) = 3x - 1$.

23. **Marketing:** A produce distributor prices oranges at \$4 per bag when the stores it supplies can sell 10,000 bags. When the demand for oranges rises to 14,000 bags, the price is reduced to \$3 per bag. Assume that the price P (in dollars) per bag is a linear function of the quantity of x bags, where $10{,}000 \le x \le 20{,}000$.

    (a) Find an equation that expresses P as a function x.

    (b) Sketch the graph of this function.

    (c) Discuss the trend. What is the price of oranges when the demand is 16,000 bags?

    (d) Suppose the price per bag is \$2, how many bags of oranges are needed?

24. **Coal Reserve:** The amount A of coal (in tons) available from a supplier t weeks after opening a new yard is given by the following piecewise-defined function:

$$A(t) = \begin{cases} -\dfrac{275}{3}t + 300 & \text{if } 0 \le t < 3 \\[2mm] -\dfrac{125}{3}t + 375 & \text{if } 3 \le t < 6 \\[2mm] -\dfrac{350}{3}t + 1100 & \text{if } 6 \le t < 9 \end{cases}$$

(a) Sketch the graph of A.

(b) How many tons were available initially? After 2 weeks? After 5 weeks? After 8 weeks?

(c) What is the domain of this function?

(d) Determine both times when 65 tons of coal were available.

25. Advertising: The graph below relates the profit (or loss) P (in thousands of dollars) to the amount of money A (in thousands of dollars) spent on advertising, up to $45,000, for a certain product.

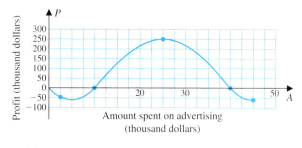

Amount spent on advertising
(thousand dollars)

(a) Explain why the graph defines a function.

(b) From the graph, approximate the domain and range.

(c) Estimate the profit when the advertising expenditure is $2500; $10,000; $25,000; $40,000.

(d) What expenditure on advertising will achieve maximum profit?

(e) Discuss the profit trend. For instance, what will happen to the profit if more than $40,000 is spent on advertising?

26. Modeling Profit: A manufacturer makes and sells x home television satellite systems per day. The daily cost function C (in dollars) is given by

$$C(x) = 225x + 450$$

and the daily revenue function R from selling x systems is

$$R(x) = -0.03x^2 + 675x.$$

(a) Express the profit P from making and selling x systems as a function of x.

(b) Graph the profit function P.

(c) Determine the profit on a day when 200 systems are made and sold.

27. Manufacturing: A radio manufacturer finds that its production cost C (in dollars) for producing x radios is given by the function

$$C(x) = 50,000 + 10,000 \sqrt[3]{x+1}.$$

It sells the radios to distributors for $10 each, and the demand is so high that all the manufactured radios are sold.

(a) Express the revenue R (in dollars) as a function of x if x radios are produced and sold.

(b) Express the profit P (in dollars) as a function of x.

(c) [G] Use a grapher to graph the function P, and then use the graph to discuss the profit trend.

28. Geometry: The area A of an equilateral triangle is given by the function

$$A = f(x) = \frac{\sqrt{3}}{4} x^2$$

where x is the side length of the triangle, and x is given by

$$x = g(P) = \left(\frac{1}{3} \right) P,$$

where P is the perimeter of the triangle.

(a) Find

$$(f \circ g)(P) .$$

(b) Interpret the resulting expression in part (a).

29. Biology: A biologist discovers that the number N of bacteria (in thousands) in a culture at a temperature T (in degrees Celsius) is given by the function

$$N(T) = -90(T + 1)^{-1} + 20.$$

The temperature T at time t (in hours) is given by the function

$$T(t) = 3t + 4, \quad \text{where} \quad 0 \le t \le 12.$$

(a) Find

$$(N \circ T)(t) .$$

(b) What does the function $N \circ T$ represent?

(c) [G] Use a grapher to graph $N \circ T$.

(d) Discuss what happens to the number of bacteria as the time elapses.

30. (a) Determine without graphing whether each function is even, odd, or neither by using function notation.

(b) [G] Demonstrate the validity of the answers in part (a) by using a grapher to graph the functions.

(i) $f(x) = 3x^2 + |x|$

(ii) $f(x) = 16x^3 + \sqrt[3]{x}$

(iii) $f(x) = \dfrac{x^3}{x^2 + 1}$

31. [G] Use a grapher to find the x intercepts of the associated function, then solve each inequality.

(a) $|5x - 3| \le |4x + 3|$

(b) $|4x + 5| > |3x - 5|$

32. Consider the square shown below, with the length of the side x and the diagonal d.

(a) Express x as a function of d. That is, write

$$x = f(d).$$

(b) Express the area A as a function of P, the perimeter. That is, write

$$A = g(P).$$

(c) Express d as a function of A, the area. That is, write

$$d = F(A).$$

33. The sum of two numbers represented by x and y is 40.

(a) Express y as a function of x.

(b) Express the product P of the two numbers as a function of x.

(c) Find the difference quotient of P when $h = 0.01$.

34. A cylindrical closed container with radius r has a volume of 36 cubic feet.

(a) Show that its surface area A is given by the function

$$A(r) = 2\pi r^2 + \frac{72}{r}.$$

(b) Find

$$A(2), A(5) \text{ and } A(8).$$

Interpret the meaning of each value. Round off to two decimal places.

(c) Find the difference quotient of A when $h = 0.01$.

◆ CHAPTER 2 TEST

1. Determine whether the correspondence described by each given diagram is a function.

(a)

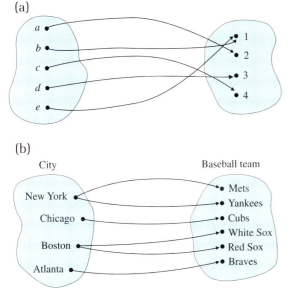

(b)

2. (a) Plot the following points on a Cartesian system: $(-2, 3)$, $(-1, 4)$, $(0, 0)$, $(3, 2)$.

(b) Suppose the points plotted in part (a) lie on the graph of an odd function f. Plot each point: $(2, f(2))$, $(1, f(1))$, $(0, f(0))$, $(-3, f(-3))$.

In problems 3 and 4:

(a) Graph the function.

(b) Find the domain and range.

(c) Indicate where the function is increasing or decreasing.

(d) Determine whether the function is even, odd, or neither.

(e) Describe any symmetry.

(f) Find the difference quotient $\dfrac{f(t + h) - f(t)}{h}$

3. $f(x) = 5x^3 - 2$

4. $f(x) = \dfrac{3}{x^2}$

5. Use the graph of $g(x) = |3x + 1| - |2x - 1|$ to solve the inequality $|3x + 1| \le |2x - 1|$.

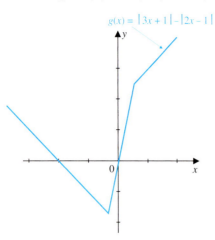

$$g(x) = |3x+1| - |2x - 1|$$

6. Sketch the graph of

$$f(x) = \begin{cases} -x & \text{if } x \le -1 \\ x^2 & \text{if } -1 < x < 1 \\ \sqrt{x} & \text{if } x \ge 1 \end{cases}$$

Find the domain and range of f.

7. Let $f(x) = 2x - 8$ and $g(x) = 1 - x^2$. Find:
(a) $(f + g)(x)$
(b) $(f \cdot g)(x)$
(c) $(f/g)(x)$
(d) $(f \circ g)(x)$
(e) $(g \circ f)(3)$
(f) $\dfrac{f(t + 3) - f(t)}{3}$

8. The figure below shows the graph of f.

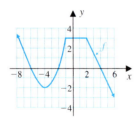

(a) Use the graph to find $f(-6)$, $f(-4)$, $f(1)$, $f(2)$, and $f(4)$.
(b) On what intervals is f increasing?
(c) On what intervals is f decreasing?
(d) On what interval is f constant?
(e) What is the domain of f?
(f) What is the range of f?

9. Use the given tables for $f(x)$ and $g(x)$ to fill in the blanks in the tables for $f[g(x)]$ and $g[f(x)]$.

x	$f(x)$	x	$g(x)$
1	3	1	4
2	4	2	1
3	2	3	3
4	1	4	2

x	$f[g(x)]$	x	$g[f(x)]$
1	—	1	—
2	—	2	—
3	—	3	—
4	—	4	—

In problems 10 and 11, use standard graphs and transformations to graph each function.

10. $g(x) = -2|x + 1| + 3$

11. $h(x) = \dfrac{1}{3}(x - 2)^3 - 4$

12. Suppose that a right triangle has a hypotenuse of length c, a leg of length a, and a leg of length 2 inches.
(a) Express c as a function f of a.
(b) Express a as a function g of c.
(c) Show that

$$(f \circ g)(c) = c \text{ and } (g \circ f)(a) = a$$

13. Medical researchers determine that the blood sugar of a person with diabetes is regulated by a mixture of time-released insulin. The graph below shows the amount of sugar A (in milligrams per deciliter) t hours after the insulin was taken.

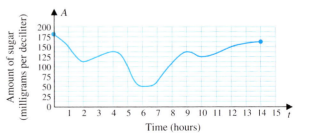

(a) What is the domain of this function?

(b) When is the blood sugar 150 milligrams per deciliter?

(c) When is the blood sugar lowest?

(d) What is the lowest blood sugar?

14. A magazine publisher has a profit of $95,000 per year when it distributes the magazine to 31,000 subscribers. When its distribution is increased by 4000 subscribers, its profit is increased by $20,000. Assume the profit P is a linear function of the number of subscribers S.

(a) Express P as a function of S.

(b) What will the profit be if the publisher distributes the magazine to 40,000 subscribers?

(c) Graph the function P.

(d) How many subscribers receive the magazine if no profit is obtained for that year?

15. Use a grapher to graph each function. Find the domain of each function.

(a) $g(x) = -\sqrt{4 - x}$

(b) $g(x) = \sqrt{x - 1} + \sqrt{x - 1}$

CHAPTER 3

Polynomial and Rational Functions

The Mackinac Bridge, one of the world's longest suspension bridges, connects the upper and lower peninsulas of Michigan. How high above the roadway is the lowest point of the cable? The answer to this question is developed in Example 11 on page 251.

In Chapter 2, we explored the general concepts associated with functions and their graphs. In this chapter, we consider *polynomial functions*—those functions whose rules are given by polynomials. In addition, we focus on showing the connection between the *x* intercepts, zeros, factors, and roots of polynomials both visually and algebraically. We also study *rational functions*—functions formed by quotients of polynomials. We will continue to emphasize graphs and applications, including mathematical modeling of real-world phenomena.

3.1 Quadratic Functions

So far we have learned about first-degree, or linear, functions whose graphs are straight lines. We begin this chapter with a study of second-degree polynomial functions, referred to as *quadratic functions.* We will investigate the properties of such functions with particular emphasis on their graphs, which are called *parabolas.*

In Section 1.6, we developed parabolas geometrically with vertices at the origin. Here we consider parabolas as graphs of equations of the forms

$$y = ax^2 + bx + c \quad \text{or} \quad x = ay^2 + by + c$$

Definition	A function of the form
Quadratic Function	$$f(x) = ax^2 + bx + c$$
	where a, b, and c are constants and $a \neq 0$, is called a **quadratic function.**

Examples of quadratic functions are:

$$h(x) = x^2,$$

$$f(x) = 3x^2 - 1,$$

and

$$g(x) = -0.5x^2 - x + 3.$$

Graphing Functions of the Form $f(x) = a(x - h)^2 + k$

Figures 1a and 1b show the graphs of the quadratic functions

$$f(x) = x^2 \quad \text{and} \quad g(x) = -x^2$$

respectively.

Figure 1

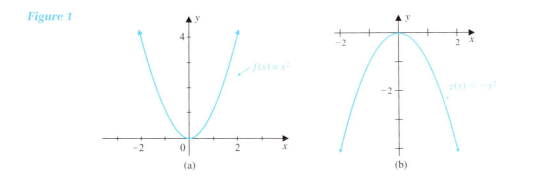

(a)   (b)

The graphs of these two functions are parabolas. In fact, the graph of any quadratic function is a *parabola,* with the following (see Section *1.6*) geometric characteristics, as displayed in Figure 2.

1. There is a vertical **axis of symmetry.**
2. The graph **opens upward** or **downward.**
3. The **vertex** of the parabola is either a *low point* on the graph (if it opens upward) or as *high point* (if it opens downward).

Figure 2

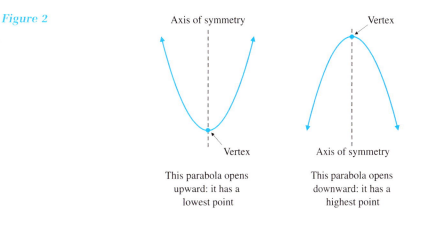

Axis of symmetry

Vertex

Vertex

Axis of symmetry

This parabola opens
upward: it has a
lowest point

This parabola opens
downward: it has a
highest point

When a quadratic function

$$f(x) = ax^2 + bx + c$$

is rewritten in the special algebraic form

$$f(x) = a(x - h)^2 + k$$

we can use the transformation techniques introduced in Section 2.4 to graph it by using the standard graph of $y = x^2$. The diagram below reviews the effects of the values of a, h, and k on transformations, and Figure 3 illustrates a transformation from the graph of $y = x^2$ to the graph of f.

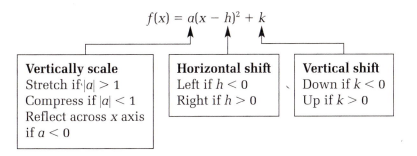

$$f(x) = a(x - h)^2 + k$$

Vertically scale
Stretch if $|a| > 1$
Compress if $|a| < 1$
Reflect across x axis
if $a < 0$

Horizontal shift
Left if $h < 0$
Right if $h > 0$

Vertical shift
Down if $k < 0$
Up if $k > 0$

Figure 3

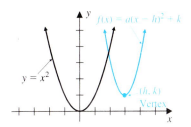

$y = x^2$

$f(x) = a(x - h)^2 + k$

(h, k)
Vertex

EXAMPLE 1   **Sketching the Graph of a Parabola by Using Transformations**

Use the graph of $y = x^2$, along with transformations, to graph the function

$$g(x) = -2(x - 1)^2 + 3.$$

Find the vertex and axis of symmetry of the parabola. Also determine the domain and range.

Solution   The graph of g is obtained by transforming the graph of $y = x^2$ as follows:

   (i)   Vertically stretch the graph of $y = x^2$ by a factor of 2.
  (ii)   Reflect the resulting graph across the x axis.
 (iii)   Shift the latter graph to the right by 1 unit.
 (iv)   Finally, shift the graph obtained in step (iii) upward by 3 units to get the graph of $g(x) = -2(x - 1)^2 + 3$ (Figure 4).

Figure 4

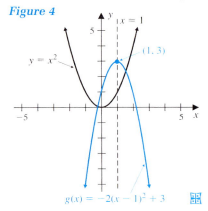

$$g(x) = -2(x - 1)^2 + 3$$

Because of the 1-unit horizontal shift to the right and the 3-unit shift upward, the vertex $(0, 0)$ of the graph of $y = x^2$ is shifted to the point $(1, 3)$ to become the vertex of the graph of g. Also, the axis of symmetry of $y = x^2$ (the y axis) is shifted 1 unit to the right to become the line $x = 1$, which is the axis of symmetry of the graph of g. From the graph, we see that the domain of g is $\mathbb{R}$ and the range includes all numbers in the interval $(-\infty, 3]$.

A quadratic function given in the *standard form*

$$f(x) = ax^2 + bx + c, \ a \neq 0$$

can be converted to the special algebraic form

$$f(x) = a(x - h)^2 + k$$

by using the technique of completing the square to produce a perfect square trinomial.

EXAMPLE 2   **Converting a Quadratic Function to the Special Algebraic Form**

Rewrite

$$f(x) = 3x^2 + 6x + 5$$

in the form

$$f(x) = a(x - h)^2 + k.$$

Find the vertex and axis of symmetry, and sketch the graph of f. Also determine the range of f.

Solution   To rewrite the quadratic function f in the special algebraic form, we proceed as follows:

$f(x) = 3x^2 + 6x + 5$	Given
$= 3(x^2 + 2x) + 5$	Group the x terms and factor out 3
$= 3\left(x^2 + 2x + \left(\dfrac{1}{2} \cdot 2\right)^2 - 1\right) + 5$	Complete the square for $x^2 + 2x$ by adding and subtracting 1 inside the parentheses
$\left.\begin{array}{l} = 3(x^2 + 2x + 1) - 3 + 5 \\ = 3(x + 1)^2 + 2 \end{array}\right\}$	Regroup, factor, and simplify
$= 3[x - (-1)]^2 + 2$	Special form

Now the graph of f can be obtained by the following transformations.
First vertically stretch the graph of

$$y = x^2$$

by a factor of 3 ($a = 3$).
Then shift the resulting curve 1 unit to the left ($h = -1$).
Finally, shift this latter curve 2 units upward ($k = 2$) (Figure 5).

Figure 5

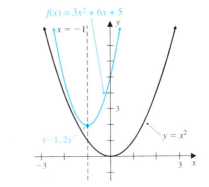

The graph of f is a parabola that opens upward with vertex at $(-1, 2)$;

its axis of symmetry is the line $x = -1$.

The range includes all numbers in the interval $[2, \infty)$.

Table 1 summarizes the information about the graph of a quadratic function

$$f(x) = ax^2 + bx + c, a \neq 0$$

that can be read from its converted special algebraic form

$$f(x) = a(x - h)^2 + k .$$

TABLE 1   Characteristics of the Graph of $f(x) = ax^2 + bx + c = a(x - h)^2 + k$

If	Parabola Opens	Vertex	Axis of Symmetry	Illustration
$a > 0$	Upward	(h, k)	$x = h$	
$a < 0$	Downward	(h, k)	$x = h$	

Up to now we have sketched graphs of quadratic functions defined by given equations. Sometimes we can reverse the process and use data from a parabolic graph to derive its equation.

EXAMPLE 3   **Determining an Equation of a Parabola**

Find an equation of the parabola, expressed in the form $f(x) = a(x - h)^2 + k$, that fits the data in Figure 6.

Solution   Figure 6 shows the vertex at the point $(-1, 9)$, and the parabola opens downward. Using the special form $f(x) = a(x - h)^2 + k$ with $h = -1$ and $k = 9$, we get

$$f(x) = a(x + 1)^2 + 9$$

To find a, notice that $(0, 8)$ is a point on the parabola, so its coordinates satisfy the equation $f(x) = a(x + 1)^2 + 9$. Thus

$$8 = a(0 + 1)^2 + 9 \quad \text{or} \quad 8 = a + 9$$

So $a = -1$.

Hence, an equation of the parabola is

$$f(x) = -(x + 1)^2 + 9.$$

By expanding the right side, we obtain another equation for the graph, namely, $f(x) = -x^2 - 2x + 8$.

Figure 6

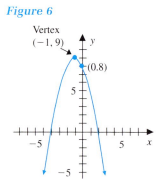

Solving Quadratic Inequalities Graphically

The real number solutions of the quadratic equation

$$ax^2 + bx + c = 0$$

correspond to the zeros of f or x intercepts of the graph of

$$f(x) = ax^2 + bx + c.$$

For instance, if a quadratic equation has two distinct real solutions, then its associated function has two real-number zeros, or, equivalently, the graph has two x intercepts (Figure 7a). If the equation has one real solution, then there is one real zero and the graph has one x intercept (Figure 7b), and if the equation has only imaginary solutions, then the associated function has no real zeros and the graph has no x intercepts (Figure 7c).

Figure 7

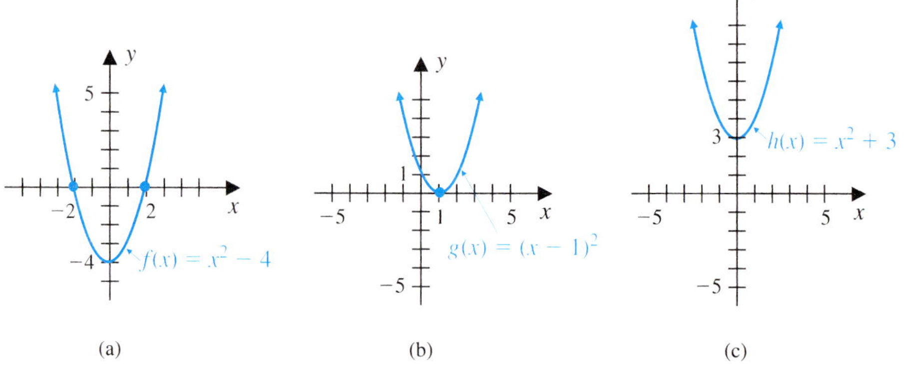

(a) $f(x) = x^2 - 4$

(b) $g(x) = (x - 1)^2$

(c) $h(x) = x^2 + 3$

As we learned in Section 2.3, we can use the locations of the x intercepts along with the graph of a function to solve a quadratic inequality by determining where the graph is above or below the x axis, as the next example shows.

EXAMPLE 4   **Solving Quadratic Inequalities Graphically**

Use the graph of $f(x) = 3x^2 - 5x + 2$ to solve each inequality.

(a) $3x^2 - 5x + 2 \geq 0$ (b) $3x^2 - 5x + 2 < 0$.

Solution   We begin by locating the x intercepts of the graph of f by finding the zeros of f; that is, we set $f(x) = 0$ and then solve the resulting equation. In this case, we have

$$3x^2 - 5x + 2 = (3x - 2)(x - 1) = 0$$

Thus $x = \dfrac{2}{3}$ or $x = 1$.

Figure 8

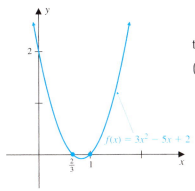

So the x intercepts are $2/3$ and 1. Figure 8 shows the graph of f and the locations of the x intercepts.

(a) By reading the graph in Figure 8, we see that $y = f(x) = 3x^2 - 5x + 2 \geq 0$ whenever (x, y) is a point on the graph of f that is above or on the x axis. So

$$3x^2 - 5x + 2 \geq 0 \text{ for all values of } x \text{ such that } x \leq \frac{2}{3} \quad \text{or} \quad x \geq 1.$$

Hence, the solution set of the inequality includes all values of x either in the interval $(-\infty, 2/3]$ or in the interval $[1, \infty)$.

(b) We observe in Figure 8 that $y = f(x) = 3x^2 - 5x + 2 < 0$ when the graph of f is below the x axis. Thus, the solution set of $3x^2 - 5x + 2 < 0$ includes all values of x such that $2/3 < x < 1$, that is, all numbers in the interval $(2/3, 1)$.

Determining Maximum and Minimum Values

We know that the graph of a quadratic function is a parabola that opens either upward or downward. If the graph opens upward, the vertex—which occurs at the lowest point on the graph—is called the **minimum point** of the function. See Figure 5 on page 207, where the minimum point of $f(x) = 3x^2 + 6x + 5$ is $(-1, 2)$.

Analogously, if the graph opens downward, the vertex, or highest point on the graph, is called the **maximum point.** See Figure 6 on page 208, where the maximum point of $f(x) = -x^2 - 2x + 8$ is $(-1, 9)$.

The minimum or maximum point is called the **extreme point** of the graph, and its y coordinate is called the **extreme value** (either the **minimum** or the **maximum value**) of the function. Thus the minimum value of $f(x) = 3x^2 + 6x + 5$ is 2, and the maximum value of $f(x) = -x^2 - 2x + 8$ is 9.

By employing the technique of completing the square, we can rewrite

$$f(x) = ax^2 + bx + c$$

as

$$f(x) = a\left(x + \frac{b}{2a}\right)^2 + \left(c - \frac{b^2}{4a}\right) \quad \text{(See problem 18).}$$

Comparing the latter form with the special form

$$f(x) = a(x - h)^2 + k$$

we see that $h = \dfrac{-b}{2a}$, so the vertex or extreme point is

$$(h, k) = \left(-\frac{b}{2a}, f\left(-\frac{b}{2a}\right)\right).$$

This result provides us with an easy way of finding the extreme value of a quadratic function (Table 2).

TABLE 2   Extreme Value of $f(x) = ax^2 + bx + c$

If	Parabola Opens	Extreme Point or Vertex	Extreme Value	Illustration
$a > 0$	Upward	$\left(-\dfrac{b}{2a}, f\left(-\dfrac{b}{2a}\right)\right)$	$f\left(-\dfrac{b}{2a}\right)$ (Minimum)	Extreme point $\left(-\frac{b}{2a}, f\left(-\frac{b}{2a}\right)\right)$ (minimum)
$a < 0$	Downward	$\left(-\dfrac{b}{2a}, f\left(-\dfrac{b}{2a}\right)\right)$	$f\left(-\dfrac{b}{2a}\right)$ (Maximum)	Extreme point $\left(-\frac{b}{2a}, f\left(-\frac{b}{2a}\right)\right)$ (maximum)

EXAMPLE 5   **Finding the Extreme Value of a Quadratic Function**

(a) Find the extreme value of

$$f(x) = -2x^2 + 12x - 16,$$

and determine whether the value is a maximum or minimum.

(b) Graph f, then interpret the graph by relating it to the result in part (a).

Solution   (a) We use the results in Table 2 as follows.

The function

$$f(x) = -2x^2 + 12x - 16$$

is in the standard form

$$f(x) = ax^2 + bx + c,$$

where $a = -2$ and $b = 12$.

Since a is negative, there is a maximum value when

$$x = -\frac{b}{2a} = -\frac{12}{-4} = 3$$

Figure 9

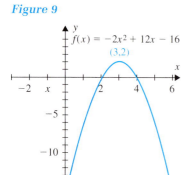

The maximum value of f is given by

$$f\left(-\frac{b}{2a}\right) = f(3)$$

$$= -2(3)^2 + 12(3) - 16$$

$$= 2$$

(b) Figure 9 shows the graph of f. From the result of part (a), we see that the function attains a maximum value of 2 at $x = 3$.

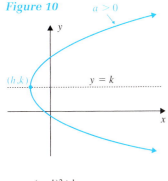

Figure 10

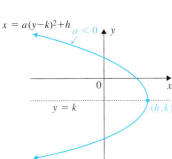

$$x = a(y-k)^2 + h$$

Graphing Horizontal Parabolas

We have seen that the graphs of quadratic functions of the form

$$y = ax^2 + bx + c$$

are parabolas that open upward if $a > 0$, or downward if $a < 0$. By interchanging the role of x and y in the equation $y = ax^2 + bx + c$, we obtain an equation of the form

$$x = ay^2 + by + c$$

This latter equation can be written in the form

$$x = a(y - k)^2 + h$$

by completing the square. Because of symmetry, the graphs of such equations are also parabolas, which open to the right if $a > 0$ (Figure 10a), or to the left if $a < 0$ (Figure 10b). The vertex is (h, k) and the axis of symmetry is the line $y = k$ (Figure 10). Parabolas whose vertical at the orgin were discussed on page 110.

Because of the vertical line test, these equations do not represent functions.

EXAMPLE 6   **Graphing Horizontal Parabolas**

Sketch the graph of each parabola. Indicate the vertex and the axis of symmetry.

(a) $x = 8y^2$ (b) $x = -y^2 - 2y + 15$

Solution   (a) The graph of the equation $x = 8y^2$ is a parabola that opens to the right, because $a = 8 > 0$. The vertex is at $(0, 0)$ and the axis of symmetry is the x axis or $y = 0$ (Figure 11a).

(b) By completing the square, we rewrite the equation in the form

$$x = a(y - k)^2 + h$$

as follows

$$x = -(y^2 + 2y + 1) + 15 + 1$$
$$= -(y^2 + 2y + 1) + 16$$
$$= -(y + 1)^2 + 16.$$

Thus, the vertex is at $(16, -1)$, and the parabola opens to the left. The axis of symmetry is the line $y = -1$ (Figure 11b).

Figure 11

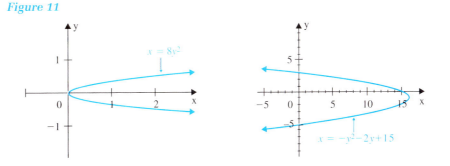

Solving Applied Problems

The solutions of many applied problems depend on finding the extreme value of a function used to model the situation. In Example 7 we construct a quadratic function to model a problem in which we want to find the maximum value.

EXAMPLE 7

Enclosing a Rectangular Corral of Maximum Area

A rancher wishes to enclose a rectangular corral. To save fencing costs, a river is used along one side of the corral and the other three sides are formed by 3000 feet of fence. Assume that x represents the width (in feet) of two of the fenced sides, y the length (in feet) of the third fenced side, and A the area of the corral (Figure 12).

(a) Express the area A as a function of x.

(b) What are the restrictions on x?

Figure 12

(c) Find the dimensions of the corral that enclose the maximum area. What is the maximum area?

(d) Graph the function defined in part (a) with the restrictions from part (b). Discuss what happens as the width increases from 0 to 1500 feet.

Solution

(a) Since the rancher plans to use 3000 feet of fence for three sides, the sketch in Figure 12 indicates that

$$2x + y = 3000$$

so

$$y = 3000 - 2x.$$

Thus the area A of the corral is given by

$$A = xy \qquad \text{Area = (Width) · (Length)}$$
$$= x(3000 - 2x) \qquad \text{Substitute for } y$$
$$= 3000x - 2x^2 \qquad \text{Multiply}$$
$$= -2x^2 + 3000x \qquad \text{Write in standard form}$$

So the area A is a quadratic function of the width x.

(b) Since the sum of the widths, $2x$, cannot be longer than the total amount of available fencing, it follows that

$$0 < 2x < 3000$$

$$0 < x < 1500 \qquad \text{Divide each part by 2}$$

(c) To find the maximum value of A, we first observe that

$$A = -2x^2 + 3000x = ax^2 + bx + c \quad \text{with } a = -2 \text{ and } b = 3000$$

So by using the results in Table 2, we find that the value of x for which the maximum value of A occurs is given by

$$x = -\frac{b}{2a}$$

$$= -\frac{3000}{2(-2)} = 750.$$

The corresponding value of y is given by

$$y = 3000 - 2x = 3000 - 2(750) = 1500.$$

Therefore the maximum value of A is given by

$$A = (750)(1500) = 1{,}125{,}000.$$

So a maximum area of 1,125,000 square feet is obtained when the corral is 750 feet wide and 1500 feet long.

(d) Next we sketch a graph of the quadratic function derived in part (a), taking into account the restrictions on x determined in part (b). The resulting graph (Figure 13) reveals that as the width x increases from 0 to 1500, the corresponding area A first increases from 0 to attain a maximum value of 1,125,000 square feet and then decreases to a value of 0 when $x = 1500$.

Figure 13

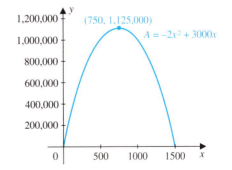

EXAMPLE 8   🅖 Modeling Population Growth

The U.S. Census Bureau has projected the population of the United States for the years from 2000 through 2100, according to the data in Table 3.

(a) Plot a scattergram of the data. Locate the t values on the horizontal axis and the P values on the vertical axis.

(b) Use a grapher and quadratic regression to find the equation of the form

$$P = at^2 + bt + c$$

that best fits the data. Use three decimal places for a, b, and c.

(c) Sketch the graph of the best fit curve from part (b).

(d) Use the regression model to predict the population in the years 2015, 2075, and 2110. Round off the answers to the nearest thousand.

TABLE 3

Year	2000	2010	2020	2030	2040	2050	2060	2070	2080	2090	2100
t	0	10	20	30	40	50	60	70	80	90	100
U.S. Population (P) (nearest thousand)	275,306	299,862	324,927	351,070	377,350	403,687	432,011	463,639	497,830	533,605	570,954

(Source: U.S. Census Bureau, January 13, 2000.)

Solution

(a) The scattergram for the given data is shown in Figure 14a, where P is used for the vertical axis and t is used for the horizontal axis.

Figure 14

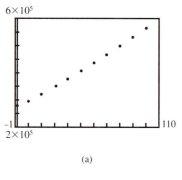

(a)

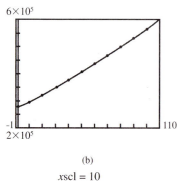

(b)

$x$scl = 10
$y$scl = 50,000

(b) By using a grapher, we find the quadratic regression model to be

$$P = 7.722t^2 + 2147.517t + 277436.245,\ t \geq 0$$

where the coefficients have been rounded to three decimal places.

(c) The graph of the regression equation in part (b) is shown in Figure 14b along with the scattergram of the given data.

(d) To predict the population in the year 2015, we substitute $t = 15$ into the regression equation to get 311,386. So the approximate population is predicted to be 311,386,000 in the year 2015.

 Similarly, in the year 2075 we substitute $t = 75$ into the equation to get the approximate population of 481,936,000.

 Finally, in the year 2110, $t = 110$ so that the approximate population is 607,099,000. 🔳

 At times, the same data can be modeled with different regression types of equations. For instance, in Chapter 4 we will use an exponential function as the regression model for the population data given in Table 3.

PROBLEM SET 3.1

Mastering the Concepts

1. Match each quadratic function with its graph in Figure 15.
(a) $f(x) = -(x - 2)^2$
(b) $g(x) = (x - 1)^2 - 2$
(c) $h(x) = -(x + 2)^2 - 3$
(d) $F(x) = (x + 1)^2 - 1$

Figure 15

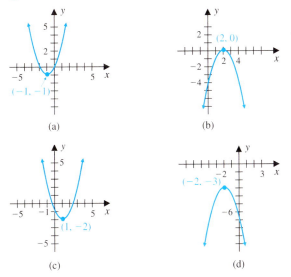

2. Describe the steps that can be used to transform the graph of $f(x) = x^2$ into the graph of g. Sketch the graph of g.
(a) $g(x) = 3x^2 + 4$
(b) $g(x) = (x + 1)^2 - 1$
(c) $g(x) = -(x + 1)^2 + 2$

In problems 3–8, use the graph of $y = x^2$ along with transformations to graph each function. Find the vertex and axis of symmetry of each parabola. Also determine the domain and range.

3. $f(x) = (x - 3)^2 + 2$
4. $g(x) = -(x + 1)^2 + 1$
5. $h(x) = -2(x + 2)^2 - 3$
6. $F(x) = 3(x - 1)^2 + 2$
7. $F(x) = \dfrac{1}{2}(x + 3)^2 + 1$
8. $G(x) = -\dfrac{1}{3}(x + 2)^2 - 1$

In problems 9–17, rewrite each function f in the form

$$f(x) = a(x - h)^2 + k$$

Find the vertex and axis of symmetry, and sketch the graph of each function. Also determine the range.

9. $f(x) = x^2 - 4x - 5$
10. $f(x) = x^2 - 10x + 20$
11. $f(x) = -x^2 - 6x - 8$
12. $f(x) = -x^2 + 5x + 11$
13. $f(x) = 3x^2 - 5x - 21$
14. $f(x) = 2x^2 - 11x + 5$
15. $f(x) = -5x^2 + 3x + 4$
16. $f(x) = -3x^2 - 6x + 5$
17. $f(x) = -\dfrac{1}{2}x^2 - 2x + 1$

18. By completing the square, show that $f(x) = ax^2 + bx + c$ can be written as

$$f(x) = a\left(x + \frac{b}{2a}\right)^2 + \left(c - \frac{b^2}{4a}\right)$$

In problems 19 and 20, find equations for the given parabolas.

19. (a) (b)

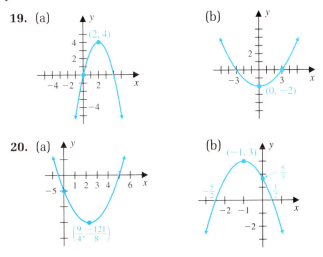

20. (a) (b)

In problems 21–24, find the equation in the form

$$f(x) = a(x - h)^2 + k$$

for the quadratic function f whose graph satisfies the given conditions. Also sketch the graph of f.

21. Vertex at $(2, -2)$ and contains the point $(0, 0)$
22. Vertex at $(-1, -9)$ and contains the point $(-4, 0)$
23. Vertex at $(0, 1)$ and contains the point $(2, -3)$
24. Vertex at $(0, 3)$ and contains the point $(-3, 0)$

In problems 25–30, use the locations of the x intercepts and the graph of the given function f to solve the associated inequality. Express the solution in interval notation.

25. $f(x) = x^2 + 5x - 14$; $x^2 + 5x - 14 > 0$
26. $f(x) = -x^2 + x + 20$; $-x^2 + x + 20 \geq 0$

27. $f(x) = -2x^2 + 5x - 3$ $-2x^2 + 5x \le 3$
28. $f(x) = 2x^2 - x - 1$ $2x^2 - x \ge 1$
29. $f(x) = 6x^2 + 7x - 20$ $6x^2 + 7x - 20 \ge 0$
30. $f(x) = 2x^2 - 9x - 5$ $2x^2 - 9x - 5 < 0$

In problems 31–36, find the extreme value algebraically. Determine whether the value is a maximum or a minimum.

31. $f(x) = x^2 + 8x + 12$
32. $g(x) = 10x^2 + x - 2$
33. $h(x) = -x^2 + 4x - 5$
34. $h(x) = -9x^2 - 6x + 8$
35. $g(x) = 2x^2 + x - 1$
36. $h(x) = -2x^2 - 9x + 5$

In problems 37–44, sketch the graph of each parabola. Indicate which way the parabola opens, the vertex, and the axis of symmetry.

37. $x = (y - 4)^2 - 3$ **38.** $x = -(y - 2)^2 + 1$
39. $x = -5(y + 3)^2 - 6$ **40.** $x = (y + 1)^2 + 5$
41. $x = -y^2 + 6y - 9$ **42.** $x = -y^2 - 5y - 4$
43. $x = 3y^2 - 12y - 36$ **44.** $x = 2y^2 - 4y - 4$

Applying the Concepts

45. Minimum Product: The difference of two real numbers is 6.
  (a) Express the product P as a function of one of the numbers.
  (b) Find the numbers that produce a minimum value of P. What is the minimum value?

46. Geometry: The sum of the height y (in centimeters) and the base x (in centimeters) of a triangle is 24 centimeters.
  (a) Express the area A of the triangle as a function of x.
  (b) Find the base that produces the maximum area of the triangle. What is the maximum area?

In problems 47–58, round off the answers to two decimal places when necessary.

47. Maximum Profit: Suppose that the profit P (in thousands of dollars) of a manufacturing company is related to the number x (in hundreds) of full-time employees by the quadratic function

$$P = -1.25x^2 + 156.25x \qquad 0 \le x \le 100$$

  (a) What is the profit if the number of employees is 500? 1000? 1500?

  (b) Find the maximum profit and the corresponding number of employees.
  (c) Graph the function. Discuss the profit trend as the number of employees increases.

48. Minimum Cost: The cost C (in dollars) to produce x tons of an alloy used for engine blocks is modeled by the function

$$C = 2.97x^2 - 603x + 62,836 \qquad 0 \le x \le 110.$$

  (a) How much does it cost if the company produces 50 tons? 100 tons? 110 tons?
  (b) Find the minimum cost and the corresponding number of tons.
  (c) Graph the function. Discuss the cost as the number of tons increases.

49. Maximum Height of a Baseball: The trajectory of a baseball hit by a player at home plate is approximated by the parabola whose function is given by

$$h(x) = -0.004x^2 + x + 3.5.$$

In this model, home plate is located at the origin, $h(x)$ (in feet) is the height of the ball above ground, and x is the horizontal distance (in feet) the ball has traveled from home plate.
  (a) How high is the ball after it has traveled horizontally 50 feet? 100 feet? 200 feet?
  (b) Graph the function. Approximate the maximum height and how far the ball travels horizontally to reach this height.
  (c) If the ball is not touched in flight, how far will it have traveled horizontally when it hits the ground?

50. Maximum Height of a Rocket: A model rocket is launched and then it accelerates until the propellant burns out, after which it coasts upward to its highest point. The height h of the rocket (in meters) above ground level t seconds after the burnout is modeled by the function

$$h = -4.9t^2 + 343t + 2010.$$

  (a) Find the height of the rocket above ground level 10 seconds, 20 seconds, and 30 seconds after the burnout.
  (b) Graph the function and approximate the maximum height attained by the rocket. How many seconds after the burnout does it take to reach the maximum height?
  (c) How long after the burnout will it take for the rocket to reach ground level?

51. Construction: A lookout tower is to be constructed on the peak of a hill. A portion of a cross section of the crown of the hill is approximately parabolic in shape. If a coordinate system is superimposed on this cross section as shown in Figure 16,

Figure 16

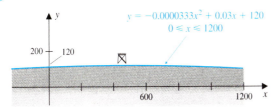

then the parabolic portion is modeled by the function

$$y = -0.0000333x^2 + 0.03x + 120$$
$$\text{where } 0 \le x \le 1200,$$

where x and y are measured in feet, and y is the number of feet above sea level. (The parabolic shape is barely visible in Figure 16 due to the scale of the drawing.)

(a) Find the height of the hill above sea level when x is 200 feet, 500 feet, and 1000 feet.

(b) Graph the function and approximate the coordinates of the peak of the hill.

(c) If the observation booth of the tower is to be 150 feet above sea level, how tall should the tower be?

52. Maximum Current of a Circuit: In an electrical circuit, the power P (in watts) delivered to the load is modeled by the function

$$P = EI - RI^2$$

where E is the voltage (in volts), R is the resistance (in ohms), and I is the current (in amperes). For a circuit of 120 volts and a resistance of 12 ohms, find the current that produces the maximum power, and find the maximum power.

53. Constructing a Playground: A recreation department plans to build a rectangular playground enclosed by 1000 feet of fencing. Assume that A represents the area of the enclosure and that x and y, respectively, represent the width and length (in feet).

(a) Express the area A as a function of the width x.

(b) What are the restrictions on x?

(c) Find the dimensions of the playground that encloses the maximum area. What is this area?

(d) Graph the function defined in part (a). Discuss what happens to the area as the width increases.

54. Shape of a Suspension Bridge: A suspension bridge has a main span of 1200 meters between the towers that support the parabolic main cables, and these towers extend 160 meters above the road surface (Figure 17).

Figure 17

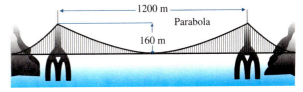

(a) Assuming that the main cable is tangent to the road at the center of the bridge and that the road is horizontal, find an equation of the main parabolic cable with respect to an xy coordinate system (with a vertical y axis, the x axis running along the road surface, and the origin at the center of the bridge).

(b) Graph the equation in part (a).

(c) Find the distance between the roadway and the cable at a point 300 meters from the center of the bridge.

55. Enclosure: A rancher has 450 feet of fencing to enclose three sides of a rectangular pen. A barn will be used to form the fourth side of the pen. What dimensions of the pen would enclose the maximum area? What is the maximum area?

56. Enclosure: A breeder of horses wants to fence in two grazing areas along a river, including one interior fence separation perpendicular to the river as shown in Figure 18. The river will serve as one side, and 600 yards of fencing are available. What is the largest area that can be enclosed? What are the exterior dimensions that yield the maximum area?

Figure 18

57. Architecture: A Norman-style window has the shape of a rectangle surmounted by a semicircular arch (Figure 19). Find the height y of the rectangle and the radius x of the

Figure 19

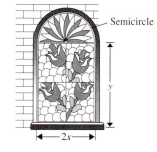

semicircle of the Norman window with maximum area whose perimeter is $8 + 2\pi$ feet. (Use $\pi = 3.14$.)

58. **Agriculture:** An orange grower has determined that if 120 orange trees are planted per acre, each will yield approximately 60 small boxes of oranges per tree over the growing season. The grower plans to add or delete trees in the orchard. The state agricultural service advises that because of crowding, each additional tree will reduce the average yield per tree by about 2 boxes over the growing season. How many trees per acre should be planted to maximize the total yield of oranges, and what is the maximum yield of the crop?

59. The following table gives discrete data points for pairs of values of the form (x, y).

x	y
0	0
0.5	8.5
1	26
2	30
3	16
4	0

(a) Plot a scattergram of this data. Locate the x values on the horizontal axis and the y values on the vertical axis.

(b) Use a grapher and quadratic regression to find the equation of the form

$$y = ax^2 + bx + c$$

that best fits the data. Round off a, b, and c to three decimal places.

(c) Sketch the graph of the best fit parabola from part (b).

(d) Use the regression equation to estimate the value of y if $x = 2.5$, 5, or 10. Round off the answers to one decimal place.

60. **Property Crime:** A study shows the number N (in millions) of property crimes in the United

States for every other year from 1996 through 2006 is given in the following table.

Year	t	N Crimes (in millions)
1996	0	10.6
1998	2	11.7
2000	4	12.4
2002	6	12.7
2004	8	12.5
2006	10	12.2

(a) Plot a scattergram of this data. Locate the t values on the horizontal axis and the N values on the vertical axis.

(b) Use a grapher and quadratic regression to find a model of the form

$$N = at^2 + bt + c$$

that best fits the data. Round off a, b, and c to three decimal places.

(c) Sketch the graph of the best fit curve from part (b).

(d) Use the regression model to predict the number of property crimes in millions for the years 2011 and 2016. Round off the answers to one decimal place.

61. **Modeling Income:** The net income N (in millions of dollars) of a company during the period from 2001 through 2005 is given by the following table.

Year	t	Income N (in millions of dollars)
2001	0	24.8
2002	1	15.6
2003	2	16.2
2004	3	17.4
2005	4	32.2

(a) Plot a scattergram of this data. Locate the t values on the horizontal axis and the N values on the vertical axis.

(b) Use a grapher and quadratic regression to find a model of the form

$$N = at^2 + bt + c$$

that best fits the data. Round off a, b, and c to three decimal places.

(c) Sketch the graph of the best fit curve from part (b).

(d) Use the regression equation to predict the net income of the company in millions of dollars for the years 2010 and 2015. Round off to one decimal place.

62. **G Fuel Economy:** The data in the following table relates the fuel economy F (in miles per gallon) of a car to different average speeds V (in miles per hour) between 40 and 80 miles per hour.

Vehicle Speed V (mph)	Fuel Economy F (mpg)
40	26.1
45	26.5
50	27.5
55	29.5
60	29.0
65	25.8
70	25.5
75	24.5
80	23.8

(a) Plot a scattergram of this data. Locate the V values on the horizontal axis and the F values on the vertical axis.

(b) Use a grapher and quadratic regression to find a model of the form

$$F = aV^2 + bV + c$$

that best fits the data. Round off a, b, and c to three decimal places.

(c) Sketch the graph of the best fit curve from part (b).

(d) Use the regression equation to estimate the fuel economy if the car speed is 52 mph, 63 mph, or 85 mph. Round off the answers to one decimal place.

Developing and Extending the Concepts

63. Decide how many x intercepts the graph of the function

$$f(x) = ax^2 + bx + c, a \neq 0$$

has if the *discriminant,* defined as

$$D = b^2 - 4ac$$

of the corresponding quadratic equation

$$ax^2 + bx + c = 0,$$

satisfies the given condition. Give examples of functions and their graphs to support your assertions.

(a) $D > 0$ (b) $D = 0$ (c) $D < 0$

64. Suppose that

$$f(x) = x^2 - kx + 16.$$

Determine a value for k such that the graph of f has

(a) No x intercept

(b) One x intercept

(c) Two x intercepts

In problems 65–67, use the results from problem 63 to determine how many x intercepts the graph of the function f has. Then graph the function to demonstrate the result.

65. $f(x) = -x^2 - 6x + 7$
66. $f(x) = 4x^2 + 12x + 9$
67. $f(x) = -4x^2 + 28x - 49$
68. **Calculus:** In calculus, *Simpson's rule* is used in numerical integration. Simpson's rule is based on the fact that the points $(-h, y_0)$, $(0, y_1)$ and (h, y_2) lie on the graph of a quadratic function $y = ax^2 + bx + c$, whose graph is given in Figure 20. Suppose that the shaded area A in Figure 20 is given by

$$A = \left(\frac{h}{3}\right)(2ah^2 + 6c).$$

Show that

$$A = \left(\frac{h}{3}\right)(y_0 + 4y_1 + y_2).$$

Figure 20

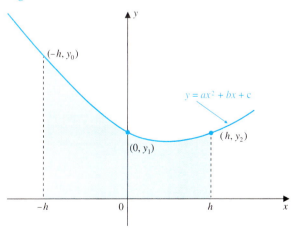

Objectives

1. Graph Functions of the Form $f(x) = a(x - h)^n + k$
2. Establish Basic Properties of Polynomial Functions
3. Relate Multiple Zeros to Graphs
4. Solve Applied Problems

3.2 Polynomial Functions of Higher Degree

Table 1 lists the type of polynomial functions that we have investigated so for.

TABLE 1

Type of Function	Form	Graph
Degree 0 (constant)	$f(x) = k \quad (k \neq 0)$	Horizontal line
Degree 1 (linear)	$f(x) = mx + b \quad (m \neq 0)$	Straight line
Degree 2 (quadratic)	$f(x) = ax^2 + bx + c \quad (a \neq 0)$	Parabola

In the following definition, we turn our attention to polynomial functions of degree higher than two.

Definition
Polynomial Function

A function f of the form

$$f(x) = a_n x^n + a_{n-1} x^{n-1} + \ldots + a_1 x + a_0, \quad a_n \neq 0$$

where n is a nonnegative integer, is called a **polynomial function of degree n.** The numbers $a_n, a_{n-1}, \ldots, a_1$, and a_0 are called the **coefficients** of the polynomial.

Thus $f(x) = 4x^3 - x^2 + 2x - 7$ is a polynomial function of degree 3, with coefficients 4, −1, 2, and −7.

It should be noted that $f(x) = 0$ is the **zero** polynomial function with no degree assigned to it.

Graphing Functions of the Form $f(x) = a(x - h)^n + k$

We begin by considering special kinds of polynomial functions of the form

$$f(x) = x^n, \text{ where } n \text{ is a positive integer.}$$

They are called **power functions.**

Examples of power functions are

$$f(x) = x^2, \quad g(x) = x^3, \quad h(x) = x^4 \text{ and } F(x) = x^5$$

Recall the graphs of the special functions $f(x) = x^2$ and $g(x) = x^3$, shown in Figure 1.

Figure 1

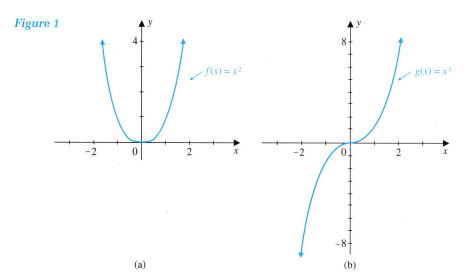

(a)   (b)

The first function

$$f(x) = x^2$$

is symmetric with respect to the y axis, the second function

$$g(x) = x^3$$

is symmetric with respect to the origin.

They will be used as models for the power functions with degrees greater than three.

In general, graphs of power functions

$$f(x) = x^n$$

fall into two categories according to whether n is an even or odd positive integer.

If $n \geq 4$ is an even integer, then the graphs of functions such as

$$f(x) = x^4 \quad \text{and} \quad g(x) = x^6$$

resemble the graph of $y = x^2$ (Figure 2a).

If $n \geq 5$ is an odd integer, then the graphs of functions such as

$$f(x) = x^5 \quad \text{and} \quad g(x) = x^7$$

resemble the graph of $y = x^3$ (Figure 2b).

Figure 2

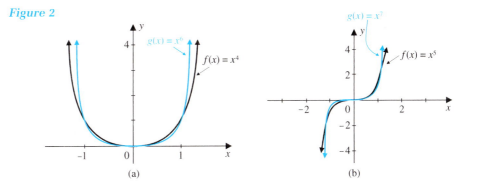

(a)   (b)

The graphs in Figures 1 and 2 suggest the general properties given in Table 2.

TABLE 2   Characteristics of Power Function Graphs

$f(x) = x^n$, n a Positive Integer		
	n is Even	**n is Odd**
Symmetry with respect to	y axis, since $f(-x) = f(x)$	Origin, since $f(-x) = -f(x)$
Shape of graph	U-shaped curve; contains points $(0, 0)$, $(-1, 1)$ and $(1, 1)$; decreasing in the interval $(-\infty, 0]$ and increasing in the interval $[0, \infty)$	Contains points $(0, 0)$, $(-1, -1)$ and $(1, 1)$; increasing in interval $(-\infty, \infty)$

Transformations of the graphs of

$$y = x^n$$

can be used to sketch graphs of functions of the form

$$f(x) = a(x - h)^n + k$$

where $n \geq 2$ is an integer.

EXAMPLE 1   **Using Transformations of a Power Function**

Use transformations of the graph of $y = x^4$ (Figure 2a) to sketch the graph of each function.

(a) $f(x) = 3x^4$ (b) $g(x) = -3x^4$

(c) $h(x) = -3(x - 2)^4$ (d) $k(x) = -3(x - 2)^4 + 1$

Solution   Each graph is obtained as follows:

(a) Vertically stretch the graph of $y = x^4$ by a factor of 3 to get the graph of

$$f(x) = 3x^4 \text{ (Figure 3a)}.$$

(b) Reflect the graph of $f(x) = 3x^4$ across the x axis to obtain the graph of

$$g(x) = -3x^4 \text{ (Figure 3b)}.$$

(c) Shift the graph of $g(x) = -3x^4$ horizontally, 2 units to the right to get the graph of

$$h(x) = -3(x - 2)^4 \text{ (Figure 3c)}.$$

(d) Shift the graph of h vertically up 1 unit to obtain the graph of

$$k(x) = -3(x - 2)^4 + 1 \text{ (Figure 3d)}.$$

Figure 3

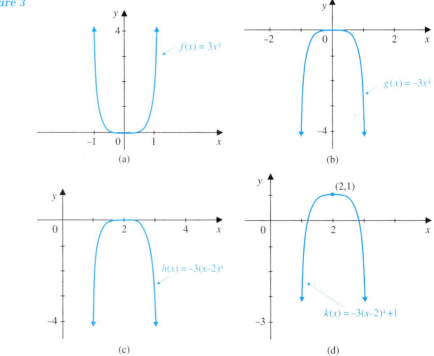

Establishing Basic Properties of Polynomial Functions

Accurately graphing a polynomial function such as

$$f(x) = 2x^3 - 3x^2 - 3x + 2$$

that is not expressed in the form

$$f(x) = a(x - h)^n + k$$

where $n \geq 3$ is an integer, often requires advanced mathematical techniques. Whether we are graphing such a function by hand or by a grapher, it helps to understand some basic properties of polynomial functions by exploring the following questions.

1. Does the graph have gaps or breaks?
2. How many turning points does the function have? Are they high or low points?
3. How many x intercepts does the function have?
4. What happens to the graph of a polynomial function $y = f(x)$ as x gets very large or very small?

The answer to the *first question* is given in the following property.

Property 1 **Continuity**	A polynomial function is **continuous** for all real numbers.

This means that a polynomial function is defined for all real numbers, and its graph is an unbroken curve, with no *jumps, gaps,* or *holes.* It can be drawn with a pencil without having to lift it from the paper.

The *second question* refers to **turning points,** which are points on the graph of a polynomial function where the graph changes direction from increasing to decreasing, or vice versa.

Figure 4 displays a graph that has four turning points, labeled T_1, T_2, T_3, and T_4. This figure also shows that the graph has *peaks* at T_1 and T_3, and *valleys* at T_2 and T_4.

Figure 4

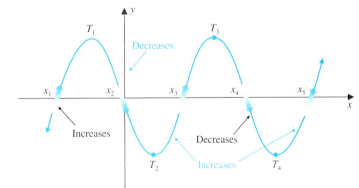

The following property gives us information about the number of turning points the graph of a polynomial function can have.

Property 2
Maximum Number
of Turning Points

> The graph of a polynomial function of degree n has at most $(n - 1)$ turning points.

The graph of a second-degree (quadratic) function has one turning point—its vertex.

This means that the total number of peaks or "high points" (called **relative maxima**) and valleys or "low points" (called **relative minima**) on the graph of a polynomial function of degree n is at most $n - 1$.

Next we turn or attention to the *third question* about the number of x intercepts. As Figure 4 shows, the graph of f has five x intercepts at x_1, x_2, x_3, x_4, and x_5. Since the zeros of f are the same as the x intercepts, there are five zeros.

In general, we have the following property.

Property 3
Maximum Number
of x intercepts
or Real Zeros

> If f is a polynomial function of degree n, then the graph of f cannot have more than n x intercepts. Accordingly, a polynomial function of degree n has at most n real zeros.

For instance, consider the number of x intercepts for the graphs of the third-degree polynomial functions

$$f(x) = x^3 + x, \quad g(x) = x^3 - 4x^2 + 4x \quad \text{and} \quad h(x) = x^3 - 7x - 1$$

in Figure 5.

The graph of f has one x intercept (Figure 5a); the graph of g has two x intercepts (Figure 5b); and the graph of h has three x intercepts (Figure 5c).

Figure 5

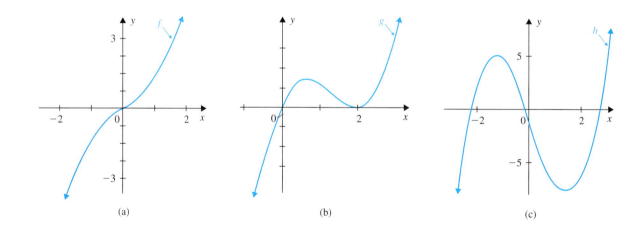

(a)   (b)   (c)

The *fourth question* regarding the *end behavior* of the graph of a polynomial function can be rephrased as follows:

What happens to the values (and graphs) of a polynomial function $y = f(x)$ as the values of $|x|$ get very large for positive or negative values of x?

Such behavior is referred to as the **limit behavior** of f.

From a numerical point of view, the data in Table 3 reveals the limit behavior of $f(x) = x^4$ in the sense that as x takes on larger and larger positive values (increases), $f(x)$ **increases without bound**. We say that $f(x)$ **approaches positive infinity as** x approaches positive infinity, and we describe this limit behavior symbolically by writing

$$f(x) \to +\infty \quad \text{as} \quad x \to +\infty.$$

TABLE 3   Limit Behavior Pattern of $f(x) = x^4$

x	-100	-10	-5	-1	0	1	5	10	100
$f(x) = x^4$	100,000,000	10,000	625	1	0	1	625	10,000	100,000,000

Similarly, we observe that $f(x)$ increases without bound as x takes on smaller and smaller negative values (decreases). So we say that $f(x)$ **approaches positive infinity as x approaches negative infinity.** Accordingly, we write

$$f(x) \to +\infty \quad \text{as} \quad x \to -\infty$$

This numerical pattern is confirmed by examining the graph of $f(x) = x^4$ in Figure 2a on page 221. As the x values get larger (or smaller), the y values on the graph get bigger and bigger without restriction.

Similarly, from a graphical point of view, the graph in Figure 1b on page 220 shows that $g(x) = x^3$ increases without bound as positive values of x increase. Also, $g(x)$ decreases without bound as negative values of x decrease. So we describe the limit behavior of g by writing, respectively,

$$g(x) \to +\infty \quad \text{as} \quad x \to +\infty \quad \text{and} \quad g(x) \to -\infty \quad \text{as} \quad x \to -\infty$$

To understand the limit behavior of a polynomial function that is not a power function, let us compare the graphs of

$$f(x) = x^3 + x^2 - 2x - 1 \text{ and } g(x) = x^3$$

given in Figures 6a and 6b, respectively.

Figure 6

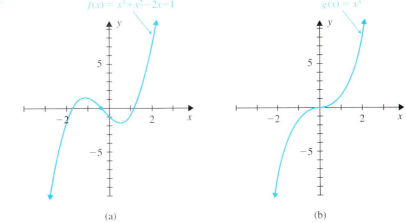

(a)   (b)

The graph of g increases without bound; that is, $g(x) \to +\infty$ as $x \to \infty$, and the same is true of the graph of f. Also, the graph of g decreases without bound; that is, $g(x) \to -\infty$ as $x \to -\infty$, and the same is true of the graph of f.

Thus, the *far left* and *far right* behavior of the polynomial function f can be discovered by using the far left and far right behavior of its leading term x^3.

In a sense, the leading term x^3 in the polynomial function f "dominates" all other terms the farther away we get from the origin. Whereas the graphs of the functions f and g are not similar near the origin, the *limit behaviors* are the same.

In general, the direction and shape of the graph of a polynomial function at the *far right* and *far left* can be determined by examining its *leading term* as explained in the following property.

Property 4
Limit Behavior

The **limit behavior** of a polynomial function

$$f(x) = a_n x^n + a_{n-1} x^{n-1} + \ldots + a_1 x + a_0 \,, \; a_n \neq 0$$

is the same limit behavior as the function defined by its *leading term, $g(x) = a_n x^n$.*

We refer to the function g as the *limit behavior model of f*. Table 4 gives two additional examples of the limit behavior property.

TABLE 4   Examples of Limit Behavior Models

Given Function	Limit Behavior Model	Limit Behavior Comparison	Graphical Display
$f(x) = 2x^3 - 3x^2 - 3x + 2$	$g(x) = 2x^3$	$g(x) \to +\infty$ as $x \to +\infty$ $g(x) \to -\infty$ as $x \to -\infty$ and so $f(x) \to +\infty$ as $x \to +\infty$ $f(x) \to -\infty$ as $x \to -\infty$	(a)
$h(x) = -2x^4 + 2x^2$	$g(x) = -2x^4$	$g(x) \to -\infty$ as $x \to +\infty$ $g(x) \to -\infty$ as $x \to -\infty$ and so $h(x) \to -\infty$ as $x \to +\infty$ $h(x) \to -\infty$ as $x \to -\infty$	(b)

To understand algebraically why f and g in the first example in Table 4 have the same limit behavior, consider

$$f(x) = 2x^3 - 3x^2 - 3x + 2.$$

Factor $f(x)$ as follows and compare with $g(x)$:

$$f(x) = 2x^3 \left(1 - \frac{3/2}{x} - \frac{3/2}{x^2} + \frac{1}{x^3} \right)$$

As $x \to +\infty$, the terms $\dfrac{3/2}{x}, \dfrac{3/2}{x^2}$, and $\dfrac{1}{x^3}$ approach zero. So that in turn

$$f(x) \text{ approaches } 2x^3(1 - 0 - 0 + 0) = 2x^3 = g(x).$$

Graphing can be done more accurately if we know the above four properties as the next example illustrates.

EXAMPLE 2   **Analyzing the Graph of a Polynomial Function**

Figure 7 shows the graph of the function

$$y = f(x) = x^3 - 4x^2 + x + 6 = (x + 1)(x - 2)(x - 3)$$

obtained by point-plotting and using the continuity and the fact that the limit behavior of f is the same as that of $y = x^3$

(a) Find the x intercepts and the zeros of f.

(b) How many turning points are there?

(c) Use the graph to solve the inequality.

$$x^3 - 4x^2 + x + 6 \leq 0$$

Figure 7

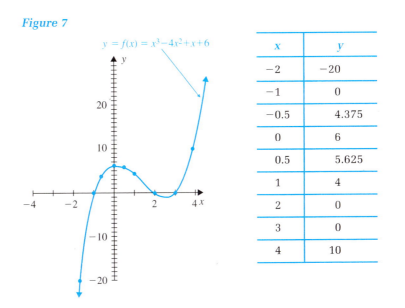

x	y
-2	-20
-1	0
-0.5	4.375
0	6
0.5	5.625
1	4
2	0
3	0
4	10

Solution   (a) Since the function is written in a factored form, it is not difficult to find the x intercepts. By setting $y = 0$ and then solving for x, we have

$$x^3 - 4x^2 + x + 6 = (x + 1)(x - 2)(x - 3) = 0$$

so that,

The factorization made it easy to find the x *intercepts. Later, we will learn how to locate* x *intercepts for polynomial functions that are difficult to factor.*

$x = -1$, $x = 2$, $x = 3$ are the three x intercepts. The zeros of f are the same as the x intercepts, namely, -1, 2, and 3.

(b) The graph has two turning points, the maximum indicated in Property 2.

(c) Next we solve the inequality

$$x^3 - 4x^2 + x + 6 \leq 0$$

From part (a), the locations of the x intercepts are -1, 2, and 3. The y values of points on the graph represent the values of the expression

$$x^3 - 4x^2 + x + 6.$$

So the portions of the graph on or below the x axis indicate the values of x that satisfy the inequality $y \leq 0$. Thus by reading the graph, we discover that the solution set of the inequality consists of all values of x such that $x \leq -1$ or $2 \leq x \leq 3$, that is, all values of x in the intervals $(-\infty, -1]$ or $[2, 3]$.

G The four properties above help us anticipate and verify the general behavior of the graph of a polynomial function even when we use a grapher.

For instance, suppose we use a grapher to graph the function

$$f(x) = x^3 - 4x^2 + x + 6$$

given in Example 2, without knowing the factored form.

By using the WINDOW specified in Figure 8a and without anticipating the behavior of the function, it would appear that the graph of f is as shown in Figure 8b.

Figure 8

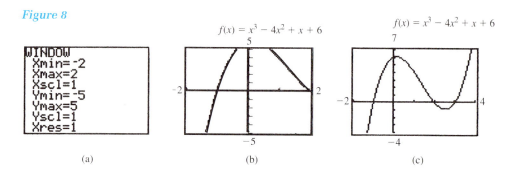

(a)   (b)   (c)

Since the graph of f is supposed to be continuous, the Ymax choice needs to be modified.

Also, the limit behavior model for f is

$$g(x) = x^3$$

(Figure 6b on page 225). Consequently, the limit behavior for f should be

$$f(x) \rightarrow +\infty \quad \text{as} \quad x \rightarrow +\infty$$

and

$$f(x) \rightarrow -\infty \quad \text{as} \quad x \rightarrow -\infty$$

Since Figure 8b does not include the appropriate limit behavior of f for large positive values of x, we need to change the Xmax choice. Figure 8c displays one possible WINDOW selection that yields a more accurate graph of f, showing both the continuity and limit behavior. More confidence in the accuracy of the graph is gained by observing that this third-degree polynomial function has the maximum number of turning points, $3 - 1 = 2$, and the maximum number of x intercepts.

Relating Multiple Zeros to Graphs

Consider the factored form of the polynomial function f in Example 2, given by

$$f(x) = x^3 - 4x^2 + x - 6 = (x + 1)(x - 2)(x - 3)$$

We note that, $f(x) = 0$ whenever $x + 1 = 0$ or $x - 2 = 0$ or $x - 3 = 0$, that is, any zero of $f(x)$ must be a zero of one of the factors of $f(x)$. Since the number of *different* linear factors of f is 3, we say the function f has zeros -1, 2, and 3, each of "multiplicity 1."

When a polynomial function f is written in factored form, the same factor, say, $x - c$, may occur more than once, in which case, c is referred to as a *repeated* or *multiple zero* of f.

For instance, $f(x) = (x - 5)^2$ has zero 5 with a multiplicity of 2.

In general, we have the following definition:

Definition **Multiplicity of a Zero**	If $f(x) = (x - c)^s \cdot Q(x)$, $Q(c) \neq 0$, and s is a positive integer, then we say that c is a **zero of multiplicity** s of the function f.

Thus the function

$$f(x) = (x + 2)^3 (x - 1)^4 (x + 1)^2$$

has three different zeros, -2, 1, and -1, with the multiplicities given in the following table.

Zero	Multiplicity
-2	3
1	4
-1	2

If c is a real zero of $f(x)$ of multiplicity s, then $f(x)$ has the factor $(x - c)^s$, and the graph of f has an x intercept at c. The general behavior of the graph of f at c depends on whether s is an odd or even integer.

For example, Figure 9 shows the graph of the function

$$f(x) = (x + 2)(x - 1)^4(x + 1)^2$$

It has zeros -2, -1 and 1, with multiplicities 1, 2, and 4 respectively. By examining the graph of f, we observe the following features.

1. The graph of f crosses the x axis at $x = -2$, a zero of odd multiplicity, 1.
2. The graph of f only touches, but does not cross the x axis at $x = -1$, a zero of even multiplicity, 2.
3. The graph of f only touches, but does not cross the x axis at $x = 1$, a zero of even multiplicity, 4.

Figure 9

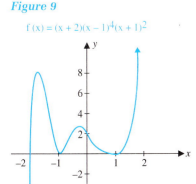

$f(x) = (x + 2)(x - 1)^4(x + 1)^2$

This illustration is generalized as follows:

Even-Odd Multiplicity Rule Graph Features	If c is a real zero of even multiplicity of a polynomial function f, then the graph of f touches the x axis at $x = c$ and bounces back. If c is a real zero of odd multiplicity of a polynomial function f, then the graph of f crosses the x axis at $x = c$.

EXAMPLE 3   G **Applying the Even-Odd Multiplicity Rule**

Use the even-odd multiplicity rule to determine where the graph of

$$f(x) = x(x - 4)^2(x - 2)^3$$

crosses the x axis and where the graph touches the x axis and bounces back.

Solution   Table 5 identifies the real zeros of f along with their multiplicities and corresponding graphical features as indicated by the even-odd multiplicity rule.

Figure 10

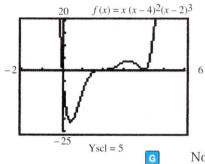

TABLE 5 $f(x) = x(x - 4)^2(x - 2)^3$

Real Zero	Multiplicity	Graphical Features
0	1 (odd)	Crosses x axis at $x = 0$
2	3 (odd)	Crosses x axis at $x = 2$
4	2 (even)	Touches x axis at $x = 4$ and bounces back

G   Note that the graph of f (Figure 10), which was obtained by using a grapher, contains the features listed in Table 5.

Solving Applied Problems

Higher-degree polynomial functions are used in mathematical modeling, as illustrated in the next example.

EXAMPLE 4   **Examining a Growth Rate Model**

A biologist studying bacteria growth in a culture determines that the growth rate $f(t)$ (in grams per hour) of bacteria during the first 30 hours of an experiment is given by the model

$$f(t) = -0.003(t - 19)(t + 5)(t + 0.75), \ 0 \le t \le 30$$

where t is the elapsed time in hours.

(a) What are the bacteria growth rates at 8 hours, 11.8 hours, 15 hours, and 29 hours? Round off the answers to two decimal places.

(b) Indicate the time when the growth rate is zero and what the growth is initially, that is, when $t = 0$. Round off the answers to two decimal places.

(c) Sketch the graph of this model.

(d) During what time interval is the growth rate positive or negative?

Solution   (a) The bacteria growth rates at 8 hours, 11.8 hours, 15 hours, and 29 hours, rounded to two decimal places, are obtained by performing the following computations.

$$f(8) = -0.003(8 - 19)(8 + 5)(8 + 0.75)$$

$$= 3.75$$

$$f(11.8) = -0.003(11.8 - 19)(11.8 + 5)(11.8 + 0.75)$$

$$= 4.55$$

$$f(15) = -0.003(15 - 19)(15 + 5)(15 + 0.75)$$

$$= 3.78$$

$$f(29) = -0.003(29 - 19)(29 + 5)(29 + 0.75)$$

$$= -30.35$$

This means that at 8 hours, at 11.8 hours, and at 15 hours, the bacteria are growing at rates of 3.75 grams per hour, 4.55 grams per hour, and 3.78 grams per hour, respectively. However, the growth rate of -30.35 grams per hour, at 29 hours indicates that the number of bacteria is declining.

(b) The growth rate is zero when $f(t) = 0$, that is, when

$$t = 19, t = -5, \text{ and } t = -0.75.$$

We only use the positive value, $t = 19$, because the given restriction indicates that $0 \le t \le 30$. Thus the growth rate is zero at 19 hours.
Notice that when $t = 0$ we get

$$f(0) = -0.003(-19)(5)(0.75)$$

$$= 0.21$$

rounded to two decimal places.
This means the initial growth rate is 0.21 gram per hour.

(c) By plotting points for $0 \le t \le 30$, including the data obtained in part (a), and connecting them with a smooth continuous curve, we obtain the graph of the model (Figure 11).

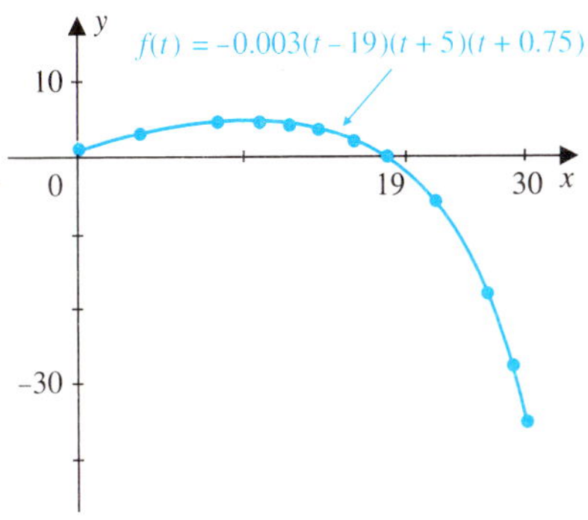

(d) The growth rate is positive whenever $f(t) > 0$. As shown in Figure 11, this occurs where the graph is above the t axis, that is, when

$$0 \le t < 19.$$

Figure 11

Similarly, the growth rate is negative when $f(t) < 0$, that is, when

$$19 < t \le 30.$$

Note that there is no systematic way to locate a maximum (high) or a minimum (low) point on the graph of a higher-degree polynomial function with the tools we have learned. However, by using programmed MIN and MAX finding features of a grapher, we are able to approximate the locations of any extreme points.

EXAMPLE 5   G **Fabricating a Box**

The liner for an outside window flower box is to be made by cutting equal squares from the corners of a rectangular sheet of perforated aluminum and turning up the sides. Assume the aluminum sheet is 40 inches by 16 inches and the length of each square is x inches.

(a) Express the volume V as a function of x.

(b) What are the restrictions on x?

(c) Use a grapher to sketch the graph of the function in part (a) incorporating the restrictions from part (b).

(d) Use a grapher to estimate the value of x that gives the maximum volume. Also, find the maximum volume. Round off the answers to one decimal place.

Solution

(a) Figure 12 displays a physical model of the situation. From the sketches we see that the height h of the liner is x inches, the width w is $16 - 2x$ inches, and the length ℓ is $40 - 2x$ inches. So the volume V of the liner is given by

Figure 12

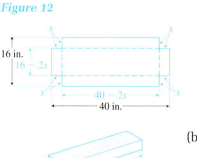

$$V = \ell wh \qquad \text{Formula from geometry}$$
$$= (40 - 2x)(16 - 2x)x \qquad \text{Substitute}$$
$$= 4x(20 - x)(8 - x) \qquad \text{Factor}$$

Thus, V is expressed as a function of x.

(b) Each physical dimension must be positive, so $x > 0$.

Also $40 - 2x > 0$ and $16 - 2x > 0$

$$40 > 2x \qquad\qquad 16 > 2x$$
$$20 > x \qquad\qquad 8 > x$$

Thus $0 < x < 8$.

(c) Figure 13 shows a viewing window of the graph of

$$V = 4x(20 - x)(8 - x) \text{ for } 0 < x < 8.$$

(d) By using the features of a grapher we find that the maximum value of V is about 1039.5 cubic inches, and it is obtained when x is approximately 3.5 inches.

Figure 13

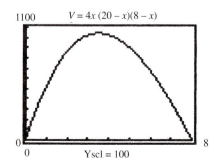

PROBLEM SET 3.2

Mastering the Concepts

In problems 1 and 2, first describe the symmetry and general shape of the graph of each function. Then sketch the graph.

1. (a) $f(x) = x^8$
    (b) $h(x) = x^9$

2. (a) $f(x) = x^{10}$
    (b) $h(x) = x^{11}$

In problems 3–14, use transformations of the graph of an appropriate power function to sketch the graph of the given function.

3. $g(x) = \dfrac{1}{3}x^4$

4. $h(x) = -5x^4$

5. $f(x) = -2x^5$

6. $g(x) = \dfrac{1}{2}x^5$

7. $h(x) = x^5 + 3$

8. $f(x) = x^6 - 7$

9. $f(x) = -\dfrac{1}{5}(x + 1)^4$

10. $h(x) = 4(x - 4)^5$

11. $g(x) = 2(x - 1)^4 + 3$

12. $f(x) = \dfrac{1}{3}(x + 2)^5 + 2$

13. $h(x) = -4(x - 2)^3 + 1$

14. $g(x) = -\dfrac{1}{2}(x + 3)^4 - 2$

In problems 15–20, use the graph of f to determine the limit behavior of $f(x)$ as:

(a) $x \to +\infty$ **(b)** $x \to -\infty$

Indicate the number of turning points on each graph, and determine whether the graph has low and high points. Also, discuss the continuity and the number of x intercepts of each graph.

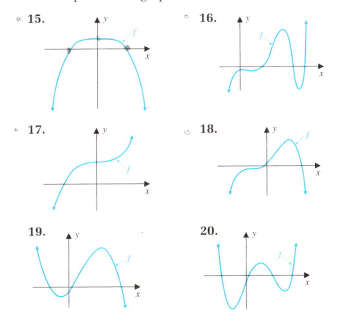

15.   **16.**

17.   **18.**

19.   **20.**

In problems 21 and 22, graph the limit behavior model function and then use the result to match the given polynomial function with its graph.

21. (a) $f(x) = -x^3 + 2x - 5$
    (b) $g(x) = x^4 - 8x^3 + 20x^2 - 16x + 8$

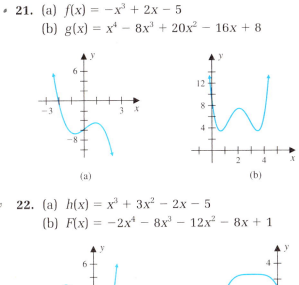

(a)   (b)

22. (a) $h(x) = x^3 + 3x^2 - 2x - 5$
    (b) $F(x) = -2x^4 - 8x^3 - 12x^2 - 8x + 1$

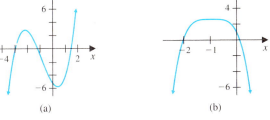

(a)   (b)

G In problems 23 and 24, graph the given function for the specified **WINDOW** selections. Explain why the selection is inappropriate.

23. $f(x) = x^3 - 6x^2 + 11x - 6$

(a)   (b)

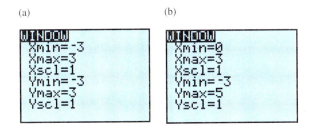

24. $h(x) = -x^4 + 2x + 3$

(a)   (b)

In problems 25–30, use the limit behavior model of the given polynomial function f to determine the limit behavior of $f(x)$:

(a) As $x \to +\infty$

(b) As $x \to -\infty$

G Then use a grapher to sketch a graph of f by selecting an appropriate viewing window. Indicate the number and types of turning points.

25. $f(x) = x^4 - x^2 - 20$
26. $f(x) = x^4 - 4x^2 - 3x + 12$
27. $f(x) = x^3 - 8x^2 + 16x$
28. $f(x) = -x^3 - 6x^2$
29. $f(x) = -2x^4 + 2x^3 - x + 3$
30. $f(x) = x(x + 1)(x - 2)$

In problems 31–36, graph each function. Find the exact locations of the x intercepts and the zeros of f of the graph, and then use the graph to solve the associated inequality.

31. $f(x) = x(x - 1)(x + 2)$;
$x(x - 1)(x + 2) > 0$
32. $h(x) = x^3 - x^2 - 2x$
$= x(x - 2)(x + 1)$;
$x^3 - x^2 - 2x \leq 0$
33. $f(x) = x^3 - x^2 - 12x$
$= x(x - 4)(x + 3)$;
$x^3 - x^2 - 12x < 0$
34. $g(x) = -2x^4 + 2x^2$
$= -2x^2(x + 1)(x - 1)$;
$-2x^4 + 2x^2 > 0$
35. $f(x) = (2x - 5)(x - 1)^2$;
$(2x - 5)(x - 1)^2 \geq 0$
36. $f(x) = 2x^4 - 3x^3 - 12x^2 + 7x + 6$
$= (2x + 1)(x - 1)(x - 3)(x + 2)$
$2x^4 - 3x^3 - 12x^2 + 7x + 6 > 0$

In problems 37–40, find all zeros of f and indicate multiplicity of each zero. Also, use the even-odd multiplicity rule to determine where the graph of f crosses the x axis, and where it touches the x axis and bounces back.

37. $f(x) = (x - 5)^2 (x + 2)^3 (x + 5)$
38. $f(x) = x(x - 4)^3 (x + 2)^2$
39. $f(x) = x(x + 3)^2 (x - 4)^3$
40. $f(x) = x(x + 2)^2 (x - 1)^3$

Applying the Concepts

41. Average Temperature: The average hourly temperature in the month of January, in degrees

Fahrenheit, in a midwestern city is modeled by the function

$f(t) = -0.1(t + 6)(t - 2)(t - 11)$ $0 \leq t \leq 12$

where $t = 0$ represents 8:00 AM and $t = 12$ represents 8:00 PM.

(a) What is the average temperature at 11:00 AM, 3:00 PM, and 6:00 PM? Round off the answers to two decimal places.

(b) Indicate when the temperature is zero.

(c) Sketch the graph of this model.

(d) During what hours of the day is the temperature above zero or below zero?

42. Population Rate: The population growth rate $f(t)$ (in thousands per year) of a certain region during a 9-year period is given by the model

$f(t) = -0.04(t - 7)(t + 9)(t + 0.8)$ $0 \leq t \leq 9$

where t is the elapsed time in years.

(a) What is the population growth rate at 5 years and 9 years? Round off the answers to two decimal places.

(b) Indicate when the growth rate is zero.

(c) Sketch the graph of this model.

(d) During what time intervals is the growth rate positive or negative?

In problems 43–46, round off the answers to two decimal places.

43. G **Unemployment Rate:** A study conducted by a state unemployment commission determined the percentage of the workforce expected to be unemployed in the state over a 4-year period beginning in January 2000. The unemployment R (in percent) over that period is approximated by the model

$R = 0.0002t^3 - 0.0147t^2 + 0.216t + 5.9$

where t is the number of months, with $t = 0$ representing the beginning of the 4-year period.

(a) What are the restrictions on t?

(b) Use a grapher to graph this function over the restricted interval.

(c) Use a grapher to determine when the unemployment is expected to be the lowest over this period. What is the lowest expected rate of unemployment?

(d) What is the highest unemployment rate expected over this period of time? When is it expected to occur?

44. G **Drug Dosage:** Data collected from a study conducted by medical researchers show the amount A of glucose (measured in tenths of a percent) in a patient's bloodstream at time t (in minutes) during a stress test. The amount A is approximated by the model

$$A = 0.1t^3 - 2.1t^2 + 12t + 2 \quad 1.5 < t \le 11$$

(a) Use a grapher to graph this function.

(b) Use a grapher to determine the maximum amount of glucose and when it occurs.

(c) What is the minimum amount of glucose, and when does it occur?

45. **G** **Package Design:** An open-top box is to be constructed by removing equal squares, each of length x inches, from the corners of a piece of cardboard that measures 20 inches by 20 inches and then folding up the sides.

(a) Draw a sketch of the situation and then express the volume V of the box as a function of x.

(b) What are the restrictions on x?

(c) Use a grapher to sketch the graph of the function from part (a) incorporating the restrictions from part (b).

(d) Use a grapher to determine the value of x for which the volume is a maximum. Also find the maximum volume.

46. **G** **U.S. Postal Package Design:** A closed box with a square end x inches on a side is to be mailed. The U.S. Postal Service will accept the box for domestic shipment only if the length L of the box plus its girth ($4x$) is at most 108 inches (Figure 14). Suppose that the length plus girth of a box is exactly 108 inches.

Figure 14

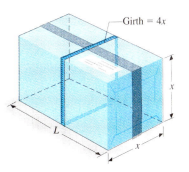

(a) Express the volume V of the box as a function of x.

(b) What are the restrictions on x?

(c) Use a grapher to sketch the graph of the function from part (a) incorporating the restrictions from part (b).

(d) Use a grapher to determine the value of x for which the volume is a maximum. Also find the maximum volume.

Developing and Extending the Concepts

47. Compare the limit behaviors of the functions

$$f(x) = ax^4 \quad \text{and} \quad g(x) = bx^4$$

if:

(a) $ab > 0$

(b) $ab < 0$

48. Compare the limit behaviors of the functions

$$f(x) = ax^3 \quad \text{and} \quad g(x) = bx^3$$

if:

(a) $ab > 0$

(b) $ab < 0$

49. Examine the graph of $y = x^3$. For $x < 0$ the shape of the curve is **concave down** (cupped downward), and for $x > 0$ the shape is **concave up** (cupped upward). The point on the graph where a curve changes concavity is said to be a **point of inflection.** For $y = x^3$, the point of inflection is (0, 0). With this in mind, graph each given function. For what values of x is the curve concave down? For what values of x is the curve concave up? Where is the point of inflection?

(a) $f(x) = 2x^3 + 1$

(b) $g(x) = -(x - 4)^3 + 1$

50. Graph

$$f(x) = x^3$$

and

$$g(x) = |x^3|$$

on the same coordinate system. How can the graph of g be obtained from the graph of f? In general, explain how the graph of

$$y = |f(x)|$$

can be geometrically obtained from the graph of the polynomial function $y = f(x)$. When does

$$|f(x)| = f(x)?$$

In problems 51–54, display a general sketch of the limit behavior of the graph of the polynomial function

$$f(x) = a(x - h)^n + k, n > 0$$

under the given conditions.

51. $a > 0$ and $n \ge 1$ is an odd integer.

52. $a < 0$ and $n \ge 1$ is an odd integer.

53. $a > 0$ and $n \ge 2$ is an even integer.

54. $a < 0$ and $n \ge 2$ is an even integer.

In problems 55 and 56, complete the following table:

$x =$	-10^3	-10^2	-10	-1	1	10	10^2	10^3
$f(x)$								
$g(x)$								
$R(x)$								

What happens to the values of $R(x)$ as

$$x \rightarrow -\infty$$

and as

$$x \rightarrow +\infty?$$

How do the data in the table show that g is the limit behavior model for f?

55. $f(x) = x^3 + 2x^2 - 5x + 1$

$$= x^3 \left(1 + \frac{2}{x} - \frac{5}{x^2} + \frac{1}{x^3} \right)$$

$$= g(x) \cdot R(x)$$

56. $f(x) = 2x^4 - x^2 + x - 7$

$$= 2x^4 \left(1 - \frac{1}{2x^2} + \frac{1}{2x^3} - \frac{7}{2x^4} \right)$$

$$= g(x) \cdot R(x)$$

[G] In problems 57 and 58, each table gives discrete points for ordered pairs (x, y).

(a) Plot a scattergram of each set of data. Locate the x values on the horizontal axis and the y values on the vertical axis.

(b) Use a grapher and cubic regression to find a model of the form

$$y = ax^3 + bx^2 + cx + d$$

that best fits the data. Round off a, b, c, and d to four decimal places.

(c) Sketch the graph of the best fit curve from part (b).

(d) Use the regression equation to estimate the value of y if x is 27, 35, or 45. Round off the answers to the nearest integer.

57.

x	y
21	33
28	41
33	45
38	49
40	51

58.

x	y
12	53
18	48
23	44
28	42
30	38

Objectives

1. Use the Remainder Theorem
2. Factor Polynomials
3. Relate Factors, Zeros, Roots, and x Intercepts
4. Determine the Number and Types of Zeros
5. Find Rational Zeros

3.3 Real Zeros of Polynomials

So far, we have introduced some properties of polynomial functions, and have seen their usefulness in analyzing graphs. To advance our ability to find the real zeros or x intercepts of these functions, we must first review some basic concepts in algebra related to long division and synthetic division (see Appendix page A10).

For instance, we can either use long division or synthetic division to divide

$$f(x) = 2x^4 - 5x^2 + 3x - 11 \text{ by } x - 2$$

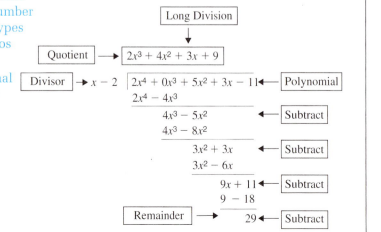

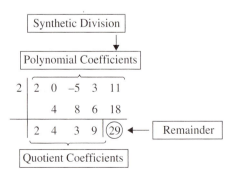

The result of this division may be written in the *multiplicative form*

$$2x^4 - 5x^2 + 3x + 11 = (x - 2)(2x^3 + 4x^2 + 3x + 9) + 29$$

| Polynomial | Divisor | Quotient | Remainder |

This observation leads to the following generalization:

The Division Algorithm Result

> If $f(x)$ and $D(x)$ are nonconstant polynomials such that the degree of $f(x)$ is greater than or equal to the degree of $D(x)$, then there exist unique polynomials $Q(x)$ and $R(x)$ such that
>
> $$f(x) = D(x) \cdot Q(x) + R(x)$$
>
> where the degree of $R(x)$ is less than the degree of $D(x)$ [$R(x)$ may be 0]. The expression $D(x)$ is called the **divisor**, $f(x)$ is the **dividend**, $Q(x)$ is the **quotient**, and $R(x)$ is the **remainder.**

As we shall see, being able to divide a polynomial function by a linear polynomial will facilitate the search for zeros of higher-degree polynomial functions.

Using the Remainder Theorem

A useful special case of the division algorithm occurs if the divisor is of the form $x - c$, where c is a real number. When this is the situation, we could use synthetic division to determine the quotient and remainder. The result is the form

$$f(x) = (x - c) \cdot Q(x) + R$$

where $Q(x)$ is the quotient and R is the remainder.

Now, if we let $x = c$, it follows that

$$f(c) = (c - c) \cdot Q(c) + R$$
$$= 0 \cdot Q(c) + R$$
$$= R$$

For example, if we divide $f(x) = 3x^3 - 2x + 1$ by $x - 2$, we obtain

$$f(x) = (x - 2)(3x^2 + 6x + 10) + 21.$$

Now if we substitute 2 for x into the latter expression we get

$$f(2) = (2 - 2)[3(2^2) + 6(2) + 10] + 21 = 0 + 21 = 21$$

Thus, we observe that $f(2) = 21$, which is the same value as the remainder when $f(x)$ is divided by $x - 2$.

In general, we have the following theorem.

The Remainder Theorem

> If a polynomial $f(x)$ of degree $n \geq 1$ is divided by $x - c$, then the remainder $R = f(c)$.

EXAMPLE 1   **Evaluating a Polynomial by Using the Remainder Theorem**

Use the remainder theorem to find $f(5)$ if $f(x) = 3x^3 - x^2 + 2x - 18$.

Solution   First, we use synthetic division to divide $3x^3 - x^2 + 2x - 18$ by $x - 5$ to get

$$
\begin{array}{r|rrrr}
5 & 3 & -1 & 2 & -18 \\
  &   & 15 & 70 & 360 \\
\hline
  & 3 & 14 & 72 & 342
\end{array}
$$

It follows that $f(5) = 342$, which is the remainder from dividing $f(x)$ by $x - 5$. ▨

Factoring Polynomials

Suppose we divide a polynomial function f by $x - c$ to obtain the result.

$$f(x) = (x - c) \cdot Q(x) + R,$$

where $Q(x)$ is the quotient and $R = f(c)$ is the remainder
If $f(c) = 0$, then the remainder is 0, so that

$$f(x) = (x - c) \cdot Q(x);$$

that is, $x - c$ is a factor of $f(x)$.
Conversely, if $x - c$ is a factor of $f(x)$, then

$$f(x) = (x - c) \cdot Q(x)$$

and by the remainder theorem, $f(c) = 0$.

Thus we have the following theorem

The Factor Theorem

> If $f(x)$ is a polynomial of degree $n \geq 1$ and $f(c) = 0$, then $x - c$ is a factor of the polynomial $f(x)$; conversely, if $x - c$ is a factor of $f(x)$, then $f(c) = 0$.

The factor theorem, in conjunction with synthetic division, enables us to determine if an expression of the form $x - c$ is a factor of a polynomial; and, if it is, we find the other factor. This idea is illustrated in the next example.

EXAMPLE 2   **Using the Factor Theorem to Determine Factors of a Polynomial**

Let $f(x) = x^3 + 5x^2 + 5x - 2$. Use the factor theorem and synthetic division to determine if $x + 2$ is a factor of $f(x)$. If it is, find the factorization.

Solution   We have $x + 2 = x - (-2)$, which has the form $x - c$ with $c = -2$. Using synthetic division to divide $f(x)$ by $x + 2$, we get

$$
\begin{array}{r|rrrr}
-2 & 1 & 5 & 5 & -2 \\
   &   & -2 & -6 & 2 \\
\hline
   & 1 & 3 & -1 & 0
\end{array}
$$

It follows that $f(-2) = 0$, so by the factor theorem, we know $x + 2$ is a factor of $f(x)$. The synthetic division indicates that the factorization is

$$x^3 + 5x^2 + 5x - 2 = (x + 2)(x^2 + 3x - 1).$$ ▨

Relating Factors, Zeros, Roots, and x Intercepts

Figure 1

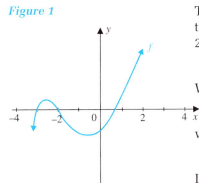

The relationship between zeros, roots, factors, and x intercepts is fundamental to the study of polynomials. For instance, by examining the factorization in Example 2, given by

$$f(x) = x^3 + 5x^2 + 5x - 2 = (x + 2)(x^2 + 3x - 1),$$

We see that -2 is a root of the equation

$$x^3 + 5x^2 + 5x - 2 = 0,$$

we also see that -2 is a zero of the function

$$f(x) = x^3 + 5x^2 + 5x - 2.$$

It follows that -2 is one of the x intercepts of the graph (Figure 1).
In general, the following statements are equivalent for any polynomial function f.

Equivalent Statements

> If f is a polynomial function and if c is a real number, then the following statements are equivalent.
>
> 1. The number c is zero of f.
> 2. The number c is a solution or root of the equation $f(x) = 0$.
> 3. $x - c$ is a factor of $f(x)$.
> 4. The number c is an x intercept for the graph of f.

EXAMPLE 3   **Relating Factors, Zeros, Roots, and x Intercepts**

Let $f(x) = x^3 + 2x^2 - 5x - 6$

(a) Use the factor theorem to show that $x + 3$ is a factor of $f(x)$, and then write $f(x)$ in complete factored form.

(b) Find the zeros of f.

(c) Find the roots of $x^3 + 2x^2 - 5x - 6 = 0$.

(d) Find the x intercepts of the graph of $f(x) = x^3 + 2x^2 - 5x - 6$.

Solution   (a) Since $x + 3 = x - (-3)$, we use $c = -3$,

to get

$$f(c) = f(-3) = (-3)^3 + 2(-3)^2 - 5(-3) - 6 = 0$$

Hence, by the factor theorem $x + 3$ is a factor of $f(x)$. Using synthetic division to divide $f(x)$ by $x + 3$, we obtain

$$
\begin{array}{r|rrrr}
-3 & 1 & 2 & -5 & -6 \\
   &   & -3 & 3 & 6 \\
\hline
   & 1 & -1 & -2 & \,0
\end{array}
$$

so that $f(x) = x^3 + 2x^2 - 5x - 6$

$$= (x + 3)(x^2 - x - 2)$$

Since $x^2 - x - 2 = (x - 2)(x + 1)$, it follows that

$$f(x) = (x + 3)(x^2 - x - 2)$$

$$= (x + 3)(x - 2)(x + 1)$$

Consequently, the complete factored form of $f(x)$ is given by

$$f(x) = (x + 3)(x - 2)(x + 1).$$

Figure 2

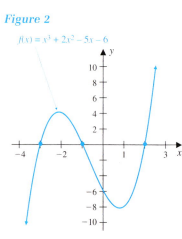

$f(x) = x^3 + 2x^2 - 5x - 6$

(b) Because of the equivalent statements, -3, -1, and 2 are the zeros of $f(x)$.

(c) Again we use the equivalent statements to conclude that the roots of the equation $f(x) = x^3 + 2x^2 - 5x - 6 = 0$ are -3, -1, and 2.

(d) Finally the equivalent statements indicate that the x intercepts of the graph of f are -3, -1, and 2 (Figure 2).

Determining the Number and Types of Zeros

In view of the equivalency between the zeros, the roots, the x intercepts, and factors of a polynomial function, we can use a graph to help predict the number and types of real zeros. For example, the graph of the function

$$f(x) = x^3 + 2x^2 - 5x - 6$$

in Figure 2 reveals that there are three real zeros, two negative and one positive. In addition, because of the equivalency, the graph in itself indirectly shows the function f has three linear factors.

To gain information about the number of real zeros for a polynomial in the real number system from an algebraic point of view, we observe that according to the factor theorem, if c_1 is a real zero of a polynomial function f of degree n, $n \geq 1$, then $f(x)$ can be factored as

$$f(x) = a(x - c_1) \cdot q_1(x)$$

where a is the leading coefficient of $f(x)$ and g, (x) of degree $n - 1$.

If c_2 is another real zero of f, then $f(x)$ can be factored as

$$f(x) = a(x - c_1)(x - c_2) \cdot q_2(x)$$

where q_2 is of degree $n - 2$.

Continuing this process, we observe that the factored form of $f(x)$ cannot have more than n linear factors. Since q_n would be of degree zero, that is a constant.

Since each linear factor produces a real zero, we are led to the following result:

Property

Maximum

Number of Zeros

If f is a polynomial function of degree n, then f cannot have more than n zeros, not necessarily all different.

Since each real zero corresponds to an x intercept, we have established the Property 3 introduced in Section 3.2, which states that the graph of a polynomial function of degree has at most n x intercepts.

EXAMPLE 4   **Finding a Polynomial with Prescribed Zeros**

Find a polynomial function f that has degree 3, zeros -1, 3, and 5, and satisfies $f(1) = 4$.

Solution   By the factor theorem, $f(x)$ has factors $x + 1$, $x - 3$, and $x - 5$.
Hence, $f(x)$ has the form:

$$f(x) = a(x + 1)(x - 3)(x - 5)$$

for some number a.

Since the above property assures us that there can be no more than three zeros,
it follows that there can only be three linear factors for $f(x)$. Because $f(1) = 4$, it
follows that

$$4 = a(1 + 1)(1 - 3)(1 - 5)$$

$$4 = 16a$$

$$a = \frac{4}{16} = \frac{1}{4}.$$

Hence,

$$f(x) = \frac{1}{4}(x + 1)(x - 3)(x - 5)$$

If we multiply this factored form, we obtain the polynomial

$$f(x) = \frac{1}{4}x^3 - \frac{7}{4}x^2 + \frac{7}{4}x + \frac{15}{4}.$$

Finding Rational Zeros

In Section 3.2, we pointed out that it is generally difficult to find the zeros of polynomial functions of higher degree. Here, we will adopt an algebraic strategy to find
rational zeros, that is, rational numbers that are zeros, if they exist for polynomial
functions with rational coefficients. We shall see that a grapher is helpful in carrying out this algebraic strategy. To understand the technique, we begin by examining the factored form of the following polynomial function.

$$f(x) = 24x^3 - 22x^2 - 5x + 6$$

$$= (4x - 3)(2x + 1)(3x - 2)$$

The zeros of f are found by solving the equation

$$f(x) = (4x - 3)(2x + 1)(3x - 2) = 0$$

to obtain the rational zeros $3/4$, $-1/2$, and $2/3$.

We observe the following relationship between the rational zeros and the
coefficients of $f(x)$:

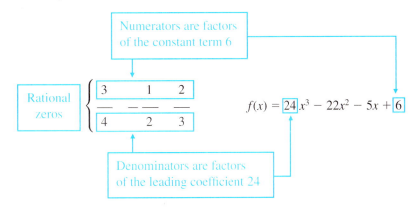

This observation leads us to the next theorem.

Rational Zero Theorem

Assume all the coefficients in the polynomial function

$$f(x) = a_n x^n + a_{n-1} x^{n-1} + \ldots + a_1 x + a_0$$

are integers, where $a_n \neq 0$ and $a_0 \neq 0$. If f has a rational zero p/q (where p and q are integers) in lowest terms, then

$$p \text{ must be a factor of } a_0 \quad \text{and} \quad q \text{ must be a factor of } a_n$$

It is important to understand the following points about the rational zero theorem:

1. A polynomial must have integer coefficients.
2. Acceptable factors for a_0 and a_n include -1 and 1.
3. The theorem does not guarantee the existence of rational zeros; indeed, there may not be any rational zero.

The rational zero theorem leads to the following strategy for finding all possible rational zeros of a polynomial function with integer coefficients.

Strategy for Finding Rational Zeros of a Polynomial Function

If p/q is a rational zero of f, then by the factor theorem, $x - \dfrac{p}{q}$ is a factor of f(x).

Step 1 List all factors of the constant term a_0 in the polynomial. (This list provides us with all possible values of p.) List all factors of the coefficient a_n of the term of the highest degree in the polynomial. (This list provides us with all possible values of q.)

Step 2 Use the lists from step 1 to determine all possible rational numbers p/q in reduced form. These rational numbers are the only possible rational zeros of f.

Step 3 Test the values produced in step 2. If none of them is a root of $f(x) = 0$, we conclude that f has no rational zeros.

As we shall see, shortcuts and variations of these steps are sometimes possible.

EXAMPLE 5   **Finding Rational Zeros**

Use the rational zero theorem to find the rational zeros of

$$f(x) = x^3 - 2x^2 - x + 2$$

and then use the results to factor $f(x)$.

Solution   Assume that p/q is a rational zero of f and p/q is reduced to lowest terms. We apply the above strategy as follows:

Step 1. The factors of $a_0 = 2$ provide the possibilities for p, which are

$$p: \pm 1, \pm 2.$$

The possible values for q are the factors of $a_3 = 1$, which are

$$q: \pm 1$$

Step 2. The possible values for p/q are

$$\frac{p}{q}: \pm 1, \pm 2.$$

We test the $\frac{p}{q}$ values by using synthetic division. If -1 is selected, then

$$
\begin{array}{r|rrrr}
-1 & 1 & -2 & -1 & 2 \\
   &   & -1 & 3 & -2 \\
\hline
   & 1 & -3 & 2 & 0
\end{array}
$$

so that -1 is a rational zero, and we obtain the factorization

$$x^3 - 2x^2 - x + 2 = (x + 1)(x^2 - 3x + 2)$$

The remaining zeros of f can be found by determining the roots of the so called *reduced equation*, $x^2 - 3x + 2 = 0$, to get

$$x^2 - 3x + 2 = 0$$

$$(x - 2)(x - 1) = 0$$

$$x = 2 \quad \text{or} \quad x = 1$$

If we had first tested 1 or 2 in Step 3, the final result would turn out to be the same.

Hence, the rational zeros of f are -1, 1, and 2. By the factor theorem,

$$f(x) = (x + 1)(x - 1)(x - 2)$$

If the rational zero theorem produces several possibilities for rational zeros (Step 2 in the above strategy), then testing each possibility can become rather tedious. At times the testing process can be shortened by first examining the graph of $y = f(x)$ to determine likely candidates for zeros. The use of the rational zero theorem in conjunction with a graph is illustrated in the next example.

EXAMPLE 6   **Finding Rational Roots of an Equation**

Find all rational roots of the equation

$$4x^3 - 13x = 6$$

Use a graph to determine likely candidates for the rational roots.

Solution   The given equation is equivalent to the equation $4x^3 - 13x - 6 = 0$. However, the solutions of this latter equation are the same as the zeros of the polynomial function

$$f(x) = 4x^3 - 13x - 6.$$

We apply the three-step strategy on page 242 to find the zeros of f. Assume p/q in reduced form, is a rational zero of f.

Step 1. The possible values of p and q are:

$$p: \pm 1, \pm 2, \pm 3, \pm 6$$

$$q: \pm 1, \pm 2, \pm 4$$

Step 2. The possible values of p/q are:

$$\frac{p}{q}: \pm 1, \ \pm 2, \ \pm 3, \ \pm 6, \ \pm \frac{1}{2}, \ \pm \frac{3}{2}, \ \pm \frac{1}{4}, \ \pm \frac{3}{4}$$

Figure 3

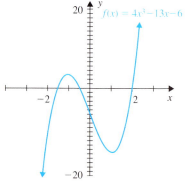

The graph of f in Figure 3 shows that there are zeros near 2, between -2 and -1, as well as between -1 and 0 so the only feasible, candidates for zeros are

$$2, \quad \frac{-3}{2}, \text{ and } -\frac{1}{2}.$$

We begin by testing 2. Using synthetic division, we get

$$
\begin{array}{r|rrrr}
2 & 4 & 0 & -13 & -6 \\
  &   & 8 & 16 & 6 \\
\hline
  & 4 & 8 & 3   & 0
\end{array}
$$

Since the reminder $f(2) = 0$, 2 is a zero of f, $x - 2$ is a factor of $f(x)$, and

$$f(x) = 4x^3 - 13x - 6 = (x - 2)(4x^2 + 8x + 3)$$

Any solution of the reduced equation $4x^2 + 8x + 3 = 0$ will also be a zero of f. Solving the reduced equation, we get

$$4x^2 + 8x + 3 = (2x + 3)(2x + 1) = 0$$

$$
\begin{array}{c|c}
2x + 3 = 0 & 2x + 1 = 0 \\[2mm]
x = -\dfrac{3}{2} & x = -\dfrac{1}{2}
\end{array}
$$

Therefore, the zeros of f, which are the same as the roots of the given equation, are $-\frac{3}{2}, -\frac{1}{2}$, and 2.

Suppose a polynomial equation $f(x) = 0$ has noninteger rational coefficients. Then we need to make an adjustment in order to apply the rational zero theorem, since it requires integer coefficients. The next example illustrates one technique for doing this.

EXAMPLE 7   **Using an Equivalent Form to Find Rational Zeros**

(a) Find the rational zeros of the polynomial function

$$f(x) = \frac{2x^4}{3} - \frac{7x^3}{3} - \frac{10x^2}{3} + 11x + 6$$

(b) Use the results from part (a) to factor $f(x)$.

Solution   (a) The zeros of f are the same as the zeros of the function

$$F(x) = \frac{1}{3}(2x^4 - 7x^3 - 10x^2 + 33x + 18)$$

obtained by factoring out $\frac{1}{3}$. Thus the zeros of f are the same as the zeros of

$$g(x) = 2x^4 - 7x^3 - 10x^2 + 33x + 18$$

Now we can apply the rational zero theorem to the function g by assuming that, $\frac{p}{q}$ in reduced form, is a rational zero of g.

The possible values for p and q are:

$$p: \pm1, \pm2, \pm3, \pm6, \pm9, \pm18$$

$$q: \pm1, \pm2$$

As a result, any rational zero of $g(x)$ must occur in the list:

$$\frac{p}{q}: \pm 1, \pm 2, \pm 3, \pm 6, \pm 9, \pm 18, \pm \frac{1}{2}, \pm \frac{3}{2}, \pm \frac{9}{2}$$

Examining the graph of $y = g(x)$ in Figure 4, we see there are zeros "near" -2 and 3, and a zero between -1 and 0. Thus, the only likely candidates are

$$-2, -\frac{1}{2}, \text{ and } 3.$$

Using synthetic division we test -2, to get

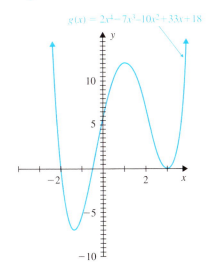

Figure 4

$g(x) = 2x^4 - 7x^3 - 10x^2 + 33x + 18$

$$
\begin{array}{r|rrrrr}
-2 & 2 & -7 & -10 & 33 & 18 \\
   &   & -4 & 22 & -24 & -18 \\
\hline
   & 2 & -11 & 12 & 9 & 0
\end{array}
$$

Thus we have

$$g(x) = (x + 2)(2x^3 - 11x^2 + 12x + 9)$$

Synthetic division is again used to test a potential zero 3, in the reduced equation $2x^3 - 11x^2 + 12x + 9 = 0$ to get

$$
\begin{array}{r|rrrr}
3 & 2 & -11 & 12 & 9 \\
  &   & 6 & -15 & -9 \\
\hline
  & 2 & -5 & -3 & 0
\end{array}
$$

So 3 is also a zero. Thus we have the factorization

$$g(x) = (x + 2)(x - 3)(2x^2 - 5x - 3).$$

All the remaining zeros must be roots of the reduced polynomial equation $2x^2 - 5x - 3 = 0$. Upon factoring this latter equation we get

$$2x^2 - 5x - 3 = (2x + 1)(x - 3) = 0$$

so that

$$x = -1/2 \text{ and } x = 3 \text{ are also zeros.}$$

Therefore, the zeros of g, and consequently of f, are

$$-2, -\frac{1}{2}, \text{ and } 3,$$

where 3 has a multiplicity of 2.

(b) Since

$$f(x) = \frac{2}{3}x^4 - \frac{7}{3}x^3 - \frac{10}{3}x^2 + 11x + 6$$

is of degree 4 and has zeros

$$-2, -1/2, \text{ and } 3 \text{ (with multiplicity 2)}$$

it follows that

$$f(x) = \frac{1}{3}(2x^4 - 7x^3 - 10x^2 + 33x + 18)$$

$$= \frac{1}{3}g(x)$$

$$= \frac{1}{3}(x + 2)(x - 3)^2(2x + 1).$$

Approximating Irrational Zeros

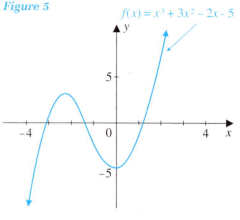

Figure 5

The technique we have learned so far for finding the zeros of polynomial functions has limitations. For instance, consider the function

$$f(x) = x^3 + 3x^2 - 2x - 5$$

The graph of f given in Figure 5 shows that there are three x intercepts. This means, of course, that there are three real number zeros for f. To determine if any of these zeros is a rational number, we apply the rational zero theorem as follows:

If $\dfrac{p}{q}$ is a rational zero, then the possibilities for

$$p \text{ are } \pm 1,\ \pm 5$$

and those for

$$q \text{ are } \pm 1$$

so that the possibilities for

$$\frac{p}{q} \text{ are } \pm 1,\ \pm 5.$$

Upon testing these values, we find that none of them is a zero for f, so we conclude that the zeros (x intercepts of the graph) must be irrational numbers. We will explore ways of approximating such values.

Numerical techniques for approximating the values of real zeros are based on the following property.

Property
Intermediate-Value
Property

Suppose that f is a polynomial function with real coefficients. If the values of $f(a)$ and $f(b)$ have opposite algebraic signs, where $a < b$, then there is at least one number c in the interval (a, b) such that c is a zero of f, that is, $f(c) = 0$.

This property is illustrated in Figure 6, which shows part of the graph of a polynomial function f that changes sign on an interval $[a, b]$. Because the graph is a continuous or unbroken curve, it must cross the x axis at least at one point, say $(c, 0)$, between the points $(a, f(a))$ and $(b, f(b))$.

In other words, there must be a real zero in the interval (a, b).

Figure 6

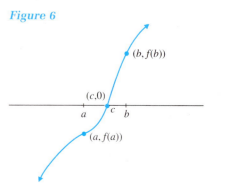

EXAMPLE 8   **Using the Intermediate-Value Property**

Use the intermediate-value property to show that

$$f(x) = x^3 + 3x^2 - 2x - 5$$

has a real zero in the interval $(1, 2)$. Notice that the graph of f shown in Figure 5 on page 246 indicates there is a zero between 1 and 2.

Solution

We find that

$$f(1) = -3 \text{ and } f(2) = 11.$$

Since

$$f(1) < 0 < f(2),$$

we conclude from the intermediate-value property that f has at least one zero in the interval $(1, 2)$.

If the intermediate-value property identifies an interval that contains a real zero of a polynomial function with rational coefficients, we can apply the rational zero theorem to find its value if it is a rational number. If the zero is an irrational number, we use an approximating method to estimate its value.

One method of approximating the irrational zeros is based on the recursive process illustrated in Figure 7, where portions of the graph of the function $y = f(x)$ are displayed.

Figure 7

Each bisection point (midpoint of the interval) is getting closer to c, where f(c) = 0.

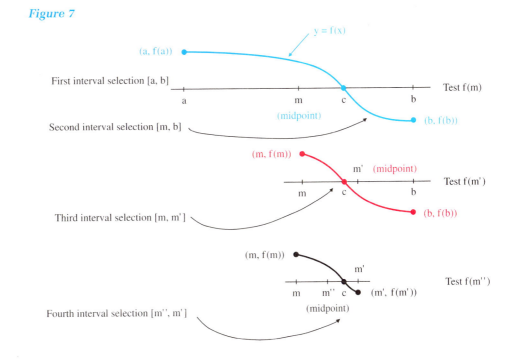

The actual algorithm for this recursive process, called the **bisection method,** procedes as follows:

The Bisection Method

> **Assume that f is a polynomial function with real coefficients.**
>
> **Step 1** Find values for a and b where $a < b$ and the values $f(a)$ and $f(b)$ change signs once, either from positive to negative or from negative to positive. (A graph of the function is helpful here.)
>
> **Step 2** Determine the midpoint m of interval $[a, b]$, namely,
>
> $$m = (a + b)/2 .$$
>
> **Step 3** From the intervals $[a, m]$ and $[m, b]$, select the interval where the values of the given function f at the endpoints have opposite algebraic signs (one is positive and the other is negative).
>
> **Step 4** Repeat steps 2 and 3 on the interval selected above. Continue this process and evaluate $f(m)$ in each repetition until the value of $f(m)$ differs from 0 by less than an amount decided on at the start.

It is common to express the desired accuracy of the approximation of a zero in terms of an absolute value inequality. For instance, if we are asked to find an approximation c for a real zero such that $|f(c)| \leq 0.0005$, then we continue the approximating process until we find a value c such that

$$-0.0005 \leq f(c) \leq 0.0005.$$

EXAMPLE 9   **Using the Bisection Method**

Use the bisection method to approximate c, the zero of the function

$$f(x) = x^3 + 3x^2 - 2x - 5$$

in the interval $(1, 2)$, to four decimal places, so that $f(c) > 0.005$.

Solution   We apply the bisection method as follows:

Step 1. From Example 8, we know that there is a zero between 1 and 2, since $f(1)$ and $f(2)$ have opposite signs.

Step 2. The midpoint of interval $[1, 2]$ is $\dfrac{(1 + 2)}{2} = 1.5$ and $f(1.5) = 2.125$

Step 3. Since $f(1) = -3$ and $f(1.5) = 2.125$, we select interval $[1, 1.5]$ to do the next bisection.

Step 4. The midpoint of interval $[1, 1.5]$ is $\dfrac{(1 + 1.5)}{2} = 1.25$ and

$$f(1.25) = -0.8594 \text{ (approx.)}$$

Next, we select interval $[1.25, 1.5]$ because $f(1.25)$ is negative and $f(1.5)$ is positive. The midpoint is 1.375 and

$$f(1.375) = 0.5215 \text{ (approx.)}$$

Then we select interval $[1.25, 1.375]$ because $f(1.25)$ is negative and $f(1.375)$ is positive. The midpoint is 1.3125 and

$$f(1.3125) = -0.1960 \text{ (approx.).}$$

Continuing this process will eventually yield more accurate approximations of the zero of f between 1 and 2. The progression of improved approximations is shown in Table 1. Since $f(1.3302) = 0.0016 < 0.005$, we conclude that 1.3302 is the desired approximation for the zero.

TABLE 1

Interval	Midpoint m (four decimal places)	$f(m)$ (four decimal places)
[1, 2]	1.5000	2.1250
[1, 1.5]	1.2500	−0.8594
[1.25, 1.5]	1.3750	0.5215
[1.25, 1.375]	1.3125	−0.1960
[1.3125, 1.375]	1.3438	0.1564
[1.3125, 1.3438]	1.3282	−0.0210
[1.3282, 1.3438]	1.3360	0.0673
[1.3282, 1.3360]	1.3321	0.0231
[1.3282, 1.3321]	1.3302	0.0016

The bisection method is easy to understand but tedious to carry out.

Another way to approximate real zeros is to use a grapher. From the graph in Figure 8, we see that there is a zero of

$$f(x) = x^3 + 3x^2 - 2x - 5$$

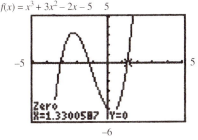

Figure 8

$f(x) = x^3 + 3x^2 - 2x - 5$

Zero
X=1.3300587 Y=0

between 1 and 2. The value of this zero is the same as the value of the x intercept between 1 and 2. To locate the x intercept we use the ROOT-finding or ZERO-finding feature on a grapher to obtain the approximation 1.3300587. Other irrational zeros of this polynomial that can also be found by this method are −3.128419 and −1.20164.

Solving Applied Problems

Many applications of mathematics involve solving polynomial equations. As we know, finding the solution of a polynomial equation is equivalent to finding the zeros of a function. The next example illustrates an application of the use of this equivalency. It also stresses the importance of carefully interpreting the results of a mathematical model.

Figure 9

EXAMPLE 10

Modeling the Dimensions of a Container

A large container for filling bags of low-nitrogen fertilizer is to be constructed by attaching an inverted cone to the bottom of a right circular cylinder of radius r. The total depth of the container is 12 feet and the radius is the same as the height of the cone, as shown in Figure 9.

(a) Express the volume V of the container as a function of r. Also, find any restrictions on r.

(b) Suppose that the container is to have a capacity of 90π cubic feet. What is the radius?

Mixed fertilizer enters container

12 ft

r

r

Individual bags of fertilizer are filled at the spigot

Solution (a) The volume V of the container is the sum of the volumes of the cylinder and the cone. Since the height of the cone is r feet, then the height of the cylinder is $12 - r$ feet. So the volume V is given by

$$V = (\text{Volume of cylinder}) + (\text{Volume of cone})$$

$$= \pi r^2 (12 - r) + \frac{1}{3}\pi r^3$$

$$\begin{cases} \text{Volume of cylinder} = \pi r^2 h; \\ \text{Volume of cone} = \dfrac{1}{3}\pi r^2 h \end{cases}$$

$$= 12\pi r^2 - \pi r^3 + \frac{1}{3}\pi r^3$$

$$= -\frac{2}{3}\pi r^3 + 12\pi r^2$$

Thus, the equation

$$V = -\left(\frac{2}{3}\right)\pi r^3 + 12\pi r^2.$$

defines V as a function of r.

Since r and V represent physical dimensions, $r > 0$ and $V > 0$.

However, since the total depth of the container is 12 feet and r is the same as the height of the cone, then r is at most 12 feet, that is, $r < 12$. So we conclude that $0 < r < 12$.

(b) Using the function found in part (a), we let $V = 90\pi$, so we need to solve the equation

$$90\pi = -\frac{2}{-3}\pi r^3 + 12\pi r^2.$$

This equation is equivalent to

$$r^3 - 18r^2 + 135 = 0.$$

Solving the latter equation is equivalent to finding the zeros of the function

$$f(r) = r^3 - 18r^2 + 135.$$

The rational zero theorem leads to the possible rational zeros:

$$\frac{p}{q}: \pm 1, \pm 3, \pm 5, \pm 9, \pm 15, \pm 27, \pm 45, \pm 135$$

However, we established in part (a) that $0 < r < 12$, so the only feasible potential zeros from this list are 1, 3, 5, and 9. After testing these possibilities, we get $f(3) = 0$, so 3 is a zero of f and $r - 3$ is a factor of $f(r)$. In fact,

$$f(r) = (r - 3)(r^2 - 15r - 45).$$

Solving the reduced equation $r - 15r - 45 = 0$ by using the quadratic formula yields two additional zeros of f:

$$\frac{15 - 9\sqrt{5}}{2} \quad \text{and} \quad \frac{15 + 9\sqrt{5}}{2}$$

Hence, the real zeros of f are

$$3, \frac{15 - 9\sqrt{5}}{2} \text{ (approx. } -2.56), \quad \text{and} \quad \frac{15 + 9\sqrt{5}}{2} \text{ (approx. } 17.56).$$

However, r cannot equal $(15 - 9\sqrt{5})/2$ because this number is negative. Furthermore, r cannot equal $(15 + 9\sqrt{5})/2$ since we have already established that $r < 12$. Consequently, the radius r must be 3 feet.

EXAMPLE 11

Figure 10

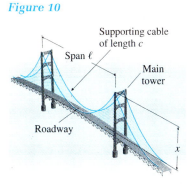

Supporting cable of length c

Span ℓ

Main tower

Roadway

x

G **Modeling a Suspension Bridge**

Figure 10 shows the main span of a suspension bridge. Assume the relationship between the length c of the supporting cable between two vertical towers and the *sag* x, the vertical distance from the top of either tower to the lowest point of the supporting cable at or above the roadway, is given by the formula

$$c = \ell + \frac{8x^2}{3\ell} - \frac{32x^4}{5\ell^3}$$

where ℓ is the horizontal distance of the *span* between the towers.

The Mackinac Bridge, one of the world's longest suspension bridges, connects the upper and lower peninsulas of Michigan. The distance between its main towers is 3800 feet, the top of each main tower is 552 feet above water level, and the roadway is 199 feet above water, as shown in Figure 11.

(a) Express the length of the supporting cable c as a function of x, the sag.

(b) Determine the restrictions on x.

(c) Use a grapher to find how high above the roadway the lowest point of the cable is, if the supporting cable is 3884 feet long. Round off the answer to one decimal place.

Solution

Figure 11

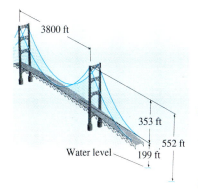

3800 ft

353 ft

552 ft

Water level

199 ft

(a) Since ℓ is given to be 3800, we substitute this value into the above formula to get

$$c = 3800 + \frac{8x^2}{3(3800)} - \frac{32x^4}{5(3800)^3}$$

$$= 3800 + \frac{x^2}{1425} - \frac{x^4}{857,375 \cdot 10^4}$$

This latter equation expresses c as a function of x.

(b) Since the lowest point of the supporting cable is at or above the roadway, then the sag x must be less than or equal to the tower's height above the roadway. Upon examining the given data, which are shown in Figure 11, it follows that $0 < x \le 353$.

(c) We begin by finding the sag x, given that the cable is 3884 feet long. To determine x, we substitute $c = 3884$ into the function in part (a) to get

$$3884 = 3800 + \frac{x^2}{1425} - \frac{x^4}{857,375 \cdot 10^4}$$

$$0 = -84 + \frac{x^2}{1425} - \frac{x^4}{857,375 \cdot 10^4}$$

Figure 12

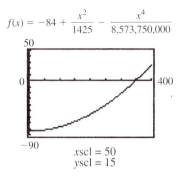

$$f(x) = -84 + \frac{x^2}{1425} - \frac{x^4}{8,573,750,000}$$

$xscl = 50$
$yscl = 15$

The solutions of this latter equation are the same as the zeros of the function f defined by

$$f(x) = -84 + \frac{x^2}{1425} - \frac{x^4}{857,375 \cdot 10^4} \neq \text{ where } 0 < x \leq 353$$

Now we use a grapher to sketch a graph of f (Figure 12). It clearly shows that there is an x intercept in the interval (0, 400). By using the zero finding feature we get an approximate value of 349.5 for the x intercept.

Thus f has a zero with value of about 349.5, making the sag approximately 349.5 feet. It follows that the lowest point of the cable is about $353 - 349.5 = 3.5$ feet above the roadway.

◈ PROBLEM SET 3.3

Mastering the Concepts

In problems 1 and 2, use the remainder theorem to find the value of each function.

1. (a) $f(-1)$ for $f(x) = 4x^3 - 2x^2 + 7x - 1$
   (b) $f(2)$ for $f(x) = 8x^3 - x^2 - x + 5$
   (c) $f(3)$ for $f(x) = 3x^4 - 2x^3 + 4x^2 + 3x + 7$
2. (a) $f(-2)$ for $f(x) = 2x^5 + 3x^4 - x^2 + x - 4$
   (b) $f(-3)$ for $f(x) = 6x^4 + 10x^2 + 7$.
   (c) $f(3/5)$ for $f(x) = 15x^4 + 5x^3 + 6x^2 - 2$
3. Use the factor theorem to determine whether each binomial is a factor of the polynomial $f(x) = x^3 - 28x - 48$. If it is, find the factorization.
   (a) $x + 2$
   (b) $x + 4$
   (c) $x + 6$
   (d) $x - 6$
4. Use the factor theorem to determine whether each binomial is a factor of the polynomial $2x^4 - 9x^3 - 34x^2 - 9x + 14$. If it is, find the factorization.
   (a) $x - 7$
   (b) $x + 2$
   (c) $x - (1/2)$
   (d) $x - 1$

In problems 5–8, use the factor theorem to show the binomial is a factor of $f(x)$, and then factor $f(x)$.

5. $f(x) = x^3 - 6x^2 + 11x - 6; x - 1$
6. $f(x) = x^3 + 6x^2 - 11x - 16; x + 1$
7. $f(x) = 2x^4 + 3x - 26; x + 2$
8. $f(x) = 6x^3 - 25x^2 - 29x + 20; x - 5$

In problems 9 and 10 use the factor theorem to show that c is a zero of f.

9. $f(x) = 4x^3 - 9x^2 - 8x - 3; c = 3$
10. $f(x) = x^3 + x^2 - 12x + 12; c = 2$

In problems 11 and 12, find a polynomial function that has degree 3 and has the given zero with the specified value of f.

11. Zeros, 1, 3, −3, and $f(2) = 5$
12. Zeros, −1, 2, 3, and $f(-2) = 4$

In problems 13–16, list all real zeros of each polynomial function. Also, specify the x intercepts, and find the real roots of the equation $f(x) = 0$.

13. $f(x) = x(x^2 - 4)(x + 3)$
14. $f(x) = x(x^2 - 1)(x^2 - 9)$
15. $f(x) = (x^2 - x - 2)(x^2 + 1)$
16. $f(x) = (x^2 + x - 6)(x^2 + 4)$

In problems 17–22, use the rational zero theorem to find the rational zeros of the function. Use the results to factor the function expression.

17. $f(x) = x^3 - x^2 - 4x + 4$
18. $g(x) = x^3 + 2x - 12$
19. $h(x) = 5x^3 - 12x^2 + 17x - 10$
20. $f(x) = x^3 - x^2 - 14x + 24$
21. $f(x) = x^5 - 5x^4 + 7x^3 + x^2 - 8x + 4$
22. $h(x) = 4x^5 - 23x^3 - 33x^2 - 17x - 3$

In problems 23–26, use the rational zero theorem to find all rational roots of each equation.

23. $x^3 + 2x^2 - 3x - 6 = 0$

24. $x^3 + 3x^2 - 6x - 18 = 0$

25. $x^4 - 24x^2 - 25 = 0$

26. $x^4 - 8x^3 - 10x^2 + 72x + 9 = 0$

In problems 27–30, use an equivalent form and the rational zero theorem to find the rational zeros of the function. Use the rational zeros to factor the function expression.

27. $f(x) = x^4 - x^3 - \dfrac{25}{4}x^2 + \dfrac{x}{4} + \dfrac{3}{2}$

28. $g(x) = \dfrac{2}{5}x^4 - x^3 - \dfrac{8}{5}x^2 + 5x - 2$

29. $g(x) = \dfrac{2}{3}x^3 + \dfrac{x^2}{15} - \dfrac{7}{15}x + \dfrac{2}{15}$

30. $f(x) = \dfrac{2}{3}x^4 - \dfrac{5}{2}x^2 + \dfrac{5}{6}x + 1$

In problems 31 and 32, use an equivalent form and the rational zero theorem to solve each equation.

31. $x^3 + \dfrac{x^2}{2} - 12x - 6 = 0$

32. $2x^3 - 3x^2 - \dfrac{16}{5}x + \dfrac{12}{5} = 0$

G In problems 33–38, use the rational zero theorem, along with the graph of the function, to find the rational zeros. Use the results to factor the function expression.

33. $h(x) = 4x^4 - 4x^3 - 7x^2 + 4x + 3$

34. $g(x) = x^4 - 9x^2 + 20$

35. $f(x) = x^4 - 3x^3 - 12x - 16$

36. $h(x) = 2x^3 + x^2 + x$

37. $f(x) = x^3 - \dfrac{7}{2}x^2 + 3x + \dfrac{5}{2}$

38. $g(x) = x^3 - \dfrac{x^2}{2} - 4x + 2$

In problems 39–44, use the intermediate-value property to show that the function f has a zero in the given interval.

39. $f(x) = x^3 - 3x^2 + 4x - 5; [1, 3]$

40. $f(x) = x^3 - 4x^2 + 3x + 1; [2, 4]$

41. $f(x) = 3x^3 - 10x + 9; [-3, -2]$

42. $f(x) = 2x^3 + 6x^2 - 8x + 2; [-5, -4]$

43. $f(x) = x^4 - 4x^3 + 10; [0.9, 2.1]$

44. $f(x) = x^4 + 6x^3 - 18x^2; [2.1, 2.2]$

In problems 45–50, use the bisection method to approximate the value of the zero c on the specified interval to four decimal places so that $|f(c)| < 0.005$.

45. $f(x) = x^3 - 3x + 1, (0, 1)$

46. $g(x) = x^4 + 2x^3 + 2x^2 - 4x - 8, (1, 2)$

47. $h(x) = x^3 - 6x^2 + 3x + 13, (-2, -1)$

48. $h(x) = -x^3 + 3x + 1, (1, 2)$

49. $h(x) = 23x - x^3, (4, 5)$

50. $g(x) = 2x^5 - 4x^4 + x^2 - 10, (2, 3)$

G In problems 51–54, use a grapher to sketch the graph of each function, and to approximate the value c of each real zero so that $|f(c)| \leq 0.0005$. Round off the answers to four decimal places, if necessary. Also, approximate the solution of the associated inequality. Round off to four decimal places.

51. $f(x) = x^3 - 3x - 2; x^3 - 3x > 2$

52. $f(x) = x^3 - 3x^2 + 5; x^3 + 5 \leq 3x^2$

53. $f(x) = x^4 - 3x^2 + 2; x^4 + 2 \leq 3x^2$

54. $f(x) = x^3 + 4x^2 - 3x - 10; x^3 + 4x^2 \leq 3x + 10$

Applying the Concepts

55. Radar Enclosure: A *radome* is a geometric solid that is formed by removing a section from a sphere so that it has a flat base, as shown in Figure 13. Such structures are used to enclose radar antennas for protection from rain, wind, and snow. The volume V of a radome of radius r and height x is given by the formula

$$V = \pi r x^2 - \dfrac{\pi}{3}x^3$$

Figure 13

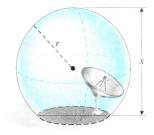

(a) Suppose that a radome has a radius of 21 feet and height of at most 40 feet. Express the volume V as function of the height x.

(b) What are the restrictions on x?

(c) Find the height x of the radome if it encloses a volume of $11,664\pi$ cubic feet.

56. Percent of Income: Suppose that the relationship between the percent of population x (expressed as a decimal) and the percent of the total income y (expressed as a decimal) of a certain community is modeled by the function

$$y = 2.7x^3 - 2.4x^2 + 0.7x$$

(a) What are the restrictions on x?

(b) What percent of the total income does 40% of the population in the community account for?

(c) What percent of the population accounts for 20% of the total income of the community?

57. Storage: A cistern to be fabricated in the shape of a right circular cylinder with a hemisphere at the bottom is to have a depth of 30 feet (Figure 14).

Figure 14

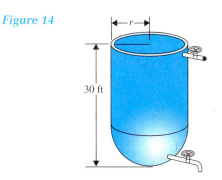

(a) Express the volume V of the cistern as a function of r, the radius of the cylinder.

(b) Determine the restrictions on r.

(c) What is the radius if the volume is 1008π cubic feet?

58. Construction: A rectangular toy chest has a square base that is 2 feet by 2 feet and a height of 1 foot. Suppose that each dimension of the chest is increased by x feet, where x is at most 2 feet.

(a) Express the volume V of the new chest as a function of x.

(b) What is the value of x if the volume of the new chest is 4.5 times the volume of the old one?

In problems 59–62, round off the answers to two decimal places.

59. G Rate of Inflation: Suppose that a typical family in the United States will save approximately s percent of its income per year if the rate of inflation is x percent, where s is modeled by the function

$$s = -0.001x^3 + 0.06x^2 - 1.2x + 12, 0 \le x \le 10.$$

Use a grapher to approximate what the rate of inflation would be if the average saving percent was 10 percent, 8 percent, or 5 percent.

60. G Hot Air Balloon: Atmospheric pressure P (in pounds per square inch) is approximated by the function

$$P = 15 - 6h + 1.2h^2 - 0.16h^3, 0 \le h \le 3.5$$

where height h is in thousands of feet above sea level. Use a grapher to approximate the height of a hot air balloon if the atmospheric pressure at that height is 10 pounds per square inch, 5 pounds per square inch, or 3.4 pounds per square inch.

61. G The Golden Gate Bridge Cable: The Golden Gate Bridge in San Francisco has twin supporting main towers extending 525 feet above the road surface. The two main towers are 4200 feet apart.

(a) Draw a sketch of the situation.

(b) Use the formula in Example 11 to express the length of the supporting cable c between the two towers as a function of the sag x.

(c) Determine the restrictions on x.

(d) Use a grapher to find how high the lowest point of the cable is above the roadway if the length of the supporting cable is 4363 feet.

62. G Radius of a Propane Tank: A steel propane gas storage tank is to be constructed in the shape of a right circular cylinder with a hemisphere attached at each end. The total length of the tank is 12 feet (Figure 15).

Figure 15

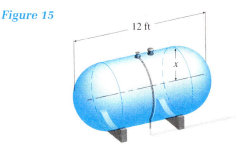

(a) Express the volume V of the tank as a function of the radius x (in feet) of the cylinder.

(b) What are the restrictions on x?

(c) Use a grapher to determine the radius so that the resulting volume is 260 cubic feet.

Developing and Extending the Concepts

In problems 63–68, indicate whether the statement is true or false. If the statement is false, give an example to disprove it; if it is true, justify it with an explanation.

63. If the graph of a polynomial function f has three x intercepts, then the degree of $f(x)$ is 3.

64. The numbers -1, 0, 2, and 3 cannot all be zeros of a polynomial function of degree 3.

65. If $\dfrac{2}{3}$ is a rational zero of the polynomial function

$$g(x) = a_n x^n + a_{n-1} x^{n-1} + \ldots + a_1 x + a_0$$

and the coefficients are all integers, then a_0 is an even integer.

66. If a polynomial function f has three rational zeros, then the graph of f has exactly three x intercepts.

67. Suppose that f is a polynomial function with real coefficients. If the values of $f(a)$ and $f(b)$ are both positive, where $a < b$, then f has no zero between a and b.

68. Suppose that f is a polynomial function with real coefficients. If the values of $f(a)$ and $f(b)$ have opposite algebraic signs, where $a < b$, then there is exactly one zero between a and b.

Objectives

1. Graph Functions of the Form

$$R(x) = \frac{a}{(x-h)^n} + k$$

2. Locate Vertical and Horizontal Asymptotes
3. Graph Nonreduced Rational Functions
4. Solve Applied Problems

3.4 Rational Functions

Just as rational numbers are defined in terms of quotients of integers, *rational functions* are defined in terms of quotients of polynomials.

A function R of the form

$$R(x) = \frac{p(x)}{q(x)}$$

where $p(x)$ and $q(x)$ are polynomials and $q(x) \neq 0$, is called a **rational function.**

Examples of rational functions are

$$R(x) = \frac{3}{x}, \; R(x) = \frac{2x}{x^2 - 4}, \; R(x) = \frac{3x - 1}{x^2 + 4}, \text{ and } R(x) = \frac{x^2 - 4}{x - 2}.$$

The **domain** of a rational function $R(x) = p(x)/q(x)$ consists of all real numbers x for which $q(x) \neq 0$ since division by 0 is not defined.

Table 1 identifies the domains of the above functions.

TABLE 1

Function	Domain
1. $R(x) = \dfrac{3}{x}$	All real numbers except $x = 0$.
2. $R(x) = \dfrac{2x}{x^2 - 4}$	All real numbers except $x = -2$ and $x = 2$ because $x^2 - 4 = 0$ when $x = -2$ or 2.
3. $R(x) = \dfrac{3x - 1}{x^2 + 4}$	All real numbers since $x^2 + 4 \neq 0$ for any real number x.
4. $R(x) = \dfrac{x^2 - 4}{x - 2}$	All real numbers except $x = 2$ because $x - 2 = 0$ when $x = 2$.

Graphing Functions of the Form $R(x) = \dfrac{a}{(x-h)^n} + k$

Our investigation of the graphs of rational functions begins with a study of functions of the form

$$R(x) = \frac{1}{x^n}.$$

This type of function is not defined when $x = 0$.

Understanding the numerical behavior of $R(x)$ for values of x "near" 0, is useful in sketching the graph of R.

To illustrate this behavior, consider the rational function

$$f(x) = \frac{1}{x}.$$

The domain of f includes all real numbers except $x = 0$. So the graph has no y intercept. The data in Table 2 suggest that as x takes on positive values closer and closer to 0 (that is, as x *approaches 0 from the right*), the positive function values increase without bound.

TABLE 2

x	1000	100	10	1	0.1	0.01	0.001	0.0001	0
$f(x) = 1/x$	0.001	0.01	0.1	1	10	100	1000	10,000	$f(0)$ is not defined

We describe this behavior symbolically by writing

$$f(x) \to +\infty \quad \text{as} \quad x \to 0^+.$$

Correspondingly, the graph of

$$f(x) = 1/x.$$

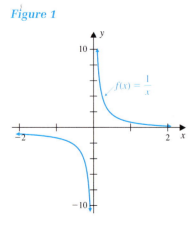
Figure 1

(Figure 1) gets "higher and higher" and approaches the y axis, but never touches it, as x approaches 0 *from the right.*

Similarly, as Table 3 suggests, the negative function values $f(x)$ decrease without bound as negative x values get closer and closer to 0 (that is, as x *approaches 0 from the left*), and we write

$$f(x) \to -\infty \quad \text{as} \quad x \to 0^-.$$

TABLE 3 As $x \to 0^-, f(x) \to -\infty$

x	-1000	-100	-10	-1	-0.1	-0.01	-0.001	-0.0001
$f(x) = 1/x$	-0.001	-0.01	-0.1	-1	-10	-100	-1000	$-10,000$

This means that as x approaches 0 from the left, the graph of $f(x) = 1/x$ gets "lower and lower" and approaches the y axis, but never touches it, as shown in Figure 1.

Because of the behavior as x approaches 0 from either the right or the left, we say that the graph of $f(x) = 1/x$ approaches the y axis *asymptotically* and the y axis (which has the equation $x = 0$) is referred to as a *vertical asymptote.*

More formally, we have the following definition:

Definition

Vertical Asymptote

> The line $x = k$, where k is constant, is called a **vertical asymptote** of the graph of a function f if
>
> $$f(x) \to -\infty \quad \text{or} \quad f(x) \to +\infty$$
>
> either as $x \to k^-$ (from the left) or as $x \to k^+$ (from the right).

Figure 2 contains illustrations of some graphs that exhibit vertical asymptotic behavior.

Figure 2

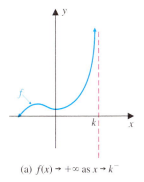

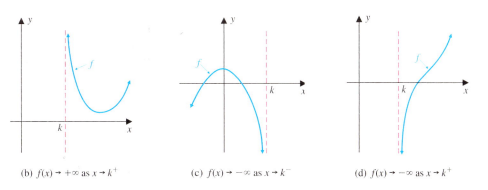

(a) $f(x) \to +\infty$ as $x \to k^-$ (b) $f(x) \to +\infty$ as $x \to k^+$ (c) $f(x) \to -\infty$ as $x \to k^-$ (d) $f(x) \to -\infty$ as $x \to k^+$

Let us reexamine the function $f(x) = 1/x$ on page 256 to determine its limit behavior. The graph in Figure 1 indicates that the function

$$f(x) = \frac{1}{x}$$

has the following limit behavior:

$$f(x) \to 0 \ \text{(through positive values) as } x \to +\infty$$
$$f(x) \to 0 \ \text{(through negative values) as } x \to -\infty$$

In other words, the graph gets closer and closer to the x axis either as

$$x \to +\infty \text{ or as } x \to -\infty.$$

Because of this limit behavior, the x axis (which has the equation $y = 0$) is referred to as a *horizontal asymptote* of the graph of f.

In general, we have the following definition:

Definition

Horizontal Asymptote

> The line $y = k$, where k is a constant, is called a **horizontal asymptote** for the graph of a function f if
>
> $$f(x) \to k \text{ as } x \to -\infty$$
>
> or
>
> $$f(x) \to k \text{ as } x \to +\infty.$$

Figure 3 contains illustrations of some graphs that exhibit horizontal asymptotic behavior.

Figure 3

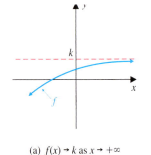

(a) $f(x) \to k$ as $x \to +\infty$

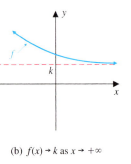

(b) $f(x) \to k$ as $x \to +\infty$

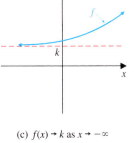

(c) $f(x) \to k$ as $x \to -\infty$

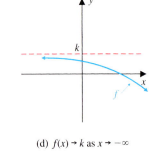

(d) $f(x) \to k$ as $x \to -\infty$

EXAMPLE 1   **Graphing $g(x) = \dfrac{1}{x^2}$**

Examine the numerical behavior of

$$g(x) = \frac{1}{x^2}$$

as $x \to 0^+$. Also, describe the limit behavior as $x \to +\infty$. Then use this information to sketch the graph of g.

Solution   The domain of g includes all real numbers except $x = 0$, so the graph has no y intercept. The data in Table 4 suggest that

$$g(x) \to +\infty \text{ as } x \to 0^+$$

so the y axis is a vertical asymptote. Also, the data in Table 4 help us to conclude that g has the following limit behavior:

$$g(x) \to 0 \text{ (through positive values) as } x \to +\infty$$

TABLE 4   As $x \to 0^+$, $g(x) \to +\infty$

x	1000	100	10	1	0.1	0.01	0.001
$g(x) = 1/x^2$	0.000001	0.0001	0.01	1	100	10,000	1,000,000

Figure 4

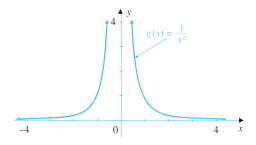

Consequently, the x axis is a horizontal asymptote of the graph of the function g.

Since $g(-x) = 1/(-x)^2 = 1/x^2 = g(x)$, it follows that g is an even function and its graph is symmetric with respect to the y axis. By using the asymptotic behavior and symmetry, we get the graph of $g(x) = 1/x^2$ (Figure 4). Note that there is no x intercept since $1/x^2 = 0$ has no solution.

In general, graphs of functions of the form

$$R(x) = \frac{1}{x^n} \text{ where } n \geq 1 \text{ is a positive integer}$$

fall into two categories according to whether n is an even or odd positive integer. If $n \geq 2$ is an even integer, then the graphs of functions such as

$$f(x) = \frac{1}{x^4} \quad \text{and} \quad g(x) = \frac{1}{x^6} \quad \text{(Figure 5a)}$$

resemble the graph of $y = 1/x^2$ shown in Figure 4 on page 258.
If $n \geq 1$ is an odd integer, then the graphs of functions such as

$$f(x) = \frac{1}{x^3} \quad \text{and} \quad g(x) = \frac{1}{x^5} \quad \text{(Figure 5b)}$$

resemble the graph of $y = 1/x$ shown in Figure 1 on page 256.

Figure 5

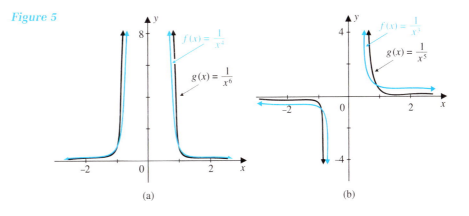

(a)   (b)

In general, the graphs of $R(x) = 1/x^n$ have the characteristics given in Table 5.

TABLE 5   Characteristics of Rational Functions $R(x) = 1/x^n$, where $n \geq 1$ is a Positive Integer

Property	n is Even	n is Odd
Vertical asymptote	y axis (equation is $x = 0$)	y axis (equation is $x = 0$)
Horizontal asymptote	x axis (equation is $y = 0$)	x axis (equation is $y = 0$)
Symmetry	y axis, since $R(-x) = R(x)$	Origin, since $R(-x) = -R(x)$
Intercepts	None	None

Transformations of graph of $R(x) = 1/x^n$ can be used to graph rational functions of the form

$$R(x) = \frac{a}{(x - h)^n} + k$$

where $n \geq 1$ is a positive integer.

EXAMPLE 2   **Using Transformations to Graph Rational Functions**

Use transformations of the graph of either $y = 1/x$ or $y = 1/x^2$ to graph each function. Identify the asymptotes.

(a) $f(x) = \dfrac{1}{x - 2}$ (b) $g(x) = \dfrac{1}{x^2} + 2$

Solution   (a) The graph of f can be obtained by shifting the graph of $y = 1/x$ horizontally 2 units to the right (Figure 6a). Note that the vertical asymptote is shifted from the y axis to the line $x = 2$, but the horizontal asymptote remains the x axis.

(b) The graph of g can be obtained by shifting the graph of $y = 1/x^2$ vertically upward 2 units (Figure 6b). Note that the vertical asymptote is still the y axis, but the horizontal asymptote is shifted 2 units up from the x axis to become the line $y = 2$.

Figure 6

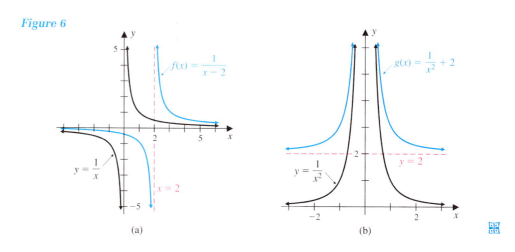

(a)   (b)

If the degree of $p(x)$ is equal to the degree of $q(x)$ for the rational function $R(x) = \dfrac{p(x)}{q(x)}$, then the graph of R has a horizontal asymptote. To find this horizontal asymptote, we may use long division to rewrite $R(x)$ in the form

$$R(x) = \frac{a}{(x - h)^n} + k$$

then proceed to sketch the graph by using transformations. The next example illustrates the technique.

EXAMPLE 3   **Graphing a Rational Function**

(a) Use long division to rewrite the given function in the form $\dfrac{a}{x - h} + k$ and then use transformations of graphs to graph the function

$$f(x) = \frac{3x + 17}{x + 5}$$

Identify the asymptotes of the graph.

(b) Find the domain and x intercepts of f.

(c) Solve the inequality $\dfrac{2}{x+5} > -3$.

Solution   (a) After dividing $3x + 17$ by $x - 5$, we rewrite $f(x)$ as

$$f(x) = \frac{2}{x+5} + 3$$

Thus the graph of f can be obtained by shifting the graph of $y = \dfrac{1}{x}$ horizontally 5 units to the left, 3 units vertically upward, and vertically scaling by a factor of 2 (Figure 7). Note that the horizontal asymptote is $y = 3$ because of the vertical shift and the vertical asymptote is $x = -5$ because of the horizontal shift.

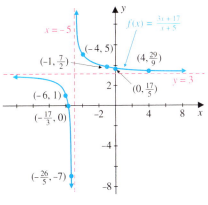

Figure 7

(b) The domain of f consists of all real numbers except -5, that is,

$$(-\infty, -5) \text{ or } (-5, \infty).$$

The x intercept of f is obtained by setting $f(x) = 0$ and then solving for x to get

$$3x + 17 = 0 \text{ or } x = -\frac{17}{3}.$$

(c) To solve $2/(x + 5) > -3$, we begin by rewriting the inequality in the equivalent form

$$\frac{2}{x+5} + 3 > 0$$

in which 0 is on one side of the inequality. Since

$$\frac{2}{x+5} + 3 = \frac{3x + 17}{x+5} = f(x)$$

we can rewrite the latter inequality as $f(x) > 0$.

Thus we need to find all values of x that yield positive values of $f(x)$, that is, we need to find all values of x that yield points on the graph of f that are above the x axis.

The graph of f in Figure 7 is above the x axis when $x < -17/3$ and when $x > -5$. Therefore, the solution of the original inequality is $x < -17/3$ or $x > -5$. In interval notation the solution includes all numbers in the interval $(-\infty, -17/3)$ or the interval $(-5, \infty)$.

Locating Vertical and Horizontal Asymptotes

To get an accurate graph of a rational function, it is helpful to first identify its asymptotes by examining the equation that defines the function. So far our investigation of the form

$$R(x) = \frac{1}{x^n}$$

indicates the following behavior:

> For any fraction $p(x)/q(x)$, as the values of the denominator $q(x)$ get closer to 0 and the values of the numerator $p(x)$ approach a value different from 0, the cumulative effect is that $p(x)/q(x)$ will either increase or decrease without bound, depending on whether the function values are positive or negative. This behavior provides the basis for determining the vertical asymptotes of a rational function from its equation.

Property 1
Locating Vertical Asymptotes

> A rational function
>
> $$R(x) = \frac{p(x)}{q(x)}$$
>
> has a vertical asymptote at the line $x = c$ if c is a zero of q and not a zero of p, that is, if $q(c) = 0$ but $p(c) \neq 0$.

For instance, Figure 8 shows the graph of the function

$$R(x) = \frac{3x^2 + 1}{x^2 - 4}, \text{which is of the form } \frac{p(x)}{q(x)}.$$

The graph of R has vertical asymptotes at the lines

$$x = -2 \text{ and } x = 2,$$

because $q(-2) = 0$ and $q(2) = 0$ but $p(-2) \neq 0$ and $p(2) \neq 0$.

Figure 8

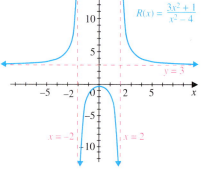

Let us turn our attention to the limit behavior of the function

$$R(x) = \frac{3x^2 + 1}{x^2 - 4}$$

as $x \to +\infty$.

By examining the graph in Figure 8, we observe the graph appears to be approaching a fixed finite value. To determine this value, we divide both the numerator and denominator of $R(x)$ by x^2 (the largest power of x together) to get

$$R(x) = \frac{3x^2 + 1}{x^2 - 4} = \frac{\dfrac{3x^2}{x^2} + \dfrac{1}{x^2}}{\dfrac{x^2}{x^2} - \dfrac{4}{x^2}} = \frac{3 + \dfrac{1}{x^2}}{1 - \dfrac{4}{x^2}}$$

When $|x|$ gets very large, the values of $\dfrac{1}{x^2}$ and $\dfrac{4}{x^2}$ approach zero; that is

$$\text{as } |x| \to +\infty, \frac{1}{x^2} \to 0 \text{ and } \frac{4}{x^2} \to 0$$

so that

$$R(x) \to \frac{3 + 0}{1 - 0} = 3, \text{ that is, 3 is the limit value.}$$

So the line $y = 3$ is a horizontal asymptote.

    In general, to determine the limit behavior of the graph of a rational function when $|x| \to \infty$, we first divide both the numerator and the denominator by x raised to its highest power. Then we examine the latter form to help locate any horizontal asymptote. The result depends on the relative degrees of the polynomials in the numerator and the denominator, as indicated in the following property:

Property 2
Locating Horizontal
Asymptotes

Assume R is a rational function (in reduced form) defined by

$$R(x) = \frac{p(x)}{q(x)} = \frac{a_m x^m + a_{m-1} x^{m-1} + \ldots + a_1 x + a_0}{b_n x^n + b_{n-1} x^{n-1} + \ldots + b_1 x + b_0}$$

where $m \geq 0$ and $n > 0$ are integers; and $a_m \neq 0$ and $b_n \neq 0$.

1. If $m < n$, then the line $y = 0$ (the x axis) is the horizontal asymptote.
2. If $m = n$, then the line $y = a_m/b_n$ is the horizontal asymptote.
3. If $m > n$, then there is no horizontal asymptote.

EXAMPLE 4

Locating a Horizontal Asymptote

Locate the horizontal asymptote of

$$g(x) = \frac{3x}{2x + 1}$$

and display this asymptotic behavior graphically.

First Solution

Because $m = n = 1$, it follows from Property 2,

that the line $y = \dfrac{3}{2}$ is a horizontal asymptote of

the graph of g.

    The graph in Figure 9 shows the following limit behaviors.

$$g(x) \to \frac{3}{2} \text{ as } x \to +\infty \text{ and } g(x) \to \frac{3}{2} \text{ as } x \to -\infty$$

Figure 9

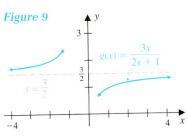

Alternate Solution

We divide both the numerator and the denominator by x (the largest power of x) to get

$$g(x) = \frac{3x}{2x + 1} = \frac{\dfrac{3x}{x}}{\dfrac{2x}{x} + \dfrac{1}{x}} = \frac{3}{2 + \dfrac{1}{x}}$$

When the values of $|x|$ get very large, the values of $\dfrac{1}{x}$ approach 0, that is

$$\text{as } |x| \rightarrow +\infty, \, 1/x \rightarrow 0$$

so

$$g(x) \rightarrow \frac{3}{2 + 0} = \frac{3}{2}$$

That is,

$$g(x) \rightarrow \frac{3}{2} \text{ as } x \rightarrow +\infty \text{ and } g(x) \rightarrow \frac{3}{2} \text{ as } x \rightarrow -\infty$$

Thus, the line $y = 3/2$ is the horizontal asymptote of the graph of g (Figure 9).

Graphing Nonreduced Rational Function

A rational function is said to be in *reduced form* if its numerator and denominator have no common factor. The next example investigates graphs of rational functions that are not in reduced form.

EXAMPLE 5   **Graphing Nonreduced Rational Functions**

Graph the given functions and find any asymptotes.

(a) $R(x) = \dfrac{x^2 - 4}{x - 2}$

(b) $h(x) = \dfrac{x^2 - 2x}{x^2 - 4x + 4}$

Solution

(a) By writing the numerator of R in factored form, we get the reduced form

$$R(x) = \frac{x^2 - 4}{x - 2} = \frac{(x - 2)(x + 2)}{x - 2} = x + 2 \quad \text{where } x \neq 2.$$

Note that the function $y = x + 2$ is *not* the same as the function R, because they have different domains. The graph of $y = x + 2$ is a straight line that includes the point $(2, 4)$ (Figure 10a), whereas the graph of R is the same straight line except that it has a *hole* at the point $(2, 4)$ (Figure 10b). In this situation, 2 is a zero of both the numerator and denominator of R. However, it turns out that there is no vertical asymptote at $x = 2$, only a hole.

(b) By writing both the numerator and denominator of $h(x)$ in factored form, we get the reduced form

$$h(x) = \frac{x^2 - 2x}{x^2 - 4x + 4}$$

$$= \frac{x(x - 2)}{(x - 2)(x - 2)} = \frac{x}{x - 2}, \quad \text{where } x \neq 2.$$

Figure 10

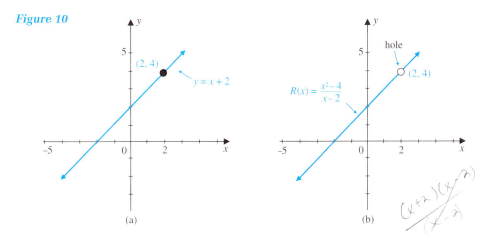

(a)

(b)

Figure 11

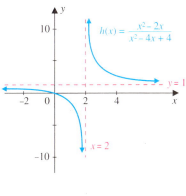

Note that the function

$$y = \frac{x}{(x-2)} = \frac{2}{x-2} + 1$$

is the same as function h. The graph is obtained by using transformations of the graph of $y = \dfrac{1}{x}$ (Figure 11). In this situation, 2 is a zero of both the numerator and denominator of h, and it turns out that there is a vertical asymptote at $x = 2$.

G The graphs of more complicated rational functions such as

$$f(x) = \frac{x^2}{x^2 - 3} \quad \text{and} \quad g(x) = \frac{3x^2 - 12}{x^2 + 2x - 3}$$

are examined extensively in calculus. For now, we can use a grapher to sketch such functions.

When a grapher is used to sketch the graph of a rational function the result may not correctly represent the behavior of the function near a vertical asymptote. The grapher may erroneously connect separate branches of the graph across a vertical asymptote. For instance, the viewing window of the graph of

$$f(x) = \frac{x - 1}{x^2 - x - 6}$$

Figure 12

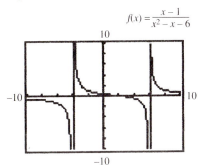

shown in Figure 12 incorrectly displays a graph that includes curves that appears to be the vertical asymptote. Actually, the grapher is erroneously connecting separate branches of the graph. Inaccuracies such as this one may occur near vertical asymptotes, depending on the selected viewing window.

Solving Applied Problems

Trends that occur in real-world situations are sometimes modeled by rational functions.

EXAMPLE 6   **Examining the Trend of Monthly Sales**

The model

$$s(t) = \frac{2000(1 + t)}{2 + t}$$

relates the monthly sales s (in dollars) of a certain new product to the time t (in months), beginning 1 month after the product has been introduced.

(a) What are the restrictions on t?

(b) Locate the horizontal asymptote for s.

(c) Use the graph of s, taking into account the restrictions on t, and the horizontal asymptote to determine the trend of the monthly sales as time elapses.

Solution   (a) We use $t \geq 1$, because the model is applicable at the beginning of the first month.

(b) Since

$$s(t) = \frac{2000(1 + t)}{2 + t} = \frac{2000t + 2000}{t + 2} = \frac{2000 + \dfrac{2000}{t}}{1 + \dfrac{2}{t}}$$

it follows that as

$$t \rightarrow +\infty, s(t) \rightarrow 2000 \,;$$

that is,

the line $s = 2000$ is a horizontal asymptote.

(c) After plotting a few points, recognizing that $s = 2000$ is a horizontal asymptote, and taking into account that $t \geq 1$, we obtain the graph shown in Figure 13. Notice that even though $t = -2$ is a vertical asymptote for the function for s (when it has no restrictions), it will not show in this graph because $t \geq 1$.

Figure 13

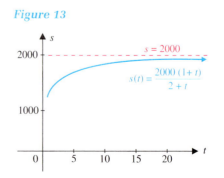

Upon examining the graph, we observe that as $t \rightarrow +\infty$ (the values of t are increasing), $s(t) \rightarrow 2000$ (from below). So as time elapses, the monthly sales trend will approach but not exceed $2000. We refer to $2000 as the *limiting value* of the monthly sales.

PROBLEM SET 3.4

Mastering the Concepts

In problems 1 and 2, graph each function. Identify the asymptotes and demonstrate the results numerically with a table of values.

1. (a) $F(x) = \dfrac{1}{x^6}$

   (b) $H(x) = \dfrac{1}{x^5}$

2. (a) $G(x) = \dfrac{1}{x^8}$

   (b) $h(x) = \dfrac{1}{x^7}$

In problems 3–6, use the given graph of each function to identify the vertical and horizontal asymptotes. Describe the asymptotic behaviors of the function.

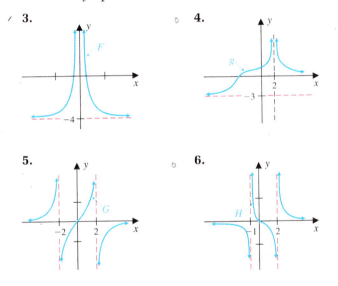

3.

4.

5.

6.

In problems 7–12, describe a sequence of transformations that will transform the graph of either $y = 1/x$ or $y = 1/x^2$ into the graph of each function. Sketch the graph of each function, and identify the asymptotes in each case. Also, find the domain and x intercepts of the graph of f.

7. (a) $f(x) = \dfrac{4}{x + 3}$

   (b) $g(x) = \dfrac{1}{x} + 3$

8. (a) $g(x) = \dfrac{1}{(x - 3)^2}$

   (b) $h(x) = \dfrac{4}{x^2} - 1$

9. (a) $f(x) = \dfrac{-3}{(x + 1)^2}$

   (b) $g(x) = \dfrac{2}{x - 1} + 3$

10. (a) $h(x) = \dfrac{3x + 2}{x}$

    (b) $f(x) = \dfrac{-2}{(x + 1)^2} - 1$

11. (a) $f(x) = \dfrac{4}{(x + 3)^2} - 1$

    (b) $g(x) = \dfrac{2x^2 + 16x + 33}{(x + 4)^2}$

    [*Hint:* Use long division first.]

12. (a) $g(x) = \dfrac{1}{x - 2} + 3$

    (b) $h(x) = \dfrac{3x + 7}{x + 2}$

    [*Hint:* Use long division first.]

In problems 13 and 14, find any vertical asymptotes. Graph each function and locate any holes in the graph.

13. (a) $f(x) = \dfrac{x^2 - 1}{x - 1}$

    (b) $g(x) = \dfrac{x^2 - 2x - 3}{x - 3}$

14. (a) $f(x) = \dfrac{x - 4}{x^2 - 16}$

    (b) $g(x) = \dfrac{x^3 - 1}{x - 1}$

In problems 15 and 16, locate the horizontal asymptote of the graph of each function, and then display this asymptotic behavior graphically.

15. (a) $f(x) = \dfrac{-2x}{x + 1}$

    (b) $g(x) = \dfrac{1 + 3x}{2x - 1}$

16. (a) $f(x) = \dfrac{x^2 + 1}{x^2 - 1}$

    (b) $g(x) = \dfrac{2x^2 + x}{x^2 - 2x + 3}$

In problems 17–20, match the function to its hand-drawn graph in Figure 14. Locate all asymptotes.

17. $f(x) = \dfrac{x-1}{x^2+x}$

18. $g(x) = \dfrac{x^2}{x^2-4}$

19. $h(x) = \dfrac{2x}{x+1}$

20. $f(x) = \dfrac{x^2-3x+4}{x+2}$

Figure 14

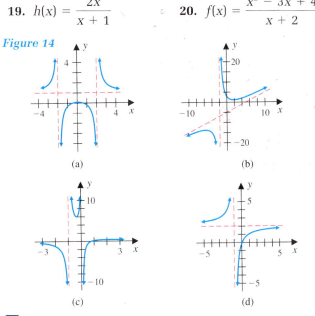

(a)

(b)

(c)

(d)

G In problems 21–28, graph each function. Locate the asymptotes (if any) for each function.

21. $f(x) = \dfrac{2x}{x+1}$

22. $g(x) = \dfrac{-3x+1}{x+4}$

23. $f(x) = \dfrac{3}{x^2-2x-3}$

24. $h(x) = \dfrac{2}{x^2-x-2}$

25. $g(x) = \dfrac{-x}{x^2-x-6}$

26. $f(x) = \dfrac{-2x}{x^2-3x-4}$

27. $f(x) = \dfrac{x^2+2}{x^2-4x+3}$

28. $g(x) = \dfrac{x^2-4}{x^2+1}$

In problems 29–36, use the graph of an associated rational function to solve each inequality.

29. $2 - \dfrac{2}{x+1} < 0$ (see problem 21)

30. $\dfrac{13}{x+4} > 3$ (see problem 22)

31. $\dfrac{3}{x^2-2x-3} \geq 0$ (see problem 23)

32. $\dfrac{2}{x^2-x-2} \leq 0$ (see problem 24)

33. $\dfrac{-x}{x^2-x-6} < 0$ (see problem 25)

34. $\dfrac{-2x}{x^2-3x-4} > 0$ (see problem 26)

35. $\dfrac{x^2+2}{x^2-4x+3} \leq 0$ (see problem 27)

36. $\dfrac{x^2-4}{x^2+1} \geq 0$ (see problem 28)

Applying the Concepts

37. Business Sales Trend: Suppose that the sale of S units of a certain product is related to the daily advertising expenditure x (in dollars) by the model

$$S(x) = \frac{5000x}{x+100}$$

(a) What are the restrictions on x?

(b) Sketch the graph of S using the restrictions from part (a). Also, locate the asymptotes.

(c) What is the sales trend as the advertising expenditures increase?

38. Government Securities: Each weekday, national newspapers such as the *New York Times* and the *Wall Street Journal* publish a curve called a *yield curve* for the U.S. government securities. Suppose that the percentage yield p on government securities that mature after t years is modeled by the function

$$p = \frac{t+0.1}{0.1t+0.03}, \quad 0.25 \leq t \leq 30$$

(a) Graph this function and locate the asymptotes.

(b) Explain the implication of the horizontal asymptote on the graph of p. That is, analyze the trend of the yield as a function of maturity.

39. Air Pollution: Suppose that the cost C (in dollars) of removing x percent of air pollutants caused by automobile emissions in a certain town is given by the model

$$C = \frac{100{,}000x}{100-1.67x}, \quad 0 < x < 60$$

(a) Graph the function C.

(b) Use the graph to analyze the trend of the cost of pollution control by examining what happens as the air pollutants approach 60%.

40. Pain Relief: Researchers have determined that the percentage p of pain relief from x grams of aspirin is given by the model

$$p = \frac{100x^2}{x^2 + 0.02} \quad 0 < x < 1$$

(a) Graph p.

(b) Use the graph to approximate the value of x when the pain relief is very close to its maximum value of 100%.

(c) Explain the implication of the horizontal asymptote of the graph of p.

41. A closed box with a square base of x centimeters per side and a height of h centimeters has a volume of 500 cubic centimeters.

(a) Write a function for the surface area A of the box in terms of x.

(b) Graph A and to find the dimensions x and h that will minimize the area needed to manufacture the box. What is the minimum surface area? Round off the answers to two decimal places.

42. A city determined that the cost C (in millions of dollars) of operating a recycling center is given by the model

$$C(p) = \frac{1.1p}{90 - p}$$

where p is the percentage of the participants from the city residents.

(a) Graph C and locate the asymptotes of the graph of C.

(b) Explain the implication of the horizontal asymptote on the graph of C as p approaches 90.

(c) Use the graph to analyze the trend of the cost if 80% of the residents are expected to participate.

Developing and Extending the Concepts

43. Let $f(x) = \frac{1}{x}$. Compute and simplify the expression

$$q = \frac{f(x) - f(2)}{x - 2}$$

Sketch the graph of q.

44. Let $f(x) = \frac{1}{x^2}$. Compute and simplify the expression

$$q = \frac{f(x) - f(1)}{x - 1}$$

Sketch the graph of q.

In problems 45 and 46, find an equation of a rational function f that satisfies the following conditions.

45. f has real zeros at -3, 1, 3, a vertical asymptote at $x = 0$ and a horizontal asymptote at $y = 2$.

46. f has no real zeros, f has a vertical asymptote at $x = 3$ and it has a horizontal asymptote at $y = 4$.

47. Explain why the following statement is true: If the line with equation $x = k$ is a vertical asymptote of the graph of the rational function $y = f(x)$, then k is not in the domain of f.

48. Explain why the following statement is false: If the line with equation

$$y = k$$

is a horizontal asymptote of the graph of the rational function

$$y = f(x),$$

then k is not in the range of f.

Hint: Consider the function

$$f(x) = \frac{2x^2 + 1}{2x^2 - 3x}.$$

To solve problems 49 and 50, consider the rational function

$$f(x) = \frac{x^2 + 1}{x}$$

By division we can rewrite f as

$$f(x) = x + \frac{1}{x}$$

since $\frac{1}{x} \to 0$ as $x \to +\infty$ or $x \to -\infty$, this means that the graph of $y = f(x)$ eventually gets closer and closer to the line $y = x$ as $x \to +\infty$ or as $x \to -\infty$. The line $y = x$ is called an **oblique asymptote** of the graph of f.

In problems 49 and 50. Use the given graph of the rational function to find the equation of the oblique asymptotes.

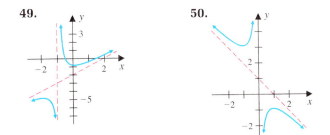

49.

50.

G In problems 51 and 52, find the oblique asymptote of the graph of each function, and then demonstrate the asymptotic behavior by using a grapher to graph the function and its asymptote.

51. (a) $f(x) = \dfrac{-4x^2 + 1}{x + 2}$

   (b) $g(x) = \dfrac{x^2 - 4x + 7}{x - 1}$

52. (a) $f(x) = \dfrac{3x^2 - 12x + 7}{x + 1}$

   (b) $g(x) = \dfrac{x^3 - 2x^2}{x^2 - 1}$

Objectives

1. Determine the Types of Zeros of a Polynomial Functions
2. Use the Variation of Signs
3. Find Bounds of Zeros of Polynomials
4. Use the Conjugate Pairs Property
5. Solve Applied Problems

3.5 Theory of Equations

So far our main concern has been to find real zeros of polynomial functions with real coefficients. In this section, we extend our discussion to include complex number zeros as well as polynomial functions with complex coefficients.

Earlier in Section 3.3, we stated the factor and remainder theorem when the number c is a real number and the polynomial function has real coefficients. The theorems are also true when c is complex and the function has complex coefficients. We also indicated that the zeros of polynomial function f are the solutions of the equation

$$f(x) = 0$$

and each real zero represents an x intercept of the graph of f. However, it is possible that a zero may be a nonreal complex number, that is, an imaginary number. For instance, as Figure 1 shows, the graph of the function

$$f(x) = x^3 - x^2 + 4x - 4 = (x - 1)(x^2 + 4)$$

has only one x intercept, and hence one real zero. But the factored form indicates that $x^2 + 4$ is also a factor of $f(x)$. By solving the equation $x^2 + 4 = 0$, we see that $x = -2i$ and $x = 2i$, are two zeros that are imaginary. The complete factored form of $f(x)$ is given by

$$f(x) = (x - 1)(x + 2i)(x - 2i)$$

Figure 1

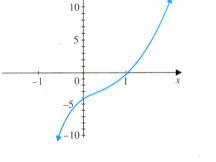

$f(x) = (x - 1)(x + 2i)(x - 2i)$

In general, to further explore the zeros of a polynomial function, we ask the following questions:

1. How many zeros of a polynomial function are real numbers, and how many are imaginary?
2. How many real zeros are positive and how many are negative?
3. In searching for the real zeros, can we determine bounds or limitations on the values of these zeros?

Here we discuss results that help answer these questions. These results form the basis of the *theory of equations.*

Determining the Types of Zeros of a Polynomial Functions

In Sections 3.2 and 3.3, we stated that it is difficult to find all the zeros of a polynomial function except in special cases. However, the next result, which is known as the fundamental theorem of algebra, proved by Karl Friedrich Gauss (1777–1855), guarantees that every polynomial function of nonzero degree has at the least one complex zero, and hence has a factor $x - c$. More formally,

The Fundamental Theorem of Algebra	Every polynomial function of degree $n \geq 1$ has at least one complex zero.

By combining the Fundamental Theorem of Algebra with the factor theorem, we obtain the following property:

The Complete Linear Factorization Property	If $f(x)$ is a polynomial of degree $n \geq 1$ with real coefficients, then there are n complex numbers $c_1, c_2, \ldots, c_n$ that form the linear factorization $$f(x) = a(x - c_1)(x - c_2)\ldots(x - c_n)$$ where a is the leading coefficient of $f(x)$.

These n factors are not necessarily distinct.

Consider the polynomial function

$$f(x) = x^4 + 9x^2$$

Since $x^4 + 9x^2 = x^2(x^2 + 9) = x^2(x + 3i)(x - 3i)$ the zeros of f are 0, $-3i$, and $3i$, and the factored form of f is

$$f(x) = x^2(x + 3i)(x - 3i).$$

Notice that 0 is a zero of multiplicity 2, whereas $-3i$ and $3i$ each have a multiplicity 1, and the number of linear factors is equal to the degree of f, which is 4.

In general, if the zeros of a polynomial function f are counted according to multiplicity, and as long as we include complex numbers as zeros, it follows that the number of linear factors is equal to the degree of f. Thus, we are led to the following result:

Number of Zeros of a Polynomial Function	If f is a polynomial function of degree $n > 0$ and if a zero of multiplicity m is counted m times, then f has precisely n zeros.

EXAMPLE 1   **Determining the Number and Types of Zeros**

Use the graph of the polynomial function f with degree 6 in Figure 2. Assume f has zeros of multiplicity 1 to determine the number of positive and negative real zeros as well as the number of imaginary zeros.

Figure 2

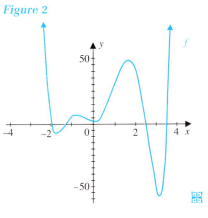

Solution   From the graph of f in Figure 2, there are four x intercepts, so the function has four real zeros: two are negative and two are positive. Since the degree of f is 6, there are six zeros, the four zeros, and $6 - 4 = 2$ imaginary zeros.

Using the Variation of Signs

Earlier we used graphs to predict the type of real zeros a polynomial function can have. Now, we turn our attention to another method that helps us to determine the number of real and imaginary zeros.

Descartes' Rule of Signs

Suppose that $f(x)$ is a polynomial with real coefficients arranged in descending powers of x. A ***variation of signs*** occurs when two successive terms have opposite signs. Missing terms (with 0 coefficients) are ignored when counting the total number of variations of signs.

(i) The **number of positive zeros** of f is either equal to the number of variations of signs for $f(x)$, or it is fewer than that number by an even positive integer.
(ii) The **number of negative zeros** of f is either equal to the number of variations of signs for $f(-x)$, or it is fewer than that number by an even positive integer.

EXAMPLE 2   **Using Descartes' Rule of Signs**

Use Descartes' rule of signs to determine the possible number of positive real zeros, negative real zeros, and imaginary zeros of

$$f(x) = 3x^5 - 2x^4 + 3x^3 + 2x^2 + 7x - 3$$

Solution   There are three variations of signs in $f(x)$:

$$f(x) = +3x^5 - 2x^4 + 3x^3 + 2x^2 + 7x - 3$$

Therefore, $f(x)$ has either three positive zeros or one positive zero.
  Since

$$f(-x) = 3(-x)^5 - 2(-x)^4 + 3(-x)^3 + 2(-x)^2 + 7(-x) - 3$$

$$= -3x^5 - 2x^4 - 3x^3 + 2x^2 - 7x - 3$$

$f(-x)$ has two variations in signs. Consequently, $f(x)$ has either two negative zeros or no negative zero.
  Since f has precisely five zeros, the actual number and types of zeros of f will be one of the four possibilities listed in Table 1.

TABLE 1 $f(x) = 3x^5 - 2x^4 + 3x^3 + 2x^2 + 7x - 3$

Possible Number of Positive Zeros		Possible Number of Negative Zeros	Possible Number of Imaginary Zeros
1.	3	2	0
2.	3	0	2
3.	1	2	2
4.	1	0	4

Finding Bounds for Zeros of Polynomial

The search for the real zeros of polynomial functions can be reduced somewhat if upper and lower bounds for the values of the zeros can be found.

By definition, a real number b is an **upper bound** for a real zero c of f if no zero is greater than b, that is, $c \leq b$. A real number a is a **lower bound** for a real zero c of f if no zero is less than a, that is, $a \leq c$.

It is possible to use synthetic division and the rational zero theorem to find upper and lower bounds for the zeros of $f(x)$. If we divide $f(x)$ by $x - c$, the third row in the division process contains the coefficients of the quotient $Q(x)$ together with the remainder R. The following property shows how the entries in the third row are used to find the bounds for the real zeros of f.

Property

Upper and Lower Bound Property

Suppose that $y = f(x)$ is a polynomial function with real coefficients and a positive leading coefficient, and that $f(x)$ is divided by $x - c$ using synthetic division.

(i) If $c > 0$ and the numbers in the last row of the synthetic division process are nonnegative (positive or 0), then c is an **upper bound** for all the real zeros of f.
(ii) If $c < 0$ and the numbers in the last row of the synthetic division process are alternately nonnegative (positive or 0) and nonpositive (negative or 0), then c is a **lower bound** for all the real zeros of f.

EXAMPLE 3

Recognizing Bounds for the Zeros of a Polynomial Function

Use the upper and lower bound property above to verify that all the real zeros of

$$f(x) = 4x^3 - 13x - 6$$

lie between -2 and 3, that is, -2 and 3 are, respectively, lower and upper bounds for the real zeros.

Solution

We first show that 3 is an upper bound for the zeros of $f(x)$ as follows:
Using synthetic division we divided

$$f(x) \text{ by } x - 3$$

to get

$$
\begin{array}{r|rrrr}
3 & 4 & 0 & -13 & -6 \\
  &   & 12 & 36 & 69 \\
\hline
  & 4 & 12 & 23 & 63 \\
\end{array}
$$

Since the numbers in the last row of the synthetic division process are positive, it follows that 3 is an upper bound for the zeros of f.

To show that -2 in a lower bound for the zeros, we divide $f(x)$ by

$$x - (-2) = x + 2$$

to obtain

$$
\begin{array}{r|rrrr}
-2 & 4 & 0 & -13 & -6 \\
   &   & -8 & 16 & -6 \\
\hline
   & 4 & -8 & 3 & -12 \\
\end{array}
$$

The numbers in the last row alternate in signs indicating that -2 is a lower bound for the zeros of f.

Upper and lower bounds are not unique numbers.

The graph of the function f (Figure 3) for Example 3 shows all three x intercepts between two numbers, -2 and 3, thus demonstrating that they are, respectively, lower and upper bounds for the zeros of f. Notice that -3 and 4 could also serve, respectively, as a lower bound and an upper bound.

Figure 3

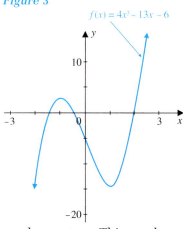

Using the Conjugate Pairs Property

The complex numbers $a + bi$ and $a - bi$ are *conjugates* of each other.

Table 2 shows examples of polynomial functions with real coefficients, where it is necessary to find the zeros of the functions in the complex number system. This can be accomplished, of course, by solving the associated equations.

TABLE 2   **Examples**

Polynomial Function f	Associated Polynomial Equation	Zeros of f
1. $f(x) = x^2 + 4$	$x^2 + 4 = 0$	$-2i, 2i$
2. $f(x) = x^2 - 6x + 25$	$x^2 - 6x + 25 = 0$	$3 - 4i, 3 + 4i$

Notice that the zeros of each function come in pairs as *complex conjugates.* That is, the zeros of $f(x) = x^2 + 4$, which are $-2i$ and $2i$, are conjugate pairs. The zeros of $f(x) = x^2 - 6x + 25$, $3 - 4i$ and $3 + 4i$, are also conjugate pairs.

These observations are generalized as follows:

Conjugate Zeros Theorem

Let f be a polynomial function with real coefficients. If $a + bi$ is a zero of f, then the complex conjugate $a - bi$ is also a zero of f.

EXAMPLE 4   **Finding Complex Zeros**

Find the zeros of

$$f(x) = x^4 - 4x^3 + 5x^2 - 4x + 4$$

given that i is a zero of f. Also, factor $f(x)$ and determine the multiplicity of each zero.

Solution   The conjugate zero theorem tells us that since i is a zero of f, then the conjugate $-i$ is also a zero of f. By the factor theorem, we have

$$f(x) = x^4 - 4x^3 + 5x^2 - 4x + 4$$
$$= (x - i)(x + i)Q(x)$$
$$= (x^2 + 1)Q(x)$$

After dividing $f(x)$ by $x^2 + 1$ using long division, we get the factorization

$$f(x) = (x^2 + 1)(x^2 - 4x + 4)$$
$$= (x - i)(x + i)(x - 2)^2$$

So i and $-i$ are each zeros of f of multiplicity 1, and 2 is a zero of f of multiplicity 2.

In the next example, we will see an example in which a complex number is a zero of a polynomial function, yet, its conjugate is not a zero.

EXAMPLE 5   **Nonpaired Zeros**

Let $f(x) = x^2 - 2x + 1 + 2i$.

Determine whether $2 - i$ and $2 + i$ are zeros of f. One of these conjugate-paired complex numbers is not a zero of f. Does this contradict the conjugate zero theorem?

Solution   If $\qquad f(x) = x^2 - 2x + (1 + 2i)$, then

$$f(2 - i) = (2 - i)^2 - 2(2 - i) + (1 + 2i)$$
$$= 4 - 4i - 1 - 4 + 2i + 1 + 2i$$
$$= 0$$

so that $2 - i$ is zero of f.

However, $f(2 + i) = (2 + i)^2 - 2(2 + i) + (1 + 2i)$

$$= 4 + 4i - 1 - 4 - 2i + 1 + 2i$$
$$= 4i \neq 0$$

and so, $2 + i$ is not a zero of f.

The fact that the conjugate pair $2 - i$ and $2 + i$ are not both zeros of f does not contradict the conjugate zeros theorem, because the coefficients of the given polynomial function are not all real numbers as required in the theorem.

EXAMPLE 6   **Using Zeros to Form a Polynomial Function**

Form a polynomial function with real coefficients, a leading coefficient of 1, the smallest possible degree, and the zeros -2, 3, and $1 - 3i$.

Solution   Since we want a polynomial of the smallest possible degree, we assume that each zero has multiplicity 1. Because we want the coefficients to be real numbers, we must include $1 + 3i$, the complex conjugate of $1 - 3i$, as a zero. By the factorization theorem we obtain

$$f(x) = [x - (-2)](x - 3)[x - (1 - 3i)][x - (1 + 3i)]$$
$$= (x^2 - x - 6)(x^2 - 2x + 10)$$
$$= x^4 - 3x^3 + 6x^2 + 2x - 60$$

It is important to understand what is meant by a *complete factorization* of a polynomial with real coefficients of degree $n > 1$. In the complex number system it is possible to write such a polynomial completely as a product of n *linear* factors. However, in the real number system, we have the following property:

Property
Complete Factorization
in the Real
Number System

In the *real number system,* any polynomial function of degree $n > 1$ with real coefficients can be written as the product of linear and/or quadratic factors with real coefficients, where the quadratic factors have no real zeros.

The examples given in Table 3 compare complete factorizations in the complex number system to those in the real number system.

TABLE 3   Examples of Complete Factorizations of Polynomials

Polynomial Function with Real Coefficients	Complete Factorization	
	Real Number System	Complex Number System
$x^2 + 4$	$x^2 + 4$	$(x + 2i)(x - 2i)$
$x^3 - 1$	$(x - 1)(x^2 + x + 1)$	$(x - 1)\left[x - \left(\dfrac{-1 + \sqrt{3}i}{2}\right)\right]\left[x - \left(\dfrac{-1 - \sqrt{3}i}{2}\right)\right]$
$x^4 - 14x^3 + 61x^2 - 54x - 130$	$(x + 1)(x - 5)(x^2 - 10x + 26)$	$(x + 1)(x - 5)[x - (5 - i)][x - (5 + i)]$

Solving Applied Problems

The next example illustrates an application of a polynomial involving complex zeros.

EXAMPLE 7   **Solving a Puzzle Involving Complex Numbers**

The Italian mathematician Girolamo Cardano (1545) is credited with the first use of complex numbers in solving the following famous problem:
Find two numbers whose sum is 10 and whose product is 40.

Solution   Let x represent the first number and $10 - x$ the second number. Since the product of the two numbers is 40, we get

$$x(10 - x) = 40$$

$$10x - x^2 = 40$$

$$x^2 - 10x + 40 = 0 \qquad \text{Write the equation in standard form}$$

$$x = 5 \pm \sqrt{15}\, i \qquad \text{Solve the quadratic equation}$$

Therefore, the two numbers are $5 - \sqrt{15}\, i$ and $5 + \sqrt{15}\, i$.

◆ PROBLEM SET 3.5

Mastering the Concepts

In problems 1 and 2, suppose that a polynomial function f with the given graph has zeros of multiplicity. Determine the number of positive real zeros, negative real zeros, and imaginary zeros of f.

1. (a) f has degree 4
  (b) f has degree 6

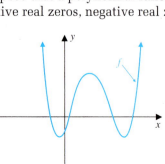

2. (a) f has degree 3
  (b) f has degree 5

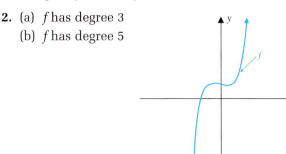

In problems 3–6, use the graph of the polynomial function to determine the possible number and types of zeros. Assume each zero has multiplicity 1.

3. $f(x) = x^4 - 10$ (Figure 4)

Figure 4

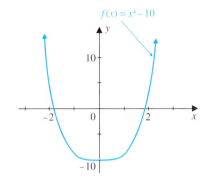

4. $g(x) = x^5 + 1$ (Figure 5)

Figure 5

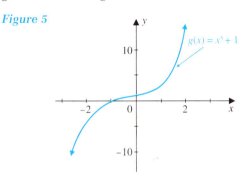

5. $g(x) = 3x^3 - 4x + 1$ (Figure 6)

Figure 6

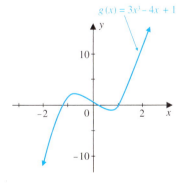

6. $h(x) = x^3 + x^2 + x + 1$ (Figure 7)

Figure 7

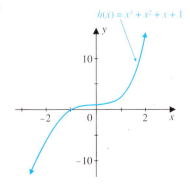

In problems 7–14, use Descartes' rule of signs to determine the possible number of positive real zeros, negative real zeros, and imaginary zeros of each given function.

7. $f(x) = x^3 - 8x - 2$

8. $g(x) = x^3 + x^2 + x + 1$

9. $g(x) = 3x^3 - 4x + 1$

10. $h(x) = -x^3 + x^2 - 2x + 7$

11. $h(x) = 7x^4 - x^3 + 11x - 1$

12. $g(x) = 2x^4 - 4x^3 - 6x^2 - 8$

13. $f(x) = x^5 - 6x - 5$

14. $h(x) = x^5 + 4x^4 - 3x^2 + x$

In problems 15–20, for each given function, use the upper and lower bound property to find upper and lower integer bounds for the real zeros.

15. $f(x) = x^3 - 3x + 1$

16. $g(x) = x^3 + 2x^2 - 6x - 18$

17. $h(x) = x^4 - 3x^3 + 4x^2 + 2x - 9$

18. $h(x) = x^4 - 24x^2 - 25$

19. $h(x) = x^5 - 3x^3 + 3x^2 + 2x - 2$

20. $g(x) = 4x^5 - 23x^3 - 33x^2 - 11x - 3$

In problems 21–24, use synthetic division to verify that the two given numbers are bounds for the real zeros of the given function f.

21. $f(x) = x^3 - 2x^2 + x + 2$
Lower bound $= -1$,
upper bound $= 2$.

22. $f(x) = 27x^3 - 27x^2 - 18x + 8$
Lower bound $= -1$,
upper bound $= 2$.

23. $f(x) = x^4 - 5x^3 - 11x^2 + 33x - 18$
Lower bound $= -4$,
upper bound $= 7$.

24. $f(x) = x^5 + 3x^3 - 9x^2 + 5$
Lower bound $= -10$,
upper bound $= 10$.

In problems 25–30, use the given zero of f to find all other complex zeros and express $f(x)$ in complete linear factored form. Also determine the multiplicity of each zero.

25. $f(x) = x^4 + 13x^2 + 36$; $2i$

26. $f(x) = 2x^5 - 7x^4 + 12x^3 - 8x^2 + 4$; $1 + i$
(multiplicity 2)

27. $f(x) = x^4 + 2x^2 + 1$; i (multiplicity 2)

28. $f(x) = x^5 - 2x^4 + 2x^3 + 8x^2 - 16x + 16$; $1 + i$

29. $f(x) = x^4 + 2x^3 - 4x - 4$; $-1 + i$

30. $f(x) = 2x^6 + x^5 + 2x^3 - 6x^2 + x - 4$;
i (multiplicity 2)

In problems 31–36, form a polynomial function with real coefficients of the specified degree, with real coefficients, a leading coefficient of 1, and the given numbers as zeros.

31. $1 + \sqrt{2}, 1 - \sqrt{2}, 3$; degree 3

32. $\sqrt{3}, -\sqrt{3}, 4$ (multiplicity 2); degree 4

33. $1 + i$; degree 2

34. $3, -3, 3i$; degree 4

35. $1 - 3i$ (multiplicity 2); degree 4

36. $1 + 5i, 2$ (multiplicity 2); degree 4

In problems 37–44, factor $f(x)$ completely:

(a) In the real number system

(b) In the complex number system

37. $f(x) = x^2 - x + 1$

38. $f(x) = 2x^2 + 4x + 6$

39. $f(x) = 2x^3 + 4x$

40. $f(x) = 4x^3 + 2x$

41. $f(x) = x^4 - 81$

42. $f(x) = x^4 + 6x^2 + 5$

43. $f(x) = 3x^3 - x^2 + 3x - 1$

44. $f(x) = x^3 + x^2 + x - 3$

45. Let $f(z) = z^3 - iz^2 + 2iz + 2$.

(a) Determine whether i and its conjugate $-i$ are zeros of f.

(b) If both are not zeros of f, does this contradict the conclusion of the conjugate zero theorem? Explain.

46. (a) [G] Use a grapher to determine the number of imaginary zeros of
$$f(z) = z^4 + 2z^2 + 1$$

(b) If possible, find all the zeros of f.

Applying the Concepts

47. Alternating Current: In 1893 Charles P. Steinmetz, an American electrical engineer, developed a theory of alternating currents based on complex numbers. He showed that *Ohm's law* for alternating currents takes the form
$$E = IZ$$
where E is the *voltage,* I is the *current,* and Z is the *impedance,* given by the equation
$$Z = R + (X_L - X_C)i$$

where X_L is the *inductive reactance* and X_C is the *capacitative reactance.* Find E if $I = 1 - i$, $R = 3$, $X_L = 2$ and $X_C = 1$.

48. Alternating Current: Use the equation $E = I[R + (X_L - X_C)i]$ to find I if $E = 3 + 2i$, $R = 4$, $X_L = 3$ and $X_C = 2$.

Developing and Extending the Concepts

49. The solutions to the equation
$$x^3 - 8 = 0$$
are the cube roots of 8.

(a) How many cube roots of 8 are there?

(b) Are the cube roots real or imaginary?

50. Show that the three cube roots of $8i$ are $\sqrt{3} + i, -\sqrt{3} + i,$ and $2i$ by cubing each of these roots to obtain $8i$.

51. The fourth roots of 16 are the solutions of the equation
$$x^4 = 16 \quad \text{or} \quad x^4 - 16 = 0.$$

(a) How many fourth roots of 16 are there?

(b) Find these roots.

52. Find all solutions of the equation
$$ix^2 - 2x + i = 0$$

53. Factor $f(x) = x^6 - 64$ completely into linear factors with complex coefficients. *Hint:* Rewrite $x^6 - 64$ as $(x^3)^2 - 8^2$.

54. Suppose that f is a polynomial function of degree n, with real coefficients where n is odd.

(a) What is the maximum number of x intercepts of f? Give an example to support your assertion.

(b) What is the minimum number of x intercepts of f? Give an example to support your assertion.

In problems 55 and 56, answer true or false. Support your assertion.

55. If the graph of a polynomial function of degree 5 has one x intercept, then the function has four imaginary zeros.

56. If 5 is an upper bound for the real zeros of a polynomial function, then no number less than 5 can be an upper bound.

◆ CHAPTER 3 REVIEW PROBLEM SET

In problems 1 and 2, use transformations of the graph of

$$y = x^2$$

to graph the given function. Find the vertex, axis of symmetry, domain, and range.

1. $f(x) = -\dfrac{1}{2}(x + 5)^2 + 3$

2. $g(x) = 3(x + 1)^2 - 2$

In problems 3 and 4, rewrite the function f in standard form $f(x) = a(x - h)^2 + k$. Sketch the graph, and locate the vertex and axis of symmetry.

3. $f(x) = x^2 + 4x - 7$

4. $f(x) = -2x^2 - 5x - 1$

In problems 5 and 6, find the standard equation $f(x) = a(x - h)^2 + k$ of the quadratic function satisfying the given conditions.

5. y intercept: -8; vertex at $(-1, -9)$

6. x intercept: 1; vertex at $\left(\dfrac{3}{2}, \dfrac{1}{2} \right)$

In problems 7 and 8, use the graph of the given quadratic function and the location of the x intercepts to solve the corresponding inequality.

7. $f(x) = 2x^2 - x - 1$;
$\quad 2x^2 - x - 1 < 0$

8. $g(x) = x^2 + 5x - 6$;
$\quad x^2 + 5x \geq 6$

In problems 9 and 10, use transformations of an appropriate power function to graph the given function.

9. $h(x) = 2(x - 1)^3 - 3$

10. $g(x) = -\dfrac{1}{3}(x - 2)^4 + 5$

In problems 11 and 12, use the graph of f to determine the limit behavior of $f(x)$ as:

11. (a) $x \to -\infty$ **12.** (b) $x \to +\infty$

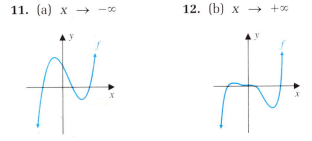

G In problems 13 and 14, use a grapher to graph the function. Determine the x intercepts of the given function, and then use the graph to discuss the limit behavior of the function and to solve the associated inequality.

13. $h(x) = x^3 - x$;
$\quad x^3 - x < 0$

14. $g(x) = x^4 - 2x^2 + 1$;
$\quad x^4 - 2x^2 + 1 \geq 0$

In problems 15 and 16, find the vertex and axis of symmetry of each parabola.

15. $x = 3y^2 - 6y + 2$ **16.** $x = 3y^2 + 2y - 7$

In problems 17 and 18, use the remainder theorem to find the value of each function if $x = c$.

17. $f(x) = 4x^3 + 2x^2 - 6x - 2$;
$\quad c = -2$

18. $g(x) = -3x^3 - 17x + 11$;
$\quad c = 4$

19. Use the factor theorem to determine whether each linear expression is a factor of f if

$$f(x) = x^4 + 3x^3 - 6x^2 - 28x - 24$$

(a) $x + 3$

(b) $x + 5$

20. Is $x + 1$ a factor of

$$f(x) = 5x^4 - 2x^3 + 11x^2 + 5x + 36?$$

Explain

In problems 21 and 22, verify that $f(c) = 0$. Then use the factor theorem to factor $f(x)$ completely.

21. $f(x) = x^3 + 6x^2 - 11x - 10$; $c = 2$

22. $f(x) = x^4 + 5x^3 - 19x^2 - 65x + 150$; $c = -5$

In problems 23 and 24, use the given graph of the function to determine the number of positive and negative zeros. Also, find lower and upper integer bounds of the real zeros. For each real zero c, find consecutive integers k and $k + 1$ such that $k < c < k + 1$.

23. $f(x) = 6x^4 - 19x^3 + 13x^2 + 4x - 4$

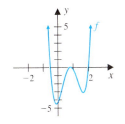

24. $g(x) = -4x^3 - 16x^2 - 9x + 9$

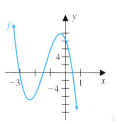

In problems 25 and 26, use the rational zero theorem to find all rational zeros of f. Use the results to factor $f(x)$ and to find other real zeros.

25. $f(x) = x^3 - 4x^2 + x + 6$

26. $f(x) = \dfrac{x^4}{2} - \dfrac{7}{2}x^3 + 9x^2 - 10x + 4$

In problems 27 and 28:

(a) Use the intermediate-value property to show that the function f has a zero in the given interval.

(b) Use the bisection method to approximate the value of the zero c, to four decimal places so that $|f(c)| < 0.005$.

27. $f(x) = x^3 - x - 5$, $[0, 2]$

28. $f(x) = x^4 - x - 1$, $[-1, 1]$

G In problems 29 and 30, use a grapher to sketch the graph of each function and to approximate all real zeros c so that

$$|f(c)| \le 0.005 .$$

Round off the answers to four decimal places.

29. $f(x) = x^4 - 9x^2 + 14$

30. $g(x) = -x^4 - 4x + 1$

In problems 31 and 32, find the complex zeros and their multiplicities.

31. $h(x) = (x^4 - 1)(x^2 + 8x + 16)$

32. $g(x) = x^3(x^2 + 9)(x^2 - 4x + 7)$

In problems 33 and 34, find the other complex zeros of f.

33. $f(x) = x^4 - 10x^3 + 37x^2 - 110x + 286$; $5 + i$ is a zero of f

34. $f(x) = x^4 - 12x^3 + 43x^2 - 22x - 78$; $5 - i$ is a zero of f

In problems 35 and 36, form a polynomial function with leading coefficient 1 that has the smallest possible degree and real coefficients with the given zeros.

35. $2, 2 - i$; degree 3   **36.** $i, 1 + i$; degree 4

In problems 37 and 38, use Descartes' rule of signs to determine the number of positive zeros, negative zeros, and imaginary zeros.

37. $f(x) = x^3 + 2x^2 - 3x - 11$

38. $h(x) = 3x^4 + 3x^3 - 13x - 6$

In problems 39 and 40, factor $f(x)$ as far as possible:

(a) In the real number system

(b) In the complex number system

39. $f(x) = x^4 - 8x^3 + 16x^2 + 8x - 17$

40. $f(x) = x^4 - 3x^2 + 6x + 8$

In problems 41 and 42, find the domain of each function.

41. $g(x) = \dfrac{3x - 1}{x^2(x^2 - 9)}$ **42.** $h(x) = \dfrac{x^3}{x^3 - 3x}$

In problems 43–46:

(a) Graph the rational function, locate the vertical and horizontal asymptotes, and find the intercepts.

(b) Use the graph to solve the inequality.

43. $f(x) = \dfrac{3}{(x - 1)^2} - 5$;

$$\dfrac{3}{(x - 1)^2} < 5$$

44. $g(x) = \dfrac{-1}{2(x + 1)} + 2$;

$$2 > \dfrac{1}{2(x + 1)}$$

45. $h(x) = \dfrac{4x + 2}{x - 7}$;

$$\dfrac{4x + 2}{x - 7} \ge 0$$

46. **G** $f(x) = \dfrac{3x^2}{x^2 - 6x + 9}$;

$$\dfrac{3x^2}{x^2 - 6x + 9} \le 0$$

47. Baseball Path: A baseball is hit so that the height h (in feet) of the ball t seconds later is given by the model

$$h(t) = -16t^2 + 64t + 5$$

(a) Use this model to find $h(0)$ and $h(1)$. What do these results mean in this situation?

(b) What restriction should be imposed on t in this model?

(c) Graph the path of the baseball and use the graph to describe the change in the height as the time increases.

(d) When does the ball reach its maximum height?

(e) What is the maximum height?

48. Revenue: The revenue R (in dollars) from a grove of oranges is given by the equation

$$R(x) = (800 - x)x$$

where x is the price (in dollars) of each bin of oranges sold.

(a) For what price of a bin of oranges is the revenue $57,600?

(b) What price of a bin of oranges will maximize the revenue?

(c) What is the maximum revenue?

49. Population Growth: A marine biologist discovers that the number N of insects found in a sample of water is given by the model

$$N(T) = -1.5T^2 + 75T + 20$$

where T is the temperature in degrees Celsius.

(a) Calculate $N(0)$, $N(10)$, and $N(25)$. Explain the meaning of each value.

(b) Find the temperatures when the number of insects is 800. Round off the answers to one decimal place.

(c) Sketch the graph of the function, and then use the graph to determine the temperature at which the population will be a maximum.

(d) Find the size of the maximum population. Round off the answer to the nearest integer.

50. ⒢ Population Growth: A biologist finds that the number N of bacteria in a culture t hours after the start of an experiment is given by the mathematical model

$$N(t) = t^3 - 6t^2 + 9t + 100$$

(a) Calculate $N(0)$, $N(10)$, and $N(25)$. Explain the meaning of each value.

(b) Use a grapher to find the time when the number of bacteria is 1000. Round off the answer to one decimal place.

51. ⒢ Ranching: A rancher wants to enclose a rectangular pen with an area of 100 square feet by fencing the entire perimeter. Suppose that the length of the pen is x feet.

(a) What restriction should be placed on x?

(b) Express the perimeter p as a function of x.

(c) If the fencing costs $6.60 per foot, express the total cost C (in dollars) as a function of x.

(d) Use a grapher to sketch the graph of C and to determine the value of x for which the cost is minimum. What is the minimum cost?

52. Chemistry: By adding x milliliters of a 10% acid solution to 10 milliliters of a 30% acid solution, a chemist obtains a mixture with an acid concentration of C%, where C (expressed in decimal form) is given by the function

$$C(x) = \frac{3 + 0.1x}{10 + x}$$

(a) What restriction should be imposed on x in this situation?

(b) Sketch the graph of C.

(c) Discuss the asymptotic behavior of the graph of C and interpret it.

(d) Determine what value of x results in a 25% acid solution. Round off the answer to one decimal place.

53. Break Even Point: Suppose that the daily cost C (in dollars) and the revenue R (in dollars) for a company that manufactures and sells x telephones are given respectively, by the models

$$C(x) = 0.004x^3 - 1.7x^2 + 16{,}000 \text{ and}$$

$$R(x) = 55x, \quad \text{where } 0 \le x \le 250$$

(a) Write the profit function P for this model.

(b) With the aid of the graph of P and the bisection method, find the production level x at which there is a break even point, that is, when the company starts to show profit. Round off the answers to two decimals.

54. ⒢ Consider the rational function

$$f(x) = \frac{2x^2 - 5x}{2x + 3}$$

(a) Sketch the graph of f. Then find all vertical asymptotes.

(b) Find the x and y intercepts of the graph of f.

(c) Use long division to find a polynomial that has the same limit behavior as the rational function f.

✦ CHAPTER 3 TEST

1. Match each function with its graph:
   (a) $f(x) = x^3 - 2x^2$
   (b) $g(x) = -x^5 + 1$
   (c) $h(x) = \dfrac{-2x + 1}{x}$

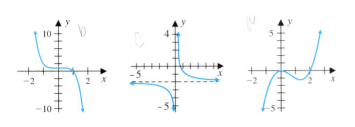

2. Let $f(x) = -2x^2 - 5x + 3$.
   (a) Rewrite $f(x)$ in the form of
   $$f(x) = a(x - h)^2 + k$$
   Use transformations of the graph of $y = x^2$ to sketch the graph.

   (b) Find the vertex and axis of symmetry.
   (c) Determine whether the graph opens upward or downward, and indicate if the vertex is a maximum or a minimum point.
   (d) Find the domain and range of f.
   (e) Solve the inequality
   $$-2x^2 - 5x + 3 \leq 0$$
   graphically. Express the solution in interval notation.

3. Describe a sequence of transformations that will transform the graph of $y = x^4$ to the graph of $f(x) = -2(x - 1)^4 + 1$.
   (a) Sketch the graph of f.
   (b) Use the graph to determine the limit behavior of $x \to -\infty$ and as $x \to +\infty$.

4. Find the zeros of $f(x) = (x^5 - x^4)(x + 1)^2$, and classify the zeros by multiplicity.

5. Let $f(x) = x^5 - 6x^4 + 11x^3 - 2x^2 - 12x + 8$.
   (a) Use the remainder theorem to find $f(-2)$.
   (b) Use the factor theorem to determine whether $x + 1$ is a factor of $f(x)$.

6. Let $f(x) = 9x^3 + 6x^2 - 5x - 2$.
   (a) Use a grapher to sketch the graph of f.
   (b) Use the graph from part (a) to determine the number and types of zeros.
   (c) Use the rational zero theorem along with the graph of f to determine the rational zeros of f.

   (d) Use the result from part (c) to express $f(x)$ in factored form.
   (e) Use the graph to find the intervals for which $9x^3 + 6x^2 - 5x - 2 < 0$.

7. Form a polynomial function of degree 4 with real coefficients that has 2, 3, and $1 + i$ as its zeros and 1 as the leading coefficient.

8. Answer true or false. For a given function
   $$f(x) \to +\infty \text{ as } x \to a.$$
   From this, we conclude that
   (a) $x = a$ is a vertical asymptote.
   (b) $y = a$ is a horizontal asymptote.
   (c) $x = -a$ is a vertical asymptote.

9. Let $f(x) = \dfrac{4x - 5}{x - 2}$.
   (a) Find the asymptotes of the graph of f.
   (b) Find the intercepts of the graph of f.
   (c) Graph f.
   (d) Use the graph to specify the domain and range of f.

10. An experimental rocket is launched from a hill 206 feet high. The height h (in feet) of the rocket from the ground t seconds after it is launched is given by the function
    $$h(t) = -16t^2 + 468t + 206$$
    (a) Calculate $h(0)$, $h(2)$, and $h(4)$. Explain the meaning of each value.
    (b) Graph h with an appropriate restriction on t.
    (c) For what value of t does the object reach its maximum height?
    (d) What is the maximum height of the object?

11. A psychologist estimates that the percent p of the information remembered by a student in an experiment t months after the student learned the material is given by the function
    $$p(t) = \dfrac{100}{1 + t}$$
    (a) What is the restriction on t?
    (b) Sketch the graph of p.
    (c) Use the graph to determine what happens to $p(t)$ as $t \to +\infty$. Explain what this behavior means in terms of the experiment.

Exponential and Logarithmic Functions

CHAPTER
4

Chapter Contents

4.1 Inverse Functions

4.2 Exponential Functions

4.3 Properties of Logarithmic Functions

4.4 Graphing Logarithmic Functions

4.5 Exponential and Logarithmic Equations and Inequalities

On December 26, 2003, there was an earthquake in Bam, Iran that caused considerable damage, and 20,000 people were killed from its effect. Its magnitude on the Richter scale was 6.5. Compare the intensity of this earthquake to the intensity of the Northern Califoria earthquake in 1989, whose magnitude was 7.1. The solution of this problem is shown in Example 10 on page 318.

City of Bam after the 2003 Earthquake in Southern Iran.

Up to now we have studied algebraic functions formed by sums, differences, products, and quotients of polynomial functions. In this chapter, we introduce two types of functions that are closely related—*exponential functions* and the *inverses* of exponential functions—*logarithmic functions*. These functions belong to a classification of functions called *transcendental* functions.

We will examine their properties, graphs, and uses in a variety of applications. As we shall see, the exponential functions are especially suitable for developing mathematical models involving population growth or decline over time.

243

4.1 Inverse Functions

Our goal in this section is to develop the general notion of inverse functions. Later in this chapter we use the idea of inverse functions to introduce logarithmic functions.

Defining Inverse Functions

Let us begin by examining the composition of two functions

$$f(x) = 5x \quad \text{and} \quad g(x) = \frac{x}{5}$$

to get $(f \circ g)(x) = f[g(x)]$

$$= f\left(\frac{x}{5}\right) = 5 \cdot \frac{x}{5}$$

$$= x$$

and $\quad (g \circ f)(x) = g[f(x)]$

$$= g(5x) = \frac{5x}{5} = x$$

$$= x$$

Thus, in this situation, for $f \circ g$, f "undoes" the output of function g, $x/5$, and produces a final output of x. The final output is the same as the original input (Figure 1a).

Similarly, for $g \circ f$, g "undoes" the output of function f and returns x as the final output (Figure 1b).

Figure 1

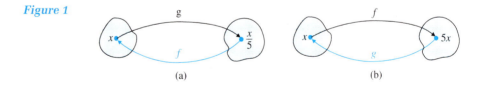

(a)   (b)

Two functions related in such a way are said to be *inverses* of each other.

Definition

Inverse Functions

> Two functions f and g are **inverses** of each other if and only if
>
> $$(f \circ g)(x) = f[g(x)] = x$$
>
> for every value of x in the domain of g, and
>
> $$(g \circ f)(x) = g[f(x)] = x$$
>
> for every value of x in the domain of f.

A function f is said to be **invertible** if such a function g exists.

EXAMPLE 1

Not every function has an inverse.

Verifying that Two Functions Are Inverses of Each Other

Show that the functions $f(x) = 3x + 2$ and $g(x) = (1/3)x - 2/3$ are inverses of each other.

Solution

From the definition, we must verify that $f[g(x)] = x$ and $g[f(x)] = x$. We have

$$f[g(x)] = f\left(\frac{1}{3}x - \frac{2}{3}\right)$$

$$= 3\left(\frac{1}{3}x - \frac{2}{3}\right) + 2 = (x - 2) + 2 = x$$

$$g[f(x)] = g(3x + 2)$$

$$= \frac{1}{3}(3x + 2) - \frac{2}{3} = \left(x + \frac{2}{3}\right) - \frac{2}{3} = x$$

for all *x*. Thus, we conclude that *f* and *g* are indeed inverses of each other.

When *f* and *g* are inverses of each other, we refer to *g* as the *inverse function of f*, and vice versa, and we write

$$g = f^{-1} \quad \text{or} \quad g^{-1} = f.$$

Care must be taken not to confuse y = f^{-1}(x), the inverse function of f, with [f (x)]$^{-1}$ = 1/f (x), the reciprocal of f (x).

Thus, in Example 1, we could write either

$$f(x) = 3x + 2 \quad \text{and} \quad f^{-1}(x) = \frac{1}{3}x - \frac{2}{3}$$

or

$$g(x) = \frac{1}{3}x - \frac{2}{3} \quad \text{and} \quad g^{-1}(x) = 3x + 2.$$

Notice that the conditions stated in the definition of an inverse function can be re-stated as follows:

$$f^{-1}[f(x)] = x \quad \text{for every } x \text{ in the domain of } f$$

$$\text{and } f[f^{-1}(x)] = x \quad \text{for every } x \text{ in the domain of } f^{-1}$$

In what follows, we shall address the following three issues related to inverse functions:

1. How are the graphs of a function and its inverse related?
2. How can we determine whether a function has an inverse?
3. How do we find the inverse of a function if it exists?

Recognizing the Symmetry of the Graphs of a Function and Its Inverse

To understand the relationship between the graphs of a function and its inverse, let us consider the two functions in Example 1

$$f(x) = 3x + 2$$

and

$$f^{-1}(x) = \frac{1}{3}x - \frac{2}{3}.$$

Tables 1a and 1b are used to graph f and f^{-1} on the same coordinate system (Figure 2). The graph of the line $y = x$ is also shown in Figure 2.

TABLE 1

f:

x	y
0	2
1	5
−1	−1
−2	−4

(a)

f^{-1}:

x	y
2	0
5	1
−1	−1
−4	−2

(b)

Figure 2

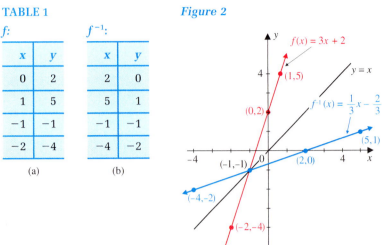

Figure 3

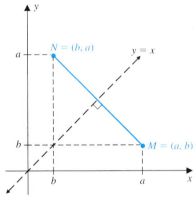

Observe from Table 1 the point $(1, 5)$ on the graph of f, and the point $(5, 1)$ on the graph of f^{-1}. Now notice the interchanging of x and y on the graphs of f and f^{-1}. That is, $f(1) = 5$ and $f^{-1}(5) = 1$. In other words, we get the point $(5, 1)$ from the point $(1, 5)$ by reflecting across the line $y = x$ (Figure 2). In general, as Figure 3 illustrates, the points $M = (a, b)$ and $N = (b, a)$ are symmetric across the line $y = x$.

This observation leads to the following result:

Property

Graphs of f and f^{-1}

> The graphs of f and f^{-1} are reflections of each other about the line $y = x$, as illustrated in Figure 4.

Figure 4

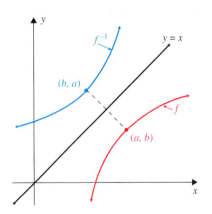

EXAMPLE 2 **Graphing f^{-1} From the Graph of f**

Use the graph of f given in Figure 5a to sketch the graph of f^{-1}. Identify the domain and range of f and f^{-1}.

Solution Because the graphs of f and f^{-1} are symmetric about the line $y = x$, the graph of f^{-1} is obtained by reflecting the graph of f about the line $y = x$ (Figure 5b).

Figure 5

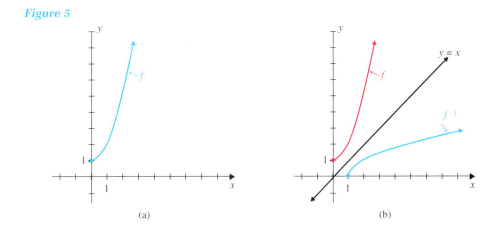

(a)   (b)

From the graphs in Figure 5, we observe that the domain of f is the interval $[0, \infty)$ and its range is the interval $[1, \infty)$. Also, we see that the domain of f^{-1} is the interval $[1, \infty)$ and its range is the interval $[0, \infty)$. In other words, the domain and range of f are, respectively, the range and domain of f^{-1}.

This example illustrates the following property:

Property **Domain and Range** **of f and f^{-1}**	The domain of f is the same as the range of f^{-1}, and the range of f is the same as the domain of f^{-1}.

Determining Whether a Function Has an Inverse

There are two ways of determining whether a given function has an inverse:

(1) graphically and (2) analytically.

1. Determining Graphically Whether a Function Has an Inverse

We can use the graph of a function to determine whether it has an inverse. Consider the graph of the function f in Figure 6. The reflected image of the graph of f about the line $y = x$ is not the graph of a function because we can draw a vertical line ℓ that intersects the graph more than once. Thus, f cannot have an inverse. Notice that the horizontal line L, obtained by reflecting ℓ across the line $y = x$, intersects the graph of f more than once.

Figure 6

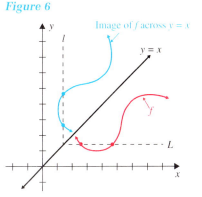

This observation provides the basis for using the graph of a function to determine whether it has an inverse.

Property **Horizontal-Line Test**	A function f has an inverse if and only if no horizontal straight line intersects its graph more than once.

EXAMPLE 3

Determining Graphically Whether a Function Has an Inverse

Use the horizontal-line test to determine whether each function has an inverse.

(a) $f(x) = \sqrt{x}$ (b) $g(x) = x^2$

Solution

The graphs of f and g are shown in Figures 7a and 7b, respectively.

(a) No horizontal line intersects the graph of f more than once (Figure 7a), so f has an inverse.

(b) Any horizontal line drawn above the x axis will intersect the graph of g twice (Figure 7b), so g does not have an inverse.

Figure 7

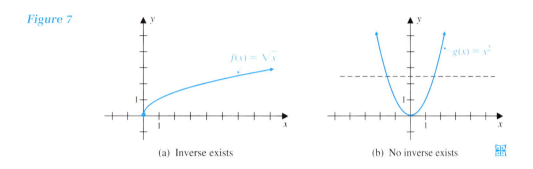

(a) Inverse exists   (b) No inverse exists

2. Determining Analytically Whether a Function Has an Inverse

Let us review two characteristics of a function $y = f(x)$ that has an inverse:

(i) The graph of the function satisfies the vertical-line test. Thus for every input value x there is one and only one output value for y.

(ii) Because the function has an inverse, the graph must also satisfy the horizontal-line test. This means that every output value y has one and only one corresponding input value for x.

It follows that a function has an inverse if and only if each distinct input value always results in a distinct output value. A function that has this property is called a *one-to-one function.*

Definition
One-to-One Function

> A function f is said to be **one-to-one** if, whenever a and b are in the domain of f and $f(a) = f(b)$, it follows that $a = b$.

EXAMPLE 4

Determining One-to-One Functions

Use the definition of a one-to-one function to determine whether each function is one-to-one. Graph each function and use the graph to demonstrate the validity of the result.

(a) $f(x) = 3x - 2$

(b) $g(x) = 4x^2$

Solution

(a) We assume that $f(a) = f(b)$ for some numbers a and b in the domain of f. According to the definition, we must show that $a = b$. We proceed as follows:

$$f(a) = f(b) \qquad \text{Assumption}$$
$$3a - 2 = 3b - 2 \qquad \text{Evaluate } f(a) \text{ and } f(b)$$
$$3a = 3b \qquad \text{Add 2 to each side}$$
$$a = b \qquad \text{Divide each side by 3}$$

Hence, f is one-to-one. So f has an inverse. The horizontal-line test confirms that f is one-to-one (Figure 8a).

(b) Here we select $x = -1$ and $x = 1$, and observe that

$$g(-1) = 4(-1)^2 = 4 \quad \text{and} \quad g(1) = 4 \cdot 1^2 = 4.$$

Hence, $g(-1) = g(1)$ but $-1 \neq 1$, so g is not one-to-one. Thus g does not have an inverse. The horizontal-line test confirms that g is not one-to-one (Figure 8b).

Figure 8

The selection of specific numbers to show that a general property does not hold, as in Example 4b, illustrates a method of proof called proof by counterexample.

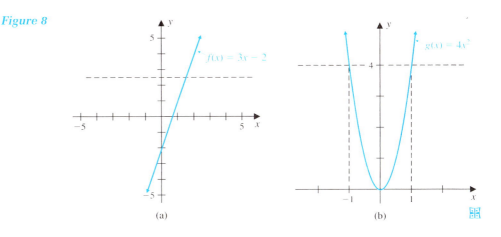

(a)

(b)

Note that functions that are either increasing or decreasing throughout their domains are one-to-one functions, and so they have inverses.

Finding the Inverse of a Function

TABLE 2

Function f

Domain	Range
3	0
1	−4
5	2
7	11
23	−6

(a)

Function f^{-1}

Domain	Range
0	3
−4	1
2	5
11	7
−6	23

(b)

We now turn our attention to answering the question:

If a function f has an inverse, how do we find the inverse function f^{-1}?

Earlier we discovered that whenever (a, b) is on the graph of f, then (b, a) is on the graph of f^{-1}. This means that f^{-1} is the function that can be formed by interchanging the roles of the domain and range values of f.

For instance, consider the function f defined by Table 2a. Since f is one-to-one, it has an inverse f^{-1}, which is formed by reversing the pairings shown in Table 2a to obtain the function f^{-1} defined by Table 2b.

If f is a function defined by the equation $y = f(x)$ and f has an inverse f^{-1}, then we have:

$$y = f(x) \qquad \text{y is expressed in terms of x}$$

$$f^{-1}(y) = f^{-1}[f(x)] \qquad \text{Apply f^{-1} to each side}$$

$$f^{-1}(y) = x \qquad \text{x is expressed in terms of y}$$

This observation provides the basis for the following procedure, which enables us to find the equation for f^{-1} if f is defined by an equation.

Procedure

Method for Finding f^{-1}

Step 1. Write the equation $y = f(x)$ that defines the one-to-one function f.

Step 2. Solve the equation $y = f(x)$ for x in terms of y, to obtain an equation for the inverse function $x = f^{-1}(y)$.

Step 3. Verify that the domain and range of f are, respectively, the same as the range and domain of f^{-1}.

EXAMPLE 5   **Finding the Inverse of a Function**

Let $f(x) = 2x - 3$.

(a) Use the above method to find $f^{-1}(x)$.

(b) Graph f and f^{-1} on the same coordinate system.

Solution   The function

$$f(x) = 2x - 3$$

is a linear function. Since the slope is $m = 2$, the graph of f is increasing for all real numbers, so f is one-to-one and thus f^{-1} exists.

(a) To find f^{-1}, we proceed as follows:

Step 1. Let $y = f(x) = 2x - 3$.

Step 2. Solve the equation for x in terms of y as follows:

$$y = 2x - 3$$

$$x = \frac{y + 3}{2}$$

So

$$f^{-1}(y) = \frac{y + 3}{2}.$$

Following the convention of using x to represent the independent variable, we rewrite the latter equation as

$$f^{-1}(x) = \frac{x + 3}{2}.$$

Step 3. The domain and range of f consist of all real numbers, which is the same for f^{-1}.

(b) The graphs of f and f^{-1} are shown in Figure 9. As expected, they are symmetric about the line $y = x$.

Figure 9

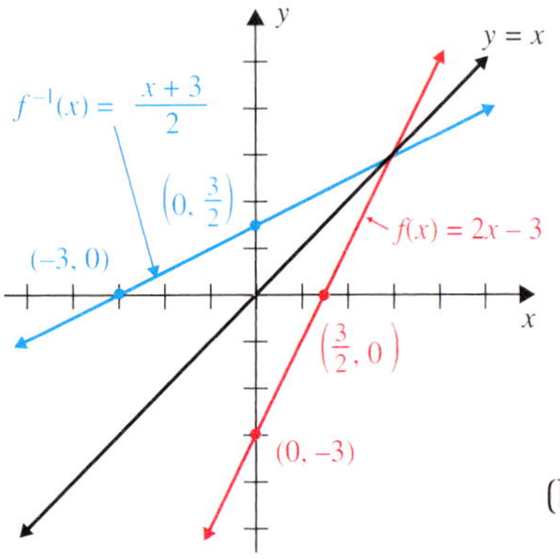

Even though two functions have the same defining equation, one may have an inverse and the other may not. The next example illustrates the situation.

EXAMPLE 6   **Comparing Two Functions with the Same Defining Equation**

Let
$$f(x) = x^2 + 1$$
and
$$g(x) = x^2 + 1, \text{ where } x \geq 0.$$

(a) Use the graphs of f and g to explain why f does not have an inverse and g does have an inverse.

(b) Find $g^{-1}(x)$.

Solution   (a) By inspecting the graph of f in Figure 10a, we see that it does not satisfy the horizontal-line test, so f does not have an inverse. On the other hand, the graph of g (Figure 10b) does satisfy the horizontal-line test, so g has an inverse function. Notice that both functions have the same defining equation, but g has a restricted domain that allows it to have an inverse.

Figure 10

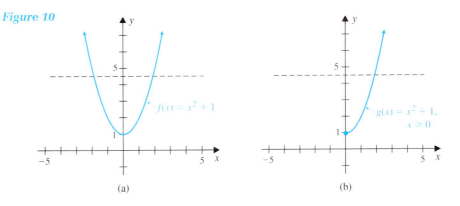

(a)   (b)

(b) We obtain an equation for the inverse function g^{-1} as follows:

Step 1. Let $y = x^2 + 1$, where $x \geq 0$
Step 2. Solve for x:
$$x^2 = y - 1$$
$$x = \sqrt{y - 1} \quad x \geq 0 \text{ is a given restriction}$$
Thus $g^{-1}(y) = \sqrt{y - 1}$

By changing the variable we get $g^{-1}(x) = \sqrt{x - 1}$.
Step 3. The domain of g and the range of g^{-1} are the same, the interval $[0, \infty)$. Also, the range of g is the same as the domain of g^{-1}, the interval $[1, \infty)$.

It should be noticed that at times we can find the inverse of the function
$$g(x) = x^2 + 1, x \geq 0$$
by inspecting the formation of its equation.

The function $g(x) = x^2 + 1$ is a composition of squaring x; and adding 1. The inverse function formed by "reversing" the composition of g by subtracting 1, then taking the square root to obtain
$$g^{-1}(x) = \sqrt{x - 1}.$$

Solving Applied Problems

The next example shows an application of inverse functions.

EXAMPLE 7   **Modeling an Insurance Premium Discount**

Insurance companies reward students who have good grades and good driving records by discounting their automobile insurance premiums. Suppose that a student who qualifies is awarded a 20% discount in premiums.

(a) Find a function f that expresses the reduced premium R in terms of the original premium P.

(b) Find the inverse function f^{-1} of part (a) and interpret it in terms of the reduced and original premiums.

Solution   (a) A 20% reduction in the original premium means the new premium is 80% of the original one. Therefore, the function f that expresses the reduced premium R in terms of the original premium P is given by

$$R = f(P) = 0.80P.$$

(b) The inverse function f^{-1} of $f(P) = 0.80P$ is obtained by solving the equation $R = 0.80P$ for P in terms of R to get

$$P = \frac{1}{0.80}R = 1.25R.$$

So

$$P = f^{-1}(R) = 1.25R.$$

This means that a 25% increase in the reduced premium is needed to obtain the original premium.

Figure 11

$$h(x) = \frac{\sqrt{x^4 + 1}}{x + 1}$$

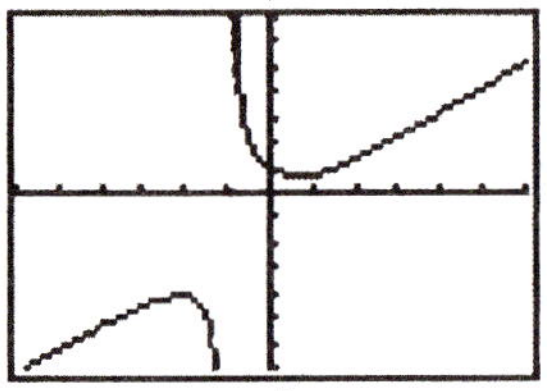

G Graphs provide us with an efficient way of determining whether a function has an inverse. If the function is difficult to graph by hand, then a grapher can help. For instance, the graph of $h(x)\dfrac{\sqrt{x^4 + 1}}{x + 1}$ (Figure 11) shows that the horizontal-line test fails. Therefore, h does not have an inverse.

◆ PROBLEM SET 4.1

Mastering the Concepts

In problems 1–6, show that the functions f and g are inverses of each other by verifying that $f[g(x)] = x$ and $g[f(x)] = x$. Sketch the graphs of f and g on the same coordinate system to show that their graphs are symmetric about the line $y = x$.

1. $f(x) = 7x - 2$;

    $g(x) = \dfrac{x}{7} + \dfrac{2}{7}$

2. $f(x) = 1 - 5x$;

    $g(x) = \dfrac{1}{5} - \dfrac{x}{5}$

3. $f(x) = x^4$,

    where $x \geq 0$;

    $g(x) = \sqrt[4]{x}$

4. $f(x) = x^3$;

    $g(x) = \sqrt[3]{x}$

5. $f(x) = \dfrac{1}{x - 2}$;

    $g(x) = \dfrac{1}{x} + 2$

6. $f(x) = \sqrt{x + 3}$;

    $g(x) = x^2 - 3$,

    where $x \geq 0$

In problems 7 and 8, use the given graph of the one-to-one function f to do the following:

(i) Sketch the graph of f^{-1} by reflecting the graph of f about the line $y = x$.

(ii) Determine and compare the domains and ranges of f and f^{-1}.

7. (a)   **(b)**

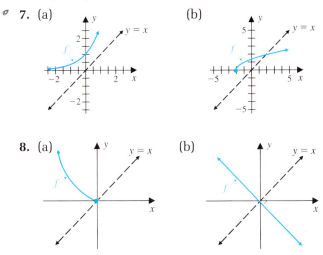

8. (a)   **(b)**

In problems 9 and 10, use the horizontal-line test to determine whether each of the functions whose graph is given has an inverse.

9. (a)   **(b)**

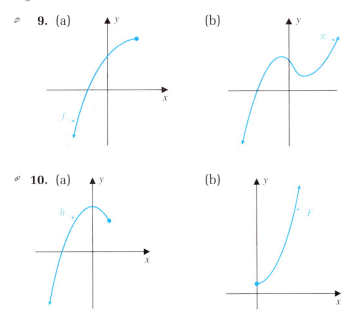

10. (a)   **(b)**

In problems 11–16, use the definition of a one-to-one function to determine whether f is one-to-one. Graph f to demonstrate the validity of the result.

11. $f(x) = -2x + 7$

12. $f(x) = -\dfrac{2}{3}x + 5$

13. $f(x) = 2x^2 + 1$

14. $f(x) = -|x|$

15. $f(x) = \sqrt{4 - x^2}$

16. $f(x) = \sqrt{5 + 3x}$

17. Write a table of values for f^{-1}, where f is given in Table 3. What are the domain and range of f? What are the domain and range of f^{-1}?

TABLE 3

x	1	2	3	4	5	6	7	8
$f(x)$	2	-3	3	5	7	100	-4	1

18. Determine whether each function has an inverse.

(a) $f(x)$ is the volume (in liters) of x kilograms of water.

(b) $f(n)$ is the number of students in a precalculus class whose birthday is on the nth day of the year.

(c) $f(t)$ is the number of customers in a department store at t minutes past 6:00 PM.

(d) $f(w)$ is the cost (in dollars) of express mailing a package that weighs w ounces.

In problems 19–32, sketch the graph of each function and use the graph to determine whether f^{-1} exists. If f^{-1} does not exist, explain why. If f^{-1} exists, use the procedure on page 290 to find $f^{-1}(x)$. Sketch the graphs of

$$f \text{ and } f^{-1}$$

on the same coordinate system.

19. $f(x) = 7x + 5$ **20.** $f(x) = 1 - 3x$

21. $f(x) = \dfrac{3}{x - 1}$ **22.** $f(x) = \dfrac{2}{x + 3}$

23. $f(x) = \sqrt{x - 3}$ **24.** $f(x) = \sqrt{3 - 2x}$

25. $f(x) = x^3 - 8$ **26.** $f(x) = x^3 + 1$

27. $f(x) = |x + 4|$ **28.** $f(x) = -2x^2 + 5$

29. $f(x) = 2 - \sqrt[3]{x}$ **30.** $f(x) = \sqrt[3]{x + 1} - 2$

31. $f(x) = -x^2,$ **32.** $f(x) = \sqrt{1 - x^2},$
$\quad\quad x \geq 0$ $\quad\quad 0 \leq x \leq 1$

In problems 33–36, the functions f and g have the same defining equation.

(a) Use the graphs of f and g to explain why f does not have an inverse but g does.

(b) Find $g^{-1}(x)$.

33. $f(x) = |x|;$
$\quad g(x) = |x|, x \leq 0$

34. $f(x) = -x^2 + 4;$
$\quad g(x) = -x^2 + 4, x \geq 0$

35. $f(x) = \sqrt{1 - x^2};$
$\quad g(x) = \sqrt{1 - x^2}, 0 \leq x \leq 1$

36. $f(x) = \sqrt{4 - x^2}$;

   $g(x) = \sqrt{4 - x^2}, -2 \le x \le 0$

37. Figure 12 shows the graph of a function f. Use symmetry and transformations to sketch the graph of each function.

   (a) $y = f^{-1}(x)$

   (b) $y = f^{-1}(x) + 1$

   (c) $y = f^{-1}(x - 1)$

   (d) $y = f^{-1}(-x)$

Figure 12

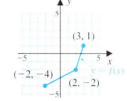

38. Figure 13 shows the graph of a function f. Use symmetry and transformations to sketch the graph of each function.

   (a) $y = f^{-1}(x)$

   (b) $y = f^{-1}(x) - 2$

   (c) $y = f^{-1}(x - 2)$

   (d) $y = -f^{-1}(x)$

Figure 13

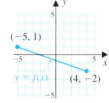

G In problems 39–42, use a grapher to sketch the graph of each function. Use the horizontal-line test to determine whether the function has an inverse.

39. $f(x) = \dfrac{x + 2}{\sqrt[3]{x} + 1}$

40. $f(x) = \dfrac{3x}{1 - \sqrt{x}}$

41. $f(x) = \dfrac{x^3 - |x|}{x}$

42. $f(x) = \dfrac{x^3 - 4x}{x^3 - 4}$

Applying the Concepts

43. Wages: To avoid layoffs, workers at a certain company accept a 15% reduction in wages.

   (a) Find a function f that expresses the reduced wages W in terms of the original wages P.

   (b) Find the inverse function of part (a) and interpret it in terms of the reduced and original wages.

44. Currency Exchange: At one time, $100 U.S. was worth 116 Euro:

   (a) Write a function f that expresses the number of dollars $f(c)$ in terms of the number of Euro c.

   (b) Write a function g that expresses the number of Euro $g(d)$ in terms of the number of U.S. dollars d.

   (c) What is the relationship between f and g? Explain.

45. Physics—Temperature Conversion: The function that converts degrees Celsius (C) to degrees Fahrenheit (F) is defined by the equation

$$F = f(C) = \frac{9}{5}C + 32.$$

   (a) Find f^{-1} and interpret the results in terms of Fahrenheit and Celsius readings.

   (b) Graph f and f^{-1} on the same coordinate system.

   (c) Find the Celsius measurements corresponding to 85°F and 113°F.

46. Business—Cost Function: A lumber yard will deliver wood for $6 per board foot plus a fixed delivery charge of $30.

   (a) Write the equation that expresses the cost function C (in dollars) in terms of x board feet of lumber delivered; that is, find $C = f(x)$.

   (b) Find f^{-1} and interpret the result in terms of cost and board feet.

   (c) Graph f and f^{-1} on the same coordinate system.

   (d) Find the number of board feet delivered if the cost is $270.

47. Sales Commission: A salesperson makes a commission of 40% of total sales plus a salary of $25,000 per year.

   (a) Determine the function that describes the total annual income y (in dollars) in terms of the total sales x (in dollars).

   (b) Find the inverse of this function and graph it.

   (c) Interpret the inverse function in terms of sales and income.

48. Dress Sizes: A size 4 dress in the United States corresponds to a European size 32 dress, and a size 12 dress in the United States corresponds to a European size 48. Assume that the relationship between the sizes in the United States and Europe is linear.

   (a) Express the European dress size D as a function f of the dress size x in the United States.

   (b) Use the result in part (a) to find the European dress sizes that correspond to sizes 6, 8, and 10 in the United States.

   (c) Explain why f has an inverse. Find f^{-1}, graph it, and interpret it in terms of dress sizes.

(d) Use the inverse function to find dress sizes in the United States to correspond to sizes 40, 52, and 60 in Europe.

Developing and Extending the Concepts

49. Use the horizontal-line test to determine whether each function is one-to-one. If the function is one-to-one, find its inverse. Determine the domains and ranges of f and f^{-1}.

(a) $f(x) = \begin{cases} \dfrac{1}{2}x - 4 & \text{if } x < 0 \\ x - 4 & \text{if } x \geq 0 \end{cases}$

(b) $f(x) = \begin{cases} x^3 & \text{if } x < 0 \\ -\sqrt{x} & \text{if } x \geq 0 \end{cases}$

50. Graph

$$f(x) = \frac{x + 1}{x}$$

and

$$g(x) = \frac{1}{1 + x}.$$

Do these functions appear to be inverses of each other? Confirm your conclusion algebraically.

51. Explain why $(f^{-1})^{-1} = f$.

52. Give examples of functions f and g to confirm that the inverse of the quotient function $(f/g)^{-1}$ does not necessarily equal the quotient function g/f. That is, give examples of functions f and g to show that $(f/g)^{-1} \neq g/f$.

53. Suppose $f(x) = 2x + 4$ and $g(x) = 3x - 1$.
(a) Determine f^{-1}, g^{-1}, and $(f \circ g)^{-1}$.
(b) Show that $(f \circ g)^{-1} \neq f^{-1} \circ g^{-1}$.
(c) Show that $(f \circ g)^{-1} = g^{-1} \circ f^{-1}$.
(d) Generalize your results to express $(g \circ f)^{-1}$ in terms of g^{-1} and f^{-1} for any two invertible functions f and g. Give examples to support your assertion.

54. Graph each function and then use symmetry with respect to the line $y = x$ to graph f^{-1} without first finding the equation for f^{-1}. Then verify the result by finding the equation for f^{-1} and graphing it.
(a) $f(x) = 3x + 2$
(b) $f(x) = 2x^3 - 1$
(c) $f(x) = 1 - 7x$
(d) $f(x) = \sqrt[3]{x} + 1$

55. **G** Match each function f with its inverse g. Then use a grapher to graph the function and its inverse on the same coordinate system.

(a) $f(x) = 5x^3 + 10$ (A) $g(x) = \dfrac{\sqrt[3]{x} - 10}{5}$

(b) $f(x) = (5x + 10)^3$ (B) $g(x) = \sqrt[3]{\dfrac{x - 10}{5}}$

(c) $f(x) = 5(x + 10)^3$ (C) $g(x) = \sqrt[3]{\dfrac{x}{5}} - 10$

(d) $f(x) = (5x)^3 + 10$ (D) $g(x) = \dfrac{\sqrt[3]{x - 10}}{5}$

Objectives

1. Define Exponential Functions
2. Graph Exponential Functions
3. Solve Applied Problems

4.2 Exponential Functions

As we will see, exponential functions are especially useful in modeling the growth or decline of such varied phenomena as population, radioactivity, and investment yields.

Defining Exponential Functions

In Section R.6, we defined b^x, where b is a positive constant, for all rational values x. For instance,

$$3^2 = 9, \quad 4^{3/2} = (\sqrt{4})^3 = 2^3 = 8, \quad \text{and} \quad 7^{1/3} = \sqrt[3]{7}.$$

It is possible to extend the notion of exponents to include all irrational numbers. However, a thorough explanation depends on concepts studied in calculus. For now, it is enough to know that if r is a rational number approximately equal to an irrational number x, then b^r is approximately equal to b^x. The better that r approximates x, the better that b^r approximates b^x. For instance, Table 1 shows a pattern of improving approximations of 4^π rounded to four decimal

TABLE 1 Approximations of 4^π

Approximation of π		Approximation of 4^π	
Improved accuracy	3.14	77.7085	Improved accuracy
	3.142	77.9242	
	3.1416	77.8810	
	3.14159	77.8799	
	⋮	⋮	

places. As the value of the exponent gets closer and closer to π, the value of 4^π gets closer and closer to a real number whose approximate decimal value is 77.8802.

With this background we are ready to define exponential functions.

Definition **Exponential Function—Base b**	A function of the form $$f(x) = b^x \quad \text{where } b > 0 \quad \text{and} \quad b \neq 1$$ is called an **exponential function of base b.** The independent variable x, which represents any real number is called the **power** or **exponent.**

In the definition, b $\neq$ 1 *because*

$$f(x) = 1^x = 1,$$

which is considered to be a constant function rather than an exponential function.

Thus

$$f(x) = 3^x \text{ is an exponential function with base 3}$$

and

$$g(x) = 5^{-x} = \left(\frac{1}{5}\right)^x \text{ is an exponential function with base } \frac{1}{5}.$$

Among the exponential functions, there is one that is encountered frequently in mathematics. This function is referred to as the *natural exponential function.*

Its base is denoted by the symbol e in honor of the Swiss mathematician Leonard Euler (1707–1783). The value of e can be approximated by evaluating the expression

$$\left(1 + \frac{1}{x}\right)^x$$

for larger and larger values of x.

Table 2 shows some of the calculations for specific values of x.

TABLE 2

x	1	10	100	1,000	10,000	100,000	1,000,000
$\left(1 + \dfrac{1}{x}\right)^x$	2	2.59374	2.70481	2.71692	2.71815	2.71827	2.71828

This leads us to the following definition:

Definition **Natural Exponential Function**	Let x be a real number, the function f defined by $$f(x) = e^x$$ is called the **natural exponential function** with base e.

This function is used extensively in formulas that model real-world phenomena. To 12 decimal places.

$$e = 2.718281828459$$

EXAMPLE 1   **Evaluating Exponential Functions**

Given the functions

$$f(x) = 3^x, \quad g(x) = e^x, \quad h(x) = 2^{-x}, \quad \text{and} \quad k(x) = e^{-x}$$

find each of the following values. Round off the answers to two decimal places.

(a) $f(\sqrt{2})$ (b) $g\left(\dfrac{3}{2}\right)$ (c) $h(2.7)$ (d) $k(\pi)$

Solution   (a) $f(\sqrt{2}) = 3^{\sqrt{2}} = 4.73$ (approx.)   (b) $g\left(\dfrac{3}{2}\right) = e^{\frac{3}{2}} = 4.48$ (approx.)

(c) $h(2.7) = 2^{-2.7} = 0.15$ (approx.)   (d) $k(\pi) = e^{-\pi} = 0.04$ (approx.)

Graphing Exponential Functions

To examine the common features of exponential functions, we use the point-plotting method to graph two such functions—one with a base greater than 1 and another with a base less then 1.

EXAMPLE 2   **Graphing Exponential Functions**

Sketch the graph of each exponential function. Determine the domain, range, and any asymptotes.

(a) $f(x) = 2^x$ (b) $g(x) = \left(\dfrac{1}{2}\right)^x$

Solution   (a) Table 3 lists some ordered pairs of numbers that satisfy $f(x) = 2^x$. After plotting these points and then connecting them with a smooth continuous curve, we obtain the graph (Figure 1a). Since 2^x is defined for any real number x, the domain of f consists of all real numbers $\mathbb{R}$. From the graph, we see that the range consists of all positive real numbers. Because of the limit behavior

Since $2^x = 0$ has no solution, there is no x intercept; that is, the graph does not intersect the x axis.

$$f(x) \to 0 \quad \text{as} \quad x \to -\infty$$

the x axis is a horizontal asymptote.

TABLE 3

x	$f(x) = 2^x$	$g(x) = \left(\dfrac{1}{2}\right)^x$
-3	$1/8$	8
-2	$1/4$	4
-1	$1/2$	2
0	1	1
1	2	$1/2$
2	4	$1/4$
3	8	$1/8$

Figure 1

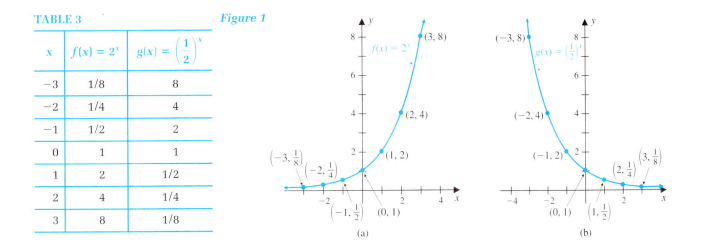

(a)   (b)

(b) Table 3 lists some ordered pairs of numbers that satisfy the equation
$g(x) = (1/2)^x$. After plotting these points and then connecting them with a
smooth continuous curve, we obtain the graph (Figure 1b). From the graph,
we see that the domain consists of all real numbers $\mathbb{R}$ and the range consists
of all positive real numbers. Since

$$g(x) \to 0 \quad \text{as} \quad x \to +\infty,$$

the x axis is a horizontal asymptote.

As illustrated in Figure 2a, graphs of functions of the form $f(x) = b^x$, where
$b > 1$, resemble the graph of $y = 2^x$.

Similarly, if $0 < b < 1$, graphs of functions of the form $f(x) = b^x$ have the same
general shape and characteristics as the graph of $y = (1/2)^x$ (Figure 2b). Notice the
similarities between the graphs in Figure 2a and in Figure 2b.

Note that $\left(\dfrac{1}{e}\right)^x = e^{-x}$

Figure 2

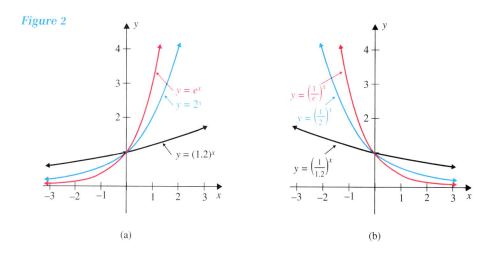

(a)   (b)

In each case the graphs contain the point $(0, 1)$. The functions are increasing
in Figure 2a and decreasing in Figure 2b. Also, all functions have the x axis as a
horizontal asymptote.

These observations lead us to the general properties given in Table 4.

TABLE 4   General Graphs and Properties of $f(x) = b^x$, $b > 0$, $b \neq 1$

Properties	General Graphs
1. The domain includes all real numbers.	
2. The range is the interval $(0, \infty)$.	
3. The function increases everywhere if $b > 1$ and decreases everywhere if $0 < b < 1$.	
4. The x axis is a horizontal asymptote.	
5. The function is one-to-one. That is, $b^u = b^v$ if and only if $u = v$.	
6. The graph contains the point $(0, 1)$.	

Known graphs of exponential functions, along with transformations, can be used
to obtain graphs of other exponential functions.

EXAMPLE 3   **Using Transformations to Graph**

(a) Describe a sequence of transformations that will transform the graph of

$$y = \left(\frac{1}{3}\right)^x \text{ into the graph of } G(x) = -\left(\frac{1}{3}\right)^x + 1.$$

Sketch the graph of G. Determine the domain and range of G, and identify any horizontal asymptote.

(b) Sketch the graph of

$$f(x) = -e^{x+3}$$

by using a sequence of transformations of the graph of $y = e^x$ and find the horizontal asymptote.

Solution   (a) As displayed in Figure 3a, the graph of G can be obtained by performing the following sequence of transformations:

(i) Reflect the graph of $y = (1/3)^x$ across the x axis to get the graph of

$$g(x) = -\left(\frac{1}{3}\right)^x.$$

(ii) Vertically shift the graph of g up 1 unit to get the graph of G.

From the graph, we see that the domain of G consists of all real numbers $\mathbb{R}$. By reading the graph of G, we find that its range is the interval $(-\infty, 1)$. Because of the 1 unit vertical shift upward, the horizontal asymptote of the graph of g (the x axis) moves 1 unit above the x axis to become the line $y = 1$, which is the horizontal asymptote of the graph of G, that is,

$$G(x) \to 1 \text{ as } x \to +\infty.$$

(b) To graph $f(x) = -e^{x+3}$, we first reflect the graph of $y = e^x$ about the x axis to get the graph of $h(x) = -e^x$. Then the graph of $f(x) = -e^{x+3} = h(x + 3)$ is obtained by shifting the graph of h to the left 3 units (Figure 3b). The horizontal asymptote of $y = e^x$ is not affected by these transformations; it is the x axis for f as well.

Figure 3

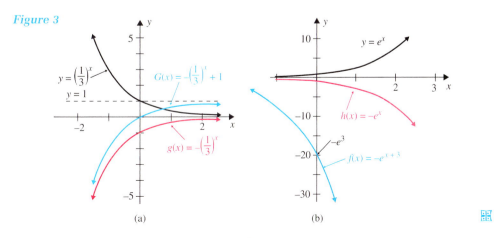

(a)   (b)

Property 5 in Table 4 indicates that $f(x) = b^x$, where $b > 0$ and $b \neq 1$, so that

$$b^u = b^v, \text{ if and only if } u = v$$

This property is used to solve certain types of exponential equations, as the next example illustrates.

EXAMPLE 4   **Using the One-to-One Property**

Solve each of the given equations by using the one-to-one property.

(a) $3^{2x+1} = 27$ (b) $4^{x-0.5} = 8$

Solution   (a) First we express each side of the equation $3^{2x+1} = 27$ in terms of base 3 to get

$$3^{2x+1} = 3^3.$$

Using the one-to-one property, we equate the exponents to get

$$2x + 1 = 3$$

so that

$$x = 1.$$

(b) In this situation, we look for a common base by proceeding as follows:

$$4^{x-0.5} = 8 \qquad \text{Given}$$

$$\left.\begin{array}{l} (2^2)^{x-0.5} = 2^3 \\ 2^{2x-1} = 2^3 \end{array}\right\} \qquad \text{Express each side in base 2}$$

$$2x - 1 = 3 \qquad \text{Equate exponents}$$

$$x = 2 \qquad \text{Solve}$$

Solving Applied Problems

Exponential functions are used to model real-world phenomena in a variety of applications, such as *compound interest, population growth and decay,* and *radio-isotope dating.*

Modeling Compound Interest

Suppose $3,000 is deposited in a saving account that pays a 3.6% annual interest rate, but that the interest is compounded or credited each quarter of the year.

The **interest rate per period** is the annual rate 3.6% = 0.036 divided by the number of compounding periods per year, 4. That is, the interest rate per period for this situation is given by $\frac{0.036}{4} = 0.009$.

If this money is left in the account for 3 years, then there are a total of 3(4) = 12 quarters, and the accumulated amount due at the end of 3 years is given by

$$S = 3000(1 + 0.009)^{12} = 3340.53 \text{ (approx.)}.$$

Thus an investment of $3000 that pays 3.6% annual interest compounded quarterly accumulates $3340.53 (including the original $3000) over 3 years.

This illustration leads to the following formula:

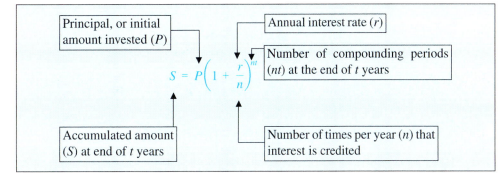

Compound Interest Formula
Discrete Case

The accumulated amount S includes the principal P.

This formula is applicable when the interest is compounded at *discrete* intervals of time such as yearly, quarterly, monthly, weekly, daily, and so forth.

Now let us examine what happens to the formula

$$S = P\left(1 + \frac{r}{n}\right)^{nt}$$

if we allow the number of annual compounding periods n to increase without bound.

First we rewrite the formula as follows:

$$S = P\left[1 + \frac{1}{\left(\frac{n}{r}\right)}\right]^{\left(\frac{n}{r}\right)rt}.$$

Next, we replace $\dfrac{n}{r}$ with x to get

$$S = P\left[\left(1 + \frac{1}{x}\right)\right]^{xrt} = P\left[\left(1 + \frac{1}{x}\right)^{x}\right]^{rt}.$$

By allowing $n \to +\infty$, correspondingly, $\dfrac{n}{r} = x \to +\infty$, so that

$$\left(1 + \frac{1}{x}\right)^{x} \to e \text{ and, in turn,}$$

$$\left[\left(1 + \frac{1}{x}\right)^{x}\right]^{rt} \to e^{rt}$$

Thus if we allow $n \to +\infty$, it turns out that correspondingly,

$$P\left(1 + \frac{r}{n}\right)^{nt} \to Pe^{rt}.$$

In other words, if the interest is compounded without interruption, the accumulated amount is found by using the following formula:

Compounding Interest Formula
Continuous Case

If P is invested at an annual rate r compounded *continuously,* then the accumulated amount S in the account at the end of t years is given by

$$S = Pe^{rt}.$$

EXAMPLE 5   **Comparing Investments**

Suppose that $5,000 is invested in a credit union paying an annual interest rate of 4% for 3 years. Find the accumulated amount if the interest is compounded:

(a) semiannually   (b) quarterly   (c) weekly

(d) daily   (e) hourly   (f) continuously

Solution   Under the given conditions, $P = 5000$, $r = 0.04$, and $t = 3$, so that if n is the number of compounding periods per year, then the accumulated amount S is given by

$$S = 5000\left(1 + \frac{0.04}{n}\right)^{3n}$$

for the discrete situations in parts (a) through (e), whereas,

$$S = 5000e^{(0.04)(3)}$$

for the continuous situation in part (f).

The accumulated amount S for each situation is computed in Table 5.

TABLE 5

Compounding Period	n	Accumulated Amount S	
(a) Semiannually	2	$5,630.81	
(b) Quarterly	4	$5,634.13	Discrete case; use
(c) Weekly	52	$5,637.22	$S = 5000\left(1 + \dfrac{0.04}{n}\right)^{3n}$
(d) Daily	365	$5,637.45	
(e) Hourly	$(24)(365) = 8760$	$5,637.48	
(f) Continuously	——	$5,637.48	Continuous case; use $S = 5000e^{(0.04)(3)}$

Notice that the largest gains appear in going from semiannually to quarterly to weekly. Subsequently, the changes in the value of S lessens as n increases. In fact, S is tending toward $5,637.48 as n gets larger and larger.

Modeling Growth and Decay

Base e is used in mathematical modeling of phenomena that exhibit continuous (uninterrupted) growth or decay as follows:

Modeling Continuous Growth or Decay

An exponential function of the form

$$P = P_0 e^{kt} \quad \text{where } k > 0$$

is used to model continuous **growth.** Its general graph is shown in Figure 4a. An exponential function of the form

$$P = P_0 e^{-kt} \quad \text{where } k > 0$$

is used to model continuous **decay.** Its general graph is shown in Figure 4b. In either case, P_0 is the initial quantity, t is the time elapsed, and k is the rate of growth or decay.

Figure 4

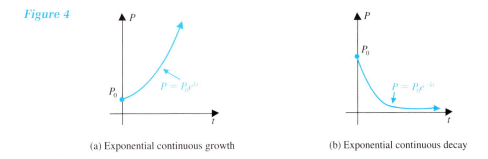

(a) Exponential continuous growth   (b) Exponential continuous decay

We say that P is growing or decaying *exponentially* at a *continuous* rate of k.

For example, suppose that the initial population of a certain city is 1.013 million, and the population P (in millions) grows exponentially at a continuous rate of about 0.58% per year. The growth model for this population t years later is given by the function

$$P(t) = 1.013e^{0.0058t}.$$

If we replace t by 1, 2, 3, and so on, we observe that the corresponding output values increase exponentially as the input values increase.

For the decay situation, the outputs decrease as the input values increase, as illustrated in the next example.

Modeling Radiocarbon Dating

Radioactive carbon-14 (C-14) is found in all living things. Once an organism dies, the C-14 begins to decay radioactively. The less C-14 found in a specimen, the older the specimen. Archaeologists are able to date objects such as relics and fossils by determining the remaining amounts of C-14. The process is called *radiocarbon dating*.

EXAMPLE 6   **Modeling Radioactive Decay of Carbon-14**

The percentage P (as a decimal) of C-14 found in a specimen t years after it begins to decay is modeled by the exponential function

$$P = e^{-0.000121t}.$$

Use the model to determine the percentage P of C-14 still present in an organism t years after it begins to decay if $t = 0$; 1000; 2000; 5000; 6000; 10,000. Round off the answers to two decimal places.

TABLE 6   Percentage of C-14 Present After t Years

t Years	P	Percentage of C-14 Present
0	1.00	100
1000	0.89	89
2000	0.79	79
5000	0.55	55
6000	0.48	48
10,000	0.30	30

Solution   We substitute each value of t into the function $P = e^{-0.000121t}$ and then use a calculator to evaluate. The results are summarized in Table 6.

[G] If the fossil has half of its C-14 remaining, it follows that $P = 0.50$ in the model in Example 6. So the solution of the equation

$$0.50 = e^{-0.000121t}$$

or, equivalently, $$0 = e^{-0.000121t} - 0.50$$

will give us the number of years of decay it took to reach this state.

Figure 5

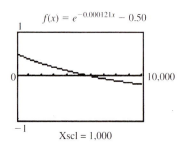

$f(x) = e^{-0.000121x} - 0.50$

Xscl = 1,000

The solution of this latter equation is the same as the x intercept of the graph of the function

$$f(x) = e^{-0.000121x} - 0.50.$$

By using a grapher to graph f (Figure 5), along with the ZERO-finding feature we find that x is approximately 5730 (to the nearest 10 years). Thus, it took about 5730 years to lose half the C-14 content.

The number of years that it takes for half the amount of a radioactive substance to decay is called its **half-life.** So in this situation we found that the half-life of C-14 is about 5730 years.

So far we have considered polynomial best fit regression models. Now we extend the idea to exponential regression models created by the least-squares method.

EXAMPLE 7   G **Modeling Population Growth**

The projected population growth in the United States for the years from 2000 through 2100 as reported by the U.S. Census Bureau is given in Table 7 (see Example in Section 3.1).

(a) Plot a scattergram of the data. Locate the t values on the horizontal axis and the P values on the vertical axis.

(b) Use a grapher and the exponential regression feature to determine the exponential model of the form

$$P = a \cdot b^t$$

that best fits the data. Round a and b to four decimal places.

(c) Sketch the graph of the function in part (b).

(d) Assuming this trend will continue, predict the population of the United States in the years 2015, 2075, and 2110. Round off the answers to the nearest thousand.

TABLE 7

Year	t	U.S. Population P (nearest thousand)
2000	0	275,306
2010	10	299,862
2020	20	324,927
2030	30	351,070
2040	40	377,350
2050	50	403,687
2060	60	432,011
2070	70	463,639
2080	80	497,830
2090	90	533,605
2100	100	570,954

(Source: U.S. Census Bureau, January 13, 2000)

Figure 6

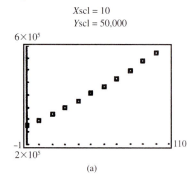

Xscl = 10
Yscl = 50,000

(a)

Solution

(a) The scattergram for the given data is shown in Figure 6a, where P is used for the vertical axis and t is used for the horizontal axis.

(b) By using a grapher we find the exponential regression model to be

$$P = 279992.9292 \cdot 1.0072^t.$$

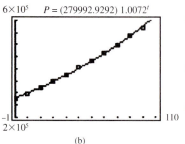

$P = (279992.9292) \, 1.0072^t$

(b)

(c) The graph of the exponential regression equation in part (b) is shown in Figure 6b along with the scattergram of the given data.

(d) To predict the population for the years 2015, 2075, and 2110, we substitute $t = 15$, 75, and 110, respectively, into the regression equation to get the population approximations:

311,805,000 for the year 2015

479,541,000 for the year 2075

and 616,419,000 for the year 2110

PROBLEM SET 4.2

In problems 1–4, use a calculator to approximate each expression, rounded to four decimal places.

1. (a) 2^π
   (b) $2^{-\sqrt{2}}$
   (c) $1000(5^{-\pi})$
   (d) $100(1.02)^{-8}$

2. (a) π^3
   (b) $5^{-\sqrt{7}}$
   (c) $2^{\sqrt[3]{5}}$
   (d) $10^x (3.14)^{\frac{1}{3}}$

3. (a) $e^{3.4}$
   (b) $e^{-\pi}$
   (c) $e^{\sqrt{2}}$
   (d) $1 - 4e^{-0.53}$

4. (a) $e^{-2.5}$
   (b) $e^{-\sqrt{3}}$
   (c) $e^{\sqrt{2 \cdot 5}}$
   (d) $3(1 + 2e^{0.9})$

In problems 5 and 6, let $f(x) = 4^{-\left(\frac{x}{2}\right)}$ and $g(x) = \left(\frac{2}{3}\right)^x$.

Find each value. Round off each answer to two decimal places.

5. (a) $f(-1)$
   (b) $f(\sqrt{3})$
   (c) $g(-2)$
   (d) $g(0.75)$

6. (a) $f(0.3)$
   (b) $f(\sqrt{5})$
   (c) $g(-\sqrt{7})$
   (d) $g(1.31)$

In problems 7 and 8, graph the functions f, g, and h on the same coordinate system. Describe the relationship among f, g, and h.

7. $f(x) = e^x$, $g(x) = 2 + e^x$, $h(x) = -2 + e^x$
8. $f(x) = 3^x$, $g(x) = 3^x - 1$, $h(x) = 3^x + 2$

In problems 9 and 10, each curve is the graph of an exponential function $y = b^x$ or b^{-x} that contains the given point. Find the value of b in each case.

9.

10.

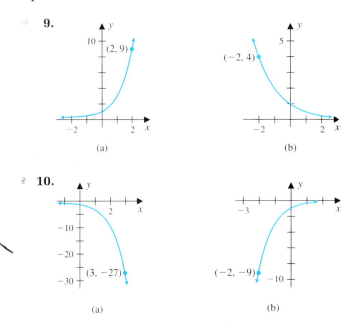

In problems 11 and 12, suppose that each curve is the graph of an exponential function of the form

$$y = A \cdot b^x$$

that contains the given pair of points. Find the values of A and b

11. (a) $P_1 = (0, 3)$ and $P_2 = (1, 9)$
    (b) $P_1 = (0, 1)$ and $P_2 = (-2, 4)$

12. (a) $P_1 = (0, 2)$ and $P_2 = (1, 6)$
    (b) $P_1 = (0, -2)$ and $P_2 = (1, -4)$

13. Each graph below is a transformation of the graph of the function $y = e^x$. Match each function with its graph.
    (a) $f(x) = e^x + 1$
    (b) $f(x) = e^{x+1}$
    (c) $f(x) = -e^x$
    (d) $f(x) = e^x - 1$

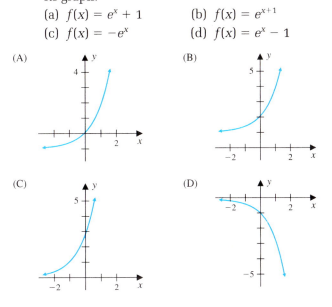

14. Each graph below is a transformation of the graph of the function $y = (0.4)^x$. Match each given function with its graph.
    (a) $g(x) = (0.4)^x - 1$
    (b) $f(x) = (0.4)^{x-1}$
    (c) $h(x) = -(0.4)^x$
    (d) $g(x) = (0.4)^{x+1}$

(A)

(B)

(C)

(D)

In problems 15–20, solve each equation for x.

15. $2^{3x} = 2^{4x-2}$ **16.** $5^{2-3x} = 5^{5x-6}$

17. $3^{x^2} = 3^{5x} + 6$ **18.** $4^x - 1 = 8^{2x}$

19. $25^{3x} = (125)^{x-1}$ **20.** $9^{x-1} = 27^{2x}$

In problems 21–28, describe a sequence of transformations that will transform the graph of a function of the form

$$y = b^x \quad \text{or} \quad y = e^x$$

to the graph of the given function. Sketch the graph in each case. Determine the domain and range, and indicate whether the function increases or decreases. Also identify any horizontal asymptote.

21. $f(x) = 5^{x+1}$ **22.** $h(x) = 4^{x-2}$

23. $f(x) = 2^x + 3$ **24.** $g(x) = 4^x - 1$

25. $f(x) = -2e^x$ **26.** $h(x) = 1 + e^x$

27. $f(x) = e^{-x} - 2$ **28.** $f(x) = 2 - 3e^{-x}$

Applying the Concepts

In problems 29–48 round off the answers to two decimal places.

29. Compound Interest: If $5000 is invested in a certificate of deposit for 5 years at an annual interest rate of 5.75%, how much money will be in the account if the interest rate is compounded:

  (a) Semiannually?   (b) Quarterly?

  (c) Monthly?   (d) Weekly?

30. Compound Interest: Suppose that $10,450 is invested in a money market account paying an annual interest rate of 4.25% compounded monthly.

  (a) Express the accumulated amount S as a function of elapsed time t in years.

  (b) After 3 years, how will the earned interest in this investment compare to the earned interest if the money is invested in a bank that pays an annual interest rate of 4.015% compounded daily?

31. Cumulative Earnings: Suppose that $5000 is invested in a money market account paying an annual interest rate of 5.75% compounded continuously.

  (a) Express the accumulated amount S as a function of the number of elapsed years t.

  (b) What is the accumulated amount after 4 years?

  (c) Will this investment earn more or less than another investment of $5000 that pays an annual interest rate of 6% compounded semiannually? Explain the difference.

32. Cumulative Interest Trend: The accumulated amount S of an investment of $10,000 paying continuously compounded interest over t years is given by

$$S = 10,000e^{0.0625t}.$$

  (a) What is the annual interest rate being paid on this investment?

  (b) What is the accumulated amount after 5 years?

  (c) Suppose $10,000 is invested in an account that pays 6.25% compounded daily. How much will accumulate after 5 years? Compare this investment to the one in part (b).

33. Population Growth: In 1999, the world population reached about 6 billion people, and it was growing at a continuous rate of about 1.3% per year.

  (a) Find the exponential growth model for the world population P (in billions) t years after 1999.

  (b) Use the model to predict the world population in the year 2009.

34. Population Growth: According to the U.S. Census Bureau, the population of the United States in 1999 was approximately 273 million people, and it was growing at a continuous rate of about 0.85% per year.

  (a) Find the exponential growth model that expresses the population P (in millions) t years after 1999.

  (b) Predict the population in the year 2050.

35. Bacteria Growth: A biologist studying bacteria in a swimming pool determines that the initial population of bacteria is 25,000 per cubic inch, and it is growing at a continuous rate given by the model

$$P = 25,000e^{0.1t}$$

where P represents the number present after t days.

  (a) What is the continuous growth rate of the bacteria?

  (b) Predict the number of bacteria per cubic inch after 5 days.

36. Bacteria Growth: The growth rate of bacteria in foods is used to determine the safe shelf-life of various products. Once the bacteria count reaches a certain level, these products are no longer safe to eat. If the daily continuous growth rate of bacteria in a certain brand of cheese is 5% of its population each day, how many times the current count level will the bacteria count be after 140 days?

37. Car Trade-in Value: Suppose that the *trade-in value V* (in dollars) of an automobile that is *t* years old is given by the formula

$$V = P(1 - r)^t$$

where *r* is the annual percentage of the depreciation and *P* is the original price of the car (in dollars). A certain make of car is purchased for $19,545, and it depreciates in value by 20% each year.

(a) Write an equation that expresses the trade-in value *V* as a function of *t*.

(b) Find the trade-in value after 1 year, after 3 years, and after 5 years. Round off answers to the nearest dollar.

38. Real Estate Appreciation: Because of inflation, the value of a home often increases with time. State and local agencies use the formula

$$S = C(1 + r)^t$$

to assess the value *S* (in dollars) of a home for property tax purposes, where *C* dollars is the cost of the home when new, *t* is the age of the home (in years), and *r* is the annual rate of inflation. Assume that the inflation rate is 3.1% and that a home originally cost $135,000.

(a) Write an equation that expresses the assessed value *S* as a function of *t*.

(b) If the inflation rate remains constant at 3.1%, what is the assessed value after 1 year? 5 years? 10 years? 15 years?

39. Recycling: In the state of Michigan, stores charge an additional 10¢ deposit for each bottle of beverage sold. Customers return these bottles for refunds, and then the stores return them to the beverage companies for recycling. Suppose that 65% of all bottles distributed will be recycled every year. If a company distributed 2.5 million bottles in 1 year, the number *N* (in millions) of recycled containers still in use after *t* years is given by the model

$$N = 2.5(0.65)^t.$$

How many recycled bottles (in millions) are still in use after 1 year? 3 years? 5 years? Round off the answers to two decimal places.

40. Electric Circuit: The electric current *I* (in amperes) flowing in a series circuit having inductance *L* henrys, resistance *R* ohms, and electromotive force *E* volts (Figure 7) is given by the model

$$I = \frac{E}{R}(1 - e^{-Rt/L})$$

Figure 7

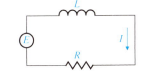

where *t* is the time (in seconds) after the current begins to flow.

(a) If *E* = 7.5 volts, *R* = 3 ohms, and *L* = 0.83 henry, express *I* as a function of *t*.

(b) Find the amount of current that flows in the circuit after 0.3 second; 0.5 second; and 1 second.

41. Environmental Science: According to the Bouguer–Lambert law, the percentage *P* of light that penetrates ordinary seawater to a depth of *d* feet is given by the function

$$P = e^{-0.044d}.$$

What is the percentage of light penetration in seawater to a depth of 4 feet? 6 feet?

42. Physics—Law of Cooling: Suppose that a heated object is placed in cooler surroundings. *Newton's law of cooling* relates the temperature *T* of the heated object to the temperature T_0 of the cooler surroundings. It states that after *t* units of time,

$$T = T_0 + Ce^{-kt}$$

where *k* and *C* are positive constants associated with the cooling object.

Suppose that a baked cake is removed from an oven and cooled to a room temperature of 70°F. The temperature *T* of the cake after *t* minutes is given by

$$T = 70 + 280e^{-0.14t}.$$

(a) What was the temperature of the cake when it was first removed from the oven?

(b) What is the temperature of the cake after 5 minutes? 10 minutes? 20 minutes?

43. Radioactive Decay: Radioactive elements decay exponentially. The decay model for a specific radioactive element is

$$A = A_0e^{-0.04463t}$$

where *A* is the amount present after *t* days and A_0 is the amount present initially. Assume there is a block of 50 grams of the element at the start. How much of the element remains after

10 days?

15 days?

30 days?

44. Financing Retirement: A method for financing a retirement is to invest an amount of money at a given rate compounded annually and then withdraw a fixed amount of money from it each year. The maximum amount of withdrawal W possible if an investment P is to last n years is given by the model

$$W = (W - Pr)(1 + r)^n.$$

If \$400,000 is invested in an account that pays 5% with the intention of making withdrawals for 15 years, how much can be withdrawn each year?

45. Population Trend: Ecologists have determined that the population N of bears in a certain protected forest area since 1996 ($t = 0$) is represented by the data in the following table.

Year	t	Population N
1996	0	225
1997	1	229
1998	2	234
1999	3	239
2000	4	244
2001	5	248
2002	6	253
2003	7	259
2004	8	264
2005	9	272

(a) Construct a scattergram that exhibits the given data. Locate the t values on the horizontal axis and the N values on the vertical axis.

(b) Use the exponential regression feature on a grapher to determine the model $N = a \cdot b^t$ that best fits the data. Round off a and b to three decimal places.

(c) Sketch the graph of the function N in part (b).

(d) Assuming that the trend of this model will continue, predict the number of bears that will inhabit the region in the year 2016.

46. G Profit Trend: Data in the following table shows the annual profit P of a company (in thousands of dollars) after t years.

(a) Construct a scattergram that exhibits the given data. Locate the t values on the horizontal axis and the P values on the vertical axis.

Year t	Profit P
0	150
1	210
2	348
3	490
4	660
5	872
6	1400

(b) Use the exponential regression feature on a grapher to determine the model $P = a \cdot b^t$ that best fits the data. Round off a and b to three decimal places.

(c) Sketch the graph of the function P in part (b).

(d) Assuming the trend of this model will continue, predict the profit after 10 years, to the nearest thousand dollars.

47. Inflation: The following data represent the amount A that \$1 will be worth t years from now, assuming that inflation runs steadily at 4% per year.

Year t	Amount A
0	\$1
1	\$0.96
2	\$0.92
3	\$0.88
4	\$0.85
5	\$0.82
6	\$0.78
7	\$0.75
8	\$0.72

(a) Construct a scattergram that exhibits the given data. Locate the t values on the horizontal axis and the A values on the vertical axis.

(b) Determine how closely the exponential model $A = (0.96)^t$ best fits the given data by comparing the function values (rounded to two decimal places) to the table values.

(c) Sketch the graph of the function A in part (b).

(d) Assuming that the trend of this model continues, predict how much \$1 now will be worth in 17 years. Round off to two decimal places.

48. [G] Depreciation: Data in the following table show the depreciation of a car that was purchased for $24,000 in 2005.

Year	2005	2006	2007	2008	2009
t	0	1	2	3	4
Value N	24,000	20,300	16,820	14,400	11,920

(a) Plot the scattergram of the data. Locate the t values on the horizontal axis and the N values on the vertical axis.

(b) Determine how closely the exponential function

$$N = 24{,}000(0.84)^t$$

best fits the given data by comparing the function values (rounded to the nearest dollar) to the table values.

(c) Sketch the graph of the function N in part (b).

(d) Assuming this trend continues, what will be the value of the car after 7 years and after 9 years? Round off the answers to the nearest dollar.

Developing and Extending the Concepts

49. Consider the exponential function

$$f(x) = b^x.$$

(a) Why does the definition exclude the value $b = 1$? How about $f(x) = 1^x$? Is this an exponential function?

(b) Why do we have the restriction $b > 0$? Is there any problem in defining and graphing the function $f(x) = (-3)^x$? Explain.

50. The *hyperbolic sine function*, denoted by *sinh*, and the *hyperbolic cosine function*, denoted by *cosh*, are defined as follows:

$$\sinh x = \frac{e^x - e^{-x}}{2} \quad \text{and} \quad \cosh x = \frac{e^x + e^{-x}}{2}$$

These functions are used in engineering. Show that $(\cosh x)^2 - (\sinh x)^2 = 1$ for every number x.

51. Let $f(x) = e^x$.

(a) Find the values of x such that

$$f(x^2) = f(7x - 12).$$

(b) Find and simplify

$$(f(x) + f(-x))^2 - (f(x) - f(-x))^2.$$

52. [G] **Civil Engineering:** When a flexible cord or cable hangs freely from its ends, it forms an arc called a *catenary*. Its equation has the form

$$y = \frac{k}{2}(e^{cx} + e^{-cx})$$

for appropriate choices of the constants c and k. The famous Gateway Arch in St. Louis, Missouri, has the shape of an *inverted catenary* (Figure 8). Suppose the arch is placed in a coordinate system so that the x axis is ground level and the y axis is the axis of symmetry of the arch. Assume its equation is given by

$$y = -63.85(e^{(x/127.7)} + e^{(-x/127.7)}) + 757.70$$

where x is in feet.

Figure 8

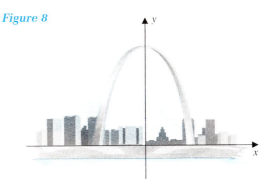

(a) Use a grapher to sketch the graph of this equation, and compare the shape of the curve to Figure 8.

(b) Locate the highest point (*apex*) of the arch.

(c) Find the distance between the bases at ground level.

53. Find the zeros of the function f if

(a) $f(x) = x^3 e^x + 8x^2 e^x - 29xe^x + 44e^x$

(b) $f(x) = 5x^3 e^{-2x} - 20x^2 e^{-2x} + 5xe^{-2x} + 30e^{-2x}$

54. Which of the given functions are equal to each other? Justify your assertion algebraically and illustrate the validity of your conclusion graphically.

(i) $f(x) = 2^{-(x-4)}$

(ii) $g(x) = (0.5)^{x-4}$

(iii) $h(x) = (0.25)^{2x-8}$

Objectives

1. Convert Exponentials to Logarithmic Forms and Vice Versa
2. Evaluate Logarithms Base 10 and Base e
3. Use the Properties of Logarithms
4. Use the Change-of-Base Formula
5. Solve Applied Problems

4.3 Properties of Logarithmic Functions

In Section 4.1, we learned that the graph of the inverse function f^{-1} is a reflection of the graph of f across the line $y = x$. In Section 4.2, we indicated that exponential functions are one-to-one, so they have inverses. Consequently, if we graph the function $f(x) = b^x$ and reflect its graph across the line $y = x$, the result is the graph of f^{-1}. This new function is given the name **logarithmic function with base b,** and it is written as $f(x) = \log_b x$.

For example, as Figure 1 shows, the graph of $f^{-1}(x) = \log_2 x$ is the reflection of the graph of $f(x) = 2^x$ across the line $y = x$. Since logarithms, in a sense, evolve as inverses of exponential functions, the algebra of logarithms is derived from the algebra of exponents, as we shall see.

Figure 1

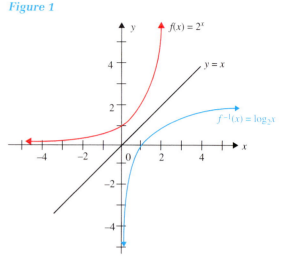

Converting Exponentials to Logarithmic Forms and Vice Versa

A logarithm is actually an exponent. For instance, consider the expression

$$49 = 7^2.$$

We refer to

2 as the *logarithm* of 49 with base 7,

and we write

$$\log_7 49 = 2.$$

The logarithm value 2 is the exponent to which 7 is raised to get 49.

In general, we have the following definition:

Definition

Logarithm

Let $b > 0$ and $b \neq 1$. The **logarithm of x with base b,** which is represented by y, is defined by

$$y = \log_b x \qquad \text{if and only if} \qquad x = b^y$$

for every $x > 0$ and for every real number y.

In words, this definition says that

the logarithm y is the exponent to which b is raised to get x.

The two equations in the above definition are equivalent and as such can be used interchangeably. The first equation is in *logarithmic form* and the second is in *exponential form*. The diagram on page 311 is helpful when changing from one form to the other:

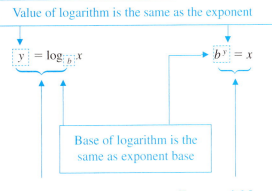

TABLE 1

x	log_2x
2^1	1
2^2	2
2^3	3
2^{-1}	-1
2^{-2}	-2
2^{-3}	-3

It is important to observe that $\log_b x$ is an exponent. For instance, the numbers in the right column of Table 1 are the logarithms (base 2) of the numbers in the left column.

EXAMPLE 1   **Converting Exponentials to Logarithmic**

Convert each exponential form equation to logarithmic form.

(a) $5^2 = 25$ (b) $\sqrt{49} = 7$ (c) $e^{-1} = \dfrac{1}{e}$

Solution   The equivalencies are listed in the following table.

	Exponential Form	Logarithmic Form
(a)	$5^2 = 25$	$\log_5 25 = 2$
(b)	$\sqrt{49} = 49^{\frac{1}{2}} = 7$	$\log_{49} 7 = \dfrac{1}{2}$
(c)	$e^{-1} = \dfrac{1}{e}$	$\log_e \dfrac{1}{2} = -1$

EXAMPLE 2   **Converting Logarithmic to Exponentials**

Convert each logarithmic form to an equivalent exponential form.

(a) $\log_2 16 = 4$ (b) $\log_9 3 = \dfrac{1}{2}$ (c) $\log_5 \dfrac{1}{25} = -2$

(d) $\log_{10} 17 = t$ (e) $\log_e(2x - 5) = y$

Solution   The following table lists the equivalency of each logarithmic form.

	Logarithmic Form	Exponential Form
(a)	$\log_2 16 = 4$	$2^4 = 16$
(b)	$\log_9 3 = \dfrac{1}{2}$	$9^{\frac{1}{2}} = 3$
(c)	$\log_5 \dfrac{1}{25} = -2$	$5^{-2} = \dfrac{1}{25}$
(d)	$\log_{10} 17 = t$	$10^t = 17$
(e)	$\log_e(2x-5) = y$	$e^y = 2x - 5$

At times, we can find the numerical value of a logarithm by converting to exponential form and then using the one-to-one property of exponents, as the next example illustrates.

EXAMPLE 3   Evaluating Logarithms

Evaluate each logarithm.

(a) $\log_4 64$ (b) $\log_3 \dfrac{1}{9}$ (c) $\log_e e^4$ (d) $\log_{\frac{1}{4}} 16$

Solution   In each case we let u equal the given expression, and write the logarithmic equation in its equivalent exponential form and then solve the resulting equation for u as shown in the following table.

	Logarithmic Form	Exponential Form	Result		
(a)	$\log_4 64 = u$	$4^u = 64$ $4^u = 4^3$	$u = 3$	or	$\log_4 64 = 3$
(b)	$\log_3 \dfrac{1}{9} = u$	$3^u = \dfrac{1}{9}$ $3^u = 3^{-2}$	$u = -2$	or	$\log_3 \dfrac{1}{9} = -2$
(c)	$\log_e e^4 = u$	$e^u = e^4$	$u = 4$	or	$\log_e e^4 = 4$
(d)	$\log_{\frac{1}{4}} 16 = u$	$\left(\dfrac{1}{4}\right)^u = 16$ $4^{-u} = 4^2$	$u = -2$	or	$\log_{\frac{1}{4}} 16 = -2$

We use the equivalency between the logarithmic and exponential forms to solve certain equations involving logarithms, as the next example shows.

EXAMPLE 4   Solving Logarithmic Equations

Solve each logarithmic equation for x.

(a) $\log_3 x = 4$ (b) $\log_5 \dfrac{1}{125} = x$

(c) $\log_x 8 = 3$ (d) $3 + \log_e 2x = t$

Solution   First we rewrite the given equation in exponential form and then solve the resulting equation.

(a) $\log_3 x = 4$ is equivalent to $3^4 = x$, so $x = 81$.

(b) $\log_5 \dfrac{1}{125} = x$ is equivalent to

$$5^x = \dfrac{1}{125}$$
$$5^x = 5^{-3} \qquad \text{Rewrite } \dfrac{1}{125} = 5^{-3}$$
$$x = -3 \qquad \text{One-to-one property}$$

(c) $\log_x 8 = 3$ is equivalent to

$$x^3 = 8$$
$$x^3 = 2^3 \qquad \text{Rewrite } 8 = 2^3$$
$$x = 2 \qquad \text{One-to-one property}$$

(d) $3 + \log_e 2x = t$ is equivalent to $\log_e 2x = t - 3$

$$2x = e^{t-3} \qquad \text{Convert to exponential form}$$

$$x = \frac{1}{2}e^{t-3} \qquad \text{Divide by 2}$$

Evaluating Logarithms—Base 10 and Base e

The base of a logarithmic function can be any positive number except 1. However, the two bases that are most widely used are 10 and e.

A logarithm with base 10 is called a **common logarithm.** Its value at x is denoted by **log x,** that is,

$$\log x = \log_{10} x.$$

A logarithm with base e is called a **natural logarithm,** and its value at x is denoted by **ln x,** that is,

$$\ln x = \log_e x.$$

Since $b^1 = b$, it follows that $\log_b b = 1$, and so

$$\log 10 = 1 \text{ and } ln\ e = 1.$$

Sometimes we can use the definition of logarithms to evaluate common and natural logarithms easily. For instance:

$$\log 1000 = 3 \qquad \text{since } 10^3 = 1000$$
$$\log 0.0001 = -4 \qquad \text{since } 10^{-4} = 0.0001$$
$$\ln e^5 = 5 \qquad \text{since } e^5 = e^5$$
$$\ln \frac{1}{e^2} = -2 \qquad \text{since } e^{-2} = \frac{1}{e^2}$$

Scientific calculators have the following function keys:

log for base 10 and an "ellen" **ln** key for the base e

to approximate values of logarithms. Table 2 lists some calculator approximations of logarithms (rounded to four decimal places) along with the equivalent exponential forms.

TABLE 2   Examples—Calculator Values of Logarithms

Approximate Logarithm Value	Exponential Form
$\log 4819 = 3.6830$	$10^{3.6830} = 4819$
$\log 0.897 = -0.0472$	$10^{-0.0472} = 0.897$
$\ln 45 = 3.8067$	$e^{3.8067} = 45$
$\ln 0.057 = -2.8647$	$e^{-2.8647} = 0.057$

Next we consider the problem of finding the value of x when the value of $\log x$ or $\ln x$ is given.

EXAMPLE 5   **Solving Logarithmic Equations**

Use a calculator to approximate the value of x to four decimal places.

(a) $\log x = 0.7235$

(b) $\ln x = 1.4674$

(c) $\log(x - 1) = -1.7$

Solution   Using a calculator and rounding the results to four decimal places, we get:

(a) $\log x = 0.7235$ is equivalent to $x = 10^{0.7235} = 5.2905$.

(b) $\ln x = 1.4674$ is equivalent to $x = e^{1.4674} = 4.3379$.

(c) $\log(x - 1) = -1.7$ is equivalent to $x - 1 = 10^{-1.7}$, so

$$x = 1 + 10^{-1.7} \quad \text{or} \quad x = 1.0200.$$

Using the Properties of Logarithms

Since logarithms are exponents, we can use the properties of exponents to derive some basic properties of logarithms.

For instance, notice that if $y = \log_b N$, then $N = b^y$. Substituting for y in the latter equation, we get

$$N = b^{\log_b N}$$

Basic Properties of Logarithms

Suppose that M, N, and b are positive real numbers, where $b \neq 1$, and r is any real number. Then we have:

(i) The Product Rule:

$$\log_b MN = \log_b M + \log_b N$$

(ii) The Quotient Rule:

$$\log_b \frac{M}{N} = \log_b M - \log_b N$$

(iii) The Logarithm of a Power:

$$\log_b N^r = r \log_b N$$

These properties can be described in words as follows:

(i) The logarithm of a product of numbers is the sum of logarithms of the numbers.

(ii) The logarithm of a quotient of numbers is the difference of the logarithms of the numbers.

(iii) The logarithm of a power of a number is the exponent times the logarithm of the number.

The proofs of these properties are based on properties of exponents. For instance, a proof of Property (i) follows.

Since

$$M = b^{\log_b M} \quad \text{and} \quad N = b^{\log_b N},$$

then

$$MN = b^{\log_b M} \cdot b^{\log_b N}$$
$$= b^{\log_b M + \log_b N}$$

By converting the equation to logarithmic form, we get

$$\log_b MN = \log_b M + \log_b N.$$

The basic properties of logarithms for the special cases of the common and natural logarithms are stated in Table 3.

TABLE 3 Basic Properties of Common and Natural Logarithms

Common Logarithms Base 10	Natural Logarithms Base e
(i) $\log MN = \log M + \log N$	(i) $\ln MN = \ln M + \ln N$
(ii) $\log \dfrac{M}{N} = \log M - \log N$	(ii) $\ln \dfrac{M}{N} = \ln M - \ln N$
(iii) $\log N^r = r \log N$	(iii) $\ln N^r = r \ln N$

In Property (iii), if the value of N is the same as the base, then we get $\log_b b^r = r \log_b b = r \cdot 1 = r$, so

$$\log_b b^r = r.$$

Specifically for $b = 10$ or $b = e$, we have

$$\log 10^r = r \quad \text{and} \quad \ln e^r = r.$$

The following identities follow directly from the definition of logarithms and their properties ($b > 0$, $b \neq 1$, and $x > 0$).

Summary of Logarithmic Identities

1. $\log_b 1 = 0$ because $b^0 = 1$
2. $\log_b b = 1$ because $b^1 = b$
3. $\log_b b^x = x$ because $b^x = b^x$
4. $b^{\log_b x} = x$ because $b^y = b^{\log_b x} = x$

The next two examples illustrate how the properties of logarithms are used to manipulate logarithmic expressions algebraically.

EXAMPLE 6

Using the Properties of Logarithms

Write each expression as a sum or difference of multiples of logarithms.

(a) $\log_8 \dfrac{x}{5}$ (b) $\log(x^7 y^{11})$ (c) $\ln\left[\dfrac{(y+7)^3}{\sqrt{y}}\right]$

Solution

We assume that all numbers whose logarithms are taken are positive.

(a) $\log_8 \dfrac{x}{5} = \log_8 x - \log_8 5$ Property (ii)

(b) $\log(x^7 y^{11}) = \log x^7 + \log y^{11}$ Property (i)

                $= 7 \log x + 11 \log y$ Property (iii)

(c) $\ln\left[\dfrac{(y+7)^3}{\sqrt{y}}\right] = \ln(y+7)^3 - \ln\sqrt{y}$ Property (ii)

                $= \ln(y+7)^3 - \ln y^{1/2}$ $\sqrt{y} = y^{1/2}$

                $= 3\ln(y+7) - \dfrac{1}{2}\ln y$ Property (iii)

EXAMPLE 7

Combining Logarithmic Expressions

Write each expression as a single logarithm.

(a) $2\ln x - 3\ln y$, where $x > 0$ and $y > 0$

(b) $\log(c^2 - cd) - \log(2c - 2d)$, where $c > d$.

Solution

(a) $2\ln x - 3\ln y = \ln x^2 - \ln y^3$ Property (iii)

                  $= \ln\left(\dfrac{x^2}{y^3}\right)$ Property (ii)

(b) $\log(c^2 - cd) - \log(2c - 2d) = \log\left[\dfrac{c^2 - cd}{2c - 2d}\right]$ Property (ii)

                  $= \log\left[\dfrac{c(c-d)}{2(c-d)}\right] = \log\left(\dfrac{c}{2}\right)$

Notice that, although the properties of the logarithms tell us how to compute the logarithm of a product or a quotient, there is no corresponding property for the logarithm of a sum or difference. That is,

$$\log_b(M + N) \neq \log_b M + \log_b N$$

and

$$\frac{\log_b M}{\log_b N} \neq \log_b\left(\frac{M}{N}\right).$$

In calculus, we encounter expressions involving exponential and logarithmic functions with base e that have to be simplified by using the properties and identities for logarithms.

EXAMPLE 8

Simplifying Expressions Involving Base e

Simplify each expression

(a) $e^{3\ln(2x+1)}$ (b) $e^{4 - 2\ln x}$

Solution

(a) $e^{3\ln(2x+1)} = e^{\ln(2x+1)^3}$ Property (iii)

$\qquad\qquad = (2x+1)^3$ $e^{\ln u} = u$ Identity 4

(b) $e^{4-2\ln x} = e^4 \cdot e^{-2\ln x}$ $e^{u+v} = e^u \cdot e^v$

$\qquad\qquad = e^4 \cdot e^{\ln x^{-2}}$ Property (iii)

$\qquad\qquad = e^4 \cdot x^{-2}$ $e^{\ln u} = u$ Identity 4

$\qquad\qquad = \dfrac{e^4}{x^2}$ $a^{-n} = \dfrac{1}{a^n}$

Using the Change-of-Base Formula

To determine logarithms with bases other than 10 or e, such as $\log_3 7$, we use the following *change-of-base formula*:

Change-of-Base Formula

$$\log_b x = \frac{\log_a x}{\log_a b}$$

where a and b are positive real numbers different from 1, and x is positive.

We can verify the formula as follows:

Let $\qquad\quad y = \log_b x$

$\qquad\qquad b^y = x$ Exponential form

$\qquad\quad \log_a b^y = \log_a x$ Take the logarithm of each side to base a

$\qquad\quad y \log_a b = \log_a x$ Property (iii)

$\qquad\qquad y = \dfrac{\log_a x}{\log_a b}$ Solve for y

$\qquad\quad \log_b x = \dfrac{\log_a x}{\log_a b}$ $y = \log_b x$

In practice, we usually choose either e or 10 for the new base, so that a calculator can be used to evaluate the necessary logarithm. Thus,

$$\log_b x = \frac{\log x}{\log b} \quad \text{or, equivalently,} \quad \log_b x = \frac{\ln x}{\ln b}.$$

EXAMPLE 9

Using the Change-of-Base Formula

Use the change-of-base formula to calculate each expression. Round off the answers to three decimal places.

(a) $\log_3 5$ $\qquad\qquad$ (b) $\log_{\sqrt{2}} \sqrt{11}$

Solution

Rounding off to three decimal places, we get:

(a) $\log_3 5 = \dfrac{\log 5}{\log 3} = 1.465$ $\qquad$ or $\qquad$ $\log_3 5 = \dfrac{\ln 5}{\ln 3} = 1.465$

(b) $\log_{\sqrt{2}} \sqrt{11} = \dfrac{\log \sqrt{11}}{\log \sqrt{2}} = 3.459$ $\qquad$ or $\qquad$ $\log_{\sqrt{2}} \sqrt{11} = \dfrac{\ln \sqrt{11}}{\ln \sqrt{2}} = 3.459$

Solving Applied Problems

Applications as diverse as measuring the strength of an earthquake, determining the acidity of a liquid, and modeling the profit from a business make use of logarithmic functions.

EXAMPLE 10   **Measuring the Strength of an Earthquake**

The **Richter scale** provides us with a way of grading the strength of an earthquake on a scale from 1 to 10—the larger the number, the more severe the earthquake. For a given earthquake, suppose that the largest seismic wave recorded on a seismograph is I and the smallest seismic wave recorded for the area is I_0. Then the **magnitude** M of the earthquake is given by

$$M = \log \frac{I}{I_0}$$

where I is called the **amplitude** of the earthquake, I_0 is called the **zero-level amplitude,** and the ratio I/I_0 is called the **intensity** of the earthquake.

(a) The 1989 Northern California earthquake had a magnitude of 7.1. Find the intensity of the earthquake and express its amplitude in terms of the zero-level amplitude.

(b) The 2003 earthquake in Bam, Iran had a magnitude of 6.5. Compare the intensity of this earthquake to the intensity of the 1989 Northern California earthquake.

Solution   (a) Substituting $M = 7.1$ into the formula $M = \log \dfrac{I}{I_0}$

we get

$$7.1 \doteq \log \frac{I}{I_0}.$$

Converting to exponential form, we find that the intensity is given by

$$\frac{I}{I_0} = 10^{7.1} \text{ so that } I = 10^{7.1}\, I_0.$$

That is, the amplitude of the earthquake was $10^{7.1}$ (approx. 12,589,254) times the zero-level amplitude.

(b) For the Bam, Iran earthquake, we have

$$6.5 = \log \frac{I}{I_0}$$

so the intensity is $\dfrac{I}{I_0} = 10^{6.5}$ (approx. 3,162,278).

Since $10^{7.1} = (10^{6.5})(10^{0.6})$ it follows that the intensity of the California earthquake was $10^{0.6}$ or approximately 4 times the intensity of the Bam, Iran earthquake.

PROBLEM SET 4.3

Mastering the Concepts

In problems 1 and 2, write each exponential equation in logarithmic form.

1. (a) $5^3 = 125$

(b) $32^{1/5} = 2$

(c) $17 = e^t$

(d) $b^x = 13z + 1$

2. (a) $\left(\dfrac{1}{2}\right)^{-3} = 8$

(b) $9^{3/2} = 27$

(c) $10^w = x - 3$

(d) $a^8 = y - 5$

In problems 3 and 4, write each logarithmic equation in exponential form.

3. (a) $\log_9 81 = 2$

(b) $\log 0.0001 = -4$

(c) $\log_c 9 = w$

(d) $\ln \dfrac{1}{2} = -1 - 3x$

4. (a) $\log_{36} 216 = \dfrac{3}{2}$

(b) $\ln s = t$

(c) $\log x = \sqrt{2}$

(d) $\log_b 7 = -3 + 8x$

In problems 5–8, find the exact value of each logarithm.

5. (a) $\log_2 \dfrac{1}{8}$

(b) $\ln e^4$

(c) $\log \sqrt[3]{10}$

(d) $\log_8 4$

6. (a) $\log_7 7$

(b) $\log_5 \dfrac{1}{625}$

(c) $\log \sqrt{10}$

(d) $\ln \sqrt[3]{e}$

7. (a) $\log_8 1$

(b) $\log_{1/2} 4$

(c) $\log_{10} 10^3$

(d) $\log_8 2$

8. (a) $\log_{0.5} 0.5$

(b) $\ln e^7$

(c) $10^{\log 3}$

(d) $e^{-\ln 4}$

In problems 9–14, solve each logarithmic equation for x by using its exponential form.

9. (a) $\log_6 x = 2$

(b) $\log_x \dfrac{1}{4} = -\dfrac{1}{2}$

10. (a) $\log_{27} x = \dfrac{1}{3}$

(b) $\log_x 4 = \dfrac{2}{3}$

11. (a) $\log_2 16 = 3x - 5$

(b) $\log_3(x + 1) = 2$

12. (a) $\log_5(5x - 1) = -2$

(b) $\log_5(x^2 - 4x) = 1$

13. (a) $y = 3 - \log x$

(b) $y = 8 + \ln(2x + 1)$

14. (a) $y = \ln(x - 1)$

(b) $y = 7 + \log(4 - x)$

In problems 15 and 16, use a calculator to approximate the value of each logarithm to four decimal places. Express the result in the equivalent exponential form.

15. (a) $\log 39.2$

(b) $\ln 961$

(c) $\log \dfrac{3}{4}$

(d) $\ln 0.23$

(e) $\ln 10$

16. (a) $\ln 39.2$

(b) $\log 961$

(c) $\ln \dfrac{3}{4}$

(d) $\log 0.23$

(e) $\log e$

In problems 17–20, use a calculator to approximate the value of x to four decimal places.

17. (a) $\ln x = 2.37$

(b) $\log x = 0.4137$

(c) $\ln x = -2.5$

(d) $\ln(2x - 1) = -3.8$

18. (a) $\log x = 2.37$

(b) $\ln x = 0.4137$

(c) $\log x = -2.5$

(d) $\log(2x - 1) = -3.8$

19. (a) $x = \dfrac{\log 3}{\log 1.71}$

(b) $x = \dfrac{\ln 3}{\ln 1.52}$

20. (a) $x = \dfrac{\ln 0.2}{-0.03}$

(b) $x = \dfrac{\log 0.7}{-0.05}$

In problems 21–26, write each expression as a sum or difference of multiples of logarithms.

21. (a) $\log_3 x(x + 1)$

(b) $\log_3 \dfrac{18}{x + 2}$

22. (a) $\log_b \dfrac{x^9}{y^7}$

(b) $\log_b x^2 y^3$

23. (a) $\log_b(x + 3)^4$

(b) $\log x\sqrt{2x + 1}$

24. (a) $\log \sqrt[6]{\dfrac{x^5}{y^2}}$

(b) $\ln \dfrac{x^5 y^2}{\sqrt[5]{z}}$

25. (a) $\ln[y(3x + 1)^{2/3}]$

(b) $\ln(x^2 + 7x)$

26. (a) $\log_5 \dfrac{x - 2}{x^2 - 4}$

(b) $\log \dfrac{5x(x^2 + 1)^2}{(x + 1)\sqrt{7x + 3}}$

In problems 27–30, use the properties of logarithms to write each expression as a single logarithm.

27. (a) $\log_5 \dfrac{5}{7} + \log_5 \dfrac{40}{25}$

(b) $\log_2 \dfrac{32}{11} + \log_2 \dfrac{121}{16} - \log_2 \dfrac{4}{5}$

28. (a) $\log(a^2 - ab) - \log(7a - 7b)$

(b) $\log\left(a + \dfrac{a}{b}\right) - \log\left(c + \dfrac{c}{b}\right)$

29. (a) $\ln(x^2 - 9) - \ln(x^2 - 6x + 9)$

(b) $\ln(3x^2 + 7x + 4) - \ln(3x^2 - 5x - 12)$

30. (a) $\log\left(\dfrac{1}{4} - \dfrac{1}{x^2}\right) - \log\left(\dfrac{1}{2} - \dfrac{1}{x}\right)$

(b) $3 \ln(x + 2) - \dfrac{1}{3} \ln x - \dfrac{1}{3} \ln(1 - 2x)$

In problems 31–34, simplify each expression.

31. (a) $e^{\ln 5}$

(b) $e^{-2 \ln 3}$

32. (a) $e^{3 + 4 \ln x}$

(b) $e^{\ln(1/x)}$

33. (a) $e^{-7 - \ln x}$

(b) $\ln e^{x^2 - 4}$

34. (a) $e^{(\ln x^2) - 1}$

(b) $e^{\ln x - 3 \ln y}$

In problems 35 and 36, compute the value to three decimal places after using the change-of-base formula to write each logarithm as a quotient of:

(i) Common logarithms   (ii) Natural logarithms

35. (a) $\log_2 5$

(b) $\log_7 2.89$

(c) $\log_4 \dfrac{3}{23}$

(d) $\log_3 e$

36. (a) $\log_8 13$

(b) $\log_4 \dfrac{1}{17}$

(c) $\log_3 0.46$

(d) $\log_{11} 7$

Applying the Concepts

37. Installment Loan Payment: An equation that relates a loan of P dollars to be paid off in n equal monthly installment payments of A dollars with an interest rate of r percent per period is given by

$\ln A - \ln P = \ln r + n \ln(1 + r) - \ln[(1 + r)^n - 1]$

where r = (Annual loan rate)/12.

(a) Solve this equation for A in terms of P, r, and n.

(b) Express A as a function of n if $18,000 is borrowed to purchase a boat at an annual rate of 10%. What are the monthly payments if the loan is to be paid off in 3 years? 4 years? 5 years?

38. Depreciation: Used-car dealers determine the value of a car by applying the formula

$$\log(1 - r) = \frac{1}{t}(\log w - \log p)$$

where p (in dollars) is the purchase price of a car when it was new, and w (in dollars) is its value t years later at an annual rate of depreciation r.

(a) Solve this equation for r in terms of t, w, and p.

(b) Express the annual rate of depreciation r as a function of t if a new car was purchased for $18,300 and sold t years later for $7500. What is the rate of depreciation if the car was sold after 3 years? 3 years and 6 months? 4 years? Round off the answers to two decimal places.

39. Population Growth—Fruit Flies: The fruit fly *Drosophila melanogaster* is used in some genetic studies in laboratories. The number N of fruit flies in a colony after t days of breeding is given by the equation

$\ln(230 - N) - \ln N = \ln 6.931 - 0.1702t.$

(a) Express N as a function of t.

(b) How many fruit flies are in the colony initially? After 10 days? After 16 days? After 30 days?

40. Physics—Speed of a Rocket: Suppose a rocket with mass M_0 (in kilograms) is moving in free space at a speed of V_0 (in kilometers per second). When the rocket is fired, its speed V is given by

$$V = V_0 + V_1(\ln M_0 - \ln M)$$

where M is the mass of the propellant that has been burned off and V_1 is the speed of the exhaust gases produced by the burn. Solve the equation for M in terms of V, V_0, V_1, and M_0.

41. Magnitude of an Earthquake: The amplitude of the earthquake in Japan in May 1983 was $10^{7.7}$ times the zero-level amplitude. Find the magnitude of the earthquake on the Richter scale.

42. Earthquake Comparisons: The 1988 Armenian earthquake had a magnitude of 6.8 on the Richter scale, and the 1985 earthquake in Mexico City had a magnitude of 7.8. How many times more powerful was the intensity of the Mexico City earthquake than the Armenian earthquake?

In problems 43 and 44, the concentration of hydrogen ions in a substance is denoted by $[H^+]$, measured in moles per liter. The **pH** of a substance is defined by the logarithmic function

$$pH = -\log [H^+].$$

This function is used to measure the *acidity* of the substance. The pH of neutral distilled water is 7. A substance with a pH of less than 7 is known as an *acid*, whereas a substance with a pH of more than 7 is called a *base*.

43. pH Scale: Find the pH of each substance and determine whether it is an acid or a base. Round off the answers to one decimal place.
  (a) Beer: $[H^+] = 3.16 \times 10^{-3}$ mole per liter
  (b) Eggs: $[H^+] = 1.6 \times 10^{-8}$ mole per liter
  (c) Milk: $[H^+] = 4 \times 10^{-7}$ mole per liter

44. pH Scale:
  (a) Express the hydrogen ion concentration $[H^+]$ as a function of pH.
  (b) Use the function in part (a) to find the hydrogen ion concentration (in moles per liter) of: calcium hydroxide, pH = 13.2; vinegar, pH = 3.1; and tomatoes, pH = 4.2. Round off the answers to three decimal places.

In problems 45 and 46, the loudness L of a sound is measured by

$$L = 10 \log \frac{I}{I_0}$$

Here, L is the number of *decibels*, I is the intensity of the sound (in watts per square meter), and I_0 is the smallest intensity that can be heard. Suppose the intensity I_0 is 10^{-12} watt per square meter.

45. Sound Intensity:
  (a) Find the number of decibels of the noise of a truck passing a pedestrian at the side of the road if the sound intensity of the truck is 10^{-3} watt per square meter.
  (b) Find the number of decibels of sound from a jet plane with sound intensity 8.3×10^2 watts per square meter. Round off the answer to two decimal places.

46. Sound Intensity:
  (a) The threshold of pain from loud sound is considered to be about 120 decibels. Find the intensity of such a sound in watts per square meter.
  (b) Find the intensity of a whisper in watts per square meter if the whisper is 25 decibels.

Developing and Extending the Concepts

47. Use $M = 10,000$, $N = 10$, $b = 10$, and $p = 3$ to show that each statement is false.
  (a) $\log_b \dfrac{M}{N} = \dfrac{\log_b M}{\log_b N}$
  (b) $\dfrac{\log_b M}{\log_b N} = \log_b M - \log_b N$

  (c) $\log_b M \cdot \log_b N = \log_b M + \log_b N$
  (d) $\log_b MN = \log_b M \cdot \log_b N$
  (e) $\log_b M^p = (\log_b M)^p$
  (f) $(\log_b M)^p = p \log_b M$

48. Show that: $\ln \left(\dfrac{\sqrt{3} + \sqrt{2}}{\sqrt{3} - \sqrt{2}} \right) = 2 \ln (\sqrt{3} + \sqrt{2})$.

In problems 49 and 50, find values for a and b that satisfy the given equation.

49. $\log(a + b) = \log a + \log b$
50. $\log(a - b) = \log a - \log b$

In problems 51 and 52, use the change-of-base formula to prove each equation.

51. $\log_b a = \dfrac{1}{\log_a b}$

52. $\dfrac{\log_b x}{\log_{ab} x} = 1 + \log_b a$

53. Find the domain and the x intercepts of the graph of

$$y = -1 + \ln(x - 2).$$

Also, find the y intercept (if any).

54. Criticize the statement:

$$1 < 2$$

$$\frac{1}{8} < \frac{2}{8} \qquad \text{Divide by 8}$$

$$\frac{1}{8} < \frac{1}{4} \qquad \text{Reduce}$$

$$\left(\frac{1}{2}\right)^3 < \left(\frac{1}{2}\right)^2$$

$$\log \left(\frac{1}{2}\right)^3 < \log \left(\frac{1}{2}\right)^2$$

$$3 \log \frac{1}{2} < 2 \log \frac{1}{2} \qquad \text{Divide by } \log \frac{1}{2}$$

$$3 < 2$$

55. Solve for x.

$$\log_3 x + \frac{1}{\log_x 3} = 4$$

56. Solve for x.

$$3 + \log_3(2x + 7) + \log_3 (2x - 1) = 5 + \log_3 x$$

Objectives

Definition

Logarithmic Function

4.4 Graphing Logarithmic Functions

As a follow up to our coverage of the algebra of logarithms in Section 4.3, we shall now study logarithmic functions in more detail. Our interest is focused on graphing these functions and exploring their applications.

Defining Logarithmic Functions

We now define a new class of functions called *logarithmic functions,* which are inverses of exponential functions, as follows:

> The inverse of the exponential function $f(x) = b^x$ is the **logarithmic function** defined by $g(x) = \log_b x$, and vice versa.

Table 1 lists examples of exponential functions along with their inverses.

TABLE 1   Examples—Inverses of Exponential Functions

Exponential Function	Inverse Function
$f(x) = 4^x$	$f^{-1}(x) = \log_4 x$
$f(x) = (1/3)^x$	$f^{-1}(x) = \log_{1/3} x$
$f(x) = 10^x$	$f^{-1}(x) = \log x$
$f(x) = e^x$	$f^{-1}(x) = \ln x$

Since $g(x) = \log_b x$ is the inverse of $f(x) = b^x$, it follows that the domain of f, which includes all real numbers, is the range of g.

By examining the definition of the logarithmic function, we observe that if

$$y = g(x) = \log_b x, \quad \text{then} \quad x = b^y \quad \text{where} \quad b > 0 \quad \text{and} \quad b \neq 1.$$

Since b^y is always positive under these conditions, it follows that x is always positive. Thus, in the real number system, we can only evaluate the logarithm of a positive number.

Table 2 lists some logarithmic functions with their domains.

TABLE 2

Function	Condition	Domain
1. $f(x) = \log_3(x - 2)$	$x - 2 > 0$	$x > 2$ or $(2, \infty)$
2. $f(x) = \log_5(4 - 7x)$	$4 - 7x > 0$	$x < \dfrac{4}{7}$ or $\left(-\infty, \dfrac{4}{7}\right)$

EXAMPLE 1

Finding the Domain

Find the domain of

$$f(x) = \ln (x^2 - x - 6).$$

Solution   To find the domain of f we use the fact that

ln $(x^2 - x - 6)$ is defined only if

$$x^2 - x - 6 = (x - 3)(x + 2) > 0.$$

Solving this inequality, we get

$$x < -2 \quad \text{or} \quad x > 3.$$

Thus the domain of f consists of all real numbers in the interval

$$(-\infty, -2) \quad \text{or} \quad (3, \infty).$$

Graphing Logarithmic Functions by Converting to Exponential Form

One approach to graphing a logarithmic function $y = \log_b x$ is to first convert its equation to the equivalent exponential form $x = b^y$. Then use the *point-plotting method* to sketch the graph of the equation $x = b^y$.

For example, to graph the logarithmic function

$$y = f(x) = \log_2 x.$$

First, we rewrite f in the form

$$x = 2^y$$

and then select values of y and their corresponding x values as shown in the table of values.

Then we graph the latter form by using point-plotting (Figure 1).

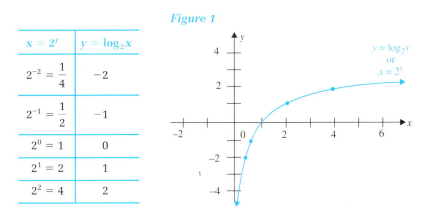

Figure 1

$x = 2^y$	$y = \log_2 x$
$2^{-2} = \dfrac{1}{4}$	-2
$2^{-1} = \dfrac{1}{2}$	-1
$2^0 = 1$	0
$2^1 = 2$	1
$2^2 = 4$	2

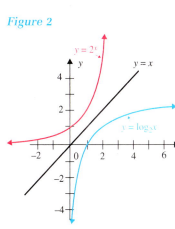

Figure 2

An alternate approach is to graph the function $y = 2^x$ and reflect the graph across the line $y = x$, as shown in Figure 2. Note that the horizontal asymptote (the x axis) for $y = 2^x$ converts to a vertical asymptote (the y axis) for $y = \log_2 x$.

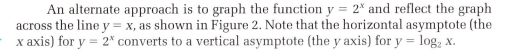

That is, $y = \log_2 x \to -\infty$ as $x \to 0^+$.

Note that, the technique, of interchanging the variables in $y = 2^x$ to obtain $f^{-1}(x) = \log_2 x$ causes the domain's end range to be interchanged. That is,

The domain of $y = 2^x$, $\mathbb{R}$, is the range of $y = \log_2{}^x$ and the range of $y = 2^x$, all positive real numbers, is the domain of $y = \log_2{}^x$.

EXAMPLE 2   **Graphing Logarithmic Functions Using Exponential Forms**

Graph each function by using its exponential form, and determine the domain and range.

(a) $f(x) = \log_4 x$ (b) $g(x) = \log_{1/4} x$

Solution   (a) To graph f we first convert $y = f(x) = \log_4 x$ to its equivalent form, $x = 4^y$. Next, we form a table of values for $x = 4^y$ by substituting values for y and then finding the corresponding x values (Table 3).

  Finally, we plot the points and connect them with a smooth curve to get the graph of f (Figure 3a).

  The domain is the interval $(0, \infty)$ and the range includes all real numbers $\mathbb{R}$.

(b) Similarly, $y = g(x) = \log_{1/4} x$ is converted to the equivalent form $x = (1/4)^y$. Then this equation is graphed by point-plotting as in part (a) (Table 3, Figure 3b).

  The domain consists of all values of x in the interval $(0, \infty)$ and the range includes all real numbers $\mathbb{R}$.

TABLE 3

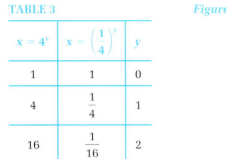

$x = 4^y$	$x = \left(\dfrac{1}{4}\right)^y$	y
1	1	0
4	$\dfrac{1}{4}$	1
16	$\dfrac{1}{16}$	2

Figure 3

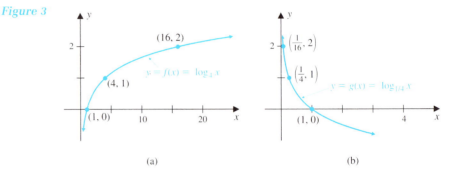

(a)   (b)

Transforming Graphs of Logarithmic Functions

The graphs in Figure 3 above serve as prototypes for the general graphs and properties of the logarithmic functions $f(x) = \log_b x$ for base $b > 1$ and for base $0 < b < 1$, respectively (Table 4).

TABLE 4   General Graphs and Properties of $f(x) = \log_b x$

Properties	General Graphs
1. The domain is the interval $(0, \infty)$. 2. The range includes all real numbers. 3. The function increases on interval $(0, \infty)$ if $b > 1$ and decreases on interval $(0, \infty)$ if $0 < b < 1$. 4. The y axis is a vertical asymptote. 5. The function is one-to-one. 6. The graph contains the point $(1, 0)$. 7. The x intercept of the graph is 1. There is no y intercept. 8. The graph is smooth and continuous.	$f(x) = \log_b x$, $b > 1$ $f(x) = \log_b x$, $0 < b < 1$

Examples of specific graphs for $b > 1$ and $0 < b < 1$ are shown in Figures 4a and 4b, respectively. We can use transformations of graphs of the functions

$$y = \log_b x$$

to graph other logarithmic functions.

Figure 4

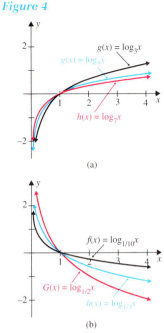

(a)

(b)

EXAMPLE 3   **Transforming Graphs of Logarithmic Functions**

Use a transformation of the graph of

$$y = \log_3 x$$

to sketch the graph of each function. Also, find the domain and range of each function.

(a) $f(x) = -\log_3 x$

(b) $g(x) = \log_3(-x)$

(c) $h(x) = \log_3(x - 1)$

Solution   (a) We can graph the function

$$f(x) = -\log_3 x$$

by reflecting the graph of $f(x) = \log_3 x$ about the x axis (Figure 5a). The domain of f is $(0, \infty)$, and the range is $\mathbb{R}$.

(b) To obtain the graph of

$$g(x) = \log_3(-x)$$

we reflect the graph of $g(x) = \log_3 x$ about the y axis (Figure 5b). The domain of g is $(-\infty, 0)$ and the range is $\mathbb{R}$.

(c) The graph of h is obtained by shifting the graph $y = \log_3 x$ one unit to the right (Figure 5c). The domain of h is $(1, \infty)$ and the graph shows that the range consists of all real numbers $\mathbb{R}$. Note that the line $x = 1$ is a vertical asymptote.

Figure 5

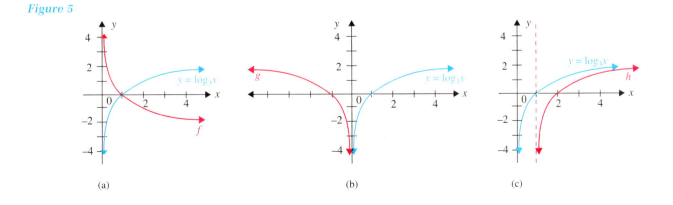

(a)   (b)   (c)

G   Graphers can be helpful in displaying the domain and the vertical asymptotes of more complicated logarithmic functions. For instance, Figure 6 shows the graph of the function

$$f(x) = \ln(x^2 - x - 6)$$

in Example 1 on page 322. The graph of f indicates the domain consists of all real numbers in the intervals $(-\infty, -2)$ or $(3, \infty)$. Also the equations of the vertical asymptotes are

$$x = -2 \quad \text{and} \quad x = 3.$$

Figure 6

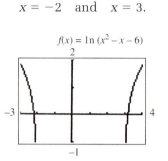

Graphers, along with the change of base formula, can be used to graph logarithmic functions with bases different than 10 or e. The next example illustrates the process.

EXAMPLE 4   G **Using a Change-of-Base to Graph**

Use the change-of-base formula and a grapher to sketch the graph of $f(x) = \log_8 x$. Also determine the domain and the vertical asymptote.

Solution   First we rewrite the function in terms of base 10 or base e. We'll select base 10. Thus

$$f(x) = \log_8 x = \frac{\log x}{\log 8}.$$

Next we use a grapher to graph $f(x) = (\log x)/(\log 8)$. A viewing window of the graph is shown in Figure 7. Since we can evaluate logarithms only for positive real numbers, it follows that the domain includes all values x in the interval $(0, \infty)$. The y axis is a vertical asymptote.

Figure 7

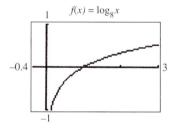

Solving Applied Problems

At times, logarithmic functions are used for best-fitting regression models, as the next example illustrates.

EXAMPLE 5 [G] **Modeling a Pricing Trend**

A manufacturer charges different prices P (in dollars) for each grapher according to the number Q of graphers purchased each year by a school, as specified in the table.

Price P (in dollars) per Calculator	Quantity Q Purchased
120	95
110	160
100	270
90	285
80	305
70	320

(a) Plot a scattergram of this data with the price P as the horizontal axis and the quantity Q as the vertical axis.

(b) Use a grapher and the logarithmic regression feature to determine the logarithmic model of the form

$$Q = a + b\ln P$$

that best fits the data. Use four decimal places for a and b.

(c) Sketch the graph of the function in part (b).

(d) Use this pricing schedule to find the number of calculators needed to be purchased for the school year if the price of each calculator is to be \$115, \$95, or \$75. Round off the answers to the nearest integer.

Solution

(a) The scattergram for the given data is shown in Figure 8, where P is used for the horizontal axis and Q is used for the vertical axis.

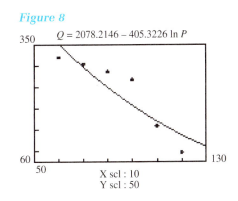

Figure 8

$Q = 2078.2146 - 405.3226 \ln P$

X scl : 10
Y scl : 50

(b) By using a grapher, we find the logarithmic regression model to be

$$Q = 2078.2146 - 405.3226 \ln P.$$

(c) The graph of the logarithmic regression equation in part (b) is shown in Figure 8, along with the scattergram of the given data.

(d) To find the number of calculators needed to be purchased if the price is to be \$115, \$95, or \$75, we substitute $P = 115$, 95, and 75, respectively, into the regression equation to obtain the following approximations:

155 for the price of \$115

232 for the price of \$95

and   328 for the price of \$75

◆ PROBLEM SET 4.4

Mastering the Concepts

In problems 1–4, find the inverse of the given function, and then use symmetry to graph both functions on the same coordinate system. Also find the domain and range of each inverse function.

 1. $f(x) = 5^x$

2. $g(x) = \left(\dfrac{1}{7}\right)^x$

3. $h(x) = \log_{1/10} x$

4. $f(x) = \log x$

5. Each curve in Figure 9 is the graph of a function of the form $y = \log_b x$. Find the base if its graph contains the given point.

Figure 9

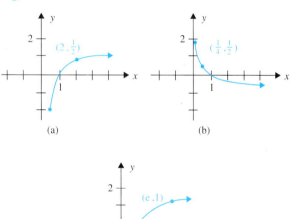

(a)

(b)

(c)

6. Simplify each expression.

(a) $\log 10^{3+2x}$

(b) $\ln\left(\dfrac{1}{e^{2x}}\right)$

In problems 7–16, find the domain of each function.

7. $f(x) = \ln(2 - x)$

8. $g(x) = \log(x + 3)$

9. $h(x) = \ln|x + 1|$

10. $F(x) = \ln\left(\dfrac{1}{\sqrt{x - 3}}\right)$

11. $f(x) = \log_5(2x + 4)$

12. $g(x) = \log\left(\dfrac{\sqrt{x + 1}}{x}\right)$

13. $g(x) = \log(x^2 - x)$

14. $h(x) = \ln(x^2 - 2x + 1)$

15. $f(x) = \log\left(\dfrac{1}{x^2 - 1}\right)$

16. $f(x) = \log_{\frac{1}{3}}\left(\dfrac{1}{x^2 - 9}\right)$

In problems 17 and 18, graph all three given logarithmic functions on the same coordinate system. Use the graphs to describe the relative steepness and limit behavior of the functions in terms of the different bases.

17. $f(x) = \log_2 x; g(x) = \log_{2.5} x; h(x) = \log_3 x$

18. $f(x) = \log_{1/3} x; g(x) = \log_{1/4} x; h(x) = \log_{1/5} x$

In problems 19–22, describe how the graph of f can be obtained from the graph of g. Find the domain of f, and write the equation of the vertical asymptote.

19. $g(x) = \log_5 x;$ $f(x) = \log_5(x - 2)$

20. $g(x) = \log_4 x;$ $f(x) = \log_4(-x)$

21. $g(x) = \ln x;$ $f(x) = -\ln x$

22. $g(x) = \log x;$ $f(x) = 2 + \log x$

In problems 23 and 24, find an equation of each curve that has been obtained by transforming the curve with the equation $y = \log_3 x$.

23.

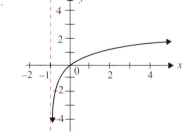

24.

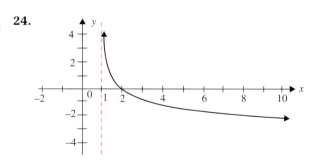

In problems 25–32, graph the given function by using transformations of the graph of $y = \log_b x$ with the same base. Determine the domain and vertical asymptote.

e 25. $f(x) = 1 + \ln x$

26. $g(x) = 3 \log x$

27. $g(x) = \dfrac{1}{3} \log_5 x - 2$

9 28. $h(x) = 1 - \log_7 x$

29. $f(x) = \log(x + 3)$

e 30. $g(x) = \log_5(x - 2)$

31. $f(x) = -2 \ln(x - e)$

32. $g(x) = \ln |1 - x|$

In problems 33–38, use the change-of-base formula along with a grapher to sketch the graph of each function. Also determine the domain and any vertical asymptotes.

33. $g(x) = \log_3 x$

34. $h(x) = \log_{1/3} x$

35. $g(x) = \log_4(5x + 1)$

36. $f(x) = \log_2(3 - 4x)$

37. $f(x) = \log_7(4x^2 - 1)$

38. $h(x) = \log_5(x^2 + 1)$

Applying the Concepts

39. **Advertising:** A company has determined that when its monthly advertising expenditure is x (in thousands of dollars), then the total monthly sales S (in thousands of dollars) is given by the model

$$S = \frac{2000 \ln(x + 5)}{\ln 10}.$$

What are the expected monthly sales (to the nearest thousand dollars) if the monthly expenditure for advertising is $10,000? $50,000? $100,000?

40. **Manufacturing:** A small auto parts supply company's weekly profit P (in dollars) is given by the function

$$P = 1828 + 914 \log x$$

where x is the number of parts produced each minute. Find the weekly profit (to the nearest dollar) if the number of parts produced each minute is 10, 50, 100, 500, and 1000.

Developing and Extending the Concepts

41. If $P_1 = -\log A$, $P_2 = -\log B$, and $0 < A < B$, then which is larger, P_1 or P_2? Explain.

42. Let $f(x) = 10 \ln \left(\dfrac{50}{50 - x} \right)$.

   (a) For what value(s) of x is:

      (i) $f(x) = 0$

      (ii) $f(x) = 20$

      Round off the answers to two decimal places.

   (b) What is the domain of f?

G In problems 43–48, use a grapher to sketch the graph of each function. Use the graph to find the domain and any vertical asymptotes. Confirm the domain algebraically.

43. $f(x) = x \ln(x - 2)$

44. $h(x) = 3[\log(3 - 2x)]^2$

45. $g(x) = \log(x^2 - 5x + 4)$

46. $f(x) = \ln(x^2 + 3x + 2)$

47. $f(x) = \log(2x + 3) + \log(2x - 3)$

48. $g(x) = \ln(x^2 - 9) - \ln(x + 3)$

49. Sketch the graph of $f(x) = 2 + \log_2 x$ and its inverse on the some coordinate system.

50. G Show that $(f \circ g)(x) = (g \circ f)(x) = x$ for $f(x) = \log_2(2x + 4)$ and $g(x) = 2^{x-1} - 2$. Also sketch the graphs of f and g on the same coordinate system.

Objectives

1. Solve Exponential Equations
2. Solve Exponential Inequalities
3. Solve Logarithmic Equations
4. Solve Applied Problems

4.5 Exponential and Logarithmic Equations and Inequalities

The properties of logarithms play an important role in solving certain equations and inequalities involving exponential and logarithmic functions such as

$$4^{2x-1} = 3, \quad e^{-3t} < 0.5 \quad \text{and} \quad \log(x + 2) - \log x = 1$$

are examples of these situations.

As we shall see, it is not always possible to solve these types of equations and inequalities algebraically. If this is the case, we rely on graphical solutions.

Solving Exponential Equations

Recall from Section 4.2, that we solved equations such as

$$2^x = 4^{2x-1}$$

by first writing each side as a power of the same base (2, in this example), equating the exponents and then solving the resulting equation. However, to solve an exponential equation such as

$$2^x = 7$$

where it is not easy to initially express each side in terms of the same base, we use the properties of logarithms, as the next example illustrates.

EXAMPLE 1   **Solving Equations Involving One Exponential Expression**

Use logarithms to solve each equation. Round off the answer to two decimal places.

(a) $4^{2x-1} = 3$

(b) $e^{-3t} = 0.5$

Algebraic Solution	Graphical Illustration

(a) Taking the common logarithm of each side of the given equation, we have

$$\log 4^{2x-1} = \log 3$$

$$(2x - 1)\log 4 = \log 3 \qquad \color{blue}{\log N^r = r \log N}$$

$$\left.\begin{array}{l} 2x - 1 = \dfrac{\log 3}{\log 4} \\[2ex] 2x = 1 + \dfrac{\log 3}{\log 4} \\[2ex] x = \dfrac{1}{2}\left(1 + \dfrac{\log 3}{\log 4}\right) \end{array}\right\} \quad \color{blue}{\text{Solve for } x}$$

$$x = 0.90 \text{ (approx.)} \qquad \color{blue}{\text{Use a calculator}}$$

(b) Taking the natural logarithm of each side of the equation $e^{-3t} = 0.5$, we have

$$\ln e^{-3t} = \ln 0.5$$

$$-3t \ln e = \ln 0.5 \qquad \color{blue}{\ln N^r = r \ln N}$$

$$t = \dfrac{-\ln 0.5}{3} \qquad \color{blue}{\ln e = 1}$$

$$t = 0.23 \text{ (approx.)} \qquad \color{blue}{\text{Use a calculator}}$$

In each case, the solution represents the x intercept of a function.

(a) Since $4^{2x-1} = 3$ is equivelent to $4^{2x-1} - 3 = 0$, the solution is the same as the x intercept of $f(x) = 4^{2x-1} - 3$.

Figure 1a shows the x intercept of the graph of f (approximately 0.90), which is the solution of $4^{2x-1} = 3$.

(b) Figure 1b shows (approximately 0.23) the t intercept of the graph of $g(t) = e^{-3t} - 0.5$, which is the same as the solution of $e^{-3t} = 0.5$.

Figure 1

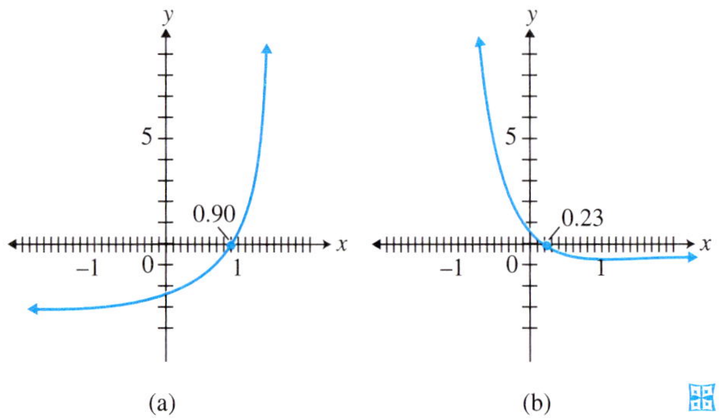

(a)   (b)

EXAMPLE 2   **Solving an Equation Involving Two Exponential Expressions**

Solve the equation

$$5^{3x-2} = 3^x$$

Round off the answer to two decimal places.

Algebraic Solution	Graphical Illustration
Taking the natural logarithm of each side, we have	Figure 2 shows the graph of the function

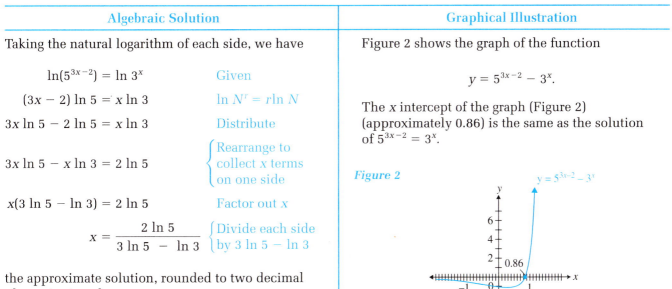

Algebraic Solution:

$\ln(5^{3x-2}) = \ln 3^x$	Given
$(3x - 2) \ln 5 = x \ln 3$	$\ln N^r = r\ln N$
$3x \ln 5 - 2 \ln 5 = x \ln 3$	Distribute
$3x \ln 5 - x \ln 3 = 2 \ln 5$	$\left\{\begin{array}{l}\text{Rearrange to}\\ \text{collect } x \text{ terms}\\ \text{on one side}\end{array}\right.$
$x(3 \ln 5 - \ln 3) = 2 \ln 5$	Factor out x
$x = \dfrac{2 \ln 5}{3 \ln 5 - \ln 3}$	$\left\{\begin{array}{l}\text{Divide each side}\\ \text{by } 3 \ln 5 - \ln 3\end{array}\right.$

the approximate solution, rounded to two decimal places, is given by

$$x = 0.86$$

Graphical Illustration:

$$y = 5^{3x-2} - 3^x.$$

The x intercept of the graph (Figure 2) (approximately 0.86) is the same as the solution of $5^{3x-2} = 3^x$.

Figure 2

G   When encountering equations such as

$$9^x - 3^x - 12 = 0 \quad \text{or} \quad e^x + e^{-x} - 6 = 0,$$

Figure 3

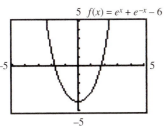

we might not recognize that this type of equation cannot be solved by using algebraic methods. However, a graphical solution is possible.
For example, by graphing the function

$$f(x) = e^x + e^{-x} - 6$$

we observe that the x intercepts of the graph are approximately, that is, -1.76 and 1.76 (Figure 3) the approximate solutions of the equation $e^x + e^{-x} - 6 = 0$.

Solving Exponential Inequalities

Earlier, we established that $f(x) = \log_b x$ is one-to-one. Therefore, if

$$u < v, \quad \text{then} \quad \log_b u < \log_b v.$$

This property is used in the next example.

EXAMPLE 3   **Solving an Inequality Involving an Exponential Expression**

Solve the inequality $2^{x-1} < 5$.

     Express the solution in interval notation and show the solution on a number line. Round off the answer to two decimal places.

Algebraic Solution	Graphical Illustration

Taking the common logarithm of each side, we have

$$\log 2^{x-1} < \log 5$$

$(x - 1) \log 2 < \log 5$ $\log N^r = r \log N$

         Divide each side by

$$x - 1 < \frac{\log 5}{\log 2}$$ log 2, which is a positive number.

$$x < 1 + \frac{\log 5}{\log 2}$$ Add 1 to each side

$x < 3.32$ (approx.)   Use a calculator

In interval notation, the solution set is $(-\infty, 3.32)$ and it is displayed in Figure 4a.

Figure 4

(a)

Figure 4b shows the graph of the function

$$y = 2^{x-1} - 5.$$

The x intercept of the graph is approximately 3.32. The solution of the inequality consists of all values of x, such that $y < 0$.

That is, $x < 3.32$ or $(-\infty, 3.32)$.

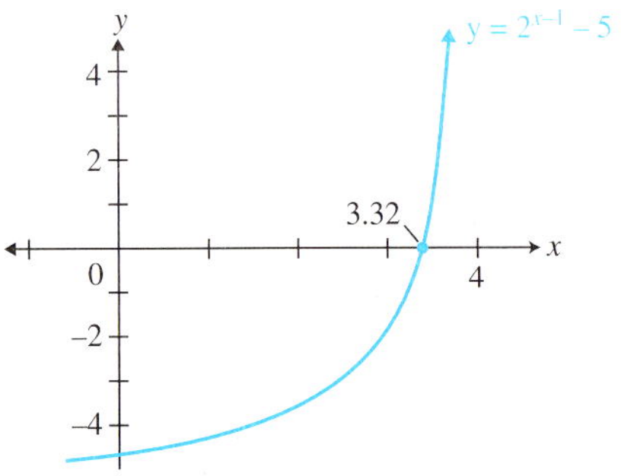

(b)

Solving Logarithmic Equations

To solve an equation involving a multiple of the sum or difference of logarithms, we first combine the logarithmic expressions into a single logarithmic expression on one side of the equation and then use the following one-to-one property

$$\log_b M = \log_b N, \quad \text{if and only if,} \quad M = N$$

For instance, to solve an equation such as

$$\log x^2 - 2 \log \sqrt{x} = 1$$

we combine the expressions on the left side as follows:

$\log x^2 - \log (\sqrt{x})^2 = 1$ Power property

$\log x^2 - \log x = 1$ Simplify

$\log \dfrac{x^2}{x} = 1$ Quotient property

$\log x = 1$ Simplify

$x = 10^1 = 10$

EXAMPLE 4   **Solving Logarithmic Equations**

Solve the equation and check for extraneous roots.

(a) $\log_3(x + 1) + \log_3(x + 3) = 1$ (b) $\log(x^2 - 4) - \log(x - 2) = 1$

Algebraic Solution	Graphical Illustration

Algebraic Solution

(a) We proceed as follows:

$\log_3(x + 1) + \log_3(x + 3) = 1$ Given

$\log_3[(x + 1)(x + 3)] = 1$ Product property

$(x + 1)(x + 3) = 3^1$ Exponential form

$\left.\begin{array}{c} x^2 + 4x + 3 = 3 \\ x^2 + 4x = 0 \end{array}\right\}$ Simplify

$x(x + 4) = 0$ Factor

$x = 0$ or $x = -4$ Solve for x

Check:

For $x = -4$

$\log_3(-4 + 1) + \log(-4 + 3) = \log_3(-3) + \log_3(-1)$,

since the logarithms of negative numbers are undefined, so -4 is an extraneous solution.

For $x = 0$ $\log_3 1 + \log_3 3 = 0 + 1 = 1$.

Thus, 0 is the only solution.

Graphical Illustration

(a) The given equation is equivalent to

$$\log_3(x + 1) + \log_3(x + 3) - 1 = 0.$$

So the solution of the equation is the same as the x intercept of the graph of

$$f(x) = \log_3(x + 1) + \log_3(x + 3) - 1.$$

Figure 5a shows the graph of the function f intercepts the x axis at the origin, which confirms that the solution is $x = 0$.

Figure 5a

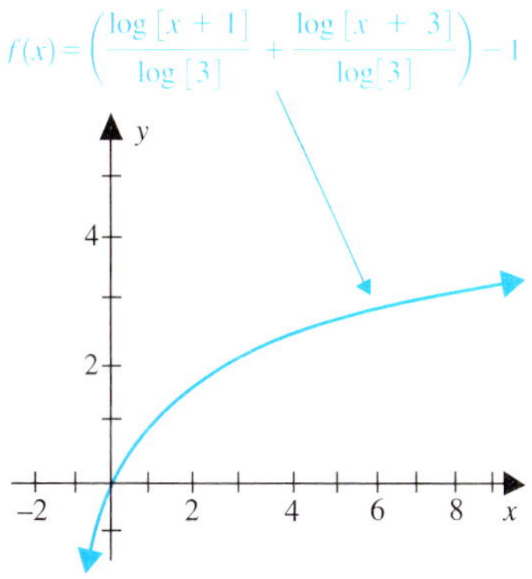

$f(x) = \left(\dfrac{\log\,[x + 1]}{\log\,[3]} + \dfrac{\log\,[x + 3]}{\log[3]}\right) - 1$

(b) We proceed as follows:

$\log(x^2 - 4) - \log(x - 2) = 1$ Given.

$\log\left[\dfrac{x^2 - 4}{x - 2}\right] = 1$ Use the quotient property

$\log\left[\dfrac{(x - 2)(x + 2)}{x - 2}\right] = 1$ Factor

$\log(x + 2) = 1$ Simplify

$x + 2 = 10$ Convert to exponential form

$x = 8$

Check:

For $x = 8$

$\log(64 - 4) - \log 6$

$= \log 60 - \log 6$

$= \log \dfrac{60}{6}$

$= \log 10$

$= 1$

Thus, 8 is the solution.

(b) Figure 5b shows the graph of

$$f(x) = \log(x^2 - 4) - \log(x - 2) - 1.$$

The x intercept of the graph is 8, which confirms that 8 is the solution of the given equation

$$\log(x^2 - 4) - \log(x - 2) - 1$$

Figure 5b

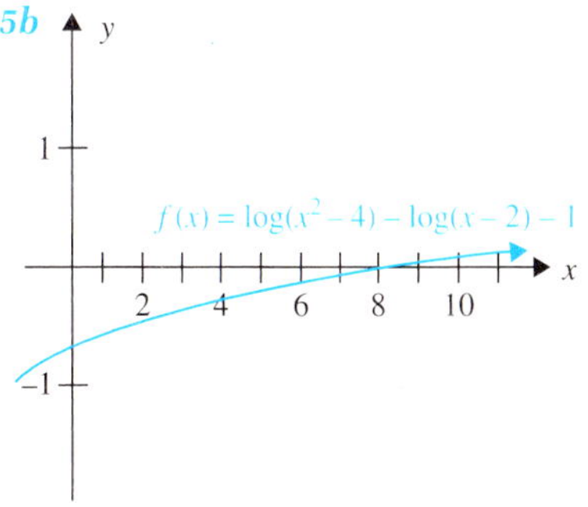

$f(x) = \log(x^2 - 4) - \log(x - 2) - 1$

Figure 6

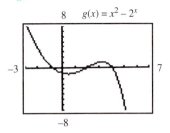

G Not all exponential and logarithmic equations and inequalities can be solved using the algebraic methods studied so far. Such problems can sometimes be solved by graphical means. For example, we solve the inequality $x^2 > 2^x$ graphically as follows:

The inequality $x^2 > 2^x$ is equivalent to $x^2 - 2^x > 0$. Figure 6 shows the graph of $g(x) = x^2 - 2^x$, using the ZERO -finding feature we get the approximate locations of the three x intercepts, namely,

$$-0.77, 2, \quad \text{and} \quad 4.$$

By reading the graph, we see that $x^2 - 2^x > 0$, where the graph of g is above the x axis—that is, whenever x is in the interval $(-\infty, -0.77)$ or in the interval $(2, 4)$. So the solution includes all values of x such that

$$x < -0.77 \quad \text{or} \quad 2 < x < 4, \quad \text{approximately.}$$

Solving Applied Problems

We know that a radioactive substance decomposes according to the decay model

$$A = A_0 e^{-kt} \quad \text{where } k > 0$$

and t represents time. Logarithms can be used to find how long it takes for a substance to decay to a given level.

EXAMPLE 5 **Using the Decay Model of an Element**

Polonium-210 is a radioactive element that decays according to the model

$$A = A_0 e^{-0.005t}$$

where A_0 is the initial weight of the sample. How long will it take for a sample of this material to decay by 30% if t represents elapsed time in days?

Solution If the sample weighs A_0 grams initially, then after decaying by 30%, its weight A is given by

$$A = A_0 - 0.30A_0$$
$$= 0.70A_0.$$

Substituting into the decay model, we get

$$0.70A_0 = A_0 e^{-0.005t}$$
$$0.70 = e^{-0.005t} \qquad \text{Divide each side by } A_0.$$
$$\ln 0.70 = \ln e^{-0.005t} \qquad \text{Take the natural logarithm of each side}$$
$$\ln 0.70 = -0.005t \ln e \qquad \ln N^r = r \ln N$$
$$\ln 0.70 = -0.005t \qquad \ln e = 1$$
$$t = \frac{\ln 0.70}{-0.005}$$
$$t = 71.33 \text{ (approx.)} \qquad \text{Use a calculator}$$

Thus it takes a little over 71 days to decay by 30%.

 **PROBLEM SET 4.5**

Mastering the Concepts

In problems 1–14, use logarithms to solve each equation. Round off the answers to four decimal places.

1. $6^x = 12$ **2.** $3^{5x} = 2$

3. $e^{-5x} = 7$ **4.** $e^{4x} = 3$

5. $7^{3x-1} = 5$ **6.** $10^{x+1} = 4$

7. $e^{x+1} = 10$ **8.** $e^{2-3x} = 8$

9. $e^{2x-1} = 5^x$ **10.** $3^{x+1} = 17^{2x}$

11. $e^{2x-1} = 10^x$ **12.** $e^x = 10^{x+1}$

13. $(1.08)^{2x+3} = (1.7)^x$ **14.** $(0.97)^{3x-1} = (3.1)^{2x}$

In problems 15–24, solve the given inequality. Express the solution in interval notation and show it on a number line. Round off the answers to two decimal places.

15. $3^x < 21$ **16.** $4^x > 3$

17. $2^{4-x} > 5$ **18.** $2^{-x} < 5$

19. $e^{3x} < 7$ **20.** $e^{-2x} > 4$

21. $3^{x+2} < 2^{1-3x}$ **22.** $e^{2x+3} < 3^{x-2}$

23. $(1.06)^x > 2$ **24.** $14.7e^{-0.21x} < 7.35$

In problems 25–34, solve each equation and check for extraneous roots.

25. $\log 2x = \log 3 + \log(x - 1)$

26. $\log_5 y + \log_5(y - 4) = 1$

27. $\ln x + \ln(x - 2) = \ln(x + 4)$

28. $2 \ln(t + 1) - \ln(t + 4) = \ln(t - 1)$

29. $\log_2(w^2 - 9) - \log_2(w + 3) = 2$

30. $\log(v + 1) - \log v = 1$

31. $\log_4 x + \log_4(6x + 10) = 1$

32. $\log_3 t + \log_3(t - 6) = \log_3 7$

33. $\log(x^2 - 144) - \log(x + 12) = 1$

34. $\log x^2 = 2 \log x$

G In problems 35–42, use a grapher to solve each equation. Round off the answers to three decimal places.

35. $2^x - 1 = 3x$ **36.** $3e^x - 7 = 5x$

37. $e^{-x^2} + x - 1 = 0$ **38.** $3^x - 4 = 4x$

39. $e^{x^2} - 3x - 5 = 0$ **40.** $3^{2x} = x^2$

41. $\ln x + x^2 = 2$ **42.** $\ln(x + 1) + x = 3$

G In problems 43–50, use a grapher to solve each inequality. Round off the answers to two decimal places.

43. $3^x < x^2$ **44.** $e^{2x} > 3 - x$

45. $\ln x > \log 2x$ **46.** $\ln(x - 3) < \log x$

47. $e^{-x} \le 3 + \ln x$ **48.** $e^{3x} - \ln x \le 5$

49. $\log(x - 1) < 3x$ **50.** $3 \ln x - x^3 > 4$

Applying the Concepts

In problems 51–60, round off each answer to two decimal places.

51. **Compound Interest:** Suppose that $10,000 is invested in a certificate of deposit at a 4.75% annual interest rate. How many years will it take for the money to double if the interest is compounded:
(a) Semiannually?
(b) Quarterly?
(c) Monthly?
(d) Continuously?

52. **Compound Interest:** Suppose that $1500 is invested at a 4.65% annual interest rate. How many years will it take for the money to triple if the interest is compounded:
(a) Semiannually?
(b) Quarterly?
(c) Monthly?
(d) Continuously?

53. **Investment Growth Model:** Suppose that P dollars are invested in a money market fund that pays a 4.85% annual interest rate compounded quarterly, and that the accumulated amount S (in dollars) after t years is a multiple of P, that is, $S = kP$.
(a) Use common logarithms to express t as a function of k.
(b) Use the function in part (a) to find out how long it takes for the investment to double, to triple, and to quadruple.

54. **Investment Growth Model:** Suppose that P dollars are invested in a credit union paying a 5.03% annual interest rate compounded continuously, and that the accumulated amount S (in dollars) after t years is a multiple of P, that is, $S = kP$.
(a) Use natural logarithms to express t as a function of k.
(b) Use the function in part (a) to find out how long it takes for the investment to double, to triple, and to quadruple.

55. **Productivity:** In an experiment conducted by a company, it was determined that the number N of days of training needed for a factory worker to produce x automobile parts per day is given by the model

$$N = 100 - 25 \ln(60 - x).$$

How many parts per day is a factory worker able to produce if 25 training days are needed?

56. Advertising: A company predicts that sales will increase during a 30-day promotional television campaign according to the logistic model

$$N = \frac{400}{1 + 300\,e^{-0.5t}}$$

where N is the number of daily sales t days after the campaign begins. How many days of advertising will result in daily sales of 300 units?

57. Deer Population: Suppose that the population P of a herd of deer newly introduced into a game preserve grows according to the model

$$P = 300 + 100 \ln(t + 1),$$

where t is the time in years.

(a) How many deer are present in the preserve initially?

(b) In how many years will the deer population be 500?

58. Radioactive Decay: Potassium-42, a radioactive element used by cardiologists as a tracer, has a half-life of about 12.5 hours.

(a) Find the decay model for this element.

(b) How long will it take for 80% of the original sample to decay?

59. Heat-Treating: A steel panel is tested for stress by heating it to 385°F and then immediately cooling it in a cooling chamber where the temperature is held at 32°F. The panel cools according to Newton's law of cooling:

$$T = 32 + 353e^{-kt}$$

where T is the temperature in degrees Fahrenheit after t minutes and k is a constant.

(a) Find k if the temperature of the panel is 198°F after 5 minutes.

(b) How long after the steel panel is placed in the cooling chamber will its temperature be 85°F?

60. Drug Dosage Model: The concentration C (in milligrams) of a certain drug after t hours in a patient's bloodstream is given by

$$C = 100e^{-0.442t}.$$

How long does it take for the drug concentration to reduce by 50%?

Developing and Extending the Concepts

61. Solve the equation

$$x^2 + 5x + 6 = 0.$$

Substitute e^t for x in the equation to get $e^{2t} + 5e^t + 6 = 0$. Explain why the latter equation has no solution.

62. Solve the equation

$$3y^2 - 28y + 9 = 0.$$

Substitute 3^x for y in the equation to get

$$3(3^{2x}) - 28(3^x) + 9 = 0.$$

63. Solve each equation.

(a) $\ln x^2 = (\ln x)^2$

(b) $\log x^3 = (\log x)^3$

64. Suppose that

$$y = 3 + 10(1 - e^{-0.2x}).$$

Solve for x in terms of y.

65. [G] Recall from Section 4.2 that the hyperbolic sine function is denoted by sinh x and defined by

$$f(x) = \sinh x = \frac{e^x - e^{-x}}{2}.$$

Use a grapher to graph f. Then use the graph to solve the equation sinh $x = 1$. Round off the answer to two decimal places.

66. [G] Is it possible to solve the equation

$$\log_3(x^4 + 1) + \log_2(x^2 + 1) = 5$$

algebraically? If so, solve it. If not, use a grapher to approximate the solution for $-3 \le x \le 3$. Round off to two decimal places.

◆ CHAPTER 4 REVIEW PROBLEM SET

1. Determine the inverse of each function. Verify that
$$f[f^{-1}(x)] = f^{-1}[f(x)] = x$$
   Graph f and f^{-1} on the same coordinate system.
   (a) $f(x) = 7 - 13x$
   (b) $f(x) = \sqrt[5]{x}$

2. Use the horizontal-line test to determine whether f has an inverse.
   (a) $f(x) = \sqrt{3x - 2} + 5$
   (b) $f(x) = x^3 + x$
   (c) $f(x) = x^{3/5} + 1$

3. Let $f(x) = 2.5^{-x}$. Find each value rounded off to three decimal places.
   (a) $f(-1)$
   (b) $f(2)$
   (c) $f(0.3)$
   (d) $f(-\sqrt{3})$
   (e) $f(1.7)$

4. Sketch the graphs of both given functions on the same coordinate system. Then use the graphs to solve each inequality.
   (a) $f(x) = 5^x$ and $g(x) = 3^x$; $5^x > 3^x$
   (b) $f(x) = \left(\dfrac{1}{4}\right)^x$ and $g(x) = \left(\dfrac{2}{5}\right)^x$; $\left(\dfrac{1}{4}\right)^x < \left(\dfrac{2}{5}\right)^x$
   (c) $f(x) = e^x$ and $g(x) = 3^x$; $e^x < 3^x$

5. Describe a sequence of transformations to change the graph of $y = 2^x$ into the graph of the given function. Graph the function. Indicate the domain, range, and whether the function is increasing or decreasing. Also, determine the horizontal asymptote.
   (a) $f(x) = 2^x + 1$
   (b) $g(x) = -2^x$
   (c) $f(x) = 3(2^x)$
   (d) $g(x) = 2^x - 3$
   (e) $f(x) = 2^{x+1} - 3$
   (f) $g(x) = -3(2^{x+1}) + 4$

6. Solve the equations by expressing each side in terms of the same base.
   (a) $2^x = 8^{x-1}$
   (b) $4^{-x} = 8^x + 2$
   (c) $3^{x^2+x} = 9$

7. Write each exponential equation in logarithmic form.
   (a) $e^{3.1} = a$
   (b) $e^{x^2} = a + b$
   (c) $5^{-2.7} = b$
   (d) $e^{a+b} = c$
   (e) $10^{a-b} = t$

8. Find the exact value of each logarithm without using a calculator.
   (a) $\log_3 9$
   (b) $\log_4 8$
   (c) $\log_6 1$
   (d) $\ln \sqrt[5]{e}$
   (e) $\ln e^{-3}$
   (f) $\log_5 0.04$
   (g) $\log_{100} 0.001$

9. Solve each equation for x.
   (a) $\log_4(x + 3) = -1$
   (b) $\log(x^2 - 6) = 1$
   (c) $\ln 2x = \ln 8 - \ln 2$
   (d) $\log_5(2x - 1) + \log_5(2x + 1) = 1$
   (e) $\ln(x + 8) - \ln x = 1$
   (f) $\ln \sqrt{x^2 + 1} = 0$
   (g) $3 \ln x + t = b$
   (h) $e^{\ln(2x+y)} = u$

10. Each of the given functions has no inverse. State one way to restrict the domain of each function so that the restricted function has an inverse.
    (a) $f(x) = |x - 2|$
    (b) $f(x) = 2x^2$

11. Solve the equation
$$\frac{e^x + e^{-x}}{e^x - e^{-x}} = u$$
    for x in terms of u.

12. Write each expression as a single logarithm and simplify.
    (a) $\ln(x^2 - 16) - \ln(x + 4)$
    (b) $\ln 3x - 2[\ln x - \ln(3 + x)]$
    (c) $\ln\left(\dfrac{e}{\sqrt[3]{x}}\right) - \ln \sqrt[3]{ex}$
    (d) $2 \log_4 x^3 + \log_4 \dfrac{2}{x} - \log_4 \dfrac{2}{x^4}$

13. Describe a sequence of transformations that will change the graph of either $y = \ln x$ or $y = \log x$ to the graph of the given function. Graph the function. Find the domain, range, and vertical asymptotes.
    (a) $f(x) = \ln(x - 3)$
    (b) $g(x) = \ln(3 - x)$
    (c) $h(x) = -\log x$

14. **G** Use a grapher to graph each function. Then find the domain of the function and any vertical asymptotes.
    (a) $f(x) = \ln(-x)$
    (b) $f(x) = e^{x^2}$
    (c) $h(x) = \ln(-x - 2)$
    (d) $h(x) = \dfrac{e^x - e^{-x}}{2}$

15. Use logarithms to solve each equation. Round off each answer to three decimal places.
    (a) $3^{x+2} = 5$
    (b) $e^{-0.5t} = 17$
    (c) $5^{x+1} = 13^{2x}$

16. **G** Determine whether each statement is true or false. Use a grapher to help support your conclusion. Assume $x > 0$.
    (a) $\ln \sqrt[5]{x} = \sqrt[5]{\ln x}$
    (b) $\ln x^2 = (\ln x)^2$
    (c) $\ln\left(\dfrac{1}{x}\right) = \dfrac{1}{\ln x}$
    (d) $\ln |x| = |\ln x|$

17. Solve each inequality. Round off the answers to two decimal places.
    (a) $e^{3x} < 4$
    (b) **G** $\ln x - e^x < -3$
    (c) $2^x \le 3^{x+3}$

18. **Bacteria Growth:** Suppose that a carton of milk contains 5000 bacteria per cubic inch at the time it was bought, and that the number doubles every day. The number N of bacteria in the carton t days after the milk was bought is given by the model
    $$N(t) = 5000 \cdot 2^t.$$
    (a) Find $N(5)$ and $N(3.5)$. Explain what each value means in this situation.
    (b) Sketch the graph of the function. Interpret the graph. Suppose that it is not safe to drink the milk when the bacteria count is 3,000,000. For how many days after the carton was bought can we safely drink the milk?

19. **Compound Interest:** If $10,000 is deposited in an account that pays 5% annual interest compounded annually, then the accumulated balance A in the account at the end of t years is given by the function
    $$A(t) = 10,000(1.05)^t.$$
    (a) Find $A(3)$, $A(7)$, and $A(10)$. Explain what each value means in this situation.
    (b) Determine when the balance will reach $35,000. Round off the answer to two decimal places.

20. **Investment:** Suppose that the interest rate allows an investment to double every 10 years. Then the accumulated value A from an initial investment of $1000 is given by
    $$A(t) = 1000 \cdot 2^t$$
    where t represents the number of 10-year periods.

    (a) Find the accumulated value of a $1000 investment after 25 years.
    (b) Determine when the investment will be worth $500,000. Round off to two decimal places.

21. **Depreciation:** Banks depreciate the value of a car according to its age. Suppose that a car is bought for $15,000 and depreciates about 25% of its value every year. The value V (in dollars) of the car t years later can be predicted by the model
    $$V(t) = 15,000(0.75)^t.$$
    (a) Find $V(2)$ and $V(3)$. Explain what each value means in this situation.
    (b) Predict when the value of the car will be $10,500. Round off the answer to two decimal places.

22. **Compound Interest:** Suppose that $1000 is put into a savings plan that yields a nominal interest rate of 5%. How much money will be accumulated in the account after 8 years if the interest is compounded:
    (a) Annually?
    (b) Semiannually?
    (c) Quarterly?
    (d) Monthly?
    (e) Continuously?

23. **Radioactive Decay:** A radioactive substance decays according to the model
    $$A = A_0 e^{-0.001t}.$$
    Thus, A grams of the substance remain after t years, where A_0 is the initial amount present.
    (a) If 100 grams of the substance are present initially, how much will be present at the end of 50 years? Round off the answer to two decimal places.
    (b) Determine when the substance decays by 70%. Round off the answer to the nearest year.

24. **Present Value:** Determine how much money must be invested now in order to have $500,000 in 5 years, if the investment during this period earns a nominal interest rate of 6% compounded annually.

25. **Compound Interest:** Suppose that $10,000 is invested at a 4.5% nominal interest rate. How many years will it take for the money to double if the interest is compounded:
    (a) Quarterly?
    (b) Continuously?

26. **Doubling Time:** Banks use the formula

$$T = \frac{\log 2}{\log(1 + r)}$$

to predict the number of years T required to double an investment at an annual interest rate of r.

(a) Find the time it takes to double an investment of $1000 at an annual interest rate of

4%, 5%, and 9%.

(b) According to the popular "rule of 70," the time T it takes for an investment to double is approximately

$$T = 70/r,$$

where r is the annual interest rate expressed as a percentage. Compare the results in part (a) to the results given by the rule of 70.

27. **Chemistry:** The pH of a solution is given by

$$pH = -\log [H^+].$$

Find the pH of each substance to two decimal places.

(a) Vinegar: $[H^+] = 1.58 \times 10^{-3}$ mole per liter

(b) Milk of magnesia: $[H^+] = 3.16 \times 10^{-11}$ mole per liter

28. **Manufacturing:** A manufacturer determines that the daily cost C (in dollars) to manufacture x VCR components is given by the function

$C(x) = 365[1 + \ln(x^2 - 225)]$ where $x > 15$

(a) Find

$$C(20) \quad \text{and} \quad C(40).$$

Explain what each value means in this situation.

(b) If the daily cost to manufacture VCR components is $3640, how many components are produced each day?

29. **Advertising:** A health club determines that the number of memberships N sold in a year is related to the number of dollars x spent on advertising in the year by the model

$$N(x) = 50 + 100 \ln\left(\frac{x}{100} + 2\right).$$

(a) How many memberships will be sold in 1 year if $1000 is spent on advertising? If $1500 is spent?

(b) If the number of memberships for a year is 400, is it worthwhile to spend $1000 on advertising?
Explain.

30. **Annuity:** If P dollars is deposited at the end of each period for n periods in an annuity that earns interest at a rate of r percent per period, the *future value A* of the annuity is given by the equation

$$n \log(1 + r) + \log P = \log(Ar + P).$$

(a) Solve this equation for A in terms of P, n, and r.

(b) Suppose that a college professor plans to invest $210 per month for retirement in a 403b tax-deferred annuity. The annuity pays an annual interest rate of 7.2%.

Express the future value A as a function of n, the number of monthly deposits.

(c) What will be the earnings of the investment after

6 months?

1 year?

5 years?

(d) **G** Use a grapher to graph the function in part (a). Then determine how long it will take (to the nearest month) an investment to become worth over $3000; over $5000; over $20,000.

31. **Business:** A group of senior citizens determines that the cost C per person (in dollars) of chartering an airplane to a nearby casino is given by the function

$$C(x) = 205 + \frac{100}{x}, 1 \le x \le 200$$

where x is the number of people.

(a) Determine

$$C^{-1}(x).$$

(b) Interpret

$$C^{-1}.$$

32. **G** **Social Science**—Spread of a Rumor: Suppose that sociologists estimate that if a person starts a rumor in a small town with a population of 20,000, the rumor will spread according to the function

$$N = 20,000[1 + 1.4e^{(-t+3)}]^{-1} - 16,700$$

where N is the number of people who have heard the rumor t hours later, where $t \ge 5$.

(a) According to this function, how many people will have heard the rumor after 5 hours? 8 hours? 10 hours?

(b) Use a grapher to graph the function. Interpret the graph, and then determine how many hours it will take for at least 2000 people to hear the rumor.

◆ **CHAPTER 4 TEST**

1. Let $f(x) = 4x - 1$.
   (a) Find $f^{-1}(x)$.
   (b) Verify that $f[f^{-1}(x)] = x$.
   (c) Sketch the graphs of f and f^{-1} on the same coordinate system.

2. Let $f(x) = 3^{x^2-x}$. Find each value. Round off the answers to three decimal places.
   (a) $f(-1)$
   (b) $f(2)$
   (c) $f(0.3)$
   (d) $f(\sqrt{5})$

3. $f(x) = \left(\dfrac{1}{5}\right)^x$.
   (a) Sketch the graph of f.
   (b) Specify the domain and range of f.
   (c) Indicate whether f is increasing or decreasing.
   (d) Identify any asymptotes for the graph of f.
   (e) Describe a sequence of transformations to change the graph of f to the graph of $g(x) = 2(1/5)^x - 3$. Then sketch the graph of g, and identify the asymptotes of the graph.

4. (a) Convert $6^u = 7$ to logarithmic form.
   (b) Convert $\log_3 x = b$ to exponential form.

5. Without using a calculator, simplify each expression.

   (a) $\ln \dfrac{1}{e^3}$ (b) $\ln \sqrt[7]{e}$

   (c) $e^{\ln \sqrt{x+3}}$ (d) $e^{\ln x^2}$

6. Let $f(x) = \log x$.
   (a) Find each value: $f(0.1)$, $f(1)$, $f(3)$, and $f(7)$. Round off each answer to four decimal places.
   (b) Sketch the graph of f, and specify its domain and range.
   (c) Does f have an inverse? Explain. If it does, find f^{-1} and sketch the graphs of f and f^{-1} on the same coordinate system.
   (d) Describe a sequence of transformations to change the graph of f to the graph of $g(x) = -\log(x + 7)$. Sketch the graph of g and find all its asymptotes.

7. Indicate whether each statement is true or false. Assume all variables are positive.

   (a) $\ln \dfrac{u}{7} = \ln u - \ln 7$

   (b) $\dfrac{\ln a}{\ln b} = \ln \dfrac{a}{b}$

   (c) $\dfrac{\log_a u}{t} = \log_a u^{1/t}$

   (d) $\log(xy)^t = t \log_y x + t$

8. Solve for x. Round off the answers to two decimal places.
   (a) $\log_3(2x - 1) = 2$
   (b) $x^5 e^{-4 \ln x} = 3$
   (c) $\log_2 x + \log_2(x + 1) = 1$
   (d) $3^{x+2} = 7$
   (e) $5^x \le 9^x$
   (f) $\ln x > 4 - x$

9. Sketch the graphs of $f(x) = \ln x + \ln 3$ and $g(x) = \ln(x + 3)$ on the same coordinate system. Use the graphs to determine the value of x for which $f(x) = g(x)$.

10. Suppose that $4000 is invested in an account paying an annual interest rate of 5%.
    (a) Find the amount in the account at the end of 2 years if the interest is compounded daily.
    (b) How long will it take for this investment to triple if interest is compounded continuously?

11. Suppose that the number N of bacteria in a certain culture is approximated by the function

    $$N(t) = N_0 \cdot 4^{0.05t}$$

    where N_0 represents the initial number and t is the number of elapsed hours.
    (a) If 6000 bacteria are present initially, how many will be present after 15 hours?
    (b) Sketch the graph of N if $N_0 = 6000$, and interpret the graph.
    (c) How many hours after the initial observation will there be 14,000 bacteria if $N_0 = 6000$? Round off the answer to two decimal places.

Trigonometric Functions

Many applications of trigonometric functions involve periodic or cyclic phenomena. For instance, tides cause ocean, sea, and lake currents to flow into and out of coastlines with varying depths that repeat over intervals of time. Trigonometric functions enable us to examine such situations. Suppose that during a 24-hour day, the average depth d (in feet) of water in a certain location, t hours after midnight, is given by the model

$$d = 8 + 4 \cos\left(\frac{4\pi}{25} t\right).$$

This model is analyzed in Example 8 on page 344.

Originally, trigonometry was limited to the numerical solutions of triangles and their applications, first in astronomy and later in surveying and navigation.

Today we recognize two points of view in the investigation of the trigonometric functions. In one approach, the unit circle is used to define the trigonometric functions for real numbers. These functions provide mathematical models for periodic phenomena encountered in electronics, engineering, the physical sciences, and the life sciences—phenomena such as sound waves, electromagnetic waves, light waves, business cycles, and biological systems. In the other approach, the trigonometric functions establish relationships between angles and sides of triangles that are used routinely in calculations made by surveyors, engineers, and navigators. We shall explore both approaches and examine the relationships between these two types of trigonometric functions and their properties.

Objectives

Figure 1

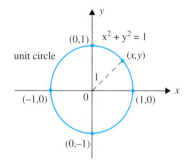

5.1 Preliminary Trigonometric Concepts

We begin by establishing some preliminary concepts necessary for the study of the trigonometric functions.

Locating Points on the Unit Circle

The graph of the equation

$$x^2 + y^2 = 1$$

is called the **unit circle.** It is a circle of radius 1 whose center is at the origin O of a rectangular (Cartesian) coordinate system (Figure 1).

By referring to Figure 2, it is possible to associate each real number with the coordinates of a point on the unit circle in the following way. First, assume that a number line L has the same scale unit as the one used for the unit circle. Next, place the number 0 on the real number line L so that it coincides with the point $(1, 0)$ on the unit circle. Then, the line L is "wrapped around" the circle, either in a *counterclockwise sense* (using the positive part of the real line L) or in a *clockwise sense* (using the negative part of the real line L). Thus, if $t_1 > 0$, it is associated with a point on the unit circle by moving $|t_1|$ units counterclockwise along the circumference of the circle, starting at the point $(1, 0)$ and ending at, say, (x_1, y_1), as illustrated in Figure 2a. If $t_2 < 0$, it is associated with a point on the circle by moving $|t_2|$ units clockwise along the circumference of the circle, starting at the point $(1, 0)$ and ending at, say, (x_2, y_2), as illustrated in Figure 2b.

Figure 2 $t_1 > 0$ associated with (x_1, y_1) $t_2 < 0$ associated with (x_2, y_2)

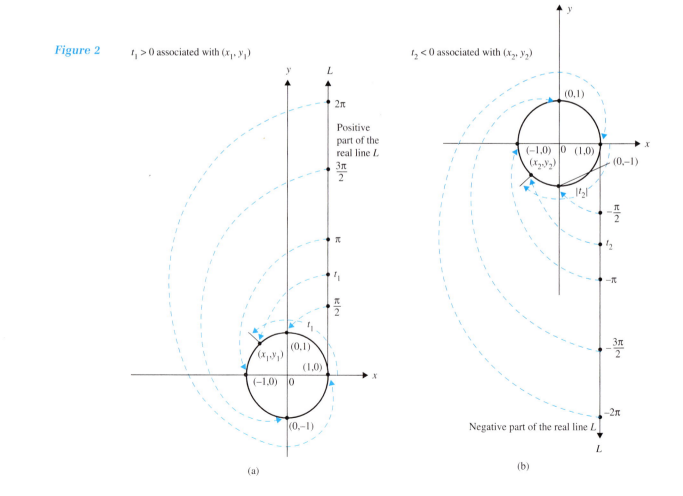

(a)   (b)

Since the circumference C of any circle is given by $C = 2\pi r$, where r represents the length of the radius of the circle, it follows that the circumference of the unit circle is 2π. If 3.14 is used as an approximation for π, the circumference of the unit circle is approximately 6.28 units. Figure 3 displays some points along the circumference of the unit circle that are obtained by the wrapping process described in Table 1.

Figure 3

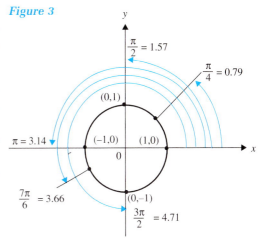

TABLE 1

Start at (1, 0) and Move Counterclockwise	Actual Distance	Approximate Distance ($\pi = 3.14$)
One-eighth of the way around the circumference	$\frac{1}{8}(2\pi) = \frac{\pi}{4}$	0.79
One-fourth of the way around the circumference	$\frac{1}{4}(2\pi) = \frac{\pi}{2}$	1.57
One-half of the way around the circumference	$\frac{1}{2}(2\pi) = \pi$	3.14
Seven-twelfths of the way around the circumference	$\frac{7}{12}(2\pi) = \frac{7\pi}{6}$	3.66
Three-fourths of the way around the circumference	$\frac{3}{4}(2\pi) = \frac{3\pi}{2}$	4.71

As a result of the wrapping process just described, it is possible to associate each real number t with a point $P = (x, y)$ on the unit circle, and we use function notation to denote this association as:

$$P(t) = (x, y).$$

Table 2 lists the designation of the points labeled in Figure 4. Thus, we observe that 0 is associated with the point $(1, 0)$; $\pi/2$ is associated with the point $(0, 1)$; π is associated with the point $(-1, 0)$; $3\pi/2$ is associated with the point $(0, -1)$; $-\pi/2$ is associated with the point $(0, -1)$; and $-\pi$ is associated with the point $(-1, 0)$.

TABLE 2

$P(t)$	$P(0)$	$P\left(\frac{\pi}{2}\right)$	$P(\pi)$	$P\left(\frac{3\pi}{2}\right)$	$P\left(-\frac{\pi}{2}\right)$	$P(-\pi)$
(x, y)	$(1, 0)$	$(0, 1)$	$(-1, 0)$	$(0, -1)$	$(0, -1)$	$(-1, 0)$

Figure 4

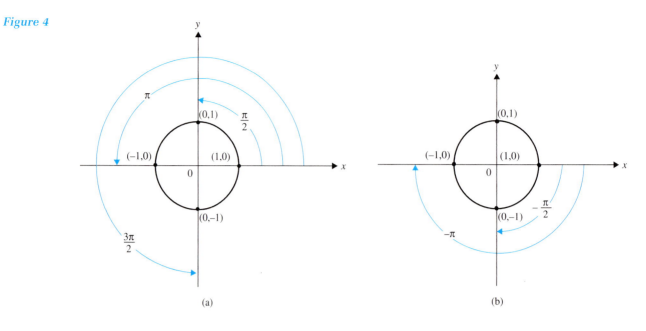

(a)   (b)

For certain values of t it is possible to find the exact values of the coordinates of $P(t) = (x, y)$ by using some of the analytical tools already studied.

For instance, the coordinates of $P(\pi/4)$ can be found as follows. Let $P(\pi/4) = (a, b)$. Since

$$P\left(\frac{\pi}{4}\right) = P\left(\frac{1}{2} \cdot \frac{\pi}{2}\right),$$

Figure 5

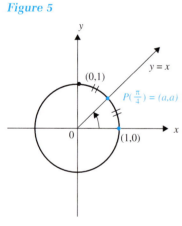

$P(\pi/4)$ is the midpoint of the arc joining the points $(1, 0)$ and $(0, 1)$ on the unit circle (Figure 5). Thus, $P(\pi/4)$ must lie on the line $y = x$, so that $a = b$; that is,

$$P\left(\frac{\pi}{4}\right) = (a, a).$$

Since the coordinates of any point on the unit circle satisfy the equation $x^2 + y^2 = 1$, we have

$$a^2 + a^2 = 1$$
$$2a^2 = 1$$
$$a^2 = \frac{1}{2}.$$

Because $a > 0$ for a point in quadrant I,

$$a = \sqrt{\frac{1}{2}} = \frac{1}{\sqrt{2}} = \frac{\sqrt{2}}{2}$$

so that

$$P\left(\frac{\pi}{4}\right) = \left(\frac{\sqrt{2}}{2}, \frac{\sqrt{2}}{2}\right).$$

Similar arguments can be used to find that the coordinates of $P(\pi/3)$ and $P(\pi/6)$ are, respectively,

$$P\left(\frac{\pi}{3}\right) = \left(\frac{1}{2}, \frac{\sqrt{3}}{2}\right) \quad \text{and} \quad P\left(\frac{\pi}{6}\right) = \left(\frac{\sqrt{3}}{2}, \frac{1}{2}\right) \quad \text{(Problem 54)}.$$

The results from the preceding discussion are summarized in Table 3 and displayed in Figure 6.

TABLE 3

$P(t)$	$P\left(\dfrac{\pi}{6}\right)$	$P\left(\dfrac{\pi}{4}\right)$	$P\left(\dfrac{\pi}{3}\right)$
(x, y)	$\left(\dfrac{\sqrt{3}}{2}, \dfrac{1}{2}\right)$	$\left(\dfrac{\sqrt{2}}{2}, \dfrac{\sqrt{2}}{2}\right)$	$\left(\dfrac{1}{2}, \dfrac{\sqrt{3}}{2}\right)$

Figure 6

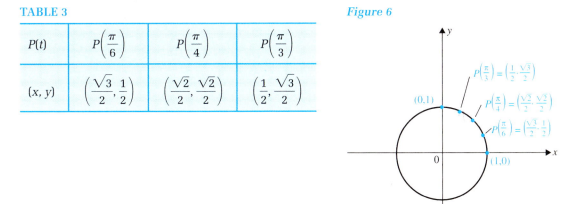

For other values of t, the tools that we have developed so far are insufficient to determine the exact location of $P(t)$. However, it is often helpful to find the quadrant containing the point.

EXAMPLE 1   **Determining the Quadrant Where $P(t)$ Lies**

Find the quadrant containing the given point (use $\pi = 3.14$).

(a) $P(2)$ (b) $P(-1)$

Figure 7

Solution   (a) Since $\pi = 3.14$ and $\pi/2 = 1.57$ (approx.), then

$$\frac{\pi}{2} < 2 < \pi$$

so that $P(2)$ is in quadrant II (Figure 7).

(b) Since $-\pi/2 = -1.57$ (approx.),

$$-\frac{\pi}{2} < -1 < 0.$$

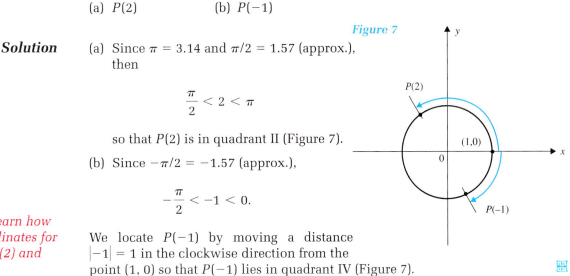

Later, we shall learn how to find the coordinates for points such as P(2) and P(−1).

We locate $P(-1)$ by moving a distance $|-1| = 1$ in the clockwise direction from the point $(1, 0)$ so that $P(-1)$ lies in quadrant IV (Figure 7).

The wrapping process enables us to associate real numbers with points on the unit circle. This concept provides the basis for the modern viewpoint of the study of trigonometry, which will be introduced in Section 5.2. Before introducing the trigonometric functions of angles, we need to review angles and their measurements (see Appendix II for review topics).

Describing and Measuring Angles

In plane geometry, recall that an **angle** is determined by rotating a ray about its endpoint (the **vertex** of the angle) from some initial position (the **initial side** of the angle) to a terminal position (the **terminal side** of the angle) (Figure 8a). In one scheme that is used to describe an angle, the middle letter represents the vertex, the first letter represents a point on the initial side, and the third letter represents a point on the terminal side. Thus we denote the angle in Figure 8a as $\angle QPR$. Angles are also denoted by lowercase Greek letters such as α in Figure 8b (α is the Greek letter *alpha*). Other Greek letters often used are β (*beta*) (Figure 8c), γ (*gamma*), θ (*theta*), and ϕ (*phi*).

Figure 8

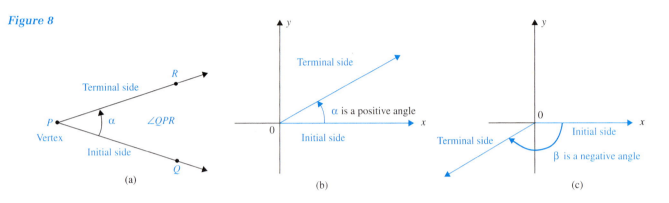

(a) (b) (c)

An angle is in **standard position** if it is placed in a rectangular coordinate system with its vertex at the origin O and with the initial side coinciding with the positive x axis. If the angle is formed by a counterclockwise rotation, the angle is considered to be **positive,** whereas if the angle is formed by a clockwise rotation, the angle is **negative.** Figure 8b displays positive angle α in standard position, and Figure 8c displays negative angle β in standard position.

Normally we measure an angle by using *degree* or *radian* measurements.

Definition
One Degree

An angle of **one degree** (1°) is an angle formed by $\dfrac{1}{360}$ of a complete counterclockwise revolution.

Figure 9 shows angles in standard position measured in degrees. We refer to an angle formed by one-half of a complete counterclockwise revolution as a **straight angle;** it has a measure of $\frac{1}{2}(360°) = 180°$ (Figure 9a). An angle of one-quarter of a counterclockwise revolution is a **right angle;** it has a measure of $\frac{1}{4}(360°) = 90°$ (Figure 9b). An angle is **acute** if its degree measure is between 0° and 90°, for example, 70° (Figure 9c). If an angle measures between 90° and 180°, it is **obtuse,** for example, 120° (Figure 9d).

Figure 9

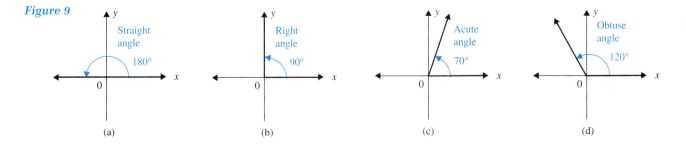

(a) (b) (c) (d)

If the terminal side of an angle in standard position lies on either the x axis or the y axis, then the angle is called **quadrantal.** For example, $-360°$, $-270°$, $-180°$, $0°$, $90°$, $180°$, $270°$, and $360°$ are quadrantal angles. Figures 9a and 9b display quadrantals $180°$ and $90°$, respectively.

Angles in standard position that have the same terminal sides are called **coterminal** angles. For example, the three angles

$$30°, \ -330°, \ \text{and} \ 750°$$

are coterminal (Figure 10).

Figure 10

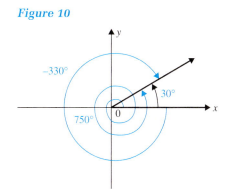

Although parts of a degree can be expressed as a decimal, such parts are sometimes given in *minutes* and *seconds*. A degree can be divided into 60 equal parts called **minutes** (′); a minute can be divided into 60 equal parts called **seconds** (″). Using these relationships, we have

$$1' = \left(\frac{1}{60}\right)^{\circ}, \quad 1'' = \left(\frac{1}{60}\right)', \quad 1'' = \left(\frac{1}{3600}\right)^{\circ}, \quad 1° = 60', \quad \text{and} \quad 1' = 60''.$$

Some calculators have keys that automatically convert decimal degree measures to degrees, minutes, and seconds and vice versa. The next example shows how such conversions can be made without such special keys.

EXAMPLE 2   **Converting Degree Measures**

(a) Express $37.45°$ in terms of degrees, minutes, and seconds.

(b) Express $23°17'37''$ in decimal degree measure to four decimal places.

Solution   (a) $37.45° = 37° + 0.45°$; and $0.45° = (0.45)(60') = 27'$.
Therefore, $37.45° = 37°27'0''$.

(b) $23°17'37'' = \left(23 + \dfrac{17}{60} + \dfrac{37}{3600}\right)^{\circ}$

$= 23.2936°$ (approx.).

Another frequently used unit of angle measurement is a *radian*.

Radian Measure
Relative to
the Unit Circle

> One **radian** is the measure of a positive angle that intercepts an arc of length 1 on a circle of radius 1 (Figure 11).

Figure 11

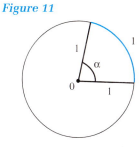

α is an angle of 1 radian

An angle of radian measure t in standard position has its initial side at the positive x axis and its terminal side at the ray containing the origin and the point $P(t)$, which is located through the use of the wrapping scheme described earlier. For example, an angle θ of one radian is generated by a counterclockwise rotation in which the point of intersection of the rotating ray with the unit circle travels 1 unit (Figure 12a). Similarly, an angle θ of radian measure $5\pi/6$ is generated by a counterclockwise rotation in which the point of intersection of the rotating ray with the unit circle travels $5\pi/6$ units (Figure 12b). The angle θ with radian measure -8.9 is generated by a clockwise rotation in which the point of intersection of the rotating ray with the unit circle travels 8.9 units (Figure 12c).

Figure 12

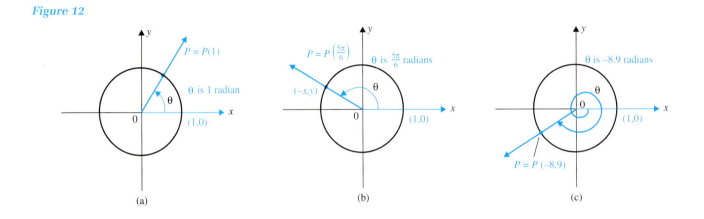

(a)   (b)   (c)

Notice that, given any angle θ, we can associate with it a real number t (the radian measure of θ). Conversely, given any real number t, we can associate with it the angle θ in standard position with terminal side containing the origin O and point $P(t)$.

Figures 13a and b show the two angles in standard position designated by θ when t is $\pi/4$ and $-5\pi/6$, respectively.

Figure 13

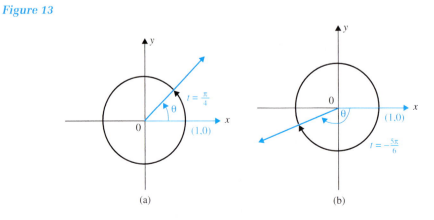

(a)   (b)

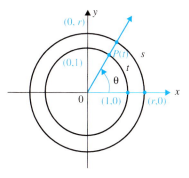

Figure 14

A **central angle** of a circle is an angle that has its vertex at the center of the circle. Figure 14 shows a central angle θ that subtends an arc of length t units on the unit circle and an arc of length s units on a concentric circle of radius r. We know that the radian measure of θ is t. From plane geometry, we know that the ratio of the arc lengths is the same as the ratio of the corresponding radii. That is,

$$\frac{t}{s} = \frac{1}{r}$$

so that we get the following results.

<u>**Radian Measure**</u>
Relative to Any Circle

$$t = \frac{s}{r}.$$

Figure 15

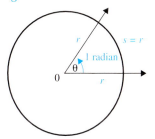

Notice that when $s = r$, the radian measure of θ equals 1; that is,

1 radian is the measure of a central angle that intercepts an arc of the circle equal in length to the radius of the circle (Figure 15).

Converting Angle Measurements

Sometimes we need to convert an angle measure from degrees to radians or vice versa. Table 4 displays corresponding degree and radian measures for some angles.

TABLE 4

Degree Measure (D)	Corresponding Radian Measure (R)
360°	2π
180°	π
90°	$\dfrac{\pi}{2}$

Graphers have a feature that automatically converts radian measure to degree measure and vice versa.

On the basis of the results in Table 4, we have the following rules for converting between radian measures and degree measures.

If an angle has radian measure $R \neq 0$ and degree measure D, then the following ratio holds:

<u>**Formulas**</u>
Angle Measurement
Conversions

$$\frac{D}{R} = \frac{180}{\pi} \quad \text{or} \quad \frac{R}{D} = \frac{\pi}{180}$$

Notice that an angle of 1 radian has a degree measure given by $(180/\pi)°$, or approximately $57.2958°$ or $57°17'45''$.

EXAMPLE 3   **Converting Degrees to Radians**

Convert each degree measure to radian measure.

(a) $-150°$ (b) $26.85°$, to four decimal places

Solution   Since $R = (\pi/180)D$, there is $\pi/180$ radian in each degree, so that

(a) $-150°$ corresponds to $\dfrac{\pi}{180}(-150) = -\dfrac{5\pi}{6}$ radians.

(b) $26.85°$ corresponds to $\dfrac{\pi}{180}(26.85) = 0.4686$ radian (approx.).

Finding Arc Lengths and Areas of Sectors

Figure 16

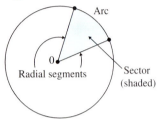

Arc

0

Radial segments

Sector (shaded)

As displayed in Figure 16, a **sector of a circle** is a region inside the circle bounded by an arc and radial line segments formed by the center of the circle 0 and the end points of the arc. Our objective is to derive formulas for the arc lengths and areas of sectors.

Earlier we established that if a central angle θ of a circle of radius r subtends an arc of length s, then the radian measure t of θ is given by the formula $t = \dfrac{s}{r}$.

By solving the formula for s, we obtain the following formula.

Formula
Arc Length

> The **arc length of a circular sector of a circle** with radius r and central angle θ is given by
>
> $$s = rt,$$
>
> where t is the measure of θ expressed in radians.

The entire interior of a circle is also a sector.

From geometry, the ratio $\dfrac{\text{Area of a Sector of a circle}}{\text{Radian measure of its central angle}}$ is constant.

Thus, if we have a sector formed from a circle of radius r with a central angle that has radian measure t (Figure 17), it follows that

$$\frac{\text{Area of Sector}}{t} = \frac{\text{Area of Circle}}{2\pi} = \frac{\pi r^2}{2\pi} = \frac{1}{2}r^2 \quad \text{or} \quad \frac{\text{Area of Sector}}{t} = \frac{1}{2}r^2$$

Figure 17

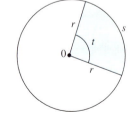

r

s

t

0

r

so that

Area of Sector $= \dfrac{1}{2}r^2t$. Since $s = rt$, we can rewrite the latter formula as

$$\text{Area of Sector} = \frac{1}{2}r \cdot (rt) = \frac{1}{2}rs.$$

Thus we have the following general result.

Formulas
Area of a Sector

> If A represents the **area of a sector of a circle** of radius r formed by a central angle θ of radian measure t and a subtended arc length s, then
>
> $$A = \frac{1}{2}r^2t \quad \text{or} \quad A = \frac{1}{2}rs.$$

EXAMPLE 4   **Finding an Arc Length and Area of a Sector**

Given a circle of radius 6 centimeters and a central angle of $\theta = 240°$ (Figure 18).

(a) Find the length s of the arc intercepted on the circle.

(b) Find the area of the sector formed by the angle.

Solution   (a) We first convert 240° to radian measure t to get

$$t = \frac{\pi}{180}\, D = \left(\frac{\pi}{180}\right)(240) = \frac{4\pi}{3} \text{ radians.}$$

Figure 18

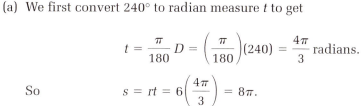

So
$$s = rt = 6\left(\frac{4\pi}{3}\right) = 8\pi.$$

Therefore, the arc length is 8π or about 25.13 centimeters.

(b) To find the area of the sector we can proceed in either of the following ways:

Using the Result from Part (a)	Alternate Solution
Substituting our result from part (a) into the formula $$A = \frac{1}{2}rs$$ we get $$A = \frac{1}{2}(6)(8\pi)$$ $$= 24\pi.$$	After converting 240° to radian measure $\frac{4\pi}{3}$, we substitute into the formula $$A = \frac{1}{2}r^2 t$$ to get $$A = \frac{1}{2}(6)^2\left(\frac{4\pi}{3}\right)$$ $$= \frac{1}{2}(36)\left(\frac{4\pi}{3}\right)$$ $$= 24\pi.$$

Either way, the area is 24π or about 75.40 square centimeters.

Solving Applied Problems—Circular Motion

Figure 19

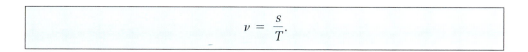

Path of a
moving object

Angles measured in radians allow us to study the motion of an object moving at a constant speed around a circle of radius r. Suppose that a moving object starts at point A and after T units of time reaches point P (Figure 19). If the arc AP has length s, then the object has moved s units of distance in T units of time. Since *distance divided by time equals speed*, we represent the *linear speed v* of the moving object as follows.

Formula
Linear Speed

$$v = \frac{s}{T}.$$

The ratio that measures the rate of change of the angle θ, where θ is measured in radians, with respect to T units of time is called the **angular speed** ω (Greek letter omega) and is given by the following formula.

Formula
Angular Speed

$$\omega = \frac{\theta}{T}.$$

Since the arc length $s = r\theta$, it follows that

$$\nu \text{ (linear speed)} = \frac{s}{T} = \frac{r\theta}{T} = r\left(\frac{\theta}{T}\right) = r\omega.$$

So the relationship between the linear speed ν and angular speed ω is given by

$$\nu = r\omega \quad \text{or} \quad \omega = \frac{\nu}{r}.$$

In other words, the linear speed of an object moving along a circular path of radius r is the product of the angular speed and the radius.

EXAMPLE 5   **Finding the Angular Speed of a Flywheel**

A belt passes over the rim of a flywheel with radius 18 centimeters (Figure 20).

(a) Find the angular speed of a point on the rim of the wheel if the belt drives the wheel at a linear speed of $\nu = 576$ centimeters per second.

(b) Find the angular speed if the belt drives the wheel at a rate of 12 rotations every 2 seconds. Round off the answer to two decimal places.

Solution

(a) Since the angular speed is given by

$$\omega = \frac{\nu}{r}, \text{ where } \nu = 576 \text{ and } r = 18,$$

it follows that

$$\omega = \frac{576}{18} = 32 \text{ radians per second.}$$

Figure 20

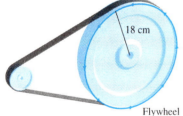

18 cm

Flywheel

(b) Since the belt drives a point on the rim of the wheel 12 times around the circle in 2 seconds, a point on the rim of the wheel rotates $12(2\pi) = 24\pi$ radians in 2 seconds. Therefore, the angular speed of the belt is

$$\omega = \frac{\theta}{T} = \frac{24\pi}{2} = 12\pi, \text{ or approximately 37.70 radians per second.}$$

PROBLEM SET 5.1

Mastering the Concepts

In problems 1 and 2, find the value of t if the point $P(t)$ is located by starting at point $(1, 0)$ and moving around the circumference of the unit circle as specified.

1. (a) One-third of the way, clockwise
   (b) Five-fourths of the way, counterclockwise

2. (a) Three-eighths of the way, clockwise
   (b) One-sixth of the way, counterclockwise

3. Use Figure 6 on page 305 and symmetry to find the coordinates of each point.

   (a) $P\left(-\dfrac{\pi}{3}\right)$ (b) $P\left(-\dfrac{\pi}{4}\right)$ (c) $P\left(-\dfrac{\pi}{6}\right)$

4. Find the coordinates of each point.

   (a) $P(4\pi)$ (b) $P\left(\dfrac{5\pi}{2}\right)$ (c) $P\left(-\dfrac{5\pi}{2}\right)$

In problems 5–8, display each point on the unit circle and find the quadrant (if any) containing each point.

5. (a) $P\left(\dfrac{\pi}{7}\right)$

   (b) $P\left(\dfrac{-3\pi}{5}\right)$

   (c) $P\left(\dfrac{-2\pi}{5}\right)$

   (d) $P\left(\dfrac{17\pi}{8}\right)$

6. (a) $P\left(\dfrac{17\pi}{6}\right)$

   (b) $P\left(\dfrac{-8\pi}{7}\right)$

   (c) $P\left(\dfrac{31\pi}{8}\right)$

   (d) $P\left(\dfrac{-9\pi}{5}\right)$

7. (a) $P(6)$
   (b) $P(1.4)$
   (c) $P(3.6)$
   (d) $P(-5.7)$

8. (a) $P(-11.2)$
   (b) $P(23.5)$
   (c) $P(-3.1)$
   (d) $P(-13.4)$

In problems 9–14, locate each angle in standard position. Specify its quadrant location, and sketch two angles coterminal with the given angle.

9. (a) $\dfrac{3\pi}{4}$

   (b) $-\dfrac{\pi}{6}$

10. (a) $-\dfrac{5\pi}{12}$

   (b) $\dfrac{7\pi}{6}$

11. (a) $-45°$
   (b) $120°$

12. (a) $210°$
   (b) $105°$

13. (a) $\dfrac{17\pi}{3}$
   (b) $-700°$

14. (a) $-\dfrac{11\pi}{3}$
   (b) $538°$

In problems 15 and 16, express each angle measure in terms of degrees, minutes, and seconds.

15. (a) $16.31°$
   (b) $-87.81°$
   (c) $-156.63°$
   (d) $89.74°$

16. (a) $64.14°$
   (b) $-12.16°$
   (c) $-213.68°$
   (d) $463.09°$

In problems 17 and 18, express each angle in terms of decimal degrees. Round off each answer to two decimal places.

17. (a) $38°18'$
   (b) $65°11'23''$
   (c) $-141°28'15''$
   (d) $244°46'15''$

18. (a) $35°41'$
   (b) $-48°15'25''$
   (c) $1°1'1''$
   (d) $-10°10'10''$

In problems 19–24, convert each angle measure to radians. Express the answer in terms of π and also as a decimal rounded off to two decimal places.

19. (a) $75°$
   (b) $-135°$

20. (a) $240°$
   (b) $-7.5°$

21. (a) $-95°$
   (b) $444°$

22. (a) $-220°$
   (b) $330°$

23. (a) $67.5°$
   (b) $30°30'36''$

24. (a) $17.45°$
   (b) $115°13'44''$

In problems 25–30, convert each radian measure to degrees. Round off each answer to two decimal places.

25. (a) $\dfrac{2\pi}{3}$

   (b) $\dfrac{43\pi}{6}$

26. (a) $\dfrac{11\pi}{6}$

   (b) $\dfrac{7\pi}{18}$

27. (a) $\dfrac{7\pi}{12}$

   (b) $-\dfrac{4\pi}{9}$

28. (a) $-\dfrac{3\pi}{8}$

   (b) $-\dfrac{\pi}{14}$

29. (a) 5
   (b) -2.3

30. (a) -1
   (b) 4.6

In problems 31–36, s represents the arc length corresponding to a central angle θ on a circle of radius r.

(a) Find the missing measurement to two decimal places.
(b) Sketch the sector that is formed in each situation and find its area. Round off the answer to two decimal places.

31. $r = 7$ inches, $\theta = \dfrac{3\pi}{14}$, $s = ?$

32. $s = 6$ feet, $\theta = \dfrac{\pi}{7}$, $r = ?$

33. $r = 1.8$ meters, $\theta = 217°$, $s = ?$

34. $r = 11$ centimeters, $\theta = \dfrac{7\pi}{5}$, $s = ?$

35. $s = 5$ yards $\theta = \dfrac{5\pi}{3}$, $r = ?$

36. $r = 7$ feet, $\theta = 110°$, $s = ?$

Applying the Concepts

In problems 37–48, round off the results to two decimal places.

37. Nautical Mile: A nautical mile is the arc length intersected on the surface of the Earth by a central angle of 1 minute (Figure 21). Assume that the radius of the Earth is 3960 miles.

    (a) How many miles are there in 1 nautical mile?

    (b) How many feet are there in 1 nautical mile?

Figure 21

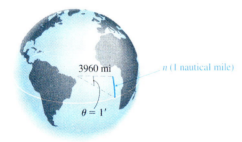

38. Pendulum Swing: Find the length of a pendulum if the tip of the pendulum traces an arc of 6 feet and the measure of the angle that this arc subtends is $\pi/5$ radian.

39. Child's Swing: The length of each chain supporting the seat of a child's swing is 8 feet. When the swing moves from its extreme forward position to its extreme backward position, the radian measure of the angle swept out by the chains is $(5\pi)/6$. How far does the seat travel in one trip between these extreme points; that is, what is the length of the arc generated by the seat through one swing between the two extreme points?

40. Revolving Door: A revolving door rotates through an angle of 2.64 radians before it is stopped. If one half of the door is 4 feet wide, through what distance does the edge of the door move?

41. Windshield Wiper: A windshield wiper of a car is 48 centimeters long, and it rotates at an angular speed of 12 degrees per second. Find the linear speed of the tip of the wiper.

42. Central Angle of a Pulley: A pulley of radius 25 centimeters uses a belt to drive another pulley of radius 15 centimeters. Find the radian measure of the angle θ_2 through which the smaller pulley turns as the larger pulley makes an angle θ_1 of 140° (Figure 22).

Figure 22

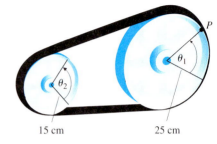

15 cm   25 cm

43. Tire Rotations: A car with tires that are 22 inches in diameter travels at 55 miles per hour.

    (a) Find the angular speed of each tire in radians per minute.

    (b) How many rotations per minute does each tire make?

44. Satellite Orbit: A satellite is orbiting the Earth in a circular orbit with a radius of 7680 kilometers.

If it makes $\dfrac{3}{4}$ of a revolution every hour, find:

    (a) Its angular speed

    (b) Its linear speed

45. Bicycle Sprocket Wheels: A bicycle's small sprocket wheel of radius 2.75 inches is connected by a chain to a large sprocket wheel of radius 4.50 inches (Figure 23). The small sprocket wheel turns at 50 rotations per minute. Find:

    (a) The linear speed (in inches per minute) of the chain connecting the sprocket wheels.

    (b) The number of rotations per minute of the large sprocket wheel.

Figure 23

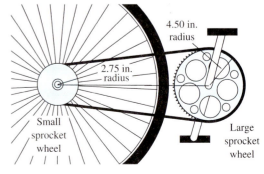

46. The Earth's Rotation: The Earth, which is approximately 93,000,000 miles from the sun, revolves about the sun in a nearly circular orbit in approximately 365 days. Find the linear speed (in miles per day) of the Earth in its orbit.

47. Waterwheel: A point on the rim of a circular waterwheel of radius 8 feet makes one complete rotation every 10 seconds. Find the angular and linear speeds of the circular motion of the moving point.

48. Steamboat: As the paddlewheel with radius 9 feet of a steamboat turns, a point on the end of the paddle blade makes one complete rotation every 8 seconds. Determine the angular and linear speeds of the circular motion of the point on the paddle blade.

Developing and Extending the Concepts

49. Figure 24 shows three circles with the same center, a common central angle θ, and radii r_1, r_2, and r_3, where $r_1 < r_2 < r_3$. A student makes the following assertion: Since the radian measure of a central angle of a circle depends on the radius of the circle, it follows that the circle with a larger radius has a bigger central angle. Explain why this assertion is false.

Figure 24

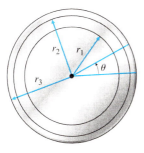

50. The following assertion is made: Suppose that angle θ_1 has radian measure t and angle θ_2 has radian measure $t + 2\pi$. When both angles are placed in standard position, they have the same terminal side. It follows that

$$\theta_1 = \theta_2.$$

Explain why this assertion is wrong.

51. Which angle is larger, angle $\alpha = 1$ or angle $\beta = 1°$? Explain your answer.

52. Suppose that you ride on a merry-go-round at a carnival. You realize that you go faster when you sit near the outside than when you sit near the inside, even though the merry-go-round travels at a constant angular speed (Figure 25). Explain this phenomenon.

Figure 25

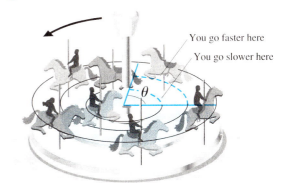

You go faster here

You go slower here

53. What is the radian measure of the smaller of the positive angles between the hands of a clock at 11:30?

54. Prove that

(a) $P\left(\dfrac{\pi}{3}\right) = \left(\dfrac{1}{2}, \dfrac{\sqrt{3}}{2}\right)$ *Hint:* Let $A = (1, 0)$,

$B = P\left(\dfrac{\pi}{3}\right) = (a, b)$, and $C = P\left(\dfrac{2\pi}{3}\right) = (-a, b)$.

Since the lengths of arcs $\overparen{AB}$ and

$\overparen{BC}$ are equal (each is of length $\dfrac{\pi}{3}$)

(Figure 26), it follows from geometry that the chords $\overline{AB}$ and $\overline{BC}$ are equal in length, so that

Figure 26

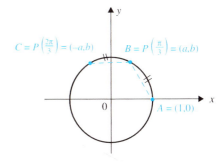

$C = P\left(\frac{2\pi}{3}\right) = (-a,b)$ $B = P\left(\frac{\pi}{3}\right) = (a,b)$

$A = (1,0)$

$|\overline{AB}| = |\overline{BC}|$ and thus $|\overline{AB}|^2 = |\overline{BC}|^2$.

Now use the distance formula.

(b) $P\left(\dfrac{\pi}{6}\right) = \left(\dfrac{\sqrt{3}}{2}, \dfrac{1}{2}\right)$

5.2 Trigonometric Functions of Real Numbers—Circular Functions

The coordinates (x, y) of point $P(t)$ on the unit circle (Figure 1) are used to define the six *trigonometric functions* or *circular functions* of t. These functions are referred to and are abbreviated as follows: (We may consider the point $P(t)$ in *any* quadrant.)

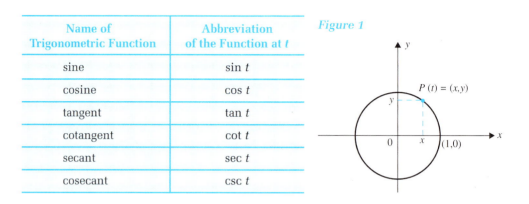

Name of Trigonometric Function	Abbreviation of the Function at t
sine	sin t
cosine	cos t
tangent	tan t
cotangent	cot t
secant	sec t
cosecant	csc t

Figure 1

Defining Trigonometric Functions of Real Numbers—Circular Functions

The six trigonometric functions, which have domains that consist of real numbers, are defined as follows:

Definition

Trigonometric Functions of a Real Number—Circular Functions

Let t be a real number and let $P(t) = (x, y)$ be the point on the unit circle associated with t (Figure 1). Then

$$\sin t = y \qquad\qquad \csc t = \frac{1}{y}, \quad y \neq 0$$

$$\cos t = x \qquad\qquad \sec t = \frac{1}{x}, \quad x \neq 0$$

$$\tan t = \frac{y}{x}, \quad x \neq 0 \qquad \cot t = \frac{x}{y}, \quad y \neq 0.$$

EXAMPLE 1   **Using the Definition of Circular Functions**

Use the definition to determine the values of the six circular functions of t when $P(t) = \left(-\frac{4}{5}, \frac{3}{5}\right)$.

Solution   Note that $\left(-\frac{4}{5}\right)^2 + \left(\frac{3}{5}\right)^2 = 1$, so that $P(t) = \left(-\frac{4}{5}, \frac{3}{5}\right)$ lies on the unit circle. It follows from the definition that

$$\sin t = y = \frac{3}{5}$$
$$\csc t = \frac{1}{y} = \frac{1}{\frac{3}{5}} = \frac{5}{3}$$

$$\cos t = x = -\frac{4}{5}$$
$$\sec t = \frac{1}{x} = \frac{1}{-\frac{4}{5}} = -\frac{5}{4}$$

$$\tan t = \frac{y}{x} = \frac{\frac{3}{5}}{-\frac{4}{5}} = -\frac{3}{4}$$
$$\cot t = \frac{x}{y} = \frac{-\frac{4}{5}}{\frac{3}{5}} = -\frac{4}{3}.$$

Evaluating Trigonometric Functions of Special Values

When $P(t)$ is a point on the coordinate axes, then the trigonometric values of t are found as illustrated in the next example.

EXAMPLE 2   **Evaluating Circular Functions**

Determine the values of the six trigonometric functions of t.

(a) $t = 0$ (b) $t = \frac{\pi}{2}$

Solution   (a) Since $P(0) = (1, 0)$ (Figure 2), it follows from the definition that the values are

Figure 2

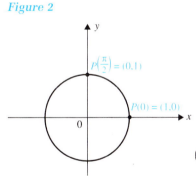

$$\sin 0 = y = 0$$
$$\csc 0 = \frac{1}{y} = \frac{1}{0} \text{ (undefined)}$$

$$\cos 0 = x = 1$$
$$\sec 0 = \frac{1}{x} = \frac{1}{1} = 1$$

$$\tan 0 = \frac{y}{x} = \frac{0}{1} = 0$$
$$\cot 0 = \frac{x}{y} = \frac{1}{0} \text{ (undefined)}.$$

(b) $P\left(\frac{\pi}{2}\right) = (0, 1)$ (Figure 2), so that from the definition it follows that the values are

$$\sin \frac{\pi}{2} = y = 1$$
$$\csc \frac{\pi}{2} = \frac{1}{y} = \frac{1}{1} = 1$$

$$\cos \frac{\pi}{2} = x = 0$$
$$\sec \frac{\pi}{2} = \frac{1}{x} = \frac{1}{0} \text{ (undefined)}$$

$$\tan \frac{\pi}{2} = \frac{y}{x} = \frac{1}{0} \text{ (undefined)}$$
$$\cot \frac{\pi}{2} = \frac{x}{y} = \frac{0}{1} = 0.$$

For *special values* of t, $\frac{\pi}{4}, \frac{\pi}{3}$, and $\frac{\pi}{6}$, we can find the trigonometric values by using results established in Section 5.1.

EXAMPLE 3   **Evaluating Circular Functions for $\dfrac{\pi}{4}$, $\dfrac{\pi}{3}$, and $\dfrac{\pi}{6}$**

Determine the six trigonometric values of t.

(a) $t = \dfrac{\pi}{4}$ (b) $t = \dfrac{\pi}{3}$ (c) $t = \dfrac{\pi}{6}$

Solution   (a) In Section 5.1, we proved that $P\left(\dfrac{\pi}{4}\right) = \left(\dfrac{\sqrt{2}}{2}, \dfrac{\sqrt{2}}{2}\right)$ (Figure 3). It follows from the definition that

Figure 3

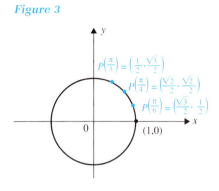

$$\sin \frac{\pi}{4} = y = \frac{\sqrt{2}}{2} \qquad \csc \frac{\pi}{4} = \frac{1}{y} = \frac{1}{\frac{\sqrt{2}}{2}} = \frac{2}{\sqrt{2}} = \sqrt{2}$$

$$\cos \frac{\pi}{4} = x = \frac{\sqrt{2}}{2} \qquad \sec \frac{\pi}{4} = \frac{1}{x} = \frac{1}{\frac{\sqrt{2}}{2}} = \frac{2}{\sqrt{2}} = \sqrt{2}$$

$$\tan \frac{\pi}{4} = \frac{y}{x} = \frac{\frac{\sqrt{2}}{2}}{\frac{\sqrt{2}}{2}} = 1 \qquad \cot \frac{\pi}{4} = \frac{x}{y} = \frac{\frac{\sqrt{2}}{2}}{\frac{\sqrt{2}}{2}} = 1.$$

(b) Since $P\left(\dfrac{\pi}{3}\right) = \left(\dfrac{1}{2}, \dfrac{\sqrt{3}}{2}\right)$ (Figure 3), it follows that

$$\sin \frac{\pi}{3} = y = \frac{\sqrt{3}}{2} \qquad \csc \frac{\pi}{3} = \frac{1}{y} = \frac{1}{\frac{\sqrt{3}}{2}} = \frac{2}{\sqrt{3}} = \frac{2\sqrt{3}}{3}$$

$$\cos \frac{\pi}{3} = x = \frac{1}{2} \qquad \sec \frac{\pi}{3} = \frac{1}{x} = \frac{1}{\frac{1}{2}} = 2$$

$$\tan \frac{\pi}{3} = \frac{y}{x} = \frac{\frac{\sqrt{3}}{2}}{\frac{1}{2}} = \sqrt{3} \qquad \cot \frac{\pi}{3} = \frac{x}{y} = \frac{\frac{1}{2}}{\frac{\sqrt{3}}{2}} = \frac{1}{\sqrt{3}} = \frac{\sqrt{3}}{3}.$$

(c) Since $P\left(\dfrac{\pi}{6}\right) = \left(\dfrac{\sqrt{3}}{2}, \dfrac{1}{2}\right)$ (Figure 3), it follows that

$$\sin \frac{\pi}{6} = y = \frac{1}{2} \qquad \csc \frac{\pi}{6} = \frac{1}{y} = \frac{1}{1/2} = 2$$

$$\cos \frac{\pi}{6} = x = \frac{\sqrt{3}}{2} \qquad \sec \frac{\pi}{6} = \frac{1}{x} = \frac{1}{\sqrt{3}/2} = \frac{2}{\sqrt{3}} = \frac{2\sqrt{3}}{3}$$

$$\tan \frac{\pi}{6} = \frac{y}{x} = \frac{\frac{1}{2}}{\frac{\sqrt{3}}{2}} = \frac{\sqrt{3}}{3} \qquad \cot \frac{\pi}{6} = \frac{x}{y} = \frac{\frac{\sqrt{3}}{2}}{\frac{1}{2}} = \sqrt{3}.$$

Table 1 lists a summary of the trigonometric function values for $\dfrac{\pi}{6}, \dfrac{\pi}{4}$ and $\dfrac{\pi}{3}$.

TABLE 1   Trigonometric Values for Special Real Numbers

Real Number t	$\sin t$	$\cos t$	$\tan t$	$\csc t$	$\sec t$	$\cot t$
$\dfrac{\pi}{6}$	$\dfrac{1}{2}$	$\dfrac{\sqrt{3}}{2}$	$\dfrac{\sqrt{3}}{3}$	2	$\dfrac{2\sqrt{3}}{3}$	$\sqrt{3}$
$\dfrac{\pi}{4}$	$\dfrac{\sqrt{2}}{2}$	$\dfrac{\sqrt{2}}{2}$	1	$\sqrt{2}$	$\sqrt{2}$	1
$\dfrac{\pi}{3}$	$\dfrac{\sqrt{3}}{2}$	$\dfrac{1}{2}$	$\sqrt{3}$	$\dfrac{2\sqrt{3}}{3}$	2	$\dfrac{\sqrt{3}}{3}$

Figure 4

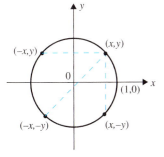

Using Symmetry to Evaluate Trigonometric Functions

Using the symmetry of the unit circle and special values we can find values of the trigonometric functions for other values of t when $P(t)$ is in quadrants other than quadrant I. Because of the symmetry of the unit circle $x^2 + y^2 = 1$, it follows that the reflection of any point on the circle across the x axis, y axis, and the origin results in another point on the circle. Thus, if (x, y) is a point on the unit circle in quadrant I, then $(x, -y)$, $(-x, y)$, and $(-x, -y)$ are also on the unit circle (Figure 4).

We know that $P(\pi/6) = (\sqrt{3}/2, 1/2)$. It follows that reflections of this point across the x axis, the y axis, and the origin are given by the points $P(11\pi/6) = (\sqrt{3}/2, -1/2)$, $P(5\pi/6) = (-\sqrt{3}/2, 1/2)$, and $P(7\pi/6) = (-\sqrt{3}/2, -1/2)$, respectively (Figure 5). With this information, the values of the trigonometric functions for $t = 5\pi/6$, $7\pi/6$, and $11\pi/6$ can be obtained by using the definition given earlier. For instance,

Figure 5

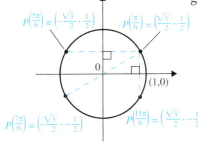

$$\sin \frac{5\pi}{6} = y = \frac{1}{2}, \cos \frac{7\pi}{6} = x = -\frac{\sqrt{3}}{2},$$

$$\tan \frac{11\pi}{6} = \frac{y}{x} = \frac{-1/2}{\sqrt{3}/2} = -\frac{1}{\sqrt{3}} = -\frac{\sqrt{3}}{3}.$$

The coordinates of the points on the unit circle corresponding to the values between 0 and 2π are listed on the back inner cover of this book.

EXAMPLE 4   **Evaluating Trigonometric Functions on a Unit Circle**

Use Figure 5 to find each value.

(a) $\sin\left(-\dfrac{\pi}{6}\right)$ (b) $\csc\left(-\dfrac{\pi}{6}\right)$

Solution   We locate $P(-\pi/6)$ by moving clockwise from the point $(1, 0)$ a distance of $|-\pi/6|$. We see that $P(-\pi/6)$ is in quadrant IV, and by using symmetry, we note that it coincides with $P(11\pi/6)$ (Figure 5), so that

$$P\left(-\frac{\pi}{6}\right) = P\left(\frac{11\pi}{6}\right) = \left(\frac{\sqrt{3}}{2}, -\frac{1}{2}\right).$$

Using the definition, we have

(a) $\sin\left(-\dfrac{\pi}{6}\right) = y = -\dfrac{1}{2}$ (b) $\csc\left(-\dfrac{\pi}{6}\right) = \dfrac{1}{y} = \dfrac{1}{-\dfrac{1}{2}} = -2.$

We can use a similar argument to find the coordinates for $P(t)$ and the values of the trigonometric functions if $t = 3\pi/4, 5\pi/4$, and $7\pi/4$ (Figure 6a), or if $t = 2\pi/3$, $4\pi/3$, and $5\pi/3$ (Figure 6b).

Figure 6

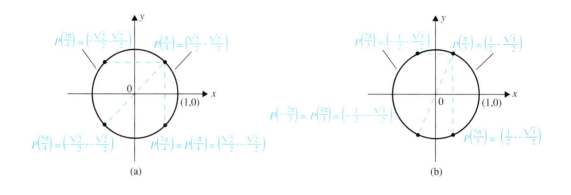

(a)   (b)

Thus,

$$\sin\left(\frac{3\pi}{4}\right) = y = \frac{\sqrt{2}}{2}$$

and

$$\cos\left(\frac{-\pi}{4}\right) = x = \frac{\sqrt{2}}{2}.$$

Also,

$$\tan\left(\frac{5\pi}{3}\right) = \frac{y}{x} = \frac{\left(-\frac{\sqrt{3}}{2}\right)}{\left(\frac{1}{2}\right)} = -\sqrt{3}$$

and

$$\cot\left(-\frac{2\pi}{3}\right) = \cot\left(\frac{4\pi}{3}\right) = \frac{x}{y} = \frac{\left(-\frac{1}{2}\right)}{\left(\frac{-\sqrt{3}}{2}\right)} = \frac{1}{\sqrt{3}} \text{ or } \frac{\sqrt{3}}{3}.$$

Approximating Trigonometric Function Values

We can approximate values of the trigonometric functions of a given real number by using a calculator. In order to perform such calculation for real numbers, the calculator must be set in *radian* mode.

Most calculators have keys for only the sine, cosine, and tangent functions. To evaluate the remaining three trigonometric functions on such calculators, we make use of the following reciprocal relationships.

Reciprocal Relationships

For a real number t, the following relationships hold:

(i) $\csc t = \dfrac{1}{\sin t}$ (ii) $\sec t = \dfrac{1}{\cos t}$ (iii) $\cot t = \dfrac{1}{\tan t}$

Figure 7

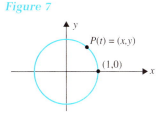

These relationships follow directly from the definitions of the trigonometric functions. By referring to Figure 7, we observe that

$$\csc t = \frac{1}{y} = \frac{1}{\sin t}$$

$$\sec t = \frac{1}{x} = \frac{1}{\cos t}$$

and

$$\cot t = \frac{x}{y} = \frac{1}{\left(\dfrac{x}{y}\right)} = \frac{1}{\tan t}.$$

The keys marked [sin⁻¹], [cos⁻¹]*, and* [tan⁻¹] *on a calculator do not represent reciprocals. They are used to calculate inverse function values.*

Thus the values of the cosecant, secant, and cotangent functions are found by using the reciprocal key [1/x] or [x⁻¹] along with the function keys for the sine, cosine, and tangent, respectively.

Before using a calculator to find function values of a real number, we must be sure that the calculator is in radian mode.

EXAMPLE 5

Using a Calculator to Approximate Values of Trigonometric Functions

Use a calculator to approximate each value to four decimal places.

(a) $\sin 1$ (b) $\cos 60$ (c) $\cot \dfrac{\pi}{8}$ (d) $\csc(-4.15)$

Solution

First we set the calculator in radian mode. Then we proceed as follows.

(a) $\sin 1 = 0.8415$ (approx.)

(b) $\cos 60 = -0.9524$ (approx.)

(c) $\cot \dfrac{\pi}{8} = \dfrac{1}{\tan \dfrac{\pi}{8}} = 2.4142$ (approx.)

(d) $\csc(-4.15) = \dfrac{1}{\sin(-4.15)} = 1.1820$ (approx.)

From the definitions of the trigonometric functions we know that if the point $P(t) = (x, y)$, then $\cos t = x$ and $\sin t = y$. By substitution it follows that

$$\boxed{P(t) = (\cos t, \sin t).}$$

For instance,

$$P\left(\frac{\pi}{4}\right) = \left(\cos \frac{\pi}{4}, \sin \frac{\pi}{4}\right) = \left(\frac{\sqrt{2}}{2}, \frac{\sqrt{2}}{2}\right)$$

$$P\left(\frac{5\pi}{6}\right) = \left(\cos \frac{5\pi}{6}, \sin \frac{5\pi}{6}\right) = \left(\frac{-\sqrt{3}}{2}, \frac{1}{2}\right)$$

and

$$P(-0.725) = (\cos(-0.725), \sin(-0.725))$$
$$= (0.7485, -0.6631) \text{ (approx.).}$$

Solving Applied Problems

EXAMPLE 6   **Solving a Motion Problem**

Figure 8a shows the shaft of a horizontal piston attached to the rim of a wheel at point A. Suppose that the radius of the wheel is 1 foot and assume that the shaft begins from a horizontal position. Determine how far the piston head moves when A moves along the rim of the wheel.

(a) 0.7 foot   (b) 0.9 foot

Round off each answer to two decimal places.

Figure 8a

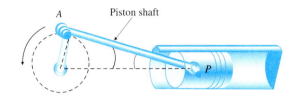

Solution   Figure 8b shows a schematic diagram of the piston. Assume that A is a point on the unit circle that starts at $(1, 0)$.

Figure 8b

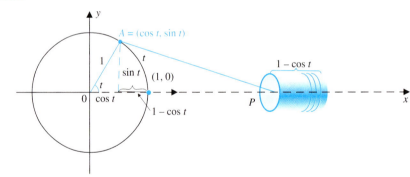

As the wheel rotates counterclockwise, the coordinates of A are given by

$$A = (\cos t, \sin t)$$

where t (in feet) is the length of the circular arc. When A has moved t feet, its x coordinate is $\cos t$, so the piston's head has moved $(1 - \cos t)$ feet.
    Consequently,

(a) if $t = 0.7$ foot, then $1 - \cos t = 1 - \cos 0.7 = 0.24$ foot (approx.);

(b) if $t = 0.9$ foot, then $1 - \cos t = 1 - \cos 0.9 = 0.38$ foot (approx.).

PROBLEM SET 5.2

Mastering the Concepts

In problems 1–6, show that each point $P(t)$ lies on the unit circle, and then find the values of the trigonometric functions of t.

1. $P(t) = \left(\dfrac{-\sqrt{3}}{2}, \dfrac{1}{2} \right)$

2. $P(t) = \left(\dfrac{-4}{5}, \dfrac{-3}{5} \right)$

3. $P(t) = \left(\dfrac{5}{13}, -\dfrac{12}{13} \right)$

4. $P(t) = \left(\dfrac{1}{\sqrt{10}}, -\dfrac{3}{\sqrt{10}} \right)$

5. $P(t) = \left(\dfrac{-3}{\sqrt{13}}, \dfrac{2}{\sqrt{13}} \right)$

6. $P(t) = \left(\dfrac{-4}{\sqrt{17}}, \dfrac{-1}{\sqrt{17}} \right)$

In problems 7–10, determine the coordinates of $P(t)$. Then find the exact values of the trigonometric functions of t.

7. (a) $t = -\dfrac{\pi}{2}$

   (b) $t = \dfrac{3\pi}{2}$

8. (a) $t = \pi$

   (b) $t = \dfrac{5\pi}{2}$

9. (a) $t = -2\pi$

   (b) $t = -\dfrac{5\pi}{2}$

10. (a) $t = \dfrac{-7\pi}{2}$

   (b) $t = -\dfrac{3\pi}{2}$

In problems 11–22, use the coordinates of $P\left(\dfrac{\pi}{6}\right)$, $P\left(\dfrac{\pi}{4}\right)$, and $P\left(\dfrac{\pi}{3}\right)$ (Figure 3), along with symmetry, to find the exact value of each expression.

11. $\sin \dfrac{3\pi}{4}$

12. $\sin\left(-\dfrac{3\pi}{4}\right)$

13. $\cos \dfrac{4\pi}{3}$

14. $\cos \dfrac{7\pi}{4}$

15. $\sin \dfrac{11\pi}{6}$

16. $\sin \dfrac{5\pi}{3}$

17. $\cos\left(-\dfrac{5\pi}{6}\right)$

18. $\cos\left(-\dfrac{7\pi}{4}\right)$

19. $\sin\left(-\dfrac{\pi}{4}\right)$

20. $\sin\left(-\dfrac{7\pi}{4}\right)$

21. $\cos \dfrac{5\pi}{4}$

22. $\cos\left(-\dfrac{5\pi}{4}\right)$

In problems 23–34, use a calculator to approximate each value to four decimal places.

23. $\sin 3$

24. $\cos \sqrt{2}$

25. $\cos(-5)$

26. $\sin(-17)$

27. $\tan 1$

28. $\cot 4$

29. $\cot(-1)$

30. $\sec 23$

31. $\csc(5.81)$

32. $\csc\left(\dfrac{7}{\pi}\right)$

33. $\sec\left(\dfrac{\pi}{7}\right)$

34. $\tan(\cos(-3.5))$

In problems 35 and 36, use $P(t) = (x, y) = (\cos t, \sin t)$ to find the coordinates of each point. Round off the answers to four decimal places.

35. (a) $P(2.7)$

   (b) $P(-4.18)$

   (c) $P(-3.58)$

   (d) $P(5.73)$

36. (a) $P(-0.432)$

   (b) $P(5.129)$

   (c) $P(0.572)$

   (d) $P(-0.636)$

Applying the Concepts

37. Piston Movement: Figure 9 shows a typical piston engine, where θ is the angle between the crankshaft $\overline{OA}$ and the connecting rod $\overline{AP}$. It can be shown that the distance $d = |\overline{OP}|$ is given by

$$d = 4\sqrt{4 - 2\cos\theta}.$$

Find the distance d (in inches), when $\theta = \pi/6$, $\pi/5$, or $\pi/4$. Round off the answers to two decimal places.

Figure 9

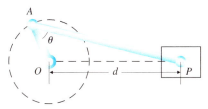

38. Speed of Rolling Pulley: If a pulley were spun in place, then a point P on the rim of the pulley would travel in a circular motion around its axle. Suppose that a pulley is moving at a constant speed v (in feet per second), then the forward speed s (in feet per second) of a fixed point on the pulley is given by the equation

$$s = v(1 + \cos\theta)$$

where θ is the angle shown in Figure 10. Find the forward speed s of a point P on the pulley moving at a rate of 45 feet per second, when $\theta = \pi/6$, $\pi/5$, or $\pi/4$. Round off the answers to two decimal places.

Figure 10

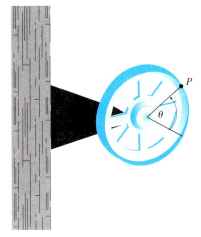

Developing and Extending the Concepts

In problems 39 and 40, find the values of the trigono-metric functions of t under the given conditions.

39. $P(t)$ is in quadrant III and lies on the line $y = 3x$.

40. $P(t)$ is in quadrant IV and lies on the line $3x + 2y = 0$.

41. Explain why $(\cos t)^2$ and $\cos t^2$ are different. Give examples to support your argument.

42. The function

$$f(t) = \frac{\sin t}{t}$$

is used in calculus. For parts (a)–(d), find each value and round off each answer to four decimal places.

(a) $f(0.1)$ (b) $f(0.01)$
(c) $f(0.001)$ (d) $f(0.0001)$
(e) Discuss the behavior of f as $t \to 0^+$.

43. In calculus it is shown that $\sin t$ can be approximated by the polynomial function

$$g(t) = t - \frac{t^3}{6} + \frac{t^5}{120} \quad \text{where } -1 \le t \le 1.$$

Compare $\sin 0.1$ and $g(0.1)$, $\sin 0.2$ and $g(0.2)$, $\sin(-0.3)$ and $g(-0.3)$, and $\sin(-1)$ and $g(-1)$. Round off the values to four decimal places.

5.3 Properties of the Trigonometric Functions

Objectives

1. Determine the Signs of Trigonometric Function Values
2. Establish the Quotient Identities
3. Establish the Even-Odd and Pythagorean Identities
4. Establish the Periodicity of Trigonometric Functions
5. Solve Applied Problems

We are able to derive several properties of the trigonometric functions by examin-ing their definitions.

Determining the Signs of Trigonometric Function Values

The algebraic sign of a trigonometric function value depends on the quadrant in which $P(t)$ lies.

For instance, as Figure 1 shows, $\sin t$ is positive for $P(t)$ in quadrants I and II, and $\sin t$ is negative in quadrants III and IV. Similarly, we can determine the alge-braic signs of the other five trigonometric functions depending on the quadrant where $P(t)$ lies. The results are listed in Table 1.

Figure 1

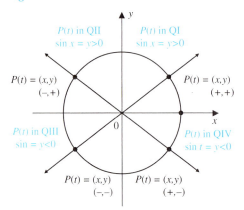

TABLE 1 Signs of the Values of the Trigonometric Functions of t

Quadrant:	Positive Values	Negative Values
I	All	None
II	$\sin t$, $\csc t$	$\cos t$, $\sec t$, $\tan t$, $\cot t$
III	$\tan t$, $\cot t$	$\sin t$, $\cos t$, $\csc t$, $\sec t$
IV	$\cos t$, $\sec t$	$\sin t$, $\csc t$, $\tan t$, $\cot t$

EXAMPLE 1 **Finding the Quadrant Location of *P*(*t*)**

Find the quadrant in which $P(t)$ lies if

$$\cot t > 0 \quad \text{and} \quad \cos t < 0.$$

Solution Using Table 1, we find that $\cot t > 0$ if $P(t)$ is in quadrant I or III, and $\cos t < 0$ if $P(t)$ is in quadrant II or III. For both conditions to be satisfied, $P(t)$ must be in quadrant III.

Establishing the Quotient Identities

Figure 2

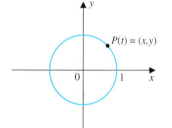

An **identity** is an equation that is true for all values of the variable for which both sides of the equation are defined. In Section 5.2 we established the reciprocal identities. Upon examining Figure 2, we discover that

$$\tan t = \frac{y}{x} = \frac{\sin t}{\cos t}$$

and

$$\cot t = \frac{x}{y} = \frac{\cos t}{\sin t}.$$

Thus we are led to the following identities.

Quotient Identities

$$\tan t = \frac{\sin t}{\cos t} \quad \text{and} \quad \cot t = \frac{\cos t}{\sin t}$$

EXAMPLE 2 **Using the Quotient Identities**

Suppose $\sin t = \dfrac{3}{5}$ and $\cos t = \dfrac{-4}{5}$.

Find $\tan t$ and $\cot t$.

Solution By the quotient identities, it follow that

$$\tan t = \frac{\sin t}{\cos t} = \frac{(3/5)}{(-4/5)} = -\frac{3}{4}$$

and

$$\cot t = \frac{\cos t}{\sin t} = \frac{(-4/5)}{(3/5)} = -\frac{4}{3}.$$

Establishing Even-Odd and Pythagorean Identities

By using the definitions of the trigonometric functions we can derive additional identities classified as the *even-odd* identities and the *Pythagorean* identities.

Suppose that

$$P(t) = (x, y).$$

Figure 3

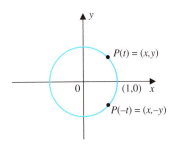

By symmetry, $P(-t) = (x, -y)$ (Figure 3).

Using the definitions of the trigonometric functions, we have

$$\cos t = x \quad \text{and} \quad \cos(-t) = x.$$

Also,

$$\sin t = y \quad \text{and} \quad \sin(-t) = -y$$

so that

$$\sin(-t) = -y = -\sin t.$$

It follows that

$$\boxed{\cos(-t) = \cos t \quad \text{and} \quad \sin(-t) = -\sin t.}$$

From the quotient identity, we get

$$\tan(-t) = \frac{\sin(-t)}{\cos(-t)} = \frac{-\sin t}{\cos t} = -\tan t.$$

So $\tan(-t) = -\tan t$.

Similar arguments can be used to establish identities for the cosecant, secant, and cotangent functions. Thus we are led to the following formulas, which are called the *even-odd* identities.

Even-Odd Identities

Function	Identity	Even or Odd
sine	$\sin(-t) = -\sin t$	odd
cosine	$\cos(-t) = \cos t$	even
tangent	$\tan(-t) = -\tan t$	odd
cosecant	$\csc(-t) = -\csc t$	odd
secant	$\sec(-t) = \sec t$	even
cotangent	$\cot(-t) = -\cot t$	odd

EXAMPLE 3 **Using the Even-Odd Identities**

Suppose that $\sin t = -3/5$ and $\cos t = 4/5$. Find

(a) $\tan(-t)$ (b) $\sec(-t)$ (c) $\csc(-t)$

Solution   (a) $\tan(-t) = -\tan t = -\dfrac{\sin t}{\cos t} = -\dfrac{\left(-\dfrac{3}{5}\right)}{\left(\dfrac{4}{5}\right)} = -\left(\dfrac{-3}{4}\right) = \dfrac{3}{4}$

(b) $\sec(-t) = \sec t = \dfrac{1}{\cos t} = \dfrac{1}{\left(\dfrac{4}{5}\right)} = \dfrac{5}{4}$

(c) $\csc(-t) = -\csc t = -\dfrac{1}{\sin t} = -\dfrac{1}{\left(-\dfrac{3}{5}\right)} = -\left(\dfrac{5}{-3}\right) = \dfrac{5}{3}$

To derive the Pythagorean identities, we begin by considering any $P(t) = (x, y)$ on the unit circle as illustrated in Figure 4.

By definition, we know that

$$\cos t = x \quad \text{and} \quad \sin t = y.$$

Figure 4

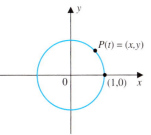

After adding the squares of these two equations to each other, we get

$$(\cos t)^2 + (\sin t)^2 = x^2 + y^2 = 1$$

because (x, y) is on the unit circle.

So we get the identity

$$(\cos t)^2 + (\sin t)^2 = 1.$$

It is customary to write powers of trigonometric expressions such as $(\sin t)^n$ and $(\cos t)^n$, respectively, as

$$(\sin t)^n = \sin^n t \quad \text{and} \quad (\cos t)^n = \cos^n t$$

except when $n = -1$.

So we write

$$\cos^2 t = (\cos t)^2 \quad \text{and} \quad \sin^2 t = (\sin t)^2.$$

Thus the identity

$$(\cos t)^2 + (\sin t)^2 = 1 \quad \text{is written as}$$

$$\cos^2 t + \sin^2 t = 1.$$

We use this latter identity to prove the other two Pythagorean identities. If we divide both sides of the identity by $\cos^2 t$ we get

$$\frac{\cos^2 t}{\cos^2 t} + \frac{\sin^2 t}{\cos^2 t} = \frac{1}{\cos^2 t}$$

$$1 + \left(\frac{\sin t}{\cos t}\right)^2 = \left(\frac{1}{\cos t}\right)^2$$

$$1 + \tan^2 t = \sec^2 t.$$

Similarly, if we divide both sides of the identity $\cos^2 t + \sin^2 t = 1$ by $\sin^2 t$ and then simplify the result, we get

$$\cot^2 t + 1 = \csc^2 t.$$

Thus we are led to the following results:

Pythagorean Identities

(i) $\cos^2 t + \sin^2 t = 1$ (ii) $1 + \tan^2 t = \sec^2 t$ (iii) $1 + \cot^2 t = \csc^2 t$

The identities can be used to compute values of the trigonometric functions as illustrated in the following examples.

EXAMPLE 4   **Evaluating Functions by Using Identities**

Use identities to find the exact values of the other five trigonometric functions of t if

$$\sin t = \frac{8}{17} \quad \text{and} \quad \cos t < 0.$$

Solution   Since $\sin t$ is positive and $\cos t$ is negative, it follows that $P(t)$ is in quadrant II. Using the Pythagorean identity $\cos^2 t + \sin^2 t = 1$, we have

$$\cos^2 t = 1 - \sin^2 t.$$

Thus

$$\cos t = \pm\sqrt{1 - \sin^2 t}.$$

However, we are given that $\cos t$ is negative, so

$$\cos t = -\sqrt{1 - \left(\frac{8}{17}\right)^2} = -\sqrt{1 - \frac{64}{289}} = -\sqrt{\frac{225}{289}} = -\frac{15}{17}.$$

Next we use the quotient and reciprocal identities to get

$$\tan t = \frac{\sin t}{\cos t} = \frac{\left(\dfrac{8}{17}\right)}{\left(-\dfrac{15}{17}\right)} = \frac{-8}{15} \qquad \cot t = \frac{\cos t}{\sin t} = \frac{\left(-\dfrac{15}{17}\right)}{\left(\dfrac{8}{17}\right)} = \frac{-15}{8}$$

$$\sec t = \frac{1}{\cos t} = \frac{1}{\left(-\dfrac{15}{17}\right)} = \frac{-17}{15} \qquad \csc t = \frac{1}{\sin t} = \frac{1}{\left(\dfrac{8}{17}\right)} = \frac{17}{8}.$$

Establishing the Periodicity of Trigonometric Functions

Because the unit circle has circumference 2π, we notice that if we add any integer multiple of 2π to t or subtract any integer multiple of 2π from t, the resulting point is at the same location as $P(t) = (x, y)$, as illustrated in Figure 5, that is,

Figure 5

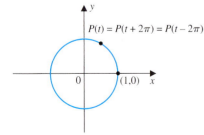

$$P(t + k \cdot 2\pi) = P(t), \ k \text{ is any integer.}$$

Correspondingly,

$$\sin t = y \quad \text{and} \quad \sin(t + k \cdot 2\pi) = y, \ k \text{ is any integer.}$$

so that

$$\boxed{\sin(t + k \cdot 2\pi) = \sin t, \text{ for any integer } k.}$$

In other words, the trigonometric function values of the sine function repeat every 2π units.

EXAMPLE 5   **Using Repetitive Behavior of Trigonometric Functions**

Use the repetitive behavior of the sine function to find the *exact* value of

$$\sin \frac{13\pi}{6}.$$

Solution   Since

Figure 6

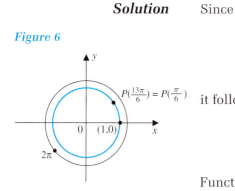

$$\frac{13\pi}{6} = \frac{\pi}{6} + \frac{12\pi}{6} = \frac{\pi}{6} + 2\pi \text{ (Figure 6),}$$

it follows that

$$\sin \frac{13\pi}{6} = \sin\left(\frac{\pi}{6} + 2\pi\right) = \sin \frac{\pi}{6} = \frac{1}{2}.$$

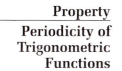

Functions with this kind of repetitive behavior are called *periodic functions.* In general, we have the following results.

Property

Periodicity of Trigonometric Functions

$$\sin t = \sin(t + k \cdot 2\pi),$$
$$\cos t = \cos(t + k \cdot 2\pi),$$
$$\tan t = \tan(t + k \cdot 2\pi),$$
$$\cot t = \cot(t + k \cdot 2\pi),$$
$$\sec t = \sec(t + k \cdot 2\pi),$$
$$\text{and} \quad \csc t = \csc(t + k \cdot 2\pi),$$

where k is any integer.

It can be shown that 2π is the *smallest positive period,* called the **fundamental period,** for the sine, cosine, cosecant, and secant functions.

However the tangent and cotangent functions have a fundamental period of π. To understand why this is the case, let us suppose that t is a given real number and

$$P(t) = (x, y).$$

Figure 7

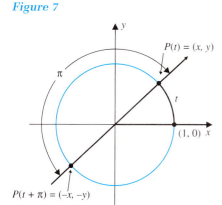

By symmetry,

$$P(t + \pi) = (-x, -y) \text{ (Figure 7).}$$

From the definition of the tangent function, we have

$$\tan(t + \pi) = \frac{-y}{-x} = \frac{y}{x} \quad \text{and} \quad \tan t = \frac{y}{x}.$$

Thus

$$\tan (t + \pi) = \tan t.$$

It follows that

$$\boxed{\tan(t + k\pi) = \tan t \quad \text{where } k \text{ is any integer.}}$$

So the tangent function has a period of π, that is, the tangent function values repeat every π units.

For example,

$$\tan \frac{10\pi}{3} = \tan\left(\frac{\pi}{3} + \frac{9\pi}{3}\right) = \tan\left(\frac{\pi}{3} + 3\pi\right) = \tan\frac{\pi}{3} = \sqrt{3}.$$

Similarly, it can be shown that the cotangent also has a period of π, that is,

$$\boxed{\cot(t + k\pi) = \cot t, \text{ where } k \text{ is any integer.}}$$

Solving Applied Problems

In calculus, algebraic expressions of the form

$$\sqrt{a^2 - u^2}, \ \sqrt{u^2 + a^2}, \quad \text{and} \quad \sqrt{u^2 - a^2},$$

where a is a positive constant, can be converted to more useful forms by making appropriate substitutions involving trigonometric functions. This technique is called **trigonometric substitution.** The idea is to rewrite radical expressions as trigonometric expressions containing no radical, as illustrated in the next example.

EXAMPLE 6   **Converting a Radical to a Trigonometric Form**

Use the substitution $u = 5 \sin t$, where t is a real number such that $P(t)$ is in quadrant I, to rewrite

$$\sqrt{25 - u^2}$$

as a trigonometric expression containing no radical.

Solution   After rewriting the Pythagorean identity

$$\cos^2 t + \sin^2 t = 1$$

as

$$\cos^2 t = 1 - \sin^2 t,$$

we proceed as follows:

$$
\begin{aligned}
\sqrt{25 - u^2} &= \sqrt{25 - (5 \sin t)^2} && \text{Substitute } u = 5 \sin t \\
&= \sqrt{25 - 25 \sin^2 t} && \text{Multiply} \\
&= \sqrt{25(1 - \sin^2 t)} && \text{Factor} \\
&= 5\sqrt{1 - \sin^2 t} && \text{Simplify} \\
&= 5\sqrt{\cos^2 t} && \text{Rewrite of Identity} \\
&= 5 \cos t && \cos t > 0 \text{ since } P(t) \text{ is in quadrant I} \quad \blacksquare
\end{aligned}
$$

 **PROBLEM SET 5.3**

Mastering the Concepts

In problems 1–8, use the reciprocal and quotient relationships to find the other four trigonometric function values of t under the given conditions.

1. $\cos t = \dfrac{12}{13}$ and $\sin t = \dfrac{5}{13}$

2. $\cos t = \dfrac{2}{5}$ and $\sin t = \dfrac{\sqrt{21}}{5}$

3. $\tan t = \dfrac{12}{5}$ and $\cos t = \dfrac{-5}{13}$

4. $\cot t = \dfrac{-2}{\sqrt{21}}$ and $\cos t = \dfrac{2}{5}$

5. $\sin t = \cos t = \dfrac{-1}{\sqrt{2}}$

6. $\tan t = 1$ and $\sin t = \dfrac{-1}{\sqrt{2}}$

7. $\sin t = 0$ and $\cos t = 1$

8. $\cos t = 0$ and $\sin t = -1$

In problems 9–12, indicate the quadrant in which $P(t)$ lies.

9. (a) $\cos t > 0$ and $\sin t < 0$
  (b) $\sec t > 0$ and $\cot t < 0$
10. (a) $\sin t > 0$ and $\cos t < 0$
  (b) $\sin t > 0$ and $\sec t < 0$
11. (a) $\tan t < 0$ and $\cos t < 0$
  (b) $\tan t < 0$ and $\cos t > 0$
12. (a) $\cos t < 0$ and $\cot t < 0$
  (b) $\sec t < 0$ and $\cot t > 0$

In problems 13–16, suppose that $\sin t = -12/13$ and $\cos t = -5/13$. Find each value.

13. $\csc(-t)$ **14.** $\tan(-t)$
15. $\cot(-t)$ **16.** $\sec(-t)$

In problems 17 and 18, verify that the equation is true for the given value of x by rounding off the values to four decimal places.

17. (a) $f(-t) = -f(t)$, where $f(t) = \sin t$ and $t = 3.752$
  (b) $f(-t) = f(t)$, where $f(t) = \cos t$ and $t = 5.384$
18. (a) $f(-t) = -f(t)$, where $f(t) = \tan t$ and $t = 4.138$
  (b) $f(-t) = f(t)$, where $f(t) = \sec t$ and $t = 2.874$

In problems 19–26, use identities to find the exact values of the other five trigonometric functions.

19. $\sin t = \dfrac{15}{17}$; t is in quadrant I

20. $\cos t = \dfrac{5}{13}$; $\tan t > 0$

21. $\cos t = \dfrac{-4}{5}$; t is in quadrant II

22. $\sin t = \dfrac{-7}{25}$; $\cot t < 0$

23. $\csc t = \dfrac{17}{8}$; t is in quadrant II

24. $\tan t = -3$; $\cos t > 0$

25. $\sec t = -\dfrac{5}{3}$; $\sin t < 0$

26. $\cot t = \dfrac{3}{4}$; $\sin t < 0$

In problems 27–30, use identities to find the approximate value of the other five trigonometric functions. Round off each answer to two decimal places.

27. $\sin t = 0.57$; $\tan t < 0$
28. $\cos t = -0.83$; $\cot t > 0$
29. $\tan t = -1.53$; $\sin t > 0$
30. $\cot t = 3.41$; $\csc t > 0$

In problems 31–41, use periodicity to find the exact values.

31. $\cos\left(\dfrac{5\pi}{6} + 2\pi\right)$ **32.** $\sin\left(\dfrac{\pi}{4} - 2\pi\right)$

33. $\sin\left(\dfrac{2\pi}{3} + 4\pi\right)$ **34.** $\cos\left(\dfrac{11\pi}{6} - 6\pi\right)$

35. $\cot\left(\dfrac{41\pi}{4}\right)$ **36.** $\tan 7\pi$

37. $\csc\left(\dfrac{-43\pi}{4}\right)$ **38.** $\sec\left(2\pi - \dfrac{\pi}{6}\right)$

39. $\tan\left(\dfrac{-65\pi}{6}\right)$ **40.** $\cot\left(\dfrac{-15\pi}{2}\right)$

Applying the Concepts

In problems 41–46, rewrite each algebraic expression as a trigonometric expression containing no radical by using the given trigonometric substitution. Assume that t is a real number such that $P(t)$ is in quadrant I.

41. $\sqrt{4 - u^2}$; $u = 2\sin t$
42. $\sqrt{u^2 - 4}$; $u = 2\sec t$
43. $\sqrt{u^2 + 9}$; $u = 3\tan t$
44. $\sqrt{9u^2 + 4}$; $u = \dfrac{2}{3}\tan t$

45. $\sqrt{4 - 9u^2}$; $u = \dfrac{2}{3} \sin t$

46. $\sqrt{u^2 - 81}$; $u = 9 \sec t$

Developing and Extending the Concepts

47. Show that the following even-odd identities are true:

(a) $\cot(-t) = -\cot t$

(b) $\sec(-t) = \sec t$

(c) $\csc(-t) = -\csc t$

48. If $f(t) = \cos t$, where t is a real number, explain why $f(t) = f(t + 2\pi) = f(t - 2\pi)$.

49. Use the Pythagorean identity

$$\cos^2 t + \sin^2 t = 1$$

to prove that

$$1 + \cot^2 t = \csc^2 t.$$

50. Explain why it is *impossible* for $\sin t > 1$ and $|\csc t| < 1$ for any real number t.

5.4 Graphs of Sine and Cosine Functions

Objectives

1. Determine the Domain and Range of the Sine and Cosine Functions
2. Graph the Sine and Cosine Functions
3. Transformations of the Sine and Cosine Graphs
4. Solve Applied Problems
5. Model Simple Harmonic Motion

Now that we have established the trigonometric functions for real numbers, we turn our attention to graphing them. We begin with the sine and cosine functions. They are used extensively in applied mathematics.

Determining the Domain and Range of the Sine and Cosine Functions

By analyzing the definitions of the sine and cosine functions we can determine their domains and ranges. Later in Section 5.5 we will do the same for the other four trigonometric functions.

Assume that $P(t) = (x, y)$ is a point on the unit circle (Figure 1), then

$$x = \cos t \quad \text{and} \quad y = \sin t.$$

Figure 1

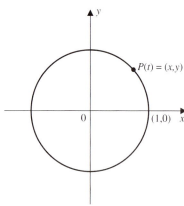

Since (x, y) is on the unit circle, it follows that *no matter what value* of t is chosen we have

$$-1 \le x \le 1 \quad \text{and} \quad -1 \le y \le 1.$$

Consequently,

$$\boxed{-1 \le \cos t \le 1 \quad \text{and} \quad -1 \le \sin t \le 1.}$$

Thus we have the following result.

Domain and Range
Sine and Cosine

For $f(t) = \sin t$ and $g(t) = \cos t$, the domain consists of all real numbers, and the range of each function is the interval $[-1, 1]$.

EXAMPLE 1   **Finding the Range of a Sine Function**

Use $-1 \le \sin t \le 1$ to find the range of $f(t) = 2 \sin t + 5$.

Solution   To find the range of f we proceed as follows:

$$-1 \le \sin t \le 1 \qquad \text{Sine has range } [-1, 1]$$

$$-2 \le 2 \sin t \le 2 \qquad \text{Multiply by 2}$$

$$3 \le 2 \sin t + 5 \le 7 \qquad \text{Add 5}$$

Thus $3 \le f(t) \le 7$, that is, the range of f is the interval $[3, 7]$.

Graphing the Sine and Cosine Functions

1. The Sine Graph

To graph the sine function, we recall two important facts:

(i) If t is a real number and $P(t) = (x, y)$ is the associated point on the unit circle, then $\sin t = y$ (Figure 2a).

(ii) As $P(t)$ moves one complete revolution counterclockwise around the unit circle, starting at $P(0) = (1, 0)$ and ending at $P(2\pi) = (1, 0)$, correspondingly, t increases from 0 to 2π along the horizontal axis in a Cartesian system (Figure 2b).

Figure 2

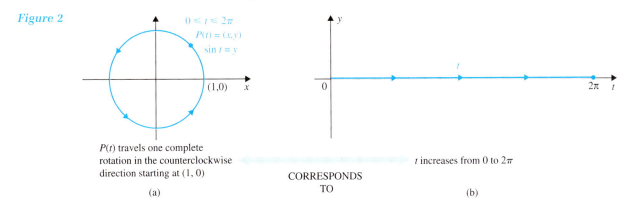

$P(t)$ travels one complete rotation in the counterclockwise direction starting at $(1, 0)$

(a)

CORRESPONDS TO

t increases from 0 to 2π

(b)

Figure 3a shows the locations of several specific points for $P(t) = (x, y)$. Since $y = \sin t$, we can obtain a rough sketch of the graph of $f(t) = \sin t$ by plotting the points (t, y) in a Cartesian system and then connecting the points with a smooth graph (Figure 3b). The graph that results when t increases from 0 to 2π, corresponding to $P(t)$ moving one counterclockwise rotation, is called **one cycle** of the sine function.

Figure 3

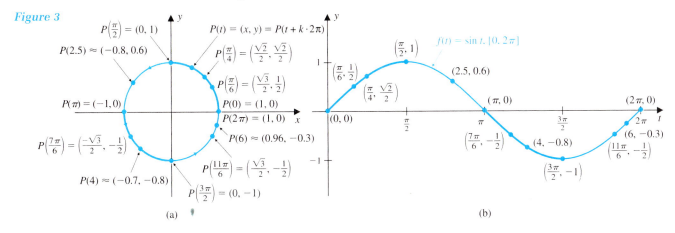

(a)

(b)

The graph in Figure 3b reveals *five key points* for the sine cycle, that is, locations of maximum and minimum points and of x intercepts. There is a maximum point at $(\pi/2, 1)$, a minimum at $(3\pi/2, -1)$, and x intercepts at $(0, 0)$, $(\pi, 0)$ and $(2\pi, 0)$.

When graphing trigonometric functions in the Cartesian plane, it is customary to use x as the independent variable instead of t, and y as the dependent variable.

Because the sine has period 2π, each time t increases or decreases by 2π units, $P(t)$ traverses the unit circle again, and correspondingly, other sine cycles are generated. So we obtain a complete graph of $y = \sin x$ by repeating the cycle in Figure 3b to the right and to the left every 2π units (Figure 4).

Upon reading the sine graph we are led to the properties listed in Table 1. Notice that the graph confirms properties already developed earlier for the sine function.

Figure 4

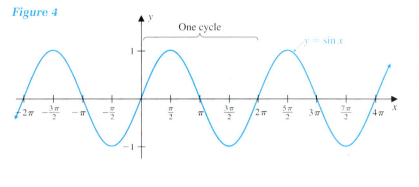

TABLE 1   Properties of the Sine Function

Property	$y = \sin x$
Domain	All real numbers
Range	Interval $[-1, 1]$; that is; $-1 \le \sin x \le 1$
Period	2π
Symmetry	Symmetric with respect to the origin
Even or odd	The function is *odd*, that is, $\sin(-x) = -\sin x$

2. The Cosine Graph

As with the sine function, we sketch one cycle of the graph of the cosine function $y = \cos x$ in the interval $[0, 2\pi]$ by first plotting some points (Table 2) and then connecting them with a smooth graph (Figure 5). The five key points are listed in Table 3.

TABLE 2

x	$\cos x$
0	1
$\dfrac{\pi}{3}$	$\dfrac{1}{2}$
$\dfrac{\pi}{2}$	0
π	-1
$\dfrac{4\pi}{3}$	$-\dfrac{1}{2}$
$\dfrac{3\pi}{2}$	0
2π	1

Figure 5

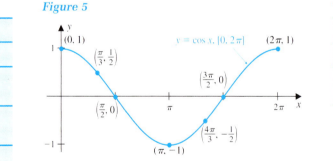

TABLE 3   Key Points for $y = \cos x$, $[0, 2\pi]$

Maximums	$(0, 1)$ and $(2\pi, 1)$
Minimum	$(\pi, -1)$
x-intercepts	$\left(\dfrac{\pi}{2}, 0\right)$ and $\left(\dfrac{3\pi}{2}, 0\right)$

Because the cosine has period 2π, we repeat the cycle in Figure 5 to the right and to the left every 2π units, to get the complete graph of $y = \cos x$ (Figure 6). From the graph, we observe the properties of the cosine function listed in Table 4. Once again, the graph confirms properties developed earlier for the cosine function.

Figure 6

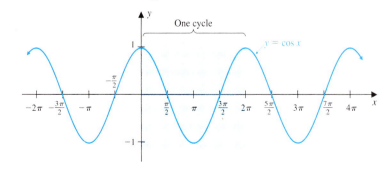

TABLE 4   Properties of the Cosine Function

Property	$y = \cos x$
Domain	All real numbers
Range	Interval $[-1, 1]$; that is, $-1 \le \cos x \le 1$
Period	2π
Symmetry	Symmetric with respect to the y axis
Even or odd	The function is *even*, that is, $\cos(-x) = \cos x$

Graphs of one cycle of $y = \sin x$ (Figure 3b) and of $y = \cos x$ (Figure 5) including their key points should be learned.

Transformations of the Sine and Cosine Graphs

Our objective here is to use transformations of the graphs of $y = \sin x$ and $y = \cos x$ to obtain graphs of functions of the more general forms.

$$F(x) = a\sin(kx + b) + c \quad \text{and} \quad G(x) = a\cos(kx + b) + c$$

where a, k, b and c are constants.

1. Amplitude: Vertical Scaling and Vertical Shifting

The next example illustrates the applications of vertical scaling and vertical shifting to the sine and cosine functions.

EXAMPLE 2   **Using Vertical Scaling and Vertical Shifting to Graph**

Use transformations of the sine and cosine graphs to sketch the graph of each function. Also determine the range and period.

(a) $F(x) = 5\sin x$

(b) $G(x) = \dfrac{1}{5}\cos x + 1$

Solution   (a) First we sketch the graph of $y = \sin x$; then we multiply each ordinate by 5. Geometrically, this means that the graph of $y = \sin x$ is vertically stretched by a factor of 5 to obtain the graph of $F(x) = 5\sin x$ (Figure 7a). From the graph, we see that the range of F is the interval $[-5, 5]$ and the period is 2π.

(b) We start by graphing $y = \cos x$.

   Next the graph of $g(x) = (1/5)\cos x$ is obtained by vertically shrinking the graph of $y = \cos x$ by a factor of $(1/5)$ (Figure 7b). Finally, we get the graph of

$$G(x) = \left(\frac{1}{5}\right)\cos x + 1$$

by shifting the graph of g up 1 unit (Figure 7b).

   The range of G is the interval

$$\left[-\frac{1}{5} + 1, \frac{1}{5} + 1\right] \quad \text{or} \quad \left[\frac{4}{5}, \frac{6}{5}\right],$$

and the period is 2π.

Figure 7

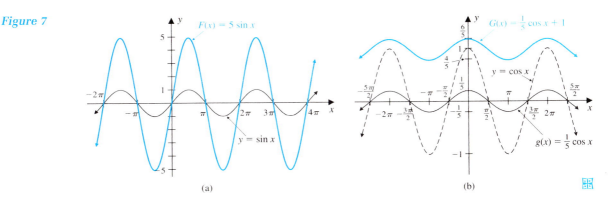

(a)   (b)

Example 2 leads us to the following generalizations.

<table>
<tr>
<td>

Property

Amplitude—Vertical Scaling and Vertical Shifting

</td>
<td>

(i) The graph of

$$f(x) = a \sin x \quad \text{or} \quad g(x) = a \cos x, \quad \text{where } a \neq 0$$

is obtained from the graph of

$$y = \sin x \quad \text{or} \quad y = \cos x,$$

respectively, by *stretching* or *compressing vertically* by a factor of $|a|$, depending on whether

$$|a| > 1 \quad \text{or} \quad |a| < 1,$$

respectively. If $a < 0$ we also *reflect* the resulting graph across the x axis.

(ii) The graph of

$$F(x) = a \sin x + c \quad \text{or} \quad G(x) = a \cos x + c$$

is obtained, respectively, by *vertically shifting* the graph of

$$f(x) = a \sin x \quad \text{or} \quad g(x) = a \cos x$$

$|c|$ units,

$$\text{upward if } c > 0$$

and

$$\text{downward if } c < 0.$$

In either case, the value of $|a|$ is called the **amplitude** of the function.

</td>
</tr>
</table>

For example, the amplitudes of

$$F(x) = 5 \sin x, G(x) = \frac{1}{5} \cos + 1, \quad \text{and} \quad H(x) = -3 \cos x + 5$$

are, respectively, $|5| = 5$, $\left| \dfrac{1}{5} \right| = \dfrac{1}{5}$, and $|-3| = 3$.

2. Period: Horizontal Scaling

Next we consider the graphs of functions of the forms

$$f(x) = \sin kx \text{ and } g(x) = \cos kx, \text{ where } k \text{ is a constant.}$$

To better understand the effect that k has on the graphs of $y = \sin x$ and $y = \cos x$, let us consider the following example in which we compare the graphs of $f(x) = \sin 2x$ and $y = \sin x$.

EXAMPLE 3   **Comparing Periods**

Sketch the graph of the function

$$f(x) = \sin 2x$$

by using the values of x in interval $[0, \pi]$. On the same coordinate system, display the graph of $y = \sin x$ in interval $[0, 2\pi]$. Compare the periods of the two functions.

Solution   Both graphs have the same amplitude, 1, but how do their periods compare? Table 5 lists values of x in interval $[0, \pi]$ along with corresponding values of $2x$ and $\sin 2x$. By using the table entries we plot points and then connect them with a smooth curve to get the graph of f (Figure 8). This graph appears to be a sine cycle with period π.

To confirm this observation, we know that

$$y = \sin \, \blacksquare \ \text{completes one cycle when } 0 \leq \blacksquare \leq 2\pi.$$

Correspondingly,

$$f(x) = \sin 2x \text{ completes one cycle}$$
$$\text{when } 0 \leq 2x \leq 2\pi.$$

That is,

$$\text{when } 0 \leq x \leq \pi.$$

So the period is π.

TABLE 5

x	$2x$	$f(x) = \sin 2x$
0	0	0
$\dfrac{\pi}{6}$	$\dfrac{\pi}{3}$	$\dfrac{\sqrt{3}}{2}$
$\dfrac{\pi}{4}$	$\dfrac{\pi}{2}$	1
$\dfrac{\pi}{2}$	π	0
$\dfrac{7\pi}{12}$	$\dfrac{7\pi}{6}$	$-\dfrac{1}{2}$
$\dfrac{3\pi}{4}$	$\dfrac{3\pi}{2}$	-1
π	2π	0

Figure 8

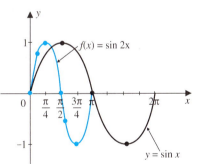

The graph of $y = \sin x$ in interval $[0, 2\pi]$ is also shown in Figure 8. The period of this function is 2π. Upon comparing the periods of the two functions, we discover that by replacing x with $2x$ in the equation $y = \sin x$, we obtain a function f whose graph is the sine curve that is horizontally compressed by one-half.

Our analysis in Example 3 is generalized as follows:
Given

$$f(x) = \sin kx, \quad \text{where } k > 0,$$

then one cycle of the graph of f occurs if

$$0 \leq kx \leq 2\pi,$$

that is, if

$$0 \leq x \leq \frac{2\pi}{k}.$$

A similar analysis applies to

$$g(x) = \cos kx, \quad \text{for} \quad k > 0.$$

Thus we have the following general property:

Property

Period—Horizontal Scaling

The graph of

$$f(x) = \sin kx \quad \text{or} \quad g(x) = \cos kx, \quad k > 0,$$

can be obtained by *horizontally scaling* the graph of

$$y = \sin x \quad \text{or} \quad y = \cos x,$$

respectively, by a factor of $1/k$. The **period** of f or g is $2\pi/k$ so that one cycle of the graph occurs in the interval $[0, 2\pi/k]$.

EXAMPLE 4   **Using Horizontal Scaling**

Use horizontal scaling to graph one cycle of $G(x) = \cos \dfrac{x}{2}$. Indicate the period of G and locate the key points.

Solution   Using the above property, we conclude that the given function

$$G(x) = \cos \frac{x}{2} = \cos \frac{1}{2} x$$

has a period given by

$$\frac{2\pi}{\left(\dfrac{1}{2}\right)} = 4\pi.$$

This means that one cycle of the graph of G can be obtained by stretching the cosine cycle horizontally by a factor of $\dfrac{1}{\left(\dfrac{1}{2}\right)} = 2$ over interval $[0, 4\pi]$ (Figure 9).

The key points, namely, the maximums, the minimum, and the x intercepts, are located by proportionally scaling their locations horizontally. They are shown in Figure 9.

Figure 9

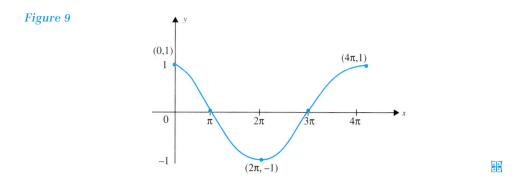

Notice that Example 3 illustrates that if $k > 1$, the cycle is *compressed* horizontally. On the other hand, if $0 < k < 1$, as was the case in Example 4, then the cycle is *stretched* horizontally.

There is no loss in generality by having k positive in the above property. If k is negative, we can rewrite the function using the even-odd properties in such a way that the coefficient of x turns out to be positive. For instance,

$f(x) = \sin(-3x) = -\sin 3x$ (Graph $y = \sin 3x$ and then reflect it across the x axis.)

and

$g(x) = \cos(-3x) = \cos 3x$.   (Same graph as the one for $y = \cos 3x$.)

3. Phase Shift: Horizontal Shifting

Now we consider graphs of functions of the forms

$$f(x) = \sin(kx + b) \text{ and } g(x) = \cos(kx + b)$$

where $k > 0$ and $b \neq 0$.

First, let us consider the situation when $k = 1$.

Recall from Section 2.4 that the graph of a function

$$y = f(x - h)$$

can be obtained by horizontally shifting the graph of $y = f(x)$ to the right if $h > 0$ and to the left if $h < 0$.

For instance, the graph of

$$f(x) = \sin\left(x + \frac{\pi}{6}\right) = \sin\left[x - \left(-\frac{\pi}{6}\right)\right]$$

Figure 10

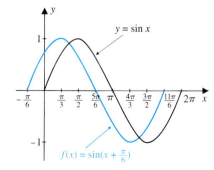

can be obtained by horizontally shifting the graph of $y = \sin x$, $\dfrac{\pi}{6}$ unit to the left. Figure 10 shows the result of shifting one cycle.

In a sense, $-\pi/6$ is the starting position of the sine cycle for f. Notice that the x values of the five key points of the graph of f are obtained by subtracting $\pi/6$ from the x values of the five key points of $y = \sin x$ to obtain

$$-\frac{\pi}{6}, \quad \frac{\pi}{2} - \frac{\pi}{6} = \frac{\pi}{3}, \quad \pi - \frac{\pi}{6} = \frac{5\pi}{6}, \quad \frac{3\pi}{2} - \frac{\pi}{6} = \frac{4\pi}{3}, \quad \text{and} \quad 2\pi - \frac{\pi}{6} = \frac{11\pi}{6}.$$

To help us understand how the graph of either

$$y = \sin x \quad \text{or} \quad y = \cos x$$

is affected when replacing

$$x \quad \text{with} \quad kx + b,$$

where $k > 0$, but $k \neq 1$, let us consider the following example.

EXAMPLE 5   **Shifting and Scaling a Graph Horizontally**

Given

$$f(x) = \sin\left(2x + \frac{\pi}{2}\right)$$

(a) Identify a starting and ending position of one cycle of the graph of f. Find the period of f.

(b) Graph one cycle by using the results in part (a)

Solution   (a) We know that $y = \sin x$ completes one cycle when $0 \leq x \leq 2\pi$. Correspondingly, $f(x) = \sin[2x + (\pi/2)]$ completes one cycle when

$$0 \leq 2x + \frac{\pi}{2} \leq 2\pi$$

that is, when

$$-\frac{\pi}{2} \leq 2x \leq 2\pi - \frac{\pi}{2} \qquad \text{Subtract } \frac{\pi}{2} \text{ from each side}$$

$$-\frac{\pi}{4} \leq x \leq \pi - \frac{\pi}{4} \qquad \text{Divide by 2}$$

$$-\frac{\pi}{4} \leq x \leq \frac{3\pi}{4} \qquad \text{Simplify}$$

$$\uparrow \qquad\qquad \uparrow$$

Starting position of one cycle	Ending position of one cycle

Here, the starting position is the solution of

$$2x + \frac{\pi}{2} = 0.$$

The length of the interval $\left[-\dfrac{\pi}{4}, \dfrac{3\pi}{4}\right]$ is given by

$$\frac{3\pi}{4} - \left(-\frac{\pi}{4}\right) = \frac{4\pi}{4} = \pi$$

which is the same as

$$\frac{2\pi}{k} = \frac{2\pi}{2} = \pi.$$

Consequently, the period of f is π.

(b) Since the graph of f generates one cycle on interval $\left[-\dfrac{\pi}{4}, \dfrac{3\pi}{4}\right]$, which has length π, its graph can be obtained by shifting the graph of $y = \sin x$ (which has period 2π) horizontally left $\dfrac{\pi}{4}$ unit and then compressing it by a factor of $\dfrac{1}{k} = \dfrac{1}{2}$ (Figure 11).

Figure 11

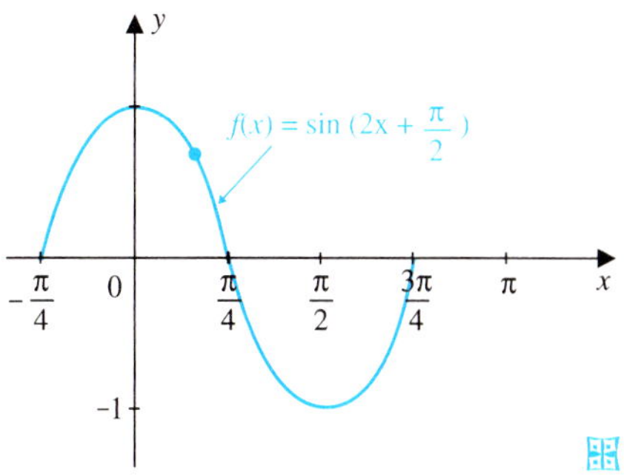

We generalize the results of Example 5 as follows.

Property

Phase Shift and Period

The graph of

$$f(x) = \sin(kx + b) \quad \text{or} \quad g(x) = \cos(kx + b), \quad \text{where} \quad k > 0$$

can be obtained by horizontally shifting the graph of

$$y = \sin x \quad \text{or} \quad y = \cos x,$$

respectively, by $|-b/k|$ units to the left if $-b/k < 0$ and to the right if $-b/k > 0$, and then horizontally scaling it by a factor of $\dfrac{1}{k}$. The resulting graph has a period of $\dfrac{2\pi}{k}$.

The number $-b/k$ is called the **phase shift** of the function.

The phase shift can be found by setting $kx + b = 0$ and solving for x to get $x = -b/k$.

In a sense, the phase shift is the starting position of one cycle.
For example,

the function $f(x) = \sin\left(2x + \dfrac{\pi}{2}\right)$ has phase shift

$$\frac{-b}{k} = \frac{\left(-\dfrac{\pi}{2}\right)}{2} = -\frac{\pi}{4} \quad \text{and period} \quad \frac{2\pi}{2} = \pi;$$

the function $g(x) = 4\cos\left(\dfrac{1}{3}x - 2\right) = 4\cos\left(\dfrac{1}{3}x + (-2)\right)$ has phase shift

$$\frac{-b}{k} = \frac{-(-2)}{\left(\dfrac{1}{3}\right)} = 6 \quad \text{and} \quad \text{period} \frac{2\pi}{\left(\dfrac{1}{3}\right)} = 6\pi.$$

Based on our previous work, let us summarize the information about sine and cosine graphs that can be extracted from function equations of the forms

$$F(x) = a\sin(kx + b) \quad \text{and} \quad G(x) = a\cos(kx + b).$$

Summary

Amplitude, Period, and Phase Shift Properties

Graphs of functions of the form

$$F(x) = a\sin(kx + b) \quad \text{or} \quad G(x) = a\cos(kx + b), \quad \text{for} \quad a \neq 0 \quad \text{and} \quad k > 0,$$

have the following properties:

(i) The amplitude is $|a|$

(ii) The period is $\dfrac{2\pi}{k}$

(iii) The phase shift is $\dfrac{-b}{k}$

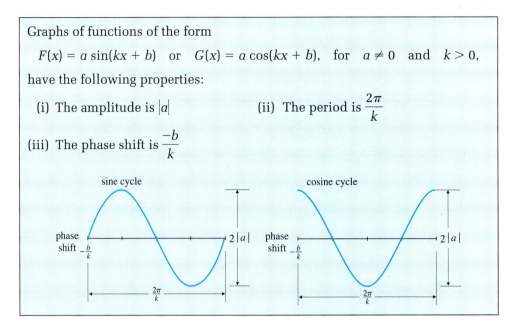

EXAMPLE 6

Using Properties to Sketch a Cosine Graph

Identify and use the phase shift, period, and amplitude to graph one cycle of the function $f(x) = 2\cos(3x - 1)$. Locate the key points, that is, the locations of minimum points, maximum points, and x intercepts.

Solution

First we rewrite the given function as

$$f(x) = 2\cos(3x - 1) = 2\cos[3x + (-1)].$$

Thus

$$a = 2, \quad k = 3, \quad \text{and} \quad b = -1$$

so that

(i) the amplitude is $|a| = |2| = 2$;

(ii) the period is $\dfrac{2\pi}{k} = \dfrac{2\pi}{3}$;

(iii) the phase shift is $\dfrac{-b}{k} = \dfrac{-(-1)}{3} = \dfrac{1}{3}$.

Figure 12

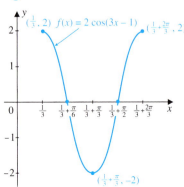

Consequently, one cycle of the graph of f, which is a cosine curve, starts at $x = 1/3$, covers an interval of length $(2\pi)/3$, to end at $x = (1/3) + (2\pi)/3$, and is vertically stretched by a factor of 2 (Figure 12).

The x values of the five key points are found by setting

$$3x - 1 = 0, \quad \frac{\pi}{2}, \quad \pi, \quad \frac{3\pi}{2}, \quad \text{and} \quad 2\pi$$

to get

$$x = \frac{1}{3}, \quad \frac{1}{3} + \frac{\pi}{6}, \quad \frac{1}{3} + \frac{\pi}{3}, \quad \frac{1}{3} + \frac{\pi}{2}, \quad \text{and} \quad \frac{1}{3} + \frac{2\pi}{3}.$$

The locations of the key points are shown in Figure 12.

It is not necessary to memorize the formulas for the period and phase shift for the functions

$$F(x) = \sin(kx + b) \quad \text{and} \quad G(x) = \cos(kx + b).$$

They are easily derived as follows:

Solve $kx + b = 0$ and $kx + b = 2\pi$ for x to get, respectively, $x = \dfrac{-b}{k}$ (phase shift) and $x = -\dfrac{b}{k} + \dfrac{2\pi}{k}$ (phase shift plus period).

For example, if $f(x) = \sin(7x - 5)$, then we solve

$$7x - 5 = 0 \text{ to get } x = \frac{5}{7}, \text{ which is the phase shift.}$$

Also, we solve

$$7x - 5 = 2\pi \text{ to get } x = \frac{5}{7} + \frac{2\pi}{7} \text{ (phase shift plus period)}$$

so that the period of f is $\dfrac{2\pi}{7}$.

The graphs of sine and cosine function are called **sine waves** or **sinusoidal curves**. The characteristics of the graph of such a curve can be used to find its equation in situations such as the one illustrated in the next example.

EXAMPLE 7   **Finding an Equation for a Sine Wave**

Given the graph of one cycle of the function $y = a \sin(kx + b)$ in Figure 13, find values for $a > 0$, $k > 0$, and b. Also find the amplitude, period, and phase shift.

Solution   The graph shows that the amplitude is 3; that is, $a = 3$. The period, which is the length of the interval containing the cycle, is given by

Figure 13

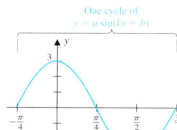

$$\frac{3\pi}{4} - \left(\frac{-\pi}{4}\right) = \pi$$

so

$$\frac{(2\pi)}{k} = \pi, \quad \text{or} \quad k = 2.$$

Since the phase shift is shown to be $-\pi/4$, it follows that $-b/k = -b/2 = -\pi/4$, or $b = \pi/2$. Thus the equation is

$$y = 3 \sin\left(2x + \frac{\pi}{2}\right).$$

Functions such as

$$f(x) = \sin 2x + \cos x, \; g(x) = 5 \sin 3x \cos 2x, \text{ and } h(x) = \sin^2 4x$$

which are *not* of the form

$$F(x) = a \sin(kx + b) \quad \text{or} \quad G(x) = a \cos(kx + b)$$

are difficult to graph by hand.

For instance, the graphs of *f, g,* and *h,* given on the interval $[0, 2\pi]$ produced by a grapher set in radian mode (for real numbers), are shown in Figures 14a, 14b, and 14c, respectively.

Figure 14

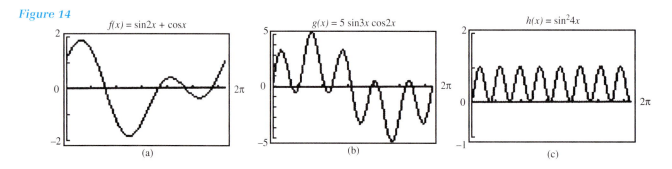

Solving Applied Problems

When sine and cosine functions are used to model periodic phenomena, their graphs help us to interpret the situation.

EXAMPLE 8

Modeling a Tide Problem

Tides cause ocean, sea, and lake currents to flow into and out of harbors, canals, and coastlines with varying depths. Suppose for a certain location, the depth of water *d* (in feet) varies with time *t* (in hours) according to the model

$$d(t) = 8 + 4 \cos\left(\frac{4\pi}{25} t\right)$$

where $t = 0$ represents midnight.

(a) Use this model to predict the depth of water at the start, at 2:00 AM and at 6:00 AM.
Round off the answers to two decimal places.

(b) Sketch the graph of this model over a 24-hour day.

(c) Find the amplitude and period. Interpret these values.

(d) Find the minimum and maximum depth of water and determine when they occur. Interpret your answers.

Solution

(a) At the start, $t = 0$, so the depth of water is given by

$$d(0) = 8 + 4 \cos 0 = 8 + 4$$
$$= 8 + 4 = 12 \text{ ft.}$$

At 2:00 AM, $t = 2$, so we have

$$d(2) = 8 + 4 \cos\left(\frac{8\pi}{25}\right)$$
$$= 10.14 \text{ ft (approx.).}$$

At 6:00 AM, $t = 6$, so we have

$$d(6) = 8 + 4 \cos\left(\frac{24\pi}{25}\right)$$

$$= 4.03 \text{ ft (approx.).}$$

(b) Figure 15 shows the graph of this model over a 24-hour day.

Figure 15

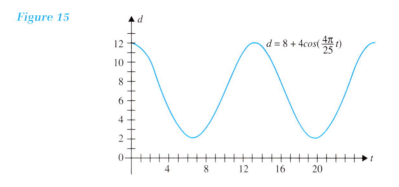

(c) The amplitude is 4, and the period is given by

$$\frac{2\pi}{\left(\dfrac{4\pi}{25}\right)} = \frac{25}{2} = 12.5 \text{ hours.}$$

The amplitude of 4 indicates that the depth varies no more than 4 feet from its average of 8 feet. Thus the water level rises to a depth of no more than $8 + 4 = 12$ feet and no less than $8 - 4 = 4$ feet. This behavior repeats every 12.5 hours.

(d) The minimum depth d occurs when the cosine is minimum, that is, when the cosine is -1. So the minimum depth is

$$8 + 4(-1) = 4 \text{ feet}$$

which occurs when

$$(4\pi/25)t = \pi \text{ and } (4\pi/25)t = 3\pi;$$

that is, when $t = \pi \cdot (25/4\pi) = 25/4 = 6.25$ and $t = 3\pi \cdot (25/4\pi) = 18.75$. This happens at 6:15 AM and then again at 6:45 PM.

The maximum depth is $8 + 4(1) = 12$ feet, when the cosine is at its maximum value of 1. This occurs when

$$(4\pi/25)t = 0, \text{ and when } (4\pi/25)t = 2\pi;$$

that is, when $t = 0$ and when $t = 2\pi \cdot (25/4\pi) = (25/2) = 12.5$.

That is, the maximum depth of 12 feet happens at the start (midnight) and again at 12:30 PM.

At times the least-squares method for finding a regression equation that best fits given data results in a sine function of the form

$$y = a \sin(kx + b) + c.$$

EXAMPLE 9   G Modeling Monthly Temperatures

The data in the given table lists average monthly morning temperatures in degrees Fahrenheit in a southern city for certain months of the year.

(a) Plot the data in the table on a scattergram. Use the horizontal axis for the t values and the vertical axis for the F values.

(b) Use a grapher to find a sine function of the form

$$F = a \sin(kt + b) + c$$

that best fits the data. Round off a, b, c, and k to three decimal places.

(c) Graph the function in part (b) of best fit on the same coordinate system as the scattergram.

(d) Predict the average monthly morning temperature in the months of

March, June, and October.

Round off the answers to one decimal place.

Month	t	Average Morning Monthly Temperature F
Jan	1	43.2
Feb	2	47.9
April	4	58.8
May	5	67.2
July	7	75.3
Aug	8	72.4
Nov	11	46.5
Dec	12	44.1

Solution   (a) By using the horizontal axis to represent the t values and the vertical axis to represent the F values, we get the scattergram shown in Figure 16a.

(b) Upon using a grapher to produce the regression equation and after rounding off to three decimal places we get the model

$$F = 15.696 \sin(0.574t - 2.305) + 59.638.$$

(c) The graph of this equation, along with the scattergram for the given data, is shown in Figure 16b.

Figure 16

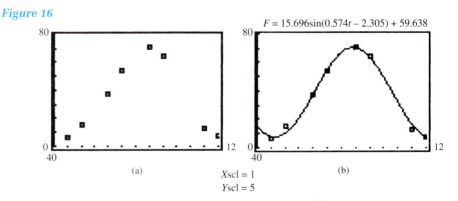

$F = 15.696\sin(0.574t - 2.305) + 59.638$

(a)

Xscl = 1
Yscl = 5

(b)

(d) To predict the average monthly morning temperature from this model for the month of March, we substitute $t = 3$ into the regression equation to get $51.0°F$ (approx.).
Similarly, for June, $t = 6$ so that $F = 73.9°$ (approx.), and for October, $t = 10$ so that $F = 55.1°$ (approx.).

Modeling Simple Harmonic Motion

In the real world there are many phenomena that involve oscillation or vibration in a uniform manner, repeating periodically in definite intervals of time. These phenomena can be modeled by *sinusoidal curves,* also called *simple harmonic curves,* which are the graphs of sine and cosine functions. Examples include alternating electrical current, sound waves, light waves, pendulums, and mass-spring systems. A mathematical model for a quantity y that is oscillating in a simple harmonic manner is given by either of the following equations:

$$y = a \cos \omega t \quad \text{or} \quad y = a \sin \omega t$$

where t represents elapsed time and $\omega > 0$ is called the **angular frequency.**
The **period T** for either $y = a \cos \omega t$ or $y = a \sin \omega t$ is given by

$$T = \frac{2\pi}{\omega}.$$

The **frequency f** of a simple harmonic motion is the number of oscillations per unit of time, so

$$f = \frac{1}{T} = \frac{\omega}{2\pi}.$$

EXAMPLE 10   **Finding the Frequency of a Simple Harmonic Model**

Suppose that alternating electric current is described by the simple harmonic model given by

$$y = 14.1 \sin 120\pi t$$

where y (in amperes) is the current at time t (in seconds). Find the amplitude, period, and frequency of the current.

Solution   In this situation, $y = a \sin \omega t$, where $a = 14.1$ and $\omega = 120\pi$. So the amplitude of the current is 14.1 and the period T is given by

$$T = \frac{2\pi}{\omega} = \frac{2\pi}{120\pi} = \frac{1}{60}$$

that is, the period is $\frac{1}{60}$ seconds. The frequency f of the current is given by

$$f = \frac{1}{T} = \frac{1}{\left(\dfrac{1}{60}\right)} = 60.$$

Therefore, the frequency is 60 cycles per second, or 60 hertz (Hz).

If a weight suspended from a spring is pulled down a distance $|a|$ and then released, it oscillates vertically in periodic motion. To model the situation, we let y (say, in centimeters) represent the *directed distance* of the weight from its rest position after time t (say, in seconds). We consider y to be a function of t. If the values of y are plotted for specific values of t, and if friction is neglected, then the resulting graph (Figure 17) has an equation of one of the following forms:

$$y = a \cos \sqrt{\frac{k}{m}}\, t \quad \text{or} \quad y = a \sin \sqrt{\frac{k}{m}}\, t$$

Figure 17

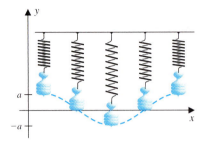

Here, k, the **stiffness coefficient,** is a constant associated with the particular spring, and m is the **mass** of the weight.

The variable y is also referred to as the **displacement.** A positive value of y indicates that it is above its rest position, and a negative value of y indicates that it is below its rest position. The farther the weight is pulled down before it is released, the greater the amplitude of motion will be. Furthermore, the stiffer the spring, the more rapidly the weight will oscillate, and thus the smaller the time period of repeating motion will be. Since the period is given by $T = 2\pi/\omega$, where $\omega = \sqrt{k/m}$, it takes $2\pi/\omega$ units of time to go through one complete oscillation.

EXAMPLE 11

Finding the Equation of Motion of an Object Suspended from a Spring

An object of mass 4 grams suspended from a spring is pulled down 2 centimeters and released at time $t = 0$ so that the object oscillates vertically.

(a) Determine which equation,

$$y = a \cos \sqrt{\frac{k}{m}}\, t \quad \text{or} \quad y = a \sin \sqrt{\frac{k}{m}}\, t,$$

is better suited to model this situation.

(b) Write an equation of motion if the stiffness coefficient k is 64 and t is the number of elapsed seconds.

(c) Find the period, frequency, and amplitude of the motion.

(d) Give a physical interpretation of the answers in part (c).

(e) Graph three cycles of the displacement y as a function of t.

Solution   (a) First, we compare the two equations available to model the situation:

Equation	$y = a \cos\sqrt{\dfrac{k}{m}}\, t$	$y = a \sin\sqrt{\dfrac{k}{m}}\, t$
Value of Equation When $t = 0$	$y = a \cos 0 = a$	$y = a \sin 0 = 0$

For the given situation, the displacement is 2 centimeters below the resting position at the <u>start</u> of the motion; that is, when $t = 0$, $y = -2$. So the equation $y = a \cos \sqrt{k/m}\, t$, with $a = -2$, is a more suitable model.

(b) Using the equation $y = a \cos \sqrt{k/m}\, t$ with $a = -2$, $k = 64$, and $m = 4$, we get the motion equation

$$y = -2 \cos\sqrt{\frac{64}{4}}\, t = -2 \cos 4t.$$

(c) The period $T = 2\pi/\omega = 2\pi/4 = \pi/2$. The frequency $f = 1/T = 2/\pi$ cycles per second, and the amplitude is $|a| = |-2| = 2$ centimeters.

(d) Since the period is $\pi/2$, it takes about 1.57 seconds for one complete oscillation. An amplitude of 2 means the highest and lowest positions are 2 centimeters above and below the resting position. A frequency of $2/\pi$ indicates that it makes about 0.64 oscillation per second.

(e) The graph in Figure 18 shows three cycles of this situation.

Figure 18

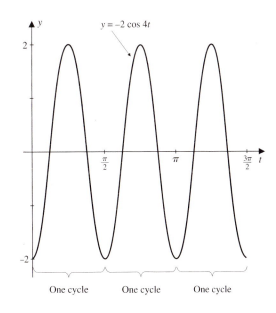

$y = -2 \cos 4t$

One cycle   One cycle   One cycle

◆ PROBLEM SET 5.4

Mastering the Concepts

In problems 1 and 2, find the range of each function. Express the answer in interval notation.

1. (a) $f(\theta) = 3 \sin \theta$

    (b) $h(\theta) = 4 - \dfrac{1}{3} \cos \theta$

    (c) $f(\theta) = \dfrac{2 \sin \theta + 5}{3}$

2. (a) $g(\theta) = -\dfrac{1}{2} \cos \theta$

    (b) $g(\theta) = \dfrac{1}{\cos \theta}$

    (c) $g(\theta) = \dfrac{3 + 2 \cos \theta}{-5}$

In problems 3–18, use the graph of the sine or cosine function and transformations to sketch the graph of each function. Indicate the range, amplitude, and period of each function.

3. (a) $y = 3 \cos x$
    (b) $y = -3 \cos x$

4. (a) $y = -5 \sin x$

    (b) $y = \dfrac{4}{3} \sin x$

5. (a) $y = \dfrac{2}{3} \cos(-x)$

    (b) $y = \dfrac{1}{2} \cos x$

6. (a) $y = -2 \cos x$
    (b) $y = 2 \cos(-x)$

7. (a) $y = \cos 2x$

    (b) $y = \cos \dfrac{x}{2}$

8. (a) $y = \sin 3x$

    (b) $y = \sin \dfrac{x}{3}$

9. (a) $y = \cos \dfrac{\pi x}{3}$

    (b) $y = \cos(-2\pi x)$

10. (a) $y = \sin \dfrac{\pi x}{3}$

    (b) $y = -\sin \dfrac{\pi x}{2}$

11. (a) $y = \dfrac{5}{3} \sin(-6x)$

    (b) $y = 5 \sin 10x$

12. (a) $y = 3 \cos \left(\dfrac{-\pi x}{6} \right)$

    (b) $y = -3 \cos 5x$

13. (a) $y = -3 \cos \dfrac{\pi x}{2}$

    (b) $y = \pi \cos \left(\dfrac{-5x}{2} \right)$

14. (a) $y = -5 \sin \dfrac{2\pi x}{3}$

    (b) $y = -\dfrac{1}{2} \sin \dfrac{3x}{2}$

15. (a) $y = 1 + \sin x$
    (b) $y = 3 - \sin x$

16. (a) $y = 2 + \cos x$

    (b) $y = -\dfrac{3}{4} + 2 \cos x$

17. (a) $y = 2 + 3 \cos \dfrac{\pi x}{4}$

    (b) $y = -2 - 3 \cos 2\pi x$

18. (a) $y = \dfrac{1}{2} - \dfrac{1}{2} \sin \dfrac{\pi x}{2}$

    (b) $y = -2 - \sin \dfrac{\pi x}{6}$

19. Match each function with one of the graphs shown in Figure 19.

$$f(x) = 0.5 \sin x; \; g(x) = \sin x; \; h(x) = 2 \sin x$$

Figure 19

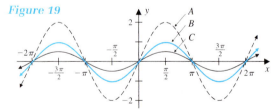

20. Match each function with one of the graphs shown in Figure 20.

$$f(x) = 2 \sin x + 1; \; g(x) = \sin x + 3;$$
$$h(x) = 0.5 \sin x - 2$$

Figure 20

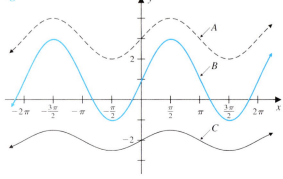

In problems 21–36, identify and use the phase shift, period, and amplitude to graph one cycle of each function.

21. $y = 2 \sin\left(x - \dfrac{\pi}{4}\right)$

22. $y = -3 \sin\left(x + \dfrac{\pi}{3}\right)$

23. $y = 4 \cos\left(x + \dfrac{\pi}{6}\right)$

24. $y = 2 \cos\left(x - \dfrac{\pi}{3}\right)$

25. $y = -2 \sin(2x - \pi)$

26. $y = -2 \cos(3x - \pi)$

27. $y = 2 \sin(\pi x - \pi)$

28. $y = 3 \cos(\pi x - \pi)$

29. $y = -3 \cos(3x + 2)$

30. $y = -3 \sin(4x - 1)$

31. $y = 2 \cos\left(\dfrac{\pi x}{2} - \dfrac{1}{4}\right)$

32. $y = -4 \sin\left(\dfrac{\pi x}{3} + \dfrac{1}{3}\right)$

33. $y = -2 \sin(5 - 3x)$

34. $y = -2 \cos(3 - 2x)$

35. $y = 2 + 3 \cos(2x - 2\pi)$

36. $y = -2 - 3 \sin(3x + \pi)$

In problems 37 and 38, each figure shows the graph of one cycle of the function

$$y = a \sin(kx + b).$$

Find values for $a > 0$, $k > 0$, and b, and write the resulting equation. Also find the amplitude, period, and phase shift.

37. (a)   (b)

38. (a)   (b)

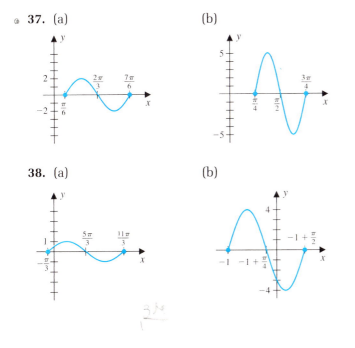

39. Write an equation for a sine function with the following properties:

amplitude $= 3$,   period $= \dfrac{\pi}{3}$,   phase shift $= 0$

Then graph one cycle.

40. Write an equation for a cosine function with the following properties:

amplitude $= 2$,   period $= \dfrac{\pi}{6}$,   phase shift $= \dfrac{\pi}{2}$

Then graph one cycle.

Applying the Concepts

In problems 41–48, round off the answers to two decimal places.

41. Blood Pressure: The blood pressure of a person is given by the model

$$y = 110 + 30 \cos \dfrac{12\pi t}{5}$$

where t is in seconds and y is in millimeters of mercury (mm Hg).

(a) Sketch the graph of this function for $0 \le t \le 2.5$.

(b) Determine the period of the function.

(c) Use the graph to determine the maximum and minimum blood pressure of the person for the given interval.

42. Electrical Circuit: The electricity supplied to homes is called alternating current (ac), and the current I (in amperes) is related to time t (in seconds) by the equation

$$I = 5 \cos 120\pi t.$$

The voltage V (in volts) that causes this current to flow is related to time t (in seconds) by the equation

$$V = 170 \cos[120\pi(t + 0.031)].$$

(a) Sketch the graphs of the current and voltage functions for $0 \le t \le 0.004$.

(b) Use the graphs to estimate the voltage at the time the current is a maximum.

(c) Use the graphs to estimate the current at the time the voltage is a maximum.

(d) Predict the first time when the voltage reaches 160 volts.

43. Height of a Point on a Waterwheel: Figure 21 shows a rotating waterwheel. At time t (in seconds) the height h (in feet) of a point P on the rim of the wheel above the surface of the water is given by the model

$$h(t) = 7 + 8 \cos\left(\frac{\pi t}{5} - \frac{2}{5}\right).$$

Figure 21

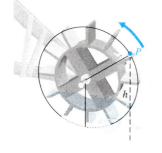

(a) Find the height of the point P on the wheel after 2 seconds, after 5 seconds, and after 10 seconds.

(b) How long does it take for the point P on the wheel to go around once?

(c) Sketch the graph of h for two cycles.

(d) Find the minimum and maximum heights of P on the wheel.

44. Sound Wave: The difference D (in dynes per square centimeter) between the atmospheric pressure and the air pressure at the eardrum produced by a sound wave at t seconds is given by the function

$$D(t) = 4 \sin\left(100\pi t + \frac{\pi}{4}\right).$$

(a) Sketch the graph of this function for $0 \le t \le 0.18$.

(b) Find the period and amplitude.

(c) Find the minimum and maximum values of $D(t)$.

45. Seasonal Business Profit: The profit P (in thousands of dollars) per month from the sales of scuba equipment for 2 years beginning in January 1, 2004, is given by the function

$$P(t) = 900 - 700 \cos \frac{\pi t}{6}$$

where t is the elapsed time (in months) and $t = 0$ corresponds to January 1, 2004.

(a) Graph the function for $0 \le t \le 24$.

(b) Find the amplitude and period, and interpret them.

(c) Determine the profit at 3 months, 9 months, 1 year, 18 months, and 2 years.

(d) What is the maximum profit, and when does it occur?

46. Tides: During a 24-hour day, the mathematical model

$$d = 3 \cos 0.52t + 9$$

is used to predict the tide. That is, it predicts how much above mean sea level the tide will raise or lower the water depth d (in feet) in a certain harbor, t hours after midnight.

(a) Sketch the graph of this function over a 24-hour day.

(b) Find the amplitude and period, and interpret them.

(c) Find the minimum and maximum depth of water in the harbor, and determine when they occur.

(d) Interpret your answers for part (c) in terms of this tidal wave.

47. Hours of Daylight: Suppose that the hours $h(t)$ of daylight for a certain city are related to the day of the year by the model

$$h(t) = 2.3 \sin\left[\frac{2\pi}{365}(t - 80)\right] + 12$$

where t is the day of the year, and $t = 1$ corresponds to January 1.

(a) Sketch the graph of this function for

$$1 \le t \le 365.$$

(b) Find the amplitude and the period.

(c) Determine the maximum and minimum number of hours of daylight and when they occur.

48. Business Profit: A company's profit fluctuates during a 12-month period according to the model

$$P(t) = 30{,}000 - 70{,}000(\sin t + 0.5 \cos 0.5t)$$

where $P(t)$ represents the profit (in dollars) after t months.

(a) Find the profit (or loss) at the beginning of the 12-month period, after 5 months, after 8 months, and after 1 year.

(b) Sketch the graph of P for

$$0 \le t \le 12.$$

(c) Interpret the graph to determine when there is a profit.

(d) What are the maximum and minimum profits over the 12-month period, and when do they occur?

49. Ⓖ **Temperature Trend:** The data in the following table lists the average monthly temperature F (in degrees Fahrenheit) for a mountainous region for certain months of the year.

Month	t	Average Monthly Temperature F
Jan	1	24
Feb	2	28.3
April	4	39.2
May	5	47.1
July	7	56.4
Aug	8	57
Oct	10	41.8
Dec	12	29.6

(a) Plot the data in the above table on a scattergram. Use the horizontal axis for the t values and the vertical axis for the F values.

(b) Use a grapher to find a sine function of the form

$$F = a \sin(kt + b) + c$$

that best fits the data. Round off a, k, b, and c to three decimal places.

(c) Graph the function in part (b) of best fit on the same coordinate system as the scattergram.

(d) Predict the average monthly temperatures in the months of March, June, and September. Round off the answers to one decimal place.

50. Ⓖ **Precipitation Trend:** The data in the following table lists the average monthly precipitation P (in inches) in a region for certain months of the year.

(a) Plot the data in the table at the top of the next column on a scattergram. Use the horizontal axis for the t values and the vertical axis for the P values.

Month	t	Average Precipitation P in inches
Jan	1	1.1
Feb	2	0.9
Mar	3	1.7
May	5	3.1
June	6	3.9
Aug	8	4.1
Sept	9	3.2
Nov	11	1.6
Dec	12	1.2

(b) Use a grapher to find a regression model of the form

$$P = a \sin(kt + b) + c$$

that best fits the data. Round off a, k, b, and c to three decimal places.

(c) Graph the function in part (b) of best fit on the same coordinate system as the scattergram.

(d) Predict the precipitation in the months of April, July, and October. Round off the answers to one decimal place.

In problems 51 and 52, a simple harmonic model is described by the given equation, where t is time (in seconds). Determine the amplitude, period, and frequency of each model.

51. (a) $y = \sin 200\pi t$
   (b) $y = 2 \cos 0.001\pi t$

52. (a) $y = 0.15 \sin 792\pi t$
   (b) $y = 0.06 \cos 120\pi t$

53. **Spring–Weight Model:** An object of mass 30 grams suspended from a spring is pulled down 50 centimeters and released at time $t = 0$ so that the mass oscillates vertically.

(a) Write an equation of motion if the stiffness coefficient $k = 8$ and t is the number of elapsed seconds.

(b) Graph two cycles of the displacement y as a function of t.

(c) Find the amplitude, period, and frequency of the motion and interpret them.

54. **Spring–Weight Model:** A weight suspended from a spring oscillates up and down according to the model

$$y = a \cos \sqrt{\frac{k}{m}}\, t.$$

If the stiffness coefficient of the spring $k = 4000$, how large a mass should be attached so that the spring will oscillate 6.4 times per second?

55. Pendulum Swing Model: A pendulum swings uniformly back and forth through an arc length of 0.05 meter, taking 2.2 seconds to move from the position directly above point A to the position directly above point B (Figure 22). Assume the motion of the pendulum is simple harmonic.

(a) Find the angular frequency ω.

(b) Find the equation for y as a function of t, where y denotes the displacement of the mass at t seconds.

Figure 22

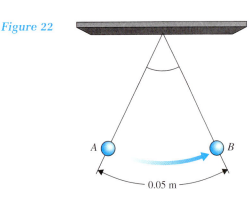

0.05 m

(c) Sketch the graph for three cycles.

56. Grandfather Clock: A grandfather clock has a pendulum of length $L = 0.99$ meter, and it swings through an arc length of 0.25 meter according to the model

$$y = a \sin \omega t.$$

(a) Find the period

$$T = 2\pi \sqrt{\frac{L}{g}},$$

where $g = 9.8$ meters per second.

(b) Find the angular frequency ω.

(c) Determine the equation for the displacement y as a function of t.

(d) Sketch the graph of the equation for three cycles.

Developing and Extending the Concepts

57. Assume

$$f(x) = \sin\left(\frac{\pi}{2} - x\right)$$

and

$$g(x) = \cos x,$$

where $-\pi \le x \le \pi$.

Compare the graphs. Are they the same? Explain.

58. Sketch the graphs of

$$f(x) = [\sin x]^2$$

and

$$g(x) = \left[\cos\left(\frac{\pi}{2} - x\right)\right]^2$$

on the interval $[0, 2\pi]$. How do the graphs compare?

59. Assume $y = \cos |x|$, where $-2\pi \le x \le 2\pi$. Compare this v graph to the graph of $f(x) = \cos x$. Are the two graphs the same or different? Explain.

60. Compare the graph of $f(x) = \sin x$ to the graph of $g(x) = |\sin x|$ for $0 \le x \le 2\pi$. Where are they the same? Where are they different? Explain.

61. **G** $f(x) = \dfrac{\sin x}{x}$

(a) Is $f(0)$ defined?

(b) Use a grapher to graph f, and then use the graph to help determine the behavior of $f(x)$ as $x \to 0$.

(c) Complete the following table to support the conclusion in part (b):

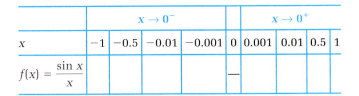

	$x \to 0^-$					$x \to 0^+$			
x	-1	-0.5	-0.01	-0.001	0	0.001	0.01	0.5	1
$f(x) = \dfrac{\sin x}{x}$					—				

62. **G** Use a grapher to graph

$$f(x) = \cos x$$

and

$$g(x) = 1 - \frac{x^2}{2} + \frac{x^4}{24}$$

for the interval

$$[-2.75, 2.75].$$

Compare the values of $f(x)$ and $g(x)$ in this interval.

G In problems 63 and 64, use a grapher to sketch the graph of each function, where

$$-2\pi \le x \le 2\pi.$$

If the function is periodic, find its period and its amplitude.

63. $y = \sin x + 2 \cos x$

64. $y = 4 \sin x + 3 \cos x$

Objectives

1. Graph the Tangent Function
2. Graph the Cotangent Function
3. Graph the Secant and Cosecant Functions

5.5 Graphs of Other Trigonometric Functions

So far, we have graphed sine and cosine functions. In this section we develop the graphs of the other four trigonometric functions by taking advantage of the fact that they can be written in terms of the sine and cosine functions.

Graphing the Tangent Function

Since the tangent has a period of π, our strategy is to graph $y = \tan x$ on interval $[-\pi/2, \pi/2]$, which has length π, and then repeat the resulting graph in both directions.

By using the graphs of $g(x) = \cos x$ and $f(x) = \sin x$ on interval $[-\pi/2, \pi/2]$ (Figures 1a and 1b), along with the quotient identity $\tan x = \sin x/\cos x$, we can discover three important features:

(1) the domain, (2) vertical asymptotes, and (3) an x intercept.

1. **Domain:** From the quotient relationship $\tan x = \sin x/\cos x$, it follows that $y = \tan x$ is undefined for those values of x where $\cos x = 0$. These values, which are the same as the x intercepts of the graph of $g(x) = \cos x$, are $x = -\pi/2$ and $x = \pi/2$ (Figure 1a). Consequently, the domain of $y = \tan x$ consists of all values of x for which $x \neq \pm\pi/2$ on interval $[-\pi/2, \pi/2]$.

2. **Asymptotes:** To help find out what happens to the values of $\tan x$ as x approaches $\pi/2$, we examine the entries in Table 1. They suggest that as x approaches $\pi/2$ from the left, $\tan x$ increases without bound; that is,

$$\tan x \to +\infty \quad \text{as} \quad x \to \left(\frac{\pi}{2}\right)^-.$$

Thus the line $x = \pi/2$ is a vertical asymptote of the graph.

Figure 1

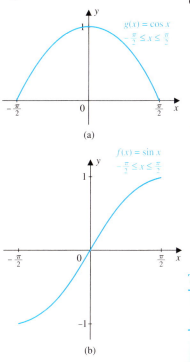

(a)

(b)

TABLE 1

				$x \to \left(\dfrac{\pi}{2}\right)^-$				
x	0	0.5	1.0	1.5	1.56	1.569	1.57	$\dfrac{\pi}{2}$
$y = \tan x$	0	0.546	1.557	14.101	92.620	556.691	1255.766	Undefined

Similarly, it can be shown that

$$\tan x \to -\infty \quad \text{as} \quad x \to \left(-\frac{\pi}{2}\right)^+$$

so $x = -\pi/2$ is also a vertical asymptote.

3. **x intercept:** The x intercept, that is, the value of x for which $\tan x = 0$ on interval $[-\pi/2, \pi/2]$, is the same as the x intercept of $f(x) = \sin x$ (Figure 1b) which is $x = 0$.

By using this information and plotting a few points, we obtain a sketch of the graph of $y = \tan x$ in the interval $(-\pi/2, \pi/2)$ (Figure 2).

Since the tangent function has a period of π, we repeat the graph in Figure 2 to the right and left every π units to get a complete graph of the tangent function (Figure 3). The periodicity of π enables us to locate all the other x intercepts and vertical asymptotes because they repeat every π units.

Figure 2

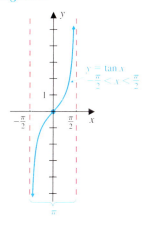

Figure 3

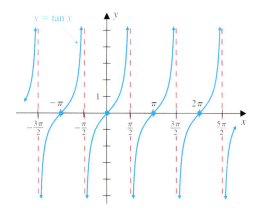

By reading the graph in Figure 3, we are led to the properties of the tangent function listed in Table 2.

TABLE 2   Properties of the Tangent Function

Property	$y = \tan x$
Domain	All real numbers except $(\pi/2) + n\pi$, where n is an integer
Range	All real numbers
Period	π
Symmetry	Symmetric with respect to the origin
Even or odd	The function is odd, that is, $\tan(-x) = -\tan x$
Asymptotes	$x = \pm\pi/2, \pm(3\pi)/2, \pm(5\pi)/2, \ldots, \pm[(\pi/2) + n\pi]$, where n is an integer
x intercepts	$0, \pm\pi, \pm2\pi, \ldots \pm n\pi$, where n is an integer
Increasing or decreasing behavior	Increases between consecutive asymptotes

By using arguments similar to those used for the sine and cosine functions in Section 5.5, we are led to the following result.

Property

Tangent Period and Phase Shift

> The period of $f(x) = a \tan (kx + b)$ is $\dfrac{\pi}{k}$,
>
> and
>
> its phase shift is $-\dfrac{b}{k}$, for $k > 0$.

If k is negative, we can use the fact that the tangent is an odd function to rewrite the given function with a positive coefficient of x, as we did for the sine and cosine functions.

For instance,

$$f(x) = \tan(-3x + 1)$$
$$= \tan[-(3x - 1)]$$
$$= -\tan(3x - 1).$$

EXAMPLE 1   **Graphing a Tangent Function**

Sketch the graph of

$$f(x) = \tan\left(x - \frac{\pi}{6}\right).$$

for one period by using a transformation of the graph of $y = \tan x$ in the interval $(-\pi/2, \pi/2)$. Specify the period, phase shift, and vertical asymptotes.

Solution   We can shift the graph of $y = \tan x$ (Figure 4a), $\pi/6$ unit to the right to get the graph of f (Figure 4b). After rewriting $f(x)$ in the form $a \tan(kx + b)$, we get

$$f(x) = 1 \tan\left[1x + \left(-\frac{\pi}{6}\right)\right].$$

So that the period is

$\pi/k = \pi/1 = \pi$, which is the same as the period for $y = \tan x$,

and the phase shift is given by

$$\frac{-b}{k} = \frac{-(-\pi/6)}{1} = \frac{\pi}{6}.$$

The left asymptote for $y = \tan x$ shifts from $x = -\pi/2$ to

$$x = -\frac{\pi}{2} + \frac{\pi}{6} = -\frac{\pi}{3}$$

for f.
The right asymptote for $y = \tan x$ shifts from $x = \pi/2$ to

$$x = \frac{\pi}{2} + \frac{\pi}{6} = \frac{2\pi}{3}$$

for f.

Figure 4

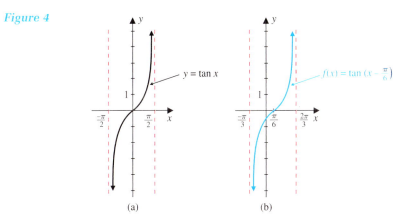

(a)   (b)

Graphing the Cotangent Function

Figure 5

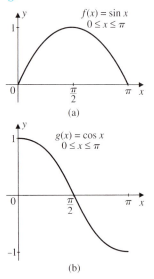

(a)

(b)

Since the cotangent has period π, we begin by graphing $y = \cot x$ on interval $[0, \pi]$. By using the quotient identity, along with the graphs of $f(x) = \sin x$ (Figure 5a) and $g(x) = \cos x$ (Figure 5b) on the interval $[0, \pi]$, we arrive at the following features.

1. **Domain:** The quotient relationship $\cot x = \cos x/\sin x$ implies that the $\cot x$ is undefined whenever $\sin x = 0$ on interval $[0, \pi]$, that is, when $x = 0$ or $x = \pi$. So the domain consists of all numbers x in interval $[0, \pi]$ except 0 and π.

2. **Asymptote:** By examining the behavior of $y = \cot x$ as $x \to 0^+$, we find that $\cot x \to +\infty$. Also, $\cot x \to -\infty$ as $x \to \pi^-$. So both $x = 0$ and $x = \pi$ are vertical asymptotes.

3. **x-intercept:** On interval $[0, \pi]$, $\cot x = 0$ whenever $\cos x = 0$, that is, if $x = \pi/2$. So $x = \pi/2$ is an x intercept.

By plotting some points and using the above information we obtain the graph of $y = \cot x$ for interval $[0, \pi]$ (Figure 6a).

To obtain the complete graph of $y = \cot x$, we use the fact that the cotangent has period π by repeating the graph in Figure 6a every π units to the right and left (Figure 6b).

Figure 6

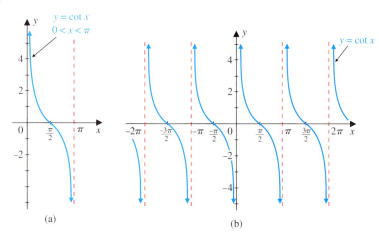

(a)

(b)

The properties of the cotangent function listed in Table 3 are read off of Figure 6b.

TABLE 3   Properties of the Cotangent Function

Property	$y = \cot x$
Domain	All real numbers except $n\pi$, where n is an integer
Range	All real numbers
Period	π
Symmetry	Symmetric with respect to the origin
Even or odd	The function is odd, that is, $$\cot(-x) = -\cot x.$$
Asymptotes	$x = 0, \pm\pi, \pm 2\pi, \ldots, \pm n\pi$, where n is an integer
x intercepts	$\pm\dfrac{\pi}{2}, \pm\dfrac{3\pi}{2}, \pm\dfrac{5\pi}{2}, \ldots, \pm\left(\dfrac{\pi}{2} + n\pi\right)$, where n is an integer
Increasing or decreasing behavior	Decreases between consecutive asymptotes

As in the case for the tangent function, graphs of cotangent functions of the form

$$f(x) = a \cot(kx + b)$$

have the following characteristics:

Property **Cotangent Period** **and Phase Shift**	The period for the function $$f(x) = a \cot(kx + b) \text{ is given by}$$ $$\frac{\pi}{k}$$ and its phase shift is $$\frac{-b}{k} \text{ for } k > 0.$$

EXAMPLE 2   **Graphing a Cotangent Function**

Sketch the graph of one period of the function

$$g(x) = \cot\left(2x - \frac{\pi}{2}\right)$$

by using transformations of the graph of one period of $y = \cot x$ in the interval $(0, \pi)$. Find the period and phase shift, and determine the asymptotes.

Solution   We begin by graphing $y = \cot x$ in the interval $(0, \pi)$ (Figure 7a). Since

$$g(x) = \cot\left(2x - \frac{\pi}{2}\right) = 1 \cot\left[2x + \left(-\frac{\pi}{2}\right)\right],$$

its period is given by

$$\frac{\pi}{k} = \frac{\pi}{2}$$

which is $\dfrac{1}{2}$ the period of $y = \cot x$.

Also, the phase shift of g is given by

$$\frac{-b}{k} = \frac{-\left(-\dfrac{\pi}{2}\right)}{2} = \frac{\pi}{4}.$$

Alternately, the phase shift can be found, as we did for sine and cosine functions, by solving

$$2x - \frac{\pi}{2} = 0 \quad \text{to get} \quad x = \frac{\pi}{4}.$$

It follows that the graph of g can be obtained by shifting the graph of $y = \cot x$ to the right $\pi/4$ unit and then horizontally compressing it by a scale factor of $\dfrac{1}{2}$ (Figure 7b).

Figure 7

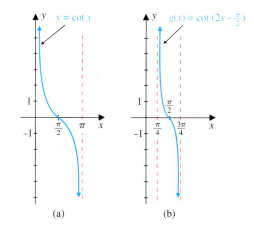

(a)   (b)

Notice that since the vertical asymptotes of $y = \cot x$ are at

$$x = 0 \text{ and } x = \pi,$$

correspondingly, the vertical asymptotes of g occur when

$$2x - \frac{\pi}{2} = 0 \text{ or } \pi, \text{ that is, when } x = \frac{\pi}{4} \text{ or } \frac{3\pi}{4}.$$

Graphing the Secant and Cosecant Functions

To graph the secant and cosecant functions, we take advantage of what we know about the graphs of the cosine and sine functions.

The graph of $y = \sec x$ can be sketched with the aid of the reciprocal identity

$$\sec x = \frac{1}{\cos x}.$$

To sketch the graph of $y = \sec x$, we first graph the cosine function $y = \cos x$. Vertical asymptotes of $y = \sec x$ occur at the x intercepts of $y = \cos x$, where $\cos x = 0$, that is, when $x = \pi/2 + n\pi$, for any integer of n.

Notice that because of the reciprocal relationship the secant increases where the cosine decreases, and it decreases where the cosine increases (Figure 8).

Figure 8

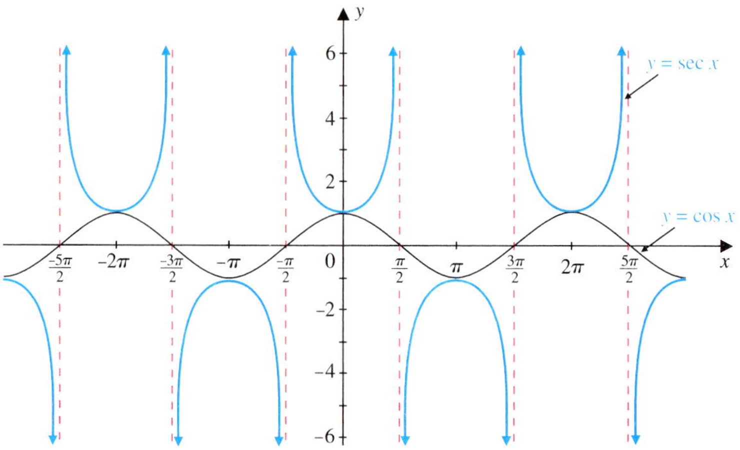

The graph shows that the range of the secant function includes all numbers y, where $y \leq -1$ or $y \geq 1$. In addition, the graph repeats itself every 2π units, confirming the fact that the function $y = \sec x$ is periodic with period 2π.

Similarly, the graph of the cosecant function $y = \csc x$ can be sketched by using

$$\csc x = \frac{1}{\sin x}$$

and the sine graph.

The vertical asymptotes occur when $\sin x = 0$, or when $x = n\pi$, for any integer n. Here the cosecant increases where the sine decreases, and it decreases where the sine increases (Figure 9).

Figure 9

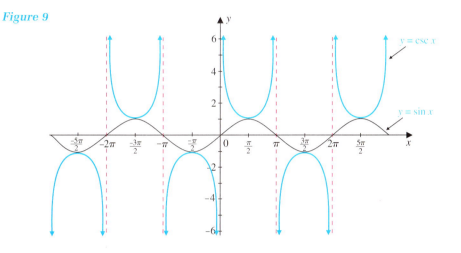

The graph reveals that the range of the cosecant function includes all numbers y, where $y \leq -1$ or $y \geq 1$. The graph repeats every 2π units, confirming that the function $y = \csc x$ is periodic with period 2π.

Table 4 summarizes properties of the secant and cosecant functions, which can be read off of the graphs in Figures 8 and 9.

TABLE 4   Properties of the Secant and Cosecant Functions

Property	$y = \sec x$	$y = \csc x$
Domain	All real numbers except $\pi/2 + n\pi$, where n is an integer	All real numbers except $n\pi$, where n is an integer
Range	All real numbers y such that $y \leq -1$ or $y \geq 1$	All real numbers y such that $y \leq -1$ or $y \geq 1$
Period	2π	2π
Symmetry	Symmetric with respect to the y axis	Symmetric with respect to the origin
Even or odd	The function is *even*, that is, $\sec(-x) = \sec x$	The function is *odd*, that is, $\csc(-x) = -\csc x$
Aymptotes	$x = (\pi/2) + n\pi$, where n is an integer	$x = n\pi$, where n is an integer

As with the sine and cosine functions, we have the following graph features.

Property

Secant and Cosecant, Period and Phase Shift

The functions

$$f(x) = a \sec(kx + b) \text{ and } g(x) = a \csc(kx + b)$$

have period

$$\frac{2\pi}{k}$$

and phase shift given by

$$\frac{-b}{k} \quad \text{for } k > 0.$$

For instance, the function

$$f(x) = 2 \csc\left(3x - \frac{\pi}{4}\right) \text{ has a period of } \frac{2\pi}{3}$$

and a phase shift of $\dfrac{-\left(\dfrac{-\pi}{4}\right)}{3} = \dfrac{\pi}{12}$.

The graphs of the secant and cosecant functions shown in Figures 8 and 9, respectively, can be used to graph other trigonometric functions, as illustrated in the next example.

EXAMPLE 3   **Graphing Secant and Cosecant Functions**

Use transformations to sketch the graph of one period of each function. Find the period, phase shift, vertical asymptotes, and the range.

(a) $f(x) = 3 \csc \dfrac{x}{2}$ (b) $g(x) = \sec(3x + \pi)$

Solution   (a) We will use one period of the graph of $y = \csc x$ in interval $(-\pi, \pi)$ (Figure 9) to sketch one period of the graph of

$$f(x) = 3 \csc \frac{x}{2} = 3 \csc\left(\frac{1}{2}x\right).$$

In the form

$$f(x) = a \csc(kx + b),$$

$$a = 3, \quad b = 0, \quad \text{and} \quad k = \frac{1}{2}.$$

So the period of f is

$$\frac{2\pi}{k} = \frac{2\pi}{\left(\dfrac{1}{2}\right)} = 4\pi = 2 \cdot 2\pi.$$

Since $b = 0$, the phase shift is 0.

Thus to get the graph of f from the graph of $y = \csc x$, which has a period of 2π, the latter graph has to be stretched horizontally by a factor of 2.

Since the vertical asymptotes for $y = \csc x$ on interval $[-\pi, \pi]$ occur when

$$x = -\pi, \quad 0, \quad \text{or} \quad \pi,$$

it follows that three vertical asymptotes for f occur when

$$x/2 = -\pi, \quad 0, \quad \text{or} \quad \pi,$$

that is, when

$$x = -2\pi, \quad 0, \quad \text{or} \quad 2\pi.$$

Because $a = 3$, we also need to stretch the graph vertically to get the graph of f. (Figure 10a). Note that the vertical stretch of multiple of 3 causes the range to become all values of y such that $y \le -3$ or $y \ge 3$.

(b)   We can obtain one period of the graph of

$$g(x) = \sec(3x + \pi)$$

from one period of the graph of $y = \sec x$ on interval $[0, 2\pi]$ (Figure 8). The period of g is given by

$$\frac{2\pi}{k} = \frac{2\pi}{3} = \frac{1}{3} \cdot 2\pi.$$

So the graph of $y = \sec x$, which has period 2π, must be horizontally shrunk by a multiple of 1/3. Also, we need to shift the graph to the left by $\pi/3$ units because the phase shift is

$$-\frac{b}{k} = -\frac{\pi}{3}.$$

Two vertical asymptotes of $y = \sec x$ occur when

$$x = \frac{\pi}{2} \quad \text{or} \quad \frac{3\pi}{2}.$$

Correspondingly, g has vertical asymptotes when

$$3x + \pi = \frac{\pi}{2} \quad \text{or} \quad \frac{3\pi}{2}$$

that is, when

$$x = -\frac{\pi}{6} \quad \text{or} \quad \frac{\pi}{6}.$$

Figure 10b shows the graph of one period of g. Since there is no vertical scaling ($a = 1$), the range of g includes all values of y such that $y \le -1$ or $y \ge 1$.

Figure 10

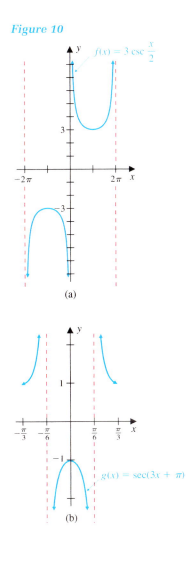

(a)

(b)

G   Most graphers do not have function keys for the cotangent, secant, and cosecant functions. To graph these functions, we use the **TAN**, **COS**, and **SIN** keys, respectively, and the reciprocal key **1/x** or **x⁻**, as we did when we evaluated these functions. For instance the graph of $y = \cot(4x - 3)$ is obtained by graphing

$$y = \frac{1}{\tan(4x - 3)}.$$

We graph $f(x) = 5 \sec 3x$ by graphing

$$f(x) = \frac{5}{\cos 3x}$$

and so forth.

Care must be taken when reading such graphs produced by a grapher. Depending on the window settings and the grapher's features, it may turn out that the graph will appear connected or continuous even though it should not because there are vertical asymptotes. Figure 11 displays such an inaccuracy for the graph of $y = \cot(4x - 3)$.

Figure 11

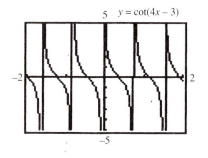

Graphers enable us to gain more insight into the characteristics of mathematical models involving trigonometric functions. For instance, Figure 12 shows the graph of the function $L = 8 \csc t + \sec t$, $0 < t < \pi/2$.

Upon reading the graph of L, we observe there is a minimum or a low point on the graph. By using the **MIN**-finding feature of a grapher, we find the low point is located at approximately $(1.11, 11.18)$.

Figure 12

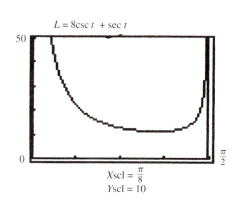

PROBLEM SET 5.5

Mastering the Concepts

In problems 1–24, use transformations of the graphs of the basic trigonometric functions to sketch the graph of one period for each function. Specify the period and phase shift.

1. $y = 3 \tan x$

2. $y = -3 \tan x$

3. $y = -\dfrac{3}{2} \cot x$

4. $y = -2 \sec x$

5. $y = 5 \csc x$

6. $y = 2 \sec \dfrac{x}{4}$

7. $y = 1 + 2 \csc x$

8. $y = 2 - \sec x$

9. $y = -2 \cot 4x$

10. $y = 2 \tan \dfrac{\pi x}{4}$

11. $y = -2 \sec \dfrac{2x}{3}$

12. $y = -3 \csc \dfrac{\pi x}{6}$

13. $y = 2 \csc 4x$

14. $y = -3 \sec \pi x$

15. $y = 2 \csc \left(x - \dfrac{\pi}{6} \right)$

16. $y = 3 \csc \left(x + \dfrac{\pi}{3} \right)$

17. $y = \tan \left(x - \dfrac{\pi}{4} \right)$

18. $y = \tan \left(x + \dfrac{2\pi}{3} \right)$

19. $y = 1.5 \cot \left(x - \dfrac{\pi}{2} \right)$

20. $y = 3 \cot \left(x + \dfrac{5\pi}{6} \right)$

21. $y = 3 \sec \left(\dfrac{x}{4} + \dfrac{\pi}{4} \right)$

22. $y = 2 \csc \left(3x - \dfrac{\pi}{2} \right)$

23. $y = -2 \cot \left(\dfrac{\pi x}{4} - \dfrac{\pi}{4} \right)$

24. $y = -2 \sec \left(\dfrac{\pi x}{6} + \dfrac{\pi}{6} \right)$

25. Use the graphs of the functions to indicate where each function listed in the table is increasing or decreasing.

	As x increases from:			
	0 to $\dfrac{\pi}{2}$	$\dfrac{\pi}{2}$ to π	π to $\dfrac{(3\pi)}{2}$	$\dfrac{(3\pi)}{2}$ to 2π
(a) $y = \tan x$				
(b) $y = \cot x$				
(c) $y = \sec x$				
(d) $y = \csc x$				

26. (a) Explain why $y = \tan x$ is not one-to-one.
  (b) Is $f(x) = \tan x$, where $-\pi/2 < x < \pi/2$, one-to-one? Use a graph to support the assertion.
  (c) Is $f(x) = \tan x$, where $0 < x < 2\pi$, one-to-one? Use a graph to support the assertion.

In problems 27–30, use the graph of the indicated function to complete each statement.

27. (a) As $x \to \left(\dfrac{\pi}{4} \right)^+$, $\tan x \to$ _____

  (b) As $x \to \left(\dfrac{\pi}{2} \right)^+$, $\tan x \to$ _____

28. (a) As $x \to 0^+$, $\cot x \to$ _____

  (b) As $x \to \left(\dfrac{\pi}{2} \right)^+$, $\sec x \to$ _____

29. (a) As $x \to \left(\dfrac{\pi}{2} \right)^-$, $\sec x \to$ _____

  (b) As $x \to \left(\dfrac{\pi}{4} \right)^-$, $\cot x \to$ _____

30. (a) As $x \to \left(\dfrac{\pi}{6} \right)^-$, $\tan x \to$ _____

  (b) As $x \to \pi^+$, $\cot \to$ _____

In problems 31–34, one period of a graph is shown. Find an equation of the given form for the graph.

31. $y = \tan kx$, $k > 0$

32. $y = a \tan kx$, $a < 0$ and $k > 0$

33. $y = a \csc(kx + b)$, $a > 0$, $b > 0$, $k > 0$

34. $y = a \sec kx$, $a > 0$ and $k > 0$

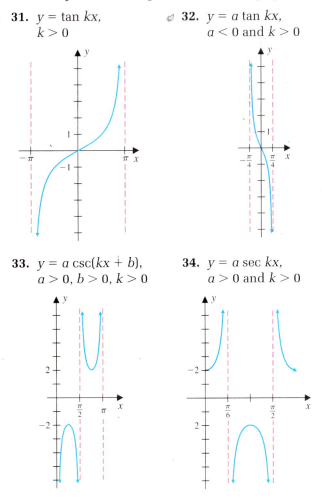

In problems 35–38, what is the distance from one vertical asymptote to the next for each function?

35. $y = \sec\left(x - \dfrac{\pi}{2}\right)$

36. $y = -2 \cot \dfrac{x}{6}$

37. $y = 2 \tan 3x$

38. $y = \csc(x + 1)$

G In problems 39 and 40,

(a) Use a grapher to graph the function defined by L in the interval $(0, \pi/2)$.

(b) Use the **MIN** -finding feature to approximate the location of the lowest point on the graph in part (a) to two decimal places.

39. $L = 2\sqrt{2}\,(\sec t + \csc t)$

40. $L = 12 \csc t + 20 \sec t$

Developing and Extending the Concepts

41. Use the graph of $y = \tan x$ to obtain the graph of

$$f(x) = |\tan x|$$

on the interval $\left(\dfrac{-\pi}{2}, \dfrac{\pi}{2}\right)$.

42. Compare the graphs of $f(x) = \sec x$ and $g(x) = \sec(-x)$. Are they the same or different? Explain.

43. Compare the graphs of $f(x) = \csc x$ and $g(x) = \csc(-x)$. Are they the same or different? Explain.

44. Sketch the graphs of

$$f(x) = \tan x$$

and

$$g(x) = \cot\left(\dfrac{\pi}{2} - x\right)$$

on the same coordinate system. How do the graphs compare?

45. Sketch the graphs of

$$f(x) = \sec x \quad \text{and} \quad g(x) = \csc\left(\dfrac{\pi}{2} - x\right)$$

on the same coordinate system. How do the graphs compare?

46. Explain why the graphs of $y = \tan x$, $f(x) = \tan(x + \pi)$, $g(x) = \tan(x - \pi)$, and $h(x) = \tan(x + 2\pi)$ are the same.

In problems 47 and 48, write an equation of the form $y = \tan kx$ or $y = \csc kx$ that satisfies the given condition.

47. (a) Tangent function of period $\dfrac{\pi}{6}$

(b) Cosecant function of period $\dfrac{5\pi}{6}$

48. (a) Tangent function of period 2

(b) Cosecant function of period 1.5

49. Show that the functions $f(x) = a \tan(kx + b)$ and $g(x) = a \cot(kx + b)$ both have period π/k and phase shift $-b/k$ for $k > 0$.

50. Show that the functions $f(x) = a \sec(kx + b)$ and $g(x) = a \csc(kx + b)$ both have period $2\pi/k$ and phase shift $-b/k$ for $k > 0$.

51. **G** Use a grapher to graph $f(x) = (\tan x)(\cos x)$ and $g(x) = \sin x$, where $-2\pi \le x \le 2\pi$. Are the graphs the same? Are the functions equal? Explain.

Objectives

1. Define Trigonometric Functions of Angles
2. Establish the Equivalency of Trigonometric Functions
3. Approximate Trigonometric Function Values of Angles
4. Solve Applied Problems

5.6 Trigonometric Functions of Angles

So far we have dealt with trigonometric functions of real numbers. Now we define the trigonometric functions of angles. We shall see how the two definitions are related.

Defining Trigonometric Functions of Angles

Here we extend the notion of trigonometric functions to include *any* angle. The angle can be measured in degrees or radians and can be positive, negative, or zero.

Definition

Trigonometric Functions of an Angle

Suppose that θ is an angle in standard position. Assume $(x, y) \neq (0, 0)$ is a point on the terminal side of θ at a distance $r = \sqrt{x^2 + y^2}$ from the origin (Figure 1).

Then the trigonometric functions of θ are defined as follows:

$$\sin \theta = \frac{y}{r} \qquad \csc \theta = \frac{r}{y}, \quad y \neq 0$$

$$\cos \theta = \frac{x}{r} \qquad \sec \theta = \frac{r}{x}, \quad x \neq 0$$

$$\tan \theta = \frac{y}{x}, \quad x \neq 0 \qquad \cot \theta = \frac{x}{y}, \quad y \neq 0$$

Figure 1

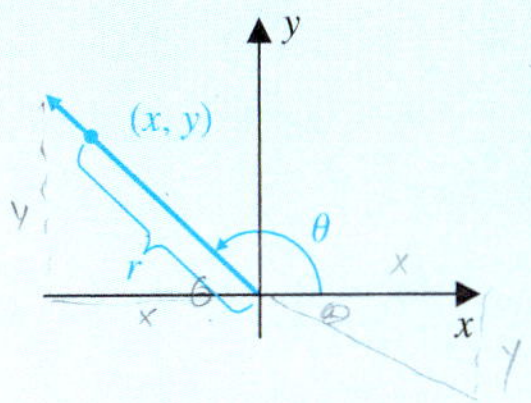

EXAMPLE 1   **Evaluating Trigonometric Functions of Angles**

Find the six trigonometric function values of each angle.

(a) α is an angle in standard position that contains the point $(5, -12)$ on its terminal side.

(b) $\beta = 180°$

Solution   (a) Figure 2a shows two possibilities for an angle α in standard position, one positive and one negative, whose terminal side contains the point $(5, -12)$. These two representations for α are coterminal.

Since $x = 5$ and $y = -12$, the distance r from the origin to $(5, -12)$ is given by

$$r = \sqrt{x^2 + y^2} = \sqrt{5^2 + (-12)^2} = \sqrt{25 + 144} = \sqrt{169} = 13$$

By the above definition, we have

$$\sin \alpha = \frac{y}{r} = \frac{-12}{13} \qquad \csc \alpha = \frac{r}{y} = \frac{13}{-12} = -\frac{13}{12}$$

$$\cos \alpha = \frac{x}{r} = \frac{5}{13} \qquad \sec \alpha = \frac{r}{x} = \frac{13}{5}$$

$$\tan \alpha = \frac{y}{x} = \frac{-12}{5} \qquad \cot \alpha = \frac{x}{y} = \frac{5}{-12} = -\frac{5}{12}.$$

Figure 2a

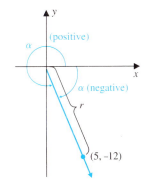

We could have chosen any point except (0, 0) on the terminal side.

(b) First, we display $\beta = 180°$ in standard position and then select a point on its terminal side. We choose the point $(-2, 0)$ (Figure 2b). For this selection,

$$x = -2 \quad \text{and} \quad y = 0,$$

and its distance from the origin is

$$r = 2.$$

By definition, we have

$$\sin 180° = \frac{y}{r} = \frac{0}{2} = 0 \qquad \csc 180° = \frac{r}{y} = \frac{2}{0} \ (\text{undefined})$$

$$\cos 180° = \frac{x}{r} = \frac{-2}{2} = -1 \qquad \sec 180° = \frac{r}{x} = \frac{2}{-2} = -1$$

$$\tan 180° = \frac{y}{x} = \frac{0}{-2} = 0 \qquad \cot 180° = \frac{x}{y} = \frac{-2}{0} \ (\text{undefined}).$$

Figure 2b

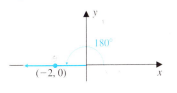

Establishing the Equivalency of Trigonometric Functions

Now we examine the connection between the trigonometric functions of real numbers or circular functions and the trigonometric functions of angles. Suppose that t is a real number and $P(t) = (x, y)$; then the trigonometric functions defined on *real numbers* (Figure 1) yields

$$\cos t = x \quad \text{and} \quad \sin t = y.$$

Figure 3

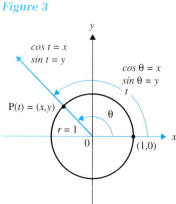

Next, we assume that θ is an angle in standard position so that point $P(t)$ is on the terminal side of θ (Figure 3). (The argument holds regardless of the quadrant where the terminal side of θ is located.) Since (x, y) is located on the unit circle where $r = 1$ (Figure 3), then, by using the definitions of trigonometric functions of angles, we have

$$\cos \theta = \frac{x}{r} = \frac{x}{1} = x \quad \text{and} \quad \sin \theta = \frac{y}{r} = \frac{y}{1} = y.$$

Notice that the real number t is the same as the radian measure of θ. Thus

$$\cos t = \cos \theta \quad \text{and} \quad \sin t = \sin \theta$$

where $\cos t$ and $\sin t$ are the values of the trigonometric functions defined on real number t, and $\cos \theta$ and $\sin \theta$ are the values of the trigonometric functions defined on angle θ with radian measure t.

Similarly, if angle θ has radian measure t, we can use the definition of the trigonometric functions of *real number t* and the definition of the trigonometric functions of *angle θ* to establish

$$\tan t = \tan \theta \qquad \cot t = \cot \theta$$
$$\sec t = \sec \theta \qquad \csc t = \csc \theta.$$

Equivalency Property

Trigonometric Functions

Consequently, we consider the trigonometric functions of angle θ to be equivalent to the trigonometric functions of a real number t, which is the radian measure of θ, and we use the terminology *trigonometric functions* regardless of whether *angles* or *real numbers* are employed.

For instance, the sine of an angle of *radian measure* $\pi/4$ is the same as the sine of the *real number* $\pi/4$. Even if an angle is given in degrees, we can associate the trigonometric functions of real numbers with the trigonometric functions of the angle by converting the degrees to radians. For example,

$$\sin 30° = \sin \frac{\pi}{6} = \frac{1}{2} \quad \text{and} \quad \cos 45° = \cos \frac{\pi}{4} = \frac{\sqrt{2}}{2}.$$

EXAMPLE 2   **Evaluating Trigonometric Functions**

Find the exact value of each expression.

(a) cos 120°   (b) sin 120°   (c) tan 240°

Solution   By using Figure 4 and the symmetry of the unit circle, we see that

Figure 4

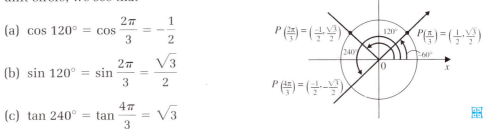

(a) $\cos 120° = \cos \dfrac{2\pi}{3} = -\dfrac{1}{2}$

(b) $\sin 120° = \sin \dfrac{2\pi}{3} = \dfrac{\sqrt{3}}{2}$

(c) $\tan 240° = \tan \dfrac{4\pi}{3} = \sqrt{3}$

Notice that the identities and properties established earlier for trigonometric functions of real numbers are also true for trigonometric functions of angles because of the above equivalancy property. In fact, with the above property in mind, we commonly use the term *trigonometric functions* regardless of whether `angles or *real numbers* are employed.

Note that the values of the trigonometric functions of angles in standard position depend only on the location of the terminal side of the angle but they do not depend on the choice of the point P on the terminal side of the angle. It follows that the trigonometric function values of coterminal angles are the same, no matter what particular point is selected on the terminal side of the angles.

For convenience, we use the symbol θ to represent an angle, the degree measure of the angle, or the radian measure of the angle. The context of our discussion will clarify which meaning of θ is being used.

Thus, if T represents any of the trigonometric functions we have the following result:

Coterminal Property

$$T(\theta) = T(\theta + 2\pi n) \text{ (if radians are used)}$$

or

$$T(\theta) = T(\theta + 360°n) \text{ (if degrees are used)}$$

for any integer n.

This property reinforces the fact that the trigonometric functions have period 2π.

EXAMPLE 3   **Using the Coterminal Property**

Use the preceding coterminal property to find the value of each expression.

(a) $\cos(-630°)$

(b) $\sin \dfrac{25\pi}{6}$

(c) $\cot\left(-\dfrac{21\pi}{4}\right)$

(d) $\sec 1380°$

Solution   (a) Since $-630°$ is coterminal with $90°$, we have

$$\cos(-630°) = \cos[-630° + 2(360°)]$$

$$= \cos 90° = \cos \frac{\pi}{2} = 0.$$

(b) Since

$$\frac{25\pi}{6} = \frac{\pi}{6} + 4\pi = \frac{\pi}{6} + 2(2\pi),$$

the angles $\dfrac{25\pi}{6}$ and $\dfrac{\pi}{6}$ are coterminal, so that

$$\sin \frac{25\pi}{6} = \sin \frac{\pi}{6} = \frac{1}{2}.$$

(c) Since

$$\frac{-21\pi}{4} = \frac{3\pi}{4} - 6\pi,$$

the angles $\dfrac{-21\pi}{4}$ and $\dfrac{3\pi}{4}$ are coterminal, so that

$$\cot\left(\frac{-21\pi}{4}\right) = \cot \frac{3\pi}{4} = -1.$$

(d) Since $1380° = 300° + 3(360°)$, the angles $1380°$ and $300°$ are coterminal, so that

$$\sec 1380° = \sec 300° = \sec \frac{5\pi}{3} = 2.$$

Approximating Trigonometric Function Values of Angles

We approximate values of trigonometric functions for angles by using a calculator and the reciprocal relationships if necessary, as we did for real numbers.

EXAMPLE 4   **Using a Calculator to Approximate Values of Trigonometric Functions**

Use a calculator to approximate each value to four decimal places.

(a) $\sin 162°$ (b) $\cot \dfrac{7\pi}{30}$ (c) $\csc(-4.6481)$

Solution   (a) After setting the calculator in degree mode, we get

$$\sin 162° = 0.3090 \text{ (approx.).}$$

(b) Here we set the calculator in radian mode. Then we use a reciprocal relationship to get

$$\cot \frac{7\pi}{30} = \frac{1}{\tan \dfrac{7\pi}{30}} = 1.1106 \text{ (approx.).}$$

(c) We set the calculator in radian mode and we use a reciprocal relationship to get

$$\csc(-4.6481) = \frac{1}{\sin(-4.6481)} = 1.0021 \text{ (approx.).}$$

Solving Applied Problems

The concepts of slopes of lines and angles are used in many applications and mathematical models. Highway engineers use slopes to measure the *steepness* of a roadway, referred to as the **grade** of the roadway, to indicate whether a roadway is *uphill* or *downhill*. For example, a (positive) 4% grade indicates a slope of 4/100 so that the roadway is uphill and it rises 4 feet vertically for every 100 feet of horizontal change. Figure 5 displays an uphill grade denoted by a *positive angle* and a downhill grade denoted by a *negative angle*.

Figure 5

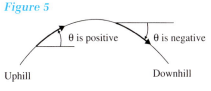

EXAMPLE 5   **Modeling a Highway Ramp**

Figure 6 shows an angle of −5° for a downhill grade of a 200-foot ramp from an elevated service drive to an expressway.

Figure 6

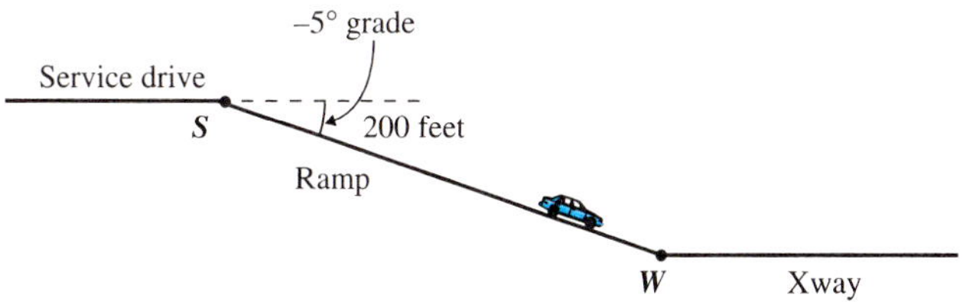

(a) Sketch a model of the situation in a Cartesian system so that the ramp entry location *S* is at the origin and the service drive is on the *x* axis.

(b) Use the model in part (a) to find the location of *W* where the ramp meets the expressway. Round off the answers to two decimal places. Interpret the location of *W* in terms of the elevation of the service drive above the expressway and the horizontal distance spanned by the ramp.

(c) What is the slope of the ramp? Round off the answer to two decimal places.

Solution   (a) Figure 7 shows the representation or model in a Cartesian system. Point *S*, the entry location of the ramp, is at (0, 0), point *W* = (*x*, *y*) is the location where the ramp meets the expressway, and the service drive is on the *x* axis. Notice the grade is shown as a negative angle since the ramp is downhill.

Figure 7

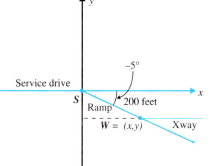

(b) We use the trigonometric functions defined for general angles to find the location of point *W* = (*x*, *y*) as follows:

$$\cos(-5°) = \frac{x}{r} = \frac{x}{200}$$

so that

$$x = 200 \cos(-5°)$$
$$= 199.24 \text{ (approx.)}$$

and

$$\sin(-5°) = \frac{y}{200}$$

so that

$$y = 200 \sin(-5°)$$

$$= -17.43 \text{ (approx.)}.$$

Thus the location of W is given by

$$W = (199.24, -17.43) \text{ (approx.)}.$$

The location of W indicates that the ramp spans a horizontal distance of approximately 199.24 feet and the service drive is approximately 17.43 feet above the expressway.

(c) The line $\overline{SW}$ defined by the ramp contains points $W = (x, y)$ and the origin $S = (0, 0)$ (Figure 7). Its slope is given by

$$\text{Slope of ramp} = \frac{y - 0}{x - 0} = \frac{y}{x}$$

Slope formula:

$$m = \frac{y_1 - y_2}{x_1 - x_2}.$$

To determine this ratio $\dfrac{y}{x}$ we can proceed in either of the following two ways:

Use the Results from Part (b)	Alternate Solution
From part (b) we found that	By definition,
$$x = 199.24 \text{ (approx.)}$$	$$\tan \theta = \frac{y}{x}$$
and	so that the
$$y = -17.43 \text{ (approx.)}$$	$$\text{slope} = \tan(-5°)$$
so that the	$$= -0.09 \text{ (approx.)}$$
$$\text{slope} = \frac{y}{x}$$	or about -9%.
$$= \frac{-17.43}{199.24}$$	
$$= -0.09 \text{ (approx.)}$$	
or about -9%.	

PROBLEM SET 5.6

Mastering the Concepts

In problems 1–10, angle θ is in standard position. Find the values of the six trigonometric functions of angle θ if the terminal side of θ contains the given point (x, y) which is r units from the origin. Sketch two possible angles, one positive and one negative, that satisfy the conditions on θ. Round off the answers to three decimal places.

1. $(x, y) = (4, 3)$
2. $(x, y) = (-3, -4)$
3. $(x, y) = (-5, 12)$
4. $(x, y) = (7, -10)$
5. $(x, y) = (1.35, -2.76)$
6. $(x, y) = (-7.89, 9.39)$

7. $(x, y) = \left(a, -\dfrac{1}{2}\right); r = 1, a < 0$

8. $(x, y) = (8, b); r = 17, b < 0$

9. $(x, y) = (a, a); a > 0$

10. $(x, y) = (-a, a); a < 0$

In problems 11–14, find the exact values of the six trigonometric functions of each angle.

11. $90°$

12. π

13. $-\dfrac{\pi}{2}$

14. $-270°$

In problems 15–22, use the known locations of $P\left(\dfrac{\pi}{6}\right)$, $P\left(\dfrac{\pi}{4}\right)$, and $P\left(\dfrac{\pi}{3}\right)$, along with symmetry, to find the exact value of each expression.

15. (a) $\sin 150°$

(b) $\cos 225°$

16. (a) $\cos\left(-\dfrac{7\pi}{6}\right)$

(b) $\cot 330°$

17. (a) $\tan 240°$

(b) $\cot \dfrac{11\pi}{6}$

18. (a) $\sec\left(-\dfrac{11\pi}{6}\right)$

(b) $\csc\left(-\dfrac{7\pi}{6}\right)$

19. (a) $\sec 315°$

(b) $\csc 300°$

20. (a) $\csc(-330°)$

(b) $\cot\left(-\dfrac{4\pi}{3}\right)$

21. (a) $\tan\left(-\dfrac{\pi}{4}\right)$

(b) $\cot 210°$

22. (a) $\sin\left(-\dfrac{7\pi}{4}\right)$

(b) $\sin \dfrac{7\pi}{6}$

In problems 23–32, use the coterminal property to find the exact value.

23. $\sin 945°$

24. $\sin(-675°)$

25. $\tan 1020°$

26. $\cos 750°$

27. $\tan\left(-\dfrac{41\pi}{4}\right)$

28. $\csc\left(-\dfrac{8\pi}{3}\right)$

29. $\cos\left(\dfrac{61\pi}{6}\right)$

30. $\cot(5\pi)$

31. $\cos 900°$

32. $\sec(-7\pi)$

In problems 33–40, use a calculator to find the approximate values of the six trigonometric functions of θ. Round off the answers to four decimal places.

33. (a) $\theta = 137°$

(b) $\theta = -95.8°$

34. (a) $\theta = 195°$

(b) $\theta = -10°$

35. (a) $\theta = 380°$

(b) $\theta = -\pi°$

36. (a) $\theta = -417°$

(b) $\theta = 215°15'5''$

37. (a) $\theta = 2$

(b) $\theta = -1.3$

38. (a) $\theta = -2$

(b) $\theta = 1.3$

39. (a) $\theta = -5.22$

(b) $\theta = \pi$

40. (a) $\theta = 5.22$

(b) $\theta = \pi^{-1}$

Applying the Concepts

41. A downhill ramp of length 185 feet from a service drive to a highway has a $-4.5°$ grade.

(a) Sketch a model of the situation in a Cartesian system so that the ramp's entry point S is at the origin and the point W where the ramp meets the highway is in quadrant IV.

(b) Use the model in part (a) to find the location W of where the ramp meets the highway. Find the elevation of the service drive above the highway and the horizontal distance spanned by the ramp.

(c) Find the slope of the ramp.

Round off all answers to two decimal places.

42. Suppose a downhill ramp connecting a service drive that is 10 feet above an expressway has a $-3°$ grade.

(a) Sketch a model of the situation in a Cartesian system so that the ramp's entry point S is at the origin and the point W where the ramp meets the expressway is in quadrant IV.

(b) Use the model in part (a) to find the location W where the ramp meets the expressway. Find the length of the ramp and the horizontal distance spanned by the ramp.

(c) Find the slope of the ramp.

Round off all answers to two decimal places.

Developing and Extending the Concepts

43. A person makes the following assertion:

Assume

$$\tan \theta = \dfrac{2}{3}.$$

Since

$$\tan \theta = \dfrac{\sin \theta}{\cos \theta}$$

it follows that

$$\sin \theta = 2 \quad \text{and} \quad \cos \theta = 3.$$

Consequently, the sine and cosine of an angle can have values larger than 1. Explain why this assertion is wrong. Give an example to support your argument.

44. Suppose that (x, y) and (x_1, y_1) are two different points in quadrant I on the terminal side of θ, as shown in Figure 8.

Figure 8

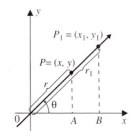

Use similar triangles to show that

$$\frac{y_1}{r_1} = \frac{y}{r} \qquad \frac{x_1}{r_1} = \frac{x}{r}$$

$$\frac{y_1}{x_1} = \frac{y}{x} \qquad \frac{r_1}{y_1} = \frac{r}{y}$$

$$\frac{r_1}{x_1} = \frac{r}{x} \qquad \frac{x_1}{y_1} = \frac{x}{y}$$

45. Show that the ratios in problem 44 also hold if angle θ is in quadrant II.

46. Show that the ratios in problem 44 also hold if angle θ is in quadrant III.

47. Show that the ratios in problem 44 also hold if angle θ is in quadrant IV.

48. A person makes the following assertion:

$$\text{If } \sin \alpha = \sin \beta, \quad \text{then} \quad \alpha = \beta.$$

Explain why this assertion is wrong. Give an example to support your argument.

49. Assume θ is an angle in standard position with terminal side lying on the line with equation $y = mx$ (Figure 9). Show that the tangent function of θ is the same as the slope of the line, that is, show that $\tan \theta = m$.

Figure 9

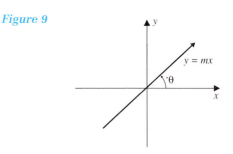

Objectives

1. Define Trigonometric Functions of Acute Angles
2. Use Right Triangles to Evaluate Trigonometric Functions for Special Angles
3. Use Reference Triangles to Evaluate Trigonometric Functions
4. Approximate Trigonometric Values for Acute Angles
5. Solve Right Triangles
6. Solve Applied Problems

5.7 Trigonometric Functions of Acute Angles and Modeling Right Triangles

The trigonometric functions enable us to establish relationships between the acute angles and lengths of the sides of right triangles. As we shall see, this enables us to solve many applied problems that are modeled by right triangles.

Defining Trigonometric Functions of Acute Angles

Suppose we place a *right triangle* (Figure 1a) on a coordinate system with one of the acute angles α in standard position (Figure 1b), then the values of the trigonometric functions of α may be expressed in terms of the lengths of the sides of the

Figure 1

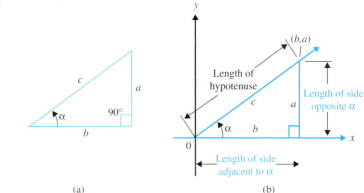

(a)   (b)

right triangle. Using the definition of the trigonometric functions of angles on page 367 with $\theta = \alpha$, $r = c$, $x = b$, and $y = a$, we express the trigonometric functions of α in terms of the lengths of the sides of the right triangle as follows:

$$\sin \alpha = \frac{a}{c} = \frac{\text{length of side opposite } \alpha}{\text{length of hypotenuse}} \qquad \csc \alpha = \frac{c}{a} = \frac{\text{length of hypotenuse}}{\text{length of side opposite } \alpha}$$

$$\cos \alpha = \frac{b}{c} = \frac{\text{length of side adjacent to } \alpha}{\text{length of hypotenuse}} \qquad \sec \alpha = \frac{c}{b} = \frac{\text{length of hypotenuse}}{\text{length of side adjacent to } \alpha}$$

$$\tan \alpha = \frac{a}{b} = \frac{\text{length of side opposite } \alpha}{\text{length of side adjacent to } \alpha} \qquad \cot \alpha = \frac{b}{a} = \frac{\text{length of side adjacent to } \alpha}{\text{length of side opposite } \alpha}$$

We can abbreviate the lengths of the side **opposite** α, the side **adjacent** to α, and the **hypotenuse** as **opp, adj,** and **hyp,** respectively (Figure 2). Thus the relationships between the trigonometric functions of an acute angle and the ratios of the lengths of the sides of a right triangle are established in the next definition.

Definition

Trigonometric Functions of an Acute Angle in a Right Triangle

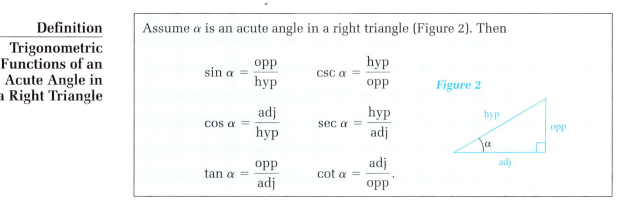

Assume α is an acute angle in a right triangle (Figure 2). Then

$$\sin \alpha = \frac{\text{opp}}{\text{hyp}} \qquad \csc \alpha = \frac{\text{hyp}}{\text{opp}}$$

$$\cos \alpha = \frac{\text{adj}}{\text{hyp}} \qquad \sec \alpha = \frac{\text{hyp}}{\text{adj}}$$

$$\tan \alpha = \frac{\text{opp}}{\text{adj}} \qquad \cot \alpha = \frac{\text{adj}}{\text{opp}}.$$

Figure 2

EXAMPLE 1

Using a Right Triangle to Evaluate Trigonometric Functions

Use the right triangle in Figure 3 to find the values of the trigonometric function of acute angle α.

Solution

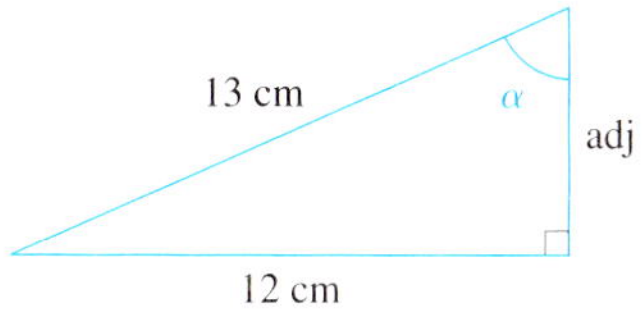

For angle α, we have opp = 12 centimeters and hyp = 13 centimeters, but the length of the adj isn't given. Using the Pythagorean theorem, we have

$$(\text{adj})^2 + (\text{opp})^2 = (\text{hyp})^2$$

Figure 3

so that

$$(\text{adj})^2 = (\text{hyp})^2 - (\text{opp})^2$$
$$= 13^2 - 12^2 = 25.$$

Therefore,

$$\text{adj} = \sqrt{25} = 5 \text{ centimeters.}$$

It follows from the right triangle relationships that

$$\sin \alpha = \frac{\text{opp}}{\text{hyp}} = \frac{12}{13} \qquad\qquad \csc \alpha = \frac{\text{hyp}}{\text{opp}} = \frac{13}{12}$$

$$\cos \alpha = \frac{\text{adj}}{\text{hyp}} = \frac{5}{13} \qquad\qquad \sec \alpha = \frac{\text{hyp}}{\text{adj}} = \frac{13}{5}$$

$$\tan \alpha = \frac{\text{opp}}{\text{adj}} = \frac{12}{5} \qquad\qquad \cot \alpha = \frac{\text{adj}}{\text{opp}} = \frac{5}{12}.$$

Using Right Triangles to Evaluate Trigonometric Functions for Special Angles

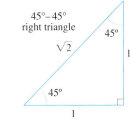

45°–45° right triangle

Our knowledge of right triangles provides us with another way to determine the exact values of the trigonometric functions for special angles 45°, 30°, and 60° (or equivalently $\pi/4$, $\pi/6$, and $\pi/3$). First, we construct an *isosceles right triangle* with two equal sides of length 1 unit (Figure 4). We know from plane geometry that both of the acute angles in this right triangle measure 45°. Using the Pythagorean theorem, we have

$$1^2 + 1^2 = (\text{hyp})^2$$

so that

$$\text{hyp} = \sqrt{2}.$$

From the right triangle trigonometric relationships, we obtain the familiar values:

$$\sin 45° = \frac{1}{\sqrt{2}} = \frac{\sqrt{2}}{2} \qquad\qquad \csc 45° = \frac{\sqrt{2}}{1} = \sqrt{2}$$

$$\cos 45° = \frac{1}{\sqrt{2}} = \frac{\sqrt{2}}{2} \qquad\qquad \sec 45° = \frac{\sqrt{2}}{1} = \sqrt{2}$$

$$\tan 45° = \frac{1}{1} = 1 \qquad\qquad \cot 45° = \frac{1}{1} = 1$$

A 30°–60° right triangle

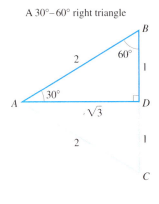

Next we consider an equilateral triangle, each of whose sides has length 2 units, such as triangle ABC in Figure 5. Recall from plane geometry that the three angles of an equilateral triangle each measure 60°. Assume that $\overline{AD}$ is the perpendicular bisector of $\overline{BC}$ (Figure 5). By using the Pythagorean theorem, we have

$$|\overline{AD}|^2 = 2^2 - 1^2 = 3 \quad \text{or} \quad |\overline{AD}| = \sqrt{3}.$$

Thus, from right triangle ADB in Figure 5, we have the known values:

$$\sin 30° = \cos 60° = \frac{1}{2} \qquad\qquad \csc 30° = \sec 60° = \frac{2}{1} = 2$$

$$\cos 30° = \sin 60° = \frac{\sqrt{3}}{2} \qquad\qquad \sec 30° = \csc 60° = \frac{2}{\sqrt{3}} = \frac{2\sqrt{3}}{3}$$

$$\tan 30° = \cot 60° = \frac{1}{\sqrt{3}} = \frac{\sqrt{3}}{3} \qquad\qquad \cot 30° = \tan 60° = \frac{\sqrt{3}}{1} = \sqrt{3}$$

Note that when we are trying to recall the values of the trigonometric functions for 45°, 30°, or 60° (or equivalently $\pi/4$, $\pi/6$, or $\pi/3$), it is useful to construct a 45°−45° right triangle (Figure 4) or a 30°−60° right triangle (triangle *ADB* in Figure 5) and then apply the right triangle trigonometric relationships.

Using Reference Triangles to Evaluate Trigonometric Functions

We often find it convenient to evaluate trigonometric functions of nonacute angles that are integer multiples of the special angles 30°, 45°, or 60° by making use of the concept of *reference triangles*. This procedure is described as follows:

Strategy for Reference Triangles to Evaluate Trigonometric Functions

Step 1 Sketch the given angle θ in standard position.

Step 2 Superimpose a reference triangle; that is, draw either a 45°−45° right triangle, or a 30°−60° right triangle in whichever quadrant will allow the hypotenuse to lie on the terminal side of θ and the adjacent side of the triangle to lie on the *x* axis.

Step 3 Use the measurements of that particular right triangle along with the quadrant location to obtain a point (x, y) on the terminal side of θ.

Step 4 Use the definitions of the trigonometric functions of angles along with the reciprocal and quotient relationships to complete the evaluation.

EXAMPLE 2

Using a Reference Triangle to Evaluate Trigonometric Functions

Use the above four-step procedure to evaluate the six trigonometric functions of $\theta = 135°$.

Solution

Refer to Figure 6

Step 1. We sketch $\theta = 135°$ in standard position.

Step 2. Then we draw a 45°−45° right triangle as a reference triangle in quadrant II so that the hypotenuse lies on the terminal side of θ and the adjacent leg lies on the negative *x* axis.

Step 3. By using the measurements of the 45°−45° right triangle, we get the point $(x, y) = (-1, 1)$ on the terminal side of θ.

It follows that $r = \sqrt{(-1)^2 + 1^2} = \sqrt{2}$.

Step 4. We get

Figure 6

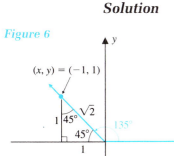

$$\sin 135° = \frac{y}{r} = \frac{1}{\sqrt{2}} \qquad \csc 135° = \frac{1}{\sin 135°} = \frac{1}{\left(\dfrac{1}{\sqrt{2}}\right)} = \sqrt{2}$$

$$\cos 135° = \frac{x}{r} = \frac{-1}{\sqrt{2}} \qquad \sec 135° = \frac{1}{\cos 135°} = \frac{1}{\left(\dfrac{-1}{\sqrt{2}}\right)} = -\sqrt{2}$$

$$\tan 135° = \frac{y}{x} = \frac{1}{-1} = -1 \qquad \cot 135° = \frac{1}{\tan 135°} = \frac{1}{-1} = -1.$$

To evaluate the trigonometric functions of a real number or of an angle given in radian measure, we can at times convert it to degree measure and then proceed as we did in Example 2. For instance, assume $\theta = (7\pi)/6$. Since $(7\pi)/6$ radians correspond to $210°$, we sketch $\theta = 210°$ in standard position along with its reference triangle as shown in Figure 7. After locating the point $(x, y) = (-\sqrt{3}, -1)$ on the terminal side of θ, we can find the trigonometric values of $(7\pi)/6$. For example,

Figure 7

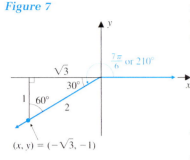

$$\sin \frac{7\pi}{6} = \sin 210° = \frac{y}{r} = \frac{-1}{2} \quad \text{and} \quad \cos \frac{7\pi}{6} = \cos 210° = \frac{x}{r} = \frac{-\sqrt{3}}{2}.$$

Approximating Trigonometric Values for Acute Angles

To approximate trigonometric function values for acute angles we proceed as we did for general angles.

For instance,

$$\sin 53.4° = 0.8028 \text{ (approx.)};$$

and

$$\tan 75°25'18'' = 3.8450 \text{ (approx.).}$$

To evaluate $\cot(\pi/8)$, we set the calculator in radian mode and then we use the reciprocal identity to get

$$\cot \frac{\pi}{8} = \frac{1}{\tan(\pi/8)}$$

$$= 2.4142 \text{ (approx.).}$$

When evaluating a trigonometric function, we *input an angle θ* and *the output V is the function value* (Figure 8a).

For acute angles, this process is *reversible or invertible* in the sense that if we specify a trigonometric function value V as input, then the output is the angle θ (Figure 8b).

Later in Section 5.8 we expand the notion of inverse trigonometric functions.

Figure 8

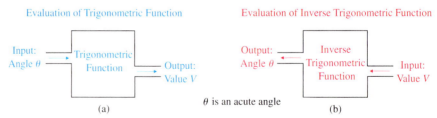

θ is an acute angle

We use the superscript $^{-1}$ notation introduced in Section 4.1 to denote this inverse function.

For instance, by using the special angles, we have

We use degree measures rather than radian measures here because we are working with right triangles.

$$\cos^{-1}\left(\frac{1}{2}\right) = 60° \quad \text{because} \quad \cos 60° = \frac{1}{2}$$

$$\sin^{-1}\left(\frac{\sqrt{2}}{2}\right) = 45° \quad \text{because} \quad \sin 45° = \frac{\sqrt{2}}{2}$$

$$\tan^{-1}(1) = 45° \quad \text{because} \quad \tan 45° = 1.$$

If an inverse value is not known, then we use a calculator to approximate its value.

For instance, after setting the calculator in degree mode and using the proper key stroke entries, we obtain the approximations

$$\sin^{-1}(0.3185) = 18.57° \quad \text{and} \quad \tan^{-1}(7.2939) = 82.19°.$$

Most calculators do not have an inverse secant key.

To calculate $\sec^{-1}(1.0313)$, we proceed as follows:

$$\sec^{-1}(1.0313) = \theta \quad \text{means} \quad \sec\theta = 1.0313$$

but since

$$\sec\theta = \frac{1}{\cos\theta}$$

it follows that

$$\frac{1}{\cos\theta} = 1.0313 \quad \text{or} \quad \frac{1}{1.0313} = \cos\theta$$

so that

$$\theta = \cos^{-1}\left(\frac{1}{1.0313}\right) = 14.15° \text{ (approx.)}.$$

Solving Right Triangles

Figure 9

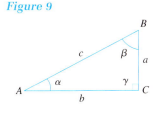

Under certain conditions, it is possible to find the measurements of unknown sides and angles of a right triangle from known sides and angles by using the trigonometric functions. This process is called *solving a right triangle.*

Consider the right triangle *ACB* shown in Figure 9. We'll adopt the convention of labeling the angles at the vertices *A*, *B*, and *C* by α, β, and γ, respectively, and the lengths of the sides opposite these angles by *a*, *b*, and *c*, respectively. Note that α and β are complementary angles and $\gamma = 90°$.

When solving a right triangle, we usually draw a sketch of the triangle approximately to scale to gain some insight into the problem.

EXAMPLE 3   **Solving Right Triangles**

Solve the right triangle *ACB* for each situation. Round off the answers to two decimal places.

(a) $a = 9$ and $\beta = 36.87°$

(b) $a = 31.42$ and $b = 26.74$

Solution   (a) Figure 10a shows a sketch of the right triangle. Here, we have to find α, *b*, and *c*. Because α and β are complementary,

$$\alpha = 90° - \beta = 90° - 36.87° = 53.13°.$$

To find *b*, we use the relationship $\tan\beta = $ opp/adj and a calculator to get

$$\tan\beta = \frac{b}{a}$$

Figure 10a

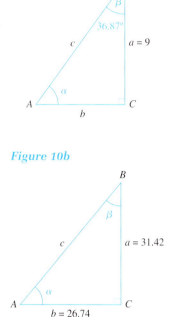

so that

$$b = a \tan \beta = 9 \tan 36.87° = 6.75 \text{ (approx.)}.$$

To find c, we use the relationship $\cos \beta = $ adj/hyp to get

$$\cos \beta = \frac{a}{c}$$

so that

$$c = \frac{a}{\cos \beta} = \frac{9}{\cos 36.87°} = 11.25 \text{ (approx.)}.$$

Figure 10b

(b) Figure 10b shows a sketch of the right triangle. Here we have to find α, β, and c. We'll start by finding α. Using the definition, it follows that

$$\tan \alpha = \frac{\text{opp}}{\text{adj}} = \frac{a}{b} = \frac{31.42}{26.74}.$$

To find α means to find the acute angle whose tangent is 31.42/26.74, so we use the inverse evaluation

$$\alpha = \tan^{-1}\left(\frac{31.42}{26.74}\right) = 49.60° \text{ (approx.)}.$$

Because α and β are complementary,

$$\beta = 90° - 49.60° = 40.40° \text{ (approx.)}.$$

Finally, to find c, we use the Pythagorean theorem to get

$$c = \sqrt{a^2 + b^2} = \sqrt{31.42^2 + 26.74^2} = 41.26 \text{ (approx.)}.$$

Solving Applied Problems

One of the major reasons for the development of trigonometry was to solve applications involving right triangles. The following examples illustrate how trigonometric functions of acute angles are used to model such situations.

EXAMPLE 4   **Finding the Height Reached by a Ladder**

If a 24-foot extension ladder leaning against a house makes a 63° angle with the ground, then how far up the side of the house does the ladder reach? Round off the answer to two decimal places.

Figure 11

Solution   The situation is shown in Figure 11. We need to determine the length $|\overline{CB}|$. By using right triangle trigonometry, we obtain

$$\sin 63° = \frac{\text{opp}}{\text{hyp}} = \frac{|\overline{CB}|}{24}$$

or

$$|\overline{CB}| = 24 \sin 63° = 21.38 \text{ (approx.)}.$$

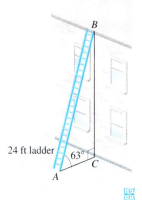

Thus, the ladder reaches about 21.38 feet up the side of the house (about 21 feet and 5 inches).

A transit is an instrument used by surveyors to determine angles.

At times, right triangle applications involve the measure of an acute angle formed by an observer's direct line of sight to an object and a horizontal line. If the object is above the horizontal line, this angle is called an **angle of elevation** (Figure 12a). If the object is below the horizontal line, the angle is called an **angle of depression** (Figure 12b).

Figure 12

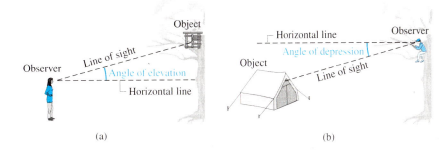

(a)   (b)

EXAMPLE 5   Using an Angle of Elevation

The Empire State Building in New York City is about 1250 feet high. A surveyor's transit is set 5 feet above ground level. The angle of elevation from the transit to a point on top of the Empire State Building is found to be 25.33°. How far is the surveyor from the building? Round off the answer to the nearest 10 feet.

Solution   The situation is modeled by the right triangle in Figure 13, where d represents the desired distance. Here we have

Figure 13

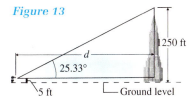

$$\tan 25.33° = \frac{\text{opp}}{\text{adj}}$$

$$= \frac{1250 - 5}{d}$$

or

$$d = \frac{1245}{\tan 25.33°}$$

$$= 2630 \text{ (approx.)}.$$

Consequently, the surveyor is about 2630 feet from the building.

EXAMPLE 6   Using Angles of Depression

An air traffic controller stands in a control tower that is 150 feet high and observes two commercial jets waiting in adjacent runways for a takeoff. The angle of depression of the first jet is 15.3°, and the angle of depression of the second is 8.2°. Assuming that the tower and the two jets lie on the same plane, find the distance between the jets. Round off the answer to the nearest 10 feet.

Solution   In Figure 14, A represents the location of the controller, C the location of the first jet, and D the location of the second jet. We wish to find $|\overline{CD}|$, which is equal to $|\overline{BD}| - |\overline{BC}|$.

Figure 14

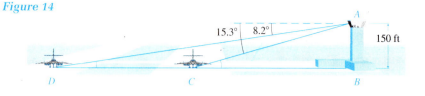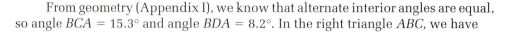

From geometry (Appendix I), we know that alternate interior angles are equal, so angle $BCA = 15.3°$ and angle $BDA = 8.2°$. In the right triangle ABC, we have

$$\tan 15.3° = \frac{|\overline{AB}|}{|\overline{BC}|} = \frac{150}{|\overline{BC}|}$$

or

$$|\overline{BC}| = \frac{150}{\tan 15.3°} = 550 \text{ (approx.)}.$$

In the right triangle ABD,

$$\tan 8.2° = \frac{|\overline{AB}|}{|\overline{BD}|} = \frac{150}{|\overline{BD}|}$$

or

$$|\overline{BD}| = \frac{150}{\tan 8.2°} = 1040 \text{ (approx.)}.$$

Hence the approximate distance between the two jets is given by

$$|\overline{CD}| = |\overline{BD}| - |\overline{BC}|$$
$$= 1040 - 550$$
$$= 490.$$

That is, the jets are approximately 490 feet apart.

◆ PROBLEM SET 5.7

Mastering the Concepts

In problems 1–6, find the values of the six trigonometric functions of the acute angle θ for each right triangle. Round off the answers to three decimal places.

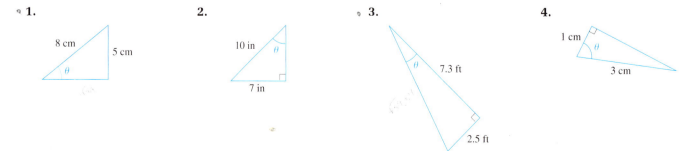

1.

8 cm   5 cm θ

2.

10 in θ 7 in

3.

θ 7.3 ft   2.5 ft

4.

1 cm θ 3 cm

5.

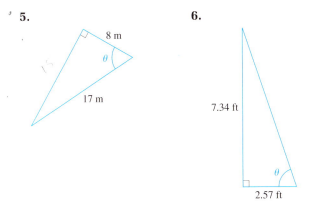

8 m

θ

17 m

6.

7.34 ft

θ

2.57 ft

7. Let θ be an acute angle of a right triangle and

$$\sin \theta = \frac{4}{5}.$$

Draw such a triangle, and then find the exact values of the other five trigonometric functions of θ.

8. Let θ be an acute angle of a right triangle for which

$$\cos \theta = \frac{12}{13}.$$

Draw such a triangle, and then find the exact values of the other five trigonometric functions of θ.

In problems 9–16, use either a 45°–45° or a 30°–60° right triangle as a reference triangle to find the exact value of each expression.

9. (a) $\sin 150°$

(b) $\cos 225°$

10. (a) $\cos\left(-\dfrac{7\pi}{6}\right)$

(b) $\cot 330°$

11. (a) $\tan 240°$

(b) $\cot \dfrac{11\pi}{6}$

12. (a) $\sec\left(-\dfrac{11\pi}{6}\right)$

(b) $\csc\left(-\dfrac{7\pi}{6}\right)$

13. (a) $\sec 315°$

(b) $\csc 300°$

14. (a) $\csc(-330°)$

(b) $\cot\left(-\dfrac{4\pi}{3}\right)$

15. (a) $\tan\left(-\dfrac{\pi}{4}\right)$

(b) $\cot 210°$

16. (a) $\sin\left(-\dfrac{7\pi}{4}\right)$

(b) $\sin \dfrac{7\pi}{6}$

In problems 17–20, use a calculator to determine each value. Round off the answer to four decimal places.

17. (a) $\sin 43°$

(c) $\cos\left(\dfrac{3\pi}{7}\right)$

(b) $\csc(61.57°)$

(d) $\sec \dfrac{3\pi}{8}$

18. (a) $\cot 88°41'$

(c) $\sec 1.536$

(b) $\sin(77°52'48'')$

(d) $\csc 1.234$

19. (a) $\tan \dfrac{\pi}{5}$

(c) $\csc 53.8°$

(b) $\cos 71°46'32''$

(d) $\sin(0.453)$

20. (a) $\cos(1.233)$

(c) $\csc 72°12'25''$

(b) $\cot 37.8°$

(d) $\tan 80.3°$

In problems 21 and 22, use a calculator to find each inverse value as an acute angle in degree measure. Round off each answer to two decimal places.

21. (a) $\sin^{-1}(0.55)$

(b) $\cos^{-1}(0.98)$

(c) $\tan^{-1}(30.12)$

(d) $\cot^{-1}(1)$

22. (a) $\cos^{-1}(0.01)$

(b) $\sin^{-1}(0.10)$

(c) $\tan^{-1}(1)$

(d) $\sec^{-1}(2)$

In problems 23–32, assume that ACB is a right triangle with $\gamma = 90°$ (Figure 15). In each case, solve the triangle.

Figure 15

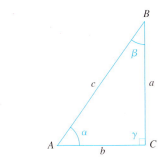

B

β

c a

α γ C

A b

Round off angles to the nearest hundredth of a degree and side lengths to two decimal places.

23. $c = 10$, $\beta = 50°$

24. $c = 9$, $\beta = 20°$

25. $b = 4$, $\beta = 43°$

26. $a = 6.5$, $\alpha = 37.2°$

27. $b = 1500$, $\alpha = 31.23°$

28. $b = 567.3$, $\alpha = 67.41°$

29. $a = 13.2$, $b = 4.1$

30. $a = 31$, $b = 4.7$

31. $\alpha = 3\beta$, $a = 1$

32. $\alpha = \dfrac{2}{3}\beta$, $a = 3$

Applying the Concepts

In problems 33–54, round off the answers to two decimal places.

33. Surveying: The lot for a new home site is uniformly pitched from the horizontal at an angle of 0.51° (Figure 16). How many inches will the ground drop if one walks 100 feet down the slope?

Figure 16

Pitched at 0.51°

34. **Surveying:** A straight sidewalk is inclined to the horizontal at an angle of 5.3°. How far must one walk on the sidewalk to change elevation by 2 meters?

35. **Foot of a Ladder:** A ladder 24 feet long is leaning against a building. The angle formed by the ladder and the ground is 63°. How far from the building is the foot of the ladder?

36. **Length of a Shadow:** A monument is 180 meters high. What is the length of the shadow cast by the monument if the angle of elevation of the sun is 58.4°?

37. **Utility Pole:** A guy wire attached to a vertical utility pole makes an angle of 71.4° with the ground. If the end of the wire attached to the ground is 15.5 feet from the pole, how high up the pole is the other end of the wire attached?

38. **Altitude of a Shuttle:** A space shuttle rises vertically. A camera 4.3 feet above the ground is located at a point 1000 feet (on the horizontal) away from the base of the launching pad. At a certain instant, the angle of elevation of the camera, focusing on the bottom of the shuttle, is 53° (Figure 17). How high above the ground is the shuttle at that instant?

Figure 17

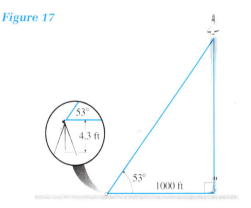

39. **Highway Patrol:** A highway patrol officer monitors traffic on a straight highway from a point *A*, which is 35 feet from the side of the highway (Figure 18). At a certain time, the patrol officer observed a speeding truck at point *B* on the highway at an angle of 25.3°. One second later, the truck was observed by the officer at point *C* at an angle of 10.3°. Find the speed of the truck in miles per hour.

Figure 18

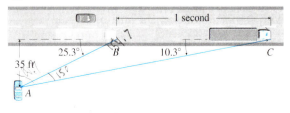

40. **Forestry:** A tree broken by the wind forms a right triangle with the ground. The broken part makes an angle of 38° with the ground, and the top of the tree is now 15 meters from the base, as shown in Figure 19. How tall was the whole tree?

Figure 19

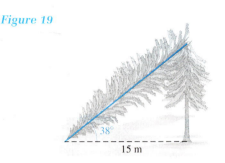

41. **Escalator:** An escalator in a department store is 50 feet long and carries customers through a vertical distance of 20 feet.
    (a) What is the angle that the escalator makes with the floor?
    (b) If it takes 25 seconds to carry a customer from the bottom of the escalator to the top, how fast (in feet per second) is the escalator moving?

42. **Surveying:** To find the length of an island, a navigator of a small plane flying at an altitude of 1540 meters determines the angles of depression of the extremities of the island to be 26.5° and 79.3° (Figure 20). What is the length of the island in kilometers?

Figure 20

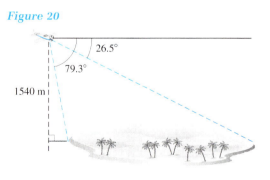

43. **Computer Monitor:** Suppose that a computer screen has a height of 7.5 inches and a width of 10 inches (Figure 21). What is the length of the diagonal *d*, and what angle does it make with the width?

Figure 21

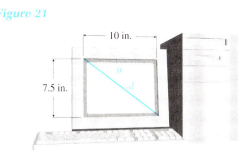

44. The Statue of Liberty: The torch of the Statue of Liberty is about 305 feet above water level. A sighting of the statue is taken from a ship 838 feet away. Find the angle of elevation from the ship to the top of the torch (Figure 22).

Figure 22

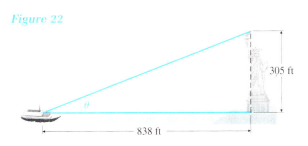

45. Angle of Elevation of the Eiffel Tower: The height of the Eiffel Tower (without the television mast added to the top) is approximately 300 meters (Figure 23). If the angle of elevation from a point on the ground to the top of the tower is 62.75°, how far from the center of the base of the tower is the point?

Figure 23

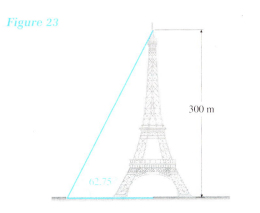

46. Light Span of a Lamp: A lamp is suspended 2 feet 10 inches above the center of a circular table. What is the radius of the table if the angle of depression from the lamp to the edge of the table is 56°15′, as shown in Figure 24?

Figure 24

47. Tower of Pisa: The Leaning Tower of Pisa was designed to stand 55 meters high when vertical. However, it was built on unstable ground and is now 5.2 meters out of perpendicular (Figure 25). Find the acute angle that the tower makes with the ground.

Figure 25

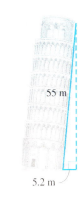

48. Length of a Ladder: A 6-foot fence stands 9 feet from a high wall. The shortest ladder that can reach the wall from outside the fence makes an angle of 41° with the horizontal (Figure 26). Find the length of the ladder.

Figure 26

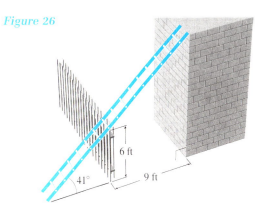

49. Height of Al Giza Pyramid: The distance measured along an edge from the original top of the Great Pyramid of Cheops at Al Giza, Egypt, to one of the corners of its square base is

219 meters. The angle of elevation from the corner of the base to the top is 42.06° (Figure 27). Find the original height of the Great Pyramid.

Figure 27

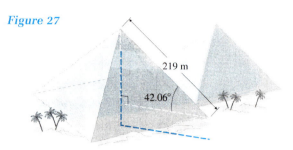

50. **Height of a Flagpole:** From the window of an apartment building 12 meters above ground, an observer determines that the angles of depression on the top and bottom of a flagpole standing on the ground (level with the base of the building) are 43° and 66°, respectively (Figure 28). Find the height of the flagpole.

Figure 28

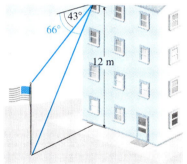

51. **Length of an Antenna:** An antenna is situated at the edge of a flat roof on top of a building that is located on level ground. From a point 100 feet from the base of the building, on the side where the antenna is placed, the angles of elevation of the top and bottom of the antenna measure 24° and 17°, respectively (Figure 29). How high is the building? How tall is the antenna?

Figure 29

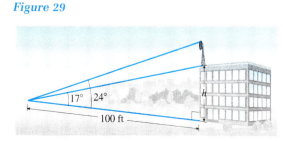

52. **Surveying:** To find the shortest distance d from a baseline l to a point T on an island offshore, a surveyor marks off a distance of 200 meters between the two points B and C on the baseline. She locates the point T on the island and then measures the angles TBC and TCB to be 68.3° and 81.5°, respectively (Figure 30). Find the distance d from the point T on the island to the nearest point on the baseline.

Figure 30

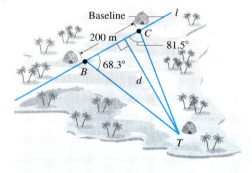

53. **Height of a Blimp:** A blimp is between two spotters who are 522 feet apart. One spotter reports that the angle of elevation of the blimp is 66°, and the other reports that it is 59°. If the blimp is directly over a line from one spotter to the other, how high is the blimp above the ground?

54. **Security:** A video security monitor is mounted at location A, 10 feet above the floor on a back wall of a drug store. The monitor scans through an angle of 51° in a vertical plane along an aisle of the store. The aisle begins at location B, 5 feet from the foot of the back wall at C and ends at D, the foot of an opposite wall (Figure 31). What is the length of the aisle; that is, what is the distance $|\overline{BD}|$?

Figure 31

Developing and Extending the Concepts

55. Use a right triangle to prove that if α and β are complementary angles, then the following equations are true. Give numerical examples to illustrate this result.

$$\sin \alpha = \cos \beta \quad \tan \alpha = \cot \beta \quad \csc \alpha = \sec \beta$$

56. Suppose that two straight lines *EC* and *DB* intersect at point *A* and right triangles *ABC* and *ADE* are formed as shown in Figure 32.

   (a) Express $\tan(90° - \theta)$ in terms of *a* and *b*.

   (b) Express $\sin \theta$ in terms of *a* and *b*.

Figure 32

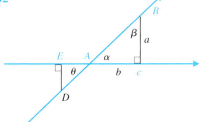

57. Two tangent lines, each of length $|T|$, are drawn from a point *P* and touch a circle of radius *r* at points *M* and *N* (Figure 33). Assume the angle *MPN* is θ. Express the length $|T|$ of each tangent line in terms of θ and *r*.

Figure 33

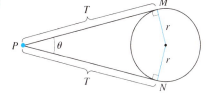

58. Figure 34 shows a regular *heptagon* (a seven-side polygon with sides of equal lengths) inscribed in a circle of radius 1. Determine the length of one side of the heptagon.

Figure 34

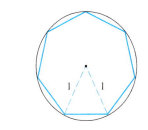

59. Explain why it is not possible to have a right triangle with a hypotenuse of 23 inches and an angle of 35° with a side opposite of length:

   (a) 11 inches

   (b) 14 inches

Objectives

1. Define the Inverse Sine, Cosine, and Tangent Functions
2. Simplify Expressions Involving Inverse Functions
3. Define the Other Inverse Trigonometric Functions
4. Solve Applied Problems

5.8 Inverse Trigonometric Functions

By examining the graphs of the six basic trigonometric functions, we observe (using the horizontal-line test) that none of them is one to one. So none of them has an inverse. However, by restricting domains, we can get trigonometric functions that do have inverses.

Defining the Inverse Sine, Cosine, and Tangent Functions

Table 1 lists three trigonometric functions with restricted domains—one for the sine, one for the cosine, and one for the tangent.

TABLE 1

Function	$f(x) = \sin x$	$g(x) = \cos x$	$h(x) = \tan x$
Restriction on x	$-\dfrac{\pi}{2} \leq x \leq \dfrac{\pi}{2}$	$0 \leq x \leq \pi$	$-\dfrac{\pi}{2} < x < \dfrac{\pi}{2}$
Graph	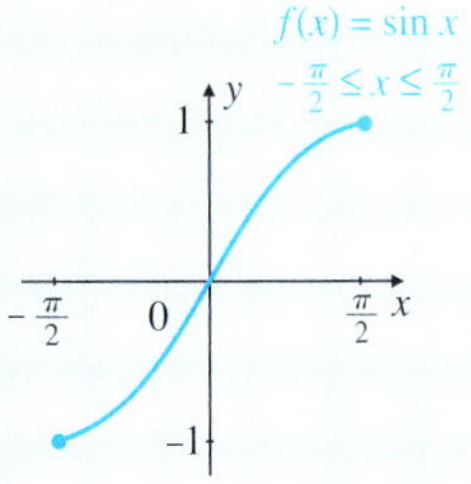	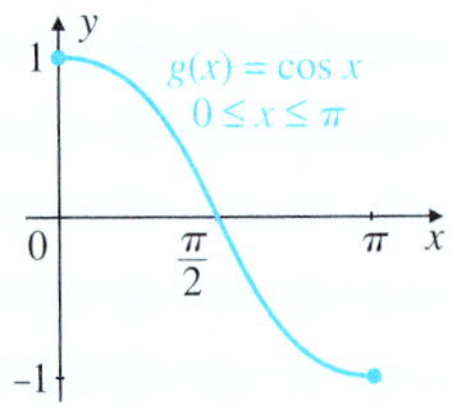	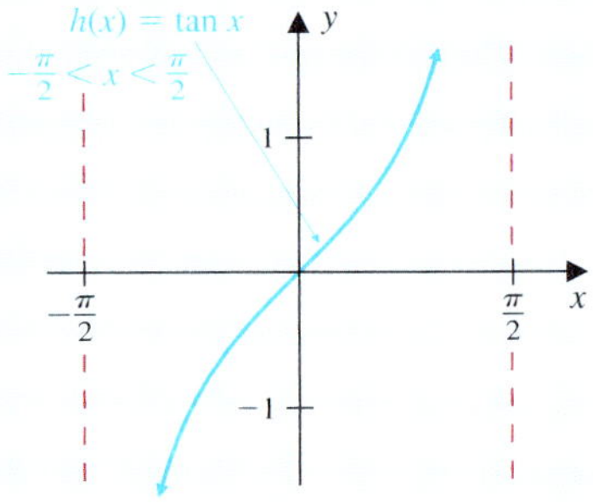

From the graphs in Table 1, we see that each of these three functions is one to one and therefore has an inverse. Let us examine each of these functions.

1. The Inverse Sine Function

When we refer to the *inverse sine function*, we mean the function that is the inverse of $f(x) = \sin x$, with the restriction $-\pi/2 \leq x \leq \pi/2$ (Table 1).

The inverse sine function is denoted by $y = \sin^{-1} x$ (read "the inverse sine of x") or by $y = \arcsin x$ (read "the arcsine of x") and is defined as follows.

Definition

The Inverse Sine Function

> The **inverse sine function,**
> $$y = \sin^{-1} x \quad \text{or} \quad y = \arcsin x$$
> is defined by
> $$\sin y = x, \quad \text{where} \quad -\pi/2 \leq y \leq \pi/2 \quad \text{and} \quad -1 \leq x \leq 1.$$

Note that even though we write $\sin^2 x$ to represent $(\sin x)^2$ we never use $\sin^{-1} x$ to represent $(\sin x)^{-1}$, since

$$(\sin x)^{-1} = \frac{1}{\sin x}$$

which is not the same as the inverse function of $\sin x$. This confusion arises because we use the "exponent" -1 in two different ways: to denote reciprocals and to denote inverse functions.

We can think of $\arcsin x$ or $\sin^{-1} x$ as the number (or the angle) between $-\pi/2$ and $\pi/2$ inclusive, whose sine is x. The special values learned earlier enable us to find exact inverse function values for certain numbers.

EXAMPLE 1 **Finding Exact Values of the Inverse Sine**

Find the exact value of each expression.

(a) $\sin^{-1}\dfrac{1}{2}$

(b) $\sin^{-1}\left(-\dfrac{\sqrt{2}}{2}\right)$

Solution By using the special values along with the restrictions on the inverse sine, we get the solutions.

(a) Suppose that $y = \sin^{-1}\dfrac{1}{2}$. This equation is equivalent to

$$\sin y = \frac{1}{2}, \text{ where } -\frac{\pi}{2} \le y \le \frac{\pi}{2}.$$

From the special values, we know that $\sin\left(\dfrac{\pi}{6}\right) = \dfrac{1}{2}$. Since $-\dfrac{\pi}{2} \le \dfrac{\pi}{6} \le \dfrac{\pi}{2}$, it

follows that $y = \dfrac{\pi}{6}$. That is, we have $\sin^{-1}\left(\dfrac{1}{2}\right) = \dfrac{\pi}{6}$.

(b) Similarly, $y = \sin^{-1}\left(-\dfrac{\sqrt{2}}{2}\right)$ is equivalent to

$$\sin y = -\frac{\sqrt{2}}{2}, \text{ where } -\frac{\pi}{2} \le y \le \frac{\pi}{2}.$$

The value $-\dfrac{\pi}{4}$ satisfies the conditions for y, so $\sin^{-1}\left(-\dfrac{\sqrt{2}}{2}\right) = -\dfrac{\pi}{4}$.

The restrictions on the inverse functions are very important to keep in mind. For instance, even though

$$\sin\left(\frac{7\pi}{4}\right) = -\frac{\sqrt{2}}{2}, \quad \sin^{-1}\left(\frac{\sqrt{2}}{2}\right) \ne \frac{7\pi}{4}$$

because $\dfrac{7\pi}{4}$ is not in the interval $\left[-\dfrac{\pi}{2}, \dfrac{\pi}{2}\right]$.

2. The Inverse Cosine Function

We now define the inverse of $g(x) = \cos x$, with x restricted to the interval $[0, \pi]$ (Table 1).

The inverse cosine function is denoted by $y = \cos^{-1} x$ or $y = \arccos x$ and defined as follows.

Definition
The Inverse Cosine Function

The **inverse cosine function,**

$$y = \cos^{-1} x \quad \text{or} \quad y = \arccos x$$

is defined by

$$\cos y = x, \quad \text{where} \quad 0 \le y \le \pi \quad \text{and} \quad -1 \le x \le 1.$$

We can think of arccos x or $\cos^{-1} x$ as the number (or the angle) between 0 and π inclusive, whose cosine is x.

EXAMPLE 2   **Finding Exact Values of the Inverse Cosine**

Find the exact value of each expression.

(a) $\cos^{-1}(-1)$

(b) $\arccos\left(-\dfrac{\sqrt{2}}{2}\right)$

Solution   The restrictions on the inverse cosine lead us to the following results:

(a) For $y = \cos^{-1}(-1)$, we know that $\cos \pi = -1$ and $0 \le \pi \le \pi$, so $y = \pi$.

(b) For $y = \arccos\left(-\dfrac{\sqrt{2}}{2}\right)$, we know that $\cos\left(\dfrac{3\pi}{4}\right) = -\dfrac{\sqrt{2}}{2}$ and $0 \le \dfrac{3\pi}{4} \le \pi$,

so $y = \dfrac{(3\pi)}{4}$.

3. The Inverse Tangent Function

The function $y = \tan^{-1} x$ or $y = \arctan x$ denotes the inverse of the function $h(x) = \tan x$, where $-\pi/2 < x < \pi/2$ (Table 1).

Definition

The Inverse Tangent Function

The values $-\pi/2$ and $\pi/2$ are excluded for the tangent function, since $\tan(-\pi/2)$ and $\tan(\pi/2)$ are not defined.

> The **inverse tangent function**
>
> $$y = \tan^{-1} x \quad \text{or} \quad y = \arctan x$$
>
> is defined by
>
> $$\tan y = x, \quad \text{where} \quad -\frac{\pi}{2} < y < \frac{\pi}{2} \text{ and } x \text{ is a real number.}$$

We can think of $\arctan x$ or $\tan^{-1} x$ as the number (or angle) between $-\pi/2$ and $\pi/2$ whose tangent is x. For instance,

$$\tan^{-1}(-1) = -\frac{\pi}{4}$$

since we know that

$$\tan\left(-\frac{\pi}{4}\right) = -1 \quad \text{and} \quad -\frac{\pi}{2} < -\frac{\pi}{4} < -\frac{\pi}{2}.$$

When the evaluation of an inverse function cannot be done with the special values, we use a calculator to approximate the value. Normally we use the radian mode for such approximations. In right triangle trigonometry we used the degree mode to evaluate the inverse functions because we worked with triangles.

EXAMPLE 3   **Approximating Values of Inverse Functions**

Use a calculator to approximate each value to two decimal places.

(a) $\sin^{-1} 0.8016$

(b) $\cos^{-1}(-0.3281)$

(c) $\arctan 0.6235$

Solution   First we set the calculator in radian mode, and then we use the appropriate inverse function key to get the following approximate values:

(a) $\sin^{-1} 0.8016 = 0.93$

(b) $\cos^{-1}(-0.3281) = 1.91$

(c) $\arctan 0.6235 = 0.56$

By reflecting each of the graphs in Table 1 on page 388 across the line $y = x$, we obtain the inverse function graphs of

$y = \sin^{-1} x$ (Figure 1a), $y = \cos^{-1} x$ (Figure 1b), and $y = \tan^{-1} x$ (Figure 1c).

Figure 1

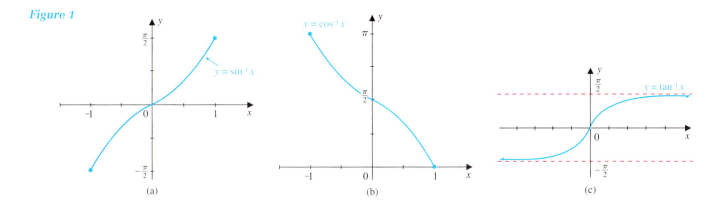

(a)   (b)   (c)

Notice that since the tangent function

$$f(x) = \tan x$$

has *vertical* asymptotes at

$$x = -\frac{\pi}{2} \quad \text{and} \quad x = \frac{\pi}{2},$$

it follows that there are *horizontal* asymptotes for the graph of

$$y = \tan^{-1} x \quad \text{at}$$
$$y = \frac{-\pi}{2} \quad \text{and} \quad y = \frac{\pi}{2} \text{ (Figure 1c)}.$$

When we rewrite an equation containing an inverse trigonometric function as an equation in terms of a trigonometric function, care must be taken to keep track of the restrictions on the variables.

EXAMPLE 4   **Rewriting an Inverse Function**

If $y = 5 + 3 \cos^{-1} 2x$, express x in terms of y. Determine the restrictions on x and y.

Solution   The given equation, $y = 5 + 3 \cos^{-1} 2x$, can be written as

$$\frac{y - 5}{3} = \cos^{-1} 2x.$$

This latter equation is equivalent to

$$\cos\left(\frac{y - 5}{3}\right) = 2x \quad \text{or} \quad \frac{1}{2}\cos\left(\frac{y - 5}{3}\right) = x$$

where

$$0 \le \frac{y - 5}{3} \le \pi \quad \text{and} \quad -1 \le 2x \le 1.$$

The restriction on y is found by solving

$$0 \leq \frac{y-5}{3} \leq \pi$$

to get:

$$0 \leq y - 5 \leq 3\pi \qquad \text{Multiply by 3}$$

$$5 \leq y \leq 3\pi + 5 \qquad \text{Add 5}$$

The restriction on x is found by solving

$$-1 \leq 2x \leq 1$$

so that

$$-\frac{1}{2} \leq x \leq \frac{1}{2} \qquad \text{Divide by 2}$$

Thus

$$x = \frac{1}{2}\cos\left(\frac{y-5}{3}\right) \quad \text{where } 5 \leq y \leq 3\pi + 5 \quad \text{and} \quad -\frac{1}{2} \leq x \leq \frac{1}{2}.$$

Simplifying Expressions Involving Inverse Functions

From Section 4.1, we learned that

$$(f \circ f^{-1})(x) = f[f^{-1}(x)] = x \quad \text{and} \quad (f^{-1} \circ f)(x) = f^{-1}[f(x)] = x.$$

When applying this property to the inverse trigonometric functions, we must take into account the restrictions on x that are delineated in the following properties.

Property
Inverse Sine
Composition

$$\sin(\sin^{-1} x) = x \quad \text{if } -1 \leq x \leq 1$$

$$\sin^{-1}(\sin x) = x \quad \text{if } -\frac{\pi}{2} \leq x \leq \frac{\pi}{2}$$

Property
Inverse Cosine
Composition

$$\cos(\cos^{-1} x) = x \quad \text{if } -1 \leq x \leq 1$$

$$\cos^{-1}(\cos x) = x \quad \text{if } 0 \leq x \leq \pi$$

Property
Inverse Tangent
Composition

$$\tan(\tan^{-1} x) = x \quad \text{for every real number } x$$

$$\tan^{-1}(\tan x) = x \quad \text{if } -\frac{\pi}{2} < x < \frac{\pi}{2}$$

EXAMPLE 5   **Using the Inverse Trigonometric Composition Properties**

Use the composition properties for the inverse trigonometric functions to find the exact value of each expression.

(a) $\sin^{-1}\left(\sin\dfrac{\pi}{6}\right)$ (b) $\cos\left(\cos^{-1}\dfrac{1}{2}\right)$ (c) $\tan^{-1}\left(\tan\dfrac{3\pi}{4}\right)$

Solution   (a) Since $-\dfrac{\pi}{2}\le\dfrac{\pi}{6}\le\dfrac{\pi}{2}$, then

$$\sin^{-1}\left(\sin\dfrac{\pi}{6}\right)=\dfrac{\pi}{6}.$$

As a check, we note that

$$\sin^{-1}\left(\sin\dfrac{\pi}{6}\right)=\sin^{-1}\dfrac{1}{2}$$

$$=\dfrac{\pi}{6}.$$

(b) Since $-1\le\dfrac{1}{2}\le 1$, then

$$\cos\left(\cos^{-1}\dfrac{1}{2}\right)=\dfrac{1}{2}.$$

(c) We first observe that

$$\tan^{-1}\left(\tan\dfrac{3\pi}{4}\right)\ne\dfrac{3\pi}{4}$$

because $\dfrac{3\pi}{4}$ does not satisfy the restriction $-\dfrac{\pi}{2}<x<\dfrac{\pi}{2}$.
Thus we proceed as follows:

$$\tan^{-1}\left(\tan\dfrac{3\pi}{4}\right)=\tan^{-1}(-1)$$

$$=-\dfrac{\pi}{4}.$$

Right triangles can help us find exact values of expressions involving inverse trigonometric functions and convert trigonometric expressions into algebraic ones.

EXAMPLE 6   **Simplifying Inverse Trigonometric Expressions**

Use right triangles to simplify each expression.

(a) $\tan\left(\cos^{-1}\dfrac{3}{4}\right)$

(b) $\sin^{-1}x+\cos^{-1}x,$ where $0<x<1$

Solution

(a) Assume that $\theta = \cos^{-1}(3/4)$.

By definition $\cos\theta = 3/4$, where $0 < \theta < \pi/2$, since the cosine is positive.

Next we construct a right triangle with one side of length 3 adjacent to angle θ and the hypotenuse of length 4 so that $\cos\theta = 3/4$ (Figure 2a).

By the Pythagorean theorem, the length of the side opposite angle θ is $\sqrt{4^2 - 3^2} = \sqrt{7}$.

Using right triangle trigonometry, we obtain

$$\tan\left(\cos^{-1}\frac{3}{4}\right) = \tan\theta$$

$$= \frac{\text{opp}}{\text{adj}} = \frac{\sqrt{7}}{3}.$$

Figure 2

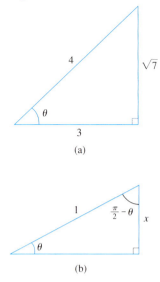

(a)

(b)

(b) Assume that $\theta = \sin^{-1}x$, so $\sin\theta = x$.

Since $0 < x < 1$, it follows that $0 < \theta < \pi/2$. We can think of θ as an acute angle in a right triangle with sides that satisfy

$$\sin\theta = \frac{\text{opp}}{\text{hyp}} = \frac{x}{1} = x \text{ (Figure 2b)}.$$

Using right triangle trigonometry, we get

$$\cos\left(\frac{\pi}{2} - \theta\right) = \frac{\text{adj}}{\text{hyp}} = \frac{x}{1} = x \quad \text{where} \quad 0 < \left(\frac{\pi}{2} - \theta\right) < \frac{\pi}{2}.$$

So

$$\cos^{-1}x = \frac{\pi}{2} - \theta.$$

It follows that

$$\sin^{-1}x + \cos^{-1}x = \theta + \left(\frac{\pi}{2} - \theta\right) = \frac{\pi}{2},$$

that is, $\sin^{-1}x + \cos^{-1}x = \dfrac{\pi}{2}$.

Defining the Other Inverse Trigonometric Functions

By restricting the domains of the cosecant, secant, and cotangent functions, we get one-to-one functions that enable us to define the other three inverse trigonometric functions.

Definition

Other Inverse Trigonometric Functions

Note that the choices of the ranges for $g(x) = \sec^{-1}x$ and $h(x) = \csc^{-1}x$ are not universally accepted and may differ in some other textbooks.

Function	Defined by
$y = \csc^{-1}x$	$\csc y = x$, where $x \leq -1$ or $x \geq 1$ and $-\pi/2 \leq y \leq \pi/2, y \neq 0$
$y = \sec^{-1}x$	$\sec y = x$, where $x \leq -1$ or $x \geq 1$ and $0 \leq y \leq \pi, y \neq \pi/2$
$y = \cot^{-1}x$	$\cot y = x$, where x is any real number and $0 < y < \pi$

The notations arccsc x, arcsec x, and arccot x, respectively, are also used for these functions.

The graphs and properties of these inverse functions are given in Table 2.

TABLE 2   Other Inverse Trigonometric Functions

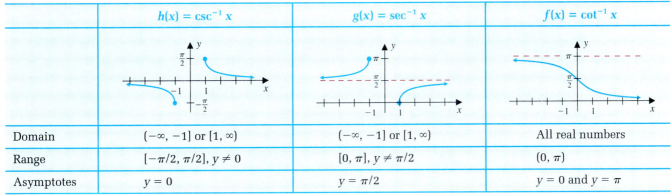

	$h(x) = \csc^{-1} x$	$g(x) = \sec^{-1} x$	$f(x) = \cot^{-1} x$
Domain	$(-\infty, -1]$ or $[1, \infty)$	$(-\infty, -1]$ or $[1, \infty)$	All real numbers
Range	$[-\pi/2, \pi/2], y \neq 0$	$[0, \pi], y \neq \pi/2$	$(0, \pi)$
Asymptotes	$y = 0$	$y = \pi/2$	$y = 0$ and $y = \pi$

EXAMPLE 7   **Evaluating Inverse Functions**

Find the exact value of each function.

(a) $\cot^{-1}\sqrt{3}$

(b) $\csc^{-1}(-2)$

Solution

(a) $\cot^{-1}\sqrt{3}$ is the number y in the interval $(0, \pi)$ that satisfies $\cot y = \sqrt{3}$, or, equivalently, $\dfrac{1}{\tan y} = \sqrt{3}$ so that $\tan y = \dfrac{1}{\sqrt{3}}$. It follows that $y = \dfrac{\pi}{6}$.

(b) $\csc^{-1}(-2)$ is the number y in the interval $[-\pi/2, \pi/2], y \neq 0$, that satisfies $\csc y = -2$ or, equivalently, $\dfrac{1}{\sin y} = -2$. That is, $\sin y = -\dfrac{1}{2}$ so that $y = -\dfrac{\pi}{6}$. 🔲

Most calculators do not have keys for the inverse cotangent, inverse secant, or inverse cosecant functions. The following formulas are used to find the approximate value:

1. If $|x| \geq 1$, then:

    (a) $\csc^{-1} x = \sin^{-1}\dfrac{1}{x}$ (b) $\sec^{-1} x = \cos^{-1}\dfrac{1}{x}$

2. If x is a real number, then:

$$\cot^{-1} x = \begin{cases} \tan^{-1}\dfrac{1}{x} & \text{if } x > 0 \\[2mm] \tan^{-1}\dfrac{1}{x} + \pi & \text{if } x < 0 \\[2mm] \dfrac{\pi}{2} & \text{if } x = 0 \end{cases}$$

EXAMPLE 8   **Approximating Values of Inverse Trigonometric Functions**

Use a calculator to approximate each value to three decimal places.

(a) $\cot^{-1} 12.3$ (b) $\sec^{-1} 1.84$ (c) $\csc^{-1}(-1.52)$

Solution   First we set the calculator in radian mode. Then we use the above formulas to get the following approximate values:

(a) $\cot^{-1} 12.3 = \tan^{-1}\left(\dfrac{1}{12.3}\right) = 0.081$

(b) $\sec^{-1} 1.84 = \cos^{-1}\left(\dfrac{1}{1.84}\right) = 0.996$

(c) $\csc^{-1}(-1.52) = \sin^{-1}\left(\dfrac{1}{-1.52}\right) = -0.718$

Solving Applied Problems

Inverse trigonometric functions are used in mathematical modeling, as the next example illustrates.

EXAMPLE 9   **Modeling a Viewing Angle**

Figure 3

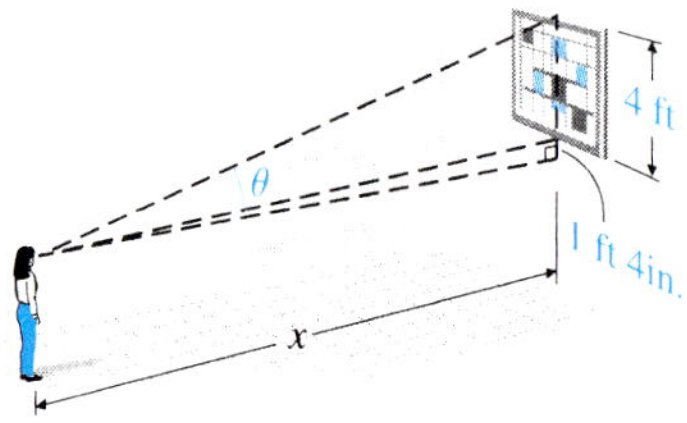

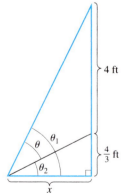

A museum plans to hang a painting for public viewing. The painting, which is 4 feet high, is mounted on a wall in such a way that its lower edge is 1 foot 4 inches above the level of the average viewer's eyes. The viewer studies the painting x feet from the wall on which the painting is mounted (Figure 3). Suppose the angle formed by the viewer's eyes and the top and bottom of the painting is θ.

(a) Use right triangle trigonometry to express the angle θ as a function of x.

(b) What is the angle θ if the viewer is standing 5 feet away from the painting? Round off the answer to two decimal places.

Solution   (a) Figure 4 helps us express θ as a function of x. (Note that 1 foot 4 inches is 4/3 feet.) We see in the figure that

Figure 4

$$\theta = \theta_1 - \theta_2$$

$$\theta = \tan^{-1}\left(\dfrac{4 + \dfrac{4}{3}}{x}\right) - \tan^{-1}\left(\dfrac{\dfrac{4}{3}}{x}\right)$$

$$\theta = \tan^{-1}\dfrac{16}{3x} - \tan^{-1}\dfrac{4}{3x}.$$

This latter equation expresses θ as a function of x.

(b) Substituting 5 for x in the equation

$$\theta = \tan^{-1}\dfrac{16}{3x} - \tan^{-1}\dfrac{4}{3x}$$

we have

$$\theta = \tan^{-1}\left(\dfrac{16}{15}\right) - \tan^{-1}\left(\dfrac{4}{15}\right) = 0.56 \text{ (approx.).}$$

Thus when the viewer stands 5 feet from the painting, the viewing angle is approximately 0.56 radian or about 31.92°.

Figure 5

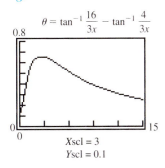

$Xscl = 3$
$Yscl = 0.1$

G Figure 5 shows a viewing window of the graph of

$$\theta = \tan^{-1}\frac{16}{3x} - \tan^{-1}\frac{4}{3x}$$

The graph indicates that there is a high or maximum point.

By using the **MAX**–finding feature, we find the maximum point of the graph of θ to be approximately (2.67, 0.64), rounded to two decimal places. This means that when x is approximately 2.67, θ attains its maximum value of about 0.64 radian or about 36.67°. Thus, when the viewer stands about 2.67 feet, or 2 feet 8 inches, from the wall, the maximum angle measure of about 37° is attained.

Standing 2 feet 8 inches from a painting 4 feet high may be too close to appreciate the artistic qualities of the painting. The maximum viewing angle does not necessarily give the best location for appreciating the art.

PROBLEM SET 5.8

Mastering the Concepts

In problems 1–10, find the exact value of each expression.

1. (a) $\sin^{-1}\dfrac{\sqrt{2}}{2}$

(b) $\sin^{-1}\left(-\dfrac{1}{2}\right)$

2. (a) $\cos^{-1}\dfrac{\sqrt{3}}{2}$

(b) $\cos^{-1}\left(-\dfrac{1}{2}\right)$

3. (a) $\tan^{-1}\dfrac{\sqrt{3}}{3}$

(b) $\tan^{-1}\left(-\dfrac{\sqrt{3}}{3}\right)$

4. (a) $\arctan(-1)$

(b) $\arccos 1$

5. (a) $\arcsin\left(-\dfrac{\sqrt{3}}{2}\right)$

(b) $\arccos\left(-\dfrac{\sqrt{3}}{2}\right)$

6. (a) $\arcsin 1$

(b) $\arctan \sqrt{3}$

7. (a) $\cos^{-1}\left(\cos\dfrac{\pi}{6}\right)$

(b) $\tan^{-1}\left(\tan\dfrac{\pi}{4}\right)$

8. (a) $\sin\left(\cos^{-1}\dfrac{1}{2}\right)$

(b) $\sin^{-1}\left[\sin\left(-\dfrac{3\pi}{2}\right)\right]$

9. (a) $\sin^{-1}\left(\sin\dfrac{5\pi}{4}\right)$

(b) $\tan[\tan^{-1}(-1)]$

10. (a) $\sec\left(\sin^{-1}\dfrac{\sqrt{3}}{2}\right)$

(b) $\csc(\tan^{-1} 1)$

In problems 11–16, find the approximate values of each expression to two decimal places.

11. (a) $\sin^{-1} 0.2182$

(b) $\arcsin(-0.7771)$

12. (a) $\cos^{-1} 0.8628$

(b) $\arccos(-0.8473)$

13. (a) $\tan^{-1} 1.072$

(b) $\arctan(-41.03)$

14. (a) $\arccos 0.4037$

(b) $\arcsin(2/\pi)$

15. (a) $\arccos(-0.7112)$

(b) $\tan^{-1} 100$

16. (a) $\sin^{-1}(1 - \sqrt{2})$

(b) $\sin^{-1}(\sqrt{3} - \sqrt{2})$

In problems 17–24, express x in terms of y. Determine the restrictions on x and y.

17. $y = \sin^{-1}(x + 1)$

18. $y = \cos^{-1}(x - 2)$

19. $y = \cos^{-1}(2x - 8)$

20. $y = 2 \sin^{-1}(3x + 7)$

21. $y = -3 \tan^{-1}(3x - 2)$

22. $y = -2 \cos^{-1}(2x - 1)$

23. $y = \dfrac{1}{3} \cos^{-1}(2x - 4)$

24. $y = 4 - 2 \sin^{-1} 4x$

In problems 25 and 26, use the inverse trigonometric composition identities, whenever possible, to find the value of each expression. Round off the answer to four decimal places when appropriate.

25. (a) $\sin(\sin^{-1} 0.3)$

(b) $\sin^{-1}(\sin 1.3)$

(c) $\cos\left[\cos^{-1}\left(-\dfrac{\pi}{6}\right)\right]$

(d) $\cos^{-1}\left(\cos\dfrac{7\pi}{6}\right)$

26. (a) $\cos(\cos^{-1} 0.8)$

(b) $\cos^{-1}(\cos \sqrt{2})$

(c) $\tan(\tan^{-1} 90)$

(d) $\tan^{-1}\left(\tan\dfrac{5\pi}{4}\right)$

In problems 27–30, find the required restrictions on x in order for the equation to be true.

27. $\sin[\sin^{-1}(3x + 2)] = 3x + 2$

28. $\cos[\cos^{-1}(5 - 2x)] = 5 - 2x$

29. $\tan^{-1}\left[\tan\left(\dfrac{x}{3} - 1\right)\right] = \dfrac{x}{3} - 1$

30. $\cos^{-1}[\cos(x^2 - 4)] = x^2 - 4$

In problems 31–36, use a right triangle whenever possible to find the exact value of each expression.

31. (a) $\cos\left(\sin^{-1}\dfrac{4}{5}\right)$

(b) $\sin\left(\cos^{-1}\dfrac{5}{13}\right)$

32. (a) $\sin\left(\tan^{-1}\dfrac{4}{3}\right)$

(b) $\cos^{-1}\left(\cos\dfrac{3\pi}{2}\right)$

33. (a) $\tan\left(\sin^{-1}\dfrac{2\sqrt{5}}{5}\right)$

(b) $\cos\left[\cos^{-1}\left(-\dfrac{2\sqrt{5}}{5}\right)\right]$

34. (a) $\sin[\tan^{-1}(-2)]$

(b) $\sec\left(\sin^{-1}\dfrac{4}{5}\right)$

35. (a) $\csc\left(\cos^{-1}\dfrac{1}{4}\right)$

(b) $\sec^{-1}\left(\sec\dfrac{\pi}{3}\right)$

36. (a) $\cot\left[\tan^{-1}\left(-\dfrac{5}{12}\right)\right]$

(b) $\csc\left(\sin^{-1}\dfrac{12}{13}\right)$

In problems 37 and 38, use a right triangle to rewrite each expression as an algebraic expression in terms of x that does not involve an inverse trigonometric function, where $x > 0$.

37. (a) $\sin(\cos^{-1} x)$

(b) $\tan(\cos^{-1} x)$

(c) $\cot(\tan^{-1} x)$

38. (a) $\cot(\sin^{-1} x)$

(b) $\cot(\cos^{-1} x)$

(c) $\csc\left(\tan^{-1}\dfrac{x}{\sqrt{2}}\right)$

In problems 39–42, find the exact value of each expression.

39. (a) $\cot^{-1} 1$

(b) $\csc^{-1}(-\sqrt{2})$

40. (a) $\cot^{-1}(-1)$

(b) $\sec^{-1} 2$

41. (a) $\sec^{-1}\sqrt{2}$

(b) $\csc^{-1}\left(-\dfrac{2}{\sqrt{3}}\right)$

42. (a) $\sec^{-1}(-\sqrt{2})$

(b) $\csc^{-1}\left(\dfrac{2}{\sqrt{3}}\right)$

In problems 43–46, find the approximate value rounded to two decimal places.

43. (a) $\cot^{-1} 0.9713$

(b) $\sec^{-1} 3.4182$

44. (a) $\csc^{-1} 1.0152$

(b) $\cot^{-1} 1.0487$

45. (a) $\sec^{-1}(-2.4182)$

(b) $\csc^{-1}(-8.8952)$

46. (a) $\sec^{-1} 8.4513$

(b) $\cot^{-1}(-0.8752)$

Applying the Concepts

47. Viewing a Billboard: A billboard is to be built adjacent to a highway so that the top and bottom will be 32 feet and 24 feet, respectively, above the eye level of a passing motorist. Suppose that the motorist passes the billboard at a distance of x feet and that the angle formed by the motorist's eye level and the top and bottom of the billboard is θ (Figure 6).

(a) Express θ as a function of x.

(b) Find the angle θ if the motorist is 100 feet from the billboard. Round off the answer to two decimal places.

Figure 6

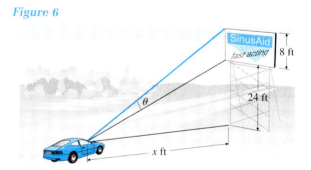

48. [G] (Refer to problem 47.)

(a) Use a grapher to graph the function found in (a).

(b) Approximate the maximum value of θ and when it occurs. What is the degree measure for this maximum angle? Round off the answers to two decimal places.

(c) Does the maximum angle give the best view of the billboard? Explain.

49. Viewing a Window: A night watchman wants the greatest possible view of a second story window of a building as he makes his rounds. The windows of the building are 4 feet high, with the lower sill 11 feet above the ground. Assume the watchman's eyes are precisely 6 feet above the ground. Suppose that the watchman stands x feet away from the building, and that the angle formed by the watchman's eyes and the top and bottom of the window is θ (Figure 7).

(a) Express θ as a function of x.

(b) Find the angle if the watchman is 20 feet from the building. Round off the answer to two decimal places.

Figure 7

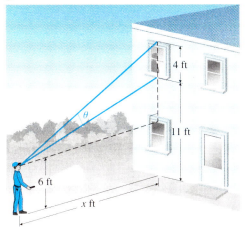

50. [G] (Refer to problem 49.)

(a) Use a grapher to graph the function found in (a).

(b) Approximate the maximum value of θ and when it occurs. What is the degree measure for this maximum angle? Round off the answers to two decimal places.

(c) Does the maximum angle give the best view of the window? Explain.

Developing and Extending the Concepts

In problems 51–56 use transformations of the graphs in Figure 1 (see page 391) to sketch the graph of each function.

51. $y = \dfrac{1}{2} \sin^{-1} x$

52. $y = -3 \cos^{-1} x$

53. $y = 3 + 2 \tan^{-1} x$

54. $y = \tan^{-1} x + \dfrac{\pi}{2}$

55. $y = 2[\sin^{-1}(x - 1)] - \pi$

56. $y = \dfrac{1}{3}[\cos^{-1}(x + 1)] + \dfrac{\pi}{2}$

57. Show that $\sin^{-1}(-x) = -\sin^{-1} x$

58. Is it true that $\tan^{-1} x = \dfrac{\sin^{-1} x}{\cos^{-1} x}$? Justify your answer.

59. Graph

$$f(x) = (\sin x)^{-1} = \frac{1}{\sin x} = \csc x \quad \text{and} \quad g(x) = \sin^{-1} x$$

for $-1 \le x \le 1$. Are the graphs the same? Explain the use of the superscript $^{-1}$ in both equations.

60. [G] Use a grapher to graph $y = \tan^{-1} x^2$. Then use the graph to describe the limit behavior of the function; that is, determine what happens to the values of $\tan^{-1} x^2$ as $x \to +\infty$ and as $x \to -\infty$.

61. [G] Use a grapher to graph $y = \sin^{-1} x + \cos^{-1} x$, where $0 \le x \le 1$. Explain how the graph demonstrates the result in Example 6b.

62. Sketch the graph of the function $y = \sin(\sin^{-1} x)$. What is the domain of this function? Explain your result.

◈ CHAPTER 5 REVIEW PROBLEM SET

1. Express each angle measure in degrees, minutes, and seconds.
   (a) $76.25°$ (b) $-61.35°$
   (c) $143.47°$ (d) $-14.47°$

2. Express each angle as a decimal. Round off the answers to four decimal places.
   (a) $4°7'$ (b) $21°19'13''$
   (c) $-45°35'25''$ (d) $15''$

3. Convert each degree measure to radian measure. Round off the answers to four decimal places.
   (a) $15°$ (b) $-17.45°$
   (c) $45°16'51''$ (d) $36°11'25''$

4. Convert each radian measure to degree measure. Round off the answers to four decimal places.
   (a) $\dfrac{17\pi}{4}$ (b) $-\dfrac{4\pi}{7}$
   (c) 5.82 (d) -7.63

5. Suppose that angle $\theta = 36°$ is in standard position. Find the measure of each of the following angles and sketch the angles.
   (a) Complement of θ
   (b) Supplement of θ
   (c) Two angles coterminal with θ, one positive and one negative.

6. Let s denote the length of the arc intercepted on a circle of radius r by a central angle θ. In each case, find the missing measurement and the area of the sector.

(a) $r = 7$ centimeters, $\theta = 75°$, $s = ?$

(b) $r = 2$ inches, $s = 5$ inches, $\theta = ?$

(c) $s = 17$ meters, $\theta = \dfrac{5\pi}{6}$, $r = ?$

7. Display the location of each point on the unit circle. Then find the coordinates of each point. Round off the answers to four decimal places.

(a) $P\left(\dfrac{2\pi}{5}\right)$

(b) $P(3.35)$

(c) $P\left(-\dfrac{3\pi}{7}\right)$

8. Find the exact values of the six trigonometric functions of t.

(a) $P(t) = \left(-\dfrac{\sqrt{3}}{2}, -\dfrac{1}{2}\right)$

(b) $P(t) = \left(\dfrac{4}{\sqrt{17}}, \dfrac{1}{\sqrt{17}}\right)$

9. Graph each function. Find the period and phase shift.

(a) $f(x) = 0.6 \tan\left(3x - \dfrac{\pi}{2}\right)$

(b) $g(x) = -1.5 \sec\left(\dfrac{x}{4} + \pi\right)$

10. Use the known graphs of the sine and cosine functions to graph one cycle of each function. Indicate the amplitude, period, and phase shift.

(a) $f(x) = 2 \sin\left(\dfrac{3x}{2} - \dfrac{\pi}{4}\right)$

(b) $g(x) = -0.2 \cos\left(2x + \dfrac{\pi}{8}\right)$

11. [G] Use a grapher to sketch the graph of each function for $0 \le x \le 4\pi$.

(a) $f(x) = 2 \cos x + 3 \cos \dfrac{x}{2}$

(b) $g(x) = \dfrac{\sin x}{x} - x$

(c) $h(x) = \dfrac{1 - \cos x}{x^2}$

12. The functions for the graphs in Figure 1 have the form $y = a \cos kx$ or $y = a \sin kx$, where $k > 0$. For each graph, specify the period and amplitude, and write a specific equation for the function.

Figure 1

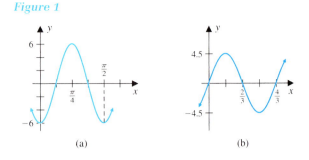

(a)   (b)

13. Sketch an angle θ in standard position and name the quadrant in which θ lies if:

(a) $\theta = -100°$

(b) $\sin \theta < 0$ and $\cos \theta < 0$

(c) $\sec \theta > 0$ and $\cot \theta < 0$

14. Find the exact values of the six trigonometric functions of θ if θ is in standard position and the terminal side of θ contains the given point

(a) $(-6, -8)$ (b) $(-3, 5)$

15. Use a calculator to find the approximate value of each expression rounded off to four decimal places.

(a) $\sin 3$ (b) $\cos(-208°)$

(c) $\tan \dfrac{3\pi}{5}$ (d) $\sec(-3.92)$

16. Use a reference triangle and periodicity to find the exact value of each expression.

(a) $\cos\left(-\dfrac{5\pi}{3}\right)$ (b) $\sec 780°$

(c) $\cot \dfrac{37\pi}{6}$ (d) $\tan(-600°)$

17. Find the exact values of the six trigonometric functions of the acute angle θ in Figure 2.

Figure 2

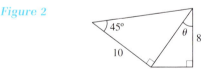

18. Assume θ is an acute angle and $\sin \theta = \dfrac{8}{17}$. Find

(a) $\cos \theta$ and $\tan \theta$

(b) $\sin(-\theta)$ and $\sec(-\theta)$

(c) θ, in both degree measure and radian measure. Round off the answers to two decimal places.

19. Suppose θ satisfies

$$180° < \theta < 270° \quad \text{and} \quad \tan \theta = a.$$

Express each of the following in terms of a.

(a) $\cot \theta$ (b) $\tan(\theta - 360°)$ (c) $\tan(\theta + 180°)$

20. Find the exact value of each expression.

(a) $\cos^{-1}\dfrac{1}{2}$

(b) $\sin^{-1}\left(-\dfrac{1}{2}\right)$

(c) $\tan^{-1}\left(\sin\dfrac{\pi}{2}\right)$

(d) $\tan\left(\sin^{-1}\dfrac{3}{5}\right)$

(e) $\sec\left[\tan^{-1}\left(-\dfrac{5}{3}\right)\right]$

21. Use a calculator to approximate the value of each expression. Give the answer in radians rounded off to two decimal places.

(a) $\sin^{-1} 0.3741$

(b) $\arccos(-0.4901)$

(c) $\sec^{-1} 9.723$

(d) $\cot^{-1} 57.29$

22. Use a right triangle to rewrite each given expression as an algebraic expression in terms of x that does not involve an inverse trigonometric function.

(a) $\sin(\tan^{-1} x)$, where $x > 0$

(b) $\tan(\sin^{-1} x)$, where $0 < x < 1$

23. In each case, solve for x in terms of y. Specify the restrictions on the values of x and y.

(a) $y = \sin^{-1}\dfrac{x}{2}$

(b) $y = \arctan(2x + 3)$

24. Solve the right triangle ABC if $\gamma = 90°$. Round off each answer to two decimal places.

(a) $b = 25$ inches and $\beta = 65°$

(b) $b = 7$ centimeters and $\alpha = 35°$

25. Engineering: A curve on a highway subtends an angle of $31°$ on a circle of radius 560 meters.

(a) How long is the curve subtended by this angle?

(b) How long will it take a car traveling 64 kilometers per hour to round the curve?

Round off the answers to two decimal places.

26. Geometry: An isosceles triangle ABC with $|\overline{AB}| = |\overline{AC}|$ and an angle $\alpha = 48°$ is inscribed in a circle. Find the radius of the circle if the side $\overline{BC}$ intercepts an arc of 6.3 centimeters on the circle. Round off the answer to two decimal places.

27. Space Satellite: A satellite traveling in a circular orbit of radius 6371 kilometers is known to have a linear speed of 30,720 kilometers per hour. Find the angular speed of the satellite in radians per hour. Round off the answer to two decimal places.

28. Simple Harmonic Motion: The position of a particle at time t is given by each equation. Find the amplitude, period, and phase shift of the path of each particle.

(a) $y = 2.7 \sin(7t - 4\pi)$

(b) $y = 4 \cos\left(3t - \dfrac{\pi}{3}\right)$

29. Hours of Daylight: Suppose the number of daylight hours N in a city at a particular time of the year is given by the model

$$N = 12 + 2.5 \sin\left[\frac{2\pi}{365}(t - 81)\right]$$

where t is the number of days, with $t = 1$ corresponding to January 1.

[G] (a) Graph the function for $0 \leq t \leq 365$.

(b) Find the maximum and minimum number of hours of daylight and the corresponding values of t rounded to two decimal places.

30. Height of a Blimp: An observer on the ground notices that the angle of elevation of a blimp is $29°$. If the observer is 2 miles away from a point directly below the blimp, what is the approximate height of the blimp? Round off the answer to two decimal places.

31. Surveying: A person standing on top of a building that is 120 feet high spots a car parked on top of a parking garage that is 7 feet high. The angle of depression is $57°$ from the person to the car. What is the approximate distance of the garage from the building? Round off the answer to two decimal places.

32. Balloonist: A balloonist 100 meters above ground level observed a car on the ground. If the horizontal distance from a point on the ground directly under the balloon to the car is 79 meters, find the angle of depression from the balloon to the car to the nearest tenth of a degree.

◆ CHAPTER 5 TEST

1. (a) Convert 520° to radian measure.
   (b) Convert $3\pi/7$ to degree measure.

2. Suppose that the sides of a central angle intercept an arc of length 15 centimeters on the circumference of a circle of radius 10 centimeters.
   (a) Find the radian measure of the central angle and the area of the sector.
   (b) Express the measure of the central angle in degrees.

3. Suppose that the point $P = (3, -7)$ is on the terminal side of angle θ in standard position.
   (a) Draw two angles, one positive and one negative, coterminal with θ.
   (b) Find the exact value of each of the six trigonometric functions of θ.

4. Suppose that $P(t) = (a, 2a)$, where $a > 0$.
   (a) Determine the value of a.
   (b) Use the results from part (a) to find $\sin t$, $\cos t$, and $\tan t$.

5. Find the exact value of each expression.
   (a) $\cos(-240°)$
   (b) $\sin \dfrac{13\pi}{4}$
   (c) $\tan \dfrac{17\pi}{3}$
   (d) $\cos^{-1}\left(-\dfrac{\sqrt{2}}{2}\right)$
   (e) $\sin\left(\cos^{-1}\dfrac{3}{5}\right)$

6. Use a calculator to approximate the value of each expression rounded off to two decimal places.
   (a) $\sin 23.17°$
   (b) $\csc 2.85$
   (c) $\cos^{-1}(-0.8413)$

7. Determine the quadrant that contains the angle θ.
   (a) $\sin \theta < 0$ and $\tan \theta > 0$
   (b) $\cos \theta > 0$ and $\csc \theta < 0$

8. Suppose that $\cos \theta = 5/17$ and θ is an acute angle. Find the exact values of the other five trigonometric functions of θ.

9. Find the amplitude, period, and phase shift, and sketch the graph of one cycle of each function.
   (a) $f(x) = 3 \sin(4x - \pi)$
   (b) $g(x) = -2 \cos\left(\dfrac{\pi x}{2} - \dfrac{\pi}{3}\right)$
   (c) $h(x) = \tan(x + \pi)$

10. The number of hours of sunshine each month for a region is given by the model

$$f(t) = 2.04 \sin\left(\dfrac{\pi}{6} t - 1.7\right) + 6.12, \ 1 \leq t \leq 12$$

where $t = 1$ represents the month of January and $t = 12$ represents the month of December.
   (a) Sketch the graph of this model.
   (b) Use the graph to determine the maximum number of hours of sunshine and in what month it occurs. Round off the answers to two decimal places.

11. A straight string of lights is to reach from the top of an 80-meter pole and make an angle of elevation of 61° with the horizontal ground. How far from the pole should the lights be fastened to the ground? Round off the answer to two decimal places.

12. One end of an 80-foot rope is attached to the top of a vertical pole standing in the center of a tent. The other end of the rope is attached to a stake 60 feet away from the base of the pole. Determine the angle of elevation of the rope with the ground.

Trigonometric Identities and Equations

Chapter Contents

Many applied problems can be modeled by trigonometric functions. For instance, suppose that an observer at point P, 111 meters away from a point Q directly beneath a hot air balloon, sights the balloon at an angle of elevation θ. At the same time, a second observer at point R, which is 74 meters closer to point Q, sights the same balloon at an angle of elevation 2θ. If the points P, R, and Q are in the same plane, find the height h of the balloon. Example 7 on page 463 gives the solution of this problem.

In Chapter 5 we introduced special trigonometric relationships called *identities.* This chapter expands the study of trigonometric identities and examines various techniques for solving trigonometric equations.

Objectives

1. Simplify Trigonometric Expressions
2. Verify Trigonometric Identities
3. Prove a Trigonometric Equation Is Not an Identity

6.1 Fundamental Identities

So far we have four types of identities: the *reciprocal relationships*, the *quotient relationships*, the *even-odd relationships*, and the *Pythagorean identities.* They are listed below and in the back inner cover.

Reciprocal Identities

1. $\csc \theta = \dfrac{1}{\sin \theta}$ 2. $\sec \theta = \dfrac{1}{\cos \theta}$ 3. $\cot \theta = \dfrac{1}{\tan \theta}$

Quotient Identities

1. $\tan \theta = \dfrac{\sin \theta}{\cos \theta}$ 2. $\cot \theta = \dfrac{\cos \theta}{\sin \theta}$

Even-Odd Identities

1. $\sin(-\theta) = -\sin \theta$ 2. $\cos(-\theta) = \cos \theta$ 3. $\tan(-\theta) = -\tan \theta$
4. $\csc(-\theta) = -\csc \theta$ 5. $\sec(-\theta) = \sec \theta$ 6. $\cot(-\theta) = -\cot \theta$

Pythagorean Identities

1. $\cos^2 \theta + \sin^2 \theta = 1$ 2. $1 + \tan^2 \theta = \sec^2 \theta$ 3. $1 + \cot^2 \theta = \csc^2 \theta$

Simplifying Trigonometric Expressions

At times it is possible to rewrite given trigonometric expressions in simpler forms by using known identities. The next two examples illustrate the process.

EXAMPLE 1   **Simplifying an Expression Using a Pythagorean Identity**

Use identities to simplify the expression $\cot^2 t \sin^2 t + \sin^2 t$.

Solution

$$\cot^2 t \sin^2 t \; + \; \sin^2 t = \frac{\cos^2 t}{\sin^2 t} \cdot \sin^2 t + \sin^2 t \qquad \text{Quotient identity}$$

$$= \cos^2 t + \sin^2 t = 1 \qquad \text{Pythagorean identity}$$

EXAMPLE 2   **Simplifying an Expression Using an Even-Odd Identity**

Use identities to simplify the expression $\dfrac{\cot \theta \cos(-\theta)}{\csc^2 \theta - 1}$.

Solution

$$\frac{\cot \theta \cos(-\theta)}{\csc^2 \theta - 1} = \frac{\cot \theta \cos \theta}{\csc^2 \theta - 1} \qquad \cos(-\theta) = \cos \theta$$

$$= \frac{\dfrac{\cos \theta}{\sin \theta} \cos \theta}{\cot^2 \theta} = \frac{\left(\dfrac{\cos^2 \theta}{\sin \theta} \right)}{\left(\dfrac{\cos^2 \theta}{\sin^2 \theta} \right)} \qquad \text{Quotient and Pythagorean identities}$$

$$= \frac{\cos^2 \theta}{\sin \theta} \cdot \frac{\sin^2 \theta}{\cos^2 \theta}$$

$$= \sin \theta \qquad \text{Simplify}$$

Verifying Trigonometric Identities

The fundamental identities play an important role in solving trigonometric equations, in performing certain operations in calculus, and in deriving other identities. In these instances we are often involved in using algebra along with the identities to manipulate and convert trigonometric expressions to other forms. This process of manipulating and converting trigonometric expressions can be practiced by *verifying trigonometric identities.*

There is no general rule for proving that a trigonometric equation is an identity; however, we normally start with one side of the equation and try to convert it to the other side by means of a sequence of algebraic manipulations and substitutions utilizing known identities. Often it is helpful to start with the side containing more terms and try to convert it to the *simpler* side. Some suggestions that may help in carrying out a proof are listed below:

Suggestions for Verifying Identities

1. Combine a sum or difference of fractions into a single fraction.
2. Reduce a fraction.
3. Factor the expression.
4. Combine like terms.
5. Multiply both the numerator and denominator by the same expression.
6. Write all trigonometric expressions in terms of sines and cosines, and then simplify.

There may be more than one way of verifying that an equation is an identity, as shown in the next example.

EXAMPLE 3   **Verifying Trigonometric Identities**

Verify that the equation

$$(\cos^2 t)(1 + \tan^2 t) = 1$$

is an identity.

Solution   Working with the left side of the equation, we get

$$(\cos^2 t)(1 + \tan^2 t) = \cos^2 t \cdot \sec^2 t \qquad \text{Pythagorean identity}$$

$$= (\cos t \cdot \sec t)^2 = \left(\cos t \cdot \frac{1}{\cos t}\right)^2 \qquad \text{Reciprocal identity}$$

$$= 1^2 = 1$$

Since we have transformed the left side into the right side, the equation is an identity.

Alternate Solution   This strategy uses a quotient identity to express the left side in terms of sin t and cos t.

$$(\cos^2 t)(1 + \tan^2 t) = \cos^2 t \left(1 + \frac{\sin^2 t}{\cos^2 t}\right) \qquad \text{Quotient identity}$$

$$= \cos^2 t + \cos^2 t \left(\frac{\sin^2 t}{\cos^2 t}\right) \qquad \text{Multiply and reduce}$$

$$= \cos^2 t + \sin^2 t = 1 \qquad \text{Pythagorean identity}$$

At times we change the form of one side of an equation before proving that it is an identity.

EXAMPLE 4   **Using Multiplication to Verify an Identity**

Verify the identity $\dfrac{\cos t}{1 - \sin t} = \dfrac{1 + \sin t}{\cos t}$.

Solution   To verify this identity, it may be tempting to multiply *both sides* by either $\cos t$ or by $1 - \sin t$. But this step is incorrect because it assumes the equation is an identity before it has been verified. Instead, we work with the left side by multiplying the numerator and denominator by the expression $1 + \sin t$ to obtain

$$\frac{\cos t}{1 - \sin t} = \frac{\cos t}{1 - \sin t} \cdot \frac{1 + \sin t}{1 + \sin t}$$

$$= \frac{(\cos t)(1 + \sin t)}{1 - \sin^2 t} \qquad \text{\color{teal}Multiply}$$

$$= \frac{(\cos t)(1 + \sin t)}{\cos^2 t} = \frac{1 + \sin t}{\cos t} \qquad \text{\color{teal}Reduce fraction}$$

Since we have transformed the left side into the right side, the equation is an identity.

In the next example we verify the identity by converting each side of the equation to the same expression.

EXAMPLE 5   **Verifying an Identity by Simplifying Each Side**

Verify that the equation $\tan(-\theta)\sin(-\theta) = \sec\theta - \cos\theta$ is an identity.

Solution   Here we use the even-odd identities to write

$$\tan(-\theta) = -\tan\theta \qquad \text{and} \qquad \sin(-\theta) = -\sin\theta$$

so it is enough to prove that

$$\tan\theta \sin\theta = \sec\theta - \cos\theta.$$

In this situation, we verify that the equation is an identity by simplifying each side until the results become identical. We proceed as follows:

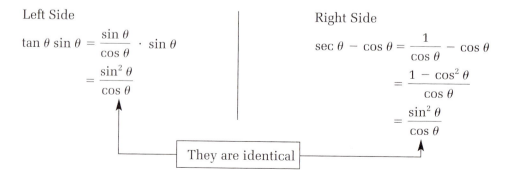

Left Side

$$\tan\theta \sin\theta = \frac{\sin\theta}{\cos\theta} \cdot \sin\theta$$

$$= \frac{\sin^2\theta}{\cos\theta}$$

Right Side

$$\sec\theta - \cos\theta = \frac{1}{\cos\theta} - \cos\theta$$

$$= \frac{1 - \cos^2\theta}{\cos\theta}$$

$$= \frac{\sin^2\theta}{\cos\theta}$$

They are identical

Therefore, the equation is an identity.

Proving a Trigonometric Equation Is Not an Identity

In order for a trigonometric equation to be an identity, the value of each side must turn out to be the same whenever we replace the variable with any number found in the domains of both sides. One *counterexample* is sufficient to show that a proposed equation is *not* an identity.

EXAMPLE 6   **Proving That an Equation Is Not an Identity**

Show that the equation

$$\sin x \cot x = \csc x - \sin x$$

is not an identity by providing a counterexample.

Solution   For the equation to be an identity, the value of each side must turn out to be the same whenever we replace x with any number found in the domains of both sides. This is not the case here. For instance, each side is defined for $x = 2$; however,

$$\sin 2 \cot 2 = -0.4161 \text{ (approx.)} \quad \text{and} \quad \csc 2 - \sin 2 = 0.1905 \text{ (approx.)}$$

are different, so we conclude that the equation is not an identity.

Viewing windows of the graphs of $y_1 = \sin x \cot x$ (Figure 1a) and $y_2 = \csc x - \sin x$ (Figure 1b) *show* that the two equations are not the same for all values in the domains of y_1 and y_2. Thus, the graphs reaffirm that the equation given in Example 6 is not an identity.

Figure 1

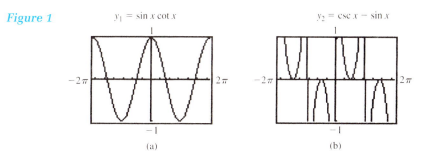

If a trigonometric expression involves radicals, then the use of absolute value is sometimes necessary, although it may be difficult to determine when to use it. In some cases we assume that the radicands are nonnegative. This means that the idenities are restricted to certain quadrants.

For example, we simplify the expression

$$\sqrt{\cos^3 x \sin x} \cdot \sqrt{\sin x}$$

for nonnegative x as follows:

$$\sqrt{\cos^3 x \sin x} \cdot \sqrt{\sin x} = \sqrt{\cos^3 x \sin^2 x}$$
$$= \sqrt{\cos^2 x \sin^2 x \cos x}$$
$$= \cos x \sin x \sqrt{\cos x}$$

a grapher can be used to confirm the results.

⬥ PROBLEM SET 6.1

Mastering the Concepts

In problems 1–8, use the fundamental identities to fill in the blank.

1. $\sin^2 3\theta = 1 - $ _____

2. $3\cos(-5\theta) = $ _____ $\cos(5\theta)$

3. $\sec^2 t - \tan^2 t = $ _____

4. $\cot^2 \theta - \csc^2 \theta = $ _____

5. $2\cos(-t)\sin(-t) = $ _____ $\cos t \sin t$

6. If $\sin t = \cos t$, then $\sin^2 t - \cos^2 t = $ _____

7. $\dfrac{4}{\sin^2 \theta} + \dfrac{4}{\cos^2 \theta} = $ _____ $\csc^2 \theta \sec^2 \theta$

8. $\dfrac{1 - \cos^2 t}{2 \cos^2 t} = $ _____ $\tan^2 t$

In problems 9 and 10, write each expression in terms of $\sin t$.

9. (a) $\tan t \cos t$
   (b) $3 + 5\cos^2 t$
   (c) $\sin t \csc t - \cos^2 t$

10. (a) $(\cos t)(\tan t + \sec t)$
    (b) $\sin^3 t + \sin t \cos^2 t$
    (c) $\sin t \cos t \cot t$

In problems 11–16, carry out each operation and simplify.

11. (a) $\cos^4 \theta + \cos^2 \theta - \cos^2 \theta \sin^2 \theta$
    (b) $\sec(-\theta)\cot(-\theta)$

12. (a) $\dfrac{4\sin^2 \theta - 1}{2\sin \theta + 1}$
    (b) $\dfrac{1 - \tan^4 t}{\sec^2 t}$

13. (a) $(\csc t - \cot t)(\sec t + 1)$
    (b) $(\sec \theta - \tan \theta)(\sin \theta + 1)$

14. (a) $\csc^4 t - \cot^4 t$
    (b) $\cot^2 t - \csc^2 t$

15. (a) $\sin t + \dfrac{\cos^2 t}{\sin t}$
    (b) $\cos \theta + \dfrac{\sin^2 \theta}{\cos \theta}$

16. (a) $\dfrac{1 - (\sin \theta - \cos \theta)^2}{\sin \theta}$
    (b) $\dfrac{1 + \cos t}{\sin t} + \dfrac{\sin t}{1 + \cos t}$

In problems 17–48, verify that the equation is an identity.

17. $\sin \theta \csc \theta = 1$

18. $\tan t \cot t = 1$

19. $\sec^2 \theta (1 - \sin^2 \theta) = 1$

20. $\cot^2 \theta (\sec^2 \theta - 1) = 1$

21. $(\sin t + \cos t)^2 + 2\sin(-t)\cos(-t) = 1$

22. $\cos(-t)\tan(-t) + \sin t = 0$

23. $(\cos t - \sin t)^2 + (\cos t + \sin t)^2 = 2$

24. $\sec^7 \theta \tan \theta - \tan \theta \sec^5 \theta = \sec^5 \theta \tan^3 \theta$

25. $\dfrac{\sin \theta}{\tan \theta} + \dfrac{\cos \theta}{\cot \theta} = \sin \theta + \cos \theta$

26. $\tan^2 \theta - \sin^2 \theta = \dfrac{\sin^4 \theta}{\cos^2 \theta}$

27. $\dfrac{\sin \theta}{\cot \theta + \csc \theta} - \dfrac{\sin \theta}{\cot \theta - \csc \theta} = 2$

28. $\dfrac{\cos \theta \cot \theta}{\cot \theta - \cos \theta} = \dfrac{\cot \theta + \cos \theta}{\cos \theta \cot \theta}$

29. $\dfrac{\sin t}{1 + \cos t} + \dfrac{1 + \cos t}{\sin t} = 2\csc t$

30. $\dfrac{\tan \theta + \cot \theta}{\tan \theta - \cot \theta} = \dfrac{\sec^2 \theta}{\tan^2 \theta - 1}$

31. $\dfrac{\cos \theta}{1 - \sin \theta} + \dfrac{\cos \theta}{1 + \sin \theta} = 2\sec \theta$

32. $\dfrac{\sin \theta}{\csc \theta - \cot \theta} = 1 + \cos \theta$

33. $\dfrac{1}{1 + \tan \theta} - \dfrac{\cot \theta}{1 + \cot \theta} = 0$

34. $(\sec \theta - \tan \theta)^2 = \dfrac{1 - \sin \theta}{1 + \sin \theta}$

35. $\dfrac{\sec \theta + 1}{\tan \theta} + \dfrac{\tan \theta}{\sec \theta + 1} = 2\csc \theta$

36. $\dfrac{1}{1 + \sin t} + \dfrac{1}{1 - \sin t} = 2\sec^2 t$

37. $\dfrac{1 - \sin \theta}{1 - \sec \theta} + \dfrac{1 + \sin \theta}{1 + \sec \theta} = 2\cot \theta (1 - \cot \theta)$

38. $\dfrac{\cos \theta + \sin^2(-\theta)\sec \theta}{\csc \theta} = \tan \theta$

39. $\dfrac{\sin t}{1 + \cos t} + \dfrac{\sin t}{1 - \cos t} = 2\csc t$

40. $\dfrac{\tan t}{\sin t(1 + \tan^2 t)} = \cos t$

41. $\dfrac{1}{\cos^2 \theta} + 1 + \dfrac{\sin^2 \theta}{\cos^2 \theta} = 2\sec^2 \theta$

42. $1 - \dfrac{\cos^2 t}{1 + \sin t} = \sin t$

43. $\dfrac{1 + \tan t}{\sin t} - \sec t = \csc t$

44. $\cot t + \dfrac{\sin t}{1 + \cos t} = \csc t$

45. $\dfrac{\tan t + \cot t}{\tan t \cot t} = \sec t \csc t$

46. $\dfrac{\sec t - \csc t}{\sec t + \csc t} = \dfrac{\tan t - 1}{\tan t + 1}$

47. $\dfrac{\cot t - \tan t}{\cot t + \tan t} = \cos^2 t - \sin^2 t$

48. $\dfrac{\sin t}{1 - \cos t} = \csc t + \cot t$

In problems 49–52, show that the given equation is *not* an identity by giving a value of t for which the given equation is false.

49. $\sin t(1 + \cot t) = 2 \sin t + \cos t$

50. $\cos t(\tan t + \cot t) = -\csc t$

51. $\sec t - \cos t = 3 \tan t \cos t$

52. $\sin t \cot t + \cos t \tan t = \sin t - \cos t$

Developing and Extending the Concepts

In problems 53–56, each equation is an identity in certain quadrants associated with t. Determine the quadrants in each case and verify the identity.

53. $\dfrac{\sin t}{\sqrt{1 - \sin^2 t}} = -\tan t$

54. $\dfrac{\cos t}{\sqrt{1 - \cos^2 t}} = \cot t$

55. $\dfrac{1}{\cos t} - \dfrac{1}{\cot t} = \sqrt{\dfrac{1 - \sin t}{1 + \sin t}}$

56. $\dfrac{\sec t - 1}{\tan t} = \sqrt{\dfrac{1 - \cos t}{1 + \cos t}}$

In problems 57–60, verify that each equation is an identity.

57. $-2 \csc(-\theta) - \dfrac{\sin \theta}{1 + \cos(-\theta)} = \dfrac{1 + \cos(-\theta)}{-\sin(-\theta)}$

58. $\dfrac{\sec(-t)}{\csc(-t)[\tan(-t) + \cot(-t)]} = \sin^2 t$

59. $\ln|\sec t + \tan t| = -\ln|\sec t - \tan t|$

60. $\dfrac{1 - \sin t \cos t}{\cos t(\sec t - \csc t)} \cdot \dfrac{\sin^2 t - \cos^2 t}{\sin^3 t + \cos^3 t} = \sin t$

61. **G** (a) Use a grapher to get two separate graphs for
$$y_1 = \cos^4 t - \sin^4 t$$
and
$$y_2 = \cos^2 t - \sin^2 t$$
for the same window settings.
(b) Compare the graphs.
(c) Is there an identity that relates y_1 and y_2? Explain.

62. **G** Use a grapher to help determine whether or not the equation
$$\ln(1 - \cos t) - 2 \ln|\sin t| = -\ln(1 + \cos t)$$
is an identity. If the equation appears to be an identity, then prove it.

Objectives

1. Use the Addition and Subtraction Cosine Identities
2. Use the Addition and Subtraction Sine Identities
3. Simplify Trigonometric Expressions and Verify Identities
4. Solve Applied Problems

6.2 Addition and Subtraction Identities

In Section 6.1 we worked with identities that express relationships among trigonometric functions of a *single variable.* In this section we develop trigonometric identities involving the sum or difference of *two variables* for

$$\cos(u - v), \quad \cos(u + v), \quad \sin(u - v), \quad \sin(u + v), \quad \text{and} \quad \tan(u - v), \quad \tan(u + v)$$

These identities are referred to as the *sum and difference identities.*

Using the Addition and Subtraction Cosine Identities

Suppose we want to express the value of $\cos(u + v)$ in terms of values of trigonometric functions of u and v. We might be *tempted* to say that

$$\cos(u + v) \text{ is the same as } \cos u + \cos v.$$

To find out whether this is true, we compare the values of the expressions

$$\cos(30° + 60°)$$

and

$$\cos 30° + \cos 60°.$$

Clearly, the results are different (see Table 1).

TABLE 1 $\cos(30° + 60°) \neq \cos 30° + \cos 60°$

$\cos(30° + 60°)$	$\cos 30° + \cos 60°$
$= \cos 90°$	$= \dfrac{\sqrt{3}}{2} + \dfrac{1}{2}$
$= 0$	$= \dfrac{\sqrt{3} + 1}{2}$

Thus, in general

$$\cos(u + v) \neq \cos u + \cos v.$$

The next result shows how it is possible to express $\cos(u + v)$ and $\cos(u - v)$ in terms of trigonometric functions of u and v.

Addition and Subtraction Identities for the Cosine

> (i) $\cos(u - v) = \cos u \cos v + \sin u \sin v$
> (ii) $\cos(u + v) = \cos u \cos v - \sin u \sin v$

The results are true for all possible values of u *and* v.

In words, the first identity states that

> *the cosine of the difference of two angles (or real numbers) equals the cosine of the first times the cosine of the second, plus the sine of the first times the sine of the second.*

The second identity can be worded in a similar way.

Proofs

(i) To prove identity (i), let u and v represent angles in standard position whose terminal sides intersect the unit circle at the points $P_1 = (x_1, y_1)$ and $P_2 = (x_2, y_2)$, respectively. Figure 1a illustrates one possible situation.
Next we place the angle $u - v$ in standard position and label the point of intersection of its terminal side and the unit circle as $P_3 = (x_3, y_3)$ (Figure 1b).

Figure 1

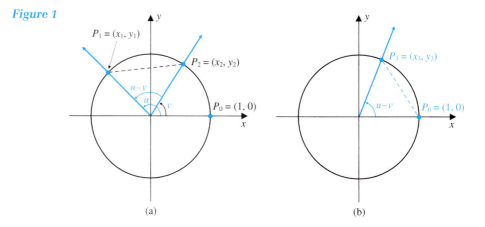

(a)   (b)

If $P_0 = (1, 0)$, then the length of arcs $\overparen{P_1 P_2}$ and $\overparen{P_0 P_3}$ are equal. It follows from geometry that the corresponding chord lengths are equal. Thus,

$$|\overline{P_0 P_3}| = |\overline{P_1 P_2}|$$

$$|\overline{P_0 P_3}|^2 = |\overline{P_1 P_2}|^2 \qquad \text{Square each side}$$

$$(x_3 - 1)^2 + y_3^2 = (x_1 - x_2)^2 + (y_1 - y_2)^2 \qquad \text{Distance formula}$$

$$(x_3^2 + y_3^2) + 1 - 2x_3 = (x_1^2 + y_1^2) + (x_2^2 + y_2^2) - 2x_1x_2 - 2y_1y_2 \qquad \left\{ \begin{array}{l} \text{Multiply and} \\ \text{rearrange} \end{array} \right.$$

$$1 + 1 - 2x_3 = 1 + 1 - 2x_1x_2 - 2y_1y_2 \qquad \left\{ \begin{array}{l} \text{Points on unit} \\ \text{circle satisfy} \\ x^2 + y^2 = 1 \end{array} \right.$$

$$-2x_3 = -2x_1x_2 - 2y_1y_2 \qquad \left\{ \begin{array}{l} \text{Subtract 2 from} \\ \text{each side} \end{array} \right.$$

[1] $$x_3 = x_1x_2 + y_1y_2 \qquad \left\{ \begin{array}{l} \text{Divide each} \\ \text{side by } -2 \end{array} \right.$$

However, we know from the definition of the trigonometric functions that

$$\cos u = x_1, \quad \cos v = x_2, \quad \cos(u - v) = x_3,$$

$$\sin u = y_1 \quad \text{and} \quad \sin v = y_2$$

Consequently, after substituting these latter expressions into the above equation labeled [1], we get

$$\cos(u - v) = \cos u \cos v + \sin u \sin v$$

(ii) To prove identity (ii), we proceed as follows:

$$\cos(u + v) = \cos[u - (-v)]$$

$$= \cos u \cos(-v) + \sin u \sin(-v) \qquad \text{Identity (i)}$$

$$= \cos u \cos v - \sin u \sin v \qquad \left\{ \begin{array}{l} \text{Cosine is even;} \\ \text{sine is odd} \end{array} \right.$$

Using the Addition and Subtraction Cosine Identities

In Examples 1 and 2 find the exact value of each expression by using the addition or subtraction cosine identities. Compare the result to the calculator value, rounded off to four decimal places.

EXAMPLE 1 $\cos 15°$

Solution Using $15° = 45° - 30°$ and the identity for $\cos(u - v)$, we get

$$\cos 15° = \cos(45° - 30°)$$

$$= \cos 45° \cos 30° + \sin 45° \sin 30°$$

$$= \frac{\sqrt{2}}{2} \cdot \frac{\sqrt{3}}{2} + \frac{\sqrt{2}}{2} \cdot \frac{1}{2}$$

$$= \frac{\sqrt{6}}{4} + \frac{\sqrt{2}}{4}$$

$$= \frac{\sqrt{6} + \sqrt{2}}{4}$$

Rounding off $(\sqrt{6} + \sqrt{2})/4$ to four decimal places gives 0.9659. Using a calculator to evaluate $\cos 15°$ and rounding off to four decimal places, we again get 0.9659.

EXAMPLE 2 $\cos \dfrac{7\pi}{12}$

Solution   Applying the identity for $\cos(u + v)$ and using

$$\frac{7\pi}{12} = \frac{\pi}{3} + \frac{\pi}{4},$$

we get

$$\cos \frac{7\pi}{12} = \cos\left(\frac{\pi}{3} + \frac{\pi}{4}\right)$$

$$= \cos \frac{\pi}{3} \cos \frac{\pi}{4} - \sin \frac{\pi}{3} \sin \frac{\pi}{4}$$

$$= \frac{1}{2} \cdot \frac{\sqrt{2}}{2} - \frac{\sqrt{3}}{2} \cdot \frac{\sqrt{2}}{2} = \frac{\sqrt{2}}{4} - \frac{\sqrt{6}}{4} = \frac{\sqrt{2} - \sqrt{6}}{4}$$

After rounding off $(\sqrt{2} - \sqrt{6})/4$ to four decimal places, we get -0.2588. Using a calculator to evaluate $\cos[7\pi/12]$ and rounding off to four decimal places, we get the same approximate value, -0.2588.

We can use the identity for $\cos(u - v)$ to establish some identities that relate the trigonometric functions to their corresponding *cofunctions*.

Cofunction Identities

If v is a real number or the radian measure of an angle, then:

(i) $\cos\left(\dfrac{\pi}{2} - v\right) = \sin v$ (ii) $\sin\left(\dfrac{\pi}{2} - v\right) = \cos v$

(iii) $\tan\left(\dfrac{\pi}{2} - v\right) = \cot v$ (iv) $\cot\left(\dfrac{\pi}{2} - v\right) = \tan v$

(v) $\sec\left(\dfrac{\pi}{2} - v\right) = \csc v$ (vi) $\csc\left(\dfrac{\pi}{2} - v\right) = \sec v$

Proof   (i) Using the subtraction identity for the cosine, we have

$$\cos\left(\frac{\pi}{2} - v\right) = \cos \frac{\pi}{2} \cos v + \sin \frac{\pi}{2} \sin v$$

$$= 0 + \sin v = \sin v.$$

This gives us the first identity.

(ii) If we replace v by $\pi/2 - v$ in the first identity, we get

$$\cos\left[\frac{\pi}{2} - \left(\frac{\pi}{2} - v\right)\right] = \sin\left(\frac{\pi}{2} - v\right)$$

$$\cos v = \sin\left(\frac{\pi}{2} - v\right)$$

that is, $\sin\left(\dfrac{\pi}{2} - v\right) = \cos v.$

(iii) We use the first two identities to obtain the cofunction identity for the tangent as follows:

$$\tan\left(\frac{\pi}{2} - v\right) = \frac{\sin\left(\dfrac{\pi}{2} - v\right)}{\cos\left(\dfrac{\pi}{2} - v\right)} = \frac{\cos v}{\sin v} = \cot v$$

The proofs of the remaining identities are similar (problem 45).

Naturally, if we measure an angle θ in degrees instead of radians, the co-function identities still hold. Thus

(i) $\cos(90° - \theta) = \sin \theta$
(ii) $\sin(90° - \theta) = \cos \theta$
(iii) $\tan(90° - \theta) = \cot \theta$
(iv) $\cot(90° - \theta) = \tan \theta$
(v) $\sec(90° - \theta) = \csc \theta$
(vi) $\csc(90° - \theta) = \sec \theta$

Using the Addition and Subtraction Sine Identities

Using the cofunction identities and the addition and subtraction identities for the cosine, we can derive identities for the sine.

Addition and Subtraction Identities for the Sine

(i) $\sin(u - v) = \sin u \cos v - \cos u \sin v$
(ii) $\sin(u + v) = \sin u \cos v + \cos u \sin v$

Proofs

(i) To prove the identity for $\sin(u - v)$, we use the cofunction identity

$$\cos\left(\frac{\pi}{2} - v\right) = \sin v.$$

Reading from right to left and replacing v with $u - v$, we get

$$\sin(u - v) = \cos\left[\frac{\pi}{2} - (u - v)\right]$$

$$= \cos\left(\frac{\pi}{2} - u + v\right)$$

$$= \cos\left[\left(\frac{\pi}{2} - u\right) + v\right].$$

So,

$$\sin(u - v) = \cos\left(\frac{\pi}{2} - u\right)\cos v - \sin\left(\frac{\pi}{2} - u\right)\sin v \qquad \begin{cases} \text{Addition identity} \\ \text{for the cosine} \end{cases}$$

$$= \sin u \cos v - \cos u \sin v \qquad \begin{cases} \text{Cofunction} \\ \text{identities} \\ \text{(i) and (ii)} \end{cases}$$

(ii) After replacing v by $-v$ in part (i), we obtain

$$\sin(u + v) = \sin[u - (-v)]$$

$$= \sin u \cos(-v) - \cos u \sin(-v) \qquad \text{Identity (i)}$$

$$= \sin u \cos v + \cos u \sin v \qquad \begin{cases} \text{Cosine is even;} \\ \text{sine is odd} \end{cases}$$

Note that by using the addition identity for the sine, we have

$$\sin\left(x + \frac{\pi}{2}\right) = \sin x \cos \frac{\pi}{2} + \cos x \sin \frac{\pi}{2}$$

$$= \sin x \cdot 0 + \cos x \cdot 1$$

$$= \cos x$$

that is, $\cos x = \sin\left(x + \dfrac{\pi}{2}\right)$.

Since $\cos x = \sin\left(x + \dfrac{\pi}{2}\right) = \sin\left[x - \left(-\dfrac{\pi}{2}\right)\right]$, it follows that the graph of $y = \cos x$ can be obtained by shifting the graph of $f(x) = \sin x$ horizontally $\pi/2$ unit to the left.

EXAMPLE 3   **Evaluating an Expression by Recognizing the Identity**

Find the exact value of the expression

$$\sin 81° \cos 21° - \cos 81° \sin 21°$$

by recognizing the applicable sine identity.

Solution   By recognizing the subtraction sine identity, from right to left, we have

$$\sin 81° \cos 21° - \cos 81° \sin 21° = \sin(81° - 21°)$$

$$= \sin 60°$$

$$= \frac{\sqrt{3}}{2}$$

EXAMPLE 4   **Finding the Exact Value of the Sine of a Difference**

Suppose that α and β are angles in standard position, where α is in quadrant I, $\cos \alpha = 4/5$, β is in quadrant II, and $\sin \beta = 3/5$.
   Use identities to determine the exact value of $\sin(\alpha - \beta)$.

Solution   Before we can apply the subtraction identity, we need to determine both $\sin \alpha$ and $\cos \beta$.
   Since the terminal side of α is in quadrant I, $\sin \alpha$ is positive, so

$$\sin \alpha = \sqrt{1 - \cos^2 \alpha}$$

$$= \sqrt{1 - (16/25)} = \frac{3}{5}.$$

Also, since the terminal side of β is in quadrant II, $\cos \beta$ is negative, so

$$\cos \beta = -\sqrt{1 - \sin^2 \beta}$$

$$= -\sqrt{1 - 9/25} = -\frac{4}{5}.$$

It follows that

$$\sin(\alpha - \beta) = \sin \alpha \cos \beta - \cos \alpha \sin \beta = \left(\frac{3}{5}\right)\left(-\frac{4}{5}\right) - \left(\frac{4}{5}\right)\left(\frac{3}{5}\right) = -\frac{24}{25}.$$

EXAMPLE 5   **Finding a Value Involving Inverse Functions**

Find the exact value of $\sin[\cos^{-1}(5/13) + \sin^{-1}(3/5)]$ by using the addition identity for the sine function and right triangle trigonometry.

Solution   First we let $\alpha = \cos^{-1}(5/13)$ and $\beta = \sin^{-1}(3/5)$, so $\cos \alpha = 5/13$, $\sin \beta = 3/5$, and α and β are acute angles.

Using right triangles, we see in Figure 2a that $\sin \alpha = 12/13$ and in Figure 2b that $\cos \beta = 4/5$.

Thus

Figure 2

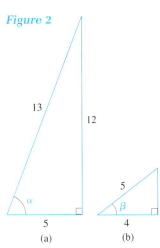

(a)   (b)

$$\sin\left(\cos^{-1}\frac{5}{13} + \sin^{-1}\frac{3}{5}\right) = \sin(\alpha + \beta)$$

$$= \sin \alpha \cos \beta + \cos \alpha \sin \beta$$

$$= \left(\frac{12}{13}\right)\left(\frac{4}{5}\right) + \left(\frac{5}{13}\right)\left(\frac{3}{5}\right)$$

$$= \frac{48}{65} + \frac{15}{65} = \frac{63}{65}$$

Using the addition and subtraction identities for the sine and cosine, we can derive addition and subtraction identities for the tangent function.

Addition and
Subtraction Identities
for the Tangent

> (i) $\tan(u - v) = \dfrac{\tan u - \tan v}{1 + \tan u \tan v}$ (ii) $\tan(u + v) = \dfrac{\tan u + \tan v}{1 - \tan u \tan v}$

Proof   (i) We prove the first identity as follows:

$$\tan(u - v) = \frac{\sin(u - v)}{\cos(u - v)}$$

$$= \frac{\sin u \cos v - \cos u \sin v}{\cos u \cos v + \sin u \sin v} \qquad \left\{\begin{array}{l}\text{Sine and cosine subtraction}\\ \text{identities}\end{array}\right.$$

$$= \frac{\dfrac{\sin u \cos v}{\cos u \cos v} - \dfrac{\cos u \sin v}{\cos u \cos v}}{\dfrac{\cos u \cos v}{\cos u \cos v} + \dfrac{\sin u \sin v}{\cos u \cos v}} \qquad \left\{\begin{array}{l}\text{Divide numerator and}\\ \text{denominator by } \cos u \cos v\end{array}\right.$$

$$= \frac{\tan u - \tan v}{1 + \tan u \tan v} \qquad \left\{\begin{array}{l}\text{Reduce fractions and use}\\ \tan t = \dfrac{\sin t}{\cos t}\end{array}\right.$$

The proof of part (ii) is left as an exercise (problem 46).

Care must be taken when applying the trigonometric identities. For example, if we replace u by $\pi/2$ in the subtraction identity for $\tan(u - v)$, we get

$$\tan\left(\frac{\pi}{2} - v\right) = \frac{\tan \dfrac{\pi}{2} - \tan v}{1 + \tan \dfrac{\pi}{2} \tan v}.$$

Since $\tan(\pi/2)$ is not defined, we might erroneously conclude that $\tan[(\pi/2) - v]$ is not defined for any value of v. However, we know from the cofunction identity that $\tan[(\pi/2) - v]$ can be written as

$$\tan\left(\frac{\pi}{2} - v\right) = \cot v.$$

This illustration emphasizes the important fact that the identities derived in trigonometry are applicable only when the values of the variables are in the domains of all the functions contained in the identities.

Simplifying Trigonometric Expressions and Verifying Identities

We can use the addition and subtraction identities to simplify certain trigonometric expressions and to verify other identities.

EXAMPLE 6   **Simplifying a Trigonometric Expression**

Write the expression $\sin 5t \cos 2t - \sin 2t \cos 5t$ in terms of the sine function.

Solution   We recognize that the expression fits the identity for $\sin(u - v)$, with $u = 5t$ and $v = 2t$. So

$$\sin 5t \cos 2t - \sin 2t \cos 5t = \sin(5t - 2t)$$

$$= \sin 3t.$$

In Section 5.4 we showed that the tangent function has a period π. In the next example, we confirm this fact by using an identity.

EXAMPLE 7   **Proving That the Tangent Function Has Period π**

Prove that $f(x) = \tan x$ has period π.

Solution   Using the addition identity for the tangent, we get

$$\tan(x + \pi) = \frac{\tan x + \tan \pi}{1 - \tan x \tan \pi}$$

$$= \frac{\tan x + 0}{1 - 0}$$

$$= \tan x.$$

Thus the tangent function has period π.

In some applications it is necessary to transform expressions from the form $P \cos u + Q \sin u$ to the form $A \cos(u - v)$.

EXAMPLE 8   **Writing $P\cos u + Q\sin u$ in the Form $A\cos(u - v)$**

Let P and Q be constants, and let v be an angle in standard position with the point (P, Q) on the terminal side of v.

(a) Show that if $A = \sqrt{P^2 + Q^2}$, then

$$P = A \cos v \quad \text{and} \quad Q = A \sin v.$$

(b) Use part (a) to show that

$$P \cos u + Q \sin u = A \cos(u - v)$$

for any angle u.

Solution

(a) Figure 3 shows angle v in standard position with the point (P, Q) on the terminal side. Notice that the distance A from the point $(0, 0)$ to (P, Q) is given by $A = \sqrt{P^2 + Q^2}$.

Using the cosine and sine definitions, we have

$$\cos v = \frac{P}{A} \quad \text{and} \quad \sin v = \frac{Q}{A}.$$

So

$$P = A \cos v \quad \text{and} \quad Q = A \sin v.$$

Figure 3

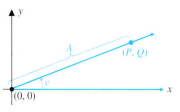

(b) Using the results from part (a), we get

$$P \cos u + Q \sin u = (A \cos v)(\cos u) + (A \sin v)(\sin u)$$

$$= A(\cos v \cos u + \sin v \sin u)$$

$$= A \cos(u - v)$$

for any angle u.

Solving Applied Problems

Applied problems from different fields such as physics, engineering, and meteorology are modeled by functions of the form

$$y = P \cos u + Q \sin u.$$

EXAMPLE 9

Modeling a Weather Pattern

The U.S. Weather Bureau conducted a study on the temperature fluctuation for a desert region. Using data collected over several years, it was determined that the average temperature T over one year (in degrees Fahrenheit) is approximated by the model

$$T = 80 - 12 \cos \frac{\pi t}{6} + 12 \sqrt{3} \sin \frac{\pi t}{6}$$

where t is the number of elapsed months, and $t = 0$ represents the month of March in each year.

(a) Write the equation representing this model in the form

$$T = 80 + A \cos(u - v).$$

(b) Find the maximum average temperature and in what month it occurs over a 12-month period starting in March.

(c) Find the minimum average temperature and in what month it occurs.

Solution   (a) We begin by rewriting the given function as

$$T = 80 + y \quad \text{where} \quad y = -12 \cos \frac{\pi t}{6} + 12\sqrt{3} \sin \frac{\pi t}{6}.$$

The expression for y is of the form

$$P \cos u + Q \sin u,$$

where $P = -12$, $Q = 12\sqrt{3}$, and $u = (\pi t)/6$.

Proceeding as we did in Example 7, we locate the point $(P, Q) = (-12, 12\sqrt{3})$ on the terminal side of an angle v in standard position (Figure 4).

Since

$$\tan v = \frac{Q}{P} = \frac{12\sqrt{3}}{-12} = -\sqrt{3}$$

and the point is in quadrant II, it follows that one possibility for angle v is $v = (2\pi)/3$.

Also

$$A = \sqrt{P^2 + Q^2}$$
$$= \sqrt{(-12)^2 + (12\sqrt{3})^2}$$
$$= 24$$

So

$$y = A \cos(u - v)$$
$$= 24 \cos\left(\frac{\pi t}{6} - \frac{2\pi}{3}\right)$$

Therefore

$$T = 80 + y$$
$$= 80 + 24 \cos\left(\frac{\pi t}{6} - \frac{2\pi}{3}\right)$$

(b) The maximum value of T occurs when the cosine value attains its maximum value of 1. So the maximum value of T is given by

$$T = 80 + 24(1) = 104°F$$

and it occurs when

$$\cos\left(\frac{\pi t}{6} - \frac{2\pi}{3}\right) = 1$$

that is, for the first time, when

$$\frac{\pi t}{6} - \frac{2\pi}{3} = 0$$

or

$$t = 4.$$

Thus the maximum average temperature occurs in July (when $t = 4$).

Figure 4

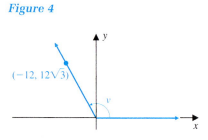

$(-12, 12\sqrt{3})$

(c) The minimum value of T occurs when the cosine value attains its minimum value of -1. This occurs when

$$\cos\left(\frac{\pi t}{6} - \frac{2\pi}{3}\right) = -1$$

that is, for the first time, when

$$\frac{\pi t}{6} - \frac{2\pi}{3} = \pi \quad \text{or} \quad \frac{\pi t}{6} = \frac{5\pi}{3}, \quad t = 10.$$

Thus the minimum value of T is given by

$$T = 80 + 24(-1)$$

$$= 56°F$$

and it occurs in the month of January (when $t = 10$).

G   Graphers are useful in performing *trend analysis.* For instance, in Example 8 we derived the result that the average temperature T (in degrees Fahrenheit) for a certain desert region is approximated by the model

$$T = 80 + 24 \cos\left(\frac{\pi t}{6} - \frac{2\pi}{3}\right)$$

where t is the number of elapsed months, and $t = 0$ represents the month of March.

A viewing window of the graph of T is shown in Figure 5 for $0 \le t \le 12$. By using the TRACE -feature, we find that starting in March ($t = 0$) the average temperature first exceeds 100°F when t is approximately 2.9 and then falls below 100°F when t is approximately 5.1. This means that the average temperature is above 100°F from about the end of May ($t = 2$) through the beginning of August ($t = 5$).

Figure 5

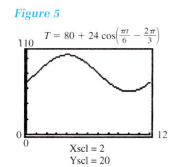

$T = 80 + 24 \cos\left(\frac{\pi t}{6} - \frac{2\pi}{3}\right)$

Xscl = 2
Yscl = 20

PROBLEM SET 6.2

Mastering the Concepts

In problems 1 and 2, use an appropriate addition or subtraction identity to find the exact value of each expression. In each case compare this value to the calculator value (rounding off to four decimal places).

1. (a) $\sin\left(\frac{\pi}{6} + \frac{3\pi}{4}\right)$ **2.** (a) $\cos\left(\frac{5\pi}{4} - \frac{\pi}{3}\right)$

   (b) $\cos(330° + 45°)$ (b) $\tan(60° - 45°)$

In problems 3 and 4, find the exact value of each expression by using an appropriate addition or subtraction identity and $(5\pi)/12 = (\pi/6) + (\pi/4)$ or $165° = 210° - 45°$. In each case compare this value to the calculator value (rounding off to four decimal places).

3. (a) $\sin\dfrac{5\pi}{12}$ **4.** (a) $\cos\dfrac{5\pi}{12}$

   (b) $\cos 165°$ (b) $\sin 165°$

   (c) $\tan\dfrac{5\pi}{12}$ (c) $\tan 165°$

In problems 5 and 6, use an appropriate addition or subtraction identity to find the exact value of each expression. In each case compare this value to the calculator value (rounding off to four decimal places).

5. (a) $\cos 105°$ **6.** (a) $\tan\dfrac{13\pi}{12}$

   (b) $\sin\dfrac{19\pi}{12}$ (b) $\cos 285°$

In problems 7 and 8, find the exact value of each expression by recognizing the applicable identity. Also, use a calculator to evaluate each expression to four decimal places. Compare the results in both approaches.

7. (a) $\sin 33° \cos 27° + \sin 27° \cos 33°$

(b) $\dfrac{\tan \dfrac{4\pi}{5} - \tan \dfrac{3\pi}{10}}{1 + \tan \dfrac{4\pi}{5} \tan \dfrac{3\pi}{10}}$

8. (a) $\cos \dfrac{5\pi}{7} \sin \dfrac{2\pi}{7} + \cos \dfrac{2\pi}{7} \sin \dfrac{5\pi}{7}$

(b) $\dfrac{\tan 17° + \tan 43°}{1 - \tan 17° \tan 43°}$

In problems 9–12, suppose that u and v are angles in standard position. Use the given information to find the value of each expression. Round off the answers in problems 11 and 12 to three decimal places.

(a) $\sin(u + v)$ (b) $\cos(u + v)$

(c) $\sin(u - v)$ (d) $\cos(u - v)$

(e) $\tan(u + v)$ (f) $\tan(u - v)$

9. $\sin u = \dfrac{12}{13}$, u is in quadrant II, $\cos v = \dfrac{-4}{5}$, and v is in quadrant II

10. $\sin u = \dfrac{3}{5}$, u is in quadrant II, $\cos v = \dfrac{3}{5}$, and v is in quadrant IV

11. $\cos u = 0.47$, u is in quadrant IV, $\sin v = -0.96$, and v is in quadrant III

12. $\tan u = 0.42$, u is in quadrant I, $\tan v = 0.59$, and v is in quadrant III

In problems 13–16, find the exact value of each expression by using the addition and subtraction identities and right triangle trigonometry.

13. $\cos\left(\sin^{-1} \dfrac{3}{5} - \cos^{-1} \dfrac{4}{5}\right)$

14. $\sin\left[\sin^{-1}\left(-\dfrac{3}{5}\right) - \cos^{-1} \dfrac{4}{5}\right]$

15. $\sin\left[\cos^{-1} \dfrac{5}{13} + \sin^{-1}\left(-\dfrac{12}{13}\right)\right]$

16. $\cos\left(\sin^{-1} \dfrac{8}{17} - \cos^{-1} \dfrac{5}{17}\right)$

In problems 17–26, rewrite each expression in terms of the sine, cosine, or tangent by using the addition and subtraction identities. Simplify each result.

17. (a) $\sin\left(\dfrac{5\pi}{2} - t\right)$ **18.** (a) $\cos\left(\dfrac{\pi}{6} - t\right)$

(b) $\cos(270° + t)$ (b) $\sin(\theta - 45°)$

19. (a) $\tan\left(t + \dfrac{\pi}{4}\right)$ **20.** (a) $\sin(\pi - t)$

(b) $\sin(45° - \theta)$ (b) $\tan(\theta + 60°)$

21. (a) $\csc(90° + \theta)$ **22.** (a) $\cot(10\pi - t)$

(b) $\sin\left(\dfrac{3\pi}{2} - t\right)$ (b) $\tan(1080° - \theta)$

23. (a) $\cos 7t \cos t - \sin 7t \sin t$

(b) $\sin 7t \cos 3t - \cos 7t \sin 3t$

24. (a) $\sin(-t) \cos 4t - \cos(-t) \sin 4t$

(b) $\cos 3\theta \cos 2\theta + \sin 3\theta \sin 2\theta$

25. (a) $\cos \dfrac{2x}{3} \cos \dfrac{x}{3} - \sin \dfrac{2x}{3} \sin \dfrac{x}{3}$

(b) $\dfrac{\tan 4t + \tan 3t}{1 - \tan 4t \tan 3t}$

26. (a) $\sin(-5t) \cos 2t - \cos(-5t) \sin 2t$

(b) $\dfrac{\tan 5t - \tan 2t}{1 + \tan 5t \tan 2t}$

In problems 27–44, verify that the equation is an identity.

27. $\sin(t - 3\pi) = -\sin t$

28. $\cos(t - 450°) = \sin t$

29. $\sin(30° + t) = \dfrac{1}{2} \cos t + \dfrac{\sqrt{3}}{2} \sin t$

30. $\sin(60° + t) - \cos(30° + t) = \sin t$

31. $\tan(45° + t) = \dfrac{1 + \tan t}{1 - \tan t}$

32. $\sin(\pi - t) - \cot t \sin\left(t - \dfrac{\pi}{2}\right) = \csc t$

33. $\sin\left(\dfrac{\pi}{4} + t\right) - \sin\left(\dfrac{\pi}{4} - t\right) = \sqrt{2} \sin t$

34. $\cos\left(\dfrac{\pi}{6} + t\right) \cos\left(\dfrac{\pi}{6} - t\right)$

$- \sin\left(\dfrac{\pi}{6} + t\right) \sin\left(\dfrac{\pi}{6} - t\right) = \dfrac{1}{2}$

35. $\sin(t + s) \sin(t - s) = \sin^2 t - \sin^2 s$

36. $\sin t \cos s - \sin\left(t + \dfrac{\pi}{2}\right) \sin(-s) = \sin(s + t)$

37. $\cos(s + t) \cos(s - t) = \cos^2 s + \cos^2 t - 1$

38. $\tan s - \tan t = \dfrac{\sin(s - t)}{\cos s \cos t}$

39. $\cot s - \tan t = \dfrac{\cos(s + t)}{\sin s \cos t}$

40. $\dfrac{1}{\tan s + \tan t} = \dfrac{\csc(s + t)}{\sec s \sec t}$

41. $\dfrac{\cos(u + v)}{\cos(u - v)} = \dfrac{1 - \tan u \tan v}{1 + \tan u \tan v}$

42. $\sin(\pi - s - t) = \sin s \cos t + \cos s \sin t$

43. $\dfrac{\sin(s + t)}{\sin(s - t)} = \dfrac{\tan s + \tan t}{\tan s - \tan t}$

44. $\cot(u + v) = \dfrac{\cot u \cot v - 1}{\cot u + \cot v}$

45. Verify that each equation is an identity.

(a) $\cot\left(\dfrac{\pi}{2} - v\right) = \tan v$

(b) $\sec\left(\dfrac{\pi}{2} - v\right) = \csc v$

(c) $\csc\left(\dfrac{\pi}{2} - v\right) = \sec v$

46. Use the identity for

$$\tan(u - v)$$

to derive the identity for

$$\tan(u + v).$$

In problems 47–50, express each function in the form

$$y = A \cos(u - v)$$

where v is the smallest positive angle. Show that the function obtained is equal to the given function.

47. $y = \sin t + \cos t$

48. $y = \sqrt{3} \cos \pi t + \sin \pi t$

49. $y = \cos 2\pi t - \sqrt{3} \sin 2\pi t$

50. $y = \cos 4t + \sqrt{3} \sin 4t$

Applying the Concepts

51. Height of a Tide: In a certain harbor, the mathematical model

$$h(t) = 0.3 \cos \dfrac{\pi t}{6} + 0.4 \sin \dfrac{\pi t}{6}$$

is used to predict the height h (in meters) of the tide above or below mean sea level t hours after midnight, over a 12-hour period.

(a) Write the equation of this model in the form

$$h(t) = A \cos(u - v).$$

Round off v to three decimal places.

(b) Find the maximum height and when it occurs. Round off the answer to one decimal place.

(c) Find the minimum height and indicate when it occurs. Round off the answer to one decimal place.

52. [G] (Refer to problem 51.)

(a) Use a grapher to graph the function found in problems 51(a) for $0 \le t \le 12$ and interpret it.

(b) Use a grapher to find when the height of the tide is more than 0.3 meter above mean sea level. Round off the answer to two decimal places.

53. Physics: An object attached to a spring vibrates vertically according to the mathematical model

$$d(t) = 4 \cos 4t + 3 \sin 4t$$

where $d(t)$ is the distance of the object from its rest position measured in centimeters, t seconds after the start of the motion.

(a) Write the equation of this model in the form

$$d(t) = A \cos(u - v).$$

Round off v to three decimal places.

(b) Determine the amplitude of the vibration of the object.

(c) Determine the period of the vibration of the object.

54. [G] (Refer to Problem 53.)

(a) Use a grapher to graph the function found in problem 52(a) and interpret it for $0 \le t \le 10$.

(b) Use a grapher to find when the object is more than 2 centimeters from its rest position. Round off the answer to two decimal places.

Developing and Extending the Concepts

55. Consider the triangle ABC in Figure 6.

(a) Express $\sin(\alpha + \beta)$ in terms of a trigonometric function of γ.

(b) Express $\cos(\alpha + \beta)$ in terms of a trigonometric function of γ.

Figure 6

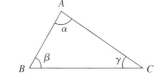

56. Figure 7 shows intersecting lines L_1 and L_2 with slopes m_1 and m_2, respectively.

(a) Show that $\tan \theta_1 = m_1$ and

$$\tan \theta_2 = m_2.$$

(b) Show that the angle $\theta = \theta_2 - \theta_1$ between L_1 and L_2 is given by

$$\tan \theta = \dfrac{m_2 - m_1}{1 + m_1 m_2}.$$

(c) Use part (b) to show that if L_1 and L_2 are perpendicular, then $m_1 m_2 = -1$.

Figure 7

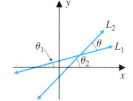

57. (a) Show that

$$\cos\left(\frac{\pi}{2} - x\right) = \cos\left(x - \frac{\pi}{2}\right).$$

(b) Explain why

$$\sin x = \cos\left(x - \frac{\pi}{2}\right).$$

(c) In view of the identity in part (b), compare the graphs of

$$f(x) = \sin x \quad \text{and} \quad g(x) = \cos\left(x - \frac{\pi}{2}\right).$$

How can the graph of $y = \cos x$ be used to obtain the graph of $f(x) = \sin x$?

58. (a) Find the value of the expression

$$\sin 1° + \sin 2° + \sin 3° + \ldots$$
$$+ \sin 357° + \sin 358° + \sin 359°$$

by using the identity

$$\sin \theta + \sin(360° - \theta) = \sin \theta - \sin \theta = 0.$$

(b) Use the identity

$$\cos \theta + \cos(360° - \theta) = 2 \cos \theta$$

to evaluate the expression

$$\cos 1° + \cos 2° + \cos 3° + \ldots$$
$$+ \cos 357° + \cos 358° + \cos 359°.$$

59. **G** Let $f(x) = \sin x$.
(a) Show that

$$\frac{f(t + h) - f(t)}{h} = \sin t \left(\frac{\cos h - 1}{h}\right) + \cos t \left(\frac{\sin h}{h}\right).$$

(b) Use a grapher to approximate the value of the expressions

$$\frac{\cos h - 1}{h} \quad \text{as} \quad h \rightarrow 0$$

and

$$\frac{\sin h}{h}$$
$$\text{as} \quad h \rightarrow 0$$

(c) Explain why

$$\frac{f(t + h) - f(t)}{h} \rightarrow \cos t \quad \text{as} \quad h \rightarrow 0$$

60. Let $g(x) = \cos x$.
(a) Show that

$$\frac{g(t + h) - g(t)}{h} = \cos t \left(\frac{\cos h - 1}{h}\right) - \sin t \left(\frac{\sin h}{h}\right)$$

(b) Use part (b) of problem 59 to show that

$$\frac{g(t + h) - g(t)}{h} \rightarrow -\sin t \quad \text{as} \quad h \rightarrow 0$$

Objectives

1. Use Double-Angle Identities
2. Use Half-Angle Identities
3. Solve Applied Problems

6.3 Double-Angle and Half-Angle Identities

In this section we'll study identities that express trigonometric functions of twice an angle and half an angle in terms of trigonometric functions of the angle. These identities follow from the addition identities. They are used in calculus.

Using Double-Angle Identities

Suppose that we want to express the value of $\sin 2\theta$ in terms of trigonometric functions of θ. We might be tempted to say that $\sin 2\theta = 2 \sin \theta$. However, this is *not* correct.

For example, assume $\theta = 30°$, then

$$\sin 2\theta = \sin[2(30°)] = \sin 60° = \frac{\sqrt{3}}{2}$$

whereas

$$2 \sin \theta = 2 \sin 30° = 2 \cdot \frac{1}{2} = 1.$$

So, in general

$$\sin 2\theta \neq 2 \sin \theta.$$

The identities that express the trigonometric functions of 2θ in terms of functions of θ are called *double-angle identities*.

Double-Angle Identities

(i) $\sin 2\theta = 2 \sin \theta \cos \theta$	(ii) $\cos 2\theta = \cos^2 \theta - \sin^2 \theta$
	$= 2 \cos^2 \theta - 1$
(iii) $\tan 2\theta = \dfrac{2 \tan \theta}{1 - \tan^2 \theta}$	$= 1 - 2 \sin^2 \theta$

Proofs

(i) To prove the first identity, we start with the addition identity for $\sin(u + v)$ and replace both u and v by θ throughout to get

$$\sin(\theta + \theta) = \sin \theta \cos \theta + \cos \theta \sin \theta$$
$$\sin 2\theta = 2 \sin \theta \cos \theta.$$

(ii) To prove the second identity, we use the identity for $\cos(u + v)$ and replace both u and v by θ to obtain

$$\cos(\theta + \theta) = \cos \theta \cos \theta - \sin \theta \sin \theta$$
$$\cos 2\theta = \cos^2 \theta - \sin^2 \theta$$

To obtain the second form for $\cos 2\theta$, we substitute the Pythagorean identity

$$\sin^2 \theta = 1 - \cos^2 \theta$$

into $\cos 2\theta = \cos^2 \theta - \sin^2 \theta$ to get

$$\cos 2\theta = \cos^2 \theta - (1 - \cos^2 \theta)$$
$$= \cos^2 \theta - 1 + \cos^2 \theta$$
$$= 2 \cos^2 \theta - 1.$$

Using the alternate form of the Pythagorean identity, $\cos^2 \theta = 1 - \sin^2 \theta$, we have

$$\cos 2\theta = \cos^2 \theta - \sin^2 \theta$$
$$= (1 - \sin^2 \theta) - \sin^2 \theta$$
$$= 1 - 2 \sin^2 \theta$$

which is the third form of the identity.

(iii) The identity for tan 2θ is obtained by replacing both u and v by θ in the addition identity for $\tan(u + v)$ to get

$$\tan(\theta + \theta) = \frac{\tan\theta + \tan\theta}{1 - \tan\theta\tan\theta}$$

$$\tan 2\theta = \frac{2\tan\theta}{1 - \tan^2\theta}$$

EXAMPLE 1   **Using Double-Angle Identities to Evaluate**

If $\sin\theta = 12/13$ and θ is an angle in quadrant I, find the exact value of each expression.

(a) $\sin 2\theta$ (b) $\cos 2\theta$ (c) $\tan 2\theta$

Solution   First we need to find $\cos\theta$. Because θ is in quadrant I, we know that $\cos\theta$ is positive, so

$$\cos\theta = \sqrt{1 - \sin^2\theta} = \sqrt{1 - (12/13)^2} = \sqrt{25/169} = 5/13.$$

Thus we get the following results:

(a) $\sin 2\theta = 2\sin\theta\cos\theta = 2\left(\dfrac{12}{13}\right)\left(\dfrac{5}{13}\right) = \dfrac{120}{169}$

(b) $\cos 2\theta = \cos^2\theta - \sin^2\theta = \left(\dfrac{5}{13}\right)^2 - \left(\dfrac{12}{13}\right)^2 = \dfrac{25}{169} - \dfrac{144}{169} = -\dfrac{119}{169}$

(c) Since

$$\tan\theta = \frac{\sin\theta}{\cos\theta} = \frac{\left(\dfrac{12}{13}\right)}{\left(\dfrac{5}{13}\right)} = \frac{12}{5}$$

it follows that

$$\tan 2\theta = \frac{2\tan\theta}{1 - \tan^2\theta} = \frac{2\left(\dfrac{12}{5}\right)}{1 - \left(\dfrac{12}{5}\right)^2} = \frac{\left(\dfrac{24}{5}\right)}{1 - \left(\dfrac{144}{25}\right)} = \frac{120}{25 - 144} = -\frac{120}{119}.$$

A more efficient way of finding tan 2θ after finding sin 2θ and cos 2θ is to proceed as follows:

$$\tan 2\theta = \frac{\sin 2\theta}{\cos 2\theta} = \frac{\left(\dfrac{120}{169}\right)}{\left(-\dfrac{119}{169}\right)} = -\frac{120}{119}.$$

EXAMPLE 2   **Rewriting a Trigonometric Expression**

Use a double-angle identity to rewrite the expression $\cos^2 3t - \sin^2 3t$ as a single trigonometric function.

Solution   We use the double-angle identity $\cos 2\theta = \cos^2 \theta - \sin^2 \theta$ by replacing θ with $3t$ to get

$$\cos^2 3t - \sin^2 3t = \cos[2(3t)] = \cos 6t.$$

EXAMPLE 3   **Finding a Value Involving Inverse Functions**

Find the exact value of $\sin\left[2 \cos^{-1}(8/17)\right]$. Compare this value to the calculator value, rounded to four decimal places.

Solution   First we let $\theta = \cos^{-1}(8/17)$, so $\cos \theta = 8/17$.

Next we draw a right triangle with an acute angle θ whose cosine is $8/17$ (Figure 1). By the pythogorean theorem, we get $b = 15$.

From right triangle trigonometry, it follows that

$$\sin \theta = 15/17.$$

Figure 1

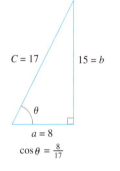

$C = 17$ $15 = b$

θ

$a = 8$

$\cos \theta = \frac{8}{17}$

Thus

$$\sin\left(2 \cos^{-1}\frac{8}{17}\right) = \sin 2\theta$$

$$= 2 \sin \theta \cos \theta$$

$$= 2\left(\frac{15}{17}\right)\left(\frac{8}{17}\right)$$

$$= \frac{240}{289}\ \text{(approx. 0.8304).}$$

The calculator value of the expression $\sin\left[2 \cos^{-1}(8/17)\right]$ also equals 0.8304, rounded to four decimal places.

EXAMPLE 4   **Verifying an Identity Using Double-Angle Identities**

Verify the identity: $\dfrac{2 \cos 2t}{\sin 2t - 2 \sin^2 t} - 1 = \cot t$.

Solution   To verify that the equation is an identity, we proceed as follows:

$$\frac{2 \cos 2t}{\sin 2t - 2 \sin^2 t} - 1 = \frac{2(\cos^2 t - \sin^2 t)}{2 \sin t \cos t - 2 \sin^2 t} - 1 \qquad \begin{cases} \cos 2t\ \text{and} \\ \sin 2t\ \text{identities} \end{cases}$$

$$= \frac{2\,(\cos t - \sin t)(\cos t + \sin t)}{2\,(\sin t)(\cos t - \sin t)} - 1 \qquad \text{Factor}$$

$$= \frac{\cos t + \sin t}{\sin t} - 1 \qquad \text{Reduce}$$

$$= \frac{\cos t}{\sin t} + \frac{\sin t}{\sin t} - 1 \qquad \frac{A + B}{C} = \frac{A}{C} + \frac{B}{C}$$

$$= \cot t + 1 - 1 = \cot t$$

Therefore, the equation is an identity.

Using Half-Angle Identities

Consider the double-angle identities

$$\cos 2t = 2\cos^2 t - 1 \quad \text{and} \quad \cos 2t = 1 - 2\sin^2 t.$$

If we solve the first for $\cos^2 t$ and the second for $\sin^2 t$ in terms of $\cos 2t$, we get the following identities:

Square Identities

$$\text{(i)} \ \cos^2 t = \frac{1 + \cos 2t}{2} \qquad \text{(ii)} \ \sin^2 t = \frac{1 - \cos 2t}{2}$$

For example, if we replace t with $3x$ in identity (i), we get

$$\cos^2 3x = \frac{1 + \cos 6x}{2}.$$

Similarly, if we replace t with $5x$ in identity (ii), we obtain

$$\sin^2 5x = \frac{1 - \cos 10x}{2}.$$

By replacing t with $\theta/2$ in the square identities, we obtain the equivalent forms

$$\text{(i)} \ \cos^2 \frac{\theta}{2} = \frac{1 + \cos \theta}{2} \quad \text{and} \quad \text{(ii)} \ \sin^2 \frac{\theta}{2} = \frac{1 - \cos \theta}{2}.$$

Next we take the square roots of both sides of the latter identities to get the following results:

Half-Angle Identities

$$\text{(i)} \ \cos \frac{\theta}{2} = \pm\sqrt{\frac{1 + \cos \theta}{2}} \quad \text{and} \quad \text{(ii)} \ \sin \frac{\theta}{2} = \pm\sqrt{\frac{1 - \cos \theta}{2}}.$$

The sign depends on the quadrant in which the terminal side of the angle (in standard position) with measure $\theta/2$ lies.

These two identities can be used to derive the identity

$$\text{(iii)} \ \tan \frac{\theta}{2} = \frac{1 - \cos \theta}{\sin \theta}$$

EXAMPLE 5 **Using a Half-Angle Identity to Evaluate**

Find the exact value of $\cos(\pi/12)$ by using the half-angle cosine identity. Compare this value to the calculator value, rounded to four decimal places.

Solution Using half-angle identity (i) for the cosine and the fact that $\cos(\pi/12)$ is positive, we get

$$\cos \frac{\pi}{12} = \cos\left[\frac{1}{2}\left(\frac{\pi}{6}\right)\right] = \sqrt{\frac{1 + \cos(\pi/6)}{2}} = \sqrt{\frac{1 + (\sqrt{3}/2)}{2}}$$

$$= \sqrt{\frac{2 + \sqrt{3}}{4}} = \frac{\sqrt{2 + \sqrt{3}}}{2} \ (\text{approx. } 0.9659).$$

The calculator value of $\cos(\pi/12)$ also equals 0.9659, to four decimal places.

EXAMPLE 6   **Evaluating Functions Involving Half-Angle Identities**

Suppose that θ is a positive obtuse angle in standard position, measured in degrees, and $\sin \theta = 3/5$. Find the exact value of each expression.

(a) $\sin \dfrac{\theta}{2}$ (b) $\cos \dfrac{\theta}{2}$ (c) $\tan \dfrac{\theta}{2}$

Solution   Since θ is obtuse, it is in quadrant II, where $\cos \theta$ is negative. So

$$\cos \theta = -\sqrt{1 - \sin^2 \theta} = -\sqrt{1 - \left(\frac{3}{5}\right)^2} = -\sqrt{\frac{16}{25}} = -\frac{4}{5}$$

Also, because $90° < \theta < 180°$, we have $45° < \theta/2 < 90°$, so $\theta/2$ is in quadrant I. Thus $\sin(\theta/2)$ and $\cos(\theta/2)$ are both positive, and we have:

(a) $\sin \dfrac{\theta}{2} = \sqrt{\dfrac{1 - \cos \theta}{2}} = \sqrt{\dfrac{1 - (-\frac{4}{5})}{2}} = \sqrt{\dfrac{9}{10}} = \dfrac{3}{\sqrt{10}} = \dfrac{3\sqrt{10}}{10}$

(b) $\cos \dfrac{\theta}{2} = \sqrt{\dfrac{1 + \cos \theta}{2}} = \sqrt{\dfrac{1 + (-\frac{4}{5})}{2}} = \sqrt{\dfrac{1}{10}} = \dfrac{1}{\sqrt{10}} = \dfrac{\sqrt{10}}{10}$

(c) $\tan \dfrac{\theta}{2} = \dfrac{\sin(\theta/2)}{\cos(\theta/2)} = \dfrac{(3\sqrt{10})/10}{\sqrt{10}/10} = 3$

Solving Applied Problems

Some applied problems can be modeled by trigonometric functions of multiple angles as the next example shows.

EXAMPLE 7   **Modeling a Balloon Sighting**

Suppose that an observer at point P, 111 meters away from point Q, directly beneath a hot air balloon at T, sights the balloon at an angle of elevation θ (Figure 2). At the same time, a second observer at point R, which is 74 meters closer to point Q, sights the same balloon at an angle of elevation 2θ. If the points P, R, and Q are in the same plane, find the height h of the balloon. Round off the answer to one decimal place.

Solution   Using the right triangles QPT and QRT in Figure 2, we have

$$\tan 2\theta = \frac{h}{37} \quad \text{and} \quad \tan \theta = \frac{h}{111}.$$

Figure 2

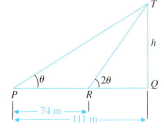

Using $\tan 2\theta = \dfrac{2 \tan \theta}{1 - \tan^2 \theta}$, we have $\dfrac{h}{37} = \dfrac{\left(\dfrac{2h}{111}\right)}{1 - \dfrac{h^2}{(111)^2}}$ Substitute for tangents

so that $1 - \dfrac{h^2}{(111)^2} = \dfrac{\left(\dfrac{2h}{111}\right)}{\left(\dfrac{h}{37}\right)} = \dfrac{2}{3} \dfrac{h^2}{(111)^2} = 1 - \dfrac{2}{3} = \dfrac{1}{3}$

$$3h^2 = (111)^2 \quad h^2 = \dfrac{\overset{37}{\cancel{(111)}}(111)}{3} \quad h = \sqrt{(37)(111)} = 37\sqrt{3}$$

Therefore, the height h, rounded off to one decimal place, is 64.1 meters.

EXAMPLE 8   **Maximizing the Distance of a Football**

A place kicker determines that if he kicks the ball at a specified angle θ he can achieve the maximum distance. If the ball is kicked at an initial velocity V_0, then the distance R (in feet) (Figure 3) is given by the model

$$R = \frac{V^2}{32} \sin 2\theta, \quad 0° \leq \theta \leq 45°.$$

(a) Find the distance R if $\theta = 37°$ and $V = 70$ feet per second. Round off the answer to two decimal places.

(b) If the football is kicked with an initial velocity of $V = 77$ feet per second, determine the value of θ that maximizes R and find the maximum value of R. Round off the answers to two decimal places.

Figure 3

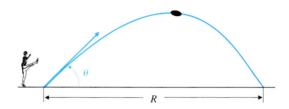

Solution   (a) Replacing V by 70 and θ by 37° in the above formula we have

$$R = \frac{(70)^2}{32} \sin 74°$$
$$= 147.19.$$

So the distance R when the ball is kicked at 37° with an initial velocity of 70 feet per second is approximately 147.19 feet or about 49.06 yards.

(b) Replacing $V = 77$ in the above equation and rounding off to two decimal places we have

$$R = 185.28 \sin 2\theta.$$

Since the amplitude of $y = \sin 2\theta$ is 1, it follows that the maximum value of

$$\sin 2\theta \text{ is } 1.$$

Consequently, the maximum value of R is attained when $\sin 2\theta = 1$, that is, when

$$2\theta = 90°$$

or

$$\theta = 45°.$$

At $\theta = 45°$, $R = 185.28 \cdot \sin 90° = 185.28$.

So the maximum value of R, about 185.28 feet or 61.76 yards, is attained if the ball is kicked at 45°.

◆ PROBLEM SET 6.3

Mastering the Concepts

1. Evaluate $\cos 2t$ and $2 \cos t$ for $t = \pi/4$. What conclusion can be made about $\cos 2t$ and $2 \cos t$?
2. Evaluate $\tan 2t$ and $2 \tan t$ for $t = \pi/4$. What conclusion can be made about $\tan 2t$ and $2 \tan t$?

In problems 3–10, use the given information to find the exact value of each expression.

(a) $\sin 2t$

(b) $\cos 2t$

(c) $\tan 2t$

3. $\sin t = \dfrac{4}{5};$ t in quadrant I

4. $\cos t = -\dfrac{12}{13};$ t in quadrant III

5. $\cos t = -\dfrac{7}{25};$ t in quadrant III

6. $\sin t = -\dfrac{5}{13};$ t in quadrant IV

7. $\tan t = -\dfrac{5}{12};$ t in quadrant II

8. $\sec t = -\dfrac{5}{2};$ t in quadrant III

9. $\cot t = -\dfrac{7}{25};$ t in quadrant IV

10. $\tan t = \dfrac{18}{5};$ t in quadrant I

In problems 11–18, use the double-angle identities to rewrite each expression in terms of a single trigonometric function.

11. $2 \sin 3t \cos 3t$

12. $1 - 2 \sin^2 3\pi x$

13. $2 \cos^2 8t - 1$

14. $4 \sin^2 7\theta \cos^2 7\theta$

15. $\cos^2 5t - \sin^2 5t$

16. $2 - 4 \sin^2 \dfrac{t}{2}$

17. $\dfrac{2 \tan 4\theta}{1 - \tan^2 4\theta}$

18. $\dfrac{\tan(t/2)}{\dfrac{1}{2} - \dfrac{1}{2} \tan^2(t/2)}$

In problems 19 and 20, use an identity and right triangle trigonometry to find the exact value of each expression. Compare each value to the calculator value, rounded to four decimal places.

19. (a) $\sin\left(2 \sin^{-1} \dfrac{1}{2}\right)$

(b) $\sin\left(2 \cos^{-1} \dfrac{\sqrt{3}}{2}\right)$

(c) $\tan\left(2 \tan^{-1} \dfrac{4}{3}\right)$

20. (a) $\cos(2 \sin^{-1} 1)$

(b) $\cos\left(2 \sin^{-1} \dfrac{\sqrt{2}}{2}\right)$

(c) $\tan\left[2 \cos^{-1}\left(-\dfrac{24}{25}\right)\right]$

In problems 21–26, use the half-angle identities to find the exact value of each expression. Compare this value to the calculator value, rounded to four decimal places.

21. (a) $\cos 22.5°$
    (b) $\sin \dfrac{5\pi}{12}$

22. (a) $\sin 75°$
    (b) $\cos \dfrac{\pi}{8}$

23. (a) $\cos \dfrac{5\pi}{12}$
    (b) $\tan \dfrac{3\pi}{8}$

24. (a) $\sin\left(-\dfrac{7\pi}{12}\right)$
    (b) $\tan \dfrac{5\pi}{8}$

25. (a) $\cos 112.5°$
    (b) $\sin 67.5°$

26. (a) $\tan \dfrac{7\pi}{12}$
    (b) $\cos 67.5°$

In problems 27–34, use the given information to find the exact values of each expression if $0 < t < 2\pi$.

(a) $\sin \dfrac{t}{2}$

(b) $\cos \dfrac{t}{2}$

(c) $\tan \dfrac{t}{2}$

27. $\sin t = \dfrac{24}{25};$ t in quadrant I

28. $\cos t = \dfrac{7}{25};$ t in quadrant IV

29. $\cos t = -\dfrac{4}{5}$; t in quadrant II

30. $\sin t = -\dfrac{5}{13}$; t in quadrant III

31. $\cot t = \dfrac{8}{15}$; t in quadrant III

32. $\sec t = \dfrac{13}{5}$; t in quadrant I

33. $\tan t = -\dfrac{8}{15}$; t in quadrant IV

34. $\cot t = \dfrac{3}{4}$; t in quadrant III

In problems 35 and 36, use the half-angle identities to rewrite each expression as a single trigonometric function.

35. (a) $\sqrt{\dfrac{1 + \cos 4t}{2}}$ **36.** (a) $\sqrt{\dfrac{1 - \cos 2\pi t}{2}}$

     (b) $\sqrt{\dfrac{1 - \cos 6t}{2}}$ (b) $-\sqrt{\dfrac{1 - \cos 2t}{1 + \cos 2t}}$

In problems 37–48, verify that the equation is an identity.

37. (a) $\sin^2 t + \cos 2t = \cos^2 t$
     (b) $\csc t \sec t = 2 \csc 2t$

38. (a) $\dfrac{\sin 2t}{2 \sin t} = \cos t$
     (b) $\cos^2 \theta (1 - \tan^2 \theta) = \cos 2\theta$

39. (a) $(\sin 2t + \cos 2t)^2 - \sin 4t = 1$
     (b) $(\sin t - \cos t)^2 + 2 \sin 2t = 1 + \sin 2t$

40. (a) $\dfrac{1 - \cos 2t}{\sin 2t} = \tan t$

     (b) $\dfrac{\sin 4\theta}{1 - \cos 4\theta} = \cot 2\theta$

41. (a) $\dfrac{\sin^2 2\theta}{\sin^2 \theta} = 4 \cos^2 \theta$

     (b) $\sin \theta \tan \dfrac{\theta}{2} \csc^2 \dfrac{\theta}{2} = 2$

42. (a) $2 \tan \dfrac{\theta}{2} \csc \theta = \sec^2 \dfrac{\theta}{2}$

     (b) $2 \cos \dfrac{\theta}{2} = (1 + \cos \theta) \sec \dfrac{\theta}{2}$

43. $\dfrac{\sin 3\theta}{\sin \theta} - \dfrac{\cos 3\theta}{\cos \theta} = 2$

44. $\dfrac{\sin 5t}{\sin t} - \dfrac{\cos 5t}{\cos t} = 4 \cos 2t$

45. $\dfrac{\cos^3 t - \sin^3 t}{\cos t - \sin t} = \dfrac{2 + \sin 2t}{2}$

46. $\dfrac{\sec^2 t}{2 - \sec^2 t} = \sec 2t$

47. $\sin^2 t \cos^2 t = \dfrac{1}{8}(1 - \cos 4t)$

48. $\sin^4 \theta = \dfrac{3}{8} - \dfrac{1}{2} \cos 2\theta + \dfrac{\cos 4\theta}{8}$

Applying the Concepts

49. Area: Suppose that triangle PBC is an isosceles triangle where each of the equal sides $\overline{PB}$ and $\overline{PC}$ has a length of 4 centimeters and θ is the angle BPC, as shown in Figure 4.

Figure 4

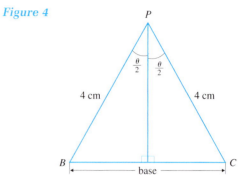

     (a) Show that the area A of the triangle is given by $A = 8 \sin \theta$.

     (b) Find the area A of the triangle when
$$\theta = 36°.$$
       Round off the answer to two decimal places.

     (c) Find the domain of the function in part (a).

     (d) Find the maximum area and the corresponding value of θ.

50. (Refer to problem 49.)

     (a) Graph the function in part (a) with the restrictions found in part (c). Interpret the graph.

     (b) Find θ when the area is 4.

51. Area: Consider an isosceles trapezoid with the dimensions shown in Figure 5.

Figure 5

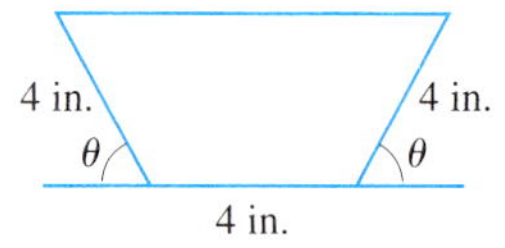

     (a) Show that the area A of the trapezoid is given by
$$A = 64 \sin \dfrac{\theta}{2} \cos^3 \dfrac{\theta}{2}.$$

     (b) Find the area of A of the trapezoid when
$$\theta = 36°.$$
       Round off the answer to two decimal places.

     (c) Find the domain of the function in part (a).

52. [G] (Refer to problem 51.)

(a) Use a grapher to graph the function in part (a) with the restrictions found in part (c). Interpret the graph.

(b) Use the graph in part (a) to find the maximum area and the value of θ when it occurs. Round off to two decimal places.

53. Household Electrical Circuit: Suppose that the voltage V (in volts) and the amperage I (in amperes) in a household electrical circuit at t seconds are given by the models

$$V = 280 \sin(120\pi t) \quad \text{and} \quad I = 7 \sin(120\pi t).$$

(a) Write the wattage $W = VI$ in the form

$$W = a \cos(k\pi t) + d$$

where a, k, and d are constants.

(b) Find the period of each of the three functions V, I, and W, and compare these periods.

(c) What is the maximum wattage?

54. Electrical Household Appliance: If a household appliance is plugged into an outlet, the wattage W used is given by the model

$$W = I^2 R$$

where $I = 12 \sin(120\pi t)$ is the amperage and R is a constant that measures the resistance (in ohms) and t is in seconds.

(a) If $R = 12$ ohms, write an expression for the wattage in the form

$$W = a \cos(240\pi t) + d$$

where a and d are constants.

(b) Find the period of the amperage I and wattage W of the appliance and compare these periods.

(c) What is the maximum wattage?

Developing and Extending the Concepts

55. Some techniques in calculus make use of the substitution $z = \tan(\theta/2)$. Use this substitution to show that:

(a) $\cos \theta = \dfrac{1 - z^2}{1 + z^2}$

(b) $\sin \theta = \dfrac{2z}{1 + z^2}$

56. Use the results from problem 55 to rewrite each expression in terms of z and simplify.

(a) $\dfrac{3}{2 \sin \theta + 3 \cos \theta}$

(b) $\dfrac{1}{3 + 2 \sin \theta}$

57. Suppose that P is a point on a unit circle, and

$$P(t) = \left(\frac{3}{5}, \frac{-4}{5} \right).$$

Find:

(a) $P(2t)$

(b) $P(t/2)$

58. (a) Suppose that $u + v = \pi/2$. Verify that $\sin(u - v) = -\cos 2u$.

(b) Suppose that $u + v = \pi$. Verify that $\sin(u - v) = -\sin 2u$.

59. (a) Express sec $2t$ in terms of sec t.

(b) Find sec $2t$ if $\tan t = 3$ and sec $t < 0$.

60. Suppose that α and β are acute angles in a right triangle. Show that

$$\sin 2\alpha = \sin 2\beta.$$

61. [G] (a) Verify that $\cos 3t = 4 \cos^3 t - 3 \cos t$.

(b) Use a grapher to demonstrate the validity of the identity in part (a) by comparing the graphs of

$$y_1 = \cos 3t \text{ and } y_2 = 4 \cos^3 t - 3 \cos t.$$

62. [G] (a) Use a grapher to graph the equation

$$y = \frac{6 \sin^2 2x}{\sin 4x}.$$

(b) Find an equation of the form $y = a \tan kx$ that has the same graph as the equation in part (a).

Objectives

1. Use Reference Angles
2. Solve Trigonometric Equations
3. Solve Applied Problems

6.4 Trigonometric Equations

We know that trigonometric identities are equations that are *true for all values* for which both sides of the equation are defined. By contrast, *conditional trigonometric equations* are equations that are *true for certain values* but false for others. Examples of such equations are

$$\sin t = 1, \quad 2 \cos \theta - 1 = 0, \quad \sin 2t = 2 \cos t \quad \text{and} \quad \tan^2 \theta + \sec \theta = 1$$

In this section, we learn how to solve conditional trigonometric equations.

Let us begin by considering the equation

$$\sin \theta = \frac{1}{2}, \text{ where } \theta \text{ represents degree measure.}$$

Suppose we want to solve this equation for θ under each of the following three conditions:

(i) $0° < \theta < 90°$ (ii) $0° < \theta < 360°$ (iii) No restriction on θ

We proceed as follows:

(i) To solve

$$\sin \theta = \frac{1}{2}, \quad 0° < \theta < 90°$$

means to find an acute angle, say θ_R, such that

$$\sin \theta_R = \frac{1}{2}.$$

Figure 1

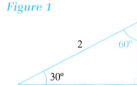

From right triangle trigonometry, we use a 30°–60° triangle (Figure 1) to observe that

$$\sin 30° = \frac{1}{2}.$$

So $\theta_R = 30°$ and the solution of the equation is 30° for the given restriction.

(ii) To solve

$$\sin \theta = \frac{1}{2}, \quad 0° < \theta < 360°$$

Figure 2

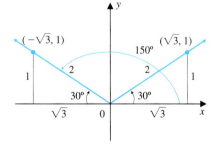

we first observe that, when placed in standard position, θ is in either quadrants I or II because the sine value is positive. To determine the exact locations of θ, we use the result in part (i), $\theta_R = 30°$, to construct reference triangles in quadrants I and II as shown in Figure 2.

Upon examining this figure, we see that both

$$\sin 30° = \frac{1}{2} \quad \text{and} \quad \sin 150° = \frac{1}{2}.$$

So, under the given restrictions, the solution of the equation includes both 30° and 150°.

(iii) Next we solve

$$\sin \theta = \frac{1}{2} \quad (\theta \text{ is unrestricted}).$$

From part (ii) we notice that there are solutions in both quadrants I and II, 30° and 150°. However, since θ is not restricted, any other angles coterminal with 30° and 150° are also solutions. Thus we conclude that the solution of this equation includes all angles of the form

$$\theta = 30° + k \cdot 360°$$

or $\theta = 150° + k \cdot 360°$,   where k is an integer.

It is important to notice the key role that the angle $\theta_R = 30°$ had in solving the equation in all three situations. In this context, θ_R is called a *reference angle*.

Using Reference Angles

Before attempting to solve other trigonometric equations, it is important to understand the general concept of reference angles.

Reference angles are defined as follows:

Definition
Reference Angle

If θ is an angle in standard position whose terminal side does not lie on either coordinate axis, then the **reference angle** θ_R for θ is defined to be the *positive acute angle* formed by the terminal side of θ and the x axis.

For instance,
the reference angle for $\theta = 115°$ is

$$\theta_R = 180° - 115° = 65° \text{ (Figure 3a)};$$

the reference angle for $\theta = 3.78$ is given by

$$\theta_R = 3.78 - \pi \quad \text{or approximately } 0.64 \text{ (Figure 3b)}.$$

Figure 3

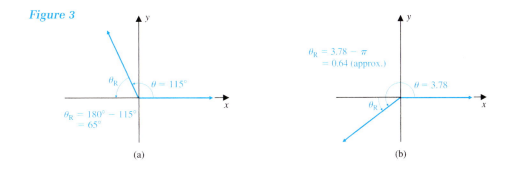

(a)   (b)

It is possible to express trigonometric function values of an angle in terms of values of its reference angle because of symmetry.
For instance, if $\theta = 115°$, its reference angle is $\theta_R = 65°$.
As shown in Figure 4, we see that

$$\sin 115° = \frac{y}{r} \qquad \text{and} \quad \cos 115° = \frac{-x}{r} = -\frac{x}{r}$$

$$= \sin 65° \qquad\qquad\qquad = -\cos 65°$$

Figure 4

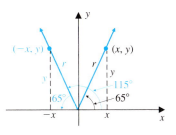

This observation leads us to the following general result

Evaluating Trigonometric Functions of Reference Angles

If $T(\theta)$ represents any of the six trigonometric functions of θ, and if θ_R is the reference angle of θ, then either

$$T(\theta) = T(\theta_R) \quad \text{or} \quad T(\theta) = -T(\theta_R)$$

depending on the function and the quadrant in which the terminal side of θ lies, when θ is in standard position.

EXAMPLE 1   Relating Trigonometric Function Evaluations to Reference Angles

For each expression, draw the angle in standard position, and then determine and display the reference angle. Rewrite the given expression in terms of the reference angle. Use a calculator to confirm the result (round off to four decimal places).

(a) $\cos 163°$ (b) $\tan 5.1416$

Solution

(a) Figure 5a, shows that the reference angle for

$$\theta = 163° \quad \text{is} \quad \theta_R = 17°.$$

Since the value of the cosine is negative in quadrant II, it follows that

$$\cos 163° = -\cos 17°.$$

By using a calculator, we find that both $\cos 163°$ and $-\cos 17°$ are approximately equal to -0.9563.

(b) Figure 5b shows that the reference angle for

$$\theta = 5.1416 \quad \text{is approximately} \quad \theta_R = 1.1416.$$

Since the tangent is negative in quadrant IV, it follows that

$$\tan 5.1416 = -\tan 1.1416.$$

Both $\tan 5.1416$ and $-\tan 1.1416$ are approximately equal to -2.1850.

Figure 5

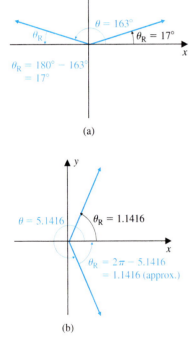

(a)

(b)

Solving Trigonometric Equations

Analytic techniques for solving trigonometric equations such as

$$\sin t = 0.8931$$

$$2 \cos^2 t - 7 \cos t = -3$$

$$\tan^4 (2\theta) = 9$$

$$3 \sin\theta = 2 \cos^2 \theta$$

$$\sin (2\theta) = \sin \theta$$

and

$$\tan \theta + 1 = \sec(3\theta)$$

involve the use of properties of the trigonometric functions such as periodicity, reference angles, signs of the values, and identities.

Commonly, trigonometric equations have restrictions that include only angles that fall within one positive rotation between 0° and 360° if degrees are used or between 0 and 2π if radians or real numbers are used.

EXAMPLE 2   **Using a Reference Angle to Solve a Trigonometric Equation**

Solve the equation $\tan t = -1.2137$, for $0 \le t < 2\pi$. Round off the answers to four decimal places.

Solution   Since the value of $\tan t$ is negative, it follows that there are two possible solutions for $0 \le t < 2\pi$, one in quadrant II, labeled t_1 and one in quadrant IV labeled t_2 as shown in Figure 6. The reference angle t_R also displayed in Figure 6 satisfies $\tan t_R = 1.2137$.

Figure 6

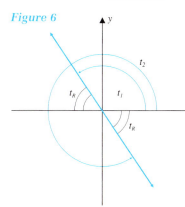

A calculator in radian mode gives us

$$t_R = \tan^{-1}(1.2137) = 0.8816 \text{ (approx.)}.$$

It follows that the two solutions are given by

$$t_1 = \pi - t_R = 2.2600 \text{ (approx.)}$$
$$\text{and } t_2 = 2\pi - t_R = 5.4016 \text{ (approx.)}.$$

EXAMPLE 3   **Using a Known Solution to Solve a Trigonometric Equation**

Use the solution of $\sin \theta = 1/2$ to solve

$$\sin(\theta - 60°) = 1/2, \quad \text{for } 0° \le \theta < 360°$$

Solution   Earlier (see page 468) we determined that if θ is not restricted the solution of $\sin \theta = 1/2$ includes all values of the form

$$\theta = 30° + k \cdot 360° \quad \text{or} \quad \theta = 150° + k \cdot 360°$$

where k is an integer.
   This means that the solutions of

$$\sin(\theta - 60°) = 1/2$$

must satisfy

$$\theta - 60° = \ldots -690°, -330°, 30°, 390°, \ldots$$
$$\text{or} \quad \theta - 60° = \ldots -210°, 150°, 510°, \ldots$$

Solving the latter equations for θ by adding 60 to each side of each equation, we get

$$\theta = \ldots -630°, -270°, 90°, 450°, \ldots$$
$$\text{or} \quad \theta = \ldots -150°, 210°, 570°, \ldots$$

However, because of the restriction $0° \le \theta < 360°$, we conclude that the solution includes $\theta = 90°$ and $\theta = 210°$ only.

EXAMPLE 4   **Solving a Trigonometric Equation by Factoring**

Solve the equation

$$\sin\theta\tan\theta - \sin\theta = 0, \quad \text{for} \quad 0° \le \theta < 360°.$$

Solution   First we factor the given equation

$$\sin\theta\tan\theta - \sin\theta = 0$$

to get

[1] $$\sin\theta(\tan\theta - 1) = 0.$$

Setting each factor equal to 0, we have two equations

$$\sin\theta = 0 \quad \text{and} \quad \tan\theta - 1 = 0.$$

For

$$\sin\theta = 0,$$

we get the solutions

$$\theta = 0° \quad \text{or} \quad \theta = 180°.$$

For $\tan\theta - 1 = 0$, we solve the equivalent equation

$$\tan\theta = 1$$

as follows:

Since the value of the tangent is a positive number, it follows that the solutions of

$$\tan\theta = 1$$

are in quadrants I and III. Also, the reference angle for $\tan\theta_R = 1$ is $\theta_R = 45°$ (Figure 7).

So the solution of $\tan\theta = 1$ in quadrant I is $\theta = 45°$, and in quadrant III is $\theta = 180° + 45° = 225°$.

Combining the results, we conclude that the solution of the original equation includes 0°, 45°, 180°, and 225°. ▦

Figure 7

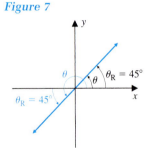

Note that we do not divide both sides of an equation by an expression containing a variable. For instance, in Example 4, if we had divided both sides of the above equation denoted by [1] by $\sin\theta$, we would have obtained the equation $\tan\theta - 1 = 0$ and two of the solutions, 0° and 180°, would have been lost.

Trigonometric equations involving multiple angles require special treatment, as illustrated in the next example.

EXAMPLE 5   **Solving a Trigonometric Equation Involving a Multiple Angle**

Solve the equation

$$\cos 3\theta = \sqrt{3}/2, \text{ for } 0° \le \theta < 360°.$$

Solution   To solve the equation, we let $\alpha = 3\theta$ in the given equation to get

$$\cos 3\theta = \cos\alpha = \frac{\sqrt{3}}{2}.$$

Since

$$0° \le \theta < 360°, \quad \text{then} \quad 0° \le 3\theta < 1080°;$$

so that

$$\cos \alpha = \frac{\sqrt{3}}{2} \quad \text{for } 0° \le \alpha < 1080°.$$

The reference angle α_R for the latter equation is given by

$$\alpha_R = \cos^{-1} \frac{\sqrt{3}}{2} = 30°.$$

Since the cosine is positive in quadrants I and IV, it follows that

$$\alpha = 30° \quad \text{and} \quad \alpha = 330°$$

Figure 8

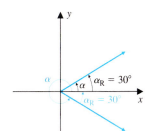

are two solutions (Figure 8).
 Next we use periodicity by adding 360° to these two angle measures to get two more solutions:

$$\alpha = 30° + 360° \quad \text{and} \quad \alpha = 330° + 360°$$
$$= 390° \qquad\qquad = 690°$$

both of which satisfy the condition $0° \le \alpha < 1080°$.
 Finally, adding 360° again, we obtain two additional solutions:

$$\alpha = 390° + 360° \quad \text{and} \quad \alpha = 690° + 360°$$
$$= 750° \qquad\qquad = 1050°.$$

We stop at 750° and 1050° because adding 360° again would result in angle measures exceeding the 1080° restriction.

Thus for $0° \le \alpha < 1080°$, we have

$$\alpha = 3\theta = 30°, 330°, 390°, 690°, 750°, \text{or } 1050°.$$

The values of θ, the original unknown, are found by dividing each value of $\alpha = 3\theta$ by 3 to obtain

$$\theta = 10°, 110°, 130°, 230°, 250°, \text{or } 350°,$$

which are the solutions of the original equation.

The next example employs an identity to solve a trigonometric equation.

EXAMPLE 6   **Solving an Equation by Using an Identity**

Solve the equation $\cos 2t = \cos t$ for t in the interval $[0, 2\pi)$.

Solution   We begin by using the double-angle identity

$$\cos 2t = 2 \cos^2 t - 1$$

to rewrite the given equation

$$\cos 2t = \cos t \quad \text{as} \quad 2 \cos^2 t - 1 = \cos t.$$

The latter equation is a quadratic equation in $\cos t$, which we can solve as follows:

$$2 \cos^2 t - 1 = \cos t$$

$$2 \cos^2 t - \cos t - 1 = 0 \qquad \text{Subtract } \cos t \text{ from each side}$$

$$(2 \cos t + 1)(\cos t - 1) = 0 \qquad \text{Factor}$$

$$
\begin{array}{lll}
2 \cos t + 1 = 0 & \quad\text{or}\quad & \cos t - 1 = 0 \\
\quad 2 \cos t = -1 & & \\
\quad \cos t = -\dfrac{1}{2} & & \cos t = 1 \\
\end{array}
$$

⎧ Set each factor equal to
⎨ zero, and solve the resulting
⎩ equations with the
restriction $0 \le t < 2\pi$

$$\text{So} \qquad t = \frac{2\pi}{3} \text{ or } \frac{4\pi}{3} \qquad\qquad t = 0$$

Therefore, the solutions are 0, $(2\pi)/3$ (approx. 2.0944) and $(4\pi)/3$ (approx. 4.1888).

Graphers enable us to solve equations containing trigonometric expressions that cannot be solved by the analytic methods that we have studied so far in algebra and trigonometry.

EXAMPLE 7 **G** **Solving an Equation Graphically**

(a) Solve the equation $\cos x = x$ graphically. Round off the result to four decimal places.

(b) Use the result from part (a) to solve the inequality $\cos x < x$.

Solution

Figure 9

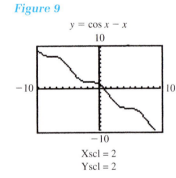

$y = \cos x - x$

(a) The equation $\cos x = x$ is equivalent to $\cos x - x = 0$.
The viewing window in Figure 9 shows that the graph of $y = \cos x - x$ has one x intercept, so there is only one solution.
By using the **ZERO** -finding feature of a grapher, we find that the solution is approximately 0.7391.

(b) The given inequality is equivalent to $\cos x - x < 0$.
The solution of this inequality incudes all values of x for which the graph of $y = \cos x - x$ is below the x axis. Since the x intercept is approximately 0.7391, it follows that the solution includes all numbers (approximately) in the interval $(0.7391, \infty)$, that is, $x > 0.7391$.

Solving Applied Problems

At times, mathematical models of real-world situations involve solving trigonometric equations.

Figure 10

EXAMPLE 8 **Modeling the Design of a Window**

(Refer to Figure 10.) A custom window is to be designed in the shape of a rectangle 2 feet wide, topped by an isosceles triangle whose equal sides make an angle θ with the width. Let y (in feet) be the height of the rectangular part of the window; and let x (in feet) be the length of each of the equal sides of the isosceles triangle. Assume the perimeter of the entire window is to be 12 feet.

(a) Express x and y, in terms of θ. Then write A, the total area (in square feet) of the window as a function of θ.

(b) What are the restrictions on θ?

(c) Find θ if y is 3 feet. Then use the value of θ, along with the results in part (a), to find x and A. Round off the answers to two decimal places.

Solution

(a) Figure 11 shows the isosceles triangle that forms the top of the window. It follows from right triangle trigonometry that

$$\sec \theta = \frac{x}{1} \quad \text{or} \quad x = \sec \theta.$$

Since the perimeter of the entire window is 12 feet, we have

$$2y + 2x + 2 = 12.$$

Substituting for x in the latter equation, we get

$$2y + 2x + 2 = 12$$
$$2y + 2 \sec \theta + 2 = 12$$
$$2y + 2 \sec \theta = 10.$$

Solving for y, we have

$$y = 5 - \sec \theta.$$

Note that the height h of the triangle in Figure 11 is given by

$$\tan \theta = h/1 = h.$$

Thus the area A of the window can be rewritten as follows:

$A = $ (area of rectangle) $+$ (area of triangle)

$\quad = $ (width)(height of rectangle) $+ \dfrac{1}{2}$(base)(height of triangle)

$\quad = 2y + \dfrac{1}{2}(2)\,h \qquad\qquad$ Substitute

$\quad = 2y + h \qquad\qquad\qquad$ Simplify

$\quad = 2(5 - \sec \theta) + \tan \theta \qquad$ Substitute

$\quad = 10 - 2 \sec \theta + \tan \theta \qquad$ Multiply

So we have

$$x = \sec \theta, \quad y = 5 - \sec \theta, \quad \text{and} \quad A = 10 - 2 \sec \theta + \tan \theta.$$

(b) Since the sum of the angles of a triangle is $180°$, it follows that

$$0° < 2\theta < 180° \quad \text{or} \quad 0° < \theta < 90°$$

that is, θ is an acute angle.

Figure 11

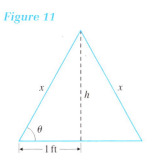

(c) If $y = 3$, then, after substituting this value into the equation $y = 5 - \sec \theta$ which was derived in part (a), we get

$$3 = 5 - \sec \theta$$

$$-2 = -\sec \theta$$

$$2 = \sec \theta, \quad \text{where } \theta \text{ is an acute angle.}$$

The solution of this equation is $\theta = 60°$.
It follows that

$$x = \sec \theta = \sec 60° = 2$$

and

$$A = 10 - 2 \sec \theta + \tan \theta$$

$$= 10 - 2 \sec 60° + \tan 60°$$

$$= 10 - 2(2) + \sqrt{3}$$

$$= 6 + \sqrt{3} = 7.73 \text{ (approx.).}$$

So, if $y = 3$, then x, the length of the two sides of the triangular top is 2 feet, and A, the total surface area, is approximately 7.73 square feet.

G In the Cartesian coordinate system, we graph trigonometric functions with real number domains (corresponding to radian measure). However, graphers also enable us to *display* the horizontal axis in terms of degrees rather than just real numbers. For instance, in Example 8, we established that the area A of the window is given by

$$A = 10 - 2 \sec \theta + \tan \theta, \quad \text{where } 0° < \theta < 90°.$$

To find the angle θ when the area $A = 7.5$, we need to solve the equation

$$7.5 = 10 - 2 \sec \theta + \tan \theta$$

Figure 12

or, equivalently,

$$0 = 2.5 - 2 \sec \theta + \tan \theta.$$

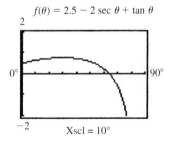

$f(\theta) = 2.5 - 2 \sec \theta + \tan \theta$

Xscl = 10°

Figure 12 shows a viewing window of the graph of

$$f(\theta) = 2.5 - 2 \sec \theta + \tan \theta \quad \text{for } 0° < \theta < 90°$$

(*set in degree mode*). By using the **ZERO** -finding feature, we find that

$$\theta \text{ is approximately } 63.83°.$$

Using the results in Example 8, we have

x (the length of the equal sides of the triangular top)

$$= \sec \theta = \sec 63.83° = 2.27 \text{ feet (approx.)}$$

and

y (the height of the rectangle) $= 5 - \sec \theta = 5 - \sec 63.83°$

$$= 2.73 \text{ feet (approx.).}$$

 **PROBLEM SET 6.4**

Mastering the Concepts

In problems 1 and 2, verify that the given values are solutions of the indicated equation.

1. (a) $2 \sin \theta - 1 = 0$; $30°, 150°$

    (b) $\tan\left(3t - \dfrac{\pi}{4}\right) = 1$; $-\dfrac{\pi}{6}, \dfrac{\pi}{6}$

2. (a) $2 \sin 2t - \sqrt{3} = 0$; $\dfrac{7\pi}{6}, \dfrac{4\pi}{3}$

    (b) $2 \sin(\theta + 30°) = -1$; $-180°, 300°$

3. Solve the equation $\sin \theta = -\sqrt{2}/2$ if:
    (a) $0° \le \theta < 90°$
    (b) $0° \le \theta < 180°$
    (c) $0° \le \theta < 360°$
    (d) θ has no restriction

4. Solve the equation $\cos t = -\sqrt{2}/2$ if:
    (a) $0 \le t < \pi/2$
    (b) $0 \le t < \pi$
    (c) $0 \le t < 2\pi$
    (d) t has no restriction

In problems 5 and 6, draw the angle in standard position. Then determine and display its reference angle. Rewrite the given expression in terms of the reference angle. Use a calculator to confirm the result (rounding off to four decimal places).

5. (a) $\sin 190°$
    (b) $\tan(-113°)$
    (c) $\sec 372°$
    (d) $\cos(-1.9)$
    (e) $\cot 7.2$

6. (a) $\cos 219°$
    (b) $\cot 114°$
    (c) $\cos(-295°)$
    (d) $\sin \dfrac{11\pi}{8}$
    (e) $\sec 5.9$

In problems 7–14, assume that $0 \le t < 2\pi$ and $0° \le \theta < 360°$. Find the exact solution of each equation.

7. (a) $\cos t = -\dfrac{1}{2}$

    (b) $\sec t = -\sqrt{2}$

8. (a) $\sin t = \dfrac{\sqrt{2}}{2}$

    (b) $3 \csc \theta = -2\sqrt{3}$

9. (a) $\tan \theta = \sqrt{3}$
    (b) $3 \cot \theta = -\sqrt{3}$

10. (a) $\csc t = \sqrt{2}$
    (b) $\cot t = -1$

11. (a) $2 \sin t - \sqrt{3} = 0$
    (b) $2 \sin \theta + \sqrt{2} = 0$

12. (a) $\tan \theta - 1 = 0$
    (b) $\cot t + \sqrt{3} = 0$

13. (a) $\sqrt{2} \cos t = -1$

    (b) $\sqrt{2} \cos\left(t - \dfrac{\pi}{4}\right) = -1$

14. (a) $\tan \theta = -1$
    (b) $\tan(\theta + 60°) = -1$

In problems 15–18, assume $0 \le t < 2\pi$. Solve each equation. Round off the answers to four decimal places.

15. (a) $\sin t = 0.8134$
    (b) $\cot t = -6.6173$

16. (a) $\cos t = -0.4176$
    (b) $\tan t = 0.6696$

17. (a) $\sec t = -\dfrac{6}{5}$

    (b) $\tan t = 10$

18. (a) $\csc t = \pi$
    (b) $\cot t = -10$

In problems 19–36, assume that $0 \le t < 2\pi$ and $0° \le \theta < 360°$. Solve each equation. Round off the answers to four decimal places, when applicable.

19. $\sin t \cos t - \sin t = 0$
20. $2 \sin \theta \cos \theta - \sqrt{2} \cos \theta = 0$
21. $2 \sin \theta \cos \theta + \sqrt{3} \cos \theta = 0$
22. $\sin t \cos t - \cos t = 0$
23. $\tan^2 t - \sqrt{3} \tan t = 0$
24. $2 \cos^2 \theta - \cos \theta = 0$
25. $\sec^2 \theta + \sec \theta = 0$
26. $\sqrt{3} \csc^2 \theta + 2 \csc \theta = 0$
27. $\tan^2 t - 3 = 0$
28. $\cot^2 t - 1 = 0$
29. $4 \sin^2 \theta - 3 = 0$
30. $\csc^2 \theta - 4 = 0$
31. $3 - 2 \cos^2 t = 3 - \cos t$
32. $1 - 2 \sin^2 t = 1 - \sin t$
33. (a) $2 \cos 3\theta = -\sqrt{2}$
    (b) $2 \cos(3\theta - 30°) = -\sqrt{2}$
34. (a) $2 \sin 2t = 1$

    (b) $2 \sin\left(2t - \dfrac{\pi}{4}\right) = 1$

35. (a) $\sec \dfrac{t}{2} = \sqrt{2}$

    (b) $\sec\left(\dfrac{t}{2} + \dfrac{\pi}{5}\right) = \sqrt{2}$

36. (a) $\tan \dfrac{\theta}{2} + \sqrt{3} = 0$

(b) $\tan \left(\dfrac{\theta}{2} - 40° \right) + \sqrt{3} = 0$

In problems 37–52, assume that

$$0 \le t < 2\pi \quad \text{and} \quad 0° \le \theta < 360°.$$

Solve each equation by using identities. Round off the answers to four decimal places, when applicable.

37. $\sin^2 t = \cos^2 t$

38. $2 \cos^2 t - 1 = 2 \sin^2 t$

39. $\cot^2 \theta = \cot \theta$

40. $2 \tan \theta - \sec^2 \theta = 0$

41. $2 - \sin t = 2 \cos^2 t$

42. $\cos^2 t - \sin^2 t - \sin t - 1 = 0$

43. $\cos \theta - \sin \theta = 1$

44. $\tan \theta + \sec \theta = 1$

45. $\sin 2t = \sqrt{2} \cos t$

46. $2 - \cos^2 t - 4 \sin^2 \dfrac{t}{2} = 0$

47. $\sin \dfrac{t}{2} + \cos t = 1$

48. $\cos 2\theta = 2 \sin^2 \theta$

49. $\cos 2\theta - 3 \sin \theta = 2$

50. $\cos 2\theta + \sin^2 \theta = 1$

51. $\cos 2t + 2 \cos^2 \left(\dfrac{t}{2} \right) - 2 = 0$

52. $\cos 2\theta - \cos \theta = 0$

Applying the Concepts

In problems 53 and 54, round off each answer to two decimal places.

53. Physics: In physics, Snell's law deals with the change in direction of a ray of light as it passes through a medium, as shown in Figure 13, where the dashed line is perpendicular to the surface of the medium:

Figure 13

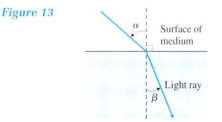

The angle α is called the **angle of incidence,** and the angle β is called the **angle of refraction. Snell's law** states that for a given medium,

$$\dfrac{\sin \alpha}{\sin \beta} = C$$

where C is a constant. The constant C is called the **index of refraction** of the medium.

(a) Find the degree measure of the angle of refraction of a light ray traveling through ice if

the angle of incidence is 45°,

and

the index of refraction is 1.31.

(b) Find the degree measure of the angle of incidence of a light ray striking a rock-salt crystal if

the angle of refraction of the ray is 35°,

and

index of refraction of rock-salt is 1.54.

54. Animal Population: Suppose that the number of rodents N in a certain population after t years is given by the mathematical model

$$N(t) = 3200 + 800 \sin \dfrac{3t}{2} \quad \text{where } 0 \le t < 5.$$

(a) How many rodents are initially in the population?

(b) Find how many years later it takes for the population to drop to 3000.

(c) Find the maximum number of rodents and when it occurs.

55. Length of a Ladder: A 5-foot fence stands 4 feet away from a high wall. A ladder on the ground leaning against the wall touches the top of the fence and makes an angle θ with the ground (Figure 14).

(a) Show that the length L of the ladder is given by

$$L = 4 \sec \theta + 5 \csc \theta.$$

(b) What are the restrictions on θ?

Figure 14

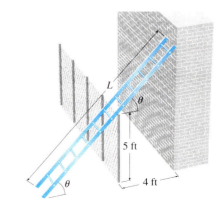

56. [G] (Refer to problem 55.) Use a grapher to find the values in degrees of θ when the length of the ladder is 13 feet. Round off to two decimal places.

Developing and Extending the Concepts

57. (a) Let θ be an angle in quadrant II in standard position and let θ_R be its reference angle. Show that the value of any trigonometric function of θ is the same as the value for the function of θ_R, except possibly for a change of algebraic sign.

   (b) Repeat part (a) for θ in quadrant III.

   (c) Repeat part (a) for θ in quadrant IV.

58. When asked to find a number in the interval

$$\left[\frac{-\pi}{2}, \frac{\pi}{2}\right]$$

that solves the equation $\dfrac{\sin x}{x} = 1$ a student responds as follows:

First multiply both sides of the given equation by x to obtain

$$x \cdot \frac{\sin x}{x} = x \cdot 1$$

$$\sin x = x.$$

Since $\sin 0 = 0$, it follows that a solution of the given equation is $x = 0$.

Do you agree with this answer? Explain.

Can you find a solution for

$$\frac{\sin x}{x} = 1$$

in the interval

$$\left[-\frac{\pi}{2}, \frac{\pi}{2}\right]?$$

In problems 59 and 60, assume $0 \le t < 2\pi$. For what values of k does the equation have:

(a) No solution?

(b) One solution?

(c) Two solutions?

(d) More than two solutions?

59. $\cos t = k$

60. $\sec t = k$

61. Find the smallest positive angle θ in radians such that

$$\frac{3.43 \times 10^6}{1 - 0.21 \cos \theta} = 3.08 \times 10^6$$

Round off the answer to two decimal places.

62. Suppose that

$$x = a \cos \theta - b \sin \theta$$
$$y = a \sin \theta + b \cos \theta$$

(a) Find the product xy in terms of a, b, and θ.

(b) Find the smallest positive angle θ so that the coefficient of ab in the result from part (a) is zero.

[G] In problems 63–68:

(a) Solve each equation graphically if $0 \le t < 2\pi$. Round off the answers to three decimal places.

(b) Use the results from part (a) to solve the corresponding inequality.

63. $15 \sin^2 t - 8 \sin t + 1 = 0$;
   $15 \sin^2 t - 8 \sin t + 1 > 0$

64. $\tan^2 t - 3 \tan t + 2 = 0$; $\tan^2 t - 3 \tan t + 2 < 0$

65. $t + \sin t = 0.32$; $t + \sin t \le 0.32$

66. $\sin t + \ln t = 2t$; $\sin t + \ln t > 2t$

67. $e^{-t} - \cos t = 0$; $e^{-t} - \cos t \ge 0$

68. $\tan t = t^2 + 1$; $\tan t \le t^2 + 1$

Objectives

1. Use Product-to-Sum Identities
2. Use Sum-to-Product Identities
3. Solve Equations by Using the Sum-to-Product Identities
4. Solve Applied Problems

6.5 Other Identities

In calculus it is necessary at times to rewrite a product of sine and cosine functions as a sum or difference of such functions, or vice versa. This can be done by using the identities in this section.

Using Product-to-Sum Identities

The addition and subtraction identities can be used to derive the following identities that express the products of sines and cosines in terms of sums or differences.

**Product-to-Sum
Identities**

> (i) $\sin u \cos v = \dfrac{1}{2}[\sin(u + v) + \sin(u - v)]$
>
> (ii) $\cos u \sin v = \dfrac{1}{2}[\sin(u + v) - \sin(u - v)]$
>
> (iii) $\cos u \cos v = \dfrac{1}{2}[\cos(u - v) + \cos(u + v)]$
>
> (iv) $\sin u \sin v = \dfrac{1}{2}[\cos(u - v) - \cos(u + v)]$

Proofs (i) We use the addition and subtraction identities for the sine to prove the first identity as follows:

$$\frac{1}{2}[\sin(u + v) + \sin(u - v)] = \frac{1}{2}\,[(\sin u \cos v + \cos u \sin v)$$

$$+ (\sin u \cos v - \cos u \sin v)]$$

$$= \frac{1}{2}[2\sin u \cos v]$$

$$= \sin u \cos v$$

The proofs of parts (ii)–(iv) are left as exercises (problems 41–43).

EXAMPLE 1 **Using a Product-to-Sum Identity to Evaluate**

(a) Find the exact value of

$$\cos \frac{\pi}{12} \sin \frac{7\pi}{12}$$

by using a product-to-sum identity.

(b) Compare the result in part (a) to the calculator value of the product, rounded to four decimal places.

Solution (a) We proceed as follows:

$$\cos \frac{\pi}{12} \sin \frac{7\pi}{12} = \frac{1}{2}\left[\sin\left(\frac{\pi}{12} + \frac{7\pi}{12}\right) - \sin\left(\frac{\pi}{12} - \frac{7\pi}{12}\right)\right] \qquad \text{Identity (ii)}$$

$$= \frac{1}{2}\left[\sin \frac{8\pi}{12} - \sin\left(-\frac{6\pi}{12}\right)\right] \qquad \text{Add/Subtract}$$

$$= \frac{1}{2}\left[\sin \frac{2\pi}{3} - \sin\left(-\frac{\pi}{2}\right)\right] \qquad \text{Reduce fractions}$$

$$= \frac{1}{2}\left[\sin \frac{2\pi}{3} + \sin \frac{\pi}{2}\right] \qquad \text{Sine is an odd function}$$

$$= \frac{1}{2}\left[\frac{\sqrt{3}}{2} + 1\right] \qquad \text{Evaluate}$$

$$= \frac{\sqrt{3}}{4} + \frac{1}{2} \qquad \text{Multiply}$$

(b) The approximate value of $\dfrac{\sqrt{3}}{4} + \dfrac{1}{2}$

to four decimal places is 0.9330.
Using a calculator to evaluate the product

$$\cos \frac{\pi}{12} \sin \frac{7\pi}{12},$$

and rounding off to four decimal places, we again get 0.9330.   ▣

EXAMPLE 2   **Rewriting a Product as a Sum**

Express $\cos 3t \cos 4t$ as a sum involving only cosines.

Solution   Using identity (iii), with $u = 3t$ and $v = 4t$, we get

$$\cos 3t \cos 4t = \frac{1}{2}[\cos(3t - 4t) + \cos(3t + 4t)]$$

$$= \frac{1}{2}[\cos(-t) + \cos 7t]$$

$$= \frac{1}{2}\cos t + \frac{1}{2}\cos 7t. \qquad \text{Cosine is an even function} \quad ▣$$

Using Sum-to-Product Identities

The sum-to-product identities enable us to rewrite the sums of sines and cosines as products.

Sum-to-Product Identities

(i) $\sin \omega + \sin t = 2 \sin\left(\dfrac{\omega + t}{2}\right) \cos\left(\dfrac{\omega - t}{2}\right)$

(ii) $\sin \omega - \sin t = 2 \cos\left(\dfrac{\omega + t}{2}\right) \sin\left(\dfrac{\omega - t}{2}\right)$

(iii) $\cos \omega + \cos t = 2 \cos\left(\dfrac{\omega + t}{2}\right) \cos\left(\dfrac{\omega - t}{2}\right)$

(iv) $\cos \omega - \cos t = -2 \sin\left(\dfrac{\omega + t}{2}\right) \sin\left(\dfrac{\omega - t}{2}\right)$

Proofs   (i) To prove the first identity, we let $\omega = u + v$ and $t = u - v$. Then

$$\omega + t = (u + v) + (u - v) = 2u.$$

So
$$u = \frac{\omega + t}{2}.$$

Also,

$$\omega - t = (u + v) - (u - v) = 2v.$$

So
$$v = \frac{\omega - t}{2}.$$

By substituting $(\omega + t)/2$ for u and $(\omega - t)/2$ for v in product-to-sum identity (i), we have

$$\sin\left(\frac{\omega + t}{2}\right)\cos\left(\frac{\omega - t}{2}\right) = \frac{1}{2}(\sin \omega + \sin t)$$

or

$$\sin \omega + \sin t = 2 \sin\left(\frac{\omega + t}{2}\right)\cos\left(\frac{\omega - t}{2}\right).$$

The proofs of parts (ii)–(iv) are left as exercises (problems 44–46).

EXAMPLE 3 **Using a Sum-to-Product Identity to Evaluate**

(a) Use a sum-to-product identity to find the exact value of
$$\sin 75° + \sin 15°.$$

(b) Compare the result from part (a) to the calculator value, rounded to four decimal places.

Solution (a) We use sum-to-product identity (i) as follows:

$$\sin 75° + \sin 15° = 2 \sin\left(\frac{75° + 15°}{2}\right)\cos\left(\frac{75° - 15°}{2}\right)$$

$$= 2 \sin 45° \cos 30°$$

$$= 2\left(\frac{\sqrt{2}}{2}\right)\left(\frac{\sqrt{3}}{2}\right) = \frac{\sqrt{6}}{2}$$

(b) Approximating $\sqrt{6}/2$ to four decimal places gives 1.2247. The calculator value of $\sin 75° + \sin 15°$, when rounded to four decimal places, also gives 1.2247.

EXAMPLE 4 **Rewriting a Difference as a Product**

Rewrite as a product $\cos 8\theta - \cos 3\theta$.

Solution Using sum-to-product identity (iv) with $\omega = 8\theta$ and $t = 3\theta$, we get

$$\cos 8\theta - \cos 3\theta = -2 \sin\left(\frac{8\theta + 3\theta}{2}\right)\sin\left(\frac{8\theta - 3\theta}{2}\right)$$

$$= -2 \sin\frac{11\theta}{2} \sin\frac{5\theta}{2}.$$

EXAMPLE 5 **Verifying an Identity**

Verify the identity

$$\frac{\cos 3t - \cos 5t}{\sin 3t + \sin 5t} = \tan t.$$

Solution   Using the sum-to-product identities for the expressions in the numerator and the denominator, we have:

$$\frac{\cos 3t - \cos 5t}{\sin 3t + \sin 5t} = \frac{-2 \sin\left(\dfrac{3t + 5t}{2}\right) \sin\left(\dfrac{3t - 5t}{2}\right)}{2 \sin\left(\dfrac{3t + 5t}{2}\right) \cos\left(\dfrac{3t - 5t}{2}\right)} \qquad \text{Identity (iv)} \\ \text{Identity (1)}$$

$$= \frac{-2 \sin 4t \sin(-t)}{2 \sin 4t \cos(-t)} \qquad \text{Simplify}$$

$$= \frac{2 \, \cancel{\sin 4t} \sin t}{2 \, \cancel{\sin 4t} \cos t} \qquad \left\{ \begin{array}{l} \text{Sine is odd and} \\ \text{cosine is even} \end{array} \right.$$

$$= \frac{\sin t}{\cos t} \qquad \text{Reduce}$$

$$= \tan t$$

Solving Equations by Using the Sum-to-Product Identities

We can employ sum-to-product identities to solve equations such as

$$\sin 3t + \sin 2t = 0$$

$$\cos 4t - \cos 3t = 0,$$

and

$$\cos 4t + \cos 2t = \cos t$$

EXAMPLE 6   **Solving a Trigonometric Equation**

Solve the equation $\sin 3t + \sin t = 0, \quad 0 \le t < 2\pi$.

Solution   Using sum-to-product identity (i), we rewrite the left side as

$$\sin 3t + \sin t = 2 \sin\left(\frac{3t + t}{2}\right) \cos\left(\frac{3t - t}{2}\right) = 2 \sin 2t \cos t$$

Next we solve the resulting equation as follows:

$$2 \sin 2t \cos t = 0 \qquad \text{Given}$$

$$\sin 2t \cos t = 0 \qquad \text{Divide each side by 2}$$

Thus there are two possibilities:

$\sin 2t = 0$	$\cos t = 0$
$2t = 0, \pi, 2\pi, 3\pi$	$t = \dfrac{\pi}{2}, \dfrac{3\pi}{2}$
$t = 0, \dfrac{\pi}{2}, \pi, \dfrac{3\pi}{2}$	

Therefore, the solution includes $0, \dfrac{\pi}{2}, \pi, \quad$ and $\quad \dfrac{3\pi}{2}$.

Solving Applied Problems

If two tones from different musical instruments are sounded simultaneously, they will interact in a phenomenon called *beats* or **heterodyning.** When the two instruments are in tune, the beats will disappear. The human ear hears beats because the sound pressure slowly rises and falls as a result of slight variations in the frequency. Consider two tones with frequencies ν_1 and ν_2 (in hertz or cycles per second) and pressures P_1 and P_2 (in pounds per square foot) given by

$$P_1 = a_1 \cos (2\pi\nu_1 t) \text{ and } P_2 = a_2 \cos(2\pi\nu_2 t)$$

where t is the time (in seconds).

The superposition

$$P = P_1 + P_2$$

can be regarded as a total pressure on the eardrum of frequency $(\nu_1 + \nu_2)/2$. If the two tones have the same intensity, then $a_1 = a_2$.

EXAMPLE 7 **Modeling Musical Beats**

Consider the two tones of sound waves with the same intensity and with frequencies 110 and 116 hertz. The pressure on an eardrum is given by the models

$$P_1 = 0.035 \cos(220\pi t) \quad \text{and} \quad P_2 = 0.035 \cos(232\pi t)$$

where P_1 and P_2 are in pounds per square foot and t is the time in seconds.

(a) Write a model for

$$P = P_1 + P_2$$

and determine the frequency of P.

(b) Use a sum-to-product identity to write

$$P = P_1 + P_2$$

as a product of trigonometric expressions.

Solution   (a) A model for $P = P_1 + P_2$ is given by

$$P = 0.035 \cos(220\pi t) + 0.035 \cos(232\pi t)$$

The frequencies ν_1 and ν_2 are 110 and 116 cycles per second and so the frequency of P is $(110 + 116)/2 = 113$ cycles per second.

(b) Using the result from part (a), we have

$$P = 0.035 \cos(220\pi t) + 0.035 \cos(232\pi t)$$

$$= 0.035\left[2 \cos\left(\frac{220\pi t + 232\pi t}{2}\right) \cos\left(\frac{220\pi t - 232\pi t}{2}\right)\right] \quad \text{Sum-to-product identity (iii)}$$

$$= 0.07 \cos 226\pi t \cos(-6\pi t) \quad \text{Simplify}$$

$$= 0.07 \cos 226\pi t \cos 6\pi t \quad \text{Cosine is even}$$

PROBLEM SET 6.5

Mastering the Concepts

In problems 1–8:

(a) Use a product-to-sum identity to find the exact value.

(b) Compare the result from part (a) to the calculator value of the product, rounded to four decimal places.

1. $\sin \dfrac{5\pi}{12} \cos \dfrac{7\pi}{12}$

2. $\sin \dfrac{\pi}{12} \cos \dfrac{5\pi}{12}$

3. $\cos 195° \sin 75°$

4. $\cos 112.5° \sin 67.5°$

5. $\cos \dfrac{17\pi}{12} \cos \dfrac{7\pi}{12}$

6. $\cos \dfrac{11\pi}{12} \cos \dfrac{5\pi}{12}$

7. $\sin 67.5° \sin 22.5°$

8. $\sin 165° \sin 75°$

In problems 9–16, rewrite each product as a sum or difference using the product-to-sum identities.

9. $\sin 4w \sin w$　　　　**10.** $\cos 3\theta \sin 9\theta$

11. $\cos 8v \cos 4v$　　　**12.** $\cos 7t \cos 3t$

13. $\cos 3t \sin 4t$　　　　**14.** $\cos 5t \cos 2t$

15. $\sin 4t \sin 5t$　　　　**16.** $\cos 3t \sin t$

In problems 17–24:

(a) Use a sum-to-product identity to find the exact value.

(b) Compare the result from part (a) to the calculator value of the expression, rounded to four decimal places.

17. $\sin \dfrac{7\pi}{18} + \sin \dfrac{\pi}{18}$　　**18.** $\sin \dfrac{13\pi}{12} + \sin \dfrac{7\pi}{12}$

19. $\sin 75° - \sin 165°$　　**20.** $\sin 105° - \sin 75°$

21. $\cos \dfrac{7\pi}{12} + \cos \dfrac{\pi}{12}$　　**22.** $\cos \dfrac{19\pi}{12} + \cos \dfrac{13\pi}{12}$

23. $\cos 75° - \cos(-15°)$

24. $\cos(-75°) - \cos 16.5°$

In problems 25–32, rewrite each sum as a product using the sum-to-product identities.

25. $\sin 3\theta + \sin \theta$　　**26.** $\sin 5x + \sin 4x$

27. $\sin 5t - \sin 7t$　　　**28.** $\sin 3v - \sin v$

29. $\cos 2t + \cos 3t$　　　**30.** $\cos 3v + \cos v$

31. $\cos 5x - \cos 6x$　　　**32.** $\cos 2x - \cos x$

In problems 33 and 34, verify that the equation is an identity.

33. $\dfrac{\sin 2t + \sin 4t}{\cos 2t - \cos 4t} = \cot t$

34. $\dfrac{\sin t + \sin 3t}{\cos t + \cos 3t} = \tan 2t$

In problems 35–38, apply the sum-to-product identities to solve each equation for $0 \le t < 2\pi$.

35. $\sin 2t - \sin 5t = 0$　　**36.** $\cos 3t + \cos 6t = 0$

37. $\cos t - \cos 3t = 0$　　**38.** $\sin 7t - \sin 3t = 0$

Solving Applied Problems

39. Musical Tones and Beats: Suppose that two tones of sound waves with the same intensity and with frequencies 64 and 72 hertz are emitted simultaneously. The pressures on an eardrum, P_1 and P_2 (in pounds per square feet), and in t seconds are given, respectively, by the models

$$P_1 = 0.5 \cos 128\pi t \quad \text{and} \quad P_2 = -0.05 \cos 144\pi t.$$

(a) Write a model for $P = P_1 + P_2$ and determine the frequency of P.

(b) Use a sum-to-product identity to write $P_1 + P_2$ as a product of two sine functions.

40. Musical Tones and Beats: Suppose that two tones of sound waves with the same intensity and with frequencies 128 and 144 hertz are emitted simultaneously. The pressures exerted on an eardrum, P_1 and P_2 (in pounds per square feet), in t seconds are given, respectively, by the models

$$P_1 = 0.25 \sin 256\pi t \quad \text{and} \quad P_2 = 0.25 \sin 288\pi t.$$

(a) Write a model for $P = P_1 + P_2$ and determine the frequency of P.

(b) Use a sum-to-product identity to write P as a product of trigonometric functions.

Developing and Extending the Concepts

In problems 41–46, prove each identity.

41. $\cos u \sin v = \dfrac{1}{2}[\sin(u + v) - \sin(u - v)]$

42. $\cos u \cos v = \dfrac{1}{2}[\cos(u - v) + \cos(u + v)]$

43. $\sin u \sin v = \dfrac{1}{2}[\cos(u - v) - \cos(u + v)]$

44. $\sin \omega - \sin t = 2 \cos\left(\dfrac{\omega + t}{2}\right) \sin\left(\dfrac{\omega - t}{2}\right)$

45. $\cos \omega + \cos t = 2 \cos\left(\dfrac{\omega + t}{2}\right) \cos\left(\dfrac{\omega - t}{2}\right)$

46. $\cos \omega - \cos t = -2 \sin\left(\dfrac{\omega + t}{2}\right) \sin\left(\dfrac{\omega - t}{2}\right)$

G In problems 47–50, use a grapher to graph each given expression and the corresponding resulting expression found in problems 25–28 separately, using the same viewing window settings. Compare the results.

47. See problem 25.

48. See problem 26.

49. See problem 27.

50. See problem 28.

◈ CHAPTER 6 REVIEW PROBLEM SET

In problems 1–4, prove that each equation is an identity.

1. (a) $\cos t \sin t \csc t \sec t = 1$

(b) $\cot \theta \cos \theta + \sin \theta = \csc \theta$

2. (a) $\sec t - \cos t = \sin t \tan t$

(b) $(\csc t + \cot t)(1 - \cos t) = \sin t$

3. (a) $\dfrac{\sin^2 t \cos t + \cos^3 t}{\cot t} = \sin t$

(b) $\cos^2 t - \sin^2 t = \dfrac{1 - \tan^2 t}{1 + \tan^2 t}$

4. (a) $\dfrac{\sin^4 t - \cos^4 t}{\sin^2 t - \cos^2 t} = 1$

(b) $\dfrac{\sec^2 t + 2 \tan t}{2 - \sec^2 t} = \dfrac{1 + \tan t}{1 - \tan t}$

In problems 5–10, simplify each expression by using identities.

5. (a) $\sin 7t \cos 2t - \cos 7t \sin 2t$

(b) $\sin(t + s) \cos s - \cos(t + s) \sin s$

6. (a) $\cos 51° \cos 39° - \sin 51° \sin 39°$

(b) $\cos 33° \cos 27° - \sin 33° \sin 27°$

7. (a) $\sin(3\pi - t)$

(b) $2 \sin^2 4\theta - 1$

8. (a) $\csc(90° - \theta) \cos \theta + \cot^2 \theta$

(b) $\cos(270° + \theta)$

9. (a) $\cos\left(\dfrac{\pi}{2} - t\right) \tan\left(\dfrac{\pi}{2} - t\right)$

(b) $\cos\left(\dfrac{\pi}{3} - t\right) - \cos\left(\dfrac{\pi}{3} + t\right)$

10. (a) $\dfrac{\tan 5t + \tan t}{1 - \tan 5t \tan t}$

(b) $\dfrac{\tan 7t - \tan 2t}{1 + \tan 7t \tan 2t}$

In problems 11–16, use an identity to find the exact value of each expression.

11. (a) $\cos\left(\dfrac{\pi}{4} + \dfrac{\pi}{6}\right)$

(b) $\cos \dfrac{\pi}{4} + \cos \dfrac{\pi}{6}$

12. (a) $\sin\left(\dfrac{3\pi}{4} - \dfrac{\pi}{3}\right)$

(b) $\sin \dfrac{3\pi}{4} - \sin \dfrac{\pi}{3}$

13. (a) $\sin 165°; \ 165° = 120° + 45°$

(b) $\cos 195°; \ 195° = 150° + 45°$

14. (a) $\cos 285°; \ 285° = 240° + 45°$

(b) $\cos 255°; \ 255° = 210° + 45°$

15. (a) $\sin 22.5°; \ 22.5° = \dfrac{1}{2}(45°)$

(b) $\cos \dfrac{11\pi}{12}; \ \dfrac{11\pi}{12} = \dfrac{1}{2}\left(\dfrac{11\pi}{6}\right)$

16. (a) $\cos 75°; \ 75° = \dfrac{1}{2}(150°)$

(b) $\sin \dfrac{7\pi}{12}; \ \dfrac{7\pi}{12} = \dfrac{1}{2}\left(\dfrac{7\pi}{6}\right)$

17. Use the fundamental identities to find the exact values of the other five trigonometric functions under the given conditions.

(a) $\sin \theta = \dfrac{3}{5}; \quad \theta$ in quadrant II

(b) $\cos \theta = \dfrac{3}{8}; \quad \theta$ in quadrant IV

18. Use the fundamental identities to simplify each expression.

(a) $\csc^2 t \tan^2 t - \tan^2 t$

(b) $\sec t - \sin t \tan t$

(c) $\sin(-\theta) \sec(-\theta)$

19. Suppose that $\sin t = 1/4$, where t is in quadrant I, and $\cos s = -5/17$, where s is in quadrant II. Find:

(a) $\sin(t - s)$

(b) $\cos(t - s)$

(c) $\sin 2t$

(d) $\cos\left(\dfrac{s}{2}\right)$

(e) $\tan 2t$

20. Suppose that $\tan t = -4/3$, where t is in quadrant II, and $\tan s = -5/12$, where s is in quadrant IV. Find:

(a) $\tan(t + s)$

(b) $\sin(t + s)$

(c) $\cos(t + s)$

(d) $\cos 2s$

(e) $\sin\left(\dfrac{t}{2}\right)$

In problems 21–28, prove that each equation is an identity.

21. $2 \csc 2\theta \cot \theta = 1 + \cot^2 \theta$

22. $\csc \theta \sin 2\theta = 2 \cos \theta$

23. $\tan\left(t + \dfrac{3\pi}{4}\right) = \dfrac{\tan t - 1}{\tan t + 1}$

24. $\dfrac{\cos\left(t - \dfrac{\pi}{4}\right)}{\cos t \sin \dfrac{\pi}{4}} = \tan t + 1$

25. $\dfrac{1 - \cos 2\theta}{\sin 2\theta} = \tan \theta$

26. $\dfrac{\cos 2\theta}{1 - \sin^2 \theta} = 2 - \sec^2 \theta$

27. $\sin \dfrac{t}{2} \cos \dfrac{t}{2} = \dfrac{\sin t}{2}$

28. $\dfrac{1 - \tan^2 \theta}{1 + \tan^2 \theta} = \cos 2\theta$

In problems 29–32, solve each equation, where $0 \le t < 2\pi$ or $0° \le \theta < 360°$. Round off part (b) of each problem to two decimal places.

29. (a) $\sin \theta = -1$

    (b) $\cos \theta = 0.3217$

30. (a) $\tan t = \sqrt{3}$

    (b) $\sec t = 1.4931$

31. (a) $5 \sin^2 t - 4 \sin t - 1 = 0$

    (b) $(2 \sin t - 1)\left(\sin t - \dfrac{1}{3}\right) = 0$

32. (a) $3 \tan^2 \theta - \sqrt{3} \tan \theta = 0$

    (b) $\tan^2 t - \tan t - 2 = 0$

In problems 33 and 34, find the exact value of each expression without using a calculator.

33. (a) $\cos 15° \cos 105°$

    (b) $\cos \dfrac{11\pi}{12} \sin \dfrac{7\pi}{12}$

34. (a) $\sin 195° \cos 15°$

    (b) $\sin \dfrac{\pi}{12} \cos \dfrac{7\pi}{12}$

In problems 35 and 36, write each expression as a sum or difference of sine or cosine functions.

35. (a) $2 \sin t \cos 2t$

    (b) $2 \cos 3t \cos 5t$

36. (a) $\sin \dfrac{5t}{2} \cos \dfrac{t}{2}$

    (b) $\sin 7t \sin 3t$

In problems 37 and 38, write each expression as a product of sine or cosine functions.

37. (a) $\sin 4t - \sin t$

    (b) $\cos 7t + \cos 3t$

38. (a) $\sin 10t + \sin 2t$

    (b) $\cos 6t - \cos 2t$

G In problems 39 and 40, use the features on a grapher to approximate the solution of each equation to two decimal places for $0 \le t < 2\pi$.

39. $\cos t = 3t$

40. $2 \sin t = 2t - 1$

In problems 41–44, write each equation in the form $y = A \cos(u - v)$.

41. $y = \sqrt{3} \sin x + \cos x$

42. $y = -\sin x - \sqrt{3} \cos x$

43. $y = -\sin x + \cos x$

44. $y = -\sin x - \cos x$

45. **Electrical Voltage:** Suppose the voltage V (in volts) in an electrical circuit is given by the model

$$V = 80 \cos\left(120\pi t - \dfrac{\pi}{4}\right)$$

where t is in seconds.

(a) Graph one cycle of the function.

(b) Determine the smallest positive value of t when the voltage is 60 volts. Round off to three decimal places.

46. G **Business Profit:** A company's profit during a 12-month period is modeled by the function

$$p(t) = 100{,}000(t - 2 \cos t)$$

where $p(t)$ (in dollars) is the profit after t months.

(a) Use a grapher to graph p for $0 \le t < 12$.

(b) Use the grapher to approximate the value of t to three decimal places, when $p(t) = \$650{,}000$.

47. Billiard Ball Motion: Figure 1 shows a billiard ball (at position A) that travels a path to hit a ball at position B. Assume the billiard ball strikes the rail of the table and rebounds so that the angles labeled θ are equal.

Figure 1

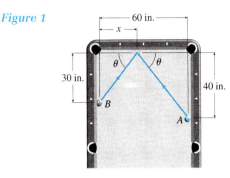

(a) Find the distance x denoted in the figure.

(b) Find $\tan \theta$, and then solve the equation for θ. Round off the answer to the nearest degree.

48. Hours of Daylight: Suppose the number of hours N of daylight on a given day of the year for a certain area is given by the model

$$N = 12 + 3 \sin\left[\frac{2\pi}{365}(t - 85)\right]$$

where t is the day of the year, and $t = 0$ is January 1.

(a) Graph the function.

(b) Determine which days of the year have 14 hours of daylight. Round off the answers to two decimal places.

◈ CHAPTER 6 TEST

1. Use
$$255° = 225° + 30°$$
to find the exact value of each expression.
(a) $\sin 255°$ (b) $\cos 255°$ (c) $\tan 255°$

2. Suppose that $\sin s = -3/5$, where s is in quadrant IV, and $\cos t = 4/5$, where t is in quadrant IV. Find the exact value of each expression.
(a) $\sin(s + t)$ (b) $\cos(s + t)$ (c) $\tan(s + t)$
(d) $\sin 2t$ (e) $\cos 2t$ (f) $\sin 4t$
(g) $\sin^2\left(\dfrac{t}{2}\right)$ (h) $\sin(s + 2t)$

3. Use $67.5° = (1/2)(135°)$ to find the exact value of each expression.
(a) $\sin 67.5°$ (b) $\cos 67.5°$ (c) $\tan 67.5°$

4. Simplify each trigonometric expression by using identities.
(a) $1 - 2\cos^2 47°$
(b) $\sin \dfrac{5\pi}{7} \cos \dfrac{5\pi}{7}$
(c) $\sin 25° \cos 35° + \sin 35° \cos 25°$
(d) $\cos \dfrac{5\pi}{7} \cos \dfrac{2\pi}{7} - \sin \dfrac{5\pi}{7} \sin \dfrac{2\pi}{7}$
(e) $\cos \dfrac{3t}{4} \cos \dfrac{t}{4} + \sin \dfrac{3t}{4} \sin \dfrac{t}{4}$

5. Solve each equation for $0° \le \theta < 360°$ or $0 \le t < 2\pi$. Round off the answers to two decimal places.
(a) $4\cos^2 t - 3 = 0$
(b) $\sin 2\theta = \dfrac{1}{2}$
(c) $(\tan t - 1)(2\tan t + 3) = 0$

6. Prove that each equation is an identity.
(a) $(\csc^2 t - 1)(1 + \tan^2 t) = \csc^2 t$
(b) $\sin 2t \sec t = 2 \sin t$
(c) $\sin\left(t + \dfrac{\pi}{6}\right) \sin\left(t - \dfrac{\pi}{6}\right) = \sin^2 t - \dfrac{1}{4}$
(d) $\dfrac{1 - \cos 2t}{\sin t} = 2 \sin t$

7. Use a grapher to solve the equation $3x + \cos x = 0$. Round off the answer to two decimal places.

8. An observer in a balloon finds that the angle of depression from a balloon to a car is θ (measured in degrees). The balloon is x feet high and the distance horizontal from a point on the ground directly under it to the car is 240 feet.
(a) Express $\cos 2\theta$ as a function of x.
(b) If $x = 296$ feet, what is the measure of θ? Round off the answer to two decimal places.

Additional Applications of Trigonometry

Chapter Contents

So far we have used trigonometry to solve right triangles. However, it can also be used to solve other types of triangles. For instance, a gear designer wishes to mount three circular gears with centers at *A, B,* and *C* at the vertices of a triangle *ABC.* The gear with center at *A* has a 2-inch radius, the gear with center at *B* has a 3-inch radius, and the gear with center *C* has a 5-inch radius. What are the measures of the angles of the triangle *ABC* to make this mounting possible? The solution to this problem is given in Example 5 on page 497.

In fields such as surveying, navigation, and engineering, it is often necessary to solve triangles that do not necessarily contain a right angle. In this chapter, we will study ways of solving such triangles. We will also use trigonometry to explore the polar coordinate system, to study vectors, and to extend our understanding of the complex number system.

Objectives

7.1 The Law of Cosines and the Law of Sines

In this section, we derive two formulas—the *law of cosines* and the *law of sines*—that enable us to find the missing parts of a triangle under certain conditions.

For convenience, we shall adopt standard notation to label the sides and angles of a triangle. Thus in triangle *ABC* (Figure 1) the angle at vertex *A* is denoted by α, the angle at *B* is denoted by β, and the angle at *C* is denoted by γ. The length of the side opposite α is *a*, the length of the side opposite β is *b*, and the length of the side opposite γ is *c*.

Figure 1

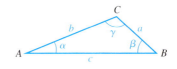

Using the Law of Cosines to Solve Triangles

We begin by establishing the *law of cosines* which is a generalization of the Pythagorean theorem.

The Law of Cosines

In any triangle *ABC*:

$$\text{(i)} \quad c^2 = a^2 + b^2 - 2ab \cos \gamma$$
$$\text{(ii)} \quad b^2 = a^2 + c^2 - 2ac \cos \beta$$
$$\text{(iii)} \quad a^2 = b^2 + c^2 - 2bc \cos \alpha$$

Proofs of Formulas

Figure 2

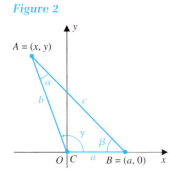

(i) To derive the law of cosines, consider triangle *ABC* in a Cartesian plane with γ in standard position and assume that vertex *A* has coordinates (x, y) (Figure 2).

By the distance formula, we have

$$c^2 = (x - a)^2 + y^2$$
$$= a^2 + (x^2 + y^2) - 2ax \qquad \text{Multiply and rearrange terms}$$
$$= a^2 + b^2 - 2ax \qquad\qquad x^2 + y^2 = b^2$$
$$= a^2 + b^2 - 2ab \cos \gamma \qquad \cos \gamma = \frac{x}{b}, \text{ so } b \cos \gamma = x$$

Formulas (ii) and (iii) are obtained by placing the other two angles in standard position and repeating the steps in the above proof.

The three formulas (i), (ii), and (iii) can be consolidated into one word form for the law of cosines as follows:

The square of the length of any side of a triangle is equal to the sum of the squares of the lengths of the other two sides, minus twice the product of these two sides and the cosine of their included angle.

The law of cosines is used to solve a triangle when given either

(i) two sides and the angle between them (SAS),

or

(ii) three sides (SSS).

EXAMPLE 1   **Using the Law of Cosines**

Use the law of cosines to find the indicated parts. Round off the answers to two decimal places.

(a) In triangle *ABC* (Figure 3), find *c* if $a = 8$, $b = 6$, and $\gamma = 60°$.

(b) Use the result in part (a) to find the measures of α and β.

Figure 3

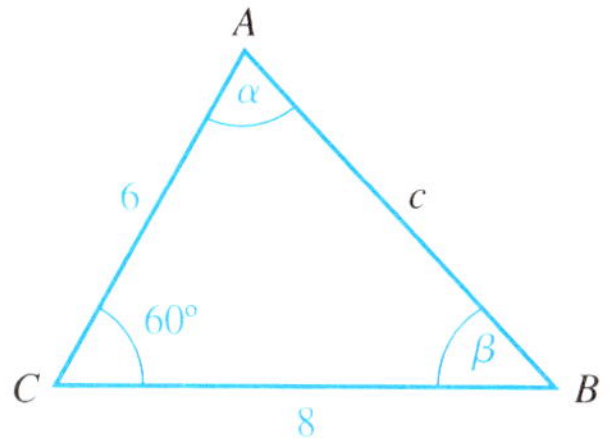

Solution

(a) We observe in Figure 3 that two sides and the included angle of the triangle are given. This corresponds to the SAS case, so the law of cosines is applicable. To find *c*, we write:

$$c^2 = a^2 + b^2 - 2ab \cos \gamma \qquad \text{Formula}$$

$$= 64 + 36 - 2(8)(6)\left(\frac{1}{2}\right) = 100 - 48 = 52 \qquad \text{Substitute}$$

So $c = \sqrt{52} = 2\sqrt{13}$ or 7.21 (approx.)   Compute

(b) We now know the lengths of all sides of triangle *ABC*:

$$a = 8, b = 6, \quad \text{and} \quad c = 2\sqrt{13}$$

This corresponds to the SSS situation, so the law of cosines can be used to find either α or β.

To find α we proceed as follows:

$$a^2 = b^2 + c^2 - 2bc \cos \alpha \qquad \text{Law of cosines}$$

$$8^2 = 6^2 + \left(2\sqrt{13}\right)^2 - 2(6)\left(2\sqrt{13}\right) \cos \alpha \qquad \text{Substitute}$$

$$\cos \alpha = \frac{36 + 52 - 64}{24\sqrt{13}} = \frac{1}{\sqrt{13}} \qquad \text{Simplify and solve for } \cos \alpha$$

Therefore

We could have solved for β first, and then found α second.

$$\alpha = \cos^{-1}\left(\frac{1}{\sqrt{13}}\right) = 73.90° \text{ (approx.)}.$$

The other angle β can be found by using the fact that the sum of the interior angles of a triangle is 180°.

So,

$$\beta = 180° - 60° - 73.90° = 46.10° \text{ (approx.)}.$$

Using the Law of Sines to Solve Triangles

Next we derive another formula used to solve triangles the *law of sines.*

The Law of Sines

In any triangle *ABC*:

$$\frac{\sin \alpha}{a} = \frac{\sin \beta}{b} = \frac{\sin \gamma}{c}$$

Proof   To derive the law of sines,

let h_1 be the height of triangle *ABC* from vertex *C* to side $\overline{AB}$

and

let h_2 be the height from vertex *B* to side $\overline{AC}$ (Figure 4).

Figure 4

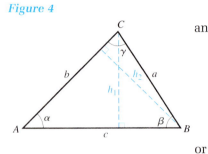

By using right triangle trigonometry, we have

$$\sin \alpha = \frac{h_1}{b} \qquad \text{and} \quad \sin \beta = \frac{h_1}{a}$$

or

$$h_1 = b \sin \alpha \quad \text{and} \qquad h_1 = a \sin \beta.$$

So

$$b \sin \alpha = a \sin \beta;$$

that is,

$$\frac{\sin \alpha}{a} = \frac{\sin \beta}{b}.$$

Similarly, we have

$$\sin \alpha = \frac{h_2}{c} \quad \text{and} \quad \sin \gamma = \frac{h_2}{a}.$$

So

$$h_2 = c \sin \alpha = a \sin \gamma,$$

or

$$\frac{\sin \alpha}{a} = \frac{\sin \gamma}{c}.$$

Thus

$$\frac{\sin \alpha}{a} = \frac{\sin \beta}{b} = \frac{\sin \gamma}{c}.$$

Note that the law of sines consists of the following three formulas

(i) $\dfrac{\sin \alpha}{a} = \dfrac{\sin \beta}{b}$ (ii) $\dfrac{\sin \alpha}{a} = \dfrac{\sin \gamma}{c}$ (iii) $\dfrac{\sin \beta}{b} = \dfrac{\sin \gamma}{c}$

and that it can also be written in the equivalent reciprocal form

$$\frac{a}{\sin \alpha} = \frac{b}{\sin \beta} = \frac{c}{\sin \gamma}.$$

The law of sines is used to solve a triangle when given

The AAS case can always be converted to the ASA case by first solving for the third angle.

   (i) Two angles and the side between them (ASA)

     or

  (ii) Two angles and a side (not the included one) (AAS)

     or

 (iii) Two sides and an angle opposite one of them (SSA)

EXAMPLE 2

Using the Law of Sines (ASA)

In triangle ABC (Figure 5), $a = 20$, $\gamma = 51°$, and $\beta = 42°$. Solve the triangle for α, b, and c. Round off the answers to two decimal places.

Figure 5

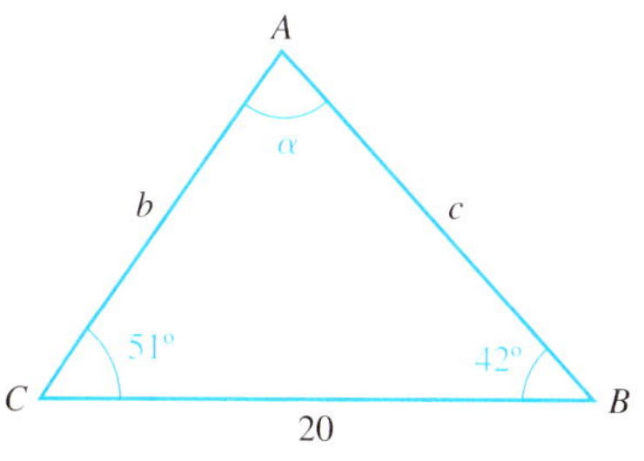

Solution

Because $\alpha + \beta + \gamma = 180°$,

$$\alpha = 180° - (\beta + \gamma) = 180° - 93° = 87°.$$

Since we have an ASA situation, the law of sines is applicable. Thus

$$\frac{b}{\sin \beta} = \frac{a}{\sin \alpha}$$

$$\frac{b}{\sin 42°} = \frac{20}{\sin 87°}.$$

So

$$b = \frac{20 \sin 42°}{\sin 87°} = 13.40 \text{ (approx.)}$$

To find c, we again use the law of sines to get

$$\frac{c}{\sin \gamma} = \frac{a}{\sin \alpha}$$

$$\frac{c}{\sin 51°} = \frac{20}{\sin 87°}.$$

So

$$c = \frac{20 \sin 51°}{\sin 87°} = 15.56 \text{ (approx.)}.$$

Solving the Ambiguous Case

Data such as those given in Example 2 lead to exactly one triangle. To decide whether or not we have enough information to determine a triangle, it is often helpful to make a sketch. For instance, if we are given two angles and the included side as shown in Figure 5, then only one triangle can be formed. However, when we are given two sides and an angle opposite one of them (SSA), a unique triangle is not always determined. In this case, there are *three* possible outcomes: no triangle; only one triangle; two different triangles. These three possible outcomes are illustrated in Table 1. For this reason, we sometimes refer to the SSA situation as the **ambiguous case** of the law of sines.

TABLE 1   Ambiguous Case (SSA) Possibilities (Given a, b, and α)

$h = b \sin \alpha$	Number of Possible Triangles	Illustration
$0 < a < h$	0	
$h < a < b$	2: $\triangle AB_1C$ and $\triangle AB_2C$	
$a \geq b$	1: $\triangle ABC$	
$a = h$	1	

Memorizing special rules to handle each of these possibilities is unnecessary. Rather, we apply the law of sines and reason our way through the problem, as illustrated in the next example.

EXAMPLE 3   **Using the Law of Sines (SSA)**

Solve each triangle (if possible). Determine all angles to the nearest hundredth of a degree and all lengths to two decimal places.

(a) $a = 5$, $b = 20$, $\alpha = 30°$ (b) $a = 5$, $b = 9$, $\alpha = 33°$

(c) $a = 20$, $b = 15$, $\alpha = 29°$

Solution   (a) Using the law of sines with $a = 5$, $b = 20$, and $\alpha = 30°$, we have

Figure 6

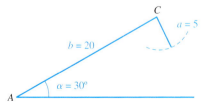

$$\frac{\sin \beta}{20} = \frac{\sin 30°}{5}$$

or

$$\sin \beta = \frac{20 \sin 30°}{5} = 2 > 1.$$

Since the sine cannot be greater than 1, there is *no possible triangle* satisfying the given conditions. Figure 6 illustrates this situation.

(b) Using the law of sines with $a = 5$, $b = 9$, and $\alpha = 33°$, we have

$$\frac{\sin \beta}{9} = \frac{\sin 33°}{5}$$

or

$$\sin \beta = \frac{9 \sin 33°}{5}.$$

One possible angle that satisfies this equation is the acute angle β_1 given by

$$\beta_1 = \sin^{-1}\left(\frac{9 \sin 33°}{5}\right)$$

$$= 78.62° \text{ (approx.)}.$$

However, using 78.62° as a reference angle, we find an obtuse angle $\beta_2 = 180° - 78.62° = 101.38°$ (approx.) that also satisfies the equation (Figure 7a).

Figure 7

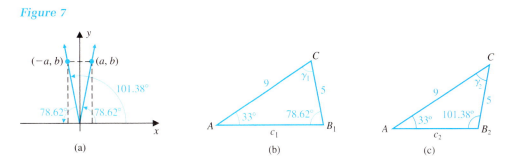

(a)   (b)   (c)

The solution $\beta_1 = 78.62°$ leads to the triangle AB_1C shown in Figure 7b. Here,

$$\gamma_1 = 180° - (33° + 78.62°) = 68.38° \text{ (approx.)}.$$

So, by the laws of sines

$$\frac{c_1}{\sin \gamma_1} = \frac{a}{\sin \alpha}$$

or

$$c_1 = \frac{a \sin \gamma_1}{\sin \alpha} = \frac{5 \sin 68.38°}{\sin 33°}$$

$$= \frac{5 \sin 68.38°}{\sin 33°}$$

$$= 8.53 \text{ (approx.)}.$$

The solution $\beta_2 = 101.38°$ leads to a second possibility, triangle AB_2C, shown in Figure 7c. For this triangle,

$$\gamma_2 = 180° - (33° + 101.38°)$$

$$= 45.62°$$

So

$$\frac{c_2}{\sin \gamma_2} = \frac{a}{\sin \alpha}$$

or

$$c_2 = \frac{a \sin \gamma_2}{\sin \alpha}$$

$$= \frac{5 \sin 45.62°}{\sin 33°}$$

$$= 6.56 \text{ (approx.)}.$$

Thus in this situation we have *two possible triangles* with the given parts.

(c) Applying the law of sines with $a = 20$, $b = 15$, and $\alpha = 29°$, we have

$$\frac{\sin \beta}{b} = \frac{\sin \alpha}{a} \quad \text{or} \quad \sin \beta = \frac{b \sin \alpha}{a} = \frac{15 \sin 29°}{20}.$$

We find that an acute angle given by

$$\beta_1 = \sin^{-1}\left(\frac{15 \sin 29°}{20}\right) = 21.32° \text{ (approx.)}$$

and an obtuse angle given by

$$\beta_2 = 180° - 21.32° = 158.68° \text{ (approx.)}.$$

both β_1 and β_2 give the same value of $\sin \beta$.
However, it is impossible to have angle $\beta_2 = 158.68°$ in a triangle that is known to have an angle of 29°, since

$$\alpha + \beta_2 = 29° + 158.68° = 187.68° > 180°.$$

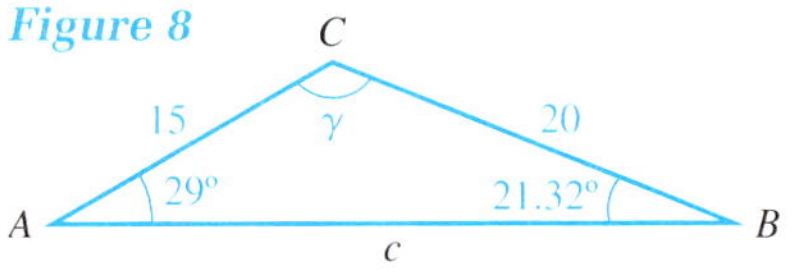

So the β_2 solution must be rejected. There is *only one possible triangle* (Figure 8) with the given parts.
The solution $\beta_1 = 21.32°$ leads to

$$\gamma = 180° - (29° + 21.32°) = 129.68° \text{ (approx.)}$$

and

$$\frac{c}{\sin 129.68°} = \frac{20}{\sin 29°}$$

or

$$c = \frac{20 \sin 129.68°}{\sin 29°} = 31.75 \text{ (approx.)}.$$

Figure 8

Solving Applied Problems

The following examples illustrate applications of the law of cosines and the law of sines in mathematical modeling involving triangles.

EXAMPLE 4

Measuring Distance Indirectly

A straight tunnel with end points A and B is to be blasted through a mountain. The distance between A and B cannot be measured directly. So, from a third point C, a surveyor finds the distance from C to A to be 573 meters and from C to B to be 819 meters (Figure 9). If the measure of angle ACB is 67°, find the length of the tunnel; that is, find $|\overline{AB}|$. Round off the answer to two decimal places.

Figure 9

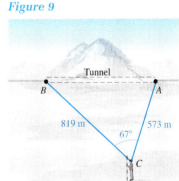

Solution

Here we have an SAS case, so we apply the law of the cosines to triangle ACB to get

$$|\overline{AB}|^2 = |\overline{CB}|^2 + |\overline{CA}|^2 - 2|\overline{CB}||\overline{CA}| \cos \gamma$$
$$= 819^2 + 573^2 - 2(819)(573) \cos 67°.$$

Hence

$$|\overline{AB}| = \sqrt{819^2 + 573^2 - 2(819)(573) \cos 67°}$$
$$= 795.21 \text{ (approx.)}$$

so the length of the tunnel is about 795.21 meters.

EXAMPLE 5   **Gear Designing**

A gear designer wishes to mount three circular gears with centers at *A*, *B*, and *C* as the vertices of a triangle *ABC*. The gear with center at *A* has a 2-inch radius, the gear with center at *B* has a 3-inch radius, and the gear with center at *C* has a 5-inch radius. What are the measures of the angles of the triangle *ABC* to make this mounting possible? Round off the answers to the nearest hundredth of a degree.

Solution   This is an SSS situation with $a = 8$, $b = 7$, and $c = 5$ (Figure 10), and so we apply the law of cosines.

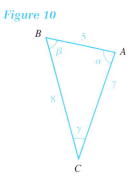

Figure 10

To find angle α, we write

$$a^2 = b^2 + c^2 - 2bc \cos \alpha \qquad \text{Law of sines}$$

$$64 = 49 + 25 - 2(7)(5) \cos \alpha \qquad \text{Substitute}$$

$$\cos \alpha = \frac{1}{7} \qquad \text{Solve for } \cos \alpha$$

So $\qquad \alpha = \cos^{-1}\left(\frac{1}{7}\right) = 81.79° \qquad$ Approximate

To find angle β, we write

$$b^2 = a^2 + c^2 - 2ac \cos \beta \qquad \text{Law of cosines}$$

$$49 = 64 + 25 - 2(8)(5) \cos \beta \qquad \text{Substitute}$$

$$\cos \beta = \frac{1}{2} \qquad \text{Solve for } \cos \beta$$

$$\beta = 60° \qquad \cos^{-1}\left(\frac{1}{2}\right) = 60°$$

To find angle γ, we use

$$\gamma = 180° - (81.79° + 60°)$$

$$= 38.21°$$

Therefore, the angles α, β, and γ are approximately, 81.79°, 60°, and 38.21°, respectively.

In applications of trigonometry to surveying and navigation, the **direction** or **bearing** of a point *Q* as viewed from a point *P* is defined as the positive acute angle θ between the ray from *P* through *Q* and the north–south line through *P* (Figure 11). Bearings are usually measured in degrees east of north, west of north, east of south, or west of south. For instance, the symbol N25°E is read as 25° east of north.

Figure 11

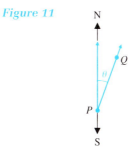

Figure 12 shows four bearings.

Figure 12

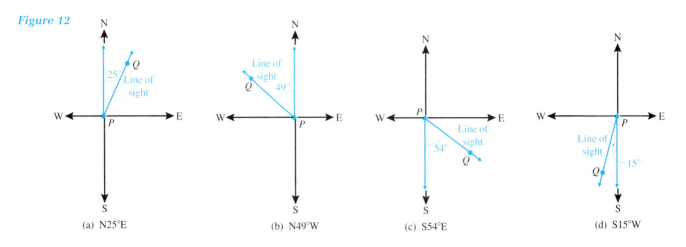

(a) N25°E   (b) N49°W   (c) S54°E   (d) S15°W

EXAMPLE 6   **Sighting a Fire from Observation Posts**

A forest ranger at observation post A sights a fire at location C in the direction N66.5°E. Another ranger at observation post B, 8 kilometers directly east of A, sights the same fire at N32.2°W. Figure 13 illustrates the situation. How far is the fire from observation post A? Round off the answer to two decimal places.

Figure 13

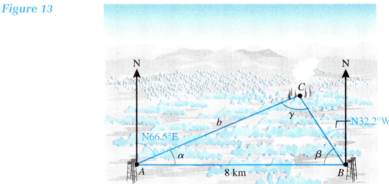

Solution   Because the interior angles α and β at vertices A and B are complementary to the known angles, we find that they measure 23.5° and 57.8°, respectively. So the angle γ at C is given by

$$\gamma = 180° - (23.5° + 57.8°) = 98.7°.$$

This ASA case can now be solved by using the law of sines to get

$$\frac{b}{\sin 57.8°} = \frac{8}{\sin 98.7°}$$

or
$$b = \frac{8 \sin 57.8°}{\sin 98.7°}$$

$$= 6.85 \text{ (approx.)}.$$

Therefore, the fire is about 6.85 kilometers from the observation post A.

EXAMPLE 7   **Modeling the Distance between Two Ships**

Two ships leave a port at the same time, traveling along straight-line paths at 13 nautical miles per hour and 20 nautical miles per hour, respectively. Assume that the angle between their directions of travel is θ, where $0° < \theta \le 180°$ and both ships maintain constant speeds.

(a) Express the distance d between them 1 hour later as a function of θ.

(b) Use the result in part (a) to find the distance between the ships if θ is 45°, 60°, or 75°. Round off the answers to two decimal places.

(c) Find angle θ if the distance between the ships is 10 nautical miles. Round off the answer to the nearest degree.

Solution   (a) After 1 hour, the ship traveling 20 nautical miles per hour will have traveled 20 nautical miles, and the ship traveling 13 nautical miles per hour will have traveled 13 nautical miles (Figure 14).
Using the law of cosines, we have

Figure 14

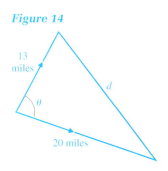

$$d^2 = 13^2 + 20^2 - 2(13)(20) \cos \theta$$

so that

$$d = \sqrt{169 + 400 - 520 \cos \theta}$$

or

$$d = \sqrt{569 - 520 \cos \theta}$$

which expresses d as a function of θ.

(b) To find the distance between the ships for the given values of θ, we substitute into the function

$$d = \sqrt{569 - 520 \cos \theta}$$

to get the following approximate values:

for $\theta = 45°$, $d = \sqrt{569 - 520 \cos 45°} = 14.19$ nautical miles;

for $\theta = 60°$, $d = \sqrt{569 - 520 \cos 60°} = 17.58$ nautical miles;

for $\theta = 75°$, $d = \sqrt{569 - 520 \cos 75°} = 20.84$ nautical miles.

(c) To find the angle θ if the distance between the ships is 10 nautical miles, we substitute 10 for d in the model to obtain the equation

$$10 = \sqrt{569 - 520 \cos \theta}.$$

Thus, $100 = 569 - 520 \cos \theta$ Square both sides

$$\cos \theta = \frac{469}{520}$$ Solve for $\cos \theta$

So, $\theta = \cos^{-1}\left(\dfrac{469}{520}\right) = 26°$ (approx.).

◆ PROBLEM SET 7.1

Mastering the Concepts

In problems 1–12, indicate whether the case is SAS or SSS, and then use the law of cosines to find the specified unknown part of triangle *ABC*. Round off angles to the nearest hundredth of a degree and side lengths to two decimal places.

1. *a;* if $b = 3$, $c = 4$, $\alpha = 30°$
2. *c;* if $a = 6$, $b = 4$, $\gamma = 59°$
3. *b;* if $a = 3$, $c = 4$, $\beta = 105°$
4. *a;* if $c = 2.4$, $b = 3.2$, $\alpha = 117°$
5. *c;* if $a = 5.8$, $b = 4.8$, $\gamma = 35.25°$
6. *b;* if $a = 6.1$, $c = 5.2$, $\beta = 5.67°$
7. *γ;* if $a = 2$, $b = 3$, $c = 4$
8. *β;* if $a = 14.3$, $b = 10.6$, $c = 8.4$
9. *β;* if $a = 23.7$, $b = 31.4$, $c = 40.6$
10. *α;* if $a = 0.6$, $b = 0.3$, $c = 0.5$
11. *α;* if $a = 81$, $b = 193$, $c = 253$
12. *γ;* if $a = 6$, $b = 5$, $c = 5.5$

In problems 13–22, indicate whether the case is

ASA or AAS,

and then use the law of sines to solve each triangle *ABC*. Round off angles to the nearest hundredth of a degree and side lengths to two decimal places.

13. $a = 20$, $\alpha = 29°$, $\beta = 136°$
14. $c = 32$, $\beta = 57°$, $\alpha = 38°$
15. $b = 30$, $\alpha = 80°$, $\gamma = 41°$
16. $a = 10.5$, $\alpha = 41°$, $\gamma = 77°$
17. $b = 19.7$, $\alpha = 42.17°$, $\gamma = 61.33°$
18. $b = 20$, $\alpha = 30°$, $\gamma = 52°$
19. $c = 5.4$, $\beta = 50.83°$, $\gamma = 70.5°$
20. $a = 13.1$, $\gamma = 100°$, $\alpha = 12.67°$
21. $b = 19$, $\alpha = 40°$, $\gamma = 62°$
22. $b = 38.8$, $\alpha = 103.45°$, $\gamma = 27.19°$

In problems 23–32, if possible, solve the ambiguous case

SSA

by using the law of sines. Round off angles to the nearest hundredth of a degree and side lengths to two decimal places. Be sure to find all possible triangles that satisfy the given conditions.

23. $b = 12$, $c = 11$, $\gamma = 81°$
24. $a = 140$, $c = 115$, $\gamma = 53.54°$
25. $a = 27$, $b = 52$, $\alpha = 70°$
26. $a = 1$, $b = 1.8$, $\alpha = 26°$
27. $a = 12.4$, $b = 8.7$, $\beta = 36.67°$
28. $a = 12.41$, $b = 81.69$, $\beta = 36.67°$
29. $a = 21.3$, $b = 18.9$, $\alpha = 65.18°$
30. $a = 263.6$, $c = 574.3$, $\alpha = 32.32°$
31. $a = 30$, $b = 44.5$, $\alpha = 33.33°$
32. $a = 10.1$, $b = 15.2$, $\alpha = 67.67°$

In problems 33–40, identify the case

SSS, SAS, ASA, AAS, or SSA

and then use the law of sines or the law of cosines to solve triangle *ABC*. Round off all angles to the nearest hundredth of a degree and side lengths to two decimal places. Be sure to find all possible triangles that satisfy the given conditions.

33. $a = 10$, $b = 50$, $\alpha = 22°$
34. $a = 14$, $\alpha = 21°$, $\beta = 35°$
35. $a = 4$, $b = 4$, $c = 6$
36. $a = 5$, $b = 10$, $c = 10$
37. $a = 14$, $\alpha = 12°$, $\beta = 97°$
38. $a = 10$, $b = 12$, $\gamma = 108°$
39. $b = 3.7$, $\alpha = 100.30°$, $\beta = 5.50°$
40. $b = 98$, $c = 39$, $\alpha = 17°$

Applying the Concepts

In problems 41–58, round off angles to the nearest hundredth of a degree and lengths to two decimal places.

41. **Surveying:** A vertical tower that is 70 feet high stands on a hill that makes an angle of 14° with the horizontal. A surveyor locates a point *C* downhill, 110 feet from the base of the tower at *A* (Figure 15). Find the distance from *C* to the top of the tower, *B;* that is, find $|\overline{CB}|$.

Figure 15

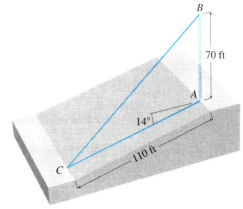

42. Distance between Jets: Two jets leave an air base at the same time and fly at the same speed along straight courses forming an angle of 112.45° with each other (Figure 16). After the jets have each flown 504 kilometers, how far apart are they?

Figure 16

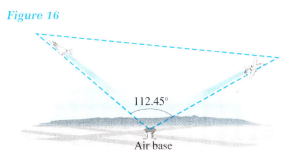

43. Surveying: Engineers are planning to construct a straight tunnel from point *A* on one side of a hill to point *B* on the other side. Since its length cannot be measured directly, point *C* is chosen by a surveyor who is 473 meters from *A* and 367 meters from *B* (Figure 17). If the measure of angle *ACB* is 46.2°, find the length of the tunnel.

Figure 17

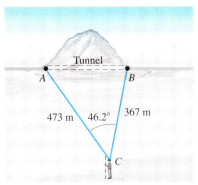

44. Surveying: To measure the length of a lake between points *A* and *B,* a point *C* on land is chosen. The distance between *C* and *A* is found to be 137 meters, and the distance between *C* and *B* is 405 meters (Figure 18). If angle *ACB* is 42.5°, how long is the lake?

Figure 18

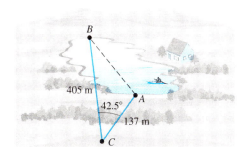

45. Joggers' Courses: Two straight roads intersect at 65° (Figure 19). A jogger on one road is 5 miles from the intersection and moving away from it at a constant rate of 7 miles per hour. At the same instant, another jogger on the other road is 4 miles from the intersection and moving away from it at a constant rate of 5 miles per hour. If the two joggers maintain constant speeds, what is the distance between them 30 minutes later?

Figure 19

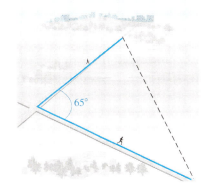

46. Navigation: A navigator of a ship plots a straight course from point *A* to point *B*. Because of an error, the ship proceeds from point *A* along a slightly different straight-line course. After traveling 3 hours on the wrong course, the error is discovered and the ship is turned through an angle of 22.5° so that it can reach point *B* (Figure 20). After 5 more hours, the ship reaches point *B*. Assuming that the ship was proceeding at a constant speed of 10 nautical miles per hour, find the time lost (to the nearest minute) because of the error.

Figure 20

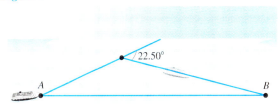

47. Lighthouse Lookout: As shown in Figure 21, two lighthouses along a straight shoreline of a lake are located 2.43 kilometers apart at points *A* and *B*. The keepers of the lighthouses at *A* and *B* sight an overturned boat at point *C* at angles 70.3° and 48.2°, respectively. A Coast Guard rescue boat at *D*, located 1.2 kilometers downshore from *B*, is to rescue the boat. How far is the Coast Guard boat from the overturned boat?

Figure 21

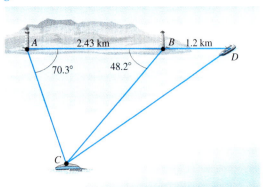

48. Baseball: Figure 22 shows a baseball diamond. It is in the form of a square that is 90 feet long on each side. The pitcher's mound at *P* is 60.5 feet from home plate, *H*, along the diagonal from home plate to second base. How far is the pitcher's mound *P* from first base, located at *F?*

Figure 22

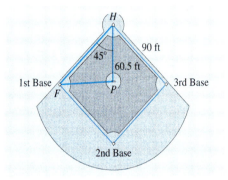

49. Fire Lookout: Two forest rangers are 1.2 kilometers apart. One ranger sights a fire at an angle of 41.43° from the line between the two observation points. The other ranger sights the same fire at an angle of 61.4° from the same line between the two observation points (Figure 23). How far is the fire from each observation point?

Figure 23

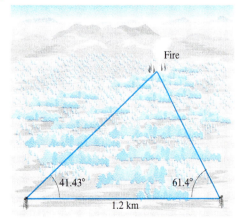

50. Distance to Top of a Tree: A biologist standing on the ground on the same horizontal plane as the base of a tree determines that the angle of elevation from a point *A* to the top of the tree is 26.2°. After walking 150 feet closer to the tree to a point *B*, she finds the angle of elevation from point *B* to the top of the tree to be 37.1°. Find the distance from point *B* to the top of the tree.

51. Distance from a Parachutist: A sports parachutist is sighted simultaneously by two observers 3 miles apart on opposite sides of the parachutist. The observed angles of elevation are 25.5° and 17.8°. Assuming that the parachutist and the two observers lie in the same vertical plane, find the distance of the parachutist from the farther observer.

52. Engineering: The crankshaft $\overline{OA}$ of an engine is 5.8 centimeters long, and the connecting rod $\overline{AP}$ is 20.3 centimeters long (Figure 24). Find angle *AOP* at the instant when angle *APO* is 12°.

Figure 24

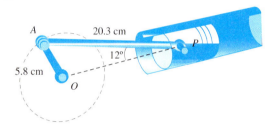

53. Streetlight Design: A streetlight is to be mounted on a brick wall. The length of brace $\overline{BC}$ is 1.3 meters, angle *BCA* is 24.8°, and angle *CAB* is 37.8° (Figure 25). Find the length of the supporting brace $\overline{AC}$.

Figure 25

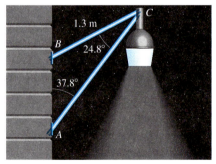

54. Communications Satellite: A communications satellite traveling in a circular orbit 1600 kilometers above the surface of the earth is located at position *A* by a tracking station *B* on the earth at a certain time (Figure 26). If the tracking antenna is aimed 35° above the horizon and if the radius of the earth is 6400 kilometers, what is the distance from the antenna to the satellite?

Figure 26

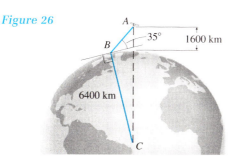

55. Navigation: A ship sails 19 nautical miles in the direction S29.3°W, and then turns onto a course S51.7°W and sails 24 nautical miles. How far is the ship from the starting point?

56. Navigation: A ship leaves a harbor at noon and sails S50°W at 18 nautical miles per hour until 2:30 PM. At that time, it changes course and sails N20°W at a reduced speed of 13 nautical miles per hour until 4:00 PM. Find:

(a) The distance of the ship from the harbor at 4:00 PM

(b) The ship's bearing from the harbor at 4:00 PM

57. Distance Model: Two straight roads intersect at an angle of 70°. A bus on one of the roads is 8 kilometers from the intersection and moving away from it at a rate of 90 kilometers per hour. At the same instant, a truck on the other road is 16 kilometers from the intersection and moving away from it at 100 kilometers per hour. Assume the bus and the truck maintain constant speeds.

(a) Express the distance d between them t hours later as a function of t.

(b) Find the distance between the two vehicles after 30 minutes, after 60 minutes, and after 75 minutes.

(c) Approximate how long it takes for the distance between the vehicles to be 50 kilometers, 100 kilometers, and 150 kilometers. Round off the answers to two decimal places.

58. Navigation—Distance Model: Two ships leave a port at the same time, traveling along straight-line paths at 12 nautical miles per hour and 16 nautical miles per hour, respectively. Assume that the angle between their directions of travel is θ, where $0° < \theta \le 180°$, and both ships maintain constant speeds.

(a) Express the distance d between them 1 hour later as a function of θ.

(b) Use the result in part (a) to find the distance between the ships if θ is 10°, 90°, or 120°.

(c) Find angle θ if the distance between the ships is 5 nautical miles, 20 nautical miles, or 25 nautical miles. Round off the answers to the nearest degree.

Developing and Extending the Concepts

59. Geometry: Figure 27 shows three circles with centers at P, Q, and R, and with radii 5, 8, and 8.4 centimeters, respectively. If the circles are tangent to one another, find the three angles of triangle PQR formed by the line segments connecting their centers.

Figure 27

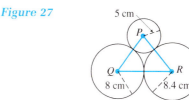

60. Geometry: Assume that a diagonal of a parallelogram is 80 inches long and that one end of it forms angles of 35° and 27°, respectively, with the two sides. Find the lengths of the sides of the parallelogram.

61. Find the lengths of the diagonals of parallelogram $ABCD$ with side $\overline{AB}$ of length 10 inches and side $\overline{BC}$ of length 15 inches, and with angle ABC of 110°.

62. Rewrite formula (i) of the law of cosines if $\gamma = 90°$. Explain the result.

63. Explain why it is not possible to have a triangle with side lengths of 7, 8, and 16. What happens if we attempt to find an angle by using the law of cosines under the SSS case?

64. Assume that triangle ABC has height $h = b \sin \alpha$ (Figure 28).

Figure 28

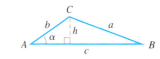

Using the law of cosines, we have

$$a^2 = b^2 + c^2 - 2bc \cos \alpha$$

or $b^2 + c^2 - a^2 - 2bc \cos \alpha = 0.$

So $c^2 - (2b \cos \alpha)c + (b^2 - a^2) = 0.$

(a) Use the quadratic formula with c as the unknown to show that

$$c = b \cos \alpha \pm \sqrt{a^2 - b^2 \sin^2 \alpha}.$$

(b) Under what conditions on a and h are there no real solutions for c? Relate the result to the ambiguous case when no triangles exist for given values of a, b, and α.

(c) If $a^2 - b^2 \sin^2 \alpha > 0$, how many possible triangles exist? Explain.

Objectives

1. Plot Points in the Polar Coordinate System
2. Convert Points from Polar to Cartesian Form and Vice Versa
3. Convert Equations from Cartesian to Polar Form and Vice Versa
4. Graph Polar Equations

7.2 Introduction to Polar Coordinates

So far, we have used the Cartesian (rectangular) coordinate system to represent points in the plane. There is another system, called the *polar coordinate system,* where we use angles and radial distances to represent points. For certain procedures, it is easier to work with equations written in terms of polar coordinates rather than Cartesian coordinates. In this section, we'll study the polar coordinate system and use trigonometry to relate it to the Cartesian coordinate system.

Plotting Points in the Polar Coordinate System

The frame of reference for the polar coordinate system consists of a fixed point O, called the **pole,** and a fixed ray with end point O, called the **polar axis** (Figure 1a). The position of a point P in this system is given by coordinates (r, θ), called **polar coordinates** of P (Figure 1b).

Figure 1

(a)   (b)

θ is *any* angle having the polar axis as its initial side and the ray from the pole through point P as its terminal side; r is the directed distance along the terminal side of θ from the pole O to P.

Figure 2 shows a sample grid for the polar coordinate system. The angle θ can be measured either in degrees or radians. It may be positive or negative, depending on whether it is generated by a counterclockwise or clockwise rotation.

Figure 2

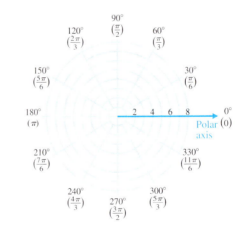

EXAMPLE 1   **Plotting Points in the Polar Coordinate System**

Plot the points with the given polar coordinates.

(a) $(3, 60°)$ (b) $(4, -30°)$ (c) $\left(4, \dfrac{3\pi}{4}\right)$

Solution   (a) To plot the point $(3, 60°)$, we first form an angle of $60°$ in such a way that the polar axis is the initial side of the angle. Then we move 3 units from the pole on the terminal side of this angle to locate the point (Figure 3).

(b)–(c) The other points in parts (b) and (c) are plotted in a similar way (Figure 3).

Figure 3

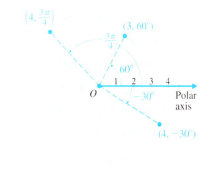

Figure 4

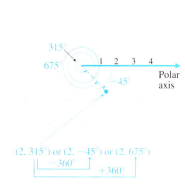

Note that in the Cartesian system each point has a *unique* number pair representation, whereas in the polar system, a point has *infinitely* many possible number pair representations because of periodicity. For example, the polar coordinates $(2, 315°)$, $(2, -45°)$, and $(2, 675°)$ all represent the same point (Figure 4).

In general, if a point P has polar coordinates (r, θ), then for any integer n,

$$(r, \theta + 2\pi n) \quad \text{or} \quad (r, \theta + 360°n)$$

are also polar coordinates of P.

The value of r for polar coordinates (r, θ) may be positive, negative, or 0. If r is negative, then the location of the point (r, θ) is the same as the point

$$(|r|, \theta + \pi) \quad \text{or} \quad (|r|, \theta + 180°)$$

For example, the point $(-3, \pi/4)$ has the same location as $(3, 5\pi/4)$ (Figure 5).

In other words, a negative value for r indicates moving $|r|$ units in the direction opposite that of the terminal side of θ in order to locate the point (r, θ). Also, if $r = 0$, then θ can be any angle.

Figure 5

Converting Points from Polar to Cartesian Form and Vice Versa

To establish the relationship between polar coordinates and Cartesian coordinates, we draw the polar axis of a polar coordinate system so that it coincides with the positive x axis of a Cartesian coordinate system.

Any point P in the plane can be located either by polar coordinates (r, θ) or by Cartesian coordinates (x, y).

For instance, suppose that P is in quadrant I and $r > 0$ (Figure 6). Using the right triangle in Figure 6, we obtain the following relationships between the coordinates:

Conversion of Coordinates

Polar or Cartesian

Figure 6

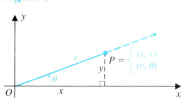

1. To convert a point P from polar coordinates (r, θ) to Cartesian coordinates (x, y), use

$$x = r \cos \theta \quad \text{and} \quad y = r \sin \theta.$$

2. To convert a point P from Cartesian coordinates (x, y) to polar coordinates (r, θ), use

$$r^2 = x^2 + y^2$$

so that if $r > 0$, then $r = \sqrt{x^2 + y^2}$; and also use

$$\tan \theta = \frac{y}{x}.$$

These relationships are true regardless of the quadrant location of P, even when $r \le 0$.

EXAMPLE 2 **Converting Polar Coordinates to Cartesian Coordinates**

Convert the given polar coordinates to Cartesian coordinates (x, y). Plot the point.

(a) $(3, 60°)$

(b) $\left(4, -\dfrac{5\pi}{6}\right)$

Solution To convert these points from polar to Cartesian coordinates, we proceed as follows:

(a) For the point $(3, 60°)$, we have

$$x = r \cos \theta = 3 \cos 60° = 3 \cdot \frac{1}{2} = \frac{3}{2}$$

$$y = r \sin \theta = 3 \sin 60° = 3 \cdot \left(\frac{\sqrt{3}}{2}\right) = \frac{3\sqrt{3}}{2}.$$

So the Cartesian coordinates are

$$\left(\frac{3}{2}, \frac{3\sqrt{3}}{2}\right) \quad \text{(Figure 7a)}.$$

(b) For the point $\left(4, -\dfrac{5\pi}{6}\right)$, we have

$$x = r \cos \theta = 4 \cos\left(-\frac{5\pi}{6}\right)$$

$$= 4\left(-\frac{\sqrt{3}}{2}\right)$$

$$= -2\sqrt{3}$$

$$y = r \sin \theta = 4 \sin\left(-\frac{5\pi}{6}\right)$$

$$= 4\left(-\frac{1}{2}\right)$$

$$= -2.$$

So the Cartesian coordinates are

$$(-2\sqrt{3}, -2) \quad \text{(Figure 7b)}.$$

Figure 7

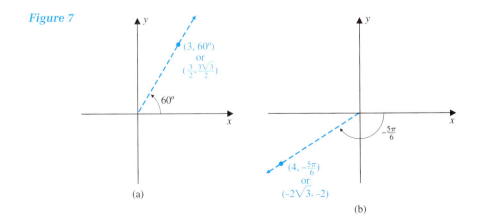

(a)

(b)

It should be noted that the conversion equation $\tan \theta = y/x$ does not define θ uniquely. To determine the polar coordinates of P, it is helpful to plot the given point because we must pay attention to the quadrant in which P lies and the restrictions on θ, as the next example illustrates.

EXAMPLE 3   **Converting Cartesian Coordinates to Polar Coordinates**

Convert the given Cartesian coordinates to polar coordinates $(r, 0)$, where $r \geq 0$ and $0° \leq \theta < 360°$.

(a) $(-1, 1)$ (b) $(5.03, -2.76)$

Round off the answers to two decimal places.

Solution   Here we use the conversion equations $r = \sqrt{x^2 + y^2}$ and $\tan \theta = y/x$, keeping in mind that the polar coordinates of a given point are not unique.

(a) For the point $(-1, 1)$ (Figure 8a), we have

$$r = \sqrt{x^2 + y^2} = \sqrt{(-1)^2 + 1^2} = \sqrt{2} \quad \text{and} \quad \tan \theta = -1.$$

Since the point $(-1, 1)$ lies in quadrant II, $90° < \theta < 180°$, so we select $\theta = 135°$ as the solution of $\tan \theta = -1$. So polar coordinates of the point are $(\sqrt{2}, 135°)$.

(b) For the point $(5.03, -2.76)$ (Figure 8b), we have

$$r = \sqrt{(5.03)^2 + (-2.76)^2} = 5.74 \text{ (approx.)} \quad \text{and} \quad \tan \theta = \frac{-2.76}{5.03}.$$

Since $\tan \theta$ is negative and θ is restricted to $0° \leq \theta < 360°$, it follows that θ is a positive angle in quadrant IV with reference angle θ_R given by

$$\theta_R = \tan^{-1} \frac{2.76}{5.03} = 28.75° \text{ (approx.)}.$$

So, θ is approximately $360° - \theta_R = 331.25°$, and the polar coordinates are $(5.74, 331.25°)$.

Figure 8

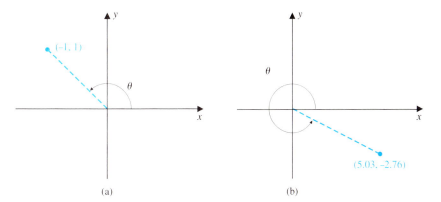

(a)   (b)

Converting Equations from Cartesian to Polar Form and Vice Versa

We know that an equation in *x* and *y* may be represented by a graph in the Cartesian coordinate system. Suppose we superimpose on the *xy* system a polar system so that the polar axis coincides with the positive *x* axis. Clearly, the graph maintains its shape. By using the conversion equations we can determine an equation for the graph in terms of polar coordinates *r* and θ.

EXAMPLE 4   **Converting a Cartesian Equation to Polar Form**

Sketch the graph of the Cartesian equation

$$(x - 2)^2 + y^2 = 4$$

and then find an equation of the graph in polar form.

Solution   We recognize the graph of this equation as a circle with radius 2 and center at (2, 0) (Figure 9). To convert the given equation to polar form, we proceed as follows:

Figure 9

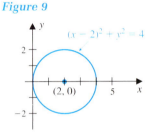

$(x - 2)^2 + y^2 = 4$	Given
$x^2 - 4x + 4 + y^2 = 4$	Multiply
$x^2 + y^2 - 4x = 0$	Simplify and rewrite
$r^2 - 4r\cos\theta = 0$	$x^2 + y^2 = r^2$ and $x = r\cos\theta$
$r^2 = 4r\cos\theta$	Add $4r\cos\theta$ to each side
$r = 4\cos\theta$	Divide each side by r

Note that the resulting equation is valid even when $r = 0$.

We can also convert an equation from polar form to Cartesian form. Once again, note that even though the equations are given in different coordinate systems, the graphs are the same.

EXAMPLE 5   **Converting a Polar Equation to Cartesian Form**

Convert the polar equation $\theta = \pi/3$ to Cartesian form, and then sketch the graph.

Solution   Since $\theta = \pi/3$, it follows from the conversion equation $\tan\theta = y/x$ that

Figure 10

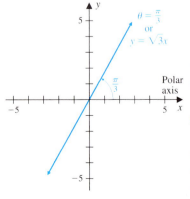

$$\tan\frac{\pi}{3} = \frac{y}{x} \quad \text{or} \quad \sqrt{3} = \frac{y}{x}.$$

That is,

$$y = \sqrt{3}x.$$

This last equation is the equation of a line with slope $\sqrt{3}$ and containing the origin (Figure 10).

Graphing Polar Equations

By the **graph of a polar equation,** we mean the set of all points (r, θ) that satisfy an equation that relates *r* and θ. In Example 5, we graphed the polar equation by first converting it to Cartesian form, and then we used the Cartesian coordinate system to sketch the graph. Polar equations can also be graphed directly by using the point-plotting method.

EXAMPLE 6   **Graphing Polar Equations Using Point-Plotting**

Use point-plotting to sketch the graph of each equation.

(a) $r = 3$ (b) $r = \theta$, where $\theta \geq 0$ (c) $r = 1 + \sin \theta$

Solution (a) The equation $r = 3$ indicates that a point is on the graph if and only if it is of the form $(3, \theta)$, where θ is any angle.

After plotting a few points of this form and then connecting them with a smooth curve, we see that the graph consists of all points that are 3 units away from the pole. The graph is the circle with center at the pole and radius 3 (Figure 11).

Figure 11

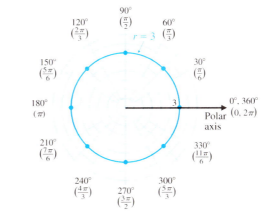

The graph of an equation of the form r = kθ is called an Archimedean spiral.

(b) Since r represents distances (real numbers) and $r = \theta$ (for $\theta \geq 0$), it is implied that θ represents real numbers or radian measures of angles. Table 1 lists some coordinates that satisfy the equation, and Figure 12 shows the graph resulting when these points are plotted and connected with a smooth curve. Note that as the angle θ is allowed to increase (rotate counterclockwise), the distances r correspondingly get longer. This results in a spiral-shaped curve.

TABLE 1

θ	0	$\dfrac{\pi}{2}$	π	$\dfrac{4\pi}{3}$	$\dfrac{3\pi}{2}$	2π
r	0	1.57 (approx.)	3.14 (approx.)	4.19 (approx.)	4.71 (approx.)	6.28 (approx.)

Figure 12

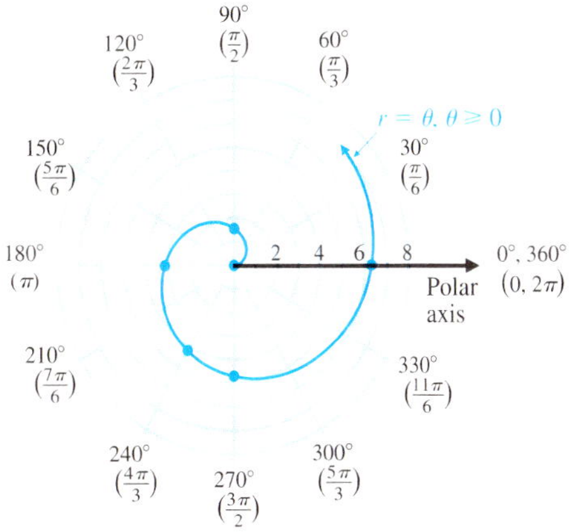

(c) To graph

$$r = 1 + \sin \theta,$$

we first build a table of coordinates that satisfy the equation (Table 2), then plot the points, and finally connect them with a smooth curve. The resulting graph is shown in Figure 13.

TABLE 2

θ	0	$\dfrac{\pi}{6}$	$\dfrac{\pi}{3}$	$\dfrac{\pi}{2}$	$\dfrac{2\pi}{3}$	$\dfrac{5\pi}{6}$	π	$\dfrac{3\pi}{2}$	2π
r	1	$\dfrac{3}{2} = 1.5$	1.87 (approx.)	2	1.87 (approx.)	$\dfrac{3}{2} = 1.5$	1	0	1

Figure 13

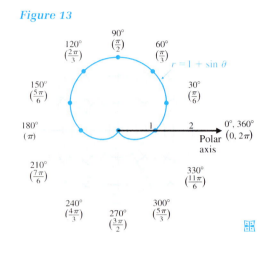

This graph is called a cardioid because of its heart shape.

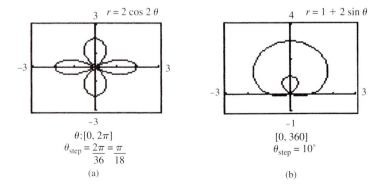

G Graphers have a polar graphing mode that enables us to sketch the graphs of polar equations of the form $r = f(\theta)$. Care must be taken when setting the range for θ, according to whether the degree mode or radian mode has been selected.

EXAMPLE 7 G **Graphing a Polar Equation**

Use a grapher to sketch the graph of each polar equation.

(a) $r = 2 \cos 2\theta$ (use radian mode)

(b) $r = 1 + 2 \sin \theta$ (use degree mode)

Solution

(a) A viewing window of the graph of $r = 2 \cos 2\theta$ is shown in Figure 14a where the radian mode was used.

(b) Figure 14b displays a viewing window of the graph of $r = 1 + 2 \sin \theta$, in which degree mode was used.

Figure 14

Note that because of the aspect ratio, graphers can produce distorted polar graphs. To improve the accuracy of a distorted graph, most graphers have features that correct the distortion. For instance, the graph of $r = 3$ is a circle, centered at the pole with radius 3 (see Example 6a). Figure 15a shows a viewing window in which the graph is distorted. After applying the distortion correction feature of the grapher, we obtain a more accurate graph (Figure 15b). Both are graphed using degree mode.

Figure 15

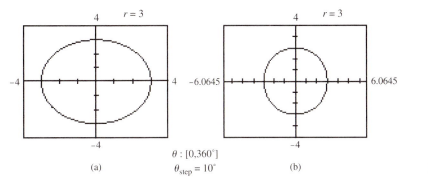

$\theta : [0, 360°]$
$\theta_{\text{step}} = 10°$

(a)   (b)

PROBLEM SET 7.2

Mastering the Concepts

In problems 1–6, plot points P and Q on the same polar coordinate system.

1. $P = (2, -45°)$; $Q = (2, 315°)$
2. $P = (3, -60°)$; $Q = (3, 300°)$
3. $P = \left(3, \dfrac{\pi}{6}\right)$; $Q = \left(-3, \dfrac{\pi}{6}\right)$
4. $P = \left(4, \dfrac{7\pi}{6}\right)$; $Q = \left(-4, \dfrac{7\pi}{6}\right)$
5. $P = \left(4, \dfrac{2\pi}{3}\right)$, $Q = \left(4, -\dfrac{2\pi}{3}\right)$
6. $P = \left(-5, \dfrac{\pi}{4}\right)$; $Q = \left(-5, -\dfrac{\pi}{4}\right)$

In problems 7 and 8, plot the given point P in the polar coordinate system, and then find three additional representations of P such that:

(i) $r > 0$ (ii) $r < 0$
7. (a) $P = (3, 100°)$ **8.** (a) $P = (4, 60°)$
   (b) $P = \left(-3, \dfrac{\pi}{4}\right)$ (b) $P = \left(4, -\dfrac{\pi}{3}\right)$

In problems 9–14, determine the Cartesian coordinates of the points with the given polar coordinates. Round off the answers to two decimal places in problems 11–14.

9. (a) $(6, 30°)$ **10.** (a) $(-8, 45°)$
   (b) $\left(4, -\dfrac{\pi}{4}\right)$ (b) $\left(4, -\dfrac{\pi}{6}\right)$

11. (a) $\left(3, -\dfrac{\pi}{7}\right)$ **12.** (a) $\left(-3, \dfrac{5\pi}{13}\right)$
   (b) $(4, -110°)$ (b) $(-6, 213°)$

13. (a) $(2, -3)$ **14.** (a) $(-2, 4)$
   (b) $\left(-3, \dfrac{5\pi}{2}\right)$ (b) $\left(4, -\dfrac{\pi}{9}\right)$

In problems 15–18, determine the polar coordinates of the points with the given Cartesian coordinates satisfying the conditions that $r > 0$ and $0 \le \theta < 2\pi$. Round off the answers to two decimal places in problems 17 and 18.

15. (a) $(-1, \sqrt{3})$ **16.** (a) $(5, 5)$
   (b) $(-6, 6\sqrt{3})$ (b) $(2\sqrt{3}, -2)$

17. (a) $(-3, 5.1)$ **18.** (a) $(4\sqrt{2}, -\sqrt{3})$
   (b) $(-6, -7)$ (b) $(-4.3, 1.4)$

In problems 19–32, sketch the graph of each Cartesian equation. Then find an equation of the graph in polar form.

19. $x = 2$ **20.** $y = 3$
21. $y = -3$ **22.** $x = -2$
23. $y = 3x - 2$ **24.** $x - y = 5$
25. $x^2 + y^2 = 25$ **26.** $x^2 + y^2 = 7$
27. $(x + 4)^2 + y^2 = 16$ **28.** $x^2 + y^2 - 6y = 0$
29. $y = 4x^2$ **30.** $xy = 4$
31. $y^2 = 8x$ **32.** $y = \sqrt{x^2 + 4}$

In problems 33–46, find an equation in terms of Cartesian coordinates that corresponds to each polar equation. Then sketch the graph.

33. $r = 3$ 34. $r = -2$

35. $r = 2 \csc \theta$ 36. $r = -4 \sec \theta$

37. $\theta = \dfrac{\pi}{4}$ 38. $\theta = \dfrac{5\pi}{3}$

39. $r = -4 \cos \theta$ 40. $r = 2 \sin \theta$

41. $r = 4 \tan \theta \sec \theta$ 42. $r^2 \sin 2\theta = 4$

43. $4r \cos \theta + 3r \sin \theta = 12$

44. $2r \sin \theta - 3r \cos \theta = 6$

45. $r = 4 \cos \theta + 2 \sin \theta$

46. $r = \cos \theta - \sin \theta$

In problems 47–54, use point-plotting to sketch the graph of each polar equation.

47. $r = 5$ 48. $r = 1$

49. $\theta = -\dfrac{\pi}{3}$ 50. $\theta = \dfrac{\pi}{6}$

51. $r = \cos \theta$ 52. $r = -2 \sin \theta$

53. $r = 1 + \cos \theta$ 54. $r = 1 - \cos \theta$

Developing and Extending the Concepts

55. Show that the graph of the polar equation

$$r = \frac{c}{a \cos \theta + b \sin \theta}$$

where a, b, and c are constants, and a and b are not both 0, is a straight line. What is the slope of the line?

56. Use the law of cosines to show that the distance d between the polar points $P = (r_1, \theta_1)$ and $Q = (r_2, \theta_2)$ satisfies the equation

$$d^2 = r_1^2 + r_2^2 - 2r_1 r_2 \cos(\theta_2 - \theta_1).$$

57. **Polar Coordinates of a Tile:** Find the polar coordinates of the corners of a tile in the shape of a regular hexagon (a polygon with six equal sides) if the center is at the pole, one vertex lies on the polar axis, and the distance from the center to any vertex is a units (Figure 16).

Figure 16

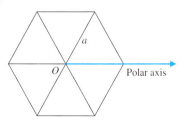

58. Give a polar equation whose graph is a circle with radius 5 and has the given point as the center in the Cartesian coordinate system.

(a) The origin   (b) $(-5, 0)$

(c) $(0, 5)$ (d) $(4, 3)$

[G] Problems 59–70 refer to the following: Some graphs of polar equations are symmetric. Table 3 indicates how to recognize certain types of symmetry: apply the symmetry tests in Table 3 to determine the symmetry (if any) of each graph. Then use a grapher to graph each polar equation to demonstrate the symmetry.

TABLE 3   Testing a Polar Equation for Symmetry
If (r, θ) Lies on the Graph:

Type of Symmetry	Then These Points Lie on the Graph
1. The polar axis	$(-r, \pi - \theta)$ or $(r, -\theta)$
2. The line $\theta = \pi/2$	$(-r, -\theta)$ or $(r, \pi - \theta)$
3. The pole	$(-r, \theta)$ or $(r, \pi + \theta)$

59. $r = 1 - \sin \theta$ (cardioid)

60. $r = 1 + 2 \cos \theta$ (limaçon)

61. $r = 2 + 3 \cos \theta$ (limaçon)

62. $r = 1 - 2 \sin \theta$ (limaçon)

63. $r = 4 + 3 \sin \theta$ (limaçon)

64. $r = 5 + 3 \cos \theta$ (limaçon)

65. $r = -2 \cos 2\theta$ (four-petaled rose)

66. $r = 4 \sin 2\theta$ (four-petaled rose)

67. $r = 2 \sin 3\theta$ (three-petaled rose)

68. $r = 2 \cos 3\theta$ (three-petaled rose)

69. $r^2 = 4 \cos 2\theta$ (lemniscate)

70. $r^2 = 4 \sin 2\theta$ (lemniscate)

71. [G] (a) Sketch the graph of each polar equation:

$$r = 2 \cos 3\theta$$

$$r = 2 \cos \left[3\left(\theta - \frac{\pi}{6} \right) \right]$$

and $$r = 2 \cos \left[3\left(\theta + \frac{\pi}{4} \right) \right]$$

(b) Describe how the graphs in part (a) are related to each other.

72. [G] (a) The graphs of $r = 2 \sin[(3\theta)/2]$ and $r = 2 \sin[(5\theta)/2]$ are called *rose curves*. Sketch a complete graph of each equation.

(b) How many petals does each graph in part (a) have?

Objectives

1. Describe Vector Operations Geometrically

2. Perform Vector Operations Analytically

3. Find Unit Vectors and Direction Angles

4. Solve Applied Problems

7.3 Vectors in the Plane

Until now, we have dealt with quantities that can be measured or represented by single real numbers, such as area, volume, time, temperature, and speed. In this section, we deal with quantities called *vectors,* which cannot be described or represented by a single real number. Vectors are important tools in applications involving force, velocity, acceleration, and displacement. We'll *see* how trigonometry is used when working with vectors. Here we restrict our study to vectors in a two dimensional setting.

Describing Vector Operations Geometrically

Geometrically, a **vector** in the plane is a line segment with a direction usually denoted by an arrowhead at one end of the segment. The end point that contains the arrowhead is called the **terminal point,** and the other end point is called the **initial point** of the vector. The length of a vector is called its **magnitude.** The vector in Figure 1, denoted by $\overrightarrow{AB}$, has initial point A and terminal point B, and its magnitude is denoted by $|\overrightarrow{AB}|$.

The **zero vector,** written **0**, is a vector whose initial and terminal points are the same. This magnitude is given by

$$|\mathbf{0}| = 0.$$

1. Geometry of Equal Vectors

Two vectors are considered to be **equal** if they have the same magnitude and the same direction, such as vectors $\overrightarrow{PQ}$ and $\overrightarrow{RS}$ in Figure 2a. Note that $\overrightarrow{AB}$ and $\overrightarrow{BA}$ in Figure 2b are not equal, even though their lengths are the same, because they have opposite directions. That is,

$$\overrightarrow{AB} \neq \overrightarrow{BA}, \quad \text{even though} \quad |\overrightarrow{AB}| = |\overrightarrow{BA}|.$$

Special notation is used for vectors so that they can be distinguished from real numbers. In this book, we normally use lowercase boldface letters to denote vectors. Hence, we speak of "vector **u**" and just write boldface **u.**

2. Geometry of Vector Addition

Two vectors **u** and **v** (Figure 3a) may be added to form the *sum* **u** + **v** as follows:

First, we shift **v** so that its initial point coincides with the terminal point of **u.** The vector having the same initial point as **u** and the same terminal point as **v** is defined to be the **sum (resultant vector) u** + **v** of **u** and **v** (Figure 3b). As Figure 3b shows, the resultant vector is a diagonal vector of the parallelogram determined by vectors **u** and **v.** This description of addition is called the **parallelogram law of addition.**

Figure 1

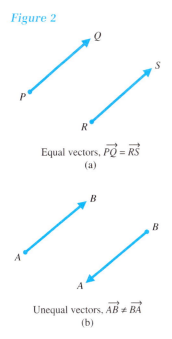

Terminal point *B*

Magnitude of $\overrightarrow{AB} = |\overrightarrow{AB}|$

Directed line segment from *A* and *B*

A   Initial point

Figure 2

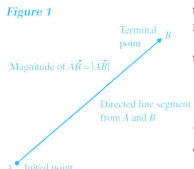

Q

S

P

R

Equal vectors, $\overrightarrow{PQ} = \overrightarrow{RS}$

(a)

B

B

A

A

Unequal vectors, $\overrightarrow{AB} \neq \overrightarrow{BA}$

(b)

*Figure 3b displays the fact that **u** + **v** = **v** + **u;** that is, vector addition is commutative.*

Figure 3

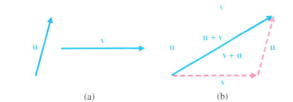

u

v

u

u + v

v + u

v

u

(a)   (b)

3. Geometry of Scalar Multiplication

Vectors can be multiplied by real numbers, called **scalars** in this context, to form other vectors. We refer to this operation as **scalar multiplication,** and we describe it as follows.

If $a \neq 0$ is a scalar (real number) and **u** is a vector, then the vector $a\mathbf{u}$ is a vector that satisfies the following conditions:

(i) **The Magnitude of au:**

$|a\mathbf{u}| = |a||\mathbf{u}|$; that is, the magnitude of $a\mathbf{u}$ is the product of the absolute value of a and the magnitude of **u.**

(ii) **The Direction of au:**

If $a > 0$, then the direction of $a\mathbf{u}$ is the same as the direction of **u.**
If $a < 0$, then the direction of $a\mathbf{u}$ is opposite the direction of **u.**

Figure 4 displays some illustrations of scalar multiplication. Note that **2u** is a vector in the same direction as **u,** with magnitude twice that of **u.** And $(-1/2)\mathbf{u}$ is a vector opposite in direction to **u,** with a length half that of **u.**

If $a = 0$, then $a\mathbf{u} = 0\mathbf{u} = \mathbf{0}$.

Figure 4

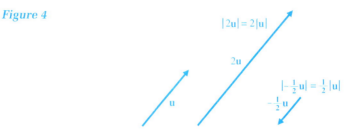

4. Geometry of Vector Subtraction

Now we can use vector addition and scalar multiplication to define vector **subtraction** as

$$\mathbf{u} - \mathbf{v} = \mathbf{u} + (-\mathbf{v})$$

where $-\mathbf{v} = (-1)\,\mathbf{v}$ (Figure 5a).

Figure 5

(a)   (b)

Observe that if we form the parallelogram defined by **u** and **v** (Figure 5b), one of the diagonal vectors is **u** + **v** and the other is **u** − **v.**

EXAMPLE 1 **Describing Vector Operations Geometrically**

Figure 6

Given vectors **u** and **v** as shown in Figure 6, display the vector 3**u** − 2**v.**

Solution To form 3**u** − 2**v,** we first form 3**u** and 2**v** (Figure 7a). Next we form the parallelogram defined by 3**u** and 2**v.** The difference 3**u** − 2**v** is the diagonal vector shown in Figure 7b.

Figure 7

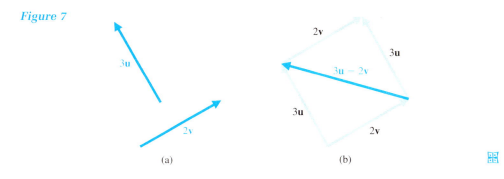

(a)   (b)

Performing Vector Operations Analytically

Figure 8

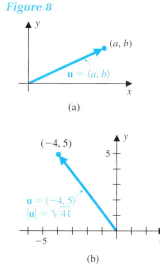

(a)

(b)

When a vector is positioned in a Cartesian plane so that its initial point is at the origin, it is called a **radius vector** or **standard position vector.** Such a vector can be uniquely represented by the coordinates of its terminal point. For instance, Figure 8a shows a radius vector **u** with terminal point (a, b). We call a the **x component** and b the **y component** of **u,** and we write **u** in the **component form**

$$\mathbf{u} = \langle a, b \rangle.$$

From the Pythagorean theorem, the **magnitude** $|\mathbf{u}|$ of vector $\mathbf{u} = \langle a, b \rangle$ is given by

$$|\mathbf{u}| = \sqrt{a^2 + b^2}.$$

For example, the vector $\mathbf{u} = \langle -4, 5 \rangle$ has x component -4, y component 5, and $|\mathbf{u}| = \sqrt{(-4)^2 + 5^2} = \sqrt{41}$ (Figure 8b).

1. Equal Vectors

If **u** and **v** are equal position vectors, they have the same initial point $(0, 0)$, so the components must also be equal; that is:

Component Rule
Equal Vectors

If $\mathbf{u} = \langle x_1, y_1 \rangle$ and $\mathbf{v} = \langle x_2, y_2 \rangle$, then

$$\mathbf{u} = \mathbf{v} \quad \text{whenever} \quad x_1 = x_2 \quad \text{and} \quad y_1 = y_2.$$

For instance, if $\mathbf{u} = \mathbf{w}$, where $\mathbf{u} = \langle 1, -3 \rangle$ and $\mathbf{w} = \langle 1, b \rangle$, then $b = -3$.

Suppose that **w** is a vector in a Cartesian plane and **w** is not a radius vector. Assume that the initial point of **w** is (x_1, y_1) and its terminal point is (x_2, y_2), as shown in Figure 9. To find the radius vector representation of **w,** we shift **w** from its given position in such a way that the initial point (x_1, y_1) is positioned at the origin to obtain the component form

$$\mathbf{w} = \langle x_2 - x_1, y_2 - y_1 \rangle.$$

Figure 9

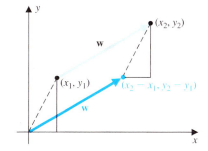

EXAMPLE 2   **Finding the Component Form**

Let $P = (3, -2)$ be the initial point and $Q = (-4, 1)$ be the terminal point of the vector

$$\mathbf{u} = \overrightarrow{PQ}.$$

Find its component form, and display $\mathbf{u}$ in both its given position and its standard position.

Solution   Here

Figure 10

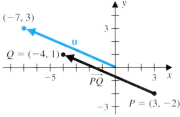

$$P = (3, -2) = (x_1, y_1) \quad \text{and} \quad Q = (-4, 1) = (x_2, y_2).$$

The x and y components of $\mathbf{u} = \overrightarrow{PQ}$ are given, respectively, by

$$x_2 - x_1 = -4 - 3 = -7 \quad \text{and} \quad y_2 - y_1 = 1 - (-2) = 3.$$

So $\mathbf{u} = \langle -7, 3 \rangle$. Figure 10 displays the vector in both its given location and its standard position.

2. Vector Addition

When vectors are written in component form, vector addition becomes very simple. Suppose that $\mathbf{u}$ and $\mathbf{v}$ are in standard position with

$$\mathbf{u} = \langle x_1, y_1 \rangle \quad \text{and} \quad \mathbf{v} = \langle x_2, y_2 \rangle.$$

As displayed in Figure 11, moving the initial point of $\mathbf{v}$ to the terminal point of $\mathbf{u}$ shows that the x component of $\mathbf{u} + \mathbf{v}$ is obtained by adding the x components of $\mathbf{u}$ and $\mathbf{v}$. Similarly, the y component of $\mathbf{u} + \mathbf{v}$ is obtained by adding the y components of $\mathbf{u}$ and $\mathbf{v}$.

Figure 11

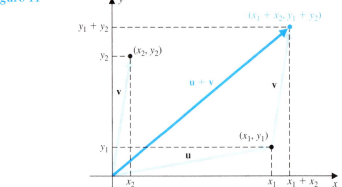

Thus we have the following property:

Component Rule **Addition of Vectors**	If $\mathbf{u} = \langle x_1, y_1 \rangle$ and $\mathbf{v} = \langle x_2, y_2 \rangle$, then $$\mathbf{u} + \mathbf{v} = \langle x_1 + x_2, y_1 + y_2 \rangle.$$

3. Scalar Multiplication

As illustrated in Figure 12, we have the following rule:

**Component Rule
Scalar Multiplication**

If $\mathbf{u} = \langle x_1, y_1 \rangle$ and c is a real number, then

$$c\mathbf{u} = c\langle x_1, y_1 \rangle = \langle cx_1, cy_1 \rangle.$$

Figure 12

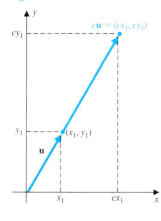

4. Vector Subtraction

Finally, we derive the subtraction rule as follows:
If $\mathbf{u} = \langle x_1, y_1 \rangle$ and $\mathbf{v} = \langle x_2, y_2 \rangle$, then $-\mathbf{v} = -1 \cdot \mathbf{v} = \langle -x_2, -y_2 \rangle$.
 So

$$\mathbf{u} - \mathbf{v} = \mathbf{u} + (-\mathbf{v})$$
$$= \langle x_1, y_1 \rangle + \langle -x_2, -y_2 \rangle$$
$$= \langle x_1 - x_2, y_1 - y_2 \rangle.$$

Thus we are led to the following property:

**Component Rule
Subtraction of Vectors**

If $\mathbf{u} = \langle x_1, y_1 \rangle$ and $\mathbf{v} = \langle x_2, y_2 \rangle$, then

$$\mathbf{u} - \mathbf{v} = \langle x_1 - x_2, y_1 - y_2 \rangle.$$

EXAMPLE 3

Performing Vector Operations

Let $\mathbf{u} = \langle 3, 4 \rangle$ and $\mathbf{v} = \langle -5, 6 \rangle$. Find:

(a) $\mathbf{u} + \mathbf{v}$ (b) $\mathbf{u} - \mathbf{v}$

(c) $4\mathbf{u} - 3\mathbf{v}$ (d) $|4\mathbf{u} - 3\mathbf{v}|$

Solution

(a) $\mathbf{u} + \mathbf{v} = \langle 3, 4 \rangle + \langle -5, 6 \rangle$
$$= \langle 3 - 5, 4 + 6 \rangle$$
$$= \langle -2, 10 \rangle$$

(b) $\mathbf{u} - \mathbf{v} = \langle 3, 4 \rangle - \langle -5, 6 \rangle$
$$= \langle 3 - (-5), 4 - 6 \rangle$$
$$= \langle 8, -2 \rangle$$

(c) $4\mathbf{u} - 3\mathbf{v} = 4\langle 3, 4 \rangle - 3\langle -5, 6 \rangle$
$$= \langle 12, 16 \rangle + \langle 15, -18 \rangle$$
$$= \langle 27, -2 \rangle$$

(d) Using the result from part (c),

$$|4\mathbf{u} - 3\mathbf{v}| = \sqrt{(27)^2 + (-2)^2} = \sqrt{729 + 4} = \sqrt{733}.$$

Finding Unit Vectors and Direction Angles

A vector of magnitude 1 is called a **unit vector.** There are two special unit vectors, called **basis vectors,** defined by

$$\mathbf{i} = \langle 1, 0 \rangle \quad \text{and} \quad \mathbf{j} = \langle 0, 1 \rangle.$$

Any vector $\mathbf{u} = \langle a, b \rangle$ can be written in terms of $\mathbf{i}$ and $\mathbf{j}$ as follows:

$$\langle a, b \rangle = \langle a, 0 \rangle + \langle 0, b \rangle = a \langle 1, 0 \rangle + b \langle 0, 1 \rangle = a\mathbf{i} + b\mathbf{j} \quad \text{(Figure 13)}$$

Figure 13

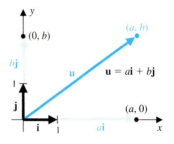

Thus we get the *component form* of the vector.

$$\mathbf{u} = \langle a, b \rangle = a\mathbf{i} + b\mathbf{j}.$$

The component rules for addition, subtraction, and scalar multiplication apply to vectors written in terms of $\mathbf{i}$ and $\mathbf{j}$. For example,

$$(3\mathbf{i} + 5\mathbf{j}) + (2\mathbf{i} - 7\mathbf{j}) = 5\mathbf{i} - 2\mathbf{j}$$

$$-2(3\mathbf{i} + 2\mathbf{j}) = -6\mathbf{i} - 4\mathbf{j}$$

$$\text{and} \quad (2\mathbf{i} - 3\mathbf{j}) - (\mathbf{i} + 5\mathbf{j}) = \mathbf{i} - 8\mathbf{j}$$

Consider the vector $\dfrac{1}{|\mathbf{u}|}\mathbf{u}$, where $\mathbf{u} \neq \mathbf{0}$. Since $\dfrac{1}{|\mathbf{u}|} > 0$, the vector $\dfrac{1}{|\mathbf{u}|}\mathbf{u}$ has the

same direction as $\mathbf{u}$. Also, the magnitude of $\dfrac{1}{|\mathbf{u}|}\mathbf{u}$ is given by

$$\left| \frac{1}{|\mathbf{u}|}\mathbf{u} \right| = \frac{1}{|\mathbf{u}|}|\mathbf{u}| = 1.$$

Thus we are led to the following notion:

Definition
Unit Vector

A **unit vector** in the direction of $\mathbf{u} = \langle a, b \rangle$, where $\mathbf{u} \neq \mathbf{0}$, is given by

$$\frac{1}{|\mathbf{u}|}\mathbf{u} = \left\langle \frac{a}{\sqrt{a^2 + b^2}}, \frac{b}{\sqrt{a^2 + b^2}} \right\rangle.$$

The process of forming this vector is referred to as **normalizing the vector u.**

EXAMPLE 4   **Normalizing a Vector**

Find a unit vector in the same direction as $\mathbf{u} = 4\mathbf{i} - 3\mathbf{j}$.

Solution   The magnitude of $\mathbf{u}$ is given by

$$|\mathbf{u}| = \sqrt{4^2 + (-3)^2} = 5.$$

Thus a unit vector in the same direction as $\mathbf{u}$ is given by

$$\frac{1}{|\mathbf{u}|}\mathbf{u} = \frac{1}{5}(4\mathbf{i} - 3\mathbf{j}) = \frac{4}{5}\mathbf{i} - \frac{3}{5}\mathbf{j}.$$

Figure 14

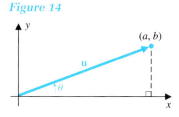

An angle θ formed by the vector $\mathbf{u} = \langle a, b \rangle$ and the positive x axis is called a **direction angle** of the vector $\mathbf{u}$ (Figure 14).

Using trigonometry, we have

$$\cos \theta = \frac{a}{|\mathbf{u}|} \quad \text{and} \quad \sin \theta = \frac{b}{|\mathbf{u}|}.$$

Rewriting these equations as $a = |\mathbf{u}| \cos \theta$ and $b = |\mathbf{u}| \sin \theta$, respectively, we get a trigonometric form of the vector.

$$\mathbf{u} = a\mathbf{i} + b\mathbf{j} = |\mathbf{u}|(\cos \theta)\mathbf{i} + |\mathbf{u}|(\sin \theta)\mathbf{j}$$

Also, if $a \neq 0$, it follows that

$$\frac{b}{a} = \frac{|\mathbf{u}| \sin \theta}{|\mathbf{u}| \cos \theta} = \frac{\sin \theta}{\cos \theta} = \tan \theta.$$

Thus a direction angle θ for $\mathbf{u} = a\mathbf{i} + b\mathbf{j}$ can be found by solving

$$\tan \theta = \frac{b}{a}$$

where θ is selected according to the quadrant location of (a, b).

For example, the direction angle of $\mathbf{u} = 5\mathbf{i} + 12\mathbf{j}$ satisfies

$$\tan \theta = \frac{b}{a} = \frac{12}{5} = 2.4.$$

Since $(5, 12)$ is in quadrant I, it follows that

$$\theta = \tan^{-1} 2.4 \quad \text{or approx. } 67.4° \quad \text{(Figure 15)}$$

is a direction angle.

Figure 15

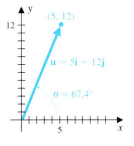

EXAMPLE 5   **Finding the Components of a Vector**

Find the component form of the vector **u** if $|\mathbf{u}| = 2$ and a direction angle $\theta = (5\pi)/4$.

Solution   The vector **u** has magnitude 2 and direction angle $\theta = (5\pi)/4$ (Figure 16). So **u** can be written in component form as

Figure 16

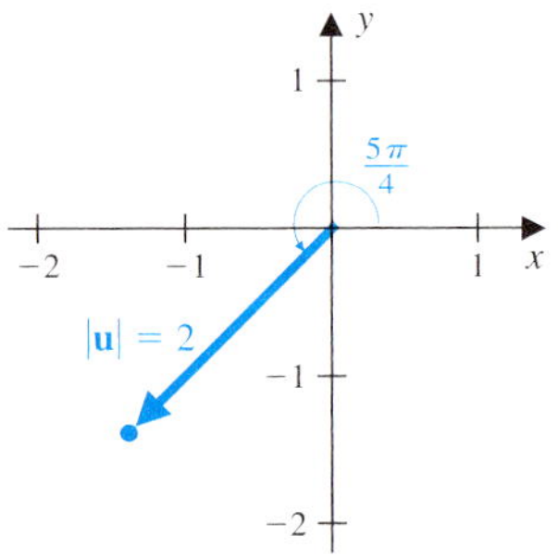

$$\mathbf{u} = a\mathbf{i} + b\mathbf{j} = |\mathbf{u}|\left(\cos\frac{5\pi}{4}\right)\mathbf{i} + |\mathbf{u}|\left(\sin\frac{5\pi}{4}\right)\mathbf{j}$$

$$= 2\left(-\frac{\sqrt{2}}{2}\right)\mathbf{i} + 2\left(-\frac{\sqrt{2}}{2}\right)\mathbf{j}$$

$$= -\sqrt{2}\,\mathbf{i} - \sqrt{2}\,\mathbf{j}.$$

Clearly,

the x component of **u** is $-\sqrt{2}$

and

the y component is also $-\sqrt{2}$.

Solving Applied Problems

Vectors are used to model applications in science and engineering.

EXAMPLE 6   **Finding the Magnitude of a Resultant Force**

A boat is pulled along a canal by two ropes on opposite sides of the canal, as shown in Figure 17. The first rope exerts a force of 350 pounds and makes an angle of 25° with respect to the axis of the boat. The other exerts a force of 200 pounds and makes an angle of 32° with respect to the axis of the boat.

Figure 17

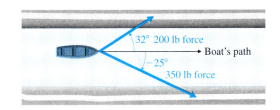

(a) What is the magnitude of the resultant force vector (to two decimal places)?

(b) What is the angle that the resultant vector makes with the axis of the boat (to the nearest hundredth of a degree)?

Solution   (a) Let $\mathbf{F}_1$ be the vector representing the 350 pound force and $\mathbf{F}_2$ be the vector representing the 200 pound force.
So we have

$$|\mathbf{F}_1| = 350$$

and

$$|\mathbf{F}_2| = 200.$$

Figure 18

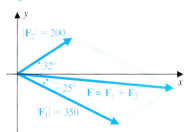

Next we set up a Cartesian coordinate system in such a way that the origin is at the point on which both $\mathbf{F}_1$ and $\mathbf{F}_2$ act. The resultant vector that is given by

$$\mathbf{F} = \mathbf{F}_1 + \mathbf{F}_2$$

describes the force on the boat (Figure 18).

Since $\mathbf{F}_1$ makes an angle of $-25°$ with the positive x axis, it follows that the components of $\mathbf{F}_1$ are given by

$$\mathbf{F}_1 = |\mathbf{F}_1| \cos(-25°)\mathbf{i} + |\mathbf{F}_1| \sin(-25°)\mathbf{j}$$
$$= 350(\cos 25°)\mathbf{i} - 350(\sin 25°)\mathbf{j}$$
$$= 317.21\mathbf{i} - 147.92\mathbf{j} \text{ (approx.)}$$

Similarly, $\mathbf{F}_2$ makes an angle of $32°$ with the positive x axis, so its components are given by

$$\mathbf{F}_2 = |\mathbf{F}_2| (\cos 32°)\mathbf{i} + |\mathbf{F}_2| (\sin 32°)\mathbf{j}$$
$$= 200(\cos 32°)\mathbf{i} + 200(\sin 32°)\mathbf{j}$$
$$= 169.61\mathbf{i} + 105.98\mathbf{j} \text{ (approx.)}.$$

The resultant vector is given by

$$\mathbf{F} = \mathbf{F}_1 + \mathbf{F}_2$$
$$= (317.21\mathbf{i} - 147.92\mathbf{j}) + (169.61\mathbf{i} + 105.98\mathbf{j})$$
$$= 486.82\mathbf{i} - 41.94\mathbf{j} \text{ (approx.)}.$$

and its magnitude is

$$|\mathbf{F}| = \sqrt{(486.82)^2 + (-41.94)^2}$$
$$= 488.62 \text{ (approx.)}.$$

Thus the resultant force has a magnitude of about 488.62 pounds.

(b) The angle θ between $\mathbf{F}$ and the axis of the boat (positive x axis) is given by

$$\tan \theta = \frac{-41.94}{486.82}$$

that is,

$$\theta = \tan^{-1}\left(\frac{-41.94}{486.82}\right) = -4.92° \text{ (approx.)}.$$

Vectors are also used in aeronautical navigation, where the following terminology is commonly used: The **heading** of an airplane is the direction in which it is pointed; its **airspeed** is its speed relative to the air. As shown in Figure 19, the vector $\mathbf{v}_1$, whose magnitude is the airspeed and whose direction is the heading, represents the **velocity of the airplane relative to the air.** The vector $\mathbf{v}_2$ is the **velocity vector** for the wind; that is, $|\mathbf{v}_2|$ is the speed of the wind, and the direction angle of $\mathbf{v}_2$ is the direction of the wind. The **course, or track,** of the airplane is the direction in which it is actually moving over the ground, and its **ground speed** is its speed relative to the ground. The magnitude of the resultant vector

$$\mathbf{v} = \mathbf{v}_1 + \mathbf{v}_2$$

is the ground speed or velocity of the airplane, and the direction of $\mathbf{v}$ is the **course.** In this context, vector $\mathbf{v}$ is called the **velocity vector** of the airplane. The angle α between the vectors $\mathbf{v}$ and $\mathbf{v}_1$ is called the **drift angle.**

Figure 19

EXAMPLE 7   **Finding the Course and Speed of an Airplane**

An airplane is headed N30°E with an airspeed of 500 miles per hour. The wind is blowing S29°E at a speed of 50 miles per hour. Find:

(a) The ground speed (to two decimal places)

(b) The course of the airplane—that is, the direction angle of the velocity vector relative to the ground (to the nearest hundredth of a degree)

(c) The drift angle (to the nearest hundredth of a degree)

Solution   Let $\mathbf{v}_1$ represent the velocity and direction of the airplane relative to the air; let $\mathbf{v}_2$ represent the velocity and direction of the wind relative to the ground; and let $\mathbf{v}$ represent the velocity vector of the airplane (Figure 20). Then

$$\mathbf{v}_1 = 500(\cos 60°)\mathbf{i} + 500(\sin 60°)\mathbf{j}$$

$$= 250.00\mathbf{i} + 433.01\mathbf{j}$$

$$\mathbf{v}_2 = 50[\cos(-61°)]\mathbf{i} + 50[\sin(-61°)]\mathbf{j}$$

$$= 24.24\mathbf{i} + (-43.73)\mathbf{j} \text{ (approx.).}$$

Figure 20

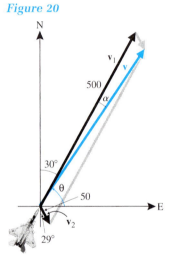

(a) Thus

$$\mathbf{v} = \mathbf{v}_1 + \mathbf{v}_2$$

$$= (250.00\mathbf{i} + 433.01\mathbf{j}) + (24.24\mathbf{i} - 43.73\mathbf{j})$$

$$= 274.24\mathbf{i} + 389.28\mathbf{j}.$$

So

$$|\mathbf{v}| = \sqrt{(274.24)^2 + (389.28)^2}$$

$$= 476.18 \text{ (approx.).}$$

Therefore, the ground speed is approximately 476.18 miles per hour.

(b) The direction angle θ of the velocity vector $\mathbf{v}$ is given by $\tan \theta = 389.28/274.24$. So one possibility is

$$\theta = \tan^{-1}\frac{389.28}{274.24} = 54.84° \text{ (approx.).}$$

(c) Upon examining Figure 20 and using the result from part (b), we find that the drift angle $\alpha = 60° - 54.84° = 5.16°$ (approx.).

 **PROBLEM SET 7.3**

Mastering the Concepts

In problems 1 and 2, let $O = (0, 0)$, $P = (-2, 1)$, $Q = (3, -2)$, $R = (4, 3)$, and $S = (-3, -5)$. Sketch each vector.

1. (a) $\mathbf{u} = \overrightarrow{OP}$
    (b) $\mathbf{v} = \overrightarrow{OQ}$

    (c) $2\mathbf{u}$

    (d) $3\mathbf{u} - 4\mathbf{v}$

2. (a) $\mathbf{u} = \overrightarrow{OS}$
    (b) $\mathbf{v} = \overrightarrow{OR}$

    (c) $-\dfrac{1}{5}\mathbf{v}$

    (d) $-2\mathbf{u} + 5\mathbf{v}$

In problems 3–8, let $\mathbf{u}$ be a vector whose initial point is P and whose terminal point is Q. Draw $\mathbf{u} = \overrightarrow{PQ}$ in standard position. Also, find the component form and the magnitude of $\mathbf{u}$.

3. $P = (8, 6)$, $Q = (3, 4)$

4. $P = (-7, -6)$, $Q = (2, 3)$

5. $P = (-2, 6)$, $Q = (3, -5)$

6. $P = (-3, 2)$, $Q = (1, -3)$
7. $P = (3, 7)$, $Q = (-3, 1)$
8. $P = (1, -3)$, $Q = (5, -1)$

In problems 9–14, use $\mathbf{u} = \langle 3, 4 \rangle$, $\mathbf{v} = \langle -2, 4 \rangle$, and $\mathbf{w} = \langle 7, 8 \rangle$ to find each expression. Display part (a) of each problem graphically.

9. (a) $\mathbf{u} + \mathbf{v}$
   (b) $|\mathbf{u} + \mathbf{v}|$
10. (a) $\mathbf{v} - \mathbf{w}$
    (b) $|\mathbf{v} - \mathbf{w}|$
11. (a) $3\mathbf{u} - 4\mathbf{w}$
    (b) $|3\mathbf{u} - 4\mathbf{w}|$
12. (a) $5\mathbf{w} + 3\mathbf{u}$
    (b) $|5\mathbf{w} + 3\mathbf{u}|$
13. (a) $-6\mathbf{v} - \mathbf{u}$
    (b) $|-6\mathbf{v} - \mathbf{u}|$
14. (a) $-(2\mathbf{u} + \mathbf{v})$
    (b) $|-(2\mathbf{u} + \mathbf{v})|$

In problems 15 and 16, express each vector in the component form of $a\mathbf{i} + b\mathbf{j}$.

15. (a) $\langle -2, 4 \rangle$
    (b) $\langle 3, 0 \rangle$
16. (a) $\langle 0, 2 \rangle$
    (b) $\langle -1, -5 \rangle$

17. Let $\mathbf{w} = \mathbf{u} + \mathbf{v}$. Find $\mathbf{v}$ in terms of $\mathbf{i}$ and $\mathbf{j}$, and sketch the vector $\mathbf{v}$ in each case.
    (a) $\mathbf{u} = 3\mathbf{i} - 2\mathbf{j}$ and $\mathbf{w} = 5\mathbf{i} + 3\mathbf{j}$
    (b) $\mathbf{u} = 2\mathbf{i} - \mathbf{j}$ and $\mathbf{w} = -2\mathbf{i} + 3\mathbf{j}$
18. Let $\mathbf{u} = \langle -2, 2 \rangle$ and $\mathbf{v} = \langle -5, 0 \rangle$. Suppose $\mathbf{w} = \mathbf{u} - \mathbf{v}$. Express $\mathbf{w}$ in terms of $\mathbf{i}$ and $\mathbf{j}$.

In problems 19 and 20, sketch the resultant vector $\mathbf{w} = \mathbf{u} + \mathbf{v}$. Find the length of $\mathbf{w}$ and the angle that $\mathbf{w}$ makes with $\mathbf{v}$.

19.   20.

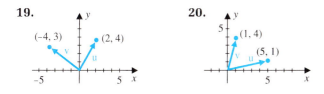

In problems 21–28, normalize each vector. Also, find the direction angle of $\mathbf{u}$ (rounded off to two decimal places).

21. $\mathbf{u} = -\sqrt{3}\mathbf{i} + \mathbf{j}$
22. $\mathbf{u} = -2\mathbf{i} - 2\sqrt{3}\mathbf{j}$
23. $\mathbf{u} = 5\mathbf{i} + 12\mathbf{j}$
24. $\mathbf{u} = 3\mathbf{i} - 3\mathbf{j}$
25. $\mathbf{u} = \sqrt{2}\mathbf{i} - \sqrt{2}\mathbf{j}$
26. $\mathbf{u} = 3\mathbf{i} + 7\mathbf{j}$
27. $\mathbf{u} = (3\mathbf{i} + 8\mathbf{j}) - (\mathbf{i} - 2\mathbf{j})$
28. $\mathbf{u} = \langle 3, -2 \rangle - 2\langle 4, 1 \rangle$

In problems 29–36, find the x and y components of the vector $\mathbf{u} = x\mathbf{i} + y\mathbf{i}$, if θ represents a direction angle of $\mathbf{u}$. Round off the answers to two decimal places.

29. $|\mathbf{u}| = 3$; $\theta = 0°$
30. $|\mathbf{u}| = 2$; $\theta = 180°$
31. $|\mathbf{u}| = 5$; $\theta = \dfrac{\pi}{4}$
32. $|\mathbf{u}| = 1$; $\theta = \dfrac{5\pi}{3}$
33. $|\mathbf{u}| = 2$; $\theta = -150°$
34. $|\mathbf{u}| = 4$; $\theta = -75°$

35. $|\mathbf{u}| = \dfrac{1}{2}$; $\theta = 100.2°$
36. $|\mathbf{u}| = 7$; $\theta = 253.3°$

Applying the Concepts

In problems 37–44, compute each scalar value to two decimal places and each angle to the nearest hundredth of a degree.

37. **Resultant Force:** Two forces $\mathbf{F}_1$ and $\mathbf{F}_2$ act on a point. If $|\mathbf{F}_1| = 40$ pounds, $|\mathbf{F}_2| = 23$ pounds, and the angle between $\mathbf{F}_1$ and $\mathbf{F}_2$ is 58°, find:
    (a) The magnitude of the resultant force $\mathbf{F}$
    (b) The angle between $\mathbf{F}_1$ and $\mathbf{F}$
38. **Resultant Force:** Two forces acting on the same object produce a resultant force of 17 pounds with an angle of 20° relative to a horizontal axis. Assume that one of the two forces is 20 pounds and has an angle of 50° relative to the horizontal axis. Find the other force and the direction angle relative to the horizontal axis.
39. **Pulling a Sled:** Two children are pulling a third child across the ice on a sled. The first child pulls a rope with a force of 5 newtons and the second child pulls a rope with a force of 8 newtons. If the angle between the ropes is 28° (Figure 21), find:
    (a) The magnitude of the resultant force
    (b) The angle the resultant force vector makes with the first child's rope

Figure 21

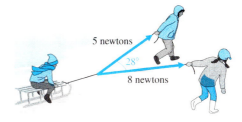

40. **Helicopter Aviation:** A helicopter is flying in the direction N50°W with an airspeed of 165 miles per hour. The wind is blowing at 35 miles per hour in the direction S60°E. Find the course and the ground speed of the helicopter.
41. **Airplane Aviation:** A commercial airplane with an airspeed of 510 miles per hour is headed N85°E. The wind is blowing N35°E at 45 miles per hour. Find:
    (a) The ground speed
    (b) The course of the airplane
    (c) The drift angle

42. **Airplane Aviation:** An airplane is flying in the direction N25°E with an airspeed of 450 kilometers per hour. Its ground speed is 500 kilometers per hour, and its course is N40°E. Find:
    (a) The speed of the wind
    (b) The direction of the wind

43. **Boat Navigation:** A boat heading S40°E at a still water speed of 40 kilometers per hour is pushed off course by a current of 32 kilometers per hour flowing in the direction S50°W. Find:
    (a) The speed of the boat
    (b) The course the boat is traveling
    (c) The drift angle

44. **Motorboat Navigation:** Suppose a motorboat that has a speed in still water of 12 miles per hour heads due west across a river that is flowing due south at a speed of 3 miles per hour.
    (a) What is the speed of the boat?
    (b) What angle does the path of the boat make with the vector representing the current?

Developing and Extending the Concepts

45. Find the terminal point of
$$\mathbf{u} = \langle 10, -4 \rangle$$
    if the initial point P is $(3, 5)$.

46. Find the initial point of
$$\mathbf{u} = \langle -4, 7 \rangle$$
    if the terminal point Q is $(2, -1)$.

Problems 47 and 48 pertain to the following properties:

Vector Addition and Scalar Multiplication Properties

Let **u**, **v**, and **w** be vectors, and let t and s be scalars; then the following relationships hold:

1. $\mathbf{u} + \mathbf{v} = \mathbf{v} + \mathbf{u}$
2. $\mathbf{u} + (\mathbf{v} + \mathbf{w}) = (\mathbf{u} + \mathbf{v}) + \mathbf{w}$
3. $\mathbf{u} + \mathbf{0} = \mathbf{u}$
4. $\mathbf{u} + (-\mathbf{u}) = \mathbf{0}$
5. $s(\mathbf{u} + \mathbf{v}) = s\mathbf{u} + s\mathbf{v}$
6. $(s + t)\mathbf{u} = s\mathbf{u} + t\mathbf{u}$
7. $(st)\mathbf{u} = s(t\mathbf{u})$
          $= t(s\mathbf{u})$
8. $1\mathbf{u} = \mathbf{u}$
    and
    $0\mathbf{u} = \mathbf{0}$

---

47. Illustrate properties 1–4 for $\mathbf{u} = \langle 1, 2 \rangle$, $\mathbf{v} = \langle -3, 2 \rangle$, and $\mathbf{w} = \langle -5, 4 \rangle$.

48. Illustrate properties 5–8 for $\mathbf{u} = \langle -2, 5 \rangle$, $\mathbf{v} = \langle 4, 3 \rangle$, $s = 4$, and $t = -3$.

The **dot product,** or **inner product,** of two vectors $\mathbf{u} = \langle a, b \rangle$ and $\mathbf{v} = \langle c, d \rangle$ is given by

$$\mathbf{u} \cdot \mathbf{v} = ac + bd.$$

The angle θ between the vectors **u** and **v** is given by

$$\cos \theta = \frac{\mathbf{u} \cdot \mathbf{v}}{|\mathbf{u}||\mathbf{v}|}, \text{ where } 0 \leq \theta \leq 180°.$$

In problems 49–54, find:
(a) The dot product of the two vectors **u** and **v**
(b) The angle θ between **u** and **v** (round off θ to the nearest tenth of a degree)

49. $\mathbf{u} = \langle -3, 2 \rangle$; $\mathbf{v} = \langle 1, -3 \rangle$
50. $\mathbf{u} = \langle 1, -3 \rangle$; $\mathbf{v} = \langle 5, -1 \rangle$
51. $\mathbf{u} = \langle -3, 6 \rangle$; $\mathbf{v} = \langle 2, -1 \rangle$
52. $\mathbf{u} = \langle -2, 6 \rangle$; $\mathbf{v} = \langle 3, -5 \rangle$
53. $\mathbf{u} = 5\mathbf{i} - 2\mathbf{j}$; $\mathbf{v} = 4\mathbf{i} + 3\mathbf{j}$
54. $\mathbf{u} = -4\mathbf{i}$; $\mathbf{v} = 4\mathbf{i} + 5\mathbf{j}$

55. Explain why two nonzero vectors **u** and **v** are **perpendicular (or orthogonal)** if and only if
$$\mathbf{u} \cdot \mathbf{v} = 0.$$
[*Hint:* Use the formulas given for problems 49–54.]

In problems 56–58, use the result from problem 55.

56. Find the values of the constant k so that **u** and **v** are orthogonal. Sketch **u** and **v** in the same coordinate system.
    (a) $\mathbf{u} = \langle k, 3 \rangle$; $\mathbf{v} = \langle -4, 2 \rangle$
    (b) $\mathbf{u} = \langle 3k, -1 \rangle$; $\mathbf{v} = \langle k, 4 \rangle$

57. Use vectors and the dot product to determine whether the triangle with vertices
$$A = (6, 1), B = (4, 3), \text{ and } C = (2, 1)$$
    is a right triangle. If it is a right triangle, which vertex angle is the right angle?

58. Suppose that vector **u** has initial point $(2, 1)$ and terminal point $(5, -4)$,
    and
$$\mathbf{v} = \langle 5, 3 \rangle.$$
    Use the dot product to determine whether **u** and **v** are orthogonal.

Objectives

7.4 Trigonometric Forms of Complex Numbers

We have already considered complex numbers and we have seen how to perform basic operations on them. Now we'll discuss how to represent these numbers in trigonometric form. Such representations make multiplication and division of complex numbers much easier to perform. In this section, we'll also see the usefulness of trigonometric forms in finding powers and roots of complex numbers.

Defining the Complex Plane

We begin by representing complex numbers as points in the plane. Each ordered pair of real numbers (a, b) can be associated with the complex number $z = a + bi$, and each complex number $z = a + bi$ can be associated with the ordered pair of real numbers (a, b).

Because of this one-to-one correspondence between the complex numbers and the ordered pairs of real numbers, we can use points in the Cartesian plane to represent the complex numbers.

The plane in which the complex numbers are represented is called the **complex plane;** the horizontal axis (x axis) is called the **real axis** and the vertical axis (y axis) is called the **imaginary axis** (Figure 1a).

For example, the ordered pairs $(2, -3)$ and $(5, 2)$ are used to represent the complex numbers $z_1 = 2 - 3i$ and $z_2 = 5 + 2i$, respectively (Figure 1b). Complex numbers of the form $z = bi$ are represented by points of the form $(0, b)$, that is, as points on the imaginary axis. For example, the complex number $3i$ is represented by the point $(0, 3)$. Complex numbers of the form $z = a$ are represented by points of the form $(a, 0)$, that is, as points on the real axis. Thus, the number 5 is represented by the point $(5, 0)$ (Figure 1c).

Figure 1

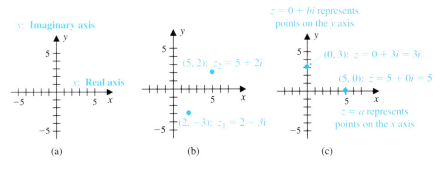

(a)   (b)   (c)

Geometrically, we interpret the *absolute value*, or *magnitude*, of a complex number $a + bi$ as the distance between the origin of the complex plane and the point (a, b). We usually denote it by the absolute value $|z| = |a + bi|$ and refer to it as the **modulus** of the complex number $a + bi$.

So, by the Pythagorean theorem, if $z = a + bi$, then $|z|$ is given by

$$|z| = \sqrt{a^2 + b^2} \quad \text{(Figure 2)}.$$

Figure 2

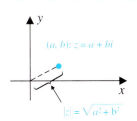

EXAMPLE 1   **Finding Absolute Values**

Let $z_1 = 4 + 3i$ and $z_2 = \sqrt{3} - i$. Find:

(a) $|z_1|$ (b) $|z_2|$ (c) $|z_1z_2|$

Solution   (a) $|z_1| = \sqrt{4^2 + 3^2} = \sqrt{16 + 9} = \sqrt{25} = 5$

(b) $|z_2| = \sqrt{(\sqrt{3})^2 + (-1)^2} = \sqrt{4} = 2$

(c) $|z_1z_2| = |(4 + 3i)(\sqrt{3} - i)|$

$$= |(4\sqrt{3} + 3) + (3\sqrt{3} - 4)i|$$

$$= \sqrt{(4\sqrt{3} + 3)^2 + (3\sqrt{3} - 4)^2}$$

$$= \sqrt{100} = 10$$

Example 1 leads us to the following general results (problem 60):

(i) $|z_1z_2| = |z_1||z_2|$ (ii) $\left|\dfrac{z_1}{z_2}\right| = \dfrac{|z_1|}{|z_2|}, \quad z_2 \neq 0$

Writing a Complex Number in Trigonometric Form

Figure 3

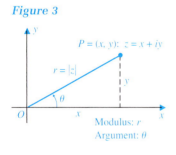

$P = (x, y): z = x + iy$

$r = |z|$

O

Modulus: r
Argument: θ

It is possible to write complex numbers in trigonometric form. Suppose that $z = x + yi$ is a nonzero complex number; then its graphical representation is the point $P = (x, y)$. Let θ be any angle in standard position whose terminal side lies on the line segment $\overline{OP}$ and $r = |z| = \sqrt{x^2 + y^2}$. The angle θ is usually chosen in the interval $[0, 2\pi)$ and is called an **argument** of z (Figure 3).

Using trigonometry, we obtain

$$\cos\theta = \frac{x}{r} \qquad \text{and} \quad \sin\theta = \frac{y}{r}$$

or $x = r\cos\theta \quad$ and $\quad y = r\sin\theta.$

Thus we can rewrite any complex number $z = x + yi$ in the form

$$z = r\cos\theta + (r\sin\theta)i = r(\cos\theta + i\sin\theta)$$

where $r = \sqrt{x^2 + y^2}$ is the modulus and θ is an argument.

This form is called the **trigonometric form,** or **polar form,** of a complex number. Notice that θ is not unique, since

$$r(\cos\theta + i\sin\theta) = r(\cos\theta_1 + i\sin\theta_1)$$

holds whenever $\theta - \theta_1$ is an integer multiple of 2π. Consequently:

Two complex numbers are equal if and only if their moduli are equal and their arguments differ by an integer multiple of 2π.

Thus if z has a modulus r and argument θ, then

$$z = x + iy = r[\cos(\theta + 2\pi k) + i\sin(\theta + 2\pi k)]$$

where k is any integer. If we use degree measure instead of radians to measure θ, then

$$z = r[\cos(\theta + 360°k) + i\sin(\theta + 360°k)]$$

where k is any integer.

EXAMPLE 2   **Writing a Complex Number in Cartesian Form**

Express the complex number

$$z = 2\left(\cos\frac{\pi}{3} + i\sin\frac{\pi}{3}\right)$$

in the Cartesian form $x + yi$.

Solution   The given complex number is in the form $z = r(\cos\theta + i\sin\theta)$, with $r = 2$ and $\theta = \pi/3$.

Figure 4

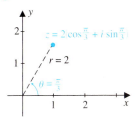

So

$$z = x + yi$$

$$= 2\cos\frac{\pi}{3} + \left(2\sin\frac{\pi}{3}\right)i$$

$$= 2\left(\frac{1}{2}\right) + 2\left(\frac{\sqrt{3}}{2}\right)i$$

$$= 1 + \sqrt{3}i \quad \text{(Figure 4)}.$$

EXAMPLE 3   **Writing a Complex Number in Trigonometric Form**

Express $z = -\sqrt{3} - i$ in trigonometric form with $0 \le \theta < 2\pi$.

Solution   First we plot the point $(-\sqrt{3}, -1)$ corresponding to $z = -\sqrt{3} - i$ (Figure 5). Then we find the modulus r and angle θ.

From Figure 5, we see that the modulus r is given by

$$r = \sqrt{(-\sqrt{3})^2 + (-1)^2} = 2.$$

Figure 5

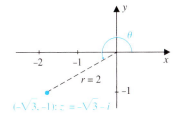

The argument θ satisfies the equation

$$\tan\theta = \frac{(-1)}{-\sqrt{3}} = \frac{1}{\sqrt{3}}$$

and its terminal side is in quadrant III. It follows from trigonometry that $\theta = (7\pi)/6$. Therefore,

$$z = 2\left(\cos\frac{7\pi}{6} + i\sin\frac{7\pi}{6}\right).$$

Finding the Powers of Complex Numbers

Trigonometric forms can be used to multiply and divide complex numbers. Let's examine the product of z_1 and z_2, where

$$z_1 = r_1(\cos\theta_1 + i\sin\theta_1) \quad \text{and} \quad z_2 = r_2(\cos\theta_2 + i\sin\theta_2)$$

We have

$$z_1 z_2 = r_1[(\cos\theta_1 + i\sin\theta_1)] \cdot r_2[(\cos\theta_2 + i\sin\theta_2)]$$

$$= r_1 r_2[(\cos\theta_1\cos\theta_2 - \sin\theta_1\sin\theta_2) + i(\cos\theta_1\sin\theta_2 + \cos\theta_2\sin\theta_1)]$$

$$= r_1 r_2[\cos(\theta_1 + \theta_2) + i\sin(\theta_1 + \theta_2)] \qquad \textcolor{teal}{\text{Sum identities for the cosine and sine}}$$

This result gives us the first of the following two formulas:

Multiplication and Division of Complex Numbers

Suppose that trigonometric forms of the complex numbers z_1 and z_2 are given by

$$z_1 = r_1(\cos\theta_1 + i\sin\theta_1) \quad \text{and} \quad z_2 = r_2(\cos\theta_2 + i\sin\theta_2)$$

Then:

(i) $z_1 z_2 = r_1 r_2[\cos(\theta_1 + \theta_2) + i\sin(\theta_1 + \theta_2)]$

(ii) $\dfrac{z_1}{z_2} = \dfrac{r_1}{r_2}[\cos(\theta_1 - \theta_2) + i\sin(\theta_1 - \theta_2)], \quad z_2 \neq 0$

The derivation of the second formula is left as an exercise (problem 62). In words, the multiplication and division formulas are stated as follows:

The product of two complex numbers in trigonometric form is obtained by multiplying their moduli and adding their arguments. Also, the quotient of two complex numbers is obtained by dividing their moduli and subtracting their arguments.

EXAMPLE 4   **Multiplying and Dividing Complex Numbers**

Let

$$z_1 = \sqrt{2}\left(\cos\frac{3\pi}{4} + i\sin\frac{3\pi}{4}\right) \quad \text{and} \quad z_2 = 4\left(\cos\frac{3\pi}{2} + i\sin\frac{3\pi}{2}\right).$$

Write each given expression in both trigonometric and Cartesian form:

(a) $z_1 z_2$ (b) $\dfrac{z_1}{z_2}$

Solution   (a) Using the multiplication formula, we have

$$z_1 z_2 = 4\sqrt{2}\left[\cos\left(\frac{3\pi}{4} + \frac{3\pi}{2}\right) + i\sin\left(\frac{3\pi}{4} + \frac{3\pi}{2}\right)\right] \qquad \textcolor{teal}{\text{Multiplication rule}}$$

$$= 4\sqrt{2}\left(\cos\frac{9\pi}{4} + i\sin\frac{9\pi}{4}\right) \qquad \textcolor{teal}{\text{Trigonometric form}}$$

$$= 4\sqrt{2}\left(\frac{\sqrt{2}}{2} + i\frac{\sqrt{2}}{2}\right) \qquad \textcolor{teal}{\text{Evaluate}}$$

$$= 4 + 4i \qquad \textcolor{teal}{\text{Cartesian form}}$$

(b) Using the division formula, we have

$$\frac{z_1}{z_2} = \frac{\sqrt{2}}{4}\left[\cos\left(\frac{3\pi}{4} - \frac{3\pi}{2}\right) + i\sin\left(\frac{3\pi}{4} - \frac{3\pi}{2}\right)\right] \qquad \text{Division rule}$$

$$= \frac{\sqrt{2}}{4}\left[\cos\left(-\frac{3\pi}{4}\right) + i\sin\left(-\frac{3\pi}{4}\right)\right] \qquad \text{Trigonometric form}$$

$$= \frac{\sqrt{2}}{4}\left(\cos\frac{3\pi}{4} - i\sin\frac{3\pi}{4}\right)$$

$$= \frac{\sqrt{2}}{4}\left(-\frac{\sqrt{2}}{2} - i\frac{\sqrt{2}}{2}\right) \qquad \text{Evaluate}$$

$$= -\frac{1}{4} - \frac{1}{4}i \qquad \text{Cartesian form}$$

Repeated use of the multiplication rule for complex numbers in trigonometric form allows us to compute powers of a complex number. For instance, if $z = r(\cos\theta + i\sin\theta)$, then

$$z^2 = z \cdot z = r \cdot r[\cos(\theta + \theta) + i\sin(\theta + \theta)] = r^2(\cos 2\theta + i\sin 2\theta).$$

Since $z^3 = z^2 \cdot z$, then

$$z^3 = r^2 \cdot r[\cos(2\theta + \theta) + i\sin(2\theta + \theta)] = r^3(\cos 3\theta + i\sin 3\theta).$$

If we repeat the process one more time, we get

$$z^4 = z^3 \cdot z = r^4(\cos 4\theta + i\sin 4\theta).$$

This pattern is expressed in the following theorem, which is attributed to Abraham DeMoivre (1667–1754).

DeMoivre's Theorem

Let $z = r(\cos\theta + i\sin\theta)$. Then for any positive integer n,
$$z^n = [r(\cos\theta + i\sin\theta)]^n = r^n(\cos n\theta + i\sin n\theta).$$

EXAMPLE 5 **Finding Powers of Complex Numbers**

Use DeMoivre's theorem to determine each of the given powers. Express the answer in Cartesian form.

(a) $[3(\cos 60° + i\sin 60°)]^4$ (b) $(1 + i)^{20}$

Solution (a) By DeMoivre's theorem,

$$[3(\cos 60° + i\sin 60°)]^4 = 3^4(\cos 240° + i\sin 240°)$$

$$= 81\left(-\frac{1}{2} - i\frac{\sqrt{3}}{2}\right) = -\frac{81}{2} - \frac{81\sqrt{3}}{2}i$$

(b) To use DeMoivre's theorem, we first convert $1 + i$ to trigonometric form:

$$1 + i = \sqrt{2}\left(\cos\frac{\pi}{4} + i\sin\frac{\pi}{4}\right)$$

So the modulus of $1 + i$ is $\sqrt{2}$ and an argument is $\pi/4$. Thus

$$(1 + i)^{20} = \left[\sqrt{2}\left(\cos\frac{\pi}{4} + i\sin\frac{\pi}{4}\right)\right]^{20} = 2^{10}(\cos 5\pi + i\sin 5\pi)$$

$$= 1024(-1 + 0i) = -1024 + 0i.$$

Finding the Roots of Complex Numbers

Recall from Section 3.5 that a polynomial equation of degree n has n roots (or solutions) in the complex number system. Hence the equation $z^4 = 16$ has four roots. One way to find these roots is to write the equation in the form

$$z^4 - 16 = 0$$

so that

$$(z^2 - 4)(z^2 + 4) = 0$$

$$(z - 2)(z + 2)(z + 2i)(z - 2i) = 0.$$

It follows that the solutions are

$$2, -2, -2i, \text{ and } 2i.$$

It was easy to find the roots for this equation because of the factorization. In more complicated situations, we can find all the roots of a complex number by using DeMoivre's theorem as follows:

Suppose we want to determine all the nth roots of a complex number w. That is, suppose we want to solve the equation

$$z^n = w \quad \text{where } n \text{ is a positive integer.}$$

Assume that trigonometric forms for w and z are given, respectively, by

$$w = R(\cos \phi + i \sin \phi) \quad \text{and} \quad z = r(\cos \theta + i \sin \theta).$$

Then applying DeMoivre's theorem to $z^n = w$ yields

$$z^n = [r(\cos \theta + i \sin \theta)]^n$$

$$= r^n(\cos n\theta + i \sin n\theta)$$

$$= R(\cos \phi + i \sin \phi).$$

So

$$r^n = R \quad \text{and} \quad n\theta = \phi + 2\pi k \quad \text{(or } n\theta = \phi + 360°k \text{ if degrees are used).}$$

Hence

$$z = r(\cos \theta + i \sin \theta) \text{ is a root of } z^n = w$$

whenever

$$r = \sqrt[n]{R} \quad \text{and} \quad \theta = \frac{\phi}{n} + \frac{2\pi k}{n} \quad \left(\text{or } \theta = \frac{\phi}{n} + \frac{360°k}{n} \text{ if degrees are used}\right)$$

where

$$k = 0, \pm 1, \pm 2, \pm 3, \ldots.$$

At first glance, it appears that there are an infinite number of roots for the equation $z^n = w$. However, there are only n roots. These n roots can be determined by letting k take on the values $0, 1, 2, 3, 4, \ldots, n - 1$. If we let $k = n$, then

$$\theta = \frac{\phi}{n} + \frac{2\pi n}{n} = \frac{\phi}{n} + 2\pi.$$

This angle has the same terminal side as ϕ/n, so we get the same value of z as when $k = 0$. Similarly, the value of θ obtained by letting $k = n + 1$ gives an angle

$$\theta = \frac{\phi}{n} + \frac{2\pi}{n} + 2\pi$$

and the resulting value of z is the same as when $k = 1$; and so on.

Thus, we are led to the following result.

nth Roots of a
Complex Number

> If $w = R(\cos \phi + i \sin \phi)$ is any nonzero complex number, and if n is any positive integer, then the distinct nth roots of w are $z_0, z_1, z_2, \ldots, z_{n-1}$, where
>
> $$z_k = \sqrt[n]{R}\left[\cos\left(\frac{\phi}{n} + \frac{2\pi k}{n}\right) + i \sin\left(\frac{\phi}{n} + \frac{2\pi k}{n}\right)\right] \qquad k = 0, 1, 2, \ldots, n - 1.$$

If we use degree measure for ϕ, the nth roots of w are given by

$$\sqrt[n]{R}\left[\cos\left(\frac{\phi}{n} + \frac{360°k}{n}\right) + i \sin\left(\frac{\phi}{n} + \frac{360°k}{n}\right)\right] \qquad k = 0, 1, 2, \ldots, n - 1.$$

EXAMPLE 6   **Finding the Fourth Roots of a Complex Number**

Find the four fourth roots of $1 + i$; that is, solve $z^4 = 1 + i$. Express the answers in both trigonometric form and Cartesian form (rounded to two decimal places). Represent the roots graphically.

Solution   First, we determine $R(\cos \phi + i \sin \phi)$, a trigonometric representation of $1 + i$.

$$\text{Here } R = \sqrt{1 + 1} = \sqrt{2} \quad \text{and} \quad \text{we use } \phi = \frac{\pi}{4}$$

So, $$1 + i = \sqrt{2}\left(\cos\frac{\pi}{4} + i \sin\frac{\pi}{4}\right).$$

Since $n = 4$, the fourth roots are given by

$$z_k = \sqrt[4]{\sqrt{2}}\left[\cos\left(\frac{\pi/4}{4} + \frac{2\pi k}{4}\right) + i \sin\left(\frac{\pi/4}{4} + \frac{2\pi k}{4}\right)\right] \quad \text{where } k = 0, 1, 2, 3.$$

Using $\sqrt[4]{\sqrt{2}} = \sqrt[8]{2}$, we substitute each value of k into this expression to obtain

$$z_0 = \sqrt[8]{2}\left(\cos\frac{\pi}{16} + i \sin\frac{\pi}{16}\right) \qquad \text{for } k = 0$$

$$z_1 = \sqrt[8]{2}\left(\cos\frac{9\pi}{16} + i \sin\frac{9\pi}{16}\right) \qquad \text{for } k = 1$$

$$z_2 = \sqrt[8]{2}\left(\cos\frac{17\pi}{16} + i \sin\frac{17\pi}{16}\right) \qquad \text{for } k = 2$$

$$z_3 = \sqrt[8]{2}\left(\cos\frac{25\pi}{16} + i \sin\frac{25\pi}{16}\right) \qquad \text{for } k = 3.$$

Figure 6

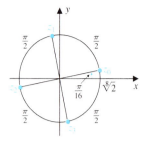

By using a calculator and rounding off the answers to two decimal places, we get the approximations

$$z_0 = 1.07 + 0.21i$$

$$z_1 = -0.21 + 1.07i$$

$$z_2 = -1.07 - 0.21i$$

$$z_3 = 0.21 - 1.07i.$$

These four fourth roots of $1 + i$ are equally spaced on the circle of radius $\sqrt[8]{2}$ centered at the origin; their arguments differ by $\pi/2$ (Figure 6).

The solutions of $z^n = 1$ are called the **nth roots of unity.** In the *real* number system, we find the solution of $x^3 = 1$ by taking the cube root of each side of the equation to get $x = 1$. There is only one real number solution. In the *complex* number system, we know that the equation $z^3 = 1$ has three possible solutions. We can now find them by using trigonometric forms of complex numbers.

EXAMPLE 7 **Finding the Cube Roots of Unity**

Find the cube roots of unity in the form $x + yi$; that is, find all three roots of the equation $z^3 = 1$.

Solution The number 1 can be written in trigonometric form as $1(\cos 0° + i \sin 0°)$, so $R = 1$, $\phi = 0°$, and $n = 3$. The three roots are given by

$$z_k = \sqrt[3]{1}\left[\cos\left(\frac{0° + 360°k}{3}\right) + i \sin\left(\frac{0° + 360°k}{3}\right)\right] \quad \text{where } k = 0, 1, 2.$$

Substituting these values for k yields the following three roots:

Figure 7

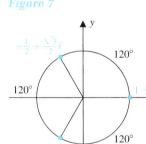

$$z_0 = 1(\cos 0° + i \sin 0°) = 1 + 0i = 1 \qquad \text{for } k = 0$$

$$z_1 = 1(\cos 120° + i \sin 120°) = -\frac{1}{2} + \frac{\sqrt{3}}{2}i \qquad \text{for } k = 1$$

$$z_2 = 1(\cos 240° + i \sin 240°) = -\frac{1}{2} - \frac{\sqrt{3}}{2}i \qquad \text{for } k = 2$$

Figure 7 illustrates the three cube roots of 1 graphically. They are equally spaced on the circle of radius 1 with center at the origin; their arguments differ by 120°.

 PROBLEM SET 7.4

Mastering the Concepts

In problems 1–4, represent each complex number in the complex plane, and find its modulus.

1. (a) $z = 2 + 7i$
   (b) $z = 5 - 2i$
3. (a) $z = 3 + 4i$
   (b) $z = -5 + 3i$

2. (a) $z = \sqrt{3} - i$
   (b) $z = -4 - 4i$
4. (a) $z = -4$
   (b) $z = 2i$

In problems 5–8, find the value of each expression, where $\bar{z} = a - bi$ is the conjugate of $z = a + bi$.

(a) $|zw|$ (b) $|z\bar{z}|$ (c) $|z||w|$ (d) $\left|\dfrac{z}{w}\right|$ (e) $\dfrac{|z|}{|w|}$

5. $z = -1 - i$; $w = 2 - 2i$
6. $z = \sqrt{3}i$; $w = -2 - 2i$

7. $z = -4 + 3i;\quad w = -2 - 5i$

8. $z = 4 + 2i;\quad w = 3 - 5i$

In problems 9–16, express each complex number in the Cartesian form $x + yi$, and graph it. Round off to two decimal places in problems 15 and 16.

9. $z = 2(\cos 30° + i \sin 30°)$

10. $z = 10\left(\cos \dfrac{3\pi}{4} + i \sin \dfrac{3\pi}{4}\right)$

11. $z = 2\left(\cos \dfrac{\pi}{2} + i \sin \dfrac{\pi}{2}\right)$

12. $z = 7\left[\cos\left(-\dfrac{3\pi}{2}\right) + i \sin\left(-\dfrac{3\pi}{2}\right)\right]$

13. $z = 8(\cos 900° + i \sin 900°)$

14. $z = 6(\cos 120° + i \sin 120°)$

15. $z = 2(\cos 10° + i \sin 10°)$

16. $z = 2[\cos(-75°) + i \sin(-75°)]$

In problems 17–22, express each complex number in trigonometric form with $0 \le \theta < 2\pi$. Round off to two decimal places when necessary.

17. (a) $z = -1 - i$
    (b) $z = -\sqrt{3} - i$

18. (a) $z = -3 + 4i$
    (b) $z = 8 + 7i$

19. (a) $z = \sqrt{2} + i$
    (b) $z = -5$

20. (a) $z = 3 + 4i$
    (b) $z = -3i$

21. (a) $z = -\dfrac{1}{2} + \dfrac{\sqrt{3}}{2}i$

    (b) $z = \dfrac{\sqrt{3}}{2} + \dfrac{1}{2}i$

22. (a) $z = \dfrac{\sqrt{2}}{2} - \dfrac{\sqrt{2}}{2}i$

    (b) $z = -\dfrac{\sqrt{2}}{2} + \dfrac{\sqrt{2}}{2}i$

In problems 23–30, find:

(a) $z_1 z_2$ (b) $\dfrac{z_1}{z_2}$

Express the answers in both trigonometric form and Cartesian form. In problems 29 and 30, round off answers to two decimal places.

23. $z_1 = 4\left(\cos \dfrac{5\pi}{6} + i \sin \dfrac{5\pi}{6}\right);$

    $z_2 = 2\left(\cos \dfrac{\pi}{3} + i \sin \dfrac{\pi}{3}\right)$

24. $z_1 = 2\left[\cos\left(-\dfrac{\pi}{6}\right) + i \sin\left(-\dfrac{\pi}{6}\right)\right];$

    $z_2 = 2\left[\cos\left(-\dfrac{5\pi}{6}\right) + i \sin\left(-\dfrac{5\pi}{6}\right)\right]$

25. $z_1 = \cos 30° + i \sin 30°;\ z_2 = \cos 60° + i \sin 60°$

26. $z_1 = 5(\cos 30° + i \sin 30°);$
    $z_2 = 6(\cos 240° + i \sin 240°)$

27. $z_1 = \sqrt{2}(\cos 45° + i \sin 45°);$
    $z_2 = \sqrt{2}(\cos 135° + i \sin 135°)$

28. $z_1 = 4\left(\cos \dfrac{3\pi}{4} + i \sin \dfrac{3\pi}{4}\right);$
    $z_2 = 2(\cos \pi + i \sin \pi)$

29. $z_1 = 2(\cos 50° + i \sin 50°);$
    $z_2 = 3(\cos 40° + i \sin 40°)$

30. $z_1 = 5(\cos 170° + i \sin 170°);$
    $z_2 = \cos 55° + i \sin 55°$

In problems 31–44, use DeMoivre's theorem to compute each of the given powers. Express the answer in Cartesian form.

31. $(\cos 30° + i \sin 30°)^7$ **32.** $(\cos 15° + i \sin 15°)^8$

33. $\left[2\left(\cos \dfrac{\pi}{6} + i \sin \dfrac{\pi}{6}\right)\right]^{10}$

34. $\left[3\left(\cos \dfrac{\pi}{18} + i \sin \dfrac{\pi}{18}\right)\right]^6$

35. $\left[2\left(\cos \dfrac{5\pi}{4} + i \sin \dfrac{5\pi}{4}\right)\right]^8$

36. $[4(\cos 36° + i \sin 36°)]^5$

37. $(5 + 5i)^6$ **38.** $(1 + \sqrt{3}i)^5$

39. $(\sqrt{3} - i)^4$ **40.** $\left(-\dfrac{1}{2} - \dfrac{\sqrt{3}}{2}i\right)^8$

41. $(\sqrt{3} + i)^{30}$ **42.** $(1 + i)^{50}$

43. $\left(\dfrac{1}{\sqrt{2}} + \dfrac{1}{\sqrt{2}}i\right)^{100}$ **44.** $\left(\dfrac{1}{2} + \dfrac{\sqrt{3}}{2}i\right)^{30}$

In problems 45–52, find all the indicated roots of each complex number in the form $x + yi$. Round off answers to two decimal places. Represent the roots graphically.

45. Square roots of i

46. Square roots of $3 - 3i$

47. Cube roots of 8

48. Cube roots of i

49. Fifth roots of $32(\cos 315° + i \sin 315°)$

50. Fifth roots of unity

51. Fourth roots of $-8 - 3\sqrt{3}i$

52. Fourth roots of -16

In problems 53–58, find all the roots of each equation in both trigonometric and Cartesian form. Round off answers to two decimal places.

53. $z^3 + 8 = 0$ **54.** $z^3 + 8i = 0$

55. $z^4 + 81 = 0$ **56.** $z^5 + 1 = 0$

57. $z^6 + 64 = 0$ **58.** $z^5 - i = 1$

Developing and Extending the Concepts

59. Prove that $|z| = \sqrt{z \cdot \bar{z}}$, where $z = a + bi$ and $\bar{z} = a - bi$ is the conjugate of z.

60. Prove each of the following statements, where $z_1 = a_1 + b_1 i$ and $z_2 = a_2 + b_2 i$:

(a) $|z_1 z_2| = |z_1||z_2|$

(b) $\left| \dfrac{z_1}{z_2} \right| = \dfrac{|z_1|}{|z_2|}$, $z_2 \neq 0$

61. Describe geometrically the set of complex numbers $z = x + yi$ that satisfy each equation. Sketch the graph.

(a) $|z| = 2$

(b) $|z - i| = 1$

62. Let

$$z_1 = r_1(\cos \theta_1 + i \sin \theta_1) \text{ and}$$
$$z_2 = r_2(\cos \theta_2 + i \sin \theta_2).$$

Show that

$$\frac{z_1}{z_2} = \frac{r_1}{r_2}[\cos(\theta_1 - \theta_2) + i \sin(\theta_1 - \theta_2)] \quad z_2 \neq 0.$$

63. Let $z = r(\cos \theta + i \sin \theta)$.

Show that

(a) $\bar{z} = r(\cos \theta - i \sin \theta)$
$$= r[\cos(-\theta) + i \sin(-\theta)]$$

(b) $z^{-1} = r^{-1}(\cos \theta - i \sin \theta)$
$$= r^{-1}[\cos(-\theta) + i \sin(-\theta)]$$

64. If $z = r(\cos \theta + i \sin \theta)$, show that

$$z^{-n} = r^{-n}(\cos n\theta - i \sin n\theta)$$

where n is a positive integer. [*Hint:* Use the result from problem 63(b).]

65. Use the result from Problem 64 to write each expression in the form $x + yi$.

(a) $\left(\dfrac{\sqrt{3}}{2} + \dfrac{1}{2}i \right)^{-5}$

(b) $(-2 + 2i)^{-3}$

66. Use DeMoivre's theorem to derive formulas for $\cos 3\theta$ in terms of $\cos \theta$, and $\sin 3\theta$ in terms of $\sin \theta$. [*Hint:* Use the identity $\cos 3\theta + i \sin 3\theta = (\cos \theta + i \sin \theta)^3$.]

CHAPTER 7 REVIEW PROBLEM SET

In problems 1 and 2, solve triangle *ABC*. Round off each angle to the nearest hundredth of a degree and each length to two decimal places.

1. (a) $a = 5$, $b = 7$, $\gamma = 30°$

(b) $c = 10$, $\alpha = 45°$, $\beta = 75°$

2. (a) $\alpha = 63.3°$, $\gamma = 81.6°$, $c = 20.7$

(b) $a = 36$, $b = 47$, $c = 41$

3. Locate the point with the given polar coordinates. Then find the corresponding Cartesian coordinates of the point.

(a) $\left(5, \dfrac{\pi}{4} \right)$ (b) $\left(2, \dfrac{\pi}{6} \right)$ (c) $(\sqrt{2}, -135°)$

4. Locate the point with the given Cartesian coordinates. Then find polar coordinates that correspond to the point, where $r > 0$ and $0° \leq \theta < 360°$.

(a) $(-3, 0)$ (b) $(-2, 2\sqrt{3})$ (c) $(-5\sqrt{3}, -5)$

5. Convert each polar equation to a corresponding equation in Cartesian form. Identify the curve and graph it.

(a) $r + 4 \sin \theta = 0$ (b) $r = -3 \cos \theta$

6. Convert each equation to a corresponding equation in polar form. Identify the curve and graph it.

(a) $3x^2 + 3y^2 = 48$ (b) $y^2 = 4x$

7. Sketch the graph of each polar equation by the point-plotting method. Then use a grapher to demonstrate the validity of the result.

(a) $r = 4 \cos \theta$ (b) $r = 1 + 3 \cos \theta$

8. **G** Use a grapher to graph each equation.

(a) $r = 2 \sin 2\theta$ (b) $r = -2 \cos 3\theta$

9. Let $\mathbf{u} = \langle 3, 4 \rangle$ and $\mathbf{v} = \langle 4, 3 \rangle$. Find each of the following vectors and represent them graphically:

(a) $3\mathbf{u}$ (b) $\mathbf{v} - \mathbf{u}$

(c) $-2\mathbf{u} + \mathbf{v}$ (d) $3\mathbf{u} + 2\mathbf{v}$

10. Determine the components of $\mathbf{u}$ if θ is a direction angle of $\mathbf{u}$.

(a) $|\mathbf{u}| = 5$ and $\theta = 30°$

(b) $|\mathbf{u}| = 6$ and $\theta = 45°$

(c) $|\mathbf{u}| = 8$ and $\theta = \dfrac{5\pi}{6}$

In problems 11 and 12, find the magnitude and direction of each vector. Round off the answers to two decimal places.

11. (a) $\mathbf{u} = 12\mathbf{i} + 5\mathbf{j}$ **12.** (a) $\mathbf{u} = 6\mathbf{i} + 3\mathbf{j}$

(b) $\mathbf{u} = 7\mathbf{i} - 2\mathbf{j}$ (b) $\mathbf{u} = \sqrt{3}\mathbf{i} - 2\mathbf{j}$

13. If **u** is a vector whose initial point is the first point given and whose terminal point is the second point given, write **u** in the form $\mathbf{u} = \langle a, b \rangle$ and find $|\mathbf{u}|$.

   (a) $(-6, 8); (4, 3)$

   (b) $(-7, 6); (3, -1)$

14. Suppose that vector **u** has a magnitude of 10 and a direction of 135°. Find the components of **u**.

In problems 15–18, express each complex number in trigonometric form. Give the exact angle in problems 15 and 16; in problems 17 and 18, round off the angle measure to two decimal places.

15. (a) $z = -1 + \sqrt{3}i$

   (b) $z = -1 + i$

16. (a) $z = 5 + 5i$

   (b) $z = 4\sqrt{3} - 12i$

17. (a) $z = 5 + 12i$

   (b) $z = 7 - 6i$

18. (a) $z = -12 + 5i$

   (b) $z = -7 - 8i$

In problems 19–22, find both $z_1 z_2$ and z_1/z_2. Write the answers in both trigonometric and Cartesian form. In problems 21 and 22, round off the answers to two decimal places.

19. $z_1 = 2(\cos \pi + i \sin \pi)$;

   $z_2 = 3\left(\cos \dfrac{\pi}{2} + i \sin \dfrac{\pi}{2}\right)$

20. $z_1 = \sqrt{2}(\cos 315° + i \sin 315°)$;

   $z_2 = 2\sqrt{2}(\cos 135° + i \sin 135°)$

21. $z_1 = 6(\cos 230° + i \sin 230°)$;

   $z_2 = 3(\cos 75° + i \sin 75°)$

22. $z_1 = \sqrt{2}\left(\cos \dfrac{\pi}{4} + i \sin \dfrac{\pi}{4}\right)$;

   $z_2 = 2\left(\cos \dfrac{2\pi}{3} + i \sin \dfrac{2\pi}{3}\right)$

In problems 23–26, use DeMoivre's theorem to write each expression in both trigonometric and Cartesian form. In problems 25 and 26, round off the answers to two decimal places.

23. (a) $(\cos 60° + i \sin 60°)^6$

   (b) $\left[\sqrt{6}\left(\cos \dfrac{\pi}{5} + i \sin \dfrac{\pi}{5}\right)\right]^5$

24. (a) $(1 + i)^{40}$

   (b) $\left(\dfrac{\sqrt{2}}{2} + \dfrac{\sqrt{2}}{2}i\right)^{100}$

25. (a) $[3(\cos 27° + i \sin 27°)]^4$

   (b) $\left[\sqrt{5}\left(\cos \dfrac{\pi}{8} + i \sin \dfrac{\pi}{8}\right)\right]^5$

26. (a) $[2(\cos 44° + i \sin 44°)]^6$

   (b) $\left[\sqrt{6}\left(\cos \dfrac{\pi}{7} + i \sin \dfrac{\pi}{7}\right)\right]^4$

In problems 27 and 28, write all the roots of each equation in trigonometric form.

27. (a) $z^3 = 8i$

   (b) $z^4 = 1 + i$

28. (a) $z^4 = 8\sqrt{2} - 8\sqrt{2}i$

   (b) $z^5 = 243i$

In problems 29–32, round off the answers to two decimal places.

29. Surveying: Two points A and B are 50 feet apart on one bank of a straight river. A point C on the bank across the river is located so that angle CAB is 70° and angle ABC is 80°. How wide is the river?

30. Geometry: A diagonal of a parallelogram is 16 inches long and forms angles of 43° and 15°, respectively, with the two sides. How long are the sides of the parallelogram?

31. Engineering: A guy wire attached to the top of a pole is 40 feet long and forms a 50° angle with the ground. How tall is the pole if it is tilted 15° from the vertical directly away from the guy wire?

32. Telephone Pole: A telephone pole AB is 30 feet tall and is located on a 10° slope from a point C on the horizontal plane (Figure 1). If angle ACB is 31°, find the length of the guy wire CB and the distance from C to the base of the pole.

Figure 1

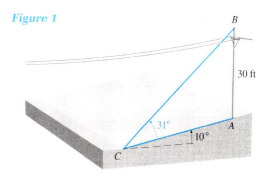

33. Engineering: Two forces, one of 63 pounds and one of 45 pounds, yield a resultant force of 75 pounds. Find the angle between the two forces (to the nearest hundredth of a degree).

34. Navigation: An airplane, which has an airspeed of 550 miles per hour, heading N30°W, is affected by a wind blowing at 45 miles per hour from S45°W. Find the velocity vector of the airplane's path. What is the speed of the airplane, and in what direction is it traveling, relative to the direction of the wind (to the nearest hundredth of a degree)?

CHAPTER 7 TEST

1. Use the law of sines to solve triangle *ABC*, with

$$a = 70, \quad \beta = 36.65°, \quad \text{and} \quad \alpha = 40°.$$

   Round off the answers to two decimal places.

2. Use the law of cosines to find *b* in triangle *ABC*, if

$$a = 5.47, \quad c = 8.94, \quad \text{and} \quad \beta = 37.2°.$$

   Round off the answer to two decimal places.

3. (a) Write the polar point $(4, 5\pi/3)$ in Cartesian form.
   (b) Write the Cartesian point $(-3, 3)$ in polar form.

4. Sketch the graph of each polar equation. Convert to rectangular form in each case.
   (a) $r = 5$
   (b) $r = 4 \sin \theta$
   (c) $\theta = \dfrac{\pi}{4}$

5. Find the polar form of each equation, and graph the equation.
   (a) $3x + 2y = 5$
   (b) $x^2 + y^2 = 16$

6. Find $4\mathbf{u} - 8\mathbf{v}$ if $\mathbf{u} = 2\mathbf{i} + 3\mathbf{j}$ and $\mathbf{v} = 4\mathbf{i} - \mathbf{j}$.

7. Find a direction angle of the vector $\mathbf{u} = 2\mathbf{i} + 3\mathbf{j}$. Round off the answer to two decimal places.

8. Let $\mathbf{u}$ and $\mathbf{v}$ be vectors such that $|\mathbf{u}| = 6$ and $|\mathbf{v}| = 8$. Assume that angles 110° and 56° are direction angles of $\mathbf{u}$ and $\mathbf{v}$, respectively. Determine $\mathbf{u} + \mathbf{v}$. Round off to two decimal places.

9. Write $z = 4 \left(\cos 315° + i \sin 315°\right)$ in Cartesian form.

10. Write $z = -1 - \sqrt{3}i$ in trigonometric form.

11. For z_1 and z_2 given below, find the following in both the Cartesian and trigonometric forms:

    (a) $z_1 z_2$ (b) $\dfrac{z_1}{z_2}$

$$z_1 = \cos 139° + i \sin 139°$$

    and

$$z_2 = 2(\cos 71° + i \sin 71°)$$

    Round off each answer to four decimal places.

12. Use DeMoivre's theorem to determine $[2(\cos 60° + i \sin 60°)]^{12}$. Express the answer in the form $a + bi$.

13. Find the three roots of $z^3 = i$. Write the answers in Cartesian form.

14. An airplane flies 40 miles due east, then changes direction and flies 60 miles N20°E. How far is the airplane from its starting point? Round off the answer to one decimal place.

CHAPTER
8

Systems of Equations and Inequalities

Chapter Contents

In the state of Michigan, most gas stations offer three grades of gasoline with the following ratings on the pumps: low grade (87 octane), middle grade (89 octane) and high grade (93 octane). However, some of these gas stations are equipped with only two large underground tanks—one for the lowest grade (87 octane) and the other for the highest grade (93 octane). To produce the middle grade (89 octane), the pump's computer mixes the other two grades from the underground tanks according to a programmed formula. How much of the 87 octane grade should be blended with the 93 octane grade to obtain 36 gallons of the 89 octane grade? The solution of this problem is given in Example 5 on page 544.

In science, business, and economics, we often encounter mathematical models that involve more than one equation and contain two or more variables. Such a set of equations is referred to as a *system of equations.* In this chapter, we study several different techniques for solving systems of equations and also *systems of inequalities.*

Objectives

1. Solve a System by Substitution or by Elimination
2. Solve a System by Gaussian Elimination
3. Solve Applied Problems

We use a brace to group together the equations in the system.

8.1 Solutions of Linear Systems by Substitution and Elimination

In this section we investigate systems that contain only linear equations. Such systems are called **linear systems.** In general, a **solution** of a linear system of equations is a collection of particular values for the variables that satisfy all equations in the system.

For example,

$$\begin{cases} 3x + 4y = 12 \\ 3x - 8y = 0 \end{cases}$$

is a linear system of two equations that contains the two variables x and y.

If we substitute 8/3 for x and 1 for y into each equation in the first system, we find that these values satisfy both equations:

$$\begin{cases} 3\left(\dfrac{8}{3}\right) + 4(1) = 8 + 4 = 12 \\ 3\left(\dfrac{8}{3}\right) - 8(1) = 8 - 8 = 0 \end{cases}$$

We say that the pair of values 8/3 and 1 for x and y, respectively, is a *solution* of the system. A solution of a system of two equations in two variables is often written as an ordered pair of numbers, so we express the solution of the first system in the form

$$\left(\frac{8}{3}, 1\right).$$

The solution of the system

$$\begin{cases} 2x - 3y + 4z = 17 \\ 4x + 5y - z = 5 \\ -x + 2y + z = -3 \end{cases}$$

is given by the ordered triple

$$(3, -1, 2).$$

Again, if we substitute 3 for x, -1 for y, and 2 for z, we find that these values satisfy each of the three equations.

Solving a System by Substitution or by Elimination

The process of finding all solutions of a system is called **solving the system.** To solve a linear system, we can use either *substitution* or *elimination.*

To solve a linear system of two equations in two variables by *substitution,* we use one of the equations to express one of the variables in terms of the other. Then we substitute this expression for the variable in the other equation to obtain an equation in one variable. This procedure is illustrated in the next example.

EXAMPLE 1 **Solving a System by Substitution**

Solve the system by substitution: $\begin{cases} 4x - y = 2 \\ 3x + y = 5 \end{cases}$ and check the solution.

Solution   We begin by solving the first equation for y in terms of x to get

$$y = 4x - 2.$$

Replacing y with $4x - 2$ in the second equation of the system, gives us

$$3x + (4x - 2) = 5, \quad 7x = 7 \quad \text{or} \quad x = 1.$$

The value of y that solves the system is obtained by substituting 1 for x in $y = 4x - 2$ to get

$$y = 4(1) - 2 = 2.$$

Thus, the solution is the pair of numbers $x = 1$ and $y = 2$, or $(1, 2)$.

Check   To check the solution, replace x by 1 and y by 2 in the original system to get

$$\begin{cases} 4(1) - 2 = 2 \\ 3(1) + 2 = 5. \end{cases}$$

Graphically, this means that the two lines with equations $4x - y = 2$ and $3x + y = 5$ intersect at the point (1, 2).

The substitution method is efficient for solving certain systems with two equations. However, it tends to become cumbersome when more than two equations are involved. An alternate method, called the *elimination method,* enables us to solve systems of linear equations in any number of variables more efficiently. The elimination method is based on the idea of *equivalent systems* of equations.

Two systems of equations are said to be **equivalent** if they have the same solutions.

For example, the systems

$$\begin{cases} 2x + y = 1 \\ -x + 3y = -4 \end{cases} \quad \text{and} \quad \begin{cases} 2x + y = 1 \\ y = -1 \end{cases}$$

are equivalent, since they have the same and only solution: $(1, -1)$.

A linear system can be converted to an equivalent form by performing one or more of the following three elementary operations:

Elementary Operations Yielding Equivalent Systems

1. Interchange the positions of two equations in the system.
2. Replace an equation in the system by a nonzero multiple of itself by multiplying each side of the equation by the same nonzero number.
3. Replace an equation in the system by the sum of a nonzero multiple of that equation and another equation in the system.

The basic strategy behind the elimination method is to use these operations to convert a system to an equivalent one that contains fewer variables and is easier to solve. For instance, consider the system

$$\begin{cases} x + y = 6 \\ x - y = 2. \end{cases}$$

If we replace the second equation by the sum of itself and the first equation, we get the equivalent system

$$\begin{cases} x + y = 6 \\ 2x \quad\;\; = 8. \end{cases}$$

This form of the system is easier to solve than the original. By inspection, we see from the second equation that $x = 4$. Substituting this value into the first equation, we get

$$4 + y = 6 \quad \text{or} \quad y = 2.$$

So the solution of the system is $(4, 2)$.

EXAMPLE 2   **Solving a System by Elimination**

Solve each system by elimination. Then represent the system graphically; that is, graph both equations in the same coordinate system. Relate the solution to the graph of the system.

(a) $\begin{cases} x - y = 1 \\ 3x - y = -1 \end{cases}$ (b) $\begin{cases} x + 3y = 4 \\ 2x + 6y = -6 \end{cases}$ (c) $\begin{cases} 3x - y = 1 \\ 6x - 2y = 2 \end{cases}$

Solution   (a) For

$$\begin{cases} x - y = 1 \\ 3x - y = -1 \end{cases}$$

we'll start by using the third elementary operation to convert to an equivalent system in which the x variable is eliminated in the second equation. This is accomplished by replacing the second equation by the sum of -3 times the first equation and the second to obtain

$$\begin{cases} x - y = 1 \\ 3x - y = -1 \end{cases} \quad \rightarrow \quad \begin{cases} x - y = 1 \\ 2y = -4 \end{cases}.$$

Next we multiply both sides of the second equation by $1/2$ to obtain an equivalent system that is simpler to solve:

$$\begin{cases} x - y = 1 \\ y = -2 \end{cases}$$

By substituting -2 for y into the first equation, we get $x - y = 1$, or $x - (-2) = 1$, that is, $x = -1$. Thus the solution is $(-1, -2)$.

The graph of the system is shown in Figure 1a. Clearly, the graphs intersect at one point. Since the coordinates of this point must satisfy both equations, it follows that the solution of the system and the point of intersection are the same: $(-1, -2)$.

(b) For

$$\begin{cases} x + 3y = 4 \\ 2x + 6y = -6 \end{cases}$$

our strategy is to produce an equivalent system by eliminating x in the second equation. To do this, we replace the second equation by the sum of -2 times the first equation and the second equation to obtain

$$\begin{cases} x + 3y = 4 \\ 2x + 6y = -6 \end{cases} \quad \rightarrow \quad \begin{cases} x + 3y = 4 \\ 0 = -14 \end{cases}.$$

This leads to a contradiction, so the system has no solution.

Figure 1

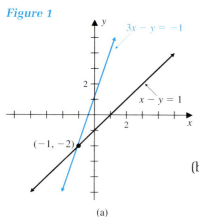

(a)

Figure 1

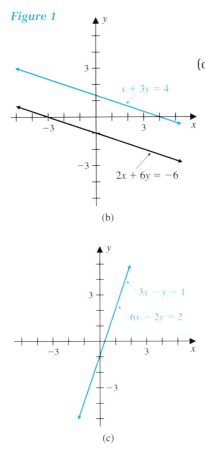

As Figure 1b shows, the two lines are parallel (the lines have the same slope $m = -1/3$), so there is no point common to the two lines. Consequently, no pair of numbers can satisfy both equations simultaneously.

(c) For

$$\begin{cases} 3x - y = 1 \\ 6x - 2y = 2 \end{cases}$$

we eliminate the x term in the second equation by replacing the second equation by the sum of -2 times the first equation and the second to get

$$\begin{cases} 3x - y = 1 \\ 6x - 2y = 2 \end{cases} \rightarrow \begin{cases} 3x - y = 1 \\ 0 = 0. \end{cases}$$

We conclude that the solution consists of all pairs of numbers (x, y) that satisfy the equation

$$3x - y = 1 \quad \text{or} \quad y = 3x - 1.$$

It follows that the solution consists of ordered pairs of the form $(x, 3x - 1)$, where x is a real number. In Figure 1c, we can see that the two equations have the same graph, so there are infinitely many solutions (one for each point on the graph).

Referring to the three systems in Example 2, the system in part (a) is said to be **consistent and independent,** with the unique solution $x = -1$ and $y = -2$ (Figure 1a). The system in part (b) is said to be **inconsistent** with no solution (Figure 1b). And the system in part (c) is said to be **consistent and dependent** with an infinite number of solutions (Figure 1c).

Solving a System by Gaussian Elimination

In Example 2a, we solved the given system by using the equivalent form

$$\begin{cases} x - y = 1 \\ \quad\quad y = -2. \end{cases}$$

This latter system is said to be in *triangular form.* Other examples of triangular form are:

$$\begin{cases} x + 2y = 7 \\ \quad\quad y = 1 \end{cases} \quad \text{and} \quad \begin{cases} x + 2y - 3z = 1 \\ \quad\quad y + 2z = 4 \\ \quad\quad\quad\quad z = 7 \end{cases}$$

Note the staggered pattern of a triangular form. The leading coefficient is 1 for each equation, and each equation after the first one has one less variable than the one above it. When a system is written in triangular form, we can find its solution by using *back-substitution.*

For instance, the solution of

$$\begin{cases} x + 2y - 3z = 1 \\ y + 2z = 4 \\ z = 7 \end{cases}$$

is found as follows:

Clearly, $z = 7$ by inspection. Substituting this value into the second equation, we get

$$y + 2(7) = 4, \quad y + 14 = 4, \quad \text{or} \quad y = -10.$$

Finally, we substitute -10 for y and 7 for z into the first equation to get

$$x + 2(-10) - 3(7) = 1, \quad x - 20 - 21 = 1, \quad x - 41 = 1, \quad \text{or} \quad x = 42.$$

Thus the solution is $(42, -10, 7)$.

The process of using elimination to convert a given linear system to an equivalent triangular one and then using back-substitution to find the solution is referred to as the **Gaussian elimination method.**

EXAMPLE 3

Solving a System by Gaussian Elimination

Subscripts are often used to denote variables in linear systems.

Solve system (A) by using the Gaussian elimination method.

$$\begin{cases} 2x_1 + x_2 - 2x_3 = 10 \\ 3x_1 + 2x_2 + 2x_3 = 1 \\ 5x_1 + 4x_2 + 3x_3 = 4 \end{cases} \tag{A}$$

Solution

First we eliminate the x_1 term in the second equation by replacing the second equation in system (A) by the sum of -3 times the first equation and 2 times the second equation to obtain the equivalent system (B):

$$\begin{cases} 2x_1 + x_2 - 2x_3 = 10 \\ x_2 + 10x_3 = -28 \\ 5x_1 + 4x_2 + 3x_3 = 4 \end{cases} \tag{B}$$

In the next step, we eliminate the x_1 term in the third equation by replacing the third equation of (B) with the sum of -5 times the first equation and 2 times the third equation to obtain the equivalent system (C):

$$\begin{cases} 2x_1 + x_2 - 2x_3 = 10 \\ x_2 + 10x_3 = -28 \\ 3x_2 + 16x_3 = -42 \end{cases} \tag{C}$$

Now we eliminate the x_2 term in the third equation by replacing the third equation of (C) by the sum of the third equation and -3 times the second equation to get the equivalent system (D):

$$\begin{cases} 2x_1 + x_2 - 2x_3 = 10 \\ x_2 + 10x_3 = -28 \\ -14x_3 = 42 \end{cases} \tag{D}$$

Next we multiply the third equation by $-1/14$ to obtain the equivalent system (E), in which the coefficient of x_3 is 1:

$$\begin{cases} 2x_1 + x_2 - 2x_3 = 10 \\ x_2 + 10x_3 = -28 \\ x_3 = -3 \end{cases} \quad \text{(E)}$$

Finally, in order to change the coefficient of x_1 in the first equation to 1, we multiply the first equation by $1/2$ to obtain the equivalent triangular system (F):

$$\begin{cases} x_1 + \frac{1}{2}x_2 - x_3 = 5 \\ x_2 + 10x_3 = -28 \\ x_3 = -3 \end{cases} \quad \text{(F)}$$

Now we use back-substitution to solve this system. Substituting $x_3 = -3$ into the second equation, we get $x_2 = 2$; then substituting $x_2 = 2$ and $x_3 = -3$ into the first equation, we get $x_1 = 1$. Thus the solution is the triple of numbers $x_1 = 1$, $x_2 = 2$, and $x_3 = -3$, or $(1, 2, -3)$.

Some graphers have a special feature that finds the coordinates of a point of intersection of two graphs. This feature can be used to approximate the values of the solution of a system of two linear equations in two variables. For example, to solve the following system graphically

$$\begin{cases} 0.3x - 2.7y = -0.82 \\ 0.2x + 5.4y = 1.83 \end{cases}$$

we graph the two equations in the some viewing window of Figure 2. By using the INTERSECT -finding feature of a grapher, we approximate the location of the point of intersection to be $(0.24, 0.33)$. So the solution of the system is approximately $x = 0.24$ and $y = 0.33$.

Figure 2

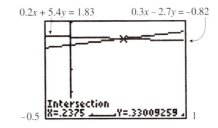

Although graphs are excellent tools for visualizing solutions of systems of equations in two variables, they do not always give the exact solution, especially if the point of intersection is not located at integer values of x and y. For this reason we normally rely on algebraic methods for solving these systems.

Solving Applied Problems

Systems of linear equations are often used to model real-life situations, as the next example shows.

EXAMPLE 4

Modeling a Physics Problem

Figure 3

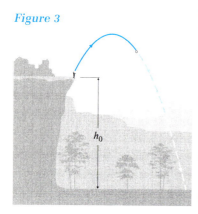

Suppose that a ball is thrown up in the air from a cliff h_0 feet above ground level with an initial upward speed of v_0 (in feet per second). Its height h (in feet) above ground level t seconds after it was thrown is given by the model

$$h = -16t^2 + v_0 t + h_0 \quad \text{(Figure 3)}.$$

(a) Suppose that the ball is 67 feet above the ground after 1 second and 61 feet above the ground after 1.5 seconds. Use a linear system to determine the initial upward speed v_0 and the height of the cliff h_0.

(b) Use the results from part (a) to determine the specific model that expresses the height h as a function of time t under the given conditions.

(c) How high above ground level is the ball after 2.5 seconds?

Solution

(a) Since the ball is 67 feet above the ground after 1 second and 61 feet after 1.5 seconds, we substitute 1 for t when $h = 67$ and 1.5 for t when $h = 61$ in the equation of the model to get the system

$$\begin{cases} 67 = -16(1)^2 + v_0(1) + h_0 \\ 61 = -16(1.5)^2 + v_0(1.5) + h_0 \end{cases} \quad \text{or} \quad \begin{cases} v_0 + h_0 = 83 \\ 1.5v_0 + h_0 = 97. \end{cases}$$

Solving this system, we get $v_0 = 28$ and $h_0 = 55$. Therefore, the initial speed is 28 feet per second and the height of the cliff is 55 feet.

(b) The general model for this situation is given by

$$h = -16t^2 + v_0 t + h_0.$$

Under the given conditions $v_0 = 28$ and $h_0 = 55$, so we get the specific model

$$h = -16t^2 + 28t + 55.$$

(c) When $t = 2.5$, $h = -16(2.5)^2 + 28(2.5) + 55 = 25$. So the height of the ball above ground level is 25 feet after 2.5 seconds.

EXAMPLE 5

Solving a Gasoline Mixture Problem

In the state of Michigan, most gas stations offer three grades of gasoline with the following ratings on the pumps: low grade (87 octane), middle grade (89 octane) and high grade (93 octane). However, some of these gas stations are equipped with only two large underground tanks—one for the lowest grade (87 octane) and the other for the highest grade (93 octane). To produce the middle grade (89 octane), the pump's computer mixes the other two grades from the underground tanks according to a programmed formula. How much of the 87 octane grade should be blended with the 93 octane grade to obtain 36 gallons of the 89 octane grade?

Solution

Let x represent the number of gallons of the 87 octane grade and y represent the number of gallons of the 93 octane grade needed to obtain a mix of 36 gallons of the 89 octane grade.

We translate the problem to the following system of equations:

$$\begin{cases} x + y = 36 & \text{\color{blue}{Total number of gallons}} \\ 87x + 93y = 36\,(89) & \text{\color{blue}{Octane mixture}} \end{cases}$$

Solving this system of equations, we get $x = 24$ and $y = 12$.

Therefore, 24 gallons of the 87 octane grade should be blended with 12 gallons of the 93 octane grade to obtain 36 gallons of the 89 octane grade.

EXAMPLE 6   **Modeling a Manufacturing Problem**

A manufacturer of radios produces three types of radios, model A, model B, and model C. The manufacturer determines that the total times allocated for production, assembly, and testing are 638 hours, 253 hours, and 159 hours, respectively. Table 1 indicates the production, assembly, and testing times required for a radio of each type. Use a linear system to determine how many of each type can be produced if all the allocated time is used up.

TABLE 1

Model	Production Hours Per radio	Assembly Hours Per radio	Testing Hours Per radio
A	1.4	0.5	0.3
B	1.6	0.6	0.4
C	1.8	0.8	0.5

Solution   Let

x represent the number of radios of model A produced,

y the number of model B, and

z the number of model C.

Now we use the data in Table 1. Since the production hours add up to 638, the assembly hours add up to 253, and the testing hours add up to 159, we obtain the system

$$\begin{cases} 1.4x + 1.6y + 1.8z = 638 \\ 0.5x + 0.6y + 0.8z = 253 \\ 0.3x + 0.4y + 0.5z = 159. \end{cases}$$

Solving this system, we get

$$x = 150, y = 110, \text{ and } z = 140.$$

Therefore, the manufacturer should be able to produce

150 radios of model A,   110 of model B,   and   140 of model C

in order to use up all the allocated times.

PROBLEM SET 8.1

Mastering the Concepts

In problems 1–12, solve each system by both substitution and elimination. Represent the system graphically, and then relate the solution to the graph of the system.

1. $\begin{cases} 3x - 2y = 1 \\ x + y = 5 \end{cases}$

2. $\begin{cases} x - y = 2 \\ 4x - 2y = 6 \end{cases}$

7. $\begin{cases} 2x - y = 5 \\ -6x + 3y = 13 \end{cases}$

8. $\begin{cases} 3x + y = 4 \\ 7x - y = 6 \end{cases}$

3. $\begin{cases} 2x + 6y = 7 \\ x + 3y = 1 \end{cases}$

4. $\begin{cases} 2x - y = 1 \\ 4x - 2y = 5 \end{cases}$

9. $\begin{cases} y = x - 5 \\ 2y = 2x - 10 \end{cases}$

10. $\begin{cases} 3x + 6y = 1 \\ x - 2y = -6 \end{cases}$

5. $\begin{cases} x + 2y = 3 \\ 2x - 3y = 1 \end{cases}$

6. $\begin{cases} 3x + 2y = 11 \\ -2x + y = 2 \end{cases}$

11. $\begin{cases} y = 1 - x \\ 5x - y = 13 \end{cases}$

12. $\begin{cases} 3x - 4y = -5 \\ x = 3 - y \end{cases}$

In problems 13–16, solve each system by elimination.

13. $\begin{cases} x + y = 5 \\ x + z = 1 \\ y + z = 2 \end{cases}$

14. $\begin{cases} x - 3y = -11 \\ 2y - 5z = 26 \\ 7x - 3z = -2 \end{cases}$

15. $\begin{cases} x + y + 2z = 11 \\ x - y + z = 3 \\ 2x + y + 3z = 17 \end{cases}$

16. $\begin{cases} x + 3y + z = 4 \\ 3x - 2y + 4z = 11 \\ 2x + y + 3z = 13 \end{cases}$

In problems 17–34, solve each system by the Gaussian elimination method.

17. $\begin{cases} x + 3y = 9 \\ x - y = 1 \end{cases}$

18. $\begin{cases} x + y = 1 \\ y - 3x = 3 \end{cases}$

19. $\begin{cases} 5x - y = 7 \\ 10x - 4y = 18 \end{cases}$

20. $\begin{cases} 2x + 4y = 3 \\ -x + y = 3 \end{cases}$

21. $\begin{cases} 3y + 2z = 5 \\ 2y = -3z + 1 \end{cases}$

22. $\begin{cases} 4x - 3y = 11 \\ 3x + 4y = 2 \end{cases}$

23. $\begin{cases} 3x + 2y = 4 \\ 5x + 3y = 7 \end{cases}$

24. $\begin{cases} 2x - 3y = 1 \\ 5x + 2y = 12 \end{cases}$

25. $\begin{cases} 2x - 7y = -5 \\ 4x + 3y = 7 \end{cases}$

26. $\begin{cases} x + y + 2z = 4 \\ x + y - 2z = 0 \\ x - y = 0 \end{cases}$

27. $\begin{cases} x + y + z = 2 \\ x + 2y - z = 4 \\ 2x + y + z = 0 \end{cases}$

28. $\begin{cases} x_1 + 2x_2 + 4x_3 = 12 \\ 2x_1 - 3x_2 + x_3 = 10 \\ 3x_1 - x_2 - 2x_3 = 1 \end{cases}$

29. $\begin{cases} x + y + z = 6 \\ x - y + 2z = 12 \\ 2x + y + z = 1 \end{cases}$

30. $\begin{cases} x + y + 2z = 4 \\ x - 5y + z = 5 \\ 3x - 4y + 7z = 24 \end{cases}$

31. $\begin{cases} x_1 + 2x_2 + 5x_3 = 4 \\ 4x_1 + x_2 + 3x_3 = 9 \\ 6x_1 + 9x_2 + x_3 = 21 \end{cases}$

32. $\begin{cases} 2x + y - 3z = 9 \\ x - 2y + 4z = 5 \\ 3x + y - 2z = 15 \end{cases}$

33. $\begin{cases} x + 3y - 2z = -21 \\ 7x - 5y + 4z = 31 \\ 2x + y + 3z = 17 \end{cases}$

34. $\begin{cases} x_1 - 5x_2 + 4x_3 = 8 \\ 3x_1 + x_2 - 2x_3 = 4 \\ 9x_1 - 3x_2 + 6x_3 = 6 \end{cases}$

[G] In problems 35 and 36, solve each system graphically by finding the location of the point of intersection by using a grapher. Round off the answers to two decimal places.

35. $\begin{cases} 0.2x + 0.3y = 0.7 \\ 0.4x - 0.5y = 0.3 \end{cases}$

36. $\begin{cases} 4.01x + 6.07y = 2.33 \\ 7.57x - 3.11y = 13.81 \end{cases}$

Applying the Concepts

In problems 37–52, use a linear system to solve each problem.

37. Grading: Suppose that three-fifths of the male students and two-thirds of the female students have A or B averages. If there are 120 students in the class and 46 have an average of C or lower, how many male and how many female students are in the class?

38. Recreation: A theater group plans to sell 500 tickets for a play. They charge $45 for each orchestra seat and $30 for each balcony seat, and they plan to collect $18,000 for a performance. How many of each type of tickets do they expect to sell?

39. Physics: An arrow is shot straight up in the air from a height of h_0 feet above the ground with an initial upward speed of v_0 (in feet per second). Its height h (in feet) above the ground t seconds after it was shot is given by

$$h = -16t^2 + v_0 t + h_0.$$

(a) If the arrow is 22 feet above the ground after 1.5 seconds and 10 feet above the ground after 2 seconds, find its initial speed and its initial height.

(b) Use the results from part (a) to find the model that expresses the height h as a function of t.

(c) How high above ground level is the arrow after 1 second? When does the arrow hit the ground? Round off the answers to two decimal places.

40. Physics: A stone is thrown vertically upward from a height of h_0 (in feet) above the ground with an initial upward speed of v_0 (in feet per second). Its height h (in feet) above the ground t seconds after it was thrown is given by

$$h = -16t^2 + v_0 t + h_0.$$

(a) If the stone is 40 feet above the ground after $1/2$ second and 48 feet above the ground after 1.5 seconds, find its initial speed v_0 and its initial height h_0.

(b) Use the results from part (a) to express h as a function of t.

(c) How high is the stone after 2 seconds? When does the stone hit the ground? Round off the answers to two decimal places.

41. Chemistry: A chemist has one solution that is 30% acid and another that is 20% acid. How much of each should be used to make 10 liters of a solution that is 22% acid?

42. Chemistry: A chemist prepares two acid solutions. The first solution is 20% acid and the second is 50% acid. How many milliliters of each solution should be mixed to obtain 12 milliliters of a 30% acid solution?

43. Aviation: An airplane flying with the wind takes 3.75 hours to fly the 2500 miles from Los Angeles to New York, but 4.4 hours to fly the same distance from New York to Los Angeles against the wind. Assuming that the airplane travels at a constant airspeed in still air and the wind blows at a constant rate, find the speed of the airplane in still air and the wind speed.

44. Water Speed: It takes a powerboat traveling upstream (against the current) 3 hours to make a 36 mile trip on a river. Returning downstream (with the current), the same trip takes 2 hours. Assuming that the boat travels at a constant speed in still water and the rate of the current is constant, find the speed of the boat in still water and the rate of the current.

45. Shoe Sales: A national sporting gear store sells running shoes and tennis shoes. A pair of tennis shoes sells for $46 and a pair of running shoes sells for $79. During a 1 day sale, a total of 260 pairs of tennis and running shoes were sold. The total receipts were $14,435. How many pairs of each kind were sold?

46. Carpooling: To encourage carpooling, a city parking lot charges $7 a day per single driver vehicle or $4 per day for each carpool vehicle with two or more people. If the city parking lot took in $649 and 115 vehicles used the lot for 1 day, how many of each kind of car used the lot that day?

47. Investment Portfolio: An investment club invested $42,000 in three accounts. During one year, the first account earned 5% simple interest, the second earned 7% simple interest, and the third earned 9% simple interest. The total interest from the three investments for the year was $2600. The interest in the first account was $200 less than the total interest from the other two accounts. How much was invested in each type of account?

48. Landscape Engineering: A landscape architect wishes to construct a sprinkler system by locating the sprinklers at the centers of three circular regions that are mutually tangent to each other. The center of each circle is labeled A, B, and C in Figure 4. If the distances $|\overline{AB}| = 15$ meters, $|\overline{BC}| = 13$ meters, and $|\overline{CA}| = 16$ meters, find the radii of the circles.

Figure 4

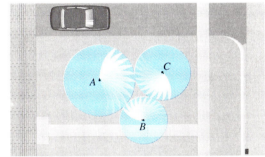

49. Airport Shuttle: An airport shuttle service has three sizes of vans. The biggest van holds 12 passengers, the next holds 10 passengers, and the smallest holds 6 passengers. The service manager has 15 vans available to accommodate 152 passengers. However, to reduce fuel costs, the manager wants to use 2 more of the available 12 passenger vans than the 6 passenger vans. How many vans of each kind should be used?

50. Sports: The total number of seats in a basketball sports arena is 18,000. The arena is divided into three sections: courtside, balcony, and end zone. There are three times as many balcony seats as courtside seats. For the conference championship game, ticket prices are $25 for courtside, $15 for balcony, and $10 for end zone. If the arena is sold out and the total receipts are $270,000, how many seats are courtside?

51. Irrigation: A grower uses three pumps to provide water to irrigate a grove. When pumps A, B, and C are used for 2 hours, they can pump 74,000 gallons. If the grower runs pump A for 4 hours and pump B for 2 hours, then 64,000 gallons of water are pumped. If pump B is used for 5 hours and pump C for 4 hours, then 120,000 gallons are pumped. How many gallons per hour can each pump handle?

52. Manufacturing: An electronics manufacturer places three wires next to each other inside a radio transmitter. The cross section of each wire is circular, and these circles are tangent to each other. Figure 5 shows the centers of these circles labeled as A, B, and C. If $|\overline{AB}| = 13$ millimeters, $|\overline{BC}| = 22$ millimeters, and $|\overline{DE}| = 52$ millimeters, find the radius of each wire.

Figure 5

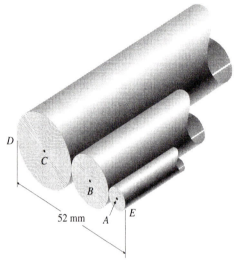

Developing and Extending the Concepts

53. Solve each system by using the substitutions

$$u = 1/x \quad \text{and} \quad v = 1/y.$$

(a) $\begin{cases} \dfrac{5}{x} - \dfrac{6}{y} = 21 \\ \dfrac{1}{x} - \dfrac{2}{y} = 5 \end{cases}$ (b) $\begin{cases} \dfrac{4}{x} + \dfrac{5}{y} = 6 \\ \dfrac{2}{x} - \dfrac{1}{y} = 10 \end{cases}$

54. Solve the system by using substitutions $u = 1/x$, $v = 1/y$, and $w = 1/z$.

$$\begin{cases} \dfrac{3}{x} + \dfrac{4}{y} - \dfrac{1}{z} = -7 \\ \dfrac{1}{x} - \dfrac{1}{y} + \dfrac{2}{z} = 0 \\ \dfrac{4}{x} + \dfrac{1}{y} - \dfrac{3}{z} = -17 \end{cases}$$

55. Give an example of a linear system that has no solution if the system has:

(a) Two variables   (b) Three variables

56. Give an example of a linear system that has infinitely many solutions if the system has:

(a) Two variables   (b) Three variables

Objectives

1. Write Linear Systems in Matrix Notation
2. Solve Linear Systems Using Row-Reduction
3. Decompose Fractions
4. Solve Applied Problems

8.2 Solutions of Linear Systems Using Augmented Matrices

In carrying out the Gaussian elimination method, we wrote the equations in the system in such a way that the same variables were aligned vertically in each step of the process. In this section we'll develop a procedure based on the Gaussian method in which we solve a system of linear equations by keeping track of the changing coefficients without writing down the variables. This procedure makes use of *matrices*.

A **matrix** is a rectangular array of numbers. The numbers in the matrix are called the *entries* or *elements* of the matrix. Traditionally, we enclose the elements of a matrix in brackets. Matrices are usually denoted by capital letters.

For example,

$$A = \begin{bmatrix} 2 & -1 & 3 \\ 5 & 4 & 7 \end{bmatrix}, \quad B = \begin{bmatrix} 1 & 3 \\ 2 & -1 \end{bmatrix}, \quad \text{and} \quad C = \begin{bmatrix} 4 & 3 & -1 \end{bmatrix}$$

are all matrices.

Each horizontal line of numbers is called a *row* of the matrix; each vertical line of numbers is called a *column*. When we refer to a matrix, we usually refer to its *size* or *dimension* by specifying the numbers of rows and the number of columns in that order.

In the given examples, matrix A is a 2 by 3 matrix (2 rows and 3 columns), which is usually written as 2×3. Matrix B is 2×2, and matrix C is 1×3.

An entry or element of a matrix can be identified by subscripts to indicate its row and column position. For instance, in matrix A, we write

$$a_{11} = 2, \quad a_{12} = -1, \quad a_{13} = 3$$
$$a_{21} = 5, \quad a_{22} = 4, \quad a_{23} = 7.$$

In general

a_{mn} is the entry in the mth row and nth column of matrix A.

Writing Linear Systems in Matrix Notation

Consider the linear system of equations

$$\begin{cases} 3x - y = 7 \\ 4x + 7y = 15. \end{cases}$$

It is customary to draw a dashed vertical line to separate the last column (the column that corresponds to the constants).

If we list the coefficients and constants of the system, we obtain the rectangular array of numbers

$$\left[\begin{array}{cc|c} 3 & -1 & 7 \\ 4 & 7 & 15 \end{array} \right].$$

This matrix is called the **augmented matrix** form of the system. The matrix of the coefficients of the variables of the system,

$$\left[\begin{array}{cc} 3 & -1 \\ 4 & 7 \end{array} \right]$$

is called the **coefficient matrix.** In this example, the first column of this matrix lists the coefficients of the x variables and the second column lists the coefficients of the y variables.

EXAMPLE 1

Writing a System in Matrix Notation

Write the system in matrix notation:

$$\begin{cases} 4x - 3y = 5 \\ x + 7y = 8 \end{cases}$$

Solution

The augmented matrix form of the system is

$$\left[\begin{array}{cc|c} 4 & -3 & 5 \\ 1 & 7 & 8 \end{array} \right].$$

If any variable does not appear in an equation, a 0 is inserted in the appropriate position in the array of coefficients.

EXAMPLE 2

Writing a System Corresponding to a Matrix

Write the system in terms of x and y corresponding to the following augmented matrix, and then solve it.

$$\begin{bmatrix} 1 & -2 & \vdots & 7 \\ 0 & 1 & \vdots & -3 \end{bmatrix}$$

Solution

Here we used a 0 to denote the coefficient of x in the second equation. So the system is

$$\begin{cases} x - 2y = 7 \\ y = -3. \end{cases}$$

Using back-substitution, we see that

$$x - 2(-3) = 7$$
$$x + 6 = 7$$
$$x = 1.$$

Therefore, the solution is $(1, -3)$.

Solving Linear Systems Using Row-Reduction

In Section 8.1 we used elementary operations to transform a linear system of equations into an equivalent system in triangular form, because the latter form is easier to solve. In this section, we perform the same process, but rather than working with the equations in the system, we work with the rows of the corresponding augmented matrix form. Each row serves as an abbreviation of the resulting equation when an elementary operation is applied.

In the following illustration we solve a system of linear equations in two ways:

1. On the left, the elementary operations are used to convert the given system to an equivalent triangular form.

2. On the right, we show the augmented matrix corresponding to each equivalent linear system and describe the elementary operation in terms of a row manipulation. A vertical arrow is used to indicate that each matrix is *row-equivalent* to the one below it in the sense that they both represent equivalent systems.

System of Equations	Augmented Matrix of System	Strategy
(A) $\begin{cases} 3x - y = 15 \\ x + 2y = -2 \end{cases}$	(A) $\begin{bmatrix} 3 & -1 & \vdots & 15 \\ 1 & 2 & \vdots & -2 \end{bmatrix}$	Given
$\downarrow$	$\downarrow$	
First interchange equations	First interchange rows	We want the first equation to have a leading coefficient of 1.
(B) $\begin{cases} x + 2y = -2 \\ 3x - y = 15 \end{cases}$	(B) $\begin{bmatrix} 1 & 2 & \vdots & -2 \\ 3 & -1 & \vdots & 15 \end{bmatrix}$	
$\downarrow$	$\downarrow$	
Next replace the second equation with the sum of -3 times the first equation and the second equation	Next we replace the second row with the sum of -3 times the first row and the second row	We want to eliminate the x term in the second equation.

(C) $\begin{cases} x + 2y = -2 \\ 0 - 7y = 21 \end{cases}$

(C) $\begin{bmatrix} 1 & 2 & \vdots & -2 \\ 0 & -7 & \vdots & 21 \end{bmatrix}$

↓

Multiply the second equation by $-\dfrac{1}{7}$

↓

Multiply the second row by $-\dfrac{1}{7}$

We want the second equation to have a leading coefficient of 1.

(D) $\begin{cases} x + 2y = -2 \\ y = -3 \end{cases}$

(D) $\begin{bmatrix} 1 & 2 & \vdots & -2 \\ 0 & 1 & \vdots & -3 \end{bmatrix}$

↓

The system is in triangular form

↓

The matrix is in triangular form

Thus the last matrix (D) is an abbreviated form of the last system form (D). Clearly,

$$y = -3.$$

To get x, we substitute -3 for y in the first equation of D to obtain

$$x + 2(-3) = -2,$$

$$\text{or} \quad x = 4.$$

So the solution is

$$(4, -3).$$

As this illustration shows, the two approaches for solving a linear system are equivalent. However, with some practice, we can work more efficiently with augmented matrices than with the actual systems of equations.

In our earlier discussion of using elimination by addition, we said that two systems were equivalent if they have the same solution. We used the elementary operations on page 539 to transform a system into an equivalent form. Analogous to this approach, we now say that two augmented matrices are **row-equivalent,** if they are augmented matrices of equivalent systems. We use the row operations listed in Table 1 to transform augmented matrices into row-equivalent matrices.

TABLE 1   Elementary Row Operations

Row Operation Performed on a Matrix	Explanation of Operation	Symbolic Representation of Operation
1. Interchange two rows	Interchange row i and row j	$R_i \leftrightarrow R_j$
2. Multiply a row by a nonzero constant	Multiply each entry in row i by a constant k to get new row i, where $k \neq 0$	$kR_i \rightarrow R_i$
3. Add a constant multiple of a row to another row	Multiply each entry of row i by the constant k and add the resulting entries to each entry in row j to get a new row j	$kR_i + R_j \rightarrow R_j$

Table 2 gives illustrations of the elementary row operations.

TABLE 2 Examples of Elementary Row Operations

Original Matrix	Row Operation	Symbolic Representation	Resulting Row-Equivalent Matrix
1. $\begin{bmatrix} 1 & 2 & -3 \\ 4 & 5 & 6 \end{bmatrix}$	Interchange row 1 and row 2	$R_1 \leftrightarrow R_2$	$\begin{bmatrix} 4 & 5 & 6 \\ 1 & 2 & -3 \end{bmatrix}$
2. $\begin{bmatrix} 1 & 2 & -3 & 5 \\ 4 & 5 & 6 & -1 \end{bmatrix}$	Multiply each entry of row 2 by -3 to get a new row 2	$-3R_2 \rightarrow R_2$	$\begin{bmatrix} 1 & 2 & -3 & 5 \\ -12 & -15 & -18 & 3 \end{bmatrix}$
3. $\begin{bmatrix} 2 & 4 & -2 \\ 3 & 1 & 5 \\ 1 & -6 & -1 \end{bmatrix}$	Multiply each entry in row 1 by 2 and add the resulting entries to the entries in row 3 to get a new row 3	$2R_1 + R_3 \rightarrow R_3$	$\begin{bmatrix} 2 & 4 & -2 \\ 3 & 1 & 5 \\ 5 & 2 & -5 \end{bmatrix}$

EXAMPLE 3

Solving a System Using Matrix Row Operations

Write the associated augmented matrix of the system and solve it using elementary row operations.

$$\begin{cases} 4x - 3y = 15 \\ x + 2y = 1 \end{cases}$$

Solution

We first write the augmented matrix of the system:

$$\left[\begin{array}{cc:c} 4 & -3 & 15 \\ 1 & 2 & 1 \end{array}\right]$$

Next we convert it into an equivalent matrix in triangular form. The easiest way to get a 1 in the upper left corner is to interchange rows 1 and 2:

$$\left[\begin{array}{cc:c} 4 & -3 & 15 \\ 1 & 2 & 1 \end{array}\right] \xrightarrow{R_1 \leftrightarrow R_2} \left[\begin{array}{cc:c} 1 & 2 & 1 \\ 4 & -3 & 15 \end{array}\right]$$

We want a 0 to be the first entry in row 2, so we multiply row 1 by -4 and then add the result to row 2:

$$\left[\begin{array}{cc:c} 1 & 2 & 1 \\ 4 & -3 & 15 \end{array}\right] \xrightarrow{-4R_1 + R_2 \rightarrow R_2} \left[\begin{array}{cc:c} 1 & 2 & 1 \\ 0 & -11 & 11 \end{array}\right]$$

Finally, to get 1 as the first nonzero entry in row 2, we multiply row 2 by $-1/11$:

$$\left[\begin{array}{cc:c} 1 & 2 & 1 \\ 0 & -11 & 11 \end{array}\right] \xrightarrow{-\frac{1}{11}R_2 \rightarrow R_2} \left[\begin{array}{cc:c} 1 & 2 & 1 \\ 0 & 1 & -1 \end{array}\right]$$

The last matrix is in the desired triangular form. By writing the associated system, we have

$$\begin{cases} x + 2y = 1 \\ y = -1 . \end{cases}$$

Using back-substitution, we replace y by -1 in the first equation to obtain

$$x + 2(-1) = 1 \quad \text{or} \quad x - 2 = 1, \quad \text{so} \quad x = 3.$$

Therefore, the solution is $(3, -1)$.

We can avoid the necessity of back-substitution in Example 3 by reducing the last matrix even further. To do this, multiply row 2 by -2 and add the result to row 1 to obtain:

$$\begin{bmatrix} 1 & 2 & \vdots & 1 \\ 0 & 1 & \vdots & -1 \end{bmatrix} \xrightarrow{\;-2R_2 + R_1 \to R_1\;} \begin{bmatrix} 1 & 0 & \vdots & 3 \\ 0 & 1 & \vdots & -1 \end{bmatrix}.$$

This final matrix is an especially simple one to interpret because the solution,

$$x = 3 \quad \text{and} \quad y = -1,$$

can be read directly from the entries in the last column.

This latter matrix is said to be in *row-reduced echelon form*.

Definition

Row-Reduced Echelon Form of a Matrix

A matrix is said to be in **row-reduced echelon form** if it satisfies the following conditions:

1. The first nonzero entry or leading term in each row is 1.
2. Any rows that consist entirely of 0s are below all rows that do not consist entirely of 0s.
3. The first nonzero entry in each row is to the right of the first nonzero entry in the preceding row.
4. In each column that contains a leading 1 of some row, all other entries are 0.

For example, the matrices

$$\begin{bmatrix} 1 & -2 & 0 \\ 0 & 0 & 1 \end{bmatrix}, \quad \begin{bmatrix} 1 & -1 & 0 & -3 \\ 0 & 0 & 1 & 4 \end{bmatrix},$$

and

$$\begin{bmatrix} 1 & -4 & 0 & 3 & 5 \\ 0 & 0 & 1 & -2 & 4 \\ 0 & 0 & 0 & 0 & 0 \end{bmatrix}$$

are in row-reduced echelon form, because each satisfies the conditions of the above definition.

However,

$$A = \begin{bmatrix} 0 & 1 \\ 0 & -2 \end{bmatrix}, \quad B = \begin{bmatrix} 0 & 0 & 0 \\ 1 & 0 & 0 \end{bmatrix},$$

and

$$C = \begin{bmatrix} 1 & 1 & 0 \\ 0 & 1 & 0 \\ 0 & 0 & 1 \end{bmatrix}$$

are not in row-reduced echelon form. In matrix A, condition 1 is violated, because the leading term in the second row is not 1. In matrix B, condition 2 is violated; and in matrix C, condition 4 is violated, because the entry in row 1, column 2 is 1 rather than 0.

EXAMPLE 4 **Interpreting Row-Reduced Echelon Matrices**

Each of the following augmented matrices is the row-reduced echelon matrix form corresponding to a linear system of equations. Solve each system.

(a) $\left[\begin{array}{cc|c} 1 & 0 & 2 \\ 0 & 1 & 3 \end{array}\right]$ (b) $\left[\begin{array}{cc|c} 1 & 0 & 5 \\ 0 & 0 & -1 \end{array}\right]$ (c) $\left[\begin{array}{ccc|c} 1 & 0 & 0 & 2 \\ 0 & 1 & 2 & 5 \\ 0 & 0 & 0 & 0 \end{array}\right]$

Solution (a) The matrix

$$\left[\begin{array}{cc|c} 1 & 0 & 2 \\ 0 & 1 & 3 \end{array}\right]$$

corresponds to the system

$$\begin{cases} x = 2 \\ y = 3 \end{cases} \text{ so the system has one solution, } (2, 3).$$

(b) The matrix

$$\left[\begin{array}{cc|c} 1 & 0 & 5 \\ 0 & 0 & -1 \end{array}\right] \text{ corresponds to the system } \begin{cases} x = 5 \\ 0 = -1. \end{cases}$$

Because of the contradiction in the second equation, we conclude that the system has no solution.

(c) The matrix

$$\left[\begin{array}{ccc|c} 1 & 0 & 0 & 2 \\ 0 & 1 & 2 & 5 \\ 0 & 0 & 0 & 0 \end{array}\right]$$

corresponds to the system

$$\begin{cases} x = 2 \\ y + 2z = 5. \end{cases}$$

This implies that there are infinitely many solutions of the form (x, y, z), where

$$x = 2, y = 5 - 2z,$$

and z is any real number.

That is, the system has infinitely many solutions of the form

$$(2, 5 - 2z, z),$$

where z is any real number.

For instance,

if we let $z = 0$, we get the solution $(2, 5, 0)$;

if $z = 1$, we get $(2, 3, 1)$,

and so forth.

We now can describe a matrix method for solving linear systems, called the *Gauss–Jordan elimination method.* 🔡

Gauss–Jordan Elimination Method

Step 1 Represent the linear system with an augmented matrix.

Step 2 Use elementary row operations to transform the original matrix to an equivalent row-reduced echelon form.

Step 3 Change the reduced matrix from step 2 to the equation form of the linear system.

Step 4 Interpret the result in step 3 to get the solution of the system, if there is one.

EXAMPLE 5 **Solving a System by Gauss–Jordan Elimination**

Solve the system below by using Gauss–Jordan elimination.

$$\begin{cases} x_1 - 2x_2 + 3x_3 = -20 \\ 2x_1 - x_2 + 2x_3 = -14 \\ 3x_1 + x_2 + 2x_3 = -11 \end{cases}$$

Solution First we write the augmented matrix of the system; then we transform the matrix into an equivalent matrix in row-reduced echelon form as follows:

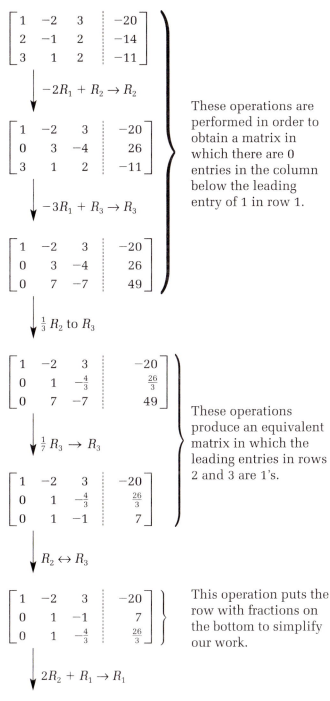

$$\begin{bmatrix} 1 & -2 & 3 & \vdots & -20 \\ 2 & -1 & 2 & \vdots & -14 \\ 3 & 1 & 2 & \vdots & -11 \end{bmatrix}$$

$\downarrow -2R_1 + R_2 \rightarrow R_2$

$$\begin{bmatrix} 1 & -2 & 3 & \vdots & -20 \\ 0 & 3 & -4 & \vdots & 26 \\ 3 & 1 & 2 & \vdots & -11 \end{bmatrix}$$

$\downarrow -3R_1 + R_3 \rightarrow R_3$

$$\begin{bmatrix} 1 & -2 & 3 & \vdots & -20 \\ 0 & 3 & -4 & \vdots & 26 \\ 0 & 7 & -7 & \vdots & 49 \end{bmatrix}$$

These operations are performed in order to obtain a matrix in which there are 0 entries in the column below the leading entry of 1 in row 1.

$\downarrow \frac{1}{3}R_2 \text{ to } R_3$

$$\begin{bmatrix} 1 & -2 & 3 & \vdots & -20 \\ 0 & 1 & -\frac{4}{3} & \vdots & \frac{26}{3} \\ 0 & 7 & -7 & \vdots & 49 \end{bmatrix}$$

$\downarrow \frac{1}{7}R_3 \rightarrow R_3$

$$\begin{bmatrix} 1 & -2 & 3 & \vdots & -20 \\ 0 & 1 & -\frac{4}{3} & \vdots & \frac{26}{3} \\ 0 & 1 & -1 & \vdots & 7 \end{bmatrix}$$

These operations produce an equivalent matrix in which the leading entries in rows 2 and 3 are 1's.

$\downarrow R_2 \leftrightarrow R_3$

$$\begin{bmatrix} 1 & -2 & 3 & \vdots & -20 \\ 0 & 1 & -1 & \vdots & 7 \\ 0 & 1 & -\frac{4}{3} & \vdots & \frac{26}{3} \end{bmatrix}$$

This operation puts the row with fractions on the bottom to simplify our work.

$\downarrow 2R_2 + R_1 \rightarrow R_1$

$$\begin{bmatrix} 1 & 0 & 1 & \vdots & -6 \\ 0 & 1 & -1 & \vdots & 7 \\ 0 & 1 & -\frac{4}{3} & \vdots & \frac{26}{3} \end{bmatrix}$$

$\downarrow -R_2 + R_3 \rightarrow R_3$

$$\begin{bmatrix} 1 & 0 & 1 & \vdots & -6 \\ 0 & 1 & -1 & \vdots & 7 \\ 0 & 0 & -\frac{1}{3} & \vdots & \frac{5}{3} \end{bmatrix}$$

These operations lead to a matrix form in which the entries in column 2 above and below the 1 in row 2 are 0s.

$\downarrow -3R_3 \rightarrow R_3$

$$\begin{bmatrix} 1 & 0 & 1 & \vdots & -6 \\ 0 & 1 & -1 & \vdots & 7 \\ 0 & 0 & 1 & \vdots & -5 \end{bmatrix}$$

This operation puts a 1 in the leading entry in row 3.

$\downarrow R_3 + R_2 \rightarrow R_2$

$$\begin{bmatrix} 1 & 0 & 1 & \vdots & -6 \\ 0 & 1 & 0 & \vdots & 2 \\ 0 & 0 & 1 & \vdots & -5 \end{bmatrix}$$

$\downarrow -R_3 + R_1 \rightarrow R_1$

$$\begin{bmatrix} 1 & 0 & 0 & \vdots & -1 \\ 0 & 1 & 0 & \vdots & 2 \\ 0 & 0 & 1 & \vdots & -5 \end{bmatrix}$$

These operations produce an equivalent matrix in which the entries in column 3 above the 1 in row 3 are 0s.

This is the row-reduced echelon form.

The last matrix corresponds to the system

$$\begin{cases} x_1 = -1 \\ x_2 = 2 \\ x_3 = -5. \end{cases}$$

So the solution is $(-1, 2, -5)$.

(Continued in next column)

Decomposing Fractions

A rational expression is said to be *proper* if the degree of its numerator is less than the degree of its denominator. Examples of proper fractions are

$$\frac{5x}{(x + 1)(x - 2)}, \quad \frac{3}{x^2 + 5}, \quad \text{and} \quad \frac{5x^2 + 7x + 3}{x^3 - 5x - 1}.$$

In calculus it is necessary at times to rewrite a rational expression as a sum (or difference) of fractions that contain linear or quadratic factors in the denominators according to the factorization of the denominator of the given rational expression. Such a representation is referred to as a **partial fraction expansion** or **decomposition.** One process of decomposing a proper fraction into partial fractions involves solving linear systems, as illustrated in the next example.

EXAMPLE 6

Decomposing a Fraction

Given that the fraction

$$\frac{2x - 2}{(x + 2)(x + 5)}$$

has a partial fraction expansion of the form

$$\frac{2x - 2}{(x + 2)(x + 5)} = \frac{A}{x + 2} + \frac{B}{x + 5}$$

find the constants A and B.

Solution

Given that

$$\frac{2x - 2}{(x + 2)(x + 5)} = \frac{A}{x + 2} + \frac{B}{x + 5}$$

we proceed to find A and B as follows:

$2x - 2 = A(x + 5) + B(x + 2)$	Multiply each side by the LCD $(x + 2)(x + 5)$
$2x - 2 = Ax + 5A + Bx + 2B$	Multiply
$2x - 2 = (A + B)x + (5A + 2B)$	Rewrite the right side

Equating the coefficients of the two sides of the latter equation results in

$$\begin{cases} A + B = 2 \\ 5A + 2B = -2. \end{cases}$$

The associated augmented matrix of the system is

$$\begin{bmatrix} 1 & 1 & \vdots & 2 \\ 5 & 2 & \vdots & -2 \end{bmatrix}.$$

The equivalent matrix in row-reduced echelon form is

$$\begin{bmatrix} 1 & 0 & \vdots & -2 \\ 0 & 1 & \vdots & 4 \end{bmatrix}$$

so $A = -2$ and $B = 4$. Thus

$$\frac{2x - 2}{(x + 2)(x + 5)} = \frac{-2}{x + 2} + \frac{4}{x + 5}.$$

Figure 1

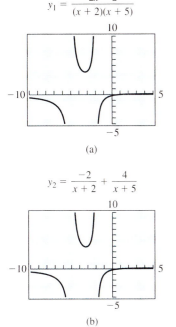

$y_1 = \dfrac{2x - 2}{(x + 2)(x + 5)}$

(a)

$y_2 = \dfrac{-2}{x + 2} + \dfrac{4}{x + 5}$

(b)

The Gauss–Jordan elimination method for solving linear systems often involves tedious numerical computations. However, some graphers are equipped with special keys that transform a matrix to an equivalent row-reduced echelon form. Also, a grapher can be used to demonstrate the validity of partial fraction decompositions graphically.

For instance, to demonstrate the validity of the result in Example 6 above graphically, we graph each side of the equation as a function of x. Figure 1a shows the graph of

$$y_1 = \frac{2x - 2}{(x + 2)(x + 5)}$$

whereas, Figure 1b shows the graph of

$$y_2 = \frac{-2}{x + 2} + \frac{4}{x + 5}$$

for the same window settings.

Figures 1a and 1b suggest that the two graphs are the same, thus demonstrating the validity of the given decomposition.

Solving Applied Problems

The following example uses row-reduction to solve a linear system of equations that models a real-life situation.

EXAMPLE 7   **Modeling a Diet Problem**

A dietician prepares a special diet that consists of iron, carbohydrates, and protein. Table 3 lists the information about the content of iron, carbohydrates, and protein in an 8 ounce glass of skim milk, 1/4 pound of lean red meat, and 2 slices of whole-grain bread. If a person on a special diet must have 10.5 milligrams of iron, 61.0 grams of carbohydrates, and 94.5 grams of protein, how many 8 ounce glasses of skim milk, 1/4 pound servings of lean red meat, and 2-slice servings of whole-grain bread will supply this diet?

TABLE 3

Food Source	Iron Milligrams Per Serving	Carbohydrates Grams Per Serving	Protein Grams Per Serving
8 ounce glass of skim milk	0.1	1.0	8.5
1/4 pound of lean red meat	3.4	20.0	22.0
2 slices of whole-grain bread	2.2	12.0	10.0

Solution   Let

x represent the number of 8 ounce glasses of skim milk,

y represent the number of 1/4 pound servings of lean red meat, and

z represent the number of 2-slice servings of whole-grain bread.

The system of linear equations that provides the special diet is given as follows:

$$\begin{cases} 0.1x + 3.4y + 2.2z = 10.5 \\ x + 20.0y + 12.0z = 61.0 \\ 8.5x + 22.0y + 10.0z = 94.5 \end{cases}$$

The augmented matrix that corresponds to this system is

$$\left[\begin{array}{ccc:c} 0.1 & 3.4 & 2.2 & 10.5 \\ 1.0 & 20.0 & 12.0 & 61.0 \\ 8.5 & 22.0 & 10.0 & 94.5 \end{array}\right].$$

By using the Gauss–Jordan elimination method, we get the equivalent row-reduced echelon matrix.

$$\left[\begin{array}{ccc:c} 1 & 0 & 0 & 5 \\ 0 & 1 & 0 & 1 \\ 0 & 0 & 1 & 3 \end{array}\right]$$

The associated system for the transformed matrix is

$$\begin{cases} x = 5 \\ y = 1 \\ z = 3. \end{cases}$$

So it takes 5 (8 ounce) glasses of skim milk,

1 serving (1/4 pound) of lean red meat,

and 3 servings (2 slices each) of whole-grain bread to supply this diet.

◆ **PROBLEM SET 8.2**

Mastering the Concepts

In problems 1–4, write the augmented matrix of each linear system.

1. $\begin{cases} x + 6y = 7 \\ 3x + 7y = 10 \end{cases}$

2. $\begin{cases} 4x + y = 7 \\ 8x - 3y = 11 \end{cases}$

3. $\begin{cases} 2x + y = 2 \\ x - y = 4 \\ 3x + z = 6 \end{cases}$

4. $\begin{cases} x_1 + 2x_2 - x_3 = 4 \\ -x_1 + 3x_2 - 2x_3 = 5 \\ 2x_1 - x_2 + x_3 = 0 \end{cases}$

In problems 5–10, each matrix is in triangular form and is the augmented matrix form of a corresponding linear system of equations. Solve each system.

5. $\begin{bmatrix} 1 & -2 & \vdots & -5 \\ 0 & 1 & \vdots & 3 \end{bmatrix}$

6. $\begin{bmatrix} 1 & 3 & \vdots & 4 \\ 0 & 1 & \vdots & 1 \end{bmatrix}$

7. $\begin{bmatrix} 1 & 2 & 3 & \vdots & 6 \\ 0 & 1 & -4 & \vdots & -3 \\ 0 & 0 & 1 & \vdots & 1 \end{bmatrix}$

8. $\begin{bmatrix} 1 & 8 & 7 & \vdots & -22 \\ 0 & 1 & 4 & \vdots & -7 \\ 0 & 0 & 1 & \vdots & -1 \end{bmatrix}$

9. $\begin{bmatrix} 1 & 0 & -1 & \vdots & 7 \\ 0 & 1 & 3 & \vdots & -3 \\ 0 & 0 & 1 & \vdots & -2 \end{bmatrix}$

10. $\begin{bmatrix} 1 & 1 & 2 & \vdots & 4 \\ 0 & 1 & 2 & \vdots & 3 \\ 0 & 0 & 1 & \vdots & 1 \end{bmatrix}$

11. Let

$$A = \begin{bmatrix} 1 & 0 & 3 \\ -3 & 1 & 4 \\ 5 & -1 & 5 \end{bmatrix}.$$

Find the matrix obtained by performing the following elementary row operations on A:
(a) Interchange the second and third rows.
(b) Multiply the third row of A by 3.
(c) Add 3 times the first row of A to the third row.

12. Let

$$A = \begin{bmatrix} 2 & 0 & 4 & 2 \\ 3 & -2 & 5 & 6 \\ -1 & 3 & 1 & 1 \end{bmatrix}.$$

Find the matrix obtained by performing the following elementary row operations on A:
(a) Interchange the second and third rows.
(b) Multiply the first row of A by 1/2.
(c) Add 2 times the third row of A to the first row.

In problems 13–16, which of the matrices are in row-reduced echelon form?

13. (a) $\begin{bmatrix} 1 & -2 & 0 & 1 \\ 0 & 0 & 1 & -4 \end{bmatrix}$

(b) $\begin{bmatrix} 1 & -1 & 0 & 4 \\ 0 & 0 & 1 & 2 \\ 0 & 0 & 0 & 0 \end{bmatrix}$

14. (a) $\begin{bmatrix} 1 & 0 & 0 & -1 \\ 0 & 1 & 0 & -3 \\ 0 & 0 & 1 & 2 \end{bmatrix}$

(b) $\begin{bmatrix} 1 & 0 & 0 \\ 0 & 0 & 1 \\ 0 & 1 & 0 \end{bmatrix}$

15. (a) $\begin{bmatrix} 1 & 0 & 1 \\ 0 & 2 & -3 \end{bmatrix}$

(b) $\begin{bmatrix} 1 & 2 & -1 & 0 & 2 \\ 0 & 0 & 0 & 1 & 4 \\ 0 & 0 & 0 & 0 & 0 \end{bmatrix}$

16. (a) $\begin{bmatrix} 1 & 2 & 0 & 2 & 5 \\ 0 & 0 & 1 & -3 & 4 \\ 0 & 0 & 0 & 0 & 0 \end{bmatrix}$

(b) $\begin{bmatrix} 1 & 0 & -2 & -2 & 3 \\ 0 & 1 & 4 & 3 & 1 \\ 0 & 0 & 1 & -1 & 5 \end{bmatrix}$

In problems 17–22, each row-reduced echelon matrix is the augmented matrix form of a corresponding linear system of equations. Solve each system.

17. $\begin{bmatrix} 1 & 0 & \vdots & 1 \\ 0 & 1 & \vdots & 1 \end{bmatrix}$

18. $\begin{bmatrix} 1 & 0 & 0 & \vdots & 0 \\ 0 & 1 & 0 & \vdots & 0 \\ 0 & 0 & 1 & \vdots & 2 \end{bmatrix}$

19. $\begin{bmatrix} 1 & 0 & -1 & \vdots & 2 \\ 0 & 1 & -2 & \vdots & 3 \end{bmatrix}$

20. $\begin{bmatrix} 1 & 0 & 0 & \vdots & 3 \\ 0 & 1 & 0 & \vdots & 0 \\ 0 & 0 & 1 & \vdots & 5 \end{bmatrix}$

21. $\begin{bmatrix} 1 & 0 & 0 & \vdots & -3 \\ 0 & 1 & 0 & \vdots & 4 \\ 0 & 0 & 0 & \vdots & 7 \end{bmatrix}$

22. $\begin{bmatrix} 1 & 0 & 0 & \vdots & 3 \\ 0 & 1 & 0 & \vdots & -2 \\ 0 & 0 & 0 & \vdots & 4 \end{bmatrix}$

In problems 23–40, solve each system by the Gauss–Jordan elimination method.

23. $\begin{cases} 3x + y = 14 \\ 2x - y = 1 \end{cases}$

24. $\begin{cases} 4x + 3y = 15 \\ 3x + 5y = 14 \end{cases}$

25. $\begin{cases} -2x + 3y = 8 \\ 2x - y = 5 \end{cases}$

26. $\begin{cases} x - 2y = 5 \\ 3x - 6y = 4 \end{cases}$

27. $\begin{cases} x + y + z = 6 \\ 3x - y + 2z = 7 \\ 2x + 3y - z = 5 \end{cases}$
28. $\begin{cases} 2x + 3y + z = 6 \\ x - 2y + 3z = -3 \\ 3x + y - z = 8 \end{cases}$

29. $\begin{cases} x + y + 2z = 4 \\ x + y - 2z = 0 \\ x - y = 0 \end{cases}$
30. $\begin{cases} x + y + z = 4 \\ x - y + 2z = 8 \\ 2x + y - z = 3 \end{cases}$

31. $\begin{cases} 2x + y - z = 7 \\ y - x = 1 \\ z - y = 1 \end{cases}$
32. $\begin{cases} x_1 + 2x_2 + 3x_3 = 0 \\ x_1 - 2x_3 = 1 \\ x_2 + x_3 = 1 \end{cases}$

33. $\begin{cases} x - 2y + z = -1 \\ 3x + y - 2z = 4 \\ y - z = 1 \end{cases}$
34. $\begin{cases} x + y - 2z = 3 \\ 3x - y + z = 5 \\ 3x + 3y - 6z = 9 \end{cases}$

35. $\begin{cases} 2x + y + z = 1 \\ 4x + 2y + 3z = 1 \\ -2x - y + z = 2 \end{cases}$
36. $\begin{cases} x + y + z = 0 \\ 2x - y - 4z = 15 \\ x - 2y - z = 7 \end{cases}$

37. $\begin{cases} 2x - 3y + z = 4 \\ x - 4y - z = 3 \end{cases}$
38. $\begin{cases} 2x + 3y - z = -2 \\ x - y + 2z = 4 \end{cases}$

39. $\begin{cases} 2x_1 - 3x_3 = 5 \\ 4x_2 + 2x_3 = -20 \\ 5x_1 + 2x_2 = -16 \end{cases}$

40. $\begin{cases} x_1 + x_2 = -1 \\ x_1 - 2x_3 = -11 \\ x_2 - 3x_3 = -15 \end{cases}$

In problems 41–50, find the partial fraction expansion.

41. $\dfrac{3x - 5}{(x - 1)(x - 2)} = \dfrac{A}{x - 1} + \dfrac{B}{x - 2}$

42. $\dfrac{x + 14}{(x + 4)(x + 12)} = \dfrac{A}{x + 4} + \dfrac{B}{x + 12}$

43. $\dfrac{-9x - 7}{(3x + 1)(x + 1)} = \dfrac{A}{3x + 1} + \dfrac{B}{x + 1}$

44. $\dfrac{13x - 24}{(3x - 5)(x - 2)} = \dfrac{A}{3x - 5} + \dfrac{B}{x - 2}$

45. $\dfrac{-21x + 11}{(x - 1)(x - 2)(x - 3)} = \dfrac{A}{x - 1} + \dfrac{B}{x - 2} + \dfrac{C}{x - 3}$

46. $\dfrac{x^2 - 3x + 4}{(x - 1)(x + 1)(x + 2)} = \dfrac{A}{x - 1} + \dfrac{B}{x + 1} + \dfrac{C}{x + 2}$

47. $\dfrac{5x^2 - 21x + 13}{(x - 3)^2(x + 2)} = \dfrac{A}{x - 3} + \dfrac{B}{(x - 3)^2} + \dfrac{C}{x + 2}$

48. $\dfrac{-x^2 + 13x - 26}{(x + 1)^2(x - 4)} = \dfrac{A}{x + 1} + \dfrac{B}{(x + 1)^2} + \dfrac{C}{x - 4}$

49. $\dfrac{6}{x(x^2 + 1)} = \dfrac{A}{x} + \dfrac{Bx + C}{x^2 + 1}$

50. $\dfrac{30x}{(x - 6)(x^2 + 4)} = \dfrac{A}{x - 6} + \dfrac{Bx + C}{x^2 + 4}$

In problems 51 and 52, use a grapher to demonstrate the validity of the partial fraction decomposition found in the problem.

51. Problem 41

52. Problem 42

Applying the Concepts

In problems 53–58, set up an appropriate linear system, and then use the Gauss–Jordan method to solve each system.

53. **Nutrition:** A nutritionist wants to create a special diet out of three substances A, B, and C. The requirements in the diet are 480 units of calcium, 190 units of iron, and 360 units of vitamins. The calcium, iron, and vitamin contents in substances A, B, and C are listed in Table 4. How many ounces of each substance should be used to prepare this diet?

TABLE 4

Substance	Calcium Units Per Ounce	Iron Units Per Ounce	Vitamin Units Per Ounce
A	30	10	15
B	20	10	10
C	10	5	20

54. **Production:** Three assembly lines, A, B, and C, can produce 8400 TV dinners per day. Together, lines A and B can produce 4900 TV dinners, whereas lines B and C together can produce 5600 TV dinners. Find the number of TV dinners each assembly line can produce alone.

55. **Horticulture:** A horticulturist wishes to mix three types of fertilizer which contain 25%, 35%, and 40% nitrogen, respectively, in order to get a mixture of 4000 pounds of $35\frac{5}{8}\%$ nitrogen. The final mixture should contain three times as much of the 40% type as the 25% type. How much of each type is in the final mixture?

56. **Fast-Food Chain:** A fast-food chain sells three types of franchises, A, B, and C. Franchise A sells for $20,000, franchise B for $25,000, and franchise C for $30,000. In one year, the company sold 18 franchises for $430,000. If the number sold of franchise A is twice the number sold of franchise C, how many of each type did the company sell that year?

57. Electrical Engineering: In electronics, the analysis of a circuit leads naturally to a system of linear equations. Suppose an electrical network contains currents I_1, I_2, and I_3 in the circuit, resistors R_1, R_2, and R_3; a 6-volt battery ($E_1 = 6$) and a 12-volt battery ($E_2 = 12$), as shown in Figure 2.

Figure 2

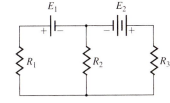

Assume the network satisfies the system given by

$$\begin{cases} I_1 - I_2 + I_3 = 0 \\ R_1I_1 + R_2I_2 \qquad = E_1 \\ \qquad R_2I_2 + R_3I_3 = E_2. \end{cases}$$

If $R_1 = 4$ ohms, $R_2 = 10$ ohms, and $R_3 = 21$ ohms, find the currents I_1, I_2, and I_3 (in amperes).

58. Irrigation Network: Figure 3 is a diagram of a system of irrigation canals that shows the flow of water into the system at junction A and out at junctions B, C, and D. The flows f_1, f_2, f_3, f_4, and f_5 (in gallons per minute) are solutions to the system

$$\begin{cases} f_1 + f_2 + f_3 \qquad = 500 \text{ at junction } A \\ f_1 \qquad\qquad - f_5 = 100 \text{ at junction } B \\ \quad f_2 - \quad f_4 + f_5 = 100 \text{ at junction } C \\ \qquad f_3 + f_4 \qquad = 300 \text{ at junction } D. \end{cases}$$

Figure 3

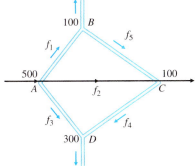

(a) Solve the system for f_1, f_2, f_3, f_4, and f_5.
(b) Discuss some choices of flows for this network.

Developing and Extending the Concepts

In problems 59–62, use matrices to solve each system. In each case, write the augmented matrix and reduce it to row-reduced echelon form.

59. $\begin{cases} x + y + 2z + 3w = 5 \\ x - 2y + z + w = 5 \\ 3x + y + z - w = 2 \\ 2x - y + z + 4w = 8 \end{cases}$

60. $\begin{cases} x_1 - x_2 + 2x_3 + x_4 = 8 \\ -x_1 + 2x_2 + 3x_3 - x_4 = 4 \\ x_1 - x_2 + x_3 - 2x_4 = -6 \\ -x_1 + x_2 + x_3 + x_4 = 7 \end{cases}$

61. $\begin{cases} 3x + y - 2z + 2w = 7 \\ 2x \qquad + z + 4w = 3 \\ \quad y + 3z + 5w = 0 \\ -x + 2y \qquad - 3w = 4 \end{cases}$

62. $\begin{cases} x_1 + 4x_2 - x_3 + x_4 = 11 \\ 3x_1 + 2x_2 + x_3 + 2x_4 = 9 \\ x_1 - 6x_2 \qquad + 3x_4 = 23 \\ x_1 + 14x_2 - 5x_3 + 2x_4 = -13 \end{cases}$

In problems 63–68, find (if possible) conditions on a, b, and c so that the system has:

(a) No solution   (b) One solution
(c) Infinite number of solutions

63. $\begin{cases} 3x_1 + x_2 - x_3 = a \\ x_1 - x_2 + 2x_3 = b \\ 5x_1 + 3x_2 - 4x_3 = c \end{cases}$

64. $\begin{cases} x_1 \qquad - x_3 = a \\ 2x_1 + x_2 - x_3 = b \\ \quad 2x_2 + 3x_3 = c \end{cases}$

65. $\begin{cases} x_1 + x_2 + x_3 = 2 \\ 2x_1 + 3x_2 + 2x_3 = 5 \\ 2x_1 + 3x_2 + (a^2 - 1)x_3 = a \end{cases}$

66. $\begin{cases} x_1 + x_2 + x_3 = 2 \\ x_1 + 2x_2 + x_3 = 3 \\ x_1 + x_2 + (c^2 - 5)x_3 = c \end{cases}$

67. Find a, b, and c, so that $(1, 2, 3)$ is a solution of the system

$$\begin{cases} ax + by + cz = 6 \\ 2bx - cy + 3az = 9 \\ -cx + 2by + 2az = 9. \end{cases}$$

68. Find the three angles of a triangle, α, β, and γ, if

$$\begin{cases} \alpha + \beta = 75° \\ \gamma - \beta = 52°. \end{cases}$$

Objectives

1. Perform Matrix Arithmetic
2. Find the Inverse of a Square Matrix
3. Solve Linear Systems Using Inverse Matrices
4. Solve Applied Problems

8.3 Solutions of Linear Systems Using Inverse Matrices

In Section 8.2 we used elementary row operations on augmented matrices to solve linear systems of equations. In this section we examine another method for solving linear systems that involves the algebra of matrices.

We say that **two matrices are equal** if and only if they are the same size and the corresponding elements in each matrix are equal. For example,

$$\begin{bmatrix} 2 & 3 & -1 \\ 5 & 8 & 3 \end{bmatrix} = \begin{bmatrix} 2 & x & y \\ a & b & c \end{bmatrix}.$$

if and only if $x = 3$, $y = -1$, $a = 5$, $b = 8$, and $c = 3$.

A matrix whose elements are all 0s is called a **zero matrix** and is denoted by **0.** For example, the zero matrix of size 2×3 is given by

$$\mathbf{0} = \begin{bmatrix} 0 & 0 & 0 \\ 0 & 0 & 0 \end{bmatrix}.$$

Performing Matrix Arithmetic

The addition (or subtraction) of two matrices can be performed *only* if the matrices are the same size. Then, to **add** (or **subtract**) the matrices, we add (or subtract) the corresponding entries.

EXAMPLE 1

Adding and Subtracting Matrices

Let

$$A = \begin{bmatrix} 3 & 5 & 1 \\ 6 & 2 & -3 \end{bmatrix} \quad \text{and} \quad B = \begin{bmatrix} 1 & 3 & 7 \\ 5 & -2 & 4 \end{bmatrix}.$$

Determine:

(a) $A + B$ (b) $A - B$

Solution

(a) $A + B = \begin{bmatrix} 3 & 5 & 1 \\ 6 & 2 & -3 \end{bmatrix} + \begin{bmatrix} 1 & 3 & 7 \\ 5 & -2 & 4 \end{bmatrix}$

$= \begin{bmatrix} 3+1 & 5+3 & 1+7 \\ 6+5 & 2+(-2) & -3+4 \end{bmatrix} = \begin{bmatrix} 4 & 8 & 8 \\ 11 & 0 & 1 \end{bmatrix}$

(b) $A - B = \begin{bmatrix} 3 & 5 & 1 \\ 6 & 2 & -3 \end{bmatrix} - \begin{bmatrix} 1 & 3 & 7 \\ 5 & -2 & 4 \end{bmatrix}$

$= \begin{bmatrix} 3-1 & 5-3 & 1-7 \\ 6-5 & 2-(-2) & -3-4 \end{bmatrix} = \begin{bmatrix} 2 & 2 & -6 \\ 1 & 4 & -7 \end{bmatrix}$

If A is a matrix and k is a real number, we define kA to be the matrix obtained by multiplying each entry of A by k. This process is referred to as **multiplication by a scalar.**

EXAMPLE 2

Multiplying a Matrix by a Real Number

Let

$$A = \begin{bmatrix} -5 & 1 \\ 3 & 2 \end{bmatrix}.$$

Find each matrix:

(a) $4A$ (b) $-2A$

Solution

(a) $4A = 4\begin{bmatrix} -5 & 1 \\ 3 & 2 \end{bmatrix} = \begin{bmatrix} 4(-5) & 4(1) \\ 4(3) & 4(2) \end{bmatrix}$

$= \begin{bmatrix} -20 & 4 \\ 12 & 8 \end{bmatrix}$

(b) $-2A = -2\begin{bmatrix} -5 & 1 \\ 3 & 2 \end{bmatrix} = \begin{bmatrix} (-2)(-5) & (-2)(1) \\ (-2)(3) & (-2)(2) \end{bmatrix}$

$= \begin{bmatrix} 10 & -2 \\ -6 & -4 \end{bmatrix}$

Because matrices are added by adding their corresponding elements, it follows from the properties of real numbers that the addition of matrices obeys both the *commutative* and *associative properties.* That is, if A, B, and C are matrices of the same size, then we have the following properties:

   1. Commutative Property: $A + B = B + A$

   2. Associative Property: $A + (B + C) = (A + B) + C$

A zero matrix acts as the identity for the operation of addition. That is, if A and $\mathbf{0}$ are of the same size, then we have:

   3. Zero Matrix Property: $A + \mathbf{0} = \mathbf{0} + A = A$

For example,

$$\begin{bmatrix} 1 & 2 \\ 3 & 8 \end{bmatrix} + \begin{bmatrix} 0 & 0 \\ 0 & 0 \end{bmatrix} = \begin{bmatrix} 1 & 2 \\ 3 & 8 \end{bmatrix}.$$

The matrix $-A$ is the same as $(-1)A$. Also, $A - B$ can be interpreted as $A - B = A + (-B)$.

If A is a matrix, the **additive inverse of** A, denoted by $-A$, is the matrix consisting of entries that are opposite in sign to the entries of A. Thus, we have:

   4. Additive Inverse Property: $A + (-A) = (-A) + A = \mathbf{0}$

The following properties involving multiplication by a scalar, and addition, can be easily verified.

Scalar Multiplication Properties

If A and B are matrices of the same size, and c and d are real numbers, then

   5. $(cd)A = c(dA)$

   6. $(c + d)A = cA + dA$

   7. $c(A + B) = cA + cB$

To understand how to *multiply two matrices,* we need some background. A matrix consisting of one row is called a **row matrix,** and a matrix consisting of a single column is called a **column matrix.**

For example,

$$\begin{bmatrix} 2 & -1 & -4 \end{bmatrix} \quad \text{and} \quad \begin{bmatrix} 1 & 0 & -1 & 5 \end{bmatrix}$$

are row matrices, whereas

$$\begin{bmatrix} 3 \\ 4 \\ -1 \end{bmatrix} \quad \text{and} \quad \begin{bmatrix} 1 \\ 0 \\ -1 \\ 5 \end{bmatrix}$$

are column matrices.

Suppose that R is a row matrix, C is a column matrix, and the number of columns in R is the same as the number of rows in C. Then the **product** of R and C is a matrix formed by multiplying the corresponding entries of the two matrices and adding the resulting products.

For instance, if

$$R = \begin{bmatrix} 2 & -1 & 3 & 4 \end{bmatrix} \quad \text{and} \quad C = \begin{bmatrix} 5 \\ 2 \\ -3 \\ 1 \end{bmatrix}$$

then

$$RC = \begin{bmatrix} 2 & -1 & 3 & 4 \end{bmatrix} \begin{bmatrix} 5 \\ 2 \\ -3 \\ 1 \end{bmatrix}$$

$$= 2(5) + (-1)(2) + 3(-3) + 4(1)$$

$$= \begin{bmatrix} 3 \end{bmatrix}.$$

Unlike addition and subtraction of matrices, the product of two matrices cannot be determined by multiplying corresponding elements.

Note that in order to multiply two matrices they do not have to be of the same size. However, in order to form the product AB of two matrices A and B, the number of columns of A must be the same as the number of rows of B. The product AB will have as many rows as A and as many columns as B. That is:

If A is an $m \times k$ matrix and B is a $k \times n$ matrix, then the product AB is an $m \times n$ matrix.

The following diagram illustrates these restrictions:

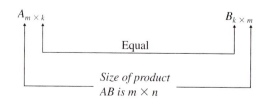

The entry in the ith row and jth column of the product AB is found by multiplying the ith row of a matrix A and the jth column of matrix B, as we did for the row and column matrices. For example, to find the product AB if

$$A = \begin{bmatrix} 1 & 3 & 2 \\ -2 & 4 & 5 \end{bmatrix} \quad \text{and} \quad B = \begin{bmatrix} -4 & 7 \\ 3 & -5 \\ 1 & 6 \end{bmatrix}.$$

we first observe that

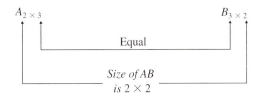

The product AB is determined as follows:

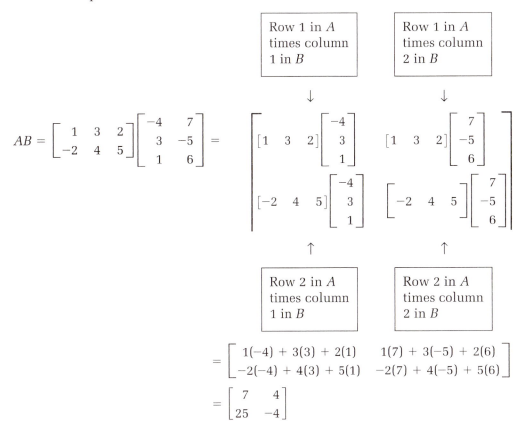

$$= \begin{bmatrix} 1(-4) + 3(3) + 2(1) & 1(7) + 3(-5) + 2(6) \\ -2(-4) + 4(3) + 5(1) & -2(7) + 4(-5) + 5(6) \end{bmatrix}$$

$$= \begin{bmatrix} 7 & 4 \\ 25 & -4 \end{bmatrix}$$

EXAMPLE 3   Multiplying Matrices

Find AB if $\quad A = \begin{bmatrix} 3 & -2 \\ 1 & 5 \\ 4 & 6 \end{bmatrix} \quad$ and $\quad B = \begin{bmatrix} 7 & -2 \\ 3 & 0 \end{bmatrix}.$

Solution

Since A is a 3×2 matrix and B is a 2×2 matrix, then AB is a 3×2 matrix. The product AB is given by

$$AB = \begin{bmatrix} 3 & -2 \\ 1 & 5 \\ 4 & 6 \end{bmatrix} \begin{bmatrix} 7 & -2 \\ 3 & 0 \end{bmatrix}$$

$$= \begin{bmatrix} 3(7) + (-2)(3) & 3(-2) + (-2)(0) \\ 1(7) + 5(3) & 1(-2) + 5(0) \\ 4(7) + 6(3) & 4(-2) + 6(0) \end{bmatrix}$$

$$= \begin{bmatrix} 15 & -6 \\ 22 & -2 \\ 46 & -8 \end{bmatrix}$$

The product AB may be defined even when BA is not. For instance, the product BA is not defined in Example 3 because the number of rows of A is not the same as the number of columns of B.

Matrix multiplication is *not* commutative; that is, it is *not* generally true that matrix product AB is the same as matrix product BA. For instance,

$$AB = \begin{bmatrix} 2 & 1 \\ 4 & 2 \end{bmatrix} \begin{bmatrix} -1 & 2 \\ 3 & 0 \end{bmatrix} = \begin{bmatrix} 1 & 4 \\ 2 & 8 \end{bmatrix}$$

whereas

$$BA = \begin{bmatrix} -1 & 2 \\ 3 & 0 \end{bmatrix} \begin{bmatrix} 2 & 1 \\ 4 & 2 \end{bmatrix} = \begin{bmatrix} 6 & 3 \\ 6 & 3 \end{bmatrix}.$$

Although matrix multiplication is not commutative, matrix multiplication obeys the associative and distributive properties. That is, if A, B, and C are matrices where multiplication is defined, and if k is a constant real number, then:

1. **Associative Properties:**

   (a) $A(BC) = (AB)C$

   (b) $(kA)B = k(AB)$

2. **Distributive Properties:**

   (a) $A(B + C) = AB + AC$

   (b) $(B + C)A = BA + CA$

Finding the Inverse of a Square Matrix

The 2×2 matrix

$$\begin{bmatrix} 1 & 0 \\ 0 & 1 \end{bmatrix}$$

and the 3×3 matrix

$$\begin{bmatrix} 1 & 0 & 0 \\ 0 & 1 & 0 \\ 0 & 0 & 1 \end{bmatrix}$$

are referred to as **identity matrices.** If I is an identity matrix and A is an $n \times n$ matrix, so that AI and IA are defined, then

$$AI = IA = A.$$

For example,

$$\begin{bmatrix} 1 & -3 \\ 2 & 4 \end{bmatrix} \begin{bmatrix} 1 & 0 \\ 0 & 1 \end{bmatrix} = \begin{bmatrix} 1 & -3 \\ 2 & 4 \end{bmatrix} \quad \text{and} \quad \begin{bmatrix} 1 & 0 \\ 0 & 1 \end{bmatrix} \begin{bmatrix} 1 & -3 \\ 2 & 4 \end{bmatrix} = \begin{bmatrix} 1 & -3 \\ 2 & 4 \end{bmatrix}.$$

An $n \times n$ matrix A is said to be **invertible** if there is an $n \times n$ matrix B such that

$$AB = BA = I.$$

The matrix B is called the **inverse matrix** of A and is denoted by A^{-1}; that is,

$$AA^{-1} = A^{-1}A = I.$$

EXAMPLE 4   **Verifying That One Matrix Is an Inverse of Another**

Show that the matrix

$$B = \begin{bmatrix} -5 & 2 \\ 3 & -1 \end{bmatrix}$$

is the inverse of the matrix

$$A = \begin{bmatrix} 1 & 2 \\ 3 & 5 \end{bmatrix}.$$

Solution   Since

$$AB = \begin{bmatrix} 1 & 2 \\ 3 & 5 \end{bmatrix} \begin{bmatrix} -5 & 2 \\ 3 & -1 \end{bmatrix} = \begin{bmatrix} 1 & 0 \\ 0 & 1 \end{bmatrix}$$

and

$$BA = \begin{bmatrix} -5 & 2 \\ 3 & -1 \end{bmatrix} \begin{bmatrix} 1 & 2 \\ 3 & 5 \end{bmatrix} = \begin{bmatrix} 1 & 0 \\ 0 & 1 \end{bmatrix}$$

it follows that $AB = I = BA$, so B is indeed the inverse of A; that is, $B = A^{-1}$. Similarly, $A = B^{-1}$.

We know from algebra that every nonzero real number a has a multiplicative inverse $a^{-1} = 1/a$ such that

$$aa^{-1} = a^{-1}a = 1.$$

The analogous statement for matrix multiplication does *not* always hold, that is, not every square nonzero matrix has an inverse.

For instance, consider the matrix

$$A = \begin{bmatrix} 1 & 1 \\ 1 & 1 \end{bmatrix}.$$

If A has an inverse, say

$$B = \begin{bmatrix} a & b \\ c & d \end{bmatrix}$$

then the product AB would have to equal the 2×2 identity matrix. However,

$$AB = \begin{bmatrix} 1 & 1 \\ 1 & 1 \end{bmatrix} \begin{bmatrix} a & b \\ c & d \end{bmatrix} = \begin{bmatrix} a+c & b+d \\ a+c & b+d \end{bmatrix}.$$

Clearly,

$$\begin{bmatrix} a+c & b+d \\ a+c & b+d \end{bmatrix} \quad \text{cannot equal} \quad \begin{bmatrix} 1 & 0 \\ 0 & 1 \end{bmatrix}$$

since it is impossible to have both $a + c = 1$ and $a + c = 0$. Thus A cannot have an inverse.

Elementary row operations can be used to find the inverse of a square matrix if it exists. For example, to find the inverse of the matrix

$$A = \begin{bmatrix} 3 & 5 \\ 1 & 2 \end{bmatrix}$$

we need to find a matrix

$$B = \begin{bmatrix} x & u \\ y & v \end{bmatrix}$$

with entries x, y, u, and v so that

$$AB = \begin{bmatrix} 3 & 5 \\ 1 & 2 \end{bmatrix} \begin{bmatrix} x & u \\ y & v \end{bmatrix} = \begin{bmatrix} 1 & 0 \\ 0 & 1 \end{bmatrix}.$$

Multiplying the left side, we get

$$\begin{bmatrix} 3x+5y & 3u+5v \\ x+2y & u+2v \end{bmatrix} = \begin{bmatrix} 1 & 0 \\ 0 & 1 \end{bmatrix}.$$

Equating the corresponding entries of the two matrices, we have

$$\begin{cases} 3x + 5y = 1 \\ x + 2y = 0 \end{cases} \quad \text{and} \quad \begin{cases} 3u + 5v = 0 \\ u + 2v = 1. \end{cases}$$

Now we can use augmented matrices and row-reduction to solve these two systems. Notice that the coefficient matrices of both systems are the same. Therefore, we can write the matrix so that the first three columns form the augmented matrix of the first system, and the first two columns and last column form the augmented matrix of the second system. That is, we write the matrix

$$\begin{bmatrix} 3 & 5 & \vdots & 1 & 0 \\ 1 & 2 & \vdots & 0 & 1 \end{bmatrix}.$$

Next we perform elementary row operations to express this matrix in a row-reduced echelon form:

$$\begin{bmatrix} 3 & 5 & \vdots & 1 & 0 \\ 1 & 2 & \vdots & 0 & 1 \end{bmatrix}$$

$$\xrightarrow{R_1 \leftrightarrow R_2} \begin{bmatrix} 1 & 2 & \vdots & 0 & 1 \\ 3 & 5 & \vdots & 1 & 0 \end{bmatrix}$$

$$\xrightarrow{-3R_1 + R_2 \to R_2} \begin{bmatrix} 1 & 2 & \vdots & 0 & 1 \\ 0 & -1 & \vdots & 1 & -3 \end{bmatrix}$$

$$\xrightarrow{(-1)R_2 \to R_2} \begin{bmatrix} 1 & 2 & \vdots & 0 & 1 \\ 0 & 1 & \vdots & -1 & 3 \end{bmatrix}$$

$$\xrightarrow{-2R_2 + R_1 \to R_1} \begin{bmatrix} 1 & 0 & \vdots & 2 & -5 \\ 0 & 1 & \vdots & -1 & 3 \end{bmatrix}$$

From the first three columns of the last matrix, we see that $x = 2$ and $y = -1$. Similarly, from the first two columns and the last column, it follows that $u = -5$ and $v = 3$. Therefore,

$$B = A^{-1} = \begin{bmatrix} 2 & -5 \\ -1 & 3 \end{bmatrix}$$

Notice that A^{-1} is the right half of the augmented matrix we found above, and that the left half is the identity matrix I.

This illustration leads us to the following general procedure for finding the inverse of a square matrix A:

Procedure

Finding the Inverse of an $n \times n$ Matrix A

1. Form the $n \times 2n$ matrix.

$$[A \vdots I]$$

$n \times n$ matrix ——⌐   ⌐—— $n \times n$ identity matrix

2. Perform elementary row operations on this matrix to transform it, if possible, to a matrix of the form.

$$[I \vdots B]$$

$n \times n$ identity matrix ——⌐

3. The matrix B is A^{-1}; that is, $A^{-1} = B$.

If the above procedure does not lead to the identity matrix on the left, then the matrix A has no inverse and we say A^{-1} does not exist.

EXAMPLE 5   **Finding the Inverse of a 3 × 3 Matrix**

Find A^{-1} where $A = \begin{bmatrix} 1 & 2 & 3 \\ 1 & 1 & 2 \\ 0 & 1 & 2 \end{bmatrix}$.

Solution   We begin by forming the matrix $[A \vdots I]$:

$$\begin{bmatrix} 1 & 2 & 3 & \vdots & 1 & 0 & 0 \\ 1 & 1 & 2 & \vdots & 0 & 1 & 0 \\ 0 & 1 & 2 & \vdots & 0 & 0 & 1 \end{bmatrix}$$

Next we perform elementary row operations on the entire matrix until the left half is transformed into the identity matrix:

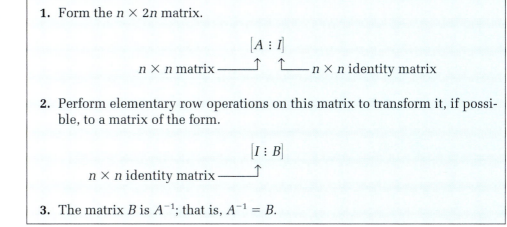

$$2R_2 + R_1 \to R_1 \quad \begin{bmatrix} 1 & 0 & 1 & : & -1 & 2 & 0 \\ 0 & -1 & -1 & : & -1 & 1 & 0 \\ 0 & 0 & 1 & : & -1 & 1 & 1 \end{bmatrix}$$

$$(-1)R_2 \to R_2 \quad \begin{bmatrix} 1 & 0 & 1 & : & -1 & 2 & 0 \\ 0 & 1 & 1 & : & 1 & -1 & 0 \\ 0 & 0 & 1 & : & -1 & 1 & 1 \end{bmatrix}$$

$$(-1)R_3 + R_2 \to R_2 \quad \begin{bmatrix} 1 & 0 & 1 & : & -1 & 2 & 0 \\ 0 & 1 & 0 & : & 2 & -2 & -1 \\ 0 & 0 & 1 & : & -1 & 1 & 1 \end{bmatrix}$$

$$(-1)R_3 + R_1 \to R_1 \quad \begin{bmatrix} 1 & 0 & 0 & : & 0 & 1 & -1 \\ 0 & 1 & 0 & : & 2 & -2 & -1 \\ 0 & 0 & 1 & : & -1 & 1 & 1 \end{bmatrix}$$

This result can be confirmed by verifying that $AA^{-1} = A^{-1}A = I.$

Therefore,
$$A^{-1} = \begin{bmatrix} 0 & 1 & -1 \\ 2 & -2 & -1 \\ -1 & 1 & 1 \end{bmatrix}.$$

Solving Linear Systems Using Inverse Matrices

Inverse matrices can also be used to solve linear systems of equations when the system has the same number of equations as variables—provided that the coefficient matrix has an inverse. Let's consider a linear system consisting of two equations in two variables x and y:

$$\begin{cases} a_1 x + a_2 y = k_1 \\ b_1 x + b_2 y = k_2 \end{cases}$$

We can express this system as a product of matrices as follows: Let A be the 2×2 coefficient matrix of the system of equations, let K be the column matrix of the constants on the right side, and let X be the column matrix of unknowns. That is, let

$$A = \begin{bmatrix} a_1 & a_2 \\ b_1 & b_2 \end{bmatrix}, \quad X = \begin{bmatrix} x \\ y \end{bmatrix}, \quad \text{and} \quad K = \begin{bmatrix} k_1 \\ k_2 \end{bmatrix}.$$

Then AX is the matrix with 2 rows and 1 column given by

$$AX = \begin{bmatrix} a_1 & a_2 \\ b_1 & b_2 \end{bmatrix} \begin{bmatrix} x \\ y \end{bmatrix} = \begin{bmatrix} a_1 x + a_2 y \\ b_1 x + b_2 y \end{bmatrix}.$$

Since $a_1 x + a_2 y = k_1$ and $b_1 x + b_2 y = k_2$, it follows that $AX = K$. The latter equation is called the **matrix equation form** of the system. To solve this equation, we multiply both sides *on the left* by A^{-1}, assuming it exists, to isolate X on the left side:

Do not change the order of multiplying A^{-1} *on each side of the equation* $AX = K$, *since* $A^{-1}(AX)$ *and* KA^{-1} *are not necessarily the same.*

$$AX = K$$
$$A^{-1}(AX) = A^{-1}K$$
$$(A^{-1}A)X = A^{-1}K$$
$$IX = A^{-1}K$$

so

$$X = A^{-1}K.$$

The entries in the matrix product $A^{-1}K$ make up the solution of the original system.

EXAMPLE 6

Solving a 2 × 2 Linear System by Using the Inverse of a Matrix

Solve the system by using the inverse of the coefficient matrix.

$$\begin{cases} 3x + 5y = 11 \\ x + 2y = 4 \end{cases}$$

Solution

The given system of equations is equivalent to the matrix equation $AX = K$, where

$$A = \begin{bmatrix} 3 & 5 \\ 1 & 2 \end{bmatrix}, \quad X = \begin{bmatrix} x \\ y \end{bmatrix}, \quad \text{and} \quad K = \begin{bmatrix} 11 \\ 4 \end{bmatrix}.$$

From the illustration on pages 568 and 569, we know that A^{-1} exists and that

$$A^{-1} = \begin{bmatrix} 2 & -5 \\ -1 & 3 \end{bmatrix}.$$

It follows that the solution X is given by

$$X = A^{-1}K$$

$$= \begin{bmatrix} 2 & -5 \\ -1 & 3 \end{bmatrix} \begin{bmatrix} 11 \\ 4 \end{bmatrix} = \begin{bmatrix} 2 \\ 1 \end{bmatrix}.$$

Therefore, $x = 2$ and $y = 1$, and the solution is (2, 1).

EXAMPLE 7

Solving a 3 × 3 Linear System by Using the Inverse of a Matrix

Solve the system by using the inverse of the coefficient matrix.

$$\begin{cases} x + 2y + 3z = 4 \\ x + y + 2z = 5 \\ y + 2z = 4 \end{cases}$$

Solution

If we let

$$A = \begin{bmatrix} 1 & 2 & 3 \\ 1 & 1 & 2 \\ 0 & 1 & 2 \end{bmatrix}, \quad X = \begin{bmatrix} x \\ y \\ z \end{bmatrix}, \quad \text{and} \quad K = \begin{bmatrix} 4 \\ 5 \\ 4 \end{bmatrix}$$

we can write the system as $AX = K$. From Example 5, we know that

$$A^{-1} = \begin{bmatrix} 0 & 1 & -1 \\ 2 & -2 & -1 \\ -1 & 1 & 1 \end{bmatrix} \text{ so that } X = A^{-1}K = \begin{bmatrix} 0 & 1 & -1 \\ 2 & -2 & -1 \\ -1 & 1 & 1 \end{bmatrix} \begin{bmatrix} 4 \\ 5 \\ 4 \end{bmatrix} = \begin{bmatrix} 1 \\ -6 \\ 5 \end{bmatrix}.$$

Thus $x = 1$, $y = -6$, and $z = 5$, and the solution is (1, −6, 5).

G Most graphers can perform matrix operations, including matrix multiplication and matrix inversion. For example, to solve the following system by using a grapher

$$\begin{cases} 0.1s + 0.2t + 0.1u = 27 \\ 0.5s + 0.2t + 0.3u = 9 \\ 0.4s + 0.8t + 0.1u = 36 \end{cases}$$

we first find the coefficient matrix A and the matrix equation

$$A = \begin{bmatrix} 0.1 & 0.2 & 0.1 \\ 0.5 & 0.2 & 0.3 \\ 0.4 & 0.8 & 0.1 \end{bmatrix} \quad \text{and} \quad AX = K,$$

$$\text{where} \quad X = \begin{bmatrix} s \\ t \\ u \end{bmatrix} \quad \text{and} \quad K = \begin{bmatrix} 27 \\ 9 \\ 36 \end{bmatrix},$$

Here we took advantage of the grapher's numerical conversion feature to display the entries of A^{-1} in fraction form rather than in decimal form.

then we input matrix A and find

$$A^{-1} = \begin{bmatrix} -\dfrac{55}{6} & \dfrac{5}{2} & \dfrac{5}{3} \\ \dfrac{35}{12} & -\dfrac{5}{4} & \dfrac{5}{6} \\ \dfrac{40}{3} & 0 & \dfrac{-10}{3} \end{bmatrix}$$

Next we use matrix multplication to obtain

$$X = A^{-1}K = \begin{bmatrix} -165 \\ 97.5 \\ 240 \end{bmatrix}.$$

So $\qquad\qquad s = -165, \quad t = 97.5 \quad \text{and} \quad u = 240.$

Solving Applied Problems

The next example illustrates a way in which matrix multiplication is used to interpret data.

EXAMPLE 8   **Solving a Business Model**

A fruit market packages fruit in three different ways for gift baskets. The economy basket E has 5 apples, 3 grapefruits, and 4 pears. The standard basket S has 4 apples, 4 grapefruits, and 5 pears. The luxury basket L has 6 apples, 5 grapefruits, and 5 pears. The cost is $0.40, $0.70 and $0.50 for each apple, grapefruit, and pear, respectively. What is the cost of preparing each basket of fruit?

Solution   We arrange the given data in matrix form as follows:

Cost per fruit

Number of items in each basket type

Apple   Grapefruit   Pear

$$A = [\$0.40 \qquad \$0.70 \qquad \$0.50]$$

$$B = \begin{matrix} \text{Apples} \\ \text{Grapefruits} \\ \text{Pears} \end{matrix} \begin{bmatrix} 5 & 4 & 6 \\ 3 & 4 & 5 \\ 4 & 5 & 5 \end{bmatrix} \begin{matrix} E & S & L \end{matrix}$$

To find the cost of each basket, multiply matrix A by matrix B to get

$$AB = [0.40 \quad 0.70 \quad 0.50] \begin{bmatrix} 5 & 4 & 6 \\ 3 & 4 & 5 \\ 4 & 5 & 5 \end{bmatrix}$$

$$= [6.10 \quad 6.90 \quad 8.40].$$

Therefore, the cost of each economy, standard, and luxury basket are

$$\$6.10, \quad \$6.90, \quad \text{and} \quad \$8.40, \quad \text{respectively.}$$

PROBLEM SET 8.3

Mastering the Concepts

In problems 1–6, perform each operation on the given matrices A and B

(a) $A + B$ 　　(b) $A - B$ 　　(c) $4A$

(d) $-5B$ 　　(e) $4A - 5B$

1. $A = \begin{bmatrix} 3 & 5 \\ -1 & 4 \end{bmatrix}$; $B = \begin{bmatrix} 5 & 6 \\ -1 & 3 \end{bmatrix}$

2. $A = \begin{bmatrix} -1 & 2 \\ 3 & 0 \end{bmatrix}$; $B = \begin{bmatrix} 2 & -1 \\ 3 & -4 \end{bmatrix}$

3. $A = \begin{bmatrix} 3 & 1 & 2 \\ -1 & 3 & 1 \end{bmatrix}$; $B = \begin{bmatrix} 5 & 11 & 6 \\ 3 & 0 & -1 \end{bmatrix}$

4. $A = \begin{bmatrix} 0 & 1 & 4 & 2 \\ -5 & 6 & 1 & 3 \end{bmatrix}$; $B = \begin{bmatrix} 6 & -1 & 3 & 1 \\ -2 & 0 & 1 & 4 \end{bmatrix}$

5. $A = \begin{bmatrix} -5 & 2 & 1 \\ -2 & 1 & 2 \\ 3 & -1 & 2 \end{bmatrix}$; $B = \begin{bmatrix} -4 & 6 & 2 \\ -1 & 2 & 5 \\ 3 & -1 & 2 \end{bmatrix}$

6. $A = \begin{bmatrix} 1 & 0 & 0 & 1 & -1 \\ 2 & 1 & 3 & 5 & 7 \\ 3 & 6 & 2 & -1 & 4 \end{bmatrix}$;

$B = \begin{bmatrix} 6 & 1 & -1 & 3 & 5 \\ 7 & 1 & 2 & 8 & -1 \\ 0 & 1 & 2 & 0 & 1 \end{bmatrix}$

In problems 7–12, if possible, find:

(a) AB 　　　　　(b) BA

7. $A = \begin{bmatrix} 1 & 2 \\ -1 & 1 \end{bmatrix}$; $B = \begin{bmatrix} 5 & -1 \\ 7 & 0 \end{bmatrix}$

8. $A = \begin{bmatrix} 1 & -1 \\ 0 & 2 \end{bmatrix}$; $B = \begin{bmatrix} 3 & 1 & -1 \\ -2 & 0 & 1 \end{bmatrix}$

9. $A = \begin{bmatrix} 2 & -3 & 5 \\ -1 & 1 & 3 \end{bmatrix}$; $B = \begin{bmatrix} -3 & 1 \\ 1 & 2 \\ 0 & -5 \end{bmatrix}$

10. $A = \begin{bmatrix} 3 & 2 & 1 \\ 1 & 4 & -1 \\ 2 & 1 & -3 \end{bmatrix}$; $B = \begin{bmatrix} -3 & 0 & 0 \\ 0 & -3 & 0 \\ 0 & 0 & -3 \end{bmatrix}$

11. $A = \begin{bmatrix} 4 & 3 & -1 \\ 8 & -2 & 3 \\ 6 & 5 & 2 \end{bmatrix}$; $B = \begin{bmatrix} 1 & 0 & 0 \\ 0 & 1 & 0 \\ 0 & 0 & 1 \end{bmatrix}$

12. $A = \begin{bmatrix} -1 & 1 & 2 & 0 \\ -2 & 3 & 5 & 1 \\ 2 & -1 & 3 & 2 \end{bmatrix}$; $B = \begin{bmatrix} 1 & -3 & 4 \\ 2 & 1 & 0 \\ -1 & 0 & 2 \\ 0 & -1 & 3 \end{bmatrix}$

In problems 13–18, determine whether A and B are inverses of each other.

13. $A = \begin{bmatrix} 1 & -1 \\ 2 & 1 \end{bmatrix}$; $B = \begin{bmatrix} \dfrac{1}{3} & \dfrac{1}{3} \\ -\dfrac{2}{3} & \dfrac{1}{3} \end{bmatrix}$

14. $A = \begin{bmatrix} 2 & 3 \\ 5 & 1 \end{bmatrix}$; $B = \begin{bmatrix} -\dfrac{1}{13} & \dfrac{3}{13} \\ \dfrac{5}{13} & -\dfrac{2}{13} \end{bmatrix}$

15. $A = \begin{bmatrix} 3 & 2 \\ 2 & -3 \end{bmatrix}$; $B = \begin{bmatrix} \dfrac{3}{13} & \dfrac{2}{13} \\ \dfrac{2}{13} & -\dfrac{3}{13} \end{bmatrix}$

16. $A = \begin{bmatrix} 1 & 1 & 1 \\ 2 & -1 & -1 \\ 1 & -1 & 2 \end{bmatrix}$; $B = \begin{bmatrix} \dfrac{1}{3} & \dfrac{1}{3} & 0 \\ \dfrac{5}{9} & -\dfrac{1}{9} & -\dfrac{1}{3} \\ \dfrac{1}{9} & -\dfrac{2}{9} & \dfrac{1}{3} \end{bmatrix}$

17. $A = \begin{bmatrix} 1 & 2 & 4 \\ 2 & -3 & 1 \\ 3 & -1 & -2 \end{bmatrix}$; $B = \begin{bmatrix} \dfrac{1}{7} & 0 & \dfrac{2}{7} \\ \dfrac{1}{7} & -\dfrac{2}{7} & \dfrac{1}{7} \\ \dfrac{1}{7} & \dfrac{1}{7} & -\dfrac{1}{7} \end{bmatrix}$

18. $A = \begin{bmatrix} 1 & 1 & 1 \\ 2 & 3 & -1 \\ 3 & 5 & 1 \end{bmatrix}$; $B = \begin{bmatrix} 2 & 1 & -1 \\ -\dfrac{5}{4} & \dfrac{1}{2} & \dfrac{3}{4} \\ \dfrac{1}{4} & -\dfrac{1}{2} & \dfrac{1}{4} \end{bmatrix}$

In problems 19–26, use elementary row operations to find the inverse of the given matrix, if it exists.

19. $\begin{bmatrix} 3 & 5 \\ 1 & 2 \end{bmatrix}$

20. $\begin{bmatrix} 2 & -5 \\ -1 & 3 \end{bmatrix}$

21. $\begin{bmatrix} -1 & 1 \\ 1 & 0 \end{bmatrix}$

22. $\begin{bmatrix} 1 & 2 \\ \dfrac{1}{4} & \dfrac{3}{4} \end{bmatrix}$

23. $\begin{bmatrix} 1 & 2 & 0 \\ 0 & 2 & 3 \\ 1 & 3 & 1 \end{bmatrix}$

24. $\begin{bmatrix} 2 & 7 & 1 \\ 1 & 4 & -1 \\ 1 & 3 & 0 \end{bmatrix}$

25. $\begin{bmatrix} 7 & 2 & -6 \\ -3 & -1 & 3 \\ 2 & 1 & -2 \end{bmatrix}$

26. $\begin{bmatrix} -1 & 0 & 2 \\ 3 & 1 & 0 \\ 0 & 2 & -3 \end{bmatrix}$

In problems 27–36, use an inverse matrix to solve each system.

27. $\begin{cases} x - y = 1 & \text{(see problem 13)} \\ 2x + y = 5 \end{cases}$

28. $\begin{cases} 2x_1 + 3x_2 = 7 & \text{(see problem 14)} \\ 5x_1 + x_2 = -2 \end{cases}$

29. $\begin{cases} 3x + 2y = 8 & \text{(see problem 15)} \\ 2x - 3y = 14 \end{cases}$

30. $\begin{cases} 3x + 5y = 13 \\ x + 2y = 5 \end{cases}$

31. $\begin{cases} 2x - 5y = -3 \\ -x + 3y = 2 \end{cases}$

32. $\begin{cases} 4x + y = 17 \\ 3x + 2y = 17 \end{cases}$

33. $\begin{cases} x + y + z = 2 \\ 2x + 3y - z = 3 \\ 3x + 5y + z = 8 \end{cases}$

34. $\begin{cases} x_1 + 2x_2 + 4x_3 = 12 \\ 2x_1 - 3x_2 + x_3 = 10 \\ 3x_1 - x_2 - 2x_3 = 1 \end{cases}$

35. $\begin{cases} x + y + z = 6 \\ 2x - y - z = 0 \\ x - y + 2z = 7 \end{cases}$

36. $\begin{cases} 2x + 7y + z = 10 \\ x + 4y - z = 4 \\ x + 3y = 4 \end{cases}$

[G] In problems 37–40, write each system in the form $AX = K$. Then use a grapher to find A^{-1} and $X = A^{-1}K$, the solution of the system. In problems 38, 39, and 40, round off the entries of A^{-1} to two decimal places.

37. $\begin{cases} x + 3y + 2z + w = 4 \\ x + 2y + 4z = -3 \\ 2x + 6y + 5z + 2w = -1 \\ x + 3y + 2z + 2w = 2 \end{cases}$

38. $\begin{cases} x + 2y + z - w = 3 \\ 2x + 4y + z + w = 2 \\ 3x + 2y - z - 2w = -4 \\ 2x + 5y + 3z = 7 \end{cases}$

39. $\begin{cases} 0.2x + 0.4y - 0.6z = 0.2 \\ -0.2x + 0.1y + 0.3z + 0.2w = 0.7 \\ 0.2x + 0.9y - 0.9z + 0.1w = 1.1 \\ 0.4x + 0.3y - 0.3z = 0.7 \end{cases}$

40. $\begin{cases} 0.3x + 0.1y - 0.1z + 0.1w = 0 \\ 0.6x + 0.4y - 0.1z + 0.4w = 1.5 \\ -0.3x - 0.1y + 0.2z - 0.1w = 1 \\ 0.3x + 0.5y + 0.3w = 2.1 \end{cases}$

Applying the Concepts

41. Crop Shipments: A nursery raises two kinds of flowers that are shipped in bundles to three different florists. The number of bundles of flower k that are shipped to store m is given by entry a_{km} in the matrix

$$A = \begin{bmatrix} \overset{\text{Store}}{\underset{1}{}} & \overset{\text{Store}}{\underset{2}{}} & \overset{\text{Store}}{\underset{3}{}} \\ 200 & 85 & 80 \\ 250 & 170 & 100 \end{bmatrix} \begin{matrix} \text{Flower 1} \\ \text{Flower 2} \end{matrix}$$

The profit (in dollars) per bundle is represented by the matrix.

$$T = \begin{bmatrix} \overset{\text{Flower}}{\underset{1}{}} & \overset{\text{Flower}}{\underset{2}{}} \\ 4.50 & 5.70 \end{bmatrix}$$

(a) Find the product TA.

(b) Indicate what each entry of the product represents.

42. Inventory Levels: A store sells two brands, B_1 and B_2, of microwave ovens. The following matrices give the sales figures and costs for three months. Use matrix multiplication to determine the total dollar sales and the total dollar costs of these items for these three months.

	May	July	October
Number of brand B_1 ovens sold	16	10	14
Number of brand B_2 ovens sold	20	15	16

	Brand B_1	Brand B_2
Retail price	210	140
Dealer cost	150	90

43. Total Revenue: An automobile manufacturer produces three different new models of cars, A, B, and C, that are shipped to two different locations for distribution. The following matrices give the number of cars that are shipped as well as the price in dollars per car.

	Number shipped	
	Site 1	Site 2
Model A	6,000	7,000
Model B	9,000	10,000
Model C	4,000	8,000

$$= S$$

Price		
Model A	Model B	Model C
10,000	12,000	14,000

$$= P$$

(a) Find the product PS.

(b) Indicate what each entry of the product represents.

44. Manufacturing: A manufacturer of calculators has two plants, each producing scientific calculators and graphing calculators. The manufacturing time requirements (in hours per calculator) and the assembly and packing costs (in dollars per hour) are given by the following matrices:

Manufacturing time
(hours per unit)

$$A = \begin{bmatrix} \overset{\text{Assembly}}{} & \overset{\text{Packaging}}{} \\ 0.2 & 0.1 \\ 0.3 & 0.1 \end{bmatrix} \begin{matrix} \text{Scientific calculator} \\ \text{Graphing calculator} \end{matrix}$$

Costs
(dollars per hour)

$$B = \begin{bmatrix} \overset{\text{First}}{\underset{\text{plant}}{}} & \overset{\text{Second}}{\underset{\text{plant}}{}} \\ 5 & 6 \\ 4 & 5 \end{bmatrix} \begin{matrix} \text{Assembly} \\ \text{Packaging} \end{matrix}$$

(a) Compute AB.

(b) Interpret the meaning of each entry in AB.

Developing and Extending the Concepts

45. Let

$$A = \begin{bmatrix} 3 & 2 \\ -1 & 4 \end{bmatrix}, \qquad B = \begin{bmatrix} 0 & 1 \\ -1 & 2 \end{bmatrix},$$

and

$$C = \begin{bmatrix} 2 & -3 \\ 1 & -4 \end{bmatrix}.$$

(a) Demonstrate the commutative property:

$$A + B = B + A$$

(b) Demonstrate the associative property:

$$A + (B + C) = (A + B) + C$$

(c) Find $-A$ and $-C$. Then show that

$$A + (-A) = \mathbf{0}$$

and

$$C + (-C) = \mathbf{0},$$

the 2×2 zero matrix.

(d) Find matrix E so that

$$A + E = C.$$

(e) Demonstrate that $A(B - C) = AB - AC$.

46. Find the values of x and y if:

(a) $\begin{bmatrix} 3 & 1 \\ 5 & 7 \end{bmatrix} = \begin{bmatrix} 3 & x \\ -y & 7 \end{bmatrix}$

(b) $\begin{bmatrix} 1 & -1 \\ 2 & 6 \end{bmatrix} + \begin{bmatrix} x & 3 \\ y & 2 \end{bmatrix} = \begin{bmatrix} 8 & 2 \\ 0 & 8 \end{bmatrix}$

47. Find the values of x, y, z, and w if:

$$\begin{bmatrix} x & y \\ z & w \end{bmatrix} \begin{bmatrix} 3 & -5 \\ -1 & 2 \end{bmatrix} = \begin{bmatrix} 1 & -1 \\ 2 & 0 \end{bmatrix}$$

48. Let

$$A = \begin{bmatrix} 3 & -1 \\ 0 & -2 \end{bmatrix}.$$

Show that

$$A^2 - A - 6I = \begin{bmatrix} 0 & 0 \\ 0 & 0 \end{bmatrix}$$

where

$$I = \begin{bmatrix} 1 & 0 \\ 0 & 1 \end{bmatrix}.$$

49. Suppose that A and B are $n \times n$ matrices and

$$AB = BA.$$

(a) Show that

$$(A + B)^2 = A^2 + 2AB + B^2.$$

(b) Show that

$$(A - B)(A + B) = A^2 - B^2.$$

50. If

$$A^{-1} = \begin{bmatrix} 1 & -1 & 3 \\ 2 & 0 & 5 \\ -1 & 1 & 0 \end{bmatrix}$$

find a matrix B such that

$$AB = \begin{bmatrix} 1 & -1 & 2 \\ 0 & 1 & 1 \\ 1 & 0 & 0 \end{bmatrix}.$$

51. If

$$A^{-1} = \begin{bmatrix} 1 & -1 & 3 \\ 2 & 1 & 1 \\ 0 & 2 & -2 \end{bmatrix}$$

find A.

52. Let $A = \begin{bmatrix} 2 & 3 \\ 4 & 5 \end{bmatrix}$ and $B = \begin{bmatrix} 7 & 8 \\ 6 & 7 \end{bmatrix}$.

(a) Compute

$$A^{-1}, B^{-1}, \text{ and } B^{-1}A^{-1}.$$

(b) Compute

$$(AB)^{-1} \text{ and compare it to } B^{-1}A^{-1}.$$

(c) Can you generalize the result in part (b)? Explain.

Objectives

1. Define and Evaluate Determinants
2. Use Elementary Operations to Evaluate Determinants
3. Use Cramer's Rule to Solve Linear Systems
4. Solve Applied Problems

8.4 Solutions of Linear Systems Using Determinants

In this section, we introduce *determinants* and then use them to solve certain systems of linear equations.

Defining and Evaluating Determinants

We begin by defining a determinant of a 2 × 2 matrix.

Definition

Determinant of a 2 × 2 Matrix

The determinant of a square matrix A is also denoted by det A, that is, |A| = det A.

Let

$$A = \begin{bmatrix} a & b \\ c & d \end{bmatrix}$$

be a 2 × 2 matrix. The symbol $|A|$ denotes the **determinant** of matrix A. Its value is defined to be the number $ad - cb$, and we write

$$|A| = \begin{vmatrix} a & b \\ c & d \end{vmatrix} = ad - cb.$$

EXAMPLE 1   **Evaluating a 2 × 2 Determinant**

Evaluate: $\begin{vmatrix} 5 & 3 \\ -3 & -6 \end{vmatrix}$

Solution $\begin{vmatrix} 5 & 3 \\ -3 & -6 \end{vmatrix} = 5(-6) - (-3)(3) = -30 + 9 = -21.$

If A is a 3 × 3 matrix, its determinant value is defined in terms of 2 × 2 determinants according to the following *expansion formula:*

An Expansion Formula for the Determinant of a 3 × 3 Matrix

Another method for evaluating the determinant of a 3 × 3 matrix is given in problem 51.

> The value of the **determinant** of a 3 × 3 matrix is given by the **expansion formula:**
>
> $$\begin{vmatrix} a_1 & a_2 & a_3 \\ b_1 & b_2 & b_3 \\ c_1 & c_2 & c_3 \end{vmatrix} = a_1 \begin{vmatrix} b_2 & b_3 \\ c_2 & c_3 \end{vmatrix} - a_2 \begin{vmatrix} b_1 & b_3 \\ c_1 & c_3 \end{vmatrix} + a_3 \begin{vmatrix} b_1 & b_2 \\ c_1 & c_2 \end{vmatrix}$$
>
> $$= a_1 b_2 c_3 - a_1 c_2 b_3 - a_2 b_1 c_3 + a_2 c_1 b_3 + a_3 b_1 c_2 - a_3 c_1 b_2$$

Note that each entry in the first row is multiplied by the 2 × 2 determinant that remains when the row and column containing the multiplier are crossed out. Thus

Similar expansion formulas can be used to evaluate n × n matrices, where n > 3.

a_1 is multiplied by $\begin{vmatrix} \cancel{a_1} & a_2 & a_3 \\ \cancel{b_1} & b_2 & b_3 \\ \cancel{c_1} & c_2 & c_3 \end{vmatrix} = \begin{vmatrix} b_2 & b_3 \\ c_2 & c_3 \end{vmatrix}$

a_2 is multiplied by $\begin{vmatrix} a_1 & \cancel{a_2} & a_3 \\ b_1 & \cancel{b_2} & b_3 \\ c_1 & \cancel{c_2} & c_3 \end{vmatrix} = \begin{vmatrix} b_1 & b_3 \\ c_1 & c_3 \end{vmatrix}$

a_3 is multiplied by $\begin{vmatrix} a_1 & a_2 & \cancel{a_3} \\ b_1 & b_2 & \cancel{b_3} \\ c_1 & c_2 & \cancel{c_3} \end{vmatrix} = \begin{vmatrix} b_1 & b_2 \\ c_1 & c_2 \end{vmatrix}$

Also, note the negative sign before a_2 in the formula.

EXAMPLE 2   **Evaluating a 3 × 3 Determinant**

Evaluate the determinant by using the expansion formula.

$$\begin{vmatrix} 3 & 2 & 7 \\ -1 & 5 & 3 \\ 2 & -3 & -6 \end{vmatrix}$$

Solution $\begin{vmatrix} 3 & 2 & 7 \\ -1 & 5 & 3 \\ 2 & -3 & -6 \end{vmatrix} = 3 \begin{vmatrix} 5 & 3 \\ -3 & -6 \end{vmatrix} - 2 \begin{vmatrix} -1 & 3 \\ 2 & -6 \end{vmatrix} + 7 \begin{vmatrix} -1 & 5 \\ 2 & -3 \end{vmatrix}$

$= 3(-30 + 9) - 2(6 - 6) + 7(3 - 10)$

$= 3(-21) - 2(0) + 7(-7)$

$= -63 - 49$

$= -112.$

EXAMPLE 3   **Evaluating a 3 × 3 Determinant of a Matrix in Triangular Form**

Evaluate the determinant $\begin{vmatrix} 3 & 4 & 5 \\ 0 & 6 & -1 \\ 0 & 0 & 2 \end{vmatrix}$.

Solution

$$\begin{vmatrix} 3 & 4 & 5 \\ 0 & 6 & -1 \\ 0 & 0 & 2 \end{vmatrix} = 3\begin{vmatrix} 6 & -1 \\ 0 & 2 \end{vmatrix} - 4\begin{vmatrix} 0 & -1 \\ 0 & 2 \end{vmatrix} + 5\begin{vmatrix} 0 & 6 \\ 0 & 0 \end{vmatrix}$$

$$= 3(12 - 0) - 4(0 - 0) + 5(0 - 0)$$

$$= 3(12)$$

$$= 36$$

Note that the value of the determinant in Example 3 is the product of the entries on its *main diagonal* (the diagonal that runs from upper left to lower right). In general we have:

The determinant of any square matrix in triangular form is equal to the product of the entries on its main diagonal.

For instance, the value of the 4 × 4 determinant

$$\begin{vmatrix} 2 & 1 & -1 & 3 \\ 0 & 1 & 1 & -2 \\ 0 & 0 & 3 & 4 \\ 0 & 0 & 0 & 5 \end{vmatrix} = 2(1)(3)(5)$$

$$= 30.$$

Using Elementary Operations to Evaluate Determinants

The effect of each elementary row (or column) operation on the value of a determinant is given in Table 1.

TABLE 1   Elementary Row Operations for Determinants

Elementary Row Operation	Example
1. If two rows of a determinant are interchanged, then the algebraic sign of the determinant changes.	$\begin{vmatrix} 3 & 4 \\ 1 & 5 \end{vmatrix} = -\begin{vmatrix} 1 & 5 \\ 3 & 4 \end{vmatrix}$
2. If every entry in one row of a determinant is multiplied by a constant k, the effect is to multiply the value of the determinant by k.	$\begin{vmatrix} 3k & -2k \\ 4 & 7 \end{vmatrix} = k\begin{vmatrix} 3 & -2 \\ 4 & 7 \end{vmatrix}$
3. If one row of a determinant is a multiple of another row, the value of the determinant is 0.	$\begin{vmatrix} 3 & -4 \\ 6 & -8 \end{vmatrix} = \begin{vmatrix} 3 & -4 \\ 2(3) & 2(-4) \end{vmatrix} = 0$
4. If a constant multiple of one row of a determinant is added to any other row, the value of the determinant will not change.	$\begin{vmatrix} -3 & 5 \\ 6 & -4 \end{vmatrix} = \begin{vmatrix} -3 & 5 \\ 6+2(-3) & -4+2(5) \end{vmatrix}$ $= \begin{vmatrix} -3 & 5 \\ 0 & 6 \end{vmatrix}$ $= -18$

The properties of determinants stated above remain true when the word "row" is replaced by the word "column." In this case, we have *elementary column operations*.

For example,

$$\begin{vmatrix} 1 & 8 & 4 \\ 0 & 2 & 7 \\ 0 & 0 & 5 \end{vmatrix} = 2 \begin{vmatrix} 1 & 4 & 4 \\ 0 & 1 & 7 \\ 0 & 0 & 5 \end{vmatrix}$$

$$= 2(1)(1)(5)$$

$$= 10.$$

EXAMPLE 4   **Evaluating a 4 × 4 Determinant**

Evaluate the determinant by using elementary row operations.

$$\begin{vmatrix} 1 & 0 & -2 & 0 \\ 0 & 3 & 0 & 2 \\ 1 & 0 & 3 & 1 \\ 0 & 4 & 0 & 5 \end{vmatrix}$$

Solution

$$\begin{vmatrix} 1 & 0 & -2 & 0 \\ 0 & 3 & 0 & 2 \\ 1 & 0 & 3 & 1 \\ 0 & 4 & 0 & 5 \end{vmatrix} = \begin{vmatrix} 1 & 0 & -2 & 0 \\ 0 & 3 & 0 & 2 \\ 0 & 0 & 5 & 1 \\ 0 & 4 & 0 & 5 \end{vmatrix}$$

Replace row 3 by the sum of -1 times row 1 and row 3:
$$(-1)R_1 + R_3 \rightarrow R_3$$
(no change in value, row operation 4)

$$= - \begin{vmatrix} 1 & 0 & -2 & 0 \\ 0 & 3 & 0 & 2 \\ 0 & 4 & 0 & 5 \\ 0 & 0 & 5 & 1 \end{vmatrix}$$

Interchange rows 3 and 4:
$$R_3 \leftrightarrow R_4$$
(change sign, row operation 1)

$$= -3 \begin{vmatrix} 1 & 0 & -2 & 0 \\ 0 & 1 & 0 & \frac{2}{3} \\ 0 & 4 & 0 & 5 \\ 0 & 0 & 5 & 1 \end{vmatrix}$$

Factor 3 from row 2 (row operation 2)

$$= -3 \begin{vmatrix} 1 & 0 & -2 & 0 \\ 0 & 1 & 0 & \frac{2}{3} \\ 0 & 0 & 0 & \frac{7}{3} \\ 0 & 0 & 5 & 1 \end{vmatrix}$$

Replace row 3 by the sum of -4 times row 2 and row 3:
$$-4R_2 + R_3 \rightarrow R_3$$
(no change, row operation 4)

$$= 3 \begin{vmatrix} 1 & 0 & -2 & 0 \\ 0 & 1 & 0 & \frac{2}{3} \\ 0 & 0 & 5 & 1 \\ 0 & 0 & 0 & \frac{7}{3} \end{vmatrix}$$

Interchange rows 3 and 4:
$$R_3 \leftrightarrow R_4$$
(change sign, row operation 1)

$$= 3(1)(1)(5)\left(\frac{7}{3}\right) = 35$$

Multiply diagonal entries (triangular matrix)

Using Cramer's Rule to Solve Linear Systems

Determinants can be used to solve a linear system with n linear equations and n variables by a technique known as *Cramer's rule*. Here, we study this rule for systems that have two equations and two unknowns, and for systems that have three equations and three unknowns. Cramer's rule enables us to use determinants to find the solution when the system has a unique solution. (The system is consistent and independent.)

Cramer's Rule for 2 × 2 Systems of Equations

The system

$$\begin{cases} ax + by = h \\ cx + dy = k \end{cases}$$

has a unique solution given by

$$x = \frac{D_x}{D} \quad \text{and} \quad y = \frac{D_y}{D}$$

where

$$D = \begin{vmatrix} a & b \\ c & d \end{vmatrix} \neq 0, \quad D_x = \begin{vmatrix} h & b \\ k & d \end{vmatrix}, \quad \text{and} \quad D_y = \begin{vmatrix} a & h \\ c & k \end{vmatrix}.$$

If $D = 0$, then the system is either inconsistent (it has no solution) or consistent and dependent (it has infinitely many solutions).

EXAMPLE 5 **Using Cramer's Rule to Solve a System of Two Equations**

Solve the system by using Cramer's rule.

$$\begin{cases} 5x - 2y = 8 \\ 3x + 4y = 10 \end{cases}$$

Solution The determinant of the coefficient matrix is given by

$$D = \begin{vmatrix} 5 & -2 \\ 3 & 4 \end{vmatrix} = 20 + 6 = 26.$$

Because $D \neq 0$, we can solve the system by applying Cramer's rule as follows:

$$D_x = \begin{vmatrix} 8 & -2 \\ 10 & 4 \end{vmatrix} = 32 + 20 = 52$$

and

$$D_y = \begin{vmatrix} 5 & 8 \\ 3 & 10 \end{vmatrix} = 50 - 24 = 26.$$

So

$$x = \frac{D_x}{D} = \frac{52}{26} = 2 \quad \text{and} \quad y = \frac{D_y}{D} = \frac{26}{26} = 1.$$

That is, the solution is (2, 1).

Cramer's Rule for
3 × 3 Systems
of Equations

The solution to the system

$$\begin{cases} a_1x + a_2y + a_3z = k_1 \\ b_1x + b_2y + b_3z = k_2 \\ c_1x + c_2y + c_3z = k_3 \end{cases}$$

is given by

$$x = \frac{D_x}{D}, \quad y = \frac{D_y}{D}, \quad \text{and} \quad z = \frac{D_z}{D}$$

where

$$D = \begin{vmatrix} a_1 & a_2 & a_3 \\ b_1 & b_2 & b_3 \\ c_1 & c_2 & c_3 \end{vmatrix} \neq 0 \qquad D_x = \begin{vmatrix} k_1 & a_2 & a_3 \\ k_2 & b_2 & b_3 \\ k_3 & c_2 & c_3 \end{vmatrix}$$

$$D_y = \begin{vmatrix} a_1 & k_1 & a_3 \\ b_1 & k_2 & b_3 \\ c_1 & k_3 & c_3 \end{vmatrix} \qquad D_z = \begin{vmatrix} a_1 & a_2 & k_1 \\ b_1 & b_2 & k_2 \\ c_1 & c_2 & k_3 \end{vmatrix}.$$

EXAMPLE 6

Using Cramer's Rule to Solve a System of Three Equations and Three Unknowns

Solve the system with Cramer's rule.

$$\begin{cases} x + y + z = 4 \\ 14x + 13y + 15z = 55 \\ 2x - y = 0 \end{cases}$$

Solution

Using Cramer's rule, we have

$$D = \begin{vmatrix} 1 & 1 & 1 \\ 14 & 13 & 15 \\ 2 & -1 & 0 \end{vmatrix} = 5 \qquad D_x = \begin{vmatrix} 4 & 1 & 1 \\ 55 & 13 & 15 \\ 0 & -1 & 0 \end{vmatrix} = 5$$

$$D_y = \begin{vmatrix} 1 & 4 & 1 \\ 14 & 55 & 15 \\ 2 & 0 & 0 \end{vmatrix} = 10 \qquad D_z = \begin{vmatrix} 1 & 1 & 4 \\ 14 & 13 & 55 \\ 2 & -1 & 0 \end{vmatrix} = 5.$$

So

$$x = \frac{D_x}{D} = \frac{5}{5} = 1, \quad y = \frac{D_y}{D} = \frac{10}{5} = 2, \quad \text{and} \quad z = \frac{D_z}{D} = \frac{5}{5} = 1.$$

Thus the solution is (1, 2, 1).

When systems of linear equations have decimal coefficients, we can use a grapher to evaluate the determinants used in Cramer's rule. For example, to solve the following system by Cramer's rule

$$\begin{cases} 4.32x - 2.01y + 0.63z = 3.7938 \\ 1.71x + 6.02y + 7.10z = 6.2912 \\ 2.81x - 6.28y + 0.25z = 1.5377 \end{cases}$$

we find the determinant of the following coefficient matrix:

$$D = \begin{vmatrix} 4.32 & -2.01 & 0.63 \\ 1.71 & 6.02 & 7.10 \\ 2.81 & -6.28 & 0.25 \end{vmatrix}$$

to get

$$D = 142.456875.$$

Since $D \neq 0$, we proceed with the calculations of D_x, D_y, and D_z. We obtain

$$D_x = \begin{vmatrix} 3.7938 & -2.01 & 0.63 \\ 6.2912 & 6.02 & 7.10 \\ 1.5377 & -6.28 & 0.25 \end{vmatrix} = 125.36205,$$

$$D_y = \begin{vmatrix} 4.32 & 3.7938 & 0.63 \\ 1.71 & 6.2912 & 7.10 \\ 2.81 & 1.5377 & 0.25 \end{vmatrix} = 24.21766875, \quad \text{and}$$

$$D_z = \begin{vmatrix} 4.32 & -2.10 & 3.7938 \\ 1.71 & 6.02 & 6.2912 \\ 2.81 & -6.28 & 1.5377 \end{vmatrix} = 74.1477513.$$

By Cramer's rule, we get the approximate values

$$x = \frac{D_x}{D} = 0.88, \quad y = \frac{D_y}{D} = 0.17, \quad \text{and} \quad z = \frac{D_z}{D} = 0.52.$$

Solving Applied Problems

Cramer's rule is a valuable tool in advanced theoretical and applied mathematics.

EXAMPLE 7   **Writing an Equation of a Line in Determinant Form**

Find an equation of the nonvertical line containing the points (x_1, y_1) and (x_2, y_2), then express this formula in determinant form.

Solution   The slope m of the line is given by

$$m = \frac{y_2 - y_1}{x_2 - x_1}.$$

Using a point–slope form for the equation of a line, we get

$$y - y_1 = m(x - x_1)$$

or

$$y - y_1 = \frac{y_2 - y_1}{x_2 - x_1}(x - x_1).$$

To clear the fraction, we multiply each side of this latter equation by $x_2 - x_1$ to obtain

$$(x_2 - x_1)(y - y_1) = (y_2 - y_1)(x - x_1).$$

Multiplying, we have

$$x_2 y - x_2 y_1 - x_1 y + x_1 y_1 = y_2 x - y_2 x_1 - y_1 x + y_1 x_1$$

or

$$x_2 y - x_2 y_1 - x_1 y + x_1 y_1 - y_2 x + y_2 x_1 + y_1 x - y_1 x_1 = 0.$$

That is,

$$x(y_1 - y_2) - y(x_1 - x_2) + (x_1 y_2 - x_2 y_1) = 0.$$

Each of the three parts of this equation defines a 2×2 determinant, which enables us to rewrite this equation as

$$x \begin{vmatrix} y_1 & 1 \\ y_2 & 1 \end{vmatrix} - y \begin{vmatrix} x_1 & 1 \\ x_2 & 1 \end{vmatrix} + \begin{vmatrix} x_1 & y_1 \\ x_2 & y_2 \end{vmatrix} = 0$$

But the left side of this equation is an expansion of the 3×3 determinant

$$\begin{vmatrix} x & y & 1 \\ x_1 & y_1 & 1 \\ x_2 & y_2 & 1 \end{vmatrix}.$$

So we can rewrite the equation of the line as

$$\begin{vmatrix} x & y & 1 \\ x_1 & y_1 & 1 \\ x_2 & y_2 & 1 \end{vmatrix} = 0.$$

For instance, if the points $(2, 3)$ and $(-1, 4)$ are on a line, then the determinant form of the equation of the line is given by

$$\begin{vmatrix} x & y & 1 \\ 2 & 3 & 1 \\ -1 & 4 & 1 \end{vmatrix} = x(-1) - y(3) + 1(11) = 0.$$

So the equation of the line becomes

$$-x - 3y + 11 = 0 \quad \text{or} \quad y = -\frac{1}{3}x + \frac{11}{3}.$$

◈ PROBLEM SET 8.4

Mastering the Concepts

In problems 1–10, evaluate each determinant by using an expansion formula.

1. (a) $\begin{vmatrix} -1 & 3 \\ -7 & 4 \end{vmatrix}$

   (b) $\begin{vmatrix} 2 & 3 \\ 9 & 4 \end{vmatrix}$

2. (a) $\begin{vmatrix} 2 & -1 \\ 3 & 2 \end{vmatrix}$

   (b) $\begin{vmatrix} 6 & 9 \\ 8 & 12 \end{vmatrix}$

3. (a) $\begin{vmatrix} 3 & -1 \\ 2 & -1 \end{vmatrix}$

   (b) $\begin{vmatrix} 4 & 3 \\ 3 & 5 \end{vmatrix}$

4. (a) $\begin{vmatrix} 14 & -1 \\ 15 & 1 \end{vmatrix}$

   (b) $\begin{vmatrix} -2 & 5 \\ 7 & 1 \end{vmatrix}$

5. $\begin{vmatrix} 2 & -1 & 3 \\ 9 & -7 & 4 \\ 11 & -6 & 2 \end{vmatrix}$

6. $\begin{vmatrix} 3 & -1 & 2 \\ 0 & 1 & -5 \\ 6 & 7 & 4 \end{vmatrix}$

7. $\begin{vmatrix} 2 & 2 & 2 \\ 3 & 3 & 3 \\ 4 & 4 & 4 \end{vmatrix}$

8. $\begin{vmatrix} \frac{1}{2} & 4 & 7 \\ 1 & -1 & 2 \\ 3 & 2 & 5 \end{vmatrix}$

9. $\begin{vmatrix} 1 & 0 & 2 & 0 \\ 0 & 1 & 0 & 0 \\ 1 & 0 & 3 & 0 \\ 0 & 0 & 0 & 3 \end{vmatrix}$

10. $\begin{vmatrix} -2 & 0 & 1 & 0 \\ -1 & 3 & 0 & 0 \\ 0 & 0 & 2 & 1 \\ 0 & 0 & 0 & 4 \end{vmatrix}$

In problems 11–16, show why each statement is true—not by evaluating each side, but by citing which row operations have been used.

11. $\begin{vmatrix} 4 & 5 \\ 3 & -2 \end{vmatrix} = -\begin{vmatrix} 3 & -2 \\ 4 & 5 \end{vmatrix}$

12. $\begin{vmatrix} 3 & 0 & 1 \\ 1 & 1 & 2 \\ 3 & 0 & 1 \end{vmatrix} = \begin{vmatrix} 0 & 0 & 0 \\ 1 & 1 & 2 \\ 3 & 0 & 1 \end{vmatrix}$

13. $\begin{vmatrix} 3 & -6 & 2 \\ 5 & -3 & 0 \\ 0 & 9 & 18 \end{vmatrix} = 9\begin{vmatrix} 3 & -6 & 2 \\ 5 & -3 & 0 \\ 0 & 1 & 2 \end{vmatrix}$

14. $\begin{vmatrix} 2 & 4 & 12 \\ -1 & 0 & 3 \\ 1 & 0 & 6 \end{vmatrix} = 18\begin{vmatrix} 1 & 2 & 6 \\ -1 & 0 & 3 \\ 0 & 0 & 1 \end{vmatrix}$

15. $\begin{vmatrix} 1 & 1 & 1 \\ 3 & 3 & 3 \\ 2 & 2 & 2 \end{vmatrix} = 6\begin{vmatrix} 0 & 0 & 0 \\ 0 & 0 & 0 \\ 1 & 1 & 1 \end{vmatrix}$

16. $\begin{vmatrix} 7 & -4 & 1 & 2 \\ 21 & 4 & 3 & -1 \\ -35 & 20 & -5 & -10 \\ 14 & 16 & 8 & -2 \end{vmatrix} = 0$

In problems 17–22, use the properties of determinants to evaluate.

17. $\begin{vmatrix} -1 & 0 & 2 \\ 0 & 0 & 0 \\ -1 & 5 & 1 \end{vmatrix}$

18. $\begin{vmatrix} 3 & 1 & 1 \\ -1 & 0 & 3 \\ 2 & 1 & 1 \end{vmatrix}$

19. $\begin{vmatrix} 2 & 1 & 3 \\ 1 & 2 & 1 \\ 4 & 0 & 0 \end{vmatrix}$

20. $\begin{vmatrix} 20 & 12 & 8 \\ 5 & 3 & 2 \\ 5 & 7 & 2 \end{vmatrix}$

21. $\begin{vmatrix} 1 & 0 & 2 & 3 \\ 1 & -3 & 0 & 1 \\ 0 & 3 & 1 & -1 \\ 2 & 1 & 2 & -2 \end{vmatrix}$

22. $\begin{vmatrix} -1 & 0 & 1 & 3 \\ 3 & -2 & 3 & 4 \\ 1 & 4 & -3 & 2 \\ 5 & 2 & -1 & 1 \end{vmatrix}$

In problems 23–36, use Cramer's rule to solve each system.

23. $\begin{cases} 2x - y = 0 \\ x + y = 1 \end{cases}$

24. $\begin{cases} -3x + y = 3 \\ -2x - y = -5 \end{cases}$

25. $\begin{cases} 3x + 2y = 7 \\ -2x + 7y = 12 \end{cases}$

26. $\begin{cases} 4x - y = 7 \\ -2x - 3y = 9 \end{cases}$

27. $\begin{cases} x + 4y = -4 \\ 3x - 2y = -19 \end{cases}$

28. $\begin{cases} 3x + y = 1 \\ -9x - 3y = -4 \end{cases}$

29. $\begin{cases} x + y + 2z = 4 \\ x + y - 2z = 0 \\ x - y = 0 \end{cases}$

30. $\begin{cases} 2x_1 - 3x_2 = 4 \\ x_1 - x_2 - 2x_3 = 1 \\ x_1 - x_2 - x_3 = 5 \end{cases}$

31. $\begin{cases} x + y + z = 4 \\ x - y + 2z = 8 \\ 2x + y - z = 3 \end{cases}$

32. $\begin{cases} 2x - 3y - z = 6 \\ x - 2y - 3z = -3 \\ 3x + y - z = 8 \end{cases}$

33. $\begin{cases} 2x_1 + x_2 + x_3 = 3 \\ -x_1 + 2x_2 - x_3 = 1 \\ 3x_1 + x_2 + 2x_3 = -1 \end{cases}$

34. $\begin{cases} 3x - 2y + 2z = 8 \\ x - 5y + 6z = 8 \\ 6x \quad\quad - 8z = 4 \end{cases}$ **35.** $\begin{cases} 5x + y - z = 4 \\ 9x + y - z = 1 \\ x - y + 5z = 2 \end{cases}$

36. $\begin{cases} 3x_1 + 4x_2 + 2x_3 = 1 \\ 4x_1 + 6x_2 + 2x_3 = 7 \\ 2x_1 + 3x_2 + x_3 = 11 \end{cases}$

G In problems 37 and 38, use a grapher to evaluate the determinants for solving each system by using Cramer's rule.

37. $\begin{cases} x + 2y + z - w = 2 \\ 2x + 4y + z + w = 2 \\ 3x + 2y - z - 2w = -4 \\ 2x + 5y + 3z = 7 \end{cases}$

38. $\begin{cases} 3x + y - z + w = 0 \\ 6x + 4y - z + 4w = 15 \\ -3x - y + 2z - w = 1 \\ 3x + 5y + 3w = 21 \end{cases}$

Applying the Concepts

39. Area of a Triangle: Let triangle PQR be located in a Cartesian plane with vertices $P = (x_1, y_1)$, $Q = (x_2, y_2)$, and $R = (x_3, y_3)$. Then the area of the triangle PQR is given by

$$A = \text{Absolute value of } \frac{1}{2} \begin{vmatrix} x_1 & y_1 & 1 \\ x_2 & y_2 & 1 \\ x_3 & y_3 & 1 \end{vmatrix}.$$

Sketch triangle PQR, where $P = (1, -5)$, $Q = (-3, -4)$, and $R = (6, 2)$. Then find its area by using the given formula.

40. Area of a Parallelogram: The determinant

$$\begin{vmatrix} a & b \\ c & d \end{vmatrix}$$

can be interpreted as the area A of the parallelogram having vertices $(0, 0)$, (a, b), and (c, d) as shown in Figure 1. Find the area of the parallelogram with vertices $(0, 0)$, $(5, -2)$, and $(-4, 3)$.

Figure 1

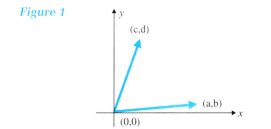

Developing and Extending the Concepts

41. Compute each determinant.

(a) $\begin{vmatrix} 1 - x & x \\ -x & 1 + x \end{vmatrix}$

(b) $\begin{vmatrix} \sqrt{6} & -2\sqrt{5} \\ 3\sqrt{5} & 4\sqrt{6} \end{vmatrix}$

(c) $\begin{vmatrix} \sqrt{5} - \sqrt{2} & 2 + \sqrt{3} \\ 2 - \sqrt{3} & \sqrt{5} + \sqrt{2} \end{vmatrix}$

42. Evaluate each determinant by using the properties of determinants.

(a) $\begin{vmatrix} p & q & r \\ p + 1 & q + 1 & r + 1 \\ p - 1 & q - 1 & r - 1 \end{vmatrix}$

(b) $\begin{vmatrix} p & q & r \\ p + q & 2q & r + q \\ 2 & 2 & 2 \end{vmatrix}$

In problems 41–44, solve for x.

43. (a) $\begin{vmatrix} x & -x \\ 5 & 3 \end{vmatrix} = 2$

(b) $\begin{vmatrix} x & 4 & 5 \\ 0 & 1 & x \\ 5 & 2 & 1 \end{vmatrix} = 7$

44. (a) $\begin{vmatrix} x & 0 & 0 \\ 3 & 1 & 2 \\ 0 & 4 & 1 \end{vmatrix} = 5$

(b) $\begin{vmatrix} 5x & 0 & 1 \\ 2x & 1 & 2 \\ 3x & 2 & 3 \end{vmatrix} = 0$

45. $\begin{vmatrix} x - 3 & -2 \\ 1 & x \end{vmatrix} = 0$

46. $\begin{vmatrix} x - 3 & 2 & 0 \\ 2 & x - 3 & 0 \\ 0 & 0 & x - 5 \end{vmatrix} = 0$

47. Find a, b, and c so that

$$\begin{vmatrix} 3 & -1 & x \\ 2 & 6 & y \\ -5 & 4 & z \end{vmatrix} = ax + by + cz.$$

48. Show that $\begin{vmatrix} 1 & x & x^2 \\ x^2 & 1 & x \\ x & x^2 & 1 \end{vmatrix} = (x^3 - 1)^2.$

49. Let $A = \begin{bmatrix} a & b \\ c & d \end{bmatrix}$.

and assume that $|A| \neq 0$. Then show that

$$A^{-1} = \frac{1}{|A|} \begin{bmatrix} d & -b \\ -c & a \end{bmatrix}.$$

Use this result to find A^{-1} if:

(a) $A = \begin{bmatrix} 5 & 3 \\ 2 & 4 \end{bmatrix}$ **(b)** $A = \begin{bmatrix} 6 & 7 \\ 7 & 8 \end{bmatrix}$

50. Use Cramer's rule to solve the system.

$$\begin{cases} x_1 & - 3x_4 = 1 \\ x_1 + 4x_2 + 2x_3 - 2x_4 = -7 \\ 2x_1 + 5x_2 \quad\quad + x_4 = 11 \\ 2x_1 - 2x_2 + 5x_3 + x_4 = 3. \end{cases}$$

Use the properties of determinants to expand each determinant by converting each to triangular form.

51. Given matrix

$$A = \begin{bmatrix} a_1 & a_2 & a_3 \\ b_1 & b_2 & b_3 \\ c_1 & c_2 & c_3 \end{bmatrix}$$

we adjoin the first two columns of A to A to form

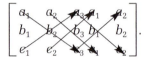

Next, we form the sum D of the products of the diagonal entries denoted by the downward arrows to get

$$D = a_1 b_2 c_3 + a_2 b_3 c_1 + a_3 b_1 c_2.$$

Similarly, we form the sum V of the products of the diagonal entries denoted by the upward arrows to get

$$V = c_1 b_2 a_3 + c_2 b_3 a_1 + c_3 b_1 a_2.$$

Explain why $|A| = D - V$.

52. Use the technique given in problem 51 to find $|A|$ if

(a) $A = \begin{bmatrix} 3 & 2 & 7 \\ -1 & 5 & 3 \\ 2 & -3 & -6 \end{bmatrix}$ (see Example 2)

(b) $A = \begin{bmatrix} 1 & 2 & 3 \\ -1 & 4 & -1 \\ 2 & 5 & -7 \end{bmatrix}$

In problems 53 and 54, use the property that the inverse of an $n \times n$ matrix A exists if and only if the value of its determinant $|A| \neq 0$ to determine whether the given matrix has an inverse.

53. (a) $\begin{bmatrix} 3 & 9 \\ 2 & 6 \end{bmatrix}$ **54.** (a) $\begin{bmatrix} 1 & -1 \\ 1 & 1 \end{bmatrix}$

(b) $\begin{bmatrix} 1 & 1 & 1 \\ 2 & -4 & -3 \\ 1 & -1 & 0 \end{bmatrix}$ (b) $\begin{bmatrix} 3 & 2 & 1 \\ 1 & -3 & 1 \\ 4 & -1 & 2 \end{bmatrix}$

Objectives

1. Graph Inequalities in Two Variables
2. Graph Systems of Inequalities
3. Use Linear Programming
4. Solve Applied Problems

8.5 Solutions of Linear Systems of Inequalities and Linear Programming

In this section, we consider systems of inequalities in two variables. Such systems are represented graphically as regions in the Cartesian plane. We also introduce *linear programming,* which is used to model a variety of situations in areas such as health, transportation, economics, engineering, and science.

Graphing Inequalities in Two Variables

Examples of *linear inequalities* in two variables are

$$y < 8x + 1, \quad 2x + y \leq 3, \quad \text{and} \quad 5x - 3y > 1.$$

As with equations in two variables, a *solution* of an inequality involving two variables is an ordered pair of numbers, which, when substituted for x and y, makes the inequality true.

For example, $(1, 2)$ is a solution of the inequality $y < 8x + 1$, since $2 < 8(1) + 1$ or $2 < 9$ is true.

The set of all ordered pairs that are solutions of an inequality is called its *solution set.* One way to identify the solution set of an inequality is to show its graph, which is the set of all points (x, y) in the xy plane whose coordinates satisfy the inequality.

To understand how to graph such an inequality, let us consider $y < 8x + 1$. We begin by graphing the linear equation $y = 8x + 1$ (Figure 1a). A dashed line is used to emphasize that points on this *boundary line* do not satisfy $y < 8x + 1$. Next we select a *test point* in one of the *two half-planes* defined by the line.

If the coordinates of the point satisfy the original inequality, then all points in that half-plane also satisfy it. If the coordinates do not satisfy the inequality, then *none* of the others do either.

Suppose we select $(0, 0)$ as the test point for $y < 8x + 1$ (Figure 1b). Substituting $x = 0$ and $y = 0$ into the inequality $y < 8x + 1$, we get $0 < 8(0) + 1$ or $0 < 1$, which is true. So all points in the half-plane containing $(0, 0)$ represent the solution set of the inequality. Consequently, the graph of $y < 8x + 1$ is the shaded region in Figure 1c.

Figure 1

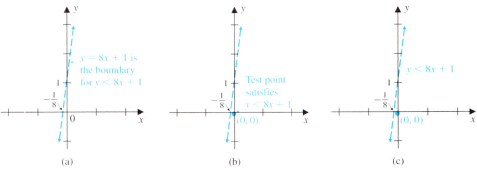

(a)   (b)   (c)

This illustration leads us to the following general procedure for graphing a linear inequality in two variables:

Procedure for Graphing a Linear Inequality in Two Variables

Step 1   Graph the associated linear equation. Use a dashed line as a boundary line if it is a *strict* inequality ($<$ or $>$) or use a solid line if it is not a strict inequality ($\leq$ or $\geq$).

Step 2   Pick a test point in either of the half-planes defined by the boundary line. Determine whether the coordinates satisfy the inequality.

Step 3   Shade the half-plane containing the test point if it satisfies the inequality; otherwise, shade the other half-plane.

The graph of the inequality consists of the shaded region from Step 3 and the line obtained in Step 1 if it is not a strict inequality.

EXAMPLE 1   **Graphing a Linear Inequality**

Graph the inequality $2x + y \leq 3$.

Solution   We use the general procedure given above to graph the inequality.

Step 1.   We begin by graphing the boundary line $2x + y = 3$.
In this case, we use a solid line to indicate that points on the line satisfy the given inequality (Figure 2a).

Figure 2

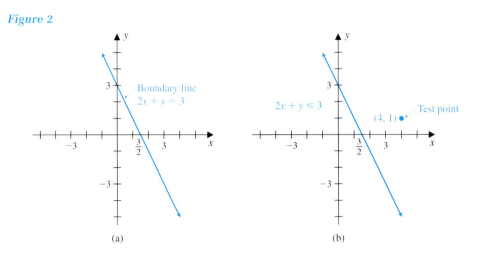

(a)   (b)

Step 2. We select a test point, say (4, 1), in the half-plane above the line (Figure 2b). Substituting $x = 4$ and $y = 1$ into the inequality, we get

$$2(4) + 1 \le 3 \text{ or } 9 \le 3, \text{ which is false.}$$

Step 3. Shade the other half-plane not containing the point (4, 1), that is, shade the half-plane below the line (Figure 2b).

The graph consists of the shaded region and the boundary line.

Graphing Systems of Inequalities

A *system of inequalities* in two variables is a collection of two or more inequalities in two variables. The solution of such a system is the set of all ordered pairs of numbers that satisfy all the inequalities simultaneously. The graph of such a system is obtained by finding the region of points common to the graphs of the individual inequalities in the system.

EXAMPLE 2 **Graphing a Linear System of Inequalities**

Graph the system of inequalities and identify where the boundary lines intersect.

$$\begin{cases} x + y \le 5 \\ -x + 2y > 4 \end{cases}$$

Solution Figure 3a shows the graph of the inequality $x + y \le 5$, whereas Figure 3b shows the graph of $-x + 2y > 4$. The graph of the system consists of all points common to the two graphs. It is shown as the shaded region in Figure 3c.

Figure 3

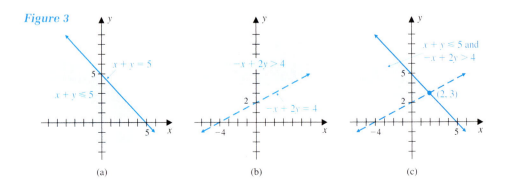

(a)   (b)   (c)

By solving the system of equations

$$\begin{cases} x + y = 5 \\ -x + 2y = 4 \end{cases}$$

we find that the boundary lines intersect at the point $(2, 3)$. Note that $(2, 3)$ is not in the graph because $(2, 3)$ does not satisfy the second inequality in the given system of inequalities.

EXAMPLE 3   **Graphing a System of Four Inequalities**

Graph the system of inequalities and identify the points of intersections of the boundary lines.

$$\begin{cases} -x + 2y \leq 6 \\ x + y \leq 4 \\ x \geq 0 \\ y \geq 0 \end{cases}$$

Solution   The graph of the first inequality consists of all points in the plane on or below the line $-x + 2y = 6$.

The graph of the second inequality consists of all points in the plane on or below the line $x + y = 4$.

The graph of $x \geq 0$ consists of all points on or to the right of the y axis whereas the graph of the inequality $y \geq 0$ includes all points in the plane on or above the x axis.

Consequently, the graph of the given system is the shaded region in the plane where all four half-planes intersect (Figure 4).

The four points of intersection of the lines are found by solving each of the linear systems

Figure 4

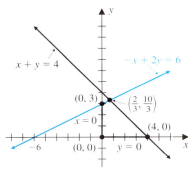

$$\begin{cases} x = 0 \\ y = 0 \end{cases} \quad \begin{cases} x + y = 4 \\ y = 0 \end{cases} \quad \begin{cases} -x + 2y = 6 \\ x + y = 4 \end{cases} \text{ and } \begin{cases} -x + 2y = 6 \\ x = 0 \end{cases}$$

to get

$(0, 0)$, $(4, 0)$, $(2/3, 10/3)$, and $(0, 3)$, respectively.

The points of intersection in Figure 4 are called *vertices* of the region.

Using Linear Programming

Systems of linear inequalities are applied in a field of mathematics called **linear programming.** To introduce this important topic, let's consider an example.

Suppose we are given

$$F = 6x + 3y$$

where x and y satisfy the system of four inequalities in Example 3. Now we ask:

Is it possible to find the maximum (largest) or the minimum (smallest) value of F under the restrictions imposed on x and y by the given system of inequalities?

In this context, we refer to the inequalities as **constraints** because they constrain the choice of x and y, and we call the solution of the linear system of inequalities the

feasible region. Our problem is to choose the point (x, y) of the feasible region so that the expression F has a maximum or minimum value. A problem of this sort is called a **linear programming problem,** and the expression F is called the **objective function.**

To determine which ordered pair of the feasible region gives the maximum or minimum value of F, we use the following general procedure:

General Procedure for Solving a Linear Programming Problem

Step 1 Find the feasible region; that is, graph the system of inequalities defined by the constraints.

Step 2 Determine the vertices (corner points) of the feasible region found in Step 1.

Step 3 Evaluate the objective function at each vertex of the feasible region.

Step 4 The maximum or minimum value of the objective function occurs at these vertices.

For our illustration, each vertex of the feasible region (see Figure 4), along with the values of F, are listed in Table 1. From the table, we see that the maximum value of F is 24, and it occurs at the point $(4, 0)$. The minimum value is 0, and it occurs at $(0, 0)$.

TABLE 1

Vertex	$F = 6x + 3y$	Value
$(0, 0)$	$F = 6(0) + 3(0)$	0 (Minimum)
$(4, 0)$	$F = 6(4) + 3(0)$	24 (Maximum)
$(2/3, 10/3)$	$F = 6(2/3) + 3(10/3)$	14
$(0, 3)$	$F = 6(0) + 3(3)$	9

EXAMPLE 4   **Solving a Linear Programming Problem**

Find the maximum and minimum values of $F = 4x + 6y$ subject to the constraints

$$\begin{cases} x - y \le 4 \\ 2x - y \ge 4 \\ x - 5y \le 2 \\ x \le 5. \end{cases}$$

Solution   Figure 5 shows the feasible region. Table 2 lists the systems and corresponding vertices that result from solving pairs of equations. Table 3 shows the values of F at these vertices.

From Table 3, we see that the maximum value of F is 56, and it occurs at $(5, 6)$. The minimum is 8, and it occurs at $(2, 0)$.

Figure 5

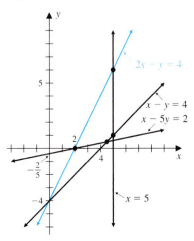

TABLE 2

System	Solution
$\begin{cases} x - 5y = 2 \\ 2x - y = 4 \end{cases}$	$(2, 0)$
$\begin{cases} x - 5y = 2 \\ x - y = 4 \end{cases}$	$\left(\dfrac{9}{2}, \dfrac{1}{2}\right)$
$\begin{cases} x - y = 4 \\ x = 5 \end{cases}$	$(5, 1)$
$\begin{cases} 2x - y = 4 \\ x = 5 \end{cases}$	$(5, 6)$

TABLE 3

Vertex	$F = 4x + 6y$	Value
$(2, 0)$	$F = 4(2) + 6(0)$	8
$(9/2, 1/2)$	$F = 4(9/2) + 6(1/2)$	21
$(5, 1)$	$F = 4(5) + 6(1)$	26
$(5, 6)$	$F = 4(5) + 6(6)$	56

G Graphers with a SHADE feature are helpful in graphing solution sets of linear inequalitites. We use a grapher to display the solution set of the system.

$$\begin{cases} x + y \geq 2 \\ -x + y \leq 2 \end{cases}$$

Figure 6

by first solving for y in terms of x in each in-equality. Then we graph the system in the resulting equivalent form.

$$\begin{cases} y \geq -x + 2 \\ y \leq x + 2 \end{cases}$$

Using the SHADE feature, the solution set is displayed in Figure 6.

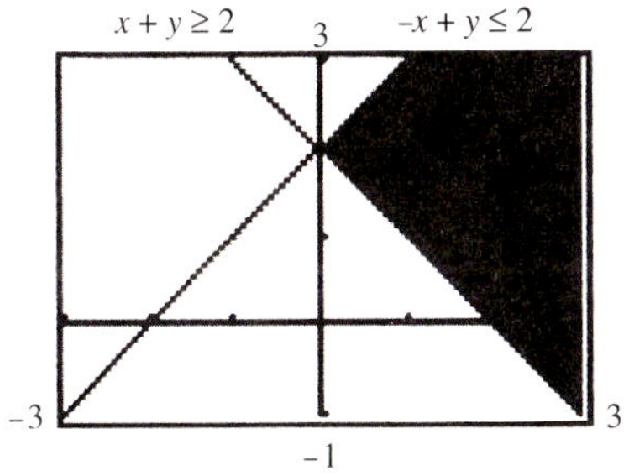

Solving Applied Problems

Systems of inequalities are used to represent applications involving restrictions on the resources available in areas such as business and manufacturing.

EXAMPLE 5 **Using a System of Inequalities to Model Manufacturing Restrictions**

An electronics company manufactures two types of calculators, a scientific calculator and a grapher. In one week, the manufacturer can produce a maximum of 450 calculators and a maximum of 360 graphers.

Suppose that the manufacturer produces at least twice as many graphers as scientific calculators.

(a) Find the constraints; that is, form a system of inequalities that represents the restrictions on the number of calculators of each type to be manufactured.

(b) Graph the system and interpret it.

Solution   (a) Let x represent the number of graphers produced in one week, and let y represent the number of scientific calculators produced in one week.

Clearly, $x \geq 0$ and $y \geq 0$. So the system of inequalities that describes the restrictions on x and y is given by

Figure 7

$$\begin{cases} x \geq 0,\ y \geq 0 \\ x \leq 360 \\ x + y \leq 450 \\ x \geq 2y. \end{cases}$$

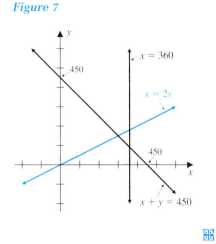

(b) The shaded region in Figure 7 shows the graph of the system of inequalities. Any point in the shaded region satisfies the restrictions. However, since the company cannot produce a fraction of a calculator, only points with nonnegative integer coordinates make sense. For example, (295.5, 143.5) is a point in the region but is not a realistic option.

The technique of linear programming uses systems of linear inequalities as restrictions to model quantities to be maximized or minimized, such as profit or cost.

EXAMPLE 6   **Using Linear Programming to Maximize Profit**

Suppose that the electronics company in Example 5 sells each grapher for $85 and each scientific calculator for $30. Assume it costs the manufacturer $55 to produce each grapher and $20 to produce each scientific calculator.

(a) Find the profit function F.

(b) Given the constraints in Example 5, find the number of calculators of each type that would have to be produced and sold in one week in order to maximize the profit.

Solution   (a) From Example 5, we are using x to represent the number of graphers produced and sold in one week, and y to represent the number of scientific calculators produced and sold. The profit for each grapher is $85 − $55 = $30, and the profit for each scientific calculator is given by $30 − $20 = $10. So the total profit is given by the objective function

$$F = 30x + 10y$$

Figure 8

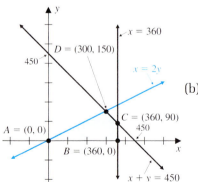

where x and y are subject to the constraints in Example 5.

(b) Figure 8 shows that the four vertices of the feasible region are

$$A = (0, 0),\ B = (360, 0),\ C = (360, 90),\ \text{and}\ D = (300, 150).$$

Table 4 lists the values of F at each vertex:

TABLE 4

Vertex	$F = 30x + 10y$	Value of F
$A = (0, 0)$	$F = 30(0) + 10(0)$	0
$B = (360, 0)$	$F = 30(360) + 10(0)$	10,800
$C = (360, 90)$	$F = 30(360) + 10(90)$	11,700
$D = (300, 150)$	$F = 30(300) + 10(150)$	10,500

From Table 4, we see that the maximum value of F is 11,700, and it occurs at the point (360, 90). Thus the maximum profit is $11,700, and it occurs when 360 graphers and 90 scientific calculators are produced and sold.

 **PROBLEM SET 8.5**

Mastering the Concepts

In problems 1–12, graph each inequality.

1. $y \geq -\dfrac{1}{2}x$

2. $3y \leq 2x + 1$

3. $2x - 3y < 0$

4. $y > 2 - x$

5. $2x - 3y < 6$

6. $2x + 3y \geq 6$

7. $x - 2y \leq 1$

8. $y + 2x \geq 5$

9. $y \leq -3$ and $x > 4$

10. $y \geq 2$ and $x \leq -1$

11. $2x + 3y < 1$

12. $3x + 4y \geq 1$

In problems 13–16, find the vertices of each shaded region.

13.

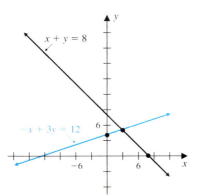

14.

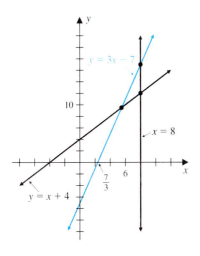

15.

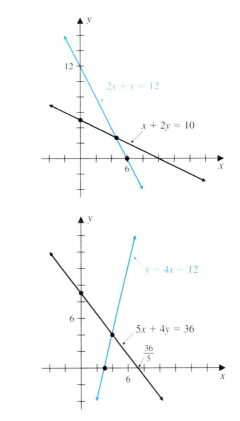

16.

In problems 17–28, graph each system of inequalities, and determine where the boundary lines intersect.

17. $\begin{cases} y \leq -5x \\ x \geq 1 \end{cases}$

18. $\begin{cases} y \geq 2x \\ x \leq -2 \end{cases}$

19. $\begin{cases} y \geq x \\ y \leq 2 \end{cases}$

20. $\begin{cases} y \leq x \\ y \geq -1 \end{cases}$

21. $\begin{cases} x + y \leq 2 \\ y - 1 > 2x \end{cases}$

22. $\begin{cases} y - x < -1 \\ 3y - x > 4 \end{cases}$

23. $\begin{cases} 5x - 2y < -10 \\ x \geq -4 \end{cases}$

24. $\begin{cases} 2x - y > 0 \\ x - 9y \leq 0 \end{cases}$

25. $\begin{cases} y \geq x \\ y \geq -x \end{cases}$

26. $\begin{cases} x + 2y \geq 8 \\ x + 4y \geq 12 \\ x \geq 2 \end{cases}$

27. $\begin{cases} x - y < 4 \\ 2x + y < 12 \\ x \ge 0, y \ge 0 \end{cases}$ **28.** $\begin{cases} 5x + 2y \ge -10 \\ -2x + 5y \le -10 \\ x \quad\quad \le 1 \end{cases}$

In problems 29–32, find the maximum and minimum value of each objective function subject to the given constraints.

29. $F = 2x + 3y$; a linearly bounded region with vertices $(0, 0), (3, 0), (4, 4), (0, 9)$

30. $F = 3x + 5y$; a linearly bounded region with vertices $(4, 1), (8, 1), (6, 5), (1, 4)$

31. $F = 7x + 2y$; the region in Figure 9

Figure 9

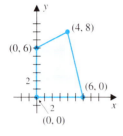

32. $F = 4x - 8y$; the region in Figure 10.

Figure 10

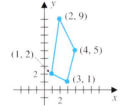

In problems 33–40, find the maximum and minimum value of each objective function subject to the given constraints.

33. $F = 3x + 5y$

$\begin{cases} x \ge 0, y \ge 0 \\ 2x + y \le 6 \end{cases}$

34. $F = 2x - y$

$\begin{cases} x + y \ge 3 \\ x \le 3 \\ y \le 3 \end{cases}$

35. $F = 2x + y$

$\begin{cases} x \ge 0, y \ge 0 \\ 4x + y \le 36 \\ 4x + 3y \le 60 \end{cases}$

36. $F = 15x + 25y$

$\begin{cases} x \ge 0, y \ge 0 \\ x + y \le 50 \\ 2x - y \le 40 \\ -3x + y \le 10 \end{cases}$

37. $F = 7x - 3y$

$\begin{cases} x \ge 0 \\ y \le 4 \\ x + y \ge 1 \\ x - y \le 1 \end{cases}$

38. $F = 5x + 2y$

$\begin{cases} x \ge 0, y \ge 0 \\ x + 3y \le 15 \\ 2x + y \le 10 \end{cases}$

39. $F = 5x + 4y$

$\begin{cases} x \ge 0, y \ge 0 \\ x + 2y \ge 3 \\ 2y \le 5 - x \end{cases}$

40. $F = x + 2y$

$\begin{cases} x \ge 0, y \ge 0 \\ 2x - 3y \ge 6 \\ 2x + y \le 14 \end{cases}$

G In problems 41–44, use a grapher with a **SHADE** feature to display each solution set.

41. $\begin{cases} x + 2y \le 4 \\ 2x - y \ge 3 \end{cases}$ **42.** $\begin{cases} 3x - 2y \le 5 \\ 2x + y \ge 8 \end{cases}$

43. $\begin{cases} 3x - y \le -3 \\ 2x + y \ge -7 \end{cases}$ **44.** $\begin{cases} x + 4y \le 13 \\ 3x - 2y \ge 4 \end{cases}$

Applying the Concepts

45. Investment: A retirement fund uses no more than $200,000 for two investments, one portion at 5% annual simple interest and the rest at 6% annual simple interest. Assume the total interest from both investments for 1 year is at least $10,600.

  (a) Write a system of linear inequalities that describes this situation.

  (b) Graph the system and interpret it.

46. Retailing: An appliance store stocks washers and dryers. The management discovered that due to demand it is necessary to have at least twice as many washers as dryers. Also, at all times, the store must have at least 10 washers and 5 dryers. Due to limitations in space, the store has room for no more than a total of 30 washers and dryers.

  (a) Write a system of inequalities to describe this situation.

  (b) Graph the system and interpret it.

47. Nutrition: A nutritionist wishes to determine a formula for the base of an instant breakfast meal. The breakfast must contain at least 12 grams of protein and 8 grams of carbohydrates. A tablespoon of protein powder made from soybeans has 2 grams of protein and 2 grams of carbohydrates. A tablespoon of protein powder made from milk solids has 2 grams of protein and 4 grams of carbohydrates.

  (a) Determine the feasible region that describes this situation.

  (b) Graph the region and interpret it.

48. Manufacturing: A compact disc manufacturer produces two models of compact disc players, model A and model B. The company allocates at least 560 hours for production and at least 210 hours for assembly. Suppose that it takes

1.4 hours to produce and 0.4 hours to assemble a unit of model *A*, and it takes 1.3 hours to produce and 0.3 hour to assemble a unit of model *B*.

(a) Determine the feasible region that describes this situation.

(b) Graph the region and interpret it.

49. **Maximizing Profit:** Because of limited storage capacity, a restaurant owner can order no more than 200 pounds of ground beef per week for making hamburgers and tacos. Each hamburger contains 1/3 pound of ground beef, whereas each taco contains 1/4 pound. The profit is 50¢ on each hamburger and 65¢ on each taco. The labor cost averages 10¢ for each hamburger and 15¢ for each taco. If the owner is willing to pay at most $300 for labor costs, how many tacos and hamburgers must be sold to maximize profit?

50. **Forestry:** Trees are harvested from a forest by a logging firm that has two kinds of crews. The first crew has 1 driver and 4 loggers. This crew can log 20 trees per day. The second crew has 2 drivers and 6 loggers, and can log 30 trees per day. If the firm employs at most 40 drivers and 150 loggers, how many of each kind of crew would log the maximum number of trees per day?

51. **Oil Refinery:** A refinery produces a combined maximum of 25,000 barrels of gasoline and diesel oil per day, of which no less than 5000 barrels must be diesel oil. If the profit is $31.50 per barrel of gasoline and $24 per barrel of diesel oil, find the maximum profit and the number of barrels of each product that must be produced to yield this maximum.

52. A manager of a racquetball club must order at least 12 racquets and at most 20 cans of balls, but cannot spend more then $450. At least 34 items must be ordered, but the manager cannot order more cans of balls than racquets. If the racquets cost $18 each and a can of balls costs $4.50 each, how many of each should be ordered if the manager intends to minimize the cost?

53. **Manufacturing:** A company manufactures two types of electric toothbrushes, one of which is cordless. The cord-type toothbrush requires 2 hours to make, and the cordless model requires 3 hours. The company has only 800 work-hours to use in manufacturing each day, and the packing department can package only 300 toothbrushes per day. If the company

sells the cord-type model for $15 and the cordless model for $22, how many of each type should it produce per day to maximize its revenue?

54. **Budgeting:** A grower has 70 acres of planting fields in which to grow flowers and eucalyptus. It costs the grower $60 per acre to grow flowers and $30 per acre to grow eucalyptus. The budget for planting is $1800. It takes 3 days to plant an acre of flowers and 4 days to plant an acre of eucalyptus. The grower has a maximum of 120 days to plant the flowers. The flowers will bring a profit of $180 per acre, and the eucalyptus will bring a profit of $100 per acre. How many acres of each crop should be planted to maximize profit?

Developing and Extending the Concepts

55. Graph the inequality
$$|x| - |y| \geq 1.$$

56. Given a linear programming problem with the objective function
$$F = 5x + 2y$$
and constraints
$$\begin{cases} 4x + 3y \leq 15 \\ x - y \geq -6 \\ x \geq 0, y \geq 0 \end{cases}$$
explain which constraints have an effect on the feasible region.

In problems 57 and 58, write a system of inequalities for each solution set shown.

57.

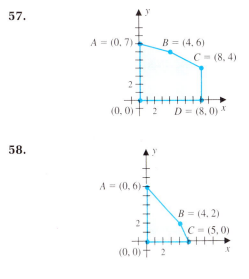

58.

8.6 Solutions of Nonlinear Systems of Equations and Inequalities

So far, the systems of equations and inequalities we have solved have been linear. In this section we consider the solution of systems in which at least one of the equations or inequalities is nonlinear. Some examples are shown below:

$$\begin{cases} x^2 + y = 6 \\ 2x - 5y = -6 \end{cases}, \qquad \begin{cases} x + y \le 16 \\ x^2 + y \ge 4 \end{cases},$$

and

$$\begin{cases} \log(x - 2) + y = 3 \\ \log\left(\dfrac{5x}{6}\right) - y = -1 \end{cases}$$

To solve these systems, we use algebraic methods and graphs.

Solving Nonlinear Systems of Equations Algebraically

At times, we are able to solve nonlinear systems of equations by using algebraic techniques such as substitution and elimination, as we did when solving linear systems.

EXAMPLE 1   **Solving a Nonlinear System by Substitution**

Solve the system below by using substitution.

$$\begin{cases} x^2 + y = 6 \\ 2x - 5y = -6 \end{cases}$$

Solution   We begin by solving the first equation for y to get $y = 6 - x^2$. Since we are trying to find values for y that satisfy both equations simultaneously, we assume that y is the same in both equations.

So we substitute the expression $6 - x^2$ for y in the second equation, $2x - 5y = -6$, to get

$$2x - 5(6 - x^2) = -6$$
$$2x - 30 + 5x^2 = -6 \qquad \text{Simplify}$$
$$5x^2 + 2x - 24 = 0$$

$$(x - 2)(5x + 12) = 0 \qquad \text{Factor}$$

$$x - 2 = 0 \quad \text{or} \quad 5x + 12 = 0 \qquad \text{Solve}$$
$$x = 2 \quad | \quad x = -\frac{12}{5}$$

If we substitute each of these two values of x into the equation

$$2x - 5y = -6 \quad \text{or} \quad y = \frac{2}{5}x + \frac{6}{5}$$

we obtain the two corresponding values for y.

Figure 1

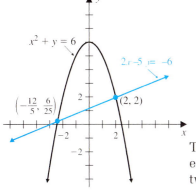

When $x = 2$,

$$y = \frac{2}{5}(2) + \frac{6}{5} = 2.$$

When $x = -\dfrac{12}{5}$,

$$y = \frac{2}{5}\left(-\frac{12}{5}\right) + \frac{6}{5} = \frac{6}{25}.$$

Thus the solutions are $(2, 2)$ and $(-12/5, 6/25)$. Note that the graphs of the two equations, $x^2 + y = 6$ (a parabola) and $2x - 5y = -6$ (a line), show that there are two points of intersection (Figure 1).

The coordinates of these two points are $(2, 2)$ and $\left(-\dfrac{12}{5}, \dfrac{6}{25}\right)$.

When solving a nonlinear system of equations, we may encounter extraneous solutions. This is where graphing the system may help.

EXAMPLE 2   **Solving a Nonlinear System That Produces Extraneous Solutions**

Graph the following equations on the same coordinate system and then solve the system

$$\begin{cases} x^2 + y^2 = 13 \\ x + y = 5. \end{cases}$$

Solution   Figure 2 shows that the circle $x^2 + y^2 = 13$ and the line $x + y = 5$ intersect at two points, so there are two solutions.

To solve this system by substitution, we first solve the equation

$$x + y = 5$$

Figure 2

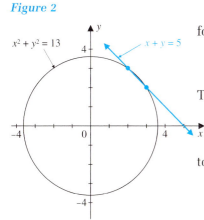

for y in terms of x to get

$$y = 5 - x.$$

Then we substitute $5 - x$ for y in the first equation,

$$x^2 + y^2 = 13,$$

to get

$$
\left.
\begin{aligned}
x^2 + (5 - x)^2 &= 13 \\
x^2 + 25 - 10x + x^2 &= 13 \\
2x^2 - 10x + 12 &= 0 \\
x^2 - 5x + 6 &= 0
\end{aligned}
\right\} \quad \text{Simplify}
$$

$$(x - 2)(x - 3) = 0 \qquad \text{Factor}$$

$$
\left.
\begin{aligned}
x - 2 = 0 \quad \text{or} \quad x - 3 &= 0 \\
x = 2 \qquad\qquad x &= 3
\end{aligned}
\right\} \quad \text{Solve}
$$

If we substitute each of these two values of x into the second equation in the system, $x + y = 5$, we obtain the corresponding values for y.

$$\text{When } x = 2, \quad 2 + y = 5 \quad \text{or} \quad y = 3.$$
$$\text{When } x = 3, \quad 3 + y = 5 \quad \text{or} \quad y = 2.$$

Therefore, the solutions of the system are $(2, 3)$ and $(3, 2)$. These ordered pairs satisfy both equations.

Let us see what happens if we substitute each of the x values, $x = 2$ and $x = 3$, into the first equation in the system, $x^2 + y^2 = 13$, instead of the second equation:

When $x = 2$:	When $x = 3$:
$4 + y^2 = 13$	$9 + y^2 = 13$
$y = \pm 3$	$y = \pm 2$

This suggests that there are four solutions: $(2, 3)$, $(2, -3)$, $(3, 2)$, $(3, -2)$. If we substitute the two suggested additional solutions, $(2, -3)$ and $(3, -2)$, into the second equation, we have

$x + y = 5$	$x + y = 5$
$2 + (-3) = 5$	$3 - 2 = 5$
$-1 = 5$ False	$1 = 5$ False

Thus neither $(2, -3)$ nor $(3, -2)$ is a solution of the original system, a fact confirmed by looking at Figure 2. The pairs $(2, -3)$ and $(3, -2)$ are extraneous solutions.

We now solve a nonlinear system of equations by using elimination. Again, a graph of the system is used to help determine how many points of intersection—and thus how many solutions—there are.

EXAMPLE 3   **Solving a Nonlinear System by Elimination**

Graph both equations on the same coordinate system and use elimination to solve the system

$$\begin{cases} x^2 + y^2 = 16 \\ x^2 + y = 4. \end{cases}$$

Solution   Figure 3 shows that the two curves $x^2 + y = 4$ and $x^2 + y^2 = 16$ intersect at three points, so there are three solutions.

To solve the system, we begin by eliminating the x^2 term as follows:

$$\begin{cases} x^2 + y^2 = 16 \\ x^2 + y = 4 \end{cases} \rightarrow \begin{cases} x^2 + y^2 = 16 \\ -x^2 - y = -4 \end{cases} \rightarrow \begin{cases} y^2 - y = 12 \\ -x^2 - y = -4 \end{cases}$$

$\uparrow$ $\uparrow$

Multiply the second   Add the second equation
equation by -1 to the first equation

Next we solve the first equation $y^2 - y = 12$ in the latter system:

$y^2 - y = 12$	Given
$y^2 - y - 12 = 0$	Subtract 12 from each side
$(y - 4)(y + 3) = 0$	Factor
$y - 4 = 0 \quad \text{or} \quad y + 3 = 0$	Solve
$y = 4 \qquad\qquad y = -3$	

Figure 3

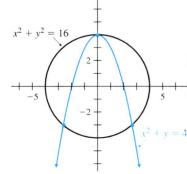

Now we substitute 4 and -3 for y into the second equation in the system, $x^2 + y = 4$, and then solve for x:

$$
\begin{array}{c|c}
\text{For } y = 4: & \text{For } y = -3: \\
x^2 + 4 = 4 & x^2 - 3 = 4 \\
x = 0 & x = \pm\sqrt{7}
\end{array}
$$

Thus $(0, 4)$, $(-\sqrt{7}, -3)$, and $(\sqrt{7}, -3)$ are the three solutions. This is consistent with what was anticipated from the graphs shown in Figure 3.

G When solving nonlinear systems of equations, the algebraic approach does not always work.

In this case, a grapher can help us approximate the solutions. For example, to approximate the solution of the system:

$$
\begin{cases}
y = 4 - x^2 \\
y = \ln x
\end{cases}
$$

to two decimal places. We observe that Figure 4 shows a viewing window of the graphs of the two equations in which they intersect at one point. Using the **INTERSECT**-finding feature of a grapher, we find the solutions to be approximately $(1.84, 0.61)$.

Figure 4

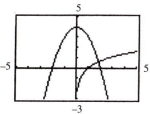

Solving Nonlinear Systems of Inequalities

Figure 5

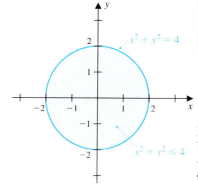

If we had selected (3, 3) as the test point, we would have concluded that the outside region is not part of the solution and so the inside region is the solution.

In Section 8.5, we graphed a linear inequality by first using the graph of the associated linear equation as a boundary line and then testing a point in either of the resulting half-planes to determine which one represents the solution. A similar technique can be used to graph nonlinear inequalities.

For instance, let us consider the inequality

$$
x^2 + y^2 \leq 4.
$$

Figure 5 shows the graph of the associated equation $x^2 + y^2 = 4$ as a solid curve because of the $\leq$ condition. This graph serves as a boundary curve that separates the plane into two regions, one region inside the circle and the other one outside. To test the inside region, we select a point, say $(0, 0)$. Since $0^2 + 0^2 = 0 \leq 4$ is true, this point and all others inside the circle are part of the graph of the given inequality $x^2 + y^2 \leq 4$. Figure 5 shows the graph, which is the shaded circular region, including the circle.

Also in Section 8.5, we found the solution of a system of linear inequalities with two variables by first graphing each of the inequalities in the same coordinate system. The points common to all graphs define the solution of the system. The same approach is used to graph the solution of nonlinear systems of inequalities with two variables.

EXAMPLE 4   **Graphing a Nonlinear System of Inequalities**

Graph the system of inequalities:
$$
\begin{cases}
y > x^2/4 \\
x + y \leq 4.
\end{cases}
$$

Solution   We begin by graphing the solution of each of the two inequalities. Then we determine where those regions overlap to get the graph of the system. Figure 6a shows the graph of the inequality $y > x^2/4$ as the shaded region above the curve $y = x^2/4$.

Figure 6

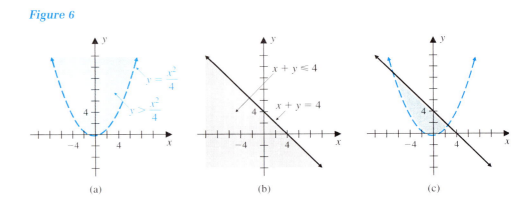

(a)   (b)   (c)

Figure 6b shows the graph of the inequality $x + y \leq 4$ as the shaded region on and below the line $x + y = 4$. Thus the graph of the solution set of the system of inequalities consists of points common to both regions, that is, the shaded region in Figure 6c along with the portion of the line that lies "inside" the parabola.

Solving Applied Problems

We conclude this section with an applied problem that is modeled by a system of nonlinear equations.

EXAMPLE 5   **Determining the Center of an Earthquake**

A seismologist determines that the center of an earthquake is located on a circle 30 miles away from a certain station. From a second station located 40 miles east and 10 miles north of the first, it is determined that the center of the earthquake is on a circle 20 miles away. Find the center of the earthquake, assuming that it is north of both stations. Round off the result to one decimal place.

Solution   First we use a coordinate system to model the situation graphically, as shown in Figure 7. Let the first station be located at the origin. Then the second station is located at the point (40, 10). According to the given conditions, the center of the earthquake lies on both the circle with center at the origin and radius 30 and on the circle with center at (40, 10) and radius 20. The equations of these two circles are, respectively, $x^2 + y^2 = 900$ and $(x - 40)^2 + (y - 10)^2 = 400$.

Figure 7

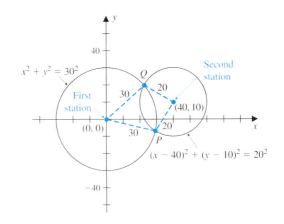

So, to determine where the center of the earthquake is located, we need to find the points of intersection of the two circles. Figure 7 suggests that there are two possible locations of the earthquake's center.

We need to solve the following system to find these locations:

$$(x - 40)^2 + (y - 10)^2 = 20^2$$
$$x^2 + y^2 = 30^2$$

Rewriting these equations, we have

$$\begin{cases} x^2 - 80x + 1600 + y^2 - 20y + 100 = 400 \\ x^2 + y^2 = 900. \end{cases}$$

Adding -1 times the second equation to the first equation, we get

$$\left. \begin{array}{r} -80x + 1600 - 20y + 100 = -500 \\ -80x + 2200 = 20y \\ y = 110 - 4x \end{array} \right\} \qquad \text{Simplify}$$

Substituting $110 - 4x$ for y into the second equation in the given system, $x^2 + y^2 = 900$, we have

$$x^2 + (110 - 4x)^2 = 900$$
$$17x^2 - 880x + 11{,}200 = 0.$$

Using the quadratic formula to solve the latter equation, we get approximately

$$x = 29.2 \text{ or } x = 22.6.$$

For $x = 29.2$, $y = 110 - 4(29.2) = -6.8$.

For $x = 22.6$, $y = 110 - 4(22.6) = 19.6$.

We reject the solution $(29.2, -6.8)$, because the center of the earthquake is located north of both stations and we accept the solution $(22.6, 19.6)$. Thus the center of the earthquake is located about 22.6 miles east and about 19.6 miles north of the first station.

PROBLEM SET 8.6

Mastering the Concepts

In problems 1–6, sketch the graphs of both equations in the system in the same coordinate system. Use the result to determine the number of solutions. Then use substitution to solve the system.

1. $\begin{cases} x^2 - 2y = 0 \\ 3x + 2y = 10 \end{cases}$ **2.** $\begin{cases} x^2 - y = -3 \\ x^2 + y^2 = 9 \end{cases}$

3. $\begin{cases} x - 5y = 0 \\ x^2 + y^2 = 1274 \end{cases}$ **4.** $\begin{cases} x - 2y = 3 \\ x^2 - 40 = -y^2 \end{cases}$

5. $\begin{cases} 2x + y = 10 \\ xy = 12 \end{cases}$ **6.** $\begin{cases} y - 2x = 3 \\ x^2 + y^2 = 16 \end{cases}$

In problems 7–12, sketch the graphs of both equations in the system in the same coordinate system. Use the result to determine the number of solutions. Then use the elimination method to solve the system. Round off to two decimal places when necessary.

7. $\begin{cases} x^2 + y = 13 \\ x^2 + y^2 = 25 \end{cases}$ **8.** $\begin{cases} x^2 + y = 1 \\ y = x^2 - 1 \end{cases}$

9. $\begin{cases} 3x - y^2 = -1 \\ x^2 + y^2 = 5 \end{cases}$ **10.** $\begin{cases} 2x + y = 1 \\ x^2 + y = 4 \end{cases}$

11. $\begin{cases} 3x + 4y = 12 \\ x^2 - y = -1 \end{cases}$ **12.** $\begin{cases} x - 2y^2 = 0 \\ x^2 + y^2 = 1 \end{cases}$

In problems 13–20, use the graphs shown for each system to determine the number of real solutions. Then solve each system by an appropriate algebraic method. Round off to two decimal places when necessary.

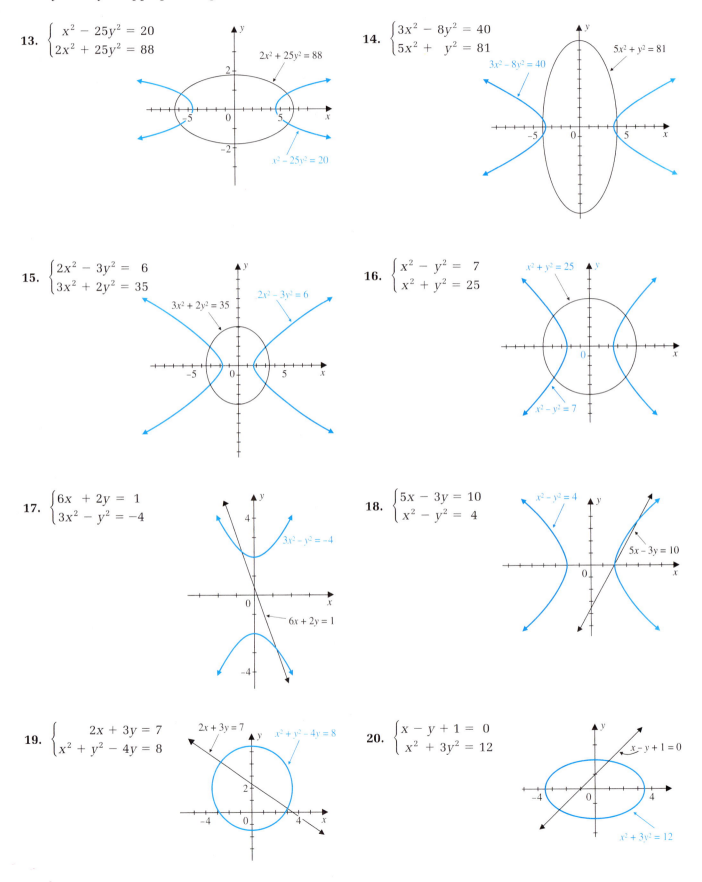

13. $\begin{cases} x^2 - 25y^2 = 20 \\ 2x^2 + 25y^2 = 88 \end{cases}$

$2x^2 + 25y^2 = 88$

$x^2 - 25y^2 = 20$

14. $\begin{cases} 3x^2 - 8y^2 = 40 \\ 5x^2 + y^2 = 81 \end{cases}$

$5x^2 + y^2 = 81$

$3x^2 - 8y^2 = 40$

15. $\begin{cases} 2x^2 - 3y^2 = 6 \\ 3x^2 + 2y^2 = 35 \end{cases}$

$3x^2 + 2y^2 = 35$

$2x^2 - 3y^2 = 6$

16. $\begin{cases} x^2 - y^2 = 7 \\ x^2 + y^2 = 25 \end{cases}$

$x^2 + y^2 = 25$

$x^2 - y^2 = 7$

17. $\begin{cases} 6x + 2y = 1 \\ 3x^2 - y^2 = -4 \end{cases}$

$3x^2 - y^2 = -4$

$6x + 2y = 1$

18. $\begin{cases} 5x - 3y = 10 \\ x^2 - y^2 = 4 \end{cases}$

$x^2 - y^2 = 4$

$5x - 3y = 10$

19. $\begin{cases} 2x + 3y = 7 \\ x^2 + y^2 - 4y = 8 \end{cases}$

$2x + 3y = 7$ $x^2 + y^2 - 4y = 8$

20. $\begin{cases} x - y + 1 = 0 \\ x^2 + 3y^2 = 12 \end{cases}$

$x - y + 1 = 0$

$x^2 + 3y^2 = 12$

In problems 21–34, solve each system.

21. $\begin{cases} u^2 + 9v^2 = 33 \\ u^2 + v^2 = 25 \end{cases}$ **22.** $\begin{cases} x^2 + 5y^2 = 70 \\ 3x^2 - 5y^2 = 30 \end{cases}$

23. $\begin{cases} 4x^2 - y^2 = 4 \\ 4x^2 + \frac{5}{3}y^2 = 36 \end{cases}$ **24.** $\begin{cases} r^2 - 2s^2 = 17 \\ 2r^2 + s^2 = 54 \end{cases}$

25. $\begin{cases} 2x^2 - 3y^2 = 20 \\ x^2 + 2y = 20 \end{cases}$ **26.** $\begin{cases} 4x^2 + 3y^2 = 43 \\ 3x^2 - y^2 = 3 \end{cases}$

27. $\begin{cases} x^2 - 2y^2 = 1 \\ x^2 + 4y^2 = 25 \end{cases}$ **28.** $\begin{cases} 2x^2 - 5y + 8 = 0 \\ x^2 - 7y^2 + 4 = 0 \end{cases}$

29. $\begin{cases} x^2 + 4y = 8 \\ x^2 + y^2 = 5 \end{cases}$ **30.** $\begin{cases} 4x^2 + 7y^2 = 32 \\ -3x^2 + 11y^2 = 41 \end{cases}$

31. $\begin{cases} x^2 + y^2 = 16 \\ x^2 - y^2 = -34 \end{cases}$ **32.** $\begin{cases} x^2 - 4y^2 = 15 \\ -x^2 + 3y^2 = 11 \end{cases}$

33. $\begin{cases} x^2 + y^2 = 25 \\ (x - 5)^2 + y^2 = 9 \end{cases}$ **34.** $\begin{cases} x^2 - y = 0 \\ x^2 + (y - 6)^2 = 36 \end{cases}$

In problems 35–44, graph the solution of each system of inequalities.

35. $\begin{cases} -x^2 + y \geq 0 \\ x + y < 1 \end{cases}$ **36.** $\begin{cases} \sqrt{x} - y > 0 \\ x - 9y \leq 0 \end{cases}$

37. $\begin{cases} x^2 + y \leq 0 \\ x + y > -2 \end{cases}$ **38.** $\begin{cases} y > x - 3 \\ y \leq \sqrt{x - 1} \\ y \geq 0 \end{cases}$

39. $\begin{cases} x \geq 0, y \geq 0 \\ y \leq 4 - x^2 \end{cases}$ **40.** $\begin{cases} x^2 + y^2 \leq 1 \\ x + y > 1 \end{cases}$

41. $\begin{cases} 4x^2 - y^2 \geq 0 \\ x + y < 9 \end{cases}$ **42.** $\begin{cases} y \geq (x - 1)^2 + 2 \\ 2x - 3y \leq -9 \end{cases}$

43. $\begin{cases} x^2 + y^2 \leq 9 \\ y - x^2 \leq 0 \end{cases}$ **44.** $\begin{cases} x^2 + y^2 > 4 \\ x^2 + y^2 < 9 \end{cases}$

G In problems 45–50, use a grapher to solve each system. Round off the answers to two decimal places.

45. $\begin{cases} y = 2 - x^2 \\ y = \log x \end{cases}$ **46.** $\begin{cases} y = e^{2x} \\ y = 1 - x \end{cases}$

47. $\begin{cases} 2^x + y = 3 \\ y = e^x \end{cases}$ **48.** $\begin{cases} y = x - 4 \\ y = x - 3^x \end{cases}$

49. $\begin{cases} y = e^{-2x} \\ y = x^2 \end{cases}$ **50.** $\begin{cases} y = \ln(x + 2) \\ y = x^4 \end{cases}$

Applying the Concepts

In problems 51–58, use a system of equations to solve each problem.

51. Computer Monitor Design: An electronics engineer designs a rectangular computer monitor screen with a 10-inch diagonal and a viewing area of 48 square inches. Find the dimensions of the screen.

52. Template Design: An artist designs a plastic template in the shape of a right triangle with one base length of 60 centimeters and an area of 1500 square centimeters. Find the lengths of the sides of the triangular template.

53. Fencing: A rancher wishes to fence two adjacent rectangular corrals with a total of 100 meters of fence. What are the dimensions of the smallest possible corral, if one corral is a square region and is 144 square meters less than the other?

54. Investment Portfolio: The annual simple interest earned on an investment is $170. If the interest rate had been 1% higher, the interest earned from this investment would have been $238. What was the amount of the investment, and what was the lower interest rate?

55. Engineering Design: An engineer designs a rectangular solar collector with a surface area of 750 square feet. If the length of the collector is 30 times its width, find its length and width.

56. Science: In an experiment with a laser beam, the path of a particle orbiting a central object is given by the equation $36x^2 + 49y^2 = 1764$, where x and y are measured (in centimeters) from the center of the object. Suppose that the laser beam follows a path along the line $y = 2x + 6$ (Figure 8). Find the coordinates of the points at which the laser will illuminate the particle. That is, find the points of intersection when the particle will pass through the beam.

Figure 8

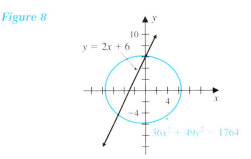

57. Construction: A 2-inch square is cut off at each corner of a rectangular piece of cardboard, and an open-topped box is formed by turning up the sides (Figure 9). Suppose that the resulting volume is 448 cubic inches, and the area of the original rectangular cardboard is 360 square inches. Find the dimensions of the original rectangular piece.

Figure 9

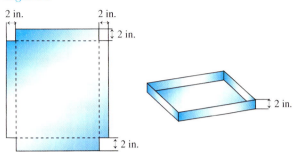

58. Treating a Polluted Lake: In order to cleanse the water in a polluted lake, biologists decided to add two different chemicals. Through laboratory tests, it was determined that the product of the amounts of the two chemicals should be 168 and the sum of their squares should be 340. How many pounds of each chemical are required for this project?

Developing and Extending the Concepts

In problems 59–64, solve each system. [*Hint:* In problem 59, let $u = 1/x^2$ and $v = 1/y^2$. Similar substitutions may be helpful in the remaining problems.]

59. $\begin{cases} \dfrac{3}{x^2} + \dfrac{2}{y^2} = 17 \\ \dfrac{4}{x^2} - \dfrac{5}{y^2} = -8 \end{cases}$

60. $\begin{cases} \dfrac{1}{x^3} - \dfrac{1}{y^3} = 63 \\ \dfrac{1}{x} - \dfrac{1}{y} = 3 \end{cases}$

61. $\begin{cases} \sqrt[3]{x^4} + \sqrt[5]{y^2} = 20 \\ \sqrt[3]{x^2} + \sqrt[5]{y} = 6 \end{cases}$

62. $\begin{cases} x + y = 72 \\ \sqrt[3]{x} + \sqrt[3]{y} = 6 \end{cases}$

63. $\begin{cases} 2x + y = 1 \\ y + z = 1 \\ x^2 + z^2 = 5 \end{cases}$

64. $\begin{cases} x^2 - 3y + 4z^2 = 31 \\ 3x + y = 5 \\ 2x - 3z = -7 \end{cases}$

65. Solve the system $\begin{cases} x_1 - x_2 + x_3 = 2 \\ 2x_2 + x_3 = 1 \\ x_1 x_2 x_3 = 0. \end{cases}$

66. Solve the system $\begin{cases} x^2 + 2y^2 - z^2 = 8 \\ x^2 - y^2 + 3z = 6 \\ y^2 - 5z = -6. \end{cases}$

In problems 67–70, solve each system algebraically.

67. $\begin{cases} 2^x - y = 0 \\ 2^{2x} - y = 2 \end{cases}$

68. $\begin{cases} \log_3(r^2 + s^2) = 2 \\ r^2 - s = 3 \end{cases}$

69. $\begin{cases} x + |y| = 1 \\ -x + 2|y| = 5 \end{cases}$

70. $\begin{cases} 2^{u+v} = 1 \\ 2^{u-v} = 4 \end{cases}$

CHAPTER 8 REVIEW PROBLEM SET

In problems 1–4, solve each system by elimination.

1. $\begin{cases} x - 2y = 3 \\ x + y = -3 \end{cases}$

2. $\begin{cases} 8r - 7s = 28 \\ 5r + 2s = -8 \end{cases}$

3. $\begin{cases} x - y = 3 \\ 2x + y = 3 \end{cases}$

4. $\begin{cases} x - y + 2z = 0 \\ 3x + y + z = 2 \\ 2x - y + 5z = 5 \end{cases}$

In problems 5–8, each row-reduced echelon matrix is the augmented matrix form of a corresponding system of linear equations. Solve each system.

5. $\begin{bmatrix} 1 & 0 & \vdots & -9 \\ 0 & 1 & \vdots & 5 \end{bmatrix}$

6. $\begin{bmatrix} 1 & 0 & 0 & \vdots & -2 \\ 0 & 1 & 3 & \vdots & 7 \\ 0 & 0 & 0 & \vdots & 0 \end{bmatrix}$

7. $\begin{bmatrix} 1 & 0 & 3 & \vdots & -1 \\ 0 & 1 & 1 & \vdots & 2 \\ 0 & 0 & 0 & \vdots & 0 \end{bmatrix}$

8. $\begin{bmatrix} 1 & 0 & -4 & \vdots & 1 \\ 0 & 0 & 0 & \vdots & 2 \end{bmatrix}$

In problems 9–12, use row-reduction to solve each system.

9. $\begin{cases} 4x - y = -4 \\ x + 2y = 6 \end{cases}$

10. $\begin{cases} 2x - 3y = 1 \\ -5x + y = 0 \end{cases}$

11. $\begin{cases} 3x - y + 2z = 1 \\ x - 2y + 4z = 2 \\ x - y + z = 0 \end{cases}$

12. $\begin{cases} 2x - y + z = 11 \\ x + y - z = -2 \\ -3x - 2y = -7 \end{cases}$

In problems 13 and 14, find the partial fraction expansion.

13. $\dfrac{x+2}{x^2-6x-7} = \dfrac{A}{x-7} + \dfrac{B}{x+1}$

14. $\dfrac{3x+1}{x^2+x} = \dfrac{A}{x} + \dfrac{B}{x+1}$

In problems 15–18, use the given matrices to perform the indicated matrix operations, if possible.

15. $A = \begin{bmatrix} 2 & 1 \\ 1 & -1 \end{bmatrix}; B = \begin{bmatrix} 5 & -2 \\ 3 & 4 \end{bmatrix}$

(a) $A + 2B$ (b) $-3A + 4B$

16. $A = \begin{bmatrix} 1 & -4 & -8 \\ 5 & 20 & 3 \end{bmatrix}; B = \begin{bmatrix} 7 & -2 & 7 \\ 3 & 5 & 3 \end{bmatrix}$

(a) $3A - 2B$ (b) $4A + 2B$

17. $A = \begin{bmatrix} 1 & -2 & 1 \\ 3 & 1 & -2 \\ 0 & 1 & -1 \end{bmatrix}; B = \begin{bmatrix} 4 & -1 \\ 3 & 1 \\ 2 & 8 \end{bmatrix}$

(a) AB (b) BA

18. $A = \begin{bmatrix} 1 & 3 \\ -1 & 2 \\ 0 & 1 \end{bmatrix}; B = \begin{bmatrix} -2 & 3 & 1 \\ 0 & 2 & -3 \\ 1 & -1 & 4 \end{bmatrix}$

(a) AB (b) BA

In problems 19 and 20, verify that $AA^{-1} = A^{-1}A = I$.

19. $A = \begin{bmatrix} 3 & -2 \\ 8 & -5 \end{bmatrix}; A^{-1} = \begin{bmatrix} -5 & 2 \\ -8 & 3 \end{bmatrix}$

20. $A = \begin{bmatrix} 1 & -1 & 1 \\ 0 & 2 & -1 \\ 2 & 3 & 0 \end{bmatrix}; A^{-1} = \begin{bmatrix} 3 & 3 & -1 \\ -2 & -2 & 1 \\ -4 & -5 & 2 \end{bmatrix}$

In problems 21 and 22, use the inverse of the matrix of coefficients to solve each system. (Refer to the inverses given in problems 19 and 20.)

21. $\begin{cases} 3x - 2y = -7 \\ 8x - 5y = -18 \end{cases}$

22. $\begin{cases} x - y + z = 7 \\ 2y - z = -7 \\ 2x + 3y = -1 \end{cases}$

In problems 23–26, evaluate each determinant.

23. $\begin{vmatrix} 1 & -1 \\ 3 & 2 \end{vmatrix}$

24. $\begin{vmatrix} -2 & 1 \\ -1 & 7 \end{vmatrix}$

25. $\begin{vmatrix} 2 & 7 & 4 \\ 3 & -1 & 5 \\ 4 & 14 & 8 \end{vmatrix}$

26. $\begin{vmatrix} 2 & -4 & 7 \\ 0 & 1 & 2 \\ 0 & 0 & 5 \end{vmatrix}$

In problems 27–30, use Cramer's rule to solve each system.

27. $\begin{cases} 3x - y = 7 \\ 2x + 3y = 12 \end{cases}$ **28.** $\begin{cases} 5x + 3y = 13 \\ 7x - 5y = 18 \end{cases}$

29. $\begin{cases} 2x - 3y + z = 10 \\ 5x + y - z = 6 \\ x - y + 2z = 6 \end{cases}$ **30.** $\begin{cases} 3x - y + 2z = 5 \\ 2x + 3y + z = 1 \\ 5x + y + 4z = 8 \end{cases}$

In problems 31 and 32, sketch the graph of the solution set of each system of inequalities.

31. $\begin{cases} x + 2y > 5 \\ 3x - y < 9 \end{cases}$ **32.** $\begin{cases} x^2 + 2y \geq 0 \\ x + 2y \leq 6 \end{cases}$

In problems 33–36, graph the region defined by each constraint system, locate each corner point, and then find the maximum and minimum values of the given linear expression over the region.

33. $F = 7x + 3y$
$\begin{cases} x \geq 0, y \geq 0 \\ 7x + 2y \leq 14 \end{cases}$

34. $F = 2x + y$
$\begin{cases} x \geq 1 \\ y \geq 2 \\ x + 2y \leq 10 \end{cases}$

35. $F = x + 5y$
$\begin{cases} x \geq 0, y \geq 0 \\ x + y \leq 4 \\ 4x + y \leq 7 \end{cases}$

36. $F = x + y$
$\begin{cases} x \geq 0, y \geq 0 \\ 3y - 2x \leq 6 \\ 3y + 4x \leq 24 \end{cases}$

G In problems 37–40, sketch the graphs of both equations in the system in the same viewing window. Use the result to determine the number of solutions. Then solve the system.

37. $\begin{cases} 3x - 4y = 25 \\ x^2 + y^2 = 25 \end{cases}$ **38.** $\begin{cases} 2x - y = 2 \\ x^2 + 2y^2 = 12 \end{cases}$

39. $\begin{cases} x + y^2 = 6 \\ x^2 + y^2 = 36 \end{cases}$

40. $\begin{cases} \log_6 x + \log_6 y = 1 \\ x^2 + y^2 = 13 \end{cases}$

41. Income Tax: An income tax service developed the following relationships among taxes for a corporation:

$$\begin{cases} x + 0.095y = 2{,}584.42 \\ 0.01x + 1.01y + 0.01z = 1{,}529.70 \\ 0.48x + 0.48y + z = 12{,}253.24 \end{cases}$$

where x denotes the state income tax owed, y denotes the city income tax owed, and z denotes the federal income tax owed (all in dollars). Solve the system to determine the taxes owed (to the nearest dollar).

In problems 42–45, use a system of equations to solve each problem.

42. Investment Portfolio: Suppose that $10,000 is invested in two funds paying 13% and 14% annual simple interest rate. If the annual return from both investments is $1400, what amount was invested at each rate?

43. Puzzle: The sum of the squares of two numbers is 113. When 5 times the square of one number is added to the square of the other, the sum is 309. What are the numbers?

44. Geometry: The perimeter of a rectangular garden is 46 meters and the area of the garden is 60 square meters. Find the length and the width of the garden.

45. Art: An artist determines the monthly cost of manufacturing a total of 190 of types A, B, and C statues is $910. Suppose that the manufacturing cost of each statue of type A is $7, of type B is $6, and of type C is $3. If statue A sells for $25, statue B for $15, and statue C for $10, how many of each type should be made to produce a monthly revenue of $2800?

 **CHAPTER 8 TEST**

1. Solve the system

$$\begin{cases} 4x - y = 3 \\ -2x + 3y = 1 \end{cases}$$

by using:
(a) The elimination method (without matrices)
(b) The elimination method using matrices

2. The row-reduced matrix

$$\begin{bmatrix} 1 & 0 & 0 & \vdots & 4 \\ 0 & 1 & 0 & \vdots & -1 \\ 0 & 0 & 1 & \vdots & 3 \end{bmatrix}$$

is the augmented matrix form of a corresponding system of linear equations. Solve the system.

3. Let

$$A = \begin{bmatrix} 3 & -1 \\ 4 & 2 \end{bmatrix} \quad \text{and} \quad B = \begin{bmatrix} 3 & -2 \\ 2 & 7 \end{bmatrix}$$

Perform each matrix operation:
(a) $A + 2B$
(b) AB
(c) $3A - 2B$

4. (a) Let $A = \begin{bmatrix} 1 & -2 \\ 1 & 1 \end{bmatrix}$.

Verify that $A^{-1} = \begin{bmatrix} \dfrac{1}{3} & \dfrac{2}{3} \\ -\dfrac{1}{3} & \dfrac{1}{3} \end{bmatrix}$.

(b) Use the inverse of the matrix of coefficients to solve the system

$$\begin{cases} x - 2y = 3 \\ x + y = -3 \end{cases}$$

5. (a) Evaluate $\begin{vmatrix} 5 & 2 \\ 2 & -3 \end{vmatrix}$.

(b) Use Cramer's rule to solve the system

$$\begin{cases} 5x + 2y = 3 \\ 2x - 3y = 5. \end{cases}$$

6. Find the partial fraction decomposition:

$$\frac{x + 4}{x^2 - 3x} = \frac{A}{x} + \frac{B}{x - 3}$$

7. Solve the system: $\begin{cases} x^2 - 2y = 0 \\ x + 2y = 6 \end{cases}$

8. Sketch the region defined by the system

$$\begin{cases} x \geq 0, y \geq 0 \\ x + y \leq 4 \\ 2x + y \leq 6. \end{cases}$$

and find the coordinates of the corner points of the boundary of the region.

9. Find the maximum and minimum values of $F = 3x + 4y + 1$ under the constraints

$$\begin{cases} x \geq 0, y \geq 0 \\ x + 2y \leq 8 \\ x + y \leq 5. \end{cases}$$

10. A plumber charges a fixed charge plus an hourly rate for service on a house call. The plumber charged $70 to repair a water tank that required 2 hours of labor and $100 to repair a water tank that took 3.5 hours of labor. Find the plumber's fixed charge and hourly rate. (Use a system of equations to solve this problem.)

11. From geometry, the sum of the interior angles of a triangle ABC is 180°. The measure of angle A is 100° less than the measure of the sum of the angles B and C, and the measure of angle C is 40° less than the measure of twice angle B. Find the measure of each angle. (Use a system of equations to solve this problem.)

Extended Topics in Analytic Geometry

Parametric equations provide us with a way of modeling the motion of a moving object. For example, suppose a space shuttle rises for its launch pad and reaches a height of 30,000 feet. Because of a computer malfunction, the booster rockets are turned off prematurely, at a moment when the shuttle is moving at a speed of 32,000 feet per second at an angle of 42° with an imaginary reference plane tangent to the earth directly below.

What is the maximum height of the space shuttle above the earth and how long will it take the space shuttle to reach its maximum height? The solution of this problem is given in Example 4 on page 639.

In this chapter we extend the study of conics which were introduced in Section 1.6. Graphing is also expanded to the use of *parametric equations.*

Objectives

1. Graph Parabolas with Vertex at $(h, k) \neq (0, 0)$
2. Graph Ellipses with Center at $(h, k) \neq (0, 0)$
3. Graph Hyperbolas with Center at $(h, k) \neq (0, 0)$

9.1 Transformations of Conics

Horizontal and vertical transformations enable us to develop additional standard equations for parabolas, ellipses, and hyperbolas to those presented in Section 1.6.

Graphing Parabolas with Vertex at $(h, k) \neq (0, 0)$

Horizontal and vertical shifting of a parabola with vertex at $(0, 0)$ and a horizontal or vertical axis of symmetry (Figure 1) to a parabola with vertex at $(h, k) \neq (0, 0)$ and $c > 0$ results in one of the following situations:

Standard Equations for Parabolas with Vertex at (h, k) and Horizontal or Vertical Axis of Symmetry

(i) $(y - k)^2 = 4c(x - h)$, opens to the right
(ii) $(y - k)^2 = -4c(x - h)$, opens to the left
(iii) $(x - h)^2 = 4c(y - k)$, opens upward
(iv) $(x - h)^2 = -4c(y - k)$, opens downward

Figure 1   *Standard equations of parabolas with vertex at the origin,* c > 0.

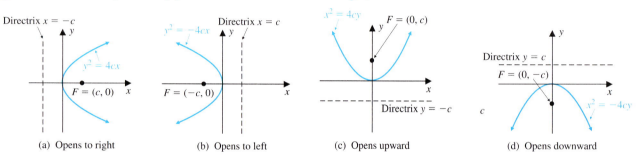

(a) Opens to right   (b) Opens to left   (c) Opens upward   (d) Opens downward

The next example illustrates the use of horizontal and vertical shifting to graph parabolas.

EXAMPLE 1   **Graphing a Parabola by Shifting**

Apply horizontal and vertical shifting to the graph of the parabola $y^2 = -24x$ to generate the graph of the equation

$$(y - 3)^2 = -24(x + 9).$$

Locate the vertex, axis of symmetry, focus, and directrix of the new parabola.

Solution   Let's compare the equations to determine how to shift the graph of

$$y^2 = -24x$$

to get the desired graph. Note that since $4c = 24$ in the equation $y^2 = -24x$, it follows that $c = 6$. Thus the focus is at $(-6, 0)$ and the directrix is at $x = 6$. The axis of symmetry is the x axis. As indicated in the following diagram,

Indicates a vertical shift 3 units up

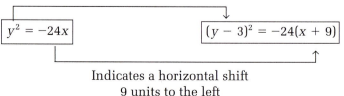

$y^2 = -24x$ $(y - 3)^2 = -24(x + 9)$

Indicates a horizontal shift
9 units to the left

Figure 2

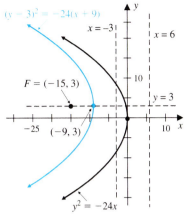

the graph of $(y - 3)^2 = -24(x + 9)$ is obtained by shifting the graph of the equation $y^2 = -24x$ to the left 9 units and up 3 units (Figure 2).

The vertex, axis of symmetry, focus, and directrix of both parabolas are given in Table 1.

TABLE 1

	$y^2 = -24x$	Shift	$(y - 3)^2 = -24(x + 9)$
Vertex	$(0, 0)$	Left 9 units, up 3 units	$(-9, 3)$
Axis of Symmetry	x axis (Line: $y = 0$)	Up 3 units	Line: $y = 0 + 3 = 3$
Focus	$(-6, 0)$	Left 9 units, up 3 units	Point: $(-6 - 9, 0 + 3) = (-15, 3)$
Directrix	Line: $x = 6$	Left 9 units	Line: $x = 6 - 9 = -3$

The equation $(y - 3)^2 = -24(x + 9)$ is the standard equation of the parabola graphed in Figure 2. However, if it is expanded algebraically, we get

$$(y - 3)^2 = -24(x + 9)$$

$$y^2 - 6y + 9 = -24x - 216$$

$$y^2 + 24x - 6y + 225 = 0.$$

This latter equation is referred to as a *general equation* of the given parabola.

The **general equation** of a parabola with an axis of symmetry that is *vertical* is given by

$$Ax^2 + Dx + Ey + F = 0.$$

If the axis of symmetry is *horizontal*, the form of the general equation is

$$Cy^2 + Dx + Ey + F = 0.$$

To convert from a general equation to the standard equation, we complete the square on the squared variable, as illustrated in the next example.

EXAMPLE 2   **Writing an Equation of a Parabola in Standard Form**

Rewrite the general equation of the parabola

$$y^2 + 2y - 8x - 3 = 0$$

in standard form by completing the square. Graph the parabola. Locate the vertex, axis of symmetry, focus, and directrix.

Solution   We rewrite the equation in standard form as follows:

$y^2 + 2y - 8x - 3 = 0$	Given
$y^2 + 2y = 8x + 3$	Isolate the y terms on one side
$y^2 + 2y + 1 = 8x + 3 + 1$	Complete the square on the y terms
$(y + 1)^2 = 8x + 4$	Factor and simplify
$(y + 1)^2 = 8\left(x + \dfrac{1}{2}\right)$	Factor out 8

We can obtain the desired graph by shifting the graph of $y^2 = 8x$ to the left 1/2 unit and down 1 unit. The graph of the resulting parabola is shown in Figure 3 along with the graph of

Figure 3

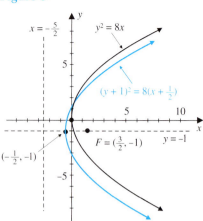

$$y^2 = 8x.$$

The vertex shifts from

$$(0, 0) \text{ to } \left(0 - \frac{1}{2}, 0 - 1 \right) = \left(-\frac{1}{2}, -1 \right).$$

The axis of symmetry shifts down 1 unit to become the line

$$y = -1.$$

Since $4c = 8$, it follows that $c = 2$, so the focus shifts from

$$(c, 0) = (2, 0) \text{ to } \left(2 - \frac{1}{2}, 0 - 1 \right) = \left(\frac{3}{2}, -1 \right).$$

Finally, the vertical directrix of $y^2 = 8x$ shifts $\frac{1}{2}$ unit to the left of directrix $x = -2$ to become the line

$$x = -2 - \frac{1}{2} = -\frac{5}{2}.$$

which is the directrix of the given parabola.

At times it is possible to use certain geometric characteristics to find the standard equation of a parabola.

EXAMPLE 3 **Finding the Standard Equation of a Parabola**

Find the standard equation of the parabola with vertex at $V = (4, -1)$ and focus at $F = (2, -1)$. Also, sketch the graph.

Solution

Let $D = (p, -1)$ be a point on the directrix of the parabola. Figure 4a shows the points $(4, -1)$, $(2, -1)$, and $(p, -1)$. Since the vertex is the midpoint of the line segment containing these points, it follows that $c = 2$, $p = 6$, and the parabola opens to the left with a horizontal axis of symmetry at $y = -1$.

Its standard equation has the form

$$(y - k)^2 = -4c(x - h)$$

with $h = 4$, $k = -1$, and $c = 2$.

After substituting these values, we get

$$[y - (-1)]^2 = -4(2)(x - 4)$$

or

$$(y + 1)^2 = -8(x - 4).$$

The graph is shown in Figure 4b.

Figure 4

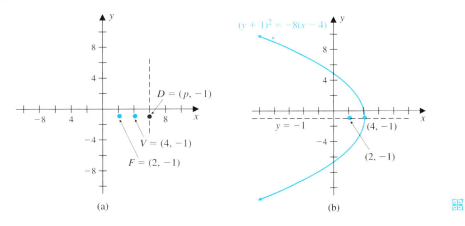

(a)   (b)

Graphing Ellipses with Center at $(h, k) \neq (0, 0)$

Horizontal and vertical shifting of an ellipse centered at the origin with a horizontal or vertical major axis (Figure 5) to an ellipse with center at (h, k) and with $a > 0$ and $b > 0$, results in one of the following two situations:

Figure 5

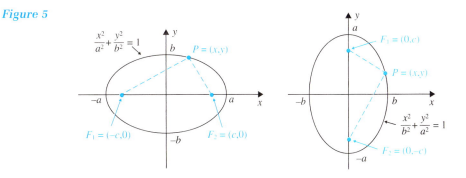

Standard equations of ellipses with center at $(0, 0)$, horizontal or vertical major axis, and foci F_1 and F_2; $a > b$, $c^2 = a^2 - b^2$.

Standard Equations for Ellipses Centered at (h, k) and Horizontal or Vertical Major Axis

(i) $\dfrac{(x - h)^2}{a^2} + \dfrac{(y - k)^2}{b^2} = 1$, if $a > b$; the major axis is horizontal

(ii) $\dfrac{(x - h)^2}{b^2} + \dfrac{(y - k)^2}{a^2} = 1$, if $b < a$; the major axis is vertical

Graphs of ellipses in either of these latter equation forms can be obtained by using shifting.

EXAMPLE 4   **Graphing an Ellipse by Shifting**

Apply horizontal and vertical shifting to the graph of the equation

$$\frac{x^2}{9} + \frac{y^2}{4} = 1$$

to obtain the graph of the equation

$$\frac{(x+3)^2}{9} + \frac{(y-5)^2}{4} = 1$$

and to find the center, vertices, and foci.

Solution   The following diagram explains how to shift the graph of the first equation to get the graph of the second:

Indicates a horizontal shift 3 units to the left

$$\frac{x^2}{9} + \frac{y^2}{4} = 1 \qquad \frac{(x+3)^2}{9} + \frac{(y-5)^2}{4} = 1$$

Indicates a vertical shift 5 units up

Figure 6

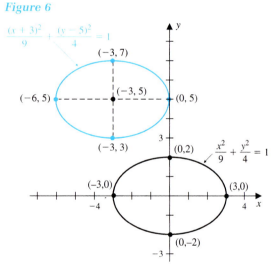

Since $c^2 = a^2 - b^2$

$$= 9 - 4 = 5$$

it follows that $(-\sqrt{5}, 0)$ and $(\sqrt{5}, 0)$ are the foci of the ellipse

$$\frac{x^2}{9} + \frac{y^2}{4} = 1$$

Figure 6 shows the graphs of both ellipses.

Comparisons of the center, vertices, and foci of both ellipses are listed in Table 2.

TABLE 2

	$\dfrac{x^2}{9} + \dfrac{y^2}{4} = 1$	Shift	$\dfrac{(x+3)^2}{9} + \dfrac{(y-5)^2}{4} = 1$
Center	$(0, 0)$	Left 3 units, up 5 units	$(-3, 5)$
Vertices	$(-3, 0)$ and $(3, 0)$,	Left 3 units, up 5 units	$(-3-3, 5) = (-6, 5)$ and $(3-3, 5) = (0, 5)$
Foci	$(-\sqrt{5}, 0)$ and $(\sqrt{5}, 0)$	Left 3 units, up 5 units	$(-3-\sqrt{5}, 5)$ and $(-3+\sqrt{5}, 5)$

The *general equation* of an ellipse with axes parallel to the coordinate axes is given by

$$Ax^2 + Cy^2 + Dx + Ey + F = 0$$

where A and C have the same sign. This equation can be rewritten in standard form by completing the square for both the x and y terms.

EXAMPLE 5 **Writing an Equation of an Ellipse in Standard Form**

Rewrite the general equation of the ellipse

$$25x^2 + 9y^2 - 100x - 54y = 44$$

in standard form by completing the squares. Find the center, vertices, and foci. Also, sketch the graph.

Solution   Here we have

$$25(x^2 - 4x \quad) + 9(y^2 - 6y \quad) = 44$$
  Factor 25 from the x terms and 9 from the y terms

$$25(x^2 - 4x + 4) + 9(y^2 - 6y + 9) = 44 + 25(4) + 9(9)$$
  Complete the squares

$$25(x - 2)^2 + 9(y - 3)^2 = 225$$
  Factor and simplify Divide both sides by 225

$$\frac{(x - 2)^2}{9} + \frac{(y - 3)^2}{25} = 1$$

So the center is at $(2, 3)$, $a = 5$, $b = 3$, and we have a vertical major axis.
Also, $c^2 = a^2 - b^2 = 25 - 9 = 16$, so $c = 4$.
Thus the foci are

$$(2, 3 - 4) = (2, -1) \quad \text{and} \quad (2, 3 + 4) = (2, 7).$$

The vertices are

$$(2, 3 - 5) = (2, -2) \quad \text{and} \quad (2, 3 + 5) = (2, 8).$$

Notice that the end points of the minor axis are

$$(2 - 3, 3) = (-1, 3) \quad \text{and} \quad (2 + 3, 5) = (5, 3).$$

The graph is displayed in Figure 7.

Figure 7

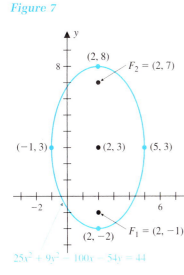

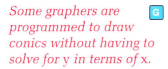

Some graphers are programmed to draw conics without having to solve for y in terms of x. 🅖

To graph an ellipse using a grapher we may have to first solve the equation for y in terms of x and then graph both of the resulting functions in the same viewing window.
For instance, to sketch the graph of the ellipse whose equation is

$$\frac{(x - 1)^2}{9} + \frac{(y + 2)^2}{4} = 1$$

we proceed as follows.

First we solve the equation for y in terms of x as follows:

$$\frac{(y+2)^2}{4} = 1 - \frac{(x-1)^2}{9}$$

$$\frac{y+2}{2} = \pm\sqrt{1 - \frac{(x-1)^2}{9}}$$

$$y+2 = \pm 2\sqrt{1 - \frac{(x-1)^2}{9}}$$

$$y = -2 \pm \frac{2\sqrt{9-(x-1)^2}}{3}$$

Figure 8

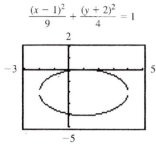

Next we graph both functions,

$$y_1 = -2 + \frac{2\sqrt{9-(x-1)^2}}{3} \quad \text{and} \quad y_2 = -2 - \frac{2\sqrt{9-(x-1)^2}}{3}$$

in the same viewing window to get the graph of the original equation (Figure 8).

Graphing Hyperbolas with Center at $(h, k) \neq (0, 0)$

In general, horizontal and vertical shifting of a hyperbola centered at $(0, 0)$ with a horizontal or vertical transverse axis (Figure 9) to a hyperbola with center (h, k) results in one of the following two situations:

Figure 9

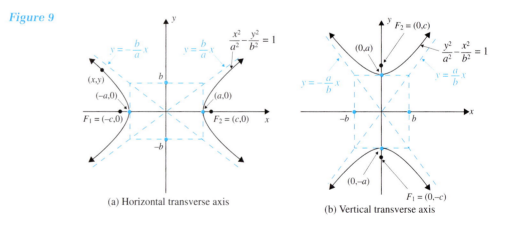

(a) Horizontal transverse axis

(b) Vertical transverse axis

Standard equations of hyperbolas with center at $(0, 0)$, horizontal or vertical transverse axis, and foci F_1 and F_2 ; $c^2 = a^2 + b^2$.

Standard Equations for Hyperbolas Centered at (h, k)

(i)	$\dfrac{(x-h)^2}{a^2} - \dfrac{(y-k)^2}{b^2} = 1$	The transverse axis, $y = k$, is horizontal.
(ii)	$\dfrac{(y-k)^2}{a^2} - \dfrac{(x-h)^2}{b^2} = 1$	The transverse axis, $x = h$, is vertical.

Shifting can be used to graph hyperbolas with equations given in either of these two forms.

EXAMPLE 6   **Graphing a Hyperbola**

Apply horizontal and vertical shifting to the graph of the equation

$$\frac{x^2}{16} - \frac{y^2}{9} = 1$$

to generate the graph of the equation

$$\frac{(x-2)^2}{16} - \frac{(y-3)^2}{9} = 1.$$

Find the center, vertices, foci, and equations of the asymptotes of the new hyperbola.

Solution   Here $a = 4$, $b = 3$, and $c^2 = 16 + 9 = 25$, so that $c = 5$ for both hyperbolas. The following diagram indicates the shifts:

Figure 10

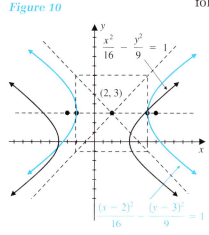

Indicates a horizontal shift 2 units to the right

$$\frac{x^2}{16} - \frac{y^2}{9} = 1 \qquad\qquad \frac{(x-2)^2}{16} - \frac{(y-3)^2}{9} = 1$$

Indicates a vertical shift 3 units up

Thus the new hyperbola is obtained by shifting the first hyperbola 2 units to the right and 3 units up. Figure 10 shows the graphs of both hyperbolas.

The center, vertices, foci, and asymptotes of both hyperbolas are listed in Table 3.

TABLE 3

	$\dfrac{x^2}{16} - \dfrac{y^2}{9} = 1$	Shift	$\dfrac{(x-2)^2}{16} - \dfrac{(y-3)^2}{9} = 1$
Center	$(0, 0)$	Right 2 units, up 3 units	$(2, 3)$
Vertices	$(-4, 0)$ and $(4, 0)$	Right 2 units, up 3 units	$(-4 + 2, 0 + 3) = (-2, 3)$ and $(4 + 2, 0 + 3) = (6, 3)$
Foci	$(-5, 0)$ and $(5, 0)$	Right 2 units, up 3 units	$(-5 + 2, 0 + 3) = (-3, 3)$ and $(5 + 2, 0 + 3) = (7, 3)$
Asymptotes	$y = \dfrac{3}{4}x$, and $y = -\dfrac{3}{4}x$	Right 2 units, up 3 units	$y - 3 = \dfrac{3}{4}(x - 2)$ or $y = \dfrac{3}{4}x + \dfrac{3}{2}$ and $y - 3 = -\dfrac{3}{4}(x - 2)$ or $y = -\dfrac{3}{4}x + \dfrac{9}{2}$

The *general equation* of a hyperbola with a horizontal or vertical transverse axis is

$$Ax^2 + Cy^2 + Dx + Ey + F = 0$$

with A and C having opposite signs. This equation can be written in standard form by completing the squares.

EXAMPLE 7   **Writing an Equation of a Hyperbola in Standard Form**

Rewrite the general equation of the hyperbola

$$4x^2 - y^2 - 8x + 2y + 7 = 0$$

in standard form by completing the squares. Find the center, vertices, foci, and equations of the asymptotes. Also, sketch the graph.

Solution   First we rewrite the equation in standard form as follows:

$$(4x^2 - 8x \quad) - (y^2 - 2y \quad) = -7 \qquad \text{Isolate variable terms}$$

$$4(x^2 - 2x + 1) - (y^2 - 2y + 1) = -7 + 4 - 1 \qquad \text{Complete the squares}$$

$$4(x - 1)^2 - (y - 1)^2 = -4 \qquad \text{Factor and simplify}$$

$$\frac{(y - 1)^2}{4} - \frac{(x - 1)^2}{1} = 1 \qquad \text{Divide each side by } -4$$

This is the standard equation of a hyperbola with center $(h, k) = (1, 1)$. In this situation, $a = 2$ and $b = 1$, so

$$c^2 = a^2 + b^2 = 4 + 1 = 5,$$

or

$$c = \sqrt{5}.$$

Since the transverse axis is vertical, the vertices are

$$(1, 1 + 2) = (1, 3) \quad \text{and} \quad (1, 1 - 2) = (1, -1).$$

The foci are

$$(1, 1 + \sqrt{5}) \qquad \text{and} \qquad (1, 1 - \sqrt{5}).$$

Figure 11

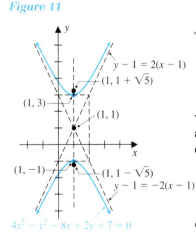

Figure 11 shows the graph of the hyperbola, obtained by drawing a rectangle with center at the point $(1, 1)$ and dimensions equal to the lengths of the transverse and conjugate axes, which are 4 and 2, respectively. The asymptotes are extensions of the diagonals of the rectangle. So their equations are

$$y - 1 = \frac{2}{1}(x - 1) \quad \text{and} \quad y - 1 = -\frac{2}{1}(x - 1)$$

or

$$y = 2x - 1 \quad \text{and} \quad y = -2x + 3.$$

✦ PROBLEM SET 9.1

Mastering the Concepts

In problems 1–6, apply horizontal and vertical shifting to the first equation of a parabola to generate the second. Find the vertex, axis of symmetry, focus, and directrix of the new parabola. Sketch the graphs of both parabolas on the same coordinate system.

1. $y^2 = 8x$; $(y - 2)^2 = 8(x - 3)$

2. $x^2 = 2y$; $(x + 3)^2 = 2(y + 2)$

3. $x^2 = -4y$; $(x + 1)^2 = -4(y + 3)$

4. $y^2 = -4x$; $(y - 3)^2 = -4(x - 1)$

5. $y^2 = -6x$; $(y + 2)^2 = -6(x - 1)$

6. $x^2 = -12y$; $(x - 2)^2 = -12(y + 1)$

In problems 7–12, write each equation in standard form. Sketch the graph of each parabola. Find the vertex, focus, axis of symmetry, and directrix.

7. $y^2 + 2x + 2y + 7 = 0$
8. $2y^2 + 2x + 6y + 9 = 0$
9. $3x^2 + 12x + 6y + 13 = 0$
10. $x^2 + 4x + 5y - 11 = 0$
11. $y^2 - 6x = 4y - 13$
12. $x^2 = \dfrac{5}{3}y + \dfrac{2}{3} + 3x$

In problems 13–22, find the standard equation of the parabola with the given characteristics. Identify the axis of symmetry and sketch the graph.

13. Vertex at $(-1, 2)$ and focus at $(3, 2)$
14. Vertex at $(3, -4)$ and focus at $(3, 0)$
15. Vertex at $(3, -5)$ and focus at $(5, -5)$
16. Vertex at $(-2, 3)$ and focus at $(-2, 8)$
17. Vertex at $(1, 3)$, contains point $(5, 7)$, and axis of symmetry parallel to the y axis
18. Vertex at $(3, -4)$, axis of symmetry parallel to the x axis, and x intercept 11
19. Focus at $(-3, 5)$ and directrix $y = -5$
20. Vertex at $(3, -5)$ and directrix $x = 2$
21. Contains the three points $(-2, 0)$, $(0, 3)$, and $(2, 0)$, and opens downward
22. Contains the three points $(0, 0)$, $(-1, 2)$, and $(-3, -2\sqrt{3})$, and opens to the left

G In problems 23–26, solve each equation for y. Then use a grapher to graph both of the resulting equations in the same viewing window to obtain a graph of the given equation.

23. (a) $(y + 1)^2 = 2(x - 1)$
    (b) $(x - 3)^2 = -6(y + 1)$
24. (a) $(y - 2)^2 = -4(x + 2)$
    (b) $(x + 4)^2 = 16(y - 2)$
25. $y^2 + 2y + 12x - 23 = 0$
26. $y^2 + 10y - x + 21 = 0$

In problems 27–34, apply horizontal and vertical shifting to the graph of the first equation to generate the graph of the second. Find the center, vertices, and foci of the new ellipse. Sketch the graphs of both equations in the same coordinate system.

27. $\dfrac{x^2}{16} + \dfrac{y^2}{9} = 1;\ \dfrac{(x - 2)^2}{16} + \dfrac{(y + 3)^2}{9} = 1$

28. $\dfrac{x^2}{25} + \dfrac{y^2}{4} = 1;\ \dfrac{(x - 2)^2}{25} + \dfrac{(y - 1)^2}{4} = 1$

29. $\dfrac{x^2}{48} + \dfrac{y^2}{64} = 1;\ \dfrac{(x - 1)^2}{48} + \dfrac{(y + 2)^2}{64} = 1$

30. $\dfrac{x^2}{10} + \dfrac{y^2}{25} = 1;\ \dfrac{(x - 3)^2}{10} + \dfrac{(y + 1)^2}{25} = 1$

31. $\dfrac{x^2}{25} + \dfrac{y^2}{16} = 1;\ \dfrac{(x + 2)^2}{25} + \dfrac{(y - 1)^2}{16} = 1$

32. $\dfrac{x^2}{9} + \dfrac{y^2}{16} = 1;\ \dfrac{(x - 2)^2}{9} + \dfrac{(y - 2)^2}{16} = 1$

33. $\dfrac{x^2}{3} + \dfrac{y^2}{15} = 1;\ \dfrac{(x + 1)^2}{3} + \dfrac{(y - 1)^2}{15} = 1$

34. $\dfrac{x^2}{15} + \dfrac{y^2}{25} = 1;\ \dfrac{(x + 1)^2}{15} + \dfrac{(y + 1)^2}{25} = 1$

In problems 35–42, rewrite each equation in standard form. Then find the center, vertices, and foci. Also, sketch the graph.

35. $4x^2 + y^2 - 16x - 6y + 21 = 0$
36. $6x^2 + 5y^2 - 60x - 10y + 65 = 0$
37. $x^2 + 4y^2 + 2x - 8y + 1 = 0$
38. $x^2 + 4y^2 - 2x - 16y + 13 = 0$
39. $9x^2 + 4y^2 + 18x - 16y - 11 = 0$
40. $9y^2 + x^2 - 18y + 2x + 9 = 0$
41. $4x^2 + 9y^2 - 32x - 36y + 64 = 0$
42. $4x^2 + 9y^2 - 24x + 36y + 36 = 0$

In problems 43–48, find the standard equation of an ellipse with the given characteristics, and sketch the graph.

43. Vertices at $(1, -2)$, $(5, -2)$, $(3, -7)$, $(3, 3)$
44. Vertices at $(0, -1)$, $(12, -1)$, $(6, -4)$, $(6, 2)$
45. Center at $(-3, 1)$; major axis parallel to the y axis and 10 units long; minor axis 2 units long
46. Center at $(0, 0)$; axes parallel to the coordinate axes; containing the points $\left(\dfrac{3\sqrt{3}}{2}, 1\right)$ and $\left(2, \dfrac{2\sqrt{5}}{3}\right)$
47. Foci at $(1, 4)$ and $(3, 4)$; major axis 4 units long
48. Foci at $(-2, 3)$ and $(-2, 7)$; minor axis 1 unit long

G In problems 49–52, solve for y in terms of x. Then use a grapher to graph both of the resulting equations in the same viewing window to obtain a graph of the given equation.

49. $0.8(x + 0.1)^2 + 1.6y^2 = 1.28$
50. $3.2(x - 0.2)^2 + 1.6y^2 = 5.12$
51. $4x^2 + y^2 - 2y = 0$
52. $4x^2 + 9y^2 - 32x = 0$

In problems 53–58, apply horizontal and vertical shifting to the graph of the first equation to generate the graph of the second. Also, find the center, vertices, foci, and asymptotes of the new graph.

53. $\dfrac{x^2}{25} - \dfrac{y^2}{4} = 1$; $\dfrac{(x-5)^2}{25} - \dfrac{(y-4)^2}{4} = 1$

54. $\dfrac{y^2}{16} - \dfrac{x^2}{25} = 1$; $\dfrac{(y-3)^2}{16} - \dfrac{(x+3)^2}{25} = 1$

55. $\dfrac{y^2}{16} - \dfrac{x^2}{16} = 1$; $\dfrac{(y+3)^2}{16} - \dfrac{(x+2)^2}{16} = 1$

56. $\dfrac{x^2}{4} - \dfrac{y^2}{20} = 1$; $\dfrac{(x+1)^2}{4} - \dfrac{(y+2)^2}{20} = 1$

57. $\dfrac{x^2}{4} - \dfrac{y^2}{9} = 1$; $\dfrac{(x-1)^2}{4} - \dfrac{(y+2)^2}{9} = 1$

58. $\dfrac{x^2}{9} - \dfrac{y^2}{\frac{9}{4}} = 1$; $\dfrac{(x+3)^2}{9} - \dfrac{(y+2)^2}{\frac{9}{4}} = 1$

In problems 59–66, rewrite each equation in standard form by completing the squares. Then find the coordinates of the center, foci, and vertices. Also, find the equations of the asymptotes and sketch the graph.

59. $4x^2 - 3y^2 - 32x + 6y + 73 = 0$

60. $4x^2 - 9y^2 - 32x + 36y + 27 = 0$

61. $9y^2 - 25x^2 + 72y - 100x + 269 = 0$

62. $9x^2 - 16y^2 - 90x - 256y = 223$

63. $4x^2 - y^2 + 8x - 2y + 6 = 0$

64. $4y^2 - x^2 + 40y - 4x + 60 = 0$

65. $4x^2 - 9y^2 - 8x + 36y - 68 = 0$

66. $9x^2 - 16y^2 - 36x - 64y + 116 = 0$

In problems 67–72, find the standard equation for the hyperbola that satisfies the given conditions and sketch the graph.

67. Center at $(3, -4)$;
horizontal transverse axis of length 6;
conjugate axis of length 4

68. Center at $(-1, 2)$;
vertical transverse axis of length 10;
conjugate axis of length 5

69. Vertices at

$$(-1, 4) \text{ and } (-1, 6);$$

foci at

$$(-1, 3) \text{ and } (-1, 7)$$

70. x intercepts at -5 and 5; asymptotes

$$y = \pm 3x$$

71. Vertices at $(6, 3)$ and $(2, 3)$;
foci at $(7, 3)$ and $(1, 3)$

72. Foci at $(-1, -2)$ and $(9, -2)$;
horizontal transverse axis of length 6

G In problems 73–76, use a grapher to graph each equation.

73. $\dfrac{(y-1)^2}{9} - \dfrac{(x+1)^2}{4} = 1$

74. $\dfrac{(x+2)^2}{25} - \dfrac{(y-1)^2}{4} = 1$

75. $x^2 - 2y^2 - 2x - 12y - 35 = 0$

76. $-3x^2 + 2y^2 + 4x + 6y - 49 = 0$

Objectives

1. Describe a Conic Using Eccentricity
2. Identify Polar Equations of Conics
3. Solve Applied Problems

9.2 Eccentricity and Polar Equation Forms for Conics

So far we have derived equations in Cartesian form for the parabola, ellipse, and hyperbola by using three different geometric definitions—one for each type of curve. By using the idea of *eccentricity*, it is possible to give a unified geometric definition for these curves.

Describing a Conic Using Eccentricity

It is traditional to use the letter e to denote the eccentricity. This e should not be confused with the natural logarithm base.

The basic idea here is to describe a conic in terms of a fixed straight line *l* called the **directrix,** a fixed point *F* not on *l* called the **focus,** and a positive constant *e* called the **eccentricity.**

A point *P* in the plane belongs to the conic if and only if

$$\frac{|\overline{PF}|}{|\overline{PD}|} = e$$

where *D* is the point on *l* and on the perpendicular line segment from *P* to *l* (Figure 1).

The value of e determines the particular conic section: *Figure 1*

1. For a *parabola*, $e = 1$ (since $|\overline{PF}| = |\overline{PD}|$).
2. For an *ellipse*, $0 < e < 1$.
3. For a *hyperbola*, $e > 1$.

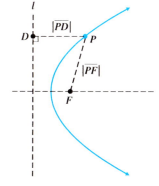

EXAMPLE 1   **Using the Eccentricity to Find an Equation of a Conic**

Find the standard equation of the conic with focus $(3, 0)$, directrix $x = 0$, and eccentricity $e = 2$. Sketch the graph.

Solution Since $e = 2$, the conic is a hyperbola. Assume that $P = (x, y)$ is a point on the graph of the conic (Figure 2). Then

Figure 2

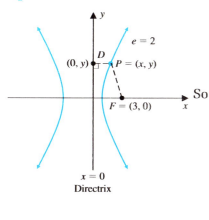

$$\frac{|\overline{PF}|}{|\overline{PD}|} = 2.$$

So

$$|\overline{PF}|^2 = 4|\overline{PD}|^2$$

$$(x - 3)^2 + y^2 = 4x^2 \qquad \text{Distance formula}$$

$$3x^2 + 6x - y^2 = 9 \qquad \text{Multiply and isolate variable terms}$$

$$3(x^2 + 2x) - y^2 = 9 \qquad \text{Factor out 3 from the } x \text{ terms}$$

$$3(x^2 + 2x + 1) - y^2 = 9 + 3 \qquad \text{Complete the square}$$

$$3(x + 1)^2 - y^2 = 12 \qquad \text{Factor and simplify}$$

$$\frac{(x + 1)^2}{4} - \frac{y^2}{12} = 1 \qquad \text{Divide by 12}$$

This latter equation is the standard equation for a hyperbola. Its graph is shown in Figure 2.

In Example 1, notice that $a = 2$, $b = 2\sqrt{3}$, and $c = 4$, so we have

$$\frac{\text{Distance between foci}}{\text{Distance between vertices}} = \frac{2c}{2a} = \frac{c}{a} = \frac{8}{4} = 2.$$

Thus the eccentricity $e = 2$ for this hyperbola is the same as c/a. In general,

> an ellipse or a hyperbola is a conic with eccentricity
> $$e = \frac{c}{a} \text{ (see Figure 3)}$$
> and a parabola is a conic with eccentricity 1.

The results in Figure 3 also hold for an ellipse or a hyperbola whose foci lie on a vertical line.

Figure 3

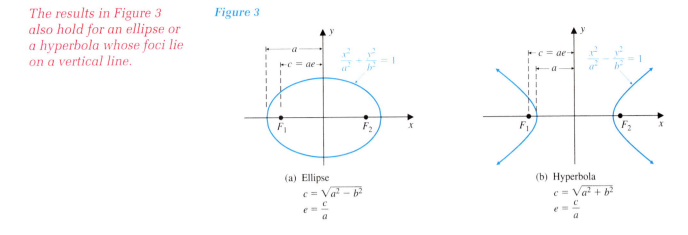

(a) Ellipse
$$c = \sqrt{a^2 - b^2}$$
$$e = \frac{c}{a}$$

(b) Hyperbola
$$c = \sqrt{a^2 + b^2}$$
$$e = \frac{c}{a}$$

EXAMPLE 2   **Finding the Eccentricity of a Conic**

Find the eccentricity of the conic

$$\frac{(x - 2)^2}{9} + \frac{(y + 1)^2}{4} = 1$$

Solution   The equation represents an ellipse with $a = 3$ and $b = 2$.

So $$c = \sqrt{a^2 - b^2} = \sqrt{9 - 4} = \sqrt{5}$$

Therefore, the eccentricity is

$$e = \frac{c}{a} = \frac{\sqrt{5}}{3} = 0.75 \text{ (approx.)}.$$

Identifying Polar Equations of Conics

Conics also have standard equations in the polar coordinate system. To derive such a polar equation, we first assume that the conic has one focus located at the pole. That is, let $P = (r, \theta)$ be a point on the graph of a conic with a focus F at the pole O, eccentricity e, and a directrix l that is a vertical line d units to the left of the focus, as displayed in Figure 4.

Figure 4

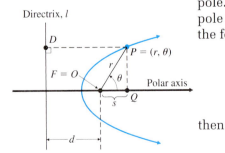

Directrix, l

If P satisfies the condition

$$\frac{|\overline{PF}|}{|\overline{PD}|} = e$$

then

(Distance from P to F) $= e \cdot$ (Distance from P to l).

That is,

$r = e(d + s)$	See Figure 4
$r = e(d + r \cos \theta)$	$\cos \theta = \dfrac{s}{r}$
$r = ed + er \cos \theta$	Multiply
$r - er \cos \theta = ed$	Isolate r terms
$r(1 - e \cos \theta) = ed$	Factor

Solving for r yields the *standard polar equation* for the conic:

$$r = \frac{ed}{1 - e \cos \theta}$$

Similar derivations lead to the standard polar equations listed in Table 1. The conics have eccentricity e, and vertical or horizontal directrix d units from the focus at the pole.

TABLE 1   Standard Polar Equations for Conics (Focus at the Pole and Eccentricity e)

Standard Polar Equation	Description of Directrix	Equation of Directrix
1. $r = \dfrac{ed}{1 + e \cos \theta}$	Vertical; that is, perpendicular to the polar axis at a distance d units to the right of the pole	$x = d$ or $r \cos \theta = d$
2. $r = \dfrac{ed}{1 - e \cos \theta}$	Vertical; that is, perpendicular to the polar axis at a distance d units to the left of the pole	$x = -d$ or $r \cos \theta = -d$
3. $r = \dfrac{ed}{1 + e \sin \theta}$	Horizontal; that is, parallel to the polar axis at a distance d units above the pole	$y = d$ or $r \sin \theta = d$
4. $r = \dfrac{ed}{1 - e \sin \theta}$	Horizontal; that is, parallel to the polar axis at a distance d units below the pole	$y = -d$ or $r \sin \theta = -d$

Upon examining Table 1, we notice that when the polar equation of the conic involves $\cos \theta$, the polar axis is an axis of symmetry. When the equation involves $\sin \theta$, the line $\theta = \pi/2$ is an axis of symmetry.

EXAMPLE 3   **Identifying and Graphing a Conic in the Polar System**

A conic is given by the polar equation

$$r = \frac{10}{3 - 2 \cos \theta}.$$

Find the eccentricity, identify the conic, and sketch the graph. Also, use the equation to locate the center and vertices.

Solution   Dividing the numerator and denominator by 3, we obtain the second standard equation in Table 1:

$$r = \frac{\dfrac{10}{3}}{1 - \dfrac{2}{3} \cos \theta} = \frac{\dfrac{2}{3}(5)}{1 - \dfrac{2}{3} \cos \theta}$$

Thus $e = 2/3$ and the conic is an ellipse with a focus at the pole and major axis along the polar axis.

The two vertices along the polar axis can be found by substituting $\theta = 0$ and $\theta = \pi$ into the original equation to obtain $r = 10$ and $r = 2$, respectively. Thus the vertices are the polar points $(10, 0)$ and $(2, \pi)$. Hence $2a = 12$, or $a = 6$.

The center of the ellipse is located at the midpoint of the line segment determined by the vertices, that is, (4, 0).

Using $e = c/a$, we obtain

$$c = ae = 6\left(\frac{2}{3}\right) = 4$$

so

$$b^2 = a^2 - c^2 = 36 - 16 = 20.$$

Figure 5

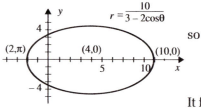

It follows that $b = \sqrt{20} = 2\sqrt{5}$ so that the end points of the minor axis are located $2\sqrt{5} \approx 4.47$ (approx.) units above and below the center. We use this information to sketch a graph of the equation (Figure 5). In a sense, the eccentricity of an ellipse measures its "roundness," and the eccentricity of a hyperbola measures the "flatness" of its branches.

To understand this notion, let us examine the graphs of

$$r = \frac{3e}{1 - e \cos \theta}$$

for changing values of e with the aid of a grapher (Figure 6). We observe that when e is close to 0, the ellipse is close to being a circle, whereas it becomes more flattened as e gets closer to 1 (Figure 6b).

When $e = 1$, the conic is a parabola (Figure 6c).

Figures 6d–f show that as $e > 1$ gets larger, the hyperbola begins to resemble a pair of parallel lines.

Figure 6

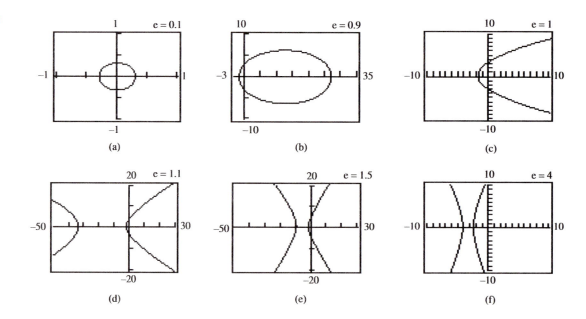

Solving Applied Problems

Mathematical models of the solar system have been revised several times to accommodate new discoveries. In the sixteenth century, the German astronomer Johannes Kepler (1571–1630) found that planets revolve around the sun in elliptical orbits, with the sun at one focus (Table 2).

Because these eccentricities are so close to 0, the orbits are almost circular.

TABLE 2 Eccentricities of Some Planetary Orbits Around the Sun

Planet	Venus	Earth	Mars	Mercury	Pluto
Eccentricity	0.01	0.02	0.09	0.21	0.25

In the next example, we use *astronomical units* to describe the elliptical orbit of Halley's Comet. By definition, 1 **astronomical unit** (or **AU**), which is the length of the semimajor axis of the earth's elliptical orbit around the sun, is about 9.26×10^7 miles.

EXAMPLE 4   **Approximating Distance in an Elliptical Path**

The orbit of Halley's Comet is an ellipse that has a major axis 36.2 AU long and a minor axis 9.1 AU wide. The center of the sun is located at one focus of the elliptical orbit.

(a) Illustrate the comet's orbit by sketching an *xy* coordinate system with the center of the orbit at the origin and the major axis on the *x* axis.

(b) Find the eccentricity of this orbit (to two decimal places).

(c) Find the minimum and maximum distances between the center of the sun and the center of the comet.

Solution   (a) The comet's orbit is sketched in Figure 7, where the center of the sun is located at one focus. The length of the major axis is $2a$ and the length of the minor axis is $2b$. The distance between the center of the ellipse and the sun is c.

(b) Here, $2a = 36.2$ and $2b = 9.1$, so $a = 18.10$ and $b = 4.55$. Thus c is given by

Figure 7

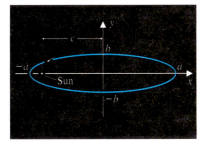

$$c = \sqrt{a - b^2}$$
$$= \sqrt{(18.1)^2 - (4.55)^2}$$
$$= 17.52 \text{ (approx.)}$$

So

$$e = \frac{c}{a} = \frac{17.52}{18.10} = 0.97 \text{ (approx.)}$$

(c) By examining Figure 7 we observe that the minimum distance between the center of the sun and the center of the comet is given by

$$a - c = 18.10 - 17.52$$
$$= 0.58 \text{ AU (approx.)}$$

or about 53,708,000 miles.
    Similarly the maximum distance between the center of the sun and the center of the comet is given by

$$a + c = 18.10 + 17.52$$
$$= 35.62 \text{ AU (approx.)}$$

or about 3,298,412,000 miles.

◆◆ PROBLEM SET 9.2

Mastering the Concepts

In problems 1–10, determine the standard equation for the conic that satisfies the given conditions and sketch the graph.

1. Focus at $(3, 0)$; directrix $y = 5/3$; eccentricity $e = 3/\sqrt{5}$
2. Focus at $(5, 0)$; directrix $x = 16/5$; eccentricity $e = 5/4$
3. Focus at $(3, 0)$; directrix $x = 25/3$; eccentricity $e = 3/5$
4. Focus at $(1, 2)$; directrix $y = -2$; eccentricity $e = 1/2$
5. Focus at $(2, 0)$; directrix $x = 4$; eccentricity $e = 1$
6. Directrix has the polar equation $r = 2 \csc \theta$ that corresponds to the focus at the pole; eccentricity $e = 3$
7. Center at $(0, 0)$; eccentricity $e = 1/2$; foci at $(-3, 0)$ and $(3, 0)$
8. Center at $(2, -1)$; eccentricity $e = 2$; foci at $(2, 1)$ and $(2, -3)$
9. Center at $(0, 0)$; eccentricity $e = 3$; containing the point $(5, 0)$
10. Eccentricity $e = 1$; containing points $(2, 3)$ and $(3, 4)$; symmetric with respect to the line $x = 1$

In problems 11–22, find the eccentricity of the given conic and identify the conic.

11. $25x^2 + 16y^2 = 400$
12. $x^2 + 2y^2 = 1$
13. $x^2 + 3y^2 = 4$
14. $y^2 - 8y + 3x + 5 = 0$
15. $9x^2 - 16y^2 + 144 = 0$
16. $4x^2 - 4y^2 + 1 = 0$
17. $16(x - 1)^2 - 9(y + 2)^2 = 144$
18. $4(x + 2)^2 - 25(y + 3)^2 = 100$
19. $3x^2 - 5x + y^2 + 22y = 1$
20. $x^2 + 16x - y + 7 = 0$
21. $x^2 + 2x - y^2 + 8y = 16$
22. $2x^2 + 12x + y^2 - 8y + 32 = 0$

In problems 23–28, find the standard polar equation of the conic with a focus at the pole, eccentricity e as given, and a directrix with the given equation.

23. $e = 1$; $r = -4 \csc \theta$
24. $e = 1/3$; $r = 4 \sec \theta$
25. $e = 2$; $r \cos \theta = -2$
26. $e = 1$; $r \sin \theta = 4$
27. $e = 2/5$; $r = 2 \csc \theta$
28. $e = 4/3$; $r = -2 \csc \theta$

In problems 29–42, find the eccentricity, identify the conic, and sketch the graph. Also, use the equation to locate the center (if applicable) and vertices in Cartesian form.

29. $r = \dfrac{5}{1 + \sin \theta}$
30. $r = \dfrac{2}{-1 + 3 \sin \theta}$
31. $r = \dfrac{2}{2 - \cos \theta}$
32. $r = \dfrac{3}{4 + 2 \sin \theta}$
33. $r = \dfrac{5}{3 + 9 \cos \theta}$
34. $r = \dfrac{10}{2 - 5 \cos \theta}$
35. $r = \dfrac{2}{2 + \cos \theta}$
36. $r = \dfrac{2}{1 + \sin\left(\theta + \dfrac{\pi}{2}\right)}$
37. $r = \dfrac{6 \csc \theta}{3 \csc \theta + 2}$
38. $r = \dfrac{8 \sec \theta}{2 \sec \theta - 1}$
39. $r = \dfrac{2}{\cos \theta - 2}$
40. $r = \dfrac{3}{2 - \sin \theta}$
41. $r = \dfrac{1}{2 - \cos (\theta - \pi)}$
42. $r = \dfrac{1}{2 - 2 \cos\left(\theta - \dfrac{\pi}{3}\right)}$

Applying the Concepts

43. **Earth's Orbit:** The orbit of the earth is an ellipse with the sun at one focus. The minimum and maximum distances from the center of the earth to the center of the sun have a ratio of about 29/30.

    (a) Find the eccentricity of this elliptical orbit.
    (b) The closest distance between the center of the earth and the center of the sun is approximately 91.3 million miles. What is the farthest distance between the earth and the sun?

44. **Kahoutek's Comet:** Kahoutek's Comet has an elliptical orbit with eccentricity $e = 0.999925$, with the sun at a focus. The minimum distance from the center of the comet to the center of the sun is about 0.13 AU.

    (a) What is the maximum distance from the center of the comet to the center of the sun (in miles)?
    (b) Find the standard equation for the orbit of the comet, with the sun at the origin and the major axis along the x axis.

(c) How far from the center of the sun is the center of the comet when a line drawn to the sun is perpendicular to the major axis?

(d) Where is the directrix of the ellipse that is nearest the sun?

45. Window Design: A designer of stained glass windows wishes to make a window in the shape of an ellipse with eccentricity $e = 0.5$. As shown in Figure 8, a rectangular pane of clear glass measuring 1 foot by 2 feet is to be positioned in the center of the elliptical window so that its four corners touch the outer frame. The remaining space of the window is to be filled with a design in stained glass.

Figure 8

(a) Sketch a coordinate system with the center of the ellipse at the origin and major axis on the x axis.

(b) Find the standard Cartesian equation of the ellipse sketched in part (a).

(c) What are the lengths of the minor and major axes of the ellipse?

(d) The designer is preparing a worktable by nailing down a rectangular wood enclosure that will allow a minimum distance of 6 inches from the edge of the window. What should be the minimum dimensions of this rectangle?

46. Mercury's Orbit: The planet Mercury travels in an elliptical orbit with eccentricity $e = 0.21$, with the sun at one focus. Suppose that the length of the major axis is about 11.6×10^7 kilometers. Use the polar equation of an ellipse,

$$r = \frac{a(1 - e^2)}{1 - e \cos \theta}$$

to find the polar equation for the elliptical orbit of Mercury.

47. [G] Use a grapher to sketch the graph of the conic

$$r = \frac{2e}{1 - e \cos \theta}$$

for the given eccentricity. What happens to the shape of the conic as e gets close to 0 and as e gets larger and larger?

(a) $e = 0.25$ (b) $e = 0.5$ (c) $e = 0.90$
(d) $e = 1$ (e) $e = 1.5$ (f) $e = 10$

48. [G] Use a grapher to sketch the graph of the conic

$$r = \frac{2e}{1 + e \sin \theta}$$

for the given eccentricity. What happens to the shape of the conic as e gets close to 0 and as e gets larger and larger?

(a) $e = 0.1$ (b) $e = 0.7$ (c) $e = 0.8$
(d) $e = 1$ (e) $e = 1.4$ (f) $e = 6$

Developing and Extending the Concepts

49. What is the relationship between the eccentricities of the following conics?

$$\frac{x^2}{a^2} + \frac{y^2}{b^2} = 1 \qquad \frac{x^2}{a^2} + \frac{y^2}{b^2} = 5$$

50. What is the relationship between the eccentricities of the following conics?

$$\frac{x^2}{a^2} - \frac{y^2}{b^2} = 2 \qquad \frac{x^2}{a^2} - \frac{y^2}{b^2} = 8$$

51. How many directrices does each type of conic have?
(a) Ellipse   (b) Hyperbola   (c) Parabola
Give an example to illustrate each case.

52. How may possible ellipses are there with eccentricity $e = 1/2$ and foci at $(0, -2)$ and $(0, 2)$? Explain and illustrate.

53. Consider the equation of the hyperbola, $x^2 - y^2 = 1$. Express this equation in polar equation form and sketch its graph.

54. What is the smaller angle between asymptotes of the following hyperbola?

$$r = \frac{1}{1 - 2 \cos \theta}$$

55. [G] What are the points of intersection of the following conics? Round off the results to two decimal places.

$$r = \frac{2}{2 + \cos \theta} \qquad r = \frac{2}{2 + \sin \theta}$$

Use a grapher to display the points of intersection.

56. [G] What point on the conic

$$r = \frac{2}{2 - \cos\left(\theta + \dfrac{\pi}{2}\right)}$$

is the closest to the origin? Use a grapher to demonstrate your result.

Objectives

1. Use Formulas for Rotation of Axes
2. Eliminate the *xy* Term
3. Identify Conics Given Their General Equations

Some choices of A, B, C, D, E, *and* F *may result in a degenerate conic. That is, the graph of the equation may be the whole plane, a pair of parallel or intersecting lines, a straight line, a point, or nothing (problems 39–42).*

9.3 Rotation of Axes

So far we have learned how to graph a conic whose general equation is given by

$$Ax^2 + Cy^2 + Dx + Ey + F = 0$$

where the axes of symmetry of the conics are parallel to the *x* or *y* axis (or both).

In this section we consider conics with second-degree equations of the general form

$$Ax^2 + Bxy + Cy^2 + Dx + Ey + F = 0, B \neq 0.$$

Because of the *cross term xy* in such equations, we will see that the graphs turn out to be *rotated* conics, in the sense that the axes of symmetry are neither horizontal nor vertical.

Using Formulas for Rotation of Axes

Suppose the axes of an *xy* coordinate system are rotated through an angle θ to form a new $\overline{x}\,\overline{y}$ coordinate system. Assume *P* is a point that has coordinates (x, y) in the original system and coordinates $(\overline{x}, \overline{y})$ in the new rotated system (Figure 1a). How are the coordinates (x, y) related to the coordinates $(\overline{x}, \overline{y})$?

Figure 1

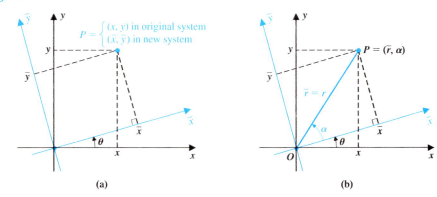

(a)   (b)

The easiest way to answer this question is to use polar coordinates. Suppose that the polar coordinates of *P* relative to the $\overline{x}\,\overline{y}$ coordinate system are $(\overline{r}, \alpha)$, where $\overline{r} = r = |\overline{OP}|$ (Figure 1b). Then the Cartesian coordinates of *P* relative to the $\overline{x}\,\overline{y}$ coordinate system are given by

$$\overline{x} = |\overline{OP}| \cos \alpha \quad \text{and} \quad \overline{y} = |\overline{OP}| \sin \alpha$$

or $\qquad\qquad \overline{x} = \overline{r} \cos \alpha \quad \text{and} \quad \overline{y} = \overline{r} \sin \alpha.$

Now we consider the same point *P* relative to the *xy* coordinate system. To find its polar coordinates, we observe in Figure 1b that one possible angle is $\theta + \alpha$. Thus Cartesian coordinates of *P* relative to the original unrotated *xy* axes are

$$x = |\overline{OP}| \cos(\alpha + \theta) \quad \text{and} \quad y = |\overline{OP}| \sin(\alpha + \theta)$$

or $\qquad\qquad x = \overline{r} \cos(\alpha + \theta) \quad \text{and} \quad y = \overline{r} \sin(\alpha + \theta).$

That is,

$$x = \bar{r}\cos(\alpha + \theta) = \bar{r}\cos\alpha\cos\theta - \bar{r}\sin\alpha\sin\theta \qquad \text{Cosine of a sum identity}$$

$$= \bar{x}\cos\theta - \bar{y}\sin\theta \qquad \text{Substitute } \bar{x} \text{ and } \bar{y}$$

and

$$y = \bar{r}\sin(\alpha + \theta) = \bar{r}\sin\alpha\cos\theta + \bar{r}\cos\alpha\sin\theta \qquad \text{Sine of a sum identity}$$

$$= \bar{y}\cos\theta + \bar{x}\sin\theta \qquad \text{Substitute } \bar{y} \text{ and } \bar{x}$$

Thus we obtain the following rotation formulas:

$$x = \bar{x}\cos\theta - \bar{y}\sin\theta$$
$$y = \bar{x}\sin\theta + \bar{y}\cos\theta.$$

These formulas express the original coordinates (x, y) in terms of the new coordinates $(\bar{x}, \bar{y})$.

The rotation formulas given above can be solved for $\bar{x}$ and $\bar{y}$ (problem 43) to obtain

$$\bar{x} = x\cos\theta + y\sin\theta$$
$$\bar{y} = -x\sin\theta + y\cos\theta.$$

These latter formulas express the new coordinates $(\bar{x}, \bar{y})$ relative to the rotated axes in terms of the original coordinates (x, y).

EXAMPLE 1   **Finding the Coordinates of a Point Relative to Rotated Axes**

Assume that a new $\bar{x}\bar{y}$ coordinate system is formed by rotating the xy coordinate system through an angle of $60°$. Find the coordinates of P with respect to the $\bar{x}\bar{y}$ coordinate system if $P = (-1, 7)$ with respect to the xy system. Display the point relative to both pairs of axes.

Solution   Here we can use the second set of rotation formulas with

$$x = -1, \quad y = 7, \quad \text{and} \quad \theta = 60° \quad \text{to get}$$

$$\begin{cases} \bar{x} = -1\cos 60° + 7\sin 60° = -\dfrac{1}{2} + \dfrac{7\sqrt{3}}{2} = \dfrac{-1 + 7\sqrt{3}}{2} \\[2mm] \bar{y} = -(-1)\sin 60° + 7\cos 60° = \dfrac{\sqrt{3}}{2} + \dfrac{7}{2} = \dfrac{\sqrt{3} + 7}{2}. \end{cases}$$

Thus

$$P = \left(\dfrac{-1 + 7\sqrt{3}}{2}, \dfrac{\sqrt{3} + 7}{2}\right) = (6.56, 4.37) \text{ (approx.)}$$

Figure 2

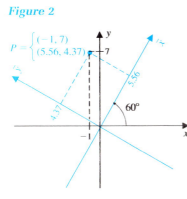

in the $\bar{x}\bar{y}$ system. The point P is shown in Figure 2 relative to both coordinate systems.

EXAMPLE 2   **Rewriting an Equation in Terms of a Rotated Coordinate System**

Assume that an xy coordinate system is rotated through a 45° angle to form an $\overline{x}\overline{y}$ coordinate system. Rewrite the equation $xy = 2$ in terms of the $\overline{x}\overline{y}$ coordinate system and sketch the graph.

Solution   For $\theta = 45°$, the rotation formulas that express x and y in terms of $\overline{x}$ and $\overline{y}$ yield

$$\begin{cases} x = \overline{x}\cos 45° - \overline{y}\sin 45° = \overline{x}\left(\dfrac{\sqrt{2}}{2}\right) - \overline{y}\left(\dfrac{\sqrt{2}}{2}\right) = \dfrac{\sqrt{2}}{2}(\overline{x} - \overline{y}) \\[2mm] y = \overline{x}\sin 45° + \overline{y}\cos 45° = \overline{x}\left(\dfrac{\sqrt{2}}{2}\right) + \overline{y}\left(\dfrac{\sqrt{2}}{2}\right) = \dfrac{\sqrt{2}}{2}(\overline{x} + \overline{y}). \end{cases}$$

A rotation of axes does not alter the shape or geometric characteristics of the conic.

Next we substitute the expressions for x and y into the given equation $xy = 2$ to get

$$\left[\frac{\sqrt{2}}{2}(\overline{x} - \overline{y})\right] \cdot \left[\frac{\sqrt{2}}{2}(\overline{x} + \overline{y})\right] = 2$$

$$\frac{1}{2}(\overline{x}^2 - \overline{y}^2) = 2 \qquad \text{\color{teal}Multiply}$$

$$\frac{\overline{x}^2}{4} - \frac{\overline{y}^2}{4} = 1 \qquad \text{\color{teal}Divide by 2}$$

This is the standard equation of a hyperbola, with

$$a = 2,\; b = 2, \quad \text{and} \quad c = \sqrt{a^2 + b^2} = \sqrt{4 + 4} = \sqrt{8} = 2\sqrt{2}.$$

Figure 3

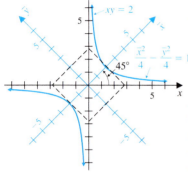

To graph the hyperbola in the $\overline{x}\overline{y}$ coordinate system, we start at the origin and find the vertices by moving 2 units in each direction along the $\overline{x}$ axis. We note that the extremities of the conjugate axis are found by moving 2 units in each direction from the origin along the $\overline{y}$ axis. Using this information, we sketch the graph (Figure 3). Note that the asymptotes for this graph are the original xy coordinate axes.

Eliminating the xy Term

Example 2 illustrates a process that enables us to convert a general equation of a conic containing an xy term to a standard equation (for the same graph) expressed in terms of a rotated $\overline{x}\overline{y}$ coordinate system where it is easier to graph. However, in the example, we were given the angle of rotation that accomplishes the conversion. Next we turn our attention to determining a way of finding an appropriate angle of rotation.

Suppose that a conic has the general equation

The given equation contains an xy term.

$$Ax^2 + Bxy + Cy^2 + Dx + Ey + F = 0, \quad B \neq 0.$$

By using the rotation formulas

$$\begin{cases} x = \overline{x}\cos\theta - \overline{y}\sin\theta \\ y = \overline{x}\sin\theta + \overline{y}\cos\theta \end{cases}$$

with an appropriate selection for θ, we can transform the given equation to an equation of the form

Note that the converted equation has no $\overline{x}\,\overline{y}$ term $(\overline{B} = 0)$.

$$\overline{A}\,\overline{x}^2 + \overline{C}\,\overline{y}^2 + \overline{D}\,\overline{x} + \overline{E}\,\overline{y} + \overline{F} = 0.$$

We refer to this process as **eliminating the *xy* term,** and we can accomplish this if we select an angle θ that satisfies

$$\cot 2\theta = \frac{A - C}{B}$$

(problem 44). For instance, in Example 2, the equation of the conic, given by $xy = 2$, has the general equation

$$xy - 2 = 0$$

where $A = 0$, $B = 1$, and $C = 0$.

Consequently, we choose θ to satisfy

$$\cot 2\theta = \frac{0 - 0}{1} = 0.$$

Normally we select the smallest positive angle for θ.

So $2\theta = 90°$, or $\theta = 45°$, will work. Note that this is precisely the angle we used in Example 2 to eliminate the *xy* term.

EXAMPLE 3

Finding an Angle of Rotation to Eliminate the *xy* Term

Find an acute angle of rotation that eliminates the *xy* term in the equation

$$21x^2 - 10\sqrt{3}\, xy + 31y^2 = 144.$$

Rewrite the equation in terms of the rotated axes $\overline{x}\,\overline{y}$, then sketch the graph showing both sets of coordinate axes.

Solution

The equation $21x^2 - 10\sqrt{3}\, xy + 31y^2 = 144$ can be written in the form

$$Ax^2 + Bxy + Cy^2 + Dx + Ey + F = 0$$

with $A = 21$, $B = -10\sqrt{3}$, $C = 31$ and $F = -144$.

So

$$\cot 2\theta = \frac{A - C}{B}$$

$$= \frac{21 - 31}{-10\sqrt{3}}$$

$$= \frac{-10}{-10\sqrt{3}} = \frac{1}{\sqrt{3}}.$$

Thus, $2\theta = 60°$, or $\theta = 30°$. The rotation formulas that will eliminate the *xy* term are given by

$$\begin{cases} x = \overline{x}\cos 30° - \overline{y}\sin 30° = \dfrac{\sqrt{3}}{2}\overline{x} - \dfrac{1}{2}\overline{y} \\[2mm] y = \overline{x}\sin 30° + \overline{y}\cos 30° = \dfrac{1}{2}\overline{x} + \dfrac{\sqrt{3}}{2}\overline{y}. \end{cases}$$

To rewrite the equation in terms of $\bar{x}$ and $\bar{y}$, we substitute these expressions for x and y into the given equation to get

$$21\left(\frac{\sqrt{3}}{2}\,\bar{x} - \frac{1}{2}\,\bar{y}\right)^2 - 10\sqrt{3}\left(\frac{\sqrt{3}}{2}\,\bar{x} - \frac{1}{2}\,\bar{y}\right)\left(\frac{1}{2}\,\bar{x} + \frac{\sqrt{3}}{2}\,\bar{y}\right)$$

$$+ 31\left(\frac{1}{2}\,\bar{x} + \frac{\sqrt{3}}{2}\,\bar{y}\right)^2 = 144.$$

Figure 4

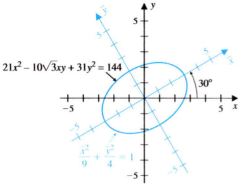

$21x^2 - 10\sqrt{3}xy + 31y^2 = 144$

$\dfrac{\bar{x}^2}{9} + \dfrac{\bar{y}^2}{4} = 1$

$30°$

After simplifying the left side of the equation, we obtain

$$16\bar{x}^2 + 36\bar{y}^2 = 144$$

or

$$\frac{\bar{x}^2}{9} + \frac{\bar{y}^2}{4} = 1.$$

This latter equation is the standard equation of an ellipse. After drawing the $\bar{x}\bar{y}$ coordinate axes by rotating the xy coordinate axes 30°, we sketch the ellipse relative to the $\bar{x}\bar{y}$ system by using $a = 3$ and $b = 2$, where the major axis lies on the $\bar{x}$ axis (Figure 4). 🔳

Inequalities of the form

$$Ax^2 + Bxy + Cy^2 + Dx + Ey + F < 0$$

or

$$Ax^2 + Bxy + Cy^2 + Dx + Ey + F > 0$$

are solved similar to the way we solved linear inequalities involving two variables in Section 8.6, as the next example illustrates.

EXAMPLE 4 **Solving an Inequality Graphically**

Use the graph of

$$21x^2 - 10\sqrt{3}\,xy + 31y^2 = 144 \quad \text{(see Example 3)}$$

to determine the region that represents the solution of

$$21x^2 - 10\sqrt{3}\,xy + 31y^2 \le 144.$$

Solution The given inequality can be rewritten as

$$21x^2 - 10\sqrt{3}\,xy + 31y^2 - 144 \le 0.$$

The graph of

$$21x^2 - 10\sqrt{3}\,xy + 31y^2 - 144 = 0$$

is shown in Figure 4. It divides the plane into two regions—the one inside the ellipse and the one outside the ellipse. To test the inside region, we select the point

Figure 5

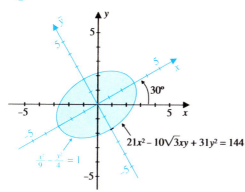

$21x^2 - 10\sqrt{3}xy + 31y^2 = 144$

$\frac{x^2}{9} - \frac{y^2}{4} = 1$

with coordinates (2, 0) and substitute $x = 2$ and $y = 0$ into the above expression to get

$$21(2)^2 - 10\sqrt{3}\,(2)(0) + 31(0)^2 - 144 = -60 < 0.$$

So (2, 0) satisfies the inequality. Thus all points inside the ellipse satisfy it.
Next we test the outside region by selecting the point (4, 0). Upon substituting the values $x = 4$ and $y = 0$ into the above expression we get

$$21(4)^2 - 10\sqrt{3}\,(4)(0) + 31(0)^2 - 144 = 192 > 0.$$

So (4, 0) does not satisfy the inequality. Thus no point outside the ellipse satisfies it.

The solution, which is shown in Figure 5, includes the curve because of the equality.

Right triangle trigonometry and trigonometric formulas can be used to eliminate the *xy* term without actually finding the angle of rotation, as the next example shows.

EXAMPLE 5   Eliminating the xy term by Using Trigonometric Identities

(a) Use the half-angle trigonometric formulas to find the values of $\sin\theta$ and $\cos\theta$ for the rotation formulas that eliminate the *xy* term in the equation

$$9x^2 + 12xy + 4y^2 + 2x - 3y = 0.$$

Then rewrite the equation in terms of the rotated $\overline{x}\,\overline{y}$ axes.

(b) Approximate the acute angle of rotation (to two decimal places) and sketch the graph, displaying both sets of the coordinate axes.

Solution   (a) In this case, $A = 9$, $B = 12$, and $C = 4$, so

$$\cot 2\theta = \frac{A - C}{B}$$

$$= \frac{9 - 4}{12} = \frac{5}{12}.$$

Figure 6

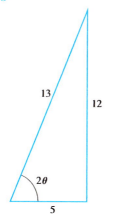

We use right triangle trigonometry and trigonometric identities to determine the appropriate rotation formulas as follows:
First we construct a right triangle that satisfies

$$\cot 2\theta = \frac{5}{12} \text{ (Figure 6)}$$

Upon examining the triangle, we see that $\cos 2\theta = 5/13$. Next we use the half-angle trigonometric formula to get

$$\sin\theta = \sqrt{\frac{1 - \cos 2\theta}{2}} = \sqrt{\frac{1 - \dfrac{5}{13}}{2}} = \sqrt{\frac{\dfrac{8}{13}}{2}} = \sqrt{\frac{4}{13}} = \frac{2}{\sqrt{13}}$$

and

$$\cos\theta = \sqrt{\frac{1 + \cos 2\theta}{2}} = \sqrt{\frac{1 + \dfrac{5}{13}}{2}} = \sqrt{\frac{\dfrac{18}{13}}{2}} = \sqrt{\frac{9}{13}} = \frac{3}{\sqrt{13}}.$$

So the rotation formulas become

$$\begin{cases} x = \bar{x}\cos\theta - \bar{y}\sin\theta = \dfrac{3}{\sqrt{13}}\bar{x} - \dfrac{2}{\sqrt{13}}\bar{y} \\[2mm] y = \bar{x}\sin\theta + \bar{y}\cos\theta = \dfrac{2}{\sqrt{13}}\bar{x} + \dfrac{3}{\sqrt{13}}\bar{y}. \end{cases}$$

To rewrite the given equation, we substitute these expressions into the original equation and simplify to get

$$\bar{y} = \sqrt{13}\,\bar{x}^2.$$

This is the equation of a parabola.

(b) Since $\cos\theta = 3/\sqrt{13}$, then

$$\theta = \cos^{-1}\frac{3}{\sqrt{13}} = 33.7° \text{ (approx.)}.$$

After rotating the xy system, we graph the parabola in the $\bar{x}\bar{y}$ system (Figure 7).

Figure 7

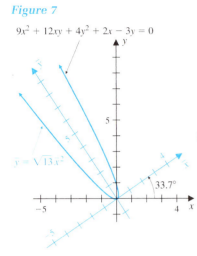

$9x^2 + 12xy + 4y^2 + 2x - 3y = 0$

$\bar{y} = \sqrt{13}\,\bar{x}^2$

33.7°

Identifying Conics Given Their General Equations

When the equation of a conic is written in the general form

$$Ax^2 + Bxy + Cy^2 + Dx + Ey + F = 0$$

it is possible to determine whether the conic is an ellipse, a parabola, or a hyperbola before graphing it by using the following property.

Property

Discriminant of a Conic

The graph of the equation

$$Ax^2 + Bxy + Cy^2 + Dx + Ey + F = 0$$

is a conic or one of its degenerate forms. We can identify the type of conic by using the **discriminant** $B^2 - 4AC$. The conic is:

1. An *ellipse* if $B^2 - 4AC < 0$
2. A *parabola* if $B^2 - 4AC = 0$
3. A *hyperbola* if $B^2 - 4AC > 0$

EXAMPLE 6   **Identifying Conics by Using the Discriminant**

Use the discriminant to identify the conic with the given equation

(a) $xy = 2$ (see Example 2)

(b) $21x^2 - 10\sqrt{3}\,xy + 31y^2 = 144$ (see Example 3)

(c) $9x^2 + 12xy + 4y^2 + 2x - 3y = 0$ (see Example 5)

Solution First we express each equation in the general form

$$Ax^2 + Bxy + Cy^2 + Dx + Ey + F = 0.$$

Then after calculating the discriminant, we use the above discriminant property. The results are given in Table 1.

TABLE 1

Given Equation	General Form $Ax^2 + Bxy + Cy^2 + Dx + Ey + F = 0$	Discriminant $B^2 - 4AC$	Conclusion
(a) $xy = 2$	$0x^2 + 1xy + 0y^2 + 0x + 0y - 2 = 0$	$1 - 0 = 1$	Hyperbola
(b) $21x^2 - 10\sqrt{3}xy + 31y^2 = 144$	$21x^2 + (-10\sqrt{3})xy + 31y^2 + 0x + 0y - 144 = 0$	$(-10\sqrt{3})^2 - 4(21)(31)$ $= -2304$	Ellipse
(c) $9x^2 + 12xy + 4y^2 + 2x - 3y = 0$	$9x^2 + 12xy + 4y^2 + 2x - 3y + 0 = 0$	$12^2 - 4(9)(4) = 0$	Parabola

G Some graphers are programmed to graph the equation of a conic given in the general form

$$Ax^2 + Bxy + Cy^2 + Dx + Ey + F = 0$$

merely by inputting the values of the coefficients A, B, C, D, E, and F.

If this feature is not available then we need to express y in terms of x and then use the grapher as we have done earlier.

For instance, consider the general equation

$$5x^2 - 8xy + 5y^2 = 10.$$

In this situation, $A = 5$, $B = -8$, and $C = 5$ in the general form

$$5x^2 + (-8)xy + 5y^2 - 10 = 0.$$

So the discriminant is given by

$$B^2 - 4AC = (-8)^2 - 4(5)(5) = -36.$$

Thus the conic is an ellipse.

To express y in terms of x, we first rewrite the equation as a standard quadratic equation by treating y as the variable and the other terms as constants:

$$5y^2 + (-8x)y + (5x^2 - 10) = 0$$

Next we use the quadratic formula to solve for y.
By letting $a = 5$, $b = -8x$, and $c = 5x^2 - 10$, we get

$$y = \frac{-b \pm \sqrt{b^2 - 4ac}}{2a}$$

$$= \frac{8x \pm \sqrt{64x^2 - 4(5)(5x^2 - 10)}}{10}.$$

Figure 8

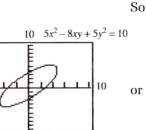

So

$$y_1 = \frac{8x - \sqrt{200 - 36x^2}}{10}$$

or

$$y_2 = \frac{8x + \sqrt{200 - 36x^2}}{10}.$$

Figure 8 shows a viewing window containing the graphs of both functions in the same coordinate system. Together they give the graph of the original equation.

PROBLEM SET 9.3

Mastering the Concepts

In problems 1–6, assume that new coordinate axes $\overline{x}\,\overline{y}$ are obtained by rotating the old coordinate axes xy through angle θ. Let P be a point with coordinates (x, y) in the old system and coordinates $(\overline{x}, \overline{y})$ in the new system. Display the point relative to both pairs of axes. In problems 5 and 6, round off each answer to two decimal places.

1. If $(x, y) = (-2, -5)$ and $\theta = 30°$, find $(\overline{x}, \overline{y})$.
2. If $(\overline{x}, \overline{y}) = (0, 3\sqrt{2})$ and $\theta = 60°$, find (x, y).
3. If $(\overline{x}, \overline{y}) = (1, -10)$ and $\theta = 45°$, find (x, y).
4. If $(x, y) = (-2, 5)$ and $\theta = 30°$, find $(\overline{x}, \overline{y})$.
5. If $(x, y) = (3.52, 5.73)$ and $\theta = 55°$, find $\overline{x}, \overline{y})$.
6. If $(\overline{x}, \overline{y}) = (-4.71, 2.13)$ and $\theta = 22°$, find (x, y).

In problems 7–14, assume an xy coordinate system is rotated through angle θ to form an $\overline{x}\,\overline{y}$ coordinate system. Rewrite the given equation in terms of the $\overline{x}\,\overline{y}$ system and sketch the graph. Display both sets of coordinate axes.

7. $xy = 1; \theta = 45°$
8. $x^2 - 4xy + y^2 - 6 = 0; \theta = \pi/4$
9. $x^2 - \sqrt{3}\,xy + 2y^2 = 4; \theta = \pi/6$
10. $5x^2 - 2xy + 5y^2 = 12; \theta = 45°$
11. $7x^2 - 6\sqrt{3}\,xy + 13y^2 = 16; \theta = 30°$
12. $7x^2 - 2\sqrt{3}\,xy + 5y^2 = 16; \theta = 30°$
13. $31x^2 + 10\sqrt{3}\,xy + 21y^2 = 144; \theta = 30°$
14. $2x + 3y = 6; \theta = \pi/6$

In problems 15–20, determine an angle θ for a rotation of axes that will eliminate the xy term from the given equation. Rewrite the equation in terms of the rotated axes $\overline{x}$ $\overline{y}$ and graph it, showing both sets of coordinate axes.

15. $5x^2 - 4xy + 5y^2 = 9$
16. $2x^2 + 4xy + 2y^2 = 14$
17. $11x^2 + 10\sqrt{3}\,xy + y^2 = 64$
18. $xy = 8$
19. $2x^2 + 4\sqrt{3}\,xy - 2y^2 = 4$
20. $3x^2 + 6\sqrt{3}\,xy + 9y^2 = 32$

In problems 21–24, use the graph of the given equation to determine the region that contains the solution of the associated inequality.

21. $xy = 2;$
    $xy < 2$ (see Example 2)
22. $2x^2 + \sqrt{3}\,xy + y^2 = 10;$
    $2x^2 + \sqrt{3}\,xy + y^2 = 10$
23. $2y^2 - \sqrt{3}\,xy + x^2 - 4 = 0;$
    $2y^2 - \sqrt{3}\,xy + x^2 - 4 \geq 0$
24. $x^2 - 4xy + y^2 - 25 = 0;$
    $x^2 - 4xy + y^2 - 25 > 0$

In problems 25–30, find the values of $\sin \theta$ and $\cos \theta$ for rotation formulas that eliminate the xy term in each equation. Write the equation in terms of the rotated $\overline{x}\,\overline{y}$ system. Determine the angle of rotation (to one decimal place) and sketch the graph, displaying both sets of coordinate axes.

25. $8x^2 - 4xy + 5y^2 = 144$
26. $x^2 - 3xy + 5y^2 - 22 = 0$
27. $5x^2 - 4xy + 8y^2 = 144$
28. $16x^2 - 24xy + 9y^2 = 80$
29. $4x^2 + 4xy + y^2 + 20x - 10y = \sqrt{5}$
30. $41x^2 - 24xy + 34y^2 = 29$

In problems 31–36, use the discriminant to determine whether the graph of the given equation is an ellipse, a parabola, or a hyperbola.

31. $x^2 - xy + y^2 = 5$
32. $x^2 + 2xy + y^2 = 16$
33. $x^2 + 24xy - 6y^2 = 30$
34. $17x^2 - 12xy + 8y^2 = 60$
35. $17x^2 - 12xy + 8y^2 = 12$
36. $24xy - 7y^2 + 36 = 0$

Developing and Extending the Concepts

37. Find an angle θ (if it exists) for which the rotation formulas give the following equations:
(a) $\bar{x} = y$ and $\bar{y} = -x$
(b) $\bar{x} = -y$ and $\bar{y} = -x$

38. **Matrix Form of a Rotation of Axes:**
(a) Multiply

$$\begin{bmatrix} \cos\theta & -\sin\theta \\ \sin\theta & \cos\theta \end{bmatrix}\begin{bmatrix} \bar{x} \\ \bar{y} \end{bmatrix}$$

and compare the result to the first set of rotation formulas on page 627.
(b) Multiply

$$\begin{bmatrix} \cos\theta & \sin\theta \\ -\sin\theta & \cos\theta \end{bmatrix}\begin{bmatrix} x \\ y \end{bmatrix}$$

and compare the result to the second set of rotation formulas on page 627.
(c) Use the matrix in part (b) to determine a matrix that will yield the $\bar{x}\bar{y}$ coordinates of a given point (x, y) if the coordinate axes are rotated $45°$.

In problems 39–42, each equation has a graph that is a degenerate conic. Verify that the graph is as described.

39. $x^2 + y^2 + 1 = 0$; the empty set (that is, no points)
40. $x^2 + y^2 - 6x + 4y + 13 = 0$; a single point
41. $2x^2 - 4xy + 2y^2 = 0$; a straight line
42. $0x^2 + 0xy + 0y^2 + 0x + 0y + 0 = 0$; the whole xy plane

43. Given the rotation formulas

$$\begin{cases} x = \bar{x}\cos\theta - \bar{y}\sin\theta \\ y = \bar{x}\sin\theta + \bar{y}\cos\theta \end{cases}$$

use the elimination method to solve for $\bar{x}$ and $\bar{y}$ in terms of x, y, $\cos\theta$, and $\sin\theta$, thus obtaining the rotation formulas

$$\begin{cases} \bar{x} = x\cos\theta + y\sin\theta \\ \bar{y} = -x\sin\theta + y\cos\theta \end{cases}$$

44. (a) Substitute the rotation formulas that express x and y (old coordinates) in terms of $\bar{x}$ and $\bar{y}$ (new coordinates) into the general quadratic form

$$Ax^2 + Bxy + Cy^2 + Dx + Ey + F = 0$$

to determine that $\bar{B}$ in the transformed equation

$$\bar{A}\bar{x}^2 + \bar{B}\bar{x}\bar{y} + \bar{C}\bar{y}^2 + \bar{D}\bar{x} + \bar{E}\bar{y} + \bar{F} = 0$$

is given by

$$\bar{B} = 2(C - A)\sin\theta\cos\theta + B(\cos^2\theta - \sin^2\theta)$$

(b) Set $\bar{B} = 0$ and verify that

$$\cot 2\theta = \frac{A - C}{B}.$$

[G] In problems 45–50, solve each equation for y in terms of x. Then use a grapher to graph the resulting equations in the same viewing window to obtain a graph of the given equation.

45. $x^2 - xy + y^2 = 4$
46. $5x^2 - 6xy + y^2 = 8$
47. $10xy - y^2 = 32$
48. $4x^2 + 4xy - y^2 = 10$
49. $2x^2 + 3xy + y^2 = 18$
50. $31x^2 + 10xy - y^2 = 14$

Objectives

1. Graph Parametric Equations by Point-Plotting
2. Eliminate the Parameter
3. Represent Cartesian Equations Parametrically
4. Solve Applied Problems

9.4 Parametric Equations

So far we have generated curves that result from graphing equations either relating x and y (in the Cartesian system) or r and θ (in the polar system). There is another way of generating curves that is used in modeling the path of a moving object. These curves are obtained from equations that express each x and y as functions of a third variable, say t (which represents time in many applications).

For instance, suppose that a particle moves in the xy plane in such a way that its x and y coordinates are functions of time t, say

$$\begin{cases} x = f(t) \\ y = g(t) \end{cases} \quad t \geq 0.$$

Then these functions taken together describe the path that the particle traces out in the *xy* plane as time *t* elapses.

In general, we have the following definition:

Definition
Parametric Equations
in the Plane

Suppose that the points (x, y) on a curve in the plane are defined by the functions

$$\begin{cases} x = f(t) \\ y = g(t) \end{cases}$$

where *t* is a real number in a given interval. The functions are called **parametric equations** for the curve, and *t* is referred to as the **parameter.**

Some examples of parametric equations for curves in the plane are

$$\begin{cases} x = t + 1 \\ y = t^2 \end{cases} \qquad \begin{cases} x = 2\cos t \\ y = 3\sin t \end{cases} \ 0 \le t \le 2\pi \qquad \begin{cases} x = t^2 \\ y = t^2 + 1 \end{cases} \ 0 \le t \le 1.$$

Graphing Parametric Equations by Point-Plotting

We can use point-plotting to graph a curve represented by parametric equations. This is done by evaluating *x* and *y* for some values of *t*, plotting the corresponding points, and then connecting them with a smooth curve, if appropriate.

By monitoring the formation of the curve as the values of *t* increase, we can see how the curve is traced out, not just its shape.

EXAMPLE 1   **Graphing Parametric Equations by Point-Plotting**

Sketch the curve represented by the parametric equations by using point-plotting.

$$\begin{cases} x = t + 1 \\ y = t^2 \end{cases} \ 0 \le t \le 5.$$

Explain how the curve is being generated as the values of *t* increase.

Solution   We begin by making a table of values of *t* and the corresponding values of *x* and *y* (Table 1). By plotting the (x, y) points in Table 1 and then connecting them with a smooth curve, we get a sketch of the graph (Figure 1).

TABLE 1

t	$x = t + 1$	$y = t^2$	(x, y)
0	1	0	$(1, 0)$
1	2	1	$(2, 1)$
2	3	4	$(3, 4)$
3	4	9	$(4, 9)$
4	5	16	$(5, 16)$
5	6	25	$(6, 25)$

Figure 1

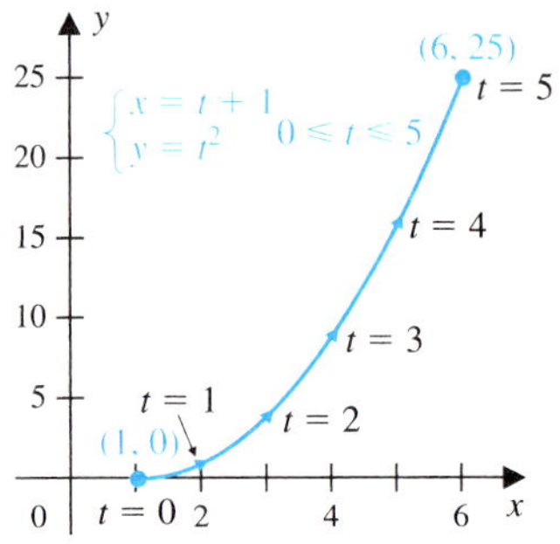

As the values of *t* increase from 0 to 5, the corresponding points on the graph trace out the curve starting at the point $(1, 0)$ and ending at the point $(6, 25)$. This *orientation* is denoted by the arrowheads in Figure 1.

Eliminating the Parameter

We can often derive an equation in terms of x and y for the curve defined by parametric equations. This process is referred to as **eliminating the parameter.**

For instance, in Example 1 where

$$\begin{cases} x = t + 1 \\ y = t^2 \end{cases} \quad 0 \le t \le 5$$

Eliminating the parameter often helps us to recognize the shape of the curve. However, it is important to examine the parametric form when determining the orientation of the curve.

we can eliminate the parameter algebraically as follows:
First solve the equation $x = t + 1$ for t to get $t = x - 1$.
Then substitute $x - 1$ for t in the second equation, $y = t^2$, to obtain

$$y = (x - 1)^2.$$

The condition $0 \le t \le 5$ requires that $0 \le x - 1 \le 5$, or $1 \le x \le 6$.
So the graph in Figure 1 is actually a portion of the parabola $y = (x - 1)^2$ where $1 \le x \le 6$.

The process of eliminating the parameter sometimes involves trigonometric identities.

EXAMPLE 2   **Eliminating the Parameter Using Trigonometry**

In many applications, t represents time, but it could denote other quantities such as an angle measure.

Eliminate the parameter and then sketch the curve of the parametric equations

$$\begin{cases} x = 2 \cos t \\ y = 3 \sin t \end{cases} \quad 0 \le t \le 2\pi.$$

Describe the orientation of a point on the curve as t varies from 0 to 2π.

Solution   First we solve the equation $x = 2 \cos t$ for $\cos t$ and the second equation for $\sin t$ to obtain

$$\cos t = \frac{x}{2} \quad \text{and} \quad \sin t = \frac{y}{3}.$$

Squaring both sides of each equation produces

$$\cos^2 t = \frac{x^2}{4} \quad \text{and} \quad \sin^2 t = \frac{y^2}{9}.$$

Figure 2

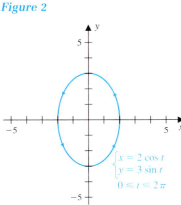

Adding these two equations, we have

$$\cos^2 t + \sin^2 t = \frac{x^2}{4} + \frac{y^2}{9}.$$

Since $\cos^2 t + \sin^2 t = 1$, we get

$$\frac{x^2}{4} + \frac{y^2}{9} = 1$$

which is an equation of an ellipse with center $(0, 0)$, semimajor axis of length 3 on the y axis, and semiminor axis of length 2 on the x axis (Figure 2).

The restriction $0 \leq t \leq 2\pi$ suggests that the orientation of a point moving along the elliptical curve, as t increases from 0 to 2π, starts at point

$$(2 \cos 0, \, 3 \sin 0) = (2, 0)$$

passes through points

$$(2 \cos(\pi/2), \, 3 \sin(\pi/2)) = (0, 3)$$

and

$$(2 \cos \pi, \, 3 \sin \pi) = (-2, 0)$$

and continues around the ellipse to end at

$$(2 \cos 2\pi, \, 3 \sin 3\pi) = (2, 0).$$

In other words, a point moves along the ellipse in a counterclockwise direction and traverses the ellipse once as t increases from 0 to 2π (Figure 2). ▨

Representing Cartesian Equations Parametrically

We now consider the reverse problem—that is, finding a parametric representation for the graph of an equation in x and y.

If a curve is described by the function $y = f(x)$, we can let $x = t$ to obtain the following parametric equations for this curve:

$$\begin{cases} x = t \\ y = f(t) \end{cases} \quad \text{where } t \text{ is in the domain of } f$$

For example, the curve defined by

$$y = 3x^2 + 7,$$

where $0 \leq x \leq 10$, can be obtained from the following parametric equations:

$$\begin{cases} x = t \\ y = 3t^2 + 7 \end{cases} \quad 0 \leq t \leq 10$$

However, parametric representations for curves are not unique. For instance, if we let $x = 2t$, then another set of parametric equations for $y = 3x^2 + 7$, where $0 \leq x \leq 10$, is given by

$$\begin{cases} x = 2t \\ y = 3(2t)^2 + 7 \end{cases} \quad \text{or} \quad \begin{cases} x = 2t \\ y = 12t^2 + 7. \end{cases}$$

Since

$$x = 2t \quad \text{and} \quad 0 \leq x \leq 10,$$

then

$$0 \leq 2t \leq 10 \quad \text{or} \quad 0 \leq t \leq 5.$$

Rewriting the Cartesian equation of a curve in terms of parametric equations may involve trigonometric functions as illustrated in the next example.

EXAMPLE 3   **Representing an Equation Parametrically by Using Trigonometry**

Represent the equation of the circle

$$x^2 + y^2 = 4$$

parametrically if the orientation of points on this curve starts at (2, 0) and traverses the circle counterclockwise once.

Solution   Using Figure 3 and trigonometry, we see that the equations

Figure 3

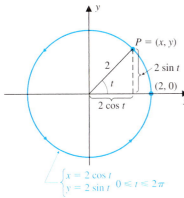

$$\begin{cases} x = 2 \cos t \\ y = 2 \sin t \end{cases} \quad 0 \le t \le 2\pi$$

give a parametric representation for every point P on the circle. We verify this result by observing that

$$x^2 + y^2 = 4 \cos^2 t + 4 \sin^2 t$$
$$= 4(\cos^2 t + \sin^2 t)$$
$$= 4.$$

The condition, $0 \le t \le 2\pi$, indicates that a point moves along the circle starting at the point (2, 0), when $t = 0$, and traverses the circle in a counterclockwise direction as t increases from 0 to 2π, ending at point (2, 0), when $t = 2\pi$.

Solving Applied Problems

Figure 4

Parametric equations are useful in modeling real-world situations, especially those in which more than one variable is dependent upon time. For example, if an object is launched from a height of s_0 feet into the air at an angle θ with the ground and with an initial speed of v_0 feet per second, then physicists have determined the position of the object after t seconds is modeled by the parametric equations

$$\begin{cases} x = (v_0 \cos \theta)t \\ y = -16t^2 + (v_0 \sin \theta)t + s_0 \end{cases} \quad t \ge 0.$$

In this model, $v_0 \cos \theta$ is the horizontal component of the initial speed and $v_0 \sin \theta$ is the vertical component of the initial speed. The constant s_0 is the vertical distance (in feet) between the ground and the point from which the object is propelled. Figure 4 shows a general graph that describes the path of the object.

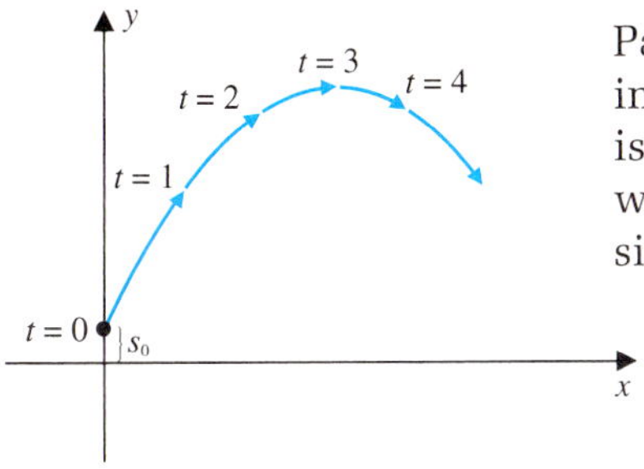

EXAMPLE 4   **Modeling the Path of a Space Shuttle**

A space shuttle rises from its launch pad and reaches a height of 30,000 feet. Because of a computer malfunction, the booster rockets are turned off prematurely, at a moment when the shuttle is moving at a speed of 32,000 feet per second at an angle of 42° with an imaginary reference plane tangent to the earth directly below.

(a) Use the model given above to determine the specific parametric equations that describe the shuttle's flight. Round off the values to the nearest thousands.

(b) Use the result in part (a) to eliminate the parameter.

(c) Find the maximum height of the space shuttle above the earth and how long it will take the space shuttle to reach its maximum height. Round off the time to the nearest 10 seconds and the height to the nearest 1,000 feet.

Solution   (a) We use the parametric equations:

$$\begin{cases} x = (v_0 \cos \theta)t \\ y = -16t^2 + (v_0 \sin \theta)t + s_0 \end{cases}$$

where $v_0 = 32{,}000$, $\theta = 42°$, and $S_0 = 30{,}000$.
The horizontal position of the space shuttle is represented by

$$x = (v_0 \cos \theta)t$$

$$= (32{,}000 \cos 42°)t = 24{,}000t \text{ (to the nearest thousands)}.$$

The vertical position is given by

$$y = -16t^2 + (32{,}000 \sin 42°)t + 30{,}000$$

$$= -16t^2 + 21{,}000t + 30{,}000 \text{ (to the nearest thousands)}.$$

Therefore, the specific parametric equations are given by

$$\begin{cases} x = 24{,}000t \\ y = -16t^2 + 21{,}000t + 30{,}000, \ t \geq 0. \end{cases}$$

(b) From part (a) we use the result $x = 24{,}000t$ to write $t = \dfrac{x}{24{,}000}$. Thus to eliminate the parameter we use substitution to get the quadratic function

$$y = -16\left(\frac{x}{24{,}000}\right)^2 + 21{,}000\left(\frac{x}{24{,}000}\right) + 30{,}000; \ x > 0$$

or

$$y = \left(\frac{-1}{36 \cdot 10^6}\right)x^2 + \frac{7}{8}x + 30{,}000; \ x \geq 0.$$

(c) Since the equation found in part (b) is a quadratic function, its graph is a parabola and its maximum occurs when

$$x = \frac{-b}{2a} = \frac{\left(-\dfrac{7}{8}\right)}{2\left(\dfrac{-1}{36 \cdot 10^6}\right)} = 15{,}750{,}000$$

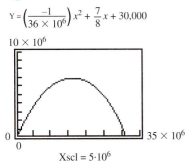

Figure 5

$Y = \left(\dfrac{-1}{36 \times 10^6}\right)x^2 + \dfrac{7}{8}x + 30{,}000$

10×10^6

35×10^6

$Xscl = 5 \cdot 10^6$
$Yscl = 10^6$

that is, when

$$t = \frac{x}{24{,}000} = \frac{15{,}750{,}000}{24{,}000} = 656.25$$

or approximately 660 seconds (about 11 minutes).
The maximum height is found by substituting $x = 15{,}750{,}000$ into the quadratic function found in part (b) to get

$$y = 6{,}921{,}000 \text{ feet (approx.)}$$

or about 1310.8 miles

G With the aid of a grapher we can graph the function found in part (b) of Example 4 (Figure 5). The graph displays the occurence of a maximum height.

Figure 6

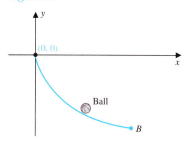

At times, parametric equations represent curves generated by physical motion in cases where it may be difficult to eliminate the parameter to find a Cartesian equation. For instance, suppose a ball at point (0, 0) is allowed to drop and then slide down a curve to point *B*, not directly under (0, 0) (Figure 6). It can be shown that the ball takes the fastest path along a curve whose parametric equations are

$$\begin{cases} x = a(t - \sin t) \\ y = a(1 - \cos t) \end{cases}$$ where *a* is a constant and *t* represents time.

When using a grapher to graph parametric equations, it is important to take into account the given restrictions on the parameter.

[G] The curve represented by these equations is called a *cycloid.* Point-plotting is impractical to graph these equations; however, it is easy to generate such a graph by using a grapher programmed to handle parametric equations.

For instance, to graph the cycloid represented by the parametric equations

$$\begin{cases} x = 2(t - \sin t) \\ y = 2(1 - \cos t) \end{cases} \quad 0 \le t \le 8\pi$$

we set the grapher in parametric mode, making certain that the restrictions on *t* are properly entered.

Figure 7 shows a viewing window of the graph of the curve. The curve is periodic with period 4π. Note that by watching certain graphers as the curve is being generated we can observe the orientation of the curve.

Figure 7

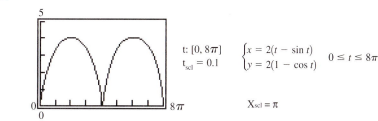

t: $[0, 8\pi]$ $\begin{cases} x = 2(t - \sin t) \\ y = 2(1 - \cos t) \end{cases}$ $0 \le t \le 8\pi$

$t_{scl} = 0.1$

$X_{scl} = \pi$

 PROBLEM SET 9.4

Mastering the Concepts

In problems 1–4, sketch the curve represented by the parametric equations by using point-plotting. Describe the orientation of a point on the curve as *t* varies over its values.

1. $\begin{cases} x = 2t + 1 \\ y = t - 1 \end{cases} \quad 0 \le t \le 4$

2. $\begin{cases} x = (1/2)t + 1 \\ y = 3t - 4 \end{cases} \quad 0 \le t \le 5$

3. $\begin{cases} x = \cos t \\ y = 4 \sin t \end{cases} \quad 0 \le t \le 2\pi$

4. $\begin{cases} x = 2 \cos t \\ y = 5 \sin t \end{cases} \quad 0 \le t \le 2\pi$

In problems 5–20, eliminate the parameter and then sketch the curve for the given parametric equations. Describe the orientation of a point on the curve as *t* varies over its values.

5. $\begin{cases} x = 1 - 3t \\ y = 2 + (5/2)t \end{cases} \quad -3 \le t \le 3$

6. $\begin{cases} x = -2 - 4t \\ y = -5 + 3t \end{cases} \quad 1 \le t \le 8$

7. $\begin{cases} x = 1 + 2t \\ y = -1 + 4t^2 \end{cases} \quad -2 \le t \le 2$

8. $\begin{cases} x = 3t^2 \\ y = 2t \end{cases} \quad -2 \le t \le 2$

9. $\begin{cases} x = t \\ y = \sqrt{1 - t^2} \end{cases}$ $-1 \le t \le 1$

10. $\begin{cases} x = -3t \\ y = \sqrt{4 + 2t} \end{cases}$ $-2 \le t \le 3$

11. $\begin{cases} x = 3 \cos t \\ y = 3 \sin t \end{cases}$ $0 \le t \le \pi$

12. $\begin{cases} x = 3 \cos t \\ y = -4 \sin t \end{cases}$ $0 \le t \le 2\pi$

13. $\begin{cases} x = -3 + \cos t \\ y = 4 - \sin t \end{cases}$ $0 \le t \le 2\pi$

14. $\begin{cases} x = 1 + \sin t \\ y = -1 + 4 \cos t \end{cases}$ $0 \le t \le 2\pi$

15. $\begin{cases} x = \csc t \\ y = \cot t \end{cases}$ $-\dfrac{\pi}{2} < t < 0$ or $0 < t < \dfrac{\pi}{2}$

16. $\begin{cases} x = \sin t \\ y = \csc t \end{cases}$ $0 < t < \pi$

17. $\begin{cases} x = e^t \\ y = e^{-2t} \end{cases}$ $t \ge 0$

18. $\begin{cases} x = t^2 \\ y = 3 \ln t \end{cases}$ $t > 0$

19. $\begin{cases} x = 2 \sec t \\ y = 3 \tan t \end{cases}$ $-\dfrac{\pi}{2} < t < \dfrac{\pi}{2}$

20. $\begin{cases} x = 2 \cos^3 t \\ y = 2 \sin^3 t \end{cases}$ $0 \le t \le 2\pi$

In problems 21–30, find parametric equations that represent the given oriented curve.

21. $y = 3x^2 + 5$; from $(0, 5)$ to $(2, 17)$
22. $y = x^3$; from $(-1, -1)$ to $(2, 8)$
23. $y = x^2 - 4x + 3$; from $(-1, 8)$ to $(1, 0)$
24. $y^2 = x$; from $(1, -1)$ to $(4, 2)$
25. Along the line segment from $(1, 2)$ to $(3, 5)$
26. Along the line segment from $(0, 2)$ to $(2, 0)$
27. Around the circle $x^2 + y^2 = 9$ once, counterclockwise starting at $(3, 0)$
28. Around the ellipse $(x^2/9) + (y^2/4) = 1$ once, counterclockwise starting at $(0, 2)$
29. Around the circle $x^2 + (y - 1)^2 = 1$ once, counterclockwise starting at $(1, 1)$
30. Along the hyperbola $x^2 - 4y^2 = 4$, where $x > 0$

Applying the Concepts

31. **Path of a Projectile:** A projectile is launched from a height of $s_0 = 10$ feet at an angle $\theta = 50°$ with the horizontal and with an initial velocity $v_0 = 200$ feet per second.
    (a) Use the model on page 639 to determine the parametric equations that describe the projectile's flight.

(b) Determine how long it takes the projectile to return to ground level (rounded off to the nearest tenth of a second). Also, find the horizontal distance the projectile has traveled to the nearest foot when it hits the ground.
(c) **G** Graph the curve.
(d) Determine the maximum height reached by the projectile to the nearest foot. When does it obtain that height?

32. **Path of a Projectile:** A projectile is launched from ground level with an initial velocity of 64 feet per second and at an angle of 45° with the horizontal.
    (a) Determine the parametric equations that model the projectile's path.
    (b) Find the time it takes for the projectile to hit the ground (to the nearest hundredth of a second). Also, find the horizontal distance the projectile has traveled when it hits the ground to the nearest foot.
    (c) **G** Graph the curve.
    (d) Determine the maximum height reached by the projectile to the nearest foot. When does it attain that height?

33. **Baseball Path:** A baseball player hits a baseball 4 feet above home plate at an angle of 25° with the horizontal and with an initial velocity of 125 feet per second.
    (a) Find parametric equations that model this path.
    (b) The outfield fence is 380 feet from home plate and the fence is 8 feet high. Determine how high the ball is when it is 380 feet from home plate. Will the ball clear the fence for a home run if it is not caught?

34. **Sports:** The parametric equations

$$\begin{cases} x = 2(t - \sin t) \\ y = -2(1 - \cos t) \end{cases}$$

describe the shape of a skateboard ramp. Locate the low point of the ramp.

Developing and Extending the Concepts

35. Find parametric equations for the curve represented by

$$2x^2 - 3xy + 2y^2 = 6$$

36. (a) Show that

$$\begin{cases} x = a \cos t + h \\ y = b \sin t + k \end{cases} \quad 0 \le t \le 2\pi$$

where $a > 0$, $b > 0$, and $a \ne b$ are parametric equations of an ellipse with center at (h, k), horizontal axis of length 2*a*, and vertical axis of length 2*b*.

(b) What is the curve if *a* = *b*?

37. Given parametric equations: $\begin{cases} x = 3 + 5t \\ y = 4 - 2t \end{cases}$

(a) Graph the parametric equations to discover that the graph is a straight line.

(b) Eliminate the parameter to find the equation of the line in terms of *x* and *y*.

(c) How does the slope of the line found in part (b) relate to the parametric equations in part (a)?

(d) Can you generalize the above results to relate the parametric equations of a line

$$\begin{cases} x = a + bt \\ y = c + dt \end{cases}$$

to the slope of the line?

38. Show that

$$\begin{cases} x = h + a \sec t \\ y = k + b \tan t \end{cases} -\frac{\pi}{2} < t < \frac{3\pi}{2}, t \neq \frac{\pi}{2}$$

where *a* > 0 and *b* > 0 are parametric equations of a hyperbola with center at (*h*, *k*), transverse axis of length 2*a*, and conjugate axis of length 2*b*.

39. Given parametric equations: $\begin{cases} x = t^2 \\ y = t^2 \end{cases}$

(a) Eliminate the parameter for the given system.

(b) Explain why the graph of the given parametric equations is different from the graph of *y* = *x*.

40. Do the following sets of parametric equations represent the same curve? Explain.

$$\begin{cases} x = \cos t \\ y = \sin t \end{cases} 0 \le t \le 2\pi$$

$$\begin{cases} x = \sin t \\ y = \cos t \end{cases} 0 \le t \le 2\pi$$

G In problems 41–46, use a grapher to graph each curve represented by the parametric equations.

41. $\begin{cases} x = 3t - 2 \\ y = 4t^2 - 3 \end{cases} 0 \le t \le 4$

42. $\begin{cases} x = t^2 - 2t \\ y = t^3 - 3t \end{cases} 0 \le t \le 4$

43. $\begin{cases} x = 2 - (1/t) \\ y = 2t + (1/t) \end{cases} 1 \le t \le 10$

44. $\begin{cases} x = 1/(t^2 - 1) \\ y = -5t + 3 \end{cases} 2 \le t \le 10$

45. $\begin{cases} x = 3t - 3 \sin t \\ y = 3 - 3 \cos t \end{cases} 0 \le t \le 10\pi$

46. $\begin{cases} x = 2 \cos^2 t \\ y = 4 \sin^2 t \end{cases} 0 \le t \le 2\pi$

CHAPTER 9 REVIEW PROBLEM SET

In problems 1–4, write the standard equation of the parabola using the given information.

1. Focus at (−4, 0); directrix *y* = 3

2. Focus at (0, 3); directrix *x* = 5/2

3. Vertex at (−2, −3); focus at (−5, −3)

4. Vertex at (1, −1); directrix *y* = 2

In problems 5–8, write the standard equation of each conic using the given information.

5. An ellipse with foci at (3, 3) and (3, −1), and containing the point (4, 0)

6. An ellipse with foci at (3, 0) and (1, 0), and a vertex at (0, 0)

7. A hyperbola with foci at (−1, 4) and (5, 4), and a vertex at (0, 4)

8. A hyperbola with vertices at (−2, 3) and (2, 3), and containing the point (8, 8)

In problems 9–14, apply horizontal and vertical shifting to the graph of the first equation to generate the graph of the second equation. Find the center (when applicable), vertices (or vertex), and foci (or focus) of the new conic. Also, sketch the graphs of both equations in the same coordinate system.

9. $y = 2x^2; y - 2 = 2(x + 1)^2$

10. $9y^2 = -16x; 9(y + 1)^2 = -16(x - 2)$

11. $\frac{x^2}{9} + \frac{y^2}{16} = 1; \frac{(x-1)^2}{9} + \frac{(y+2)^2}{16} = 1$

12. $\frac{x^2}{4} + \frac{y^2}{9} = 1; \frac{(x-4)^2}{4} + \frac{(y+2)^2}{9} = 1$

13. $\frac{y^2}{4} - \frac{x^2}{9} = 1; \frac{(y-1)^2}{4} - \frac{(x-2)^2}{9} = 1$

14. $\frac{x^2}{9} - \frac{y^2}{16} = 1; \frac{(x+2)^2}{9} - \frac{(y-1)^2}{16} = 1$

In problems 15–20, identify and sketch the graph of the conic. Find the vertices (or vertex), foci (or focus), and equations of the asymptotes when applicable.

15. $y^2 + 2y - 2x + 7 = 0$

16. $x^2 - 2x - 6y - 7 = 0$

17. $9x^2 + 4y^2 - 90x - 16y + 205 = 0$

18. $9x^2 + 4y^2 - 90x - 16y - 83 = 0$

19. $4x^2 - 9y^2 - 16x - 90y + 16 = 0$

20. $4x^2 - 9y^2 - 16x - 90y - 210 = 0$

In problems 21–24, determine the standard equation for the conic that satisfies the given conditions.

21. Focus at $(2, 0)$; directrix $x = 9/2$; eccentricity $e = 2/3$

22. Focus at $(0, 0)$; directrix $y = -2$; eccentricity $e = 3$

23. Foci at $(-10, 0)$ and $(10, 0)$; eccentricity $e = 5/4$

24. Focus at $(6, -10)$; directrix $x = 2$; eccentricity $e = 1$

In problems 25 and 26, find the eccentricity of each conic.

25. (a) $8x^2 + 3y^2 = 12$

  (b) $12(x + 3)^2 + 8y = 1$

26. (a) $18x^2 - 9y^2 = 18$

  (b) $24(x - 3)^2 - y^2 = 1$

In problems 27 and 28, find the standard polar equation of the conic that satisfies the given conditions. (Assume the focus is at the pole.)

27. (a) $e = 1$; directrix $r = (-1/2)\csc \theta$

  (b) $e = 2$; directrix $r = 4 \sec \theta$

28. (a) $e = 3$; directrix $r = 2 \csc \theta$

  (b) $e = 1/3$; directrix $r = \sec \theta$

In problems 29 and 30, find the eccentricity, identify the conic, and sketch the graph. Also, locate the center and vertices of the conic.

29. (a) $r = \dfrac{3}{3 - 2 \cos \theta}$

  (b) $r = \dfrac{5}{2 - 5 \sin \theta}$

30. (a) $r = \dfrac{2}{4 + \cos \theta}$

  (b) $r = \dfrac{2}{3 - 3 \sin \theta}$

In problems 31–34, use the discriminant to determine whether the graph of the given equation is an ellipse, a parabola, or a hyperbola. Perform a rotation to eliminate the xy term and sketch the graph.

31. $x^2 - xy + y^2 = 2$

32. $14x^2 + 24xy + 7y^2 = 1$

33. $x^2 - \sqrt{3}\, xy = 1$

34. $6x^2 - 6xy + 14y^2 = 45$

Ⓖ In problems 35 and 36, use a grapher to graph each equation.

35. $x^2 - y^2 - 6x + 8y - 3 = 0$

36. $x^2 - 4xy - 2y^2 = 4$

In problems 37 and 38, eliminate the parameter and sketch the curve for the given parametric equations. Describe the orientation.

37. $\begin{cases} x = 2t \\ y = 2t^2 - 3t \end{cases}$ $0 \le t \le 4$

38. $\begin{cases} x = 1 - \sin t \\ y = 1 + \cos t \end{cases}$ $0 \le t \le 2\pi$

◈ CHAPTER 9 TEST

In problems 1 and 2, find the vertices (or vertex) and the foci (or focus) of the given conics. If the conic is a hyperbola, find the equations of its asymptotes. Sketch the graph.

1. $4x^2 + y^2 + 16x + 7 = 0$

2. $y^2 + 6y - x + 21 = 0$

3. Determine an acute angle θ for a rotation of axes that will eliminate the xy term from $x^2 - 4xy + y^2 = 6$. Express the equation in terms of $\overline{x}$ and $\overline{y}$, and graph the rotated equation.

4. Eliminate the parameter and sketch the curve given by the parametric equations.

$$\begin{cases} x = 2 + \sin t \\ y = 1 - \cos t \end{cases} \quad 0 \le t \le 2\pi$$

5. Find the eccentricity of the ellipse whose equation is

$$2y^2 + 9x^2 = 18.$$

6. Consider the conic given in polar coordinates by the equation

$$r = \frac{4}{2 + \sin\theta}.$$

   (a) Identify the conic.
   (b) Find the eccentricity of the conic.
   (c) Find the foci of the conic.
   (d) Sketch the conic.

7. Identify the conic given by the equation

$$3x^2 + 4xy + y^2 + y - 10 = 0.$$

Graph the conic.

CHAPTER 10

Discrete Mathematics

Sequences and series, part of a field of study referred to as *discrete mathematics,* provide mathematical tools that enable us to model certain applications. For instance, consider a theater that has 25 rows of seats. The first row contains 30 seats, the second contains 32 seats, the third contains 34 seats, and so on. How many seats are there in the twentieth row and what is the total number of seats in the theater? The solution of this problem is given in Example 11 on page 665.

In this chapter, topics from *discrete mathematics* are also covered. The function concept is used to introduce *sequences* and *series. Mathematical induction,* a method of proof, and the expansions of positive integer powers of binomials are included. In addition, we introduce combinations, permutations, and probability.

10.1 Sequences and Summation Notation

Many applications of mathematics involve ordered lists of numbers called *sequences.*

Describing a Sequence

Sequences are described in various ways.

For example, the pattern given by the following positive odd integers describes a sequence:

$$1, 3, 5, 7, \ldots$$

The list of paired numbers

$$(1, 3), (2, 6), (3, 9), (4, 12), (5, 15), \ldots$$

describes a sequence in a way that lends itself to graphing.

A sequence establishes a correspondence between positive integers, which indicate order or position, and other numbers.

Definition

Sequence

> A function whose domain consists of consecutive positive integers and whose range is a set of numbers is a **sequence.**

At times, only the range of a sequence is given. For instance, the sequence of positive odd integers is simply listed as

$$1, 3, 5, 7, \ldots$$

This is an example of an *infinite sequence,* since the domain is the set of *all* positive integers. By comparison, the set of numbers

$$1^2, 2^2, 3^2, 4^2, \text{ and } 5^2 \quad \text{or} \quad 1, 4, 9, 16, \text{ and } 25$$

illustrates a *finite sequence* whose domain consists of only

$$1, 2, 3, 4, \text{ and } 5.$$

To denote the particular numbers in a sequence we often give a rule defined by a formula. For instance, if we let

$$f(n) = 3n,$$

where n is a positive integer, then this infinite sequence is given by

$$f(1), f(2), f(3), f(4), \ldots, f(n), \ldots \quad \text{or} \quad 3, 6, 9, 12, \ldots, 3n, \ldots$$

The three dots at the end are used to indicate that the sequence continues indefinitely, following the pattern defined by $f(n) = 3n$.

The numbers in the range of a sequence are called the **terms** of the sequence. The symbolism

$$a_1, a_2, a_3, \ldots, a_n, \ldots$$

is often used to denote a sequence, where a_1 is the **first term,** a_2 is the **second term,** and a_n is the **nth,** or **general term** of the sequence.

The more compact notation $\{a_n\}$, in which the general term is enclosed in braces, is also used to denote a sequence.

For example, $\{a_n\} = \{3n\}$ indicates that the sequence is generated by successively substituting the integer values $n = 1, 2, 3, \ldots$ into the formula $3n$. Thus we have

$$a_1 = 3(1) = 3 \ , \ a_2 = 3(2) = 6 \ , \ a_3 = 3(3) = 9 \ , \ a_4 = 3(4) = 12, \ldots$$

EXAMPLE 1   **Determining Terms of a Sequence**

Determine the first five terms of the sequence with the given nth term.

(a) $a_n = (-1)^n$ 　　　　　　　(b) $b_n = 3 - \dfrac{1}{n}$

Solution   To find the first five terms of each sequence, we substitute the positive integers 1, 2, 3, 4, and 5, in turn, for n in the formula for the general term.

(a) 　　　　　$a_1 = (-1)^1 = -1, a_2 = (-1)^2 = 1, a_3 = (-1)^3 = -1,$
　　　　　　　$a_4 = (-1)^4 = 1, \quad \text{and} \quad a_5 = (-1)^5 = -1$

So the first five terms of this sequence are $-1, 1, -1, 1,$ and -1.

(b) For $b_n = 3 - 1/n$, we have

$$b_1 = 3 - (1/1) = 2, b_2 = 3 - (1/2) = 5/2, b_3 = 3 - (1/3) = 8/3,$$

$$b_4 = 3 - 1/4 = 11/4, \quad \text{and} \quad b_5 = 3 - 1/5 = 14/5$$

Hence the first five terms of this sequence are 2, 5/2, 8/3, 11/4, and 14/5.

In Example 1 we used the formulas for the nth term to determine the first few terms of a sequence. It may be possible to reverse the process—that is, use the first few terms of a sequence to find its general nth term.

EXAMPLE 2   **Finding the nth Term of a Sequence**

Find a formula for the general term of the sequence:

$$5, 10, 15, 20, 25, \ldots$$

Solution   By examining the given terms and looking for a pattern, we can see by inspection that the first five terms are multiples of 5, so we can write

$$a_1 = 5 = 5 \cdot 1, \quad a_2 = 10 = 5 \cdot 2, \quad a_3 = 15 = 5 \cdot 3,$$

$$a_4 = 20 = 5 \cdot 4, \quad \text{and} \quad a_5 = 25 = 5 \cdot 5.$$

Not every sequence can be expressed in terms of a formula for its nth (general) term.

The pattern of these five terms suggests that one formula for the general term is given by $a_n = 5n$.

Using Recursive Formulas

So far we have considered sequences whose nth terms are defined by *explicit formulas*. Another way used to specify the terms of a sequence is by a formula called a **recursive formula**. This type of formula gives one or more initial terms of a sequence and then defines each subsequent one in terms of the preceding term or terms.

EXAMPLE 3   **Finding Terms of a Sequence by Using a Recursive Formula**

The sequence $\{a_n\}$ is defined by

$$a_1 = 1, a_2 = 1, \quad \text{and} \quad a_n = a_{n-1} + a_{n-2}, \quad \text{where } n \geq 3.$$

Find the first six terms of this sequence.

Solution   The first six terms of this sequence are found as follows:

$$a_1 = 1 \qquad\qquad\qquad\qquad \text{Given}$$
$$a_2 = 1 \qquad\qquad\qquad\qquad \text{Given}$$
$$\left.\begin{aligned} a_3 &= a_{3-1} + a_{3-2} = a_2 + a_1 = 1 + 1 = 2 \\ a_4 &= a_{4-1} + a_{4-2} = a_3 + a_2 = 2 + 1 = 3 \\ a_5 &= a_{5-1} + a_{5-2} = a_4 + a_3 = 3 + 2 = 5 \\ a_6 &= a_{6-1} + a_{6-2} = a_5 + a_4 = 5 + 3 = 8 \end{aligned}\right\} \quad a_n = a_{n-1} + a_{n-2}$$

So the first six terms of the sequence are

$$1, 1, 2, 3, 5, \text{ and } 8.$$

The sequence found in Example 3 is a **Fibonacci sequence**, and the numbers that appear in it are called **Fibonacci numbers.** This sequence has interesting mathematical properties and it is used in a wide variety of applications in fields such as physics and ecology.

Graphing Sequences

Since a sequence is a function, the graph of the sequence $\{a_n\}$ is the graph of the function $y = f(n)$, where $f(n) = a_n$ and n is a positive integer. As usual, we use the horizontal axis for the inputs n and the vertical axis for the outputs

$$a_n = f(n).$$

Figure 1

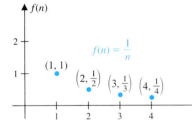

For example, the graph of the sequence $\left\{\dfrac{1}{n}\right\}$ is the graph of the function

$$f(n) = \frac{1}{n} \quad \text{where } n = 1, 2, 3, \ldots$$

The graph consists of a succession of isolated discrete or disconnected points, since the right side of the equation is defined only for positive integer values of n (Figure 1).

In the next example we contrast the graph of a sequence, which is discrete or disconnected, to the graph of a counterpart function, which is continuous or unbroken.

EXAMPLE 4   **Graphing a Sequence**

(a) Graph the sequence whose general term is given by

$$a_n = \frac{2}{n + 1}.$$

(b) Compare the graph of the function

$$f(x) = \frac{2}{x + 1}, \quad x \geq 1$$

to that of the sequence in part (a).

(c) Compare the limit behavior of the sequence $\{a_n\}$ as $n \to \infty$ to that of $f(x)$ as $x \to \infty$.

Solution (a) Since this is an infinite sequence, it is impossible to plot all points and so we will display the first five points. To find the first five terms of the sequence, we substitute

$n = 1, 2, 3, 4,$ and 5 into the general term $a_n = 2/(n + 1)$ to get

$$a_1 = \frac{2}{1 + 1} = 1, \quad a_2 = \frac{2}{2 + 1} = \frac{2}{3}, \quad a_3 = \frac{2}{3 + 1} = \frac{2}{4} = \frac{1}{2},$$

$$a_4 = \frac{2}{4 + 1} = \frac{2}{5}, \quad \text{and} \quad a_5 = \frac{2}{5 + 1} = \frac{2}{6} = \frac{1}{3}.$$

Because the sequence is defined only for positive integers values of n, the graph includes the following discrete points (Figure 2a):

$$(1, 1), \left(2, \frac{2}{3}\right), \left(3, \frac{1}{2}\right), \left(4, \frac{2}{5}\right), \text{ and } \left(5, \frac{1}{3}\right)$$

(b) By contrast, the graph of

$$f(x) = \frac{2}{(x + 1)}, \text{ for } x \geq 1$$

is formed by connecting the points with a smooth, continuous or unbroken curve (Figure 2b). Observe that the limit behavior is given by $f(x) \to 0$ as $x \to \infty$.

Figure 2

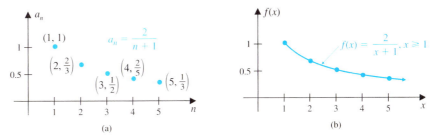

(c) Note that in both situations, as the domain values (represented by n for the sequence and by x for the function f) increase, the function values approach 0. That is, the limit behavior of

$$a_n = \frac{2}{n + 1} \quad \text{as} \quad n \to +\infty$$

is the same as the limit behavior of

$$f(x) = \frac{2}{x + 1} \quad \text{as} \quad x \to +\infty$$

So

$$\frac{2}{n + 1} \to 0 \quad \text{as} \quad n \to +\infty.$$

The limit behavior of a sequence $\{a_n\}$ as $n \to \infty$ is called the **limit of the sequence,** which is written in symbols as

$$\lim_{n \to \infty} a_n.$$

Thus, in Example 4, we write

$$\lim_{n\to\infty}\frac{2}{n+1} = 0.$$

G   Graphers can help us investigate the limit of a sequence.
For instance, consider the sequence

$$\left\{\sqrt{n^2 + 4n} - n\right\}.$$

The first five terms of the sequence rounded to two decimal places are given by

(1, 1.24), (2, 1.46), (3, 1.58), (4, 1.66) and (5, 1.71).

Figure 3a shows the plot of the first five terms of the sequence, whereas Figure 3b shows a viewing window of the associated function $f(x) = \sqrt{x^2 + 4x} - x$, $x \geq 1$, and the line $y = 2$.

Figure 3

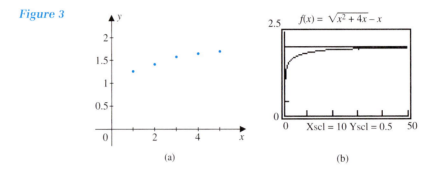

(a)   (b)

By using the [TRACE] feature on the graph of f, we monitor the y values as the x values get larger and larger. It appears that the limit behavior of f is approaching 2. So we conclude that

$$\lim_{n\to\infty}\left(\sqrt{n^2 + 4n} - n\right)$$

also seems to be approaching 2.

Using Summation Notation

The sum of the first n terms of an infinite sequence $\{a_n\}$ is given by

$$a_1 + a_2 + a_3 + \cdots + a_n.$$

The symbol Σ is a stylized version of the Greek capital letter sigma.

Frequently, such a sum is written more compactly by using the following **sigma** (Σ), or **summation notation:**

$$\sum_{k=1}^{n} a_k = a_1 + a_2 + a_3 + \cdots + a_n.$$

Note that the lower limit of a summation may also be 0 or any positive integer. It does not have to be 1.

Here, Σ indicates a sum and the letter k is called the **index of summation.** The symbols n and 1 above and below Σ are called the **upper** and **lower limits** of the summation, respectively. In this case, they indicate that k takes on integer values from 1 to n, inclusive.

Any letter can be used for the index; however, $i, j,$ and k are the most commonly used.

For instance,

$$\sum_{k=1}^{n} 5^k = \sum_{i=1}^{n} 5^i = \sum_{j=1}^{n} 5^j = 5^1 + 5^2 + 5^3 + \cdots + 5^n.$$

EXAMPLE 5   **Using Sigma Notation**

Evaluate each sum.

(a) $\displaystyle\sum_{k=1}^{3} (4k^2 - 3k)$ (b) $\displaystyle\sum_{k=2}^{5} \frac{k-1}{k+1}$

Solution   (a) Here we have

$$a_k = 4k^2 - 3k.$$

To find the sum, we substitute the integers 1, 2, and 3 for k in succession, and then add the resulting numbers:

$$\sum_{k=1}^{3} (4k^2 - 3k) = [4(1^2) - 3(1)] + [4(2^2) - 3(2)] + [4(3^2) - 3(3)]$$

$$= 1 + 10 + 27 = 38$$

(b) Here $a_k = \dfrac{k-1}{k+1}$, and the index starts at $k = 2$ and ends at $k = 5$. Thus

$$\sum_{k=1}^{5} \frac{k-1}{k+1} = \frac{2-1}{2+1} + \frac{3-1}{3+1} + \frac{4-1}{4+1} + \frac{5-1}{5+1}$$

$$= \frac{1}{3} + \frac{2}{4} + \frac{3}{5} + \frac{4}{6} = \frac{21}{10}.$$

Following are some basic properties of summations.

Basic Properties of Summation

If $\{a_n\}$ and $\{b_n\}$ are given sequences and if c represents a constant, then we have the following results:

1. **Constant Property:** $\displaystyle\sum_{k=1}^{n} c = nc$

2. **Homogeneous Property:** $\displaystyle\sum_{k=1}^{n} ca_k = c \sum_{k=1}^{n} a_k$

3. **Additive Property:** $\displaystyle\sum_{k=1}^{n} (a_k + b_k) = \sum_{k=1}^{n} a_k + \sum_{k=1}^{n} b_k$

4. **Sum of Successive Integers:**

$$\sum_{k=1}^{n} k = 1 + 2 + 3 + \cdots + n$$

$$= \frac{n(n+1)}{2}$$

5. **Sum of Successive Squares:**

$$\sum_{k=1}^{n} k^2 = 1^2 + 2^2 + 3^2 + \cdots + n^2$$

$$= \frac{n(n+1)(2n+1)}{6}$$

The first three properties can be verified by first expanding the summation and then using basic properties of algebra. For instance, Property 2 can be proved as follows:

$$\sum_{k=1}^{n} ca_k = ca_1 + ca_2 + \cdots + ca_n$$

$$= c(a_1 + a_2 + \cdots + a_n)$$

$$= c\sum_{k=1}^{n} a_k$$

Properties 4 and 5 will be proved in Section 10.2.

EXAMPLE 6

Using the Basic Properties to Find a Sum

Use the basic summation properties to evaluate

$$\sum_{k=1}^{20} (2k^2 - 3k + 4).$$

Solution

We evaluate the given sum by applying the basic summation properties as follows:

$$\sum_{k=1}^{20} (2k^2 - 3k + 4) = \sum_{k=1}^{20} 2k^2 + \sum_{k=1}^{20} (-3k) + \sum_{k=1}^{20} 4 \qquad \text{Property 3}$$

$$= 2\sum_{k=1}^{20} k^2 + (-3)\sum_{k=1}^{20} k + 20 \cdot 4 \qquad \text{Properties 1 and 2}$$

$$= 2\left[\frac{(20)(21)(41)}{6}\right] - 3\left[\frac{(20)(21)}{2}\right] + 80 \qquad \text{Properties 4 and 5}$$

$$= 5190$$

Solving Applied Problems

Sequences are used to model data involving ordered events.

EXAMPLE 7

Modeling a Savings Pattern

A newspaper carrier saved $1 the first week of delivering newspapers, $2 the second week, $4 the third week, and $7 the fourth week. For the fifth week, the amount saved is obtained by adding $4 to the amount for the preceding week; for the sixth week, the amount saved is obtained by adding $5 to the amount for the preceding week, and so on.

(a) Find the sequence that describes the savings for each week.

(b) Find a recursive formula for this sequence.

(c) If this pattern continues, predict the newspaper carrier's savings for the seventh week.

(d) Find the sum saved after 8 weeks.

Solution

(a) By examining the sequence, we observe that the second number is obtained by adding 1 to the first, the third is obtained by adding 2 to the second, the fourth by adding 3 to the third, and so on. Therefore, the sequence that describes the savings is given by

$$\underbrace{1,\quad 2,\quad 4,\quad 7,\quad 11,\quad 16,\ldots}_{+1\quad +2\quad +3\quad +4\quad +5}$$

(b) If we denote the sequence by $\{a_n\}$, then we have

$$
\begin{aligned}
a_1 &= 1\\
a_2 &= 2 = 1 \quad + 1 = a_1 + 1\\
a_3 &= 4 = 2 \quad + 2 = a_2 + 2\\
a_4 &= 7 = 4 \quad + 3 = a_3 + 3\\
a_5 &= 11 = 7 \quad + 4 = a_4 + 4\\
a_6 &= 16 = 11 + 5 = a_5 + 5\\
&\vdots \qquad\qquad \vdots\\
a_n &= a_{n-1} + (n - 1).
\end{aligned}
$$

So the recursive formula is given by

$$a_n = a_{n-1} + (n - 1), \text{ where } n > 1 \text{ and } a_1 = 1.$$

(c) If the pattern continues, then

$$
\begin{aligned}
a_7 &= a_6 + (7 - 1)\\
&= a_6 + 6\\
&= 16 + 6 = 22.
\end{aligned}
$$

So the seventh week's savings will be $22.

(d) The total of the first 8 terms of the sequence is given by the sum

$$\sum_{k=1}^{8} a_k = 1 + 2 + 4 + 7 + 11 + 16 + 22 + 29$$

$$= 92.$$

So the total amount saved after 8 weeks is $92.

PROBLEM SET 10.1

Mastering the Concepts

In problems 1–6, indicate whether the sequence is finite or infinite. Then write the general term of the sequence and express it in both subscript and function notation. Indicate the domain and range of each sequence.

1. $1, 3, 5, 7, 9$

2. $2, 4, 6, 8$

3. $1, 4, 9, 16, 25, \ldots$

4. $1, 8, 27, 64, \ldots$

5. $1, -1, 1, -1, 1, \ldots$

6. $-2, 2, -2, 2, -2, \ldots$

In problems 7–12, find the first five terms of the sequence whose general term is a_n.

7. $a_n = 2n + 3$

8. $a_n = \dfrac{1}{n^2}$

9. $a_n = \dfrac{n}{n + 2}$

10. $a_n = \dfrac{2^n}{1 + 2^{-n}}$

11. $a_n = e^{-n}$

12. $a_n = (-1)^n + 1$

In problems 13–18, find the first five terms of each infinite, recursively defined sequence.

13. $a_1 = 1; a_n = \dfrac{1}{2} a_{n-1}, n \ge 2$

14. $a_1 = 2; a_n = 2a_{n-1} - 1, n \ge 2$

15. $a_1 = -2; a_n = 3a_{n-1}, n \ge 2$

16. $a_1 = 1, a_2 = 2; a_n = a_{n-1} + a_{n-2}, n \ge 3$

17. $a_1 = 1, a_2 = 3; a_n = a_{n-2} + a_{n-1}, n > 2$

18. $a_1 = 2, a_2 = -1; a_n = a_{n-2} - a_{n-1}, n > 2$

In problems 19–24, find a formula for the general term of each sequence.

19. $3, 5, 7, \ldots$

20. $2, 5, 8, 11, \ldots$

21. $\dfrac{1}{3}, \dfrac{1}{5}, \dfrac{1}{7}, \dfrac{1}{9}, \ldots$

22. $\dfrac{1}{2}, \dfrac{3}{4}, \dfrac{5}{6}, \dfrac{7}{8}, \ldots$

23. $1, -4, 9, -16, \ldots$

24. $1, -8, 27, -64, \ldots$

In problems 25–34:

(a) Find the first five terms of each sequence. Then plot the corresponding pairs to show the pattern of the graph.

(b) Graph the associated continuous function of each sequence.

(c) Use the graph from part (b) to find the limit behavior of the sequence, that is, find

$$\lim_{n \to \infty} a_n.$$

25. $a_n = 3n + 2$

26. $a_n = 1 - 4n$

27. $a_n = 1 + \dfrac{n}{2}$

28. $a_n = \dfrac{2n + 1}{n}$

29. $a_n = 2^n$

30. $a_n = 2^{-n}$

31. $a_n = 1 - \dfrac{1}{n}$

32. $a_n = 1 + \dfrac{1}{n}$

33. $a_n = \dfrac{3n^2 + 1}{n^2}$

34. $a_n = \dfrac{3n^2 - 1}{n^2}$

In problems 35–42, expand and then evaluate each sum.

35. $\displaystyle\sum_{k=1}^{10} 5k$

36. $\displaystyle\sum_{k=0}^{4} \dfrac{2^k}{k + 1}$

37. $\displaystyle\sum_{k=0}^{4} 3^{2k}$

38. $\displaystyle\sum_{k=2}^{5} 2^{k-2}$

39. $\displaystyle\sum_{k=1}^{3} (2k + 1)^2$

40. $\displaystyle\sum_{i=2}^{6} \dfrac{1}{i(i + 1)}$

41. $\displaystyle\sum_{k=1}^{4} \dfrac{(-1)^k + 1}{k}$

42. $\displaystyle\sum_{k=1}^{5} \left(-\dfrac{1}{3}\right)^{k-1}$

In problems 43–46, use the basic properties of summation to evaluate each expression.

43. $\displaystyle\sum_{k=1}^{20} (4k + 3)$

44. $\displaystyle\sum_{k=1}^{30} (7 - 3k)$

45. $\displaystyle\sum_{k=1}^{100} (5k^2 - 3)$

46. $\displaystyle\sum_{k=1}^{10} (3k^2 - 5k + 1)$

G In problems 47–52:

(a) Find the first five terms of each sequence. Round off these values to two decimal places.

(b) Plot the corresponding pairs from part (a) to show the pattern of the graph of the sequence. Then use a grapher to graph the associated function of each sequence.

(c) Use the TRACE feature on the graph from part (b) to find the limit behavior of the sequence, that is, find $\displaystyle\lim_{n \to \infty} a_n$.

47. $a_n = \dfrac{\sin n}{n}$

48. $a_n = \dfrac{\ln(n + 1)}{n + 1}$

49. $a_n = \dfrac{1 - \cos n}{n}$

50. $a_n = \sqrt[n]{n}$

51. $a_n = \dfrac{n}{e^n}$

52. $a_n = \dfrac{n^2}{1 - e^n}$

Applying the Concepts

53. Mixing Paint: A house painter wishes to mix paint in a 1 gallon bucket from some cans of colored paint. The first contains 1/2 gallon of paint, the second can contains 1/4 gallon of paint, the third can contains 1/8 gallon of paint, and so on, so that each successive can holds half the amount of the preceding one.

(a) Find the first five terms of the sequence for the amount of colored paint in each can according to the specified order.

(b) Find a formula for this sequence.

(c) If the first five such cans were poured into the 1 gallon bucket, use summation notation to represent the total amount of paint in the bucket. Then evaluate this summation.

(d) Is the bucket full after the contents of the first five cans are poured into it? If this process continues, will the can ever be filled?

54. Telemarketing: An advertising agency begins a telemarketing campaign by making 50 phone calls on the first day. Each day after, the agency will make 10 more calls than were made the day before.
(a) Find the first five terms of this sequence.
(b) Let a_n be the total number of calls made on the nth day. Find a formula that defines a_n.
(c) Find the total number of phone calls made during the first 10 days.

55. Botany: A botanist measured the height of a plant once a month and charted its growth. Suppose that the initial height is recorded as 4 inches and the height increases 10% per month for a year.
(a) Find the first five terms of the sequence.
(b) Find a formula a_n for the height of the plant in the nth month.
(c) What is the total height of the plant after 9 months? Round off the answer to two decimal places.

56. Computer Fund: To save for a computer, a teacher made an initial deposit of $40 into a savings account. Then $15 was deposited into the account after 1 week, $20 after 2 weeks, $25 after 3 weeks, $30 after 4 weeks, and so on, until enough money was saved for the computer.
(a) Find the first eight terms of the sequence that shows the amount saved each week.
(b) Write a formula a_n for the sequence.
(c) What is the total amount saved after 8 weeks, including the initial deposit?

Developing and Extending the Concepts

57. Find a formula for the sequence defined by the recursive formula:
$$a_1 = 1 \quad \text{and} \quad a_n = a_{n-1} + 8. \ n > 1$$

58. Find the first five terms of the sequence defined by the recursive formula:
$$a_1 = 10 \quad \text{and} \quad a_n = \sqrt{a_{n-1}}, \quad n > 1$$
Round off each term to four decimal places. Graph the sequence.

59. Solve each equation for t.
(a) $\sum_{k=1}^{10} kt = 110$
(b) $\sum_{k=1}^{20} k(k - t) = 1820$

60. Consider the sequence defined recursively by
$$a_1 = 2 \quad \text{and} \quad a_n = 5 - a_{n-1}, n \geq 2.$$
(a) Find the first eight terms of the sequence $\{a_n\}$ and graph it.

(b) Find the first eight terms of the sequence
$$\left\{ \frac{a_n}{a_n - 1} \right\}$$
and graph it.

In problems 61–68, suppose that
$$S_n = \sum_{k=1}^{n} a_k$$
Then a recursive formula for a_n is given by
$$a_n = S_n - S_{n-1} \quad \text{where} \quad n \geq 2 \quad \text{and} \quad a_1 = S_1$$
Use this result to find an explicit formula for the general term a_n of each sequence.

61. $S_n = 3n + 4$ **62.** $S_n = 4 - 3n$
63. $S_n = n(n + 2)$ **64.** $S_n = 3n(n + 1)$
65. $S_n = \dfrac{n}{n + 2}$ **66.** $S_n = \dfrac{2n}{3n + 1}$
67. $S_n = 2^n$ **68.** $S_n = 2^{n+1} - 3$

In problems 69–74, find S_3 and S_6.

69. $S_n = \sum_{k=1}^{n} \dfrac{k(k + 1)}{2}$

70. $S_n = \sum_{k=1}^{n} \dfrac{k(k + 1)(2k + 1)}{6}$

71. $S_n = \sum_{k=1}^{n} \dfrac{k^2(k + 1)^2}{4}$

72. $S_n = \sum_{k=1}^{n} (2^{k+1} - 8)$

73. $S_n = \sum_{k=1}^{n} (-1)^k (k + 2)$

74. $S_n = \sum_{k=1}^{n} \dfrac{(-1)^{k+1}}{k + 2}$

In problems 75 and 76, use
$$S_n = \sum_{k=1}^{n} (a_k - a_{k+1})$$
$$= (a_1 - a_2) + (a_2 - a_3) + (a_3 - a_4) + \ldots + (a_n - a_{n+1})$$
$$= a_1 - a_{n+1}$$
to find an explicit formula for S_n in terms of n.

75. (a) $S_n = \sum_{k=1}^{n} \left(\dfrac{1}{k} - \dfrac{1}{k + 1} \right)$
(b) $S_n = \sum_{k=1}^{n} \left(\dfrac{1}{k + 1} - \dfrac{1}{k + 2} \right)$

76. (a) $S_n = \sum_{k=1}^{n} \left(\dfrac{1}{3k - 1} - \dfrac{1}{3k + 2} \right)$
(b) $S_n = \sum_{k=1}^{n} \left(\dfrac{1}{k + 2} - \dfrac{1}{k + 3} \right)$

10.2 Arithmetic and Geometric Sequences

In this section we investigate two types of sequences, *arithmetic sequences* and *geometric sequences.*

Identifying an Arithmetic Sequence and Finding Its General Term

Consider the sequence

$$2, 6, 10, 14, 18, \ldots$$

Note that the difference between successive terms is always 4. This sequence is an example of an *arithmetic sequence.*

Definition

Arithmetic Sequence

A sequence $a_1, a_2, a_3, \ldots, a_n, \ldots$ is called an **arithmetic sequence** if each term (after the first term) differs from the preceding term by a constant number. That is, there is a constant d for which

$$a_n - a_{n-1} = d \quad \text{for all integers} \quad n > 1.$$

The constant number d is called the **common difference** for the sequence.

For the arithmetic sequence given above the common difference is 4, so each term of the sequence can be obtained by adding 4 to its predecessor. That is,

$$
\begin{aligned}
a_1 &= 2 \\
a_2 &= 6 = 2 + 4 \\
a_3 &= 10 = 2 + 2 \cdot 4 \\
a_4 &= 14 = 2 + 3 \cdot 4 \\
&\ \vdots \\
a_n &= \quad 2 + (n - 1) \cdot 4.
\end{aligned}
$$

In general, any term a_n of an arithmetic sequence is the sum of the first term a_1 and a multiple of the common difference d. We can find a general formula for the nth term of an arithmetic sequence as follows:

$$
\begin{aligned}
a_1 &= a_1 \\
a_2 &= a_1 + d \\
a_3 &= a_2 + d = (a_1 + d) + d = a_1 + 2d \\
a_4 &= a_3 + d = (a_1 + 2d) + d = a_1 + 3d \\
&\ \vdots
\end{aligned}
$$

By continuing this pattern, we discover that the *general term* a_n of an arithmetic sequence with a first term a_1 and a common difference d is given by

$$a_n = a_1 + (n - 1)d.$$

Table 1 shows a few examples of arithmetic sequences.

TABLE 1   Examples of Arithmetic Sequences

Sequence	First Term, a_1	Common Difference, d	General Term, a_n
$3, 7, 11, 15, \ldots$	3	$7 - 3 = 4$	$3 + (n-1)4 = 4n - 1$
$4, 9, 14, 19, \ldots$	4	$9 - 4 = 5$	$4 + (n-1)5 = 5n - 1$
$1, -4, -9, -14, \ldots$	1	$-4 - 1 = -5$	$1 + (n-1)(-5) = -5n + 6$
$\dfrac{10}{7}, \dfrac{1}{7}, -\dfrac{8}{7}, -\dfrac{17}{7}, \ldots$	$\dfrac{10}{7}$	$\dfrac{1}{7} - \dfrac{10}{7} = -\dfrac{9}{7}$	$\dfrac{10}{7} + (n-1)\left(-\dfrac{9}{7}\right) = \dfrac{-9}{7}n + \dfrac{19}{7}$

The next example will illustrate how it is possible to determine the general term of an arithmetic sequence when given two specific, nonconsecutive terms in the sequence.

EXAMPLE 1   **Finding the General Term of an Arithmetic Sequence**

Find the general term of an arithmetic sequence if the third term is 7 and the seventh term is 15.

Solution   Using the formula

$$a_n = a_1 + (n-1)d \quad \text{twice,}$$

first with $n = 3$ and $a_3 = 7$, and then with $n = 7$ and $a_7 = 15$, we obtain the following system of linear equations in the unknown a_1 and d:

$$\begin{cases} 7 = a_1 + (3-1)d \\ 15 = a_1 + (7-1)d \end{cases} \quad \text{or} \quad \begin{cases} a_1 + 2d = 7 \\ a_1 + 6d = 15 \end{cases}$$

Solving for d and a_1 gives us $a_1 = 3$ and $d = 2$. Therefore, the general term a_n of this sequence is given by

$$a_n = 3 + (n-1)2$$
$$= 3 + 2n - 2$$
$$= 2n + 1.$$

EXAMPLE 2   **Finding n When a_n is Given**

Which term of the arithmetic sequence $7, 3, -1, -5, -9, \ldots$ is -389?

Solution   Here we have $a_n = -389$ and we need to find n.
First we note that $a_1 = 7$ and $d = 3 - 7 = -4$.
Substituting these numbers into the formula $a_n = a_1 + (n-1)d$, we get

$$a_n = a_1 + (n-1)d$$
$$-389 = 7 + (n-1)(-4)$$
$$-389 = 7 - 4n + 4$$
$$-400 = -4n$$
$$100 = n.$$

Therefore, the 100th term is -389; that is, $a_{100} = -389$.

The graph of an arithmetic sequence consists of discrete points that define a linear pattern, as the next example illustrates.

EXAMPLE 3   **Graphing an Arithmetic Sequence**

(a) Graph the given terms of an arithmetic sequence

$$1, \frac{3}{2}, 2, \frac{5}{2}, 3, \ldots$$

(b) Compare the graph in part (a) to the continuous graph of the line

$$y = \frac{1}{2}x + \frac{1}{2} \quad \text{where } x \geq 1.$$

Solution

(a) The graph of the given terms of the sequence is displayed in Figure 1a.

(b) The continuous graph of the line $y = (1/2)x + 1/2$, where $x \geq 1$, is shown in Figure 1b. For the given sequence

$$a_1 = 1 \quad \text{and} \quad d = \frac{3}{2} - 1 = \frac{1}{2}.$$

Thus the general term of the sequence is given by

$$a_n = a_1 + (n - 1)d$$

$$= 1 + (n - 1)\left(\frac{1}{2}\right)$$

$$= 1 + \frac{1}{2}n - \frac{1}{2}$$

$$= \frac{1}{2}n + \frac{1}{2}.$$

Upon comparing

$$a_n = \frac{1}{2}n + \frac{1}{2} \quad \text{to} \quad y = \frac{1}{2}x + \frac{1}{2}$$

where $x \geq 1$, we conclude that the graph of the arithmetic sequence consists of discrete points that lie on the graph of the line $y = (1/2)x + 1/2$.

Figure 1

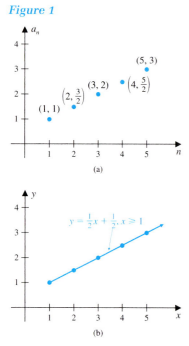

(a)

(b)

Finding the Sum of an Arithmetic Sequence

Consider the arithmetic sequence given by

$$3, 8, 13, 18, \ldots, 5n - 2, \ldots$$

By adding successive terms of this sequence, we create another sequence $\{S_n\}$, called the *sequence of partial sums,* whose terms are given by

$$S_1 = 3$$
$$S_2 = 3 + 8 = 11$$
$$S_3 = 3 + 8 + 13 = 24$$
$$S_4 = 3 + 8 + 13 + 18 = 42$$
$$\vdots \qquad \vdots$$
$$S_n = 3 + 8 + 13 + 18 + \cdots + (5n - 2).$$

In general, if

$$a_1, a_2, a_3, \ldots, a_n, \ldots$$

is an arithmetic sequence, then the partial sum S_n for the sequence $\{a_n\}$ is given by

$$S_n = a_1 + a_2 + \cdots + a_n$$

$$= \sum_{k=1}^{n} a_k.$$

We can derive a formula for S_n as follows:

If d is the common difference, by using the basic properties of summation given on page 649, we get

$$S_n = \sum_{k=1}^{n} a_k$$

$$= \sum_{k=1}^{n} [a_1 + (k - 1)d] \qquad a_k = a_1 + (k - 1)d$$

$$= \sum_{k=1}^{n} [a_1 + kd - d] \qquad \text{Multiply}$$

$$= \sum_{k=1}^{n} a_1 + \sum_{k=1}^{n} kd - \sum_{k=1}^{n} d \qquad \text{Property 3}$$

$$= na_1 + d \sum_{k=1}^{n} k - nd \qquad \text{Properties 1 and 2}$$

$$= na_1 + d\left[\frac{n(n + 1)}{2}\right] - nd \qquad \text{Property 4}$$

$$= \frac{n}{2}[2a_1 + d(n + 1) - 2d] \qquad \text{Factor out } \frac{n}{2}$$

$$= \frac{n}{2}[2a_1 + (n - 1)d] \qquad \text{Simplify}$$

Thus the partial sum S_n of an arithmetic sequence $\{a_1 + (n - 1)d\}$ is given by

$$S_n = \frac{n}{2}[2a_1 + (n - 1)d].$$

Using the fact that $a_n = a_1 + (n - 1)d$, we can rewrite this formula as follows:

$$S_n = \frac{n}{2}[2a_1 + (n - 1)d] = \frac{n}{2}[a_1 + a_1 + (n - 1)d] = \frac{n}{2}(a_1 + a_n)$$

Therefore,

$$S_n = \frac{n}{2}(a_1 + a_n).$$

EXAMPLE 4   **Finding the Partial Sum of an Arithmetic Sequence**

Find the common difference and the sum of the first 20 terms of the arithmetic sequence

$$2, 6, 10, 14, \ldots$$

Solution   Here the common difference $d = 6 - 2 = 4$, so using the formula

$$S_n = \left(\frac{n}{2}\right)[2a_1 + (n - 1)d]$$

with $a_1 = 2$, $d = 4$, and $n = 20$, we have

$$S_{20} = \frac{20}{2}[2(2) + (20 - 1)4]$$

$$= 10(4 + 76) = 800.$$

EXAMPLE 5   **Finding n, Given the Partial Sum S_n for an Arithmetic Sequence**

Find n for the arithmetic sequence $\{a_n\}$ if $a_1 = 3$, the common difference $d = 5$, and the partial sum $S_n = 255$.

Solution   Using the formula

$$S_n = \left(\frac{n}{2}\right)[2a_1 + (n - 1)d]$$

with $a_1 = 3$, $d = 5$, and $S_n = 255$, we have

$$255 = \left(\frac{n}{2}\right)[6 + (n - 1)5]$$

$$510 = n(6 + 5n - 5) \qquad \text{Multiply each side by 2}$$

$$510 = n(1 + 5n) \qquad \text{Simplify}$$

$$510 = n + 5n^2 \qquad \text{Multiply}$$

$$5n^2 + n - 510 = 0 \qquad \text{Subtract 510 from each side}$$

$$(5n + 51)(n - 10) = 0 \qquad \text{Factor}$$

So,

$$n = 10 \quad \text{or} \quad n = -\frac{51}{5}.$$

Since n must be a positive integer, $n = 10$.

Identifying a Geometric Sequence and Finding Its General Term

Consider the sequence

$$3, 6, 12, 24, \ldots$$

Note that each term after the first one can be obtained by multiplying the preceding term by 2. This is an example of a *geometric sequence*. In general, we have the following definition:

Definition
Geometric Sequence

A sequence $a_1, a_2, a_3, \ldots, a_n, \ldots$ is called a **geometric sequence** if each term (after the first) is obtained by multiplying the preceding term by a constant number. That is, there is a constant r for which

$$\frac{a_n}{a_{n-1}} = r$$

for all integers $n > 1$. The constant number r is called the **common ratio** of the geometric sequence.

Thus the geometric sequence

$$3, 6, 12, 24, \ldots, 3(2^{n-1}), \ldots$$

has the common ratio 2, that is, the ratio of any two successive terms is 2. For instance,

$$\frac{a_2}{a_1} = \frac{6}{3} = 2, \frac{a_3}{a_2} = \frac{12}{6} = 2, \ldots, \quad \text{and} \quad \frac{3(2^{n-1})}{3(2^{n-2})} = 2.$$

Consider a geometric sequence $\{a_n\}$ with common ratio r and with first term a_1. Then the nth term a_n may be expressed in terms of r and a_1 as follows:

$$a_1 = a_1$$
$$a_2 = a_1 r$$
$$a_3 = a_2 r = (a_1 r)r = a_1 r^2$$
$$a_4 = a_3 r = (a_1 r^2)r = a_1 r^3$$
$$a_5 = a_4 r = (a_1 r^3)r = a_1 r^4$$

Continuing this pattern, we are led to the following general formula.

$$a_n = a_1 r^{n-1}$$

EXAMPLE 6   **Finding the General Term of a Geometric Sequence**

Find the general term of the given geometric sequence.

(a) $\dfrac{1}{3}, \dfrac{1}{6}, \dfrac{1}{12}, \dfrac{1}{24}, \ldots$

(b) The third term is 8, the fifth term is 32, and all terms are positive.

Solution   (a) For $\dfrac{1}{3}, \dfrac{1}{6}, \dfrac{1}{12}, \dfrac{1}{24}, \ldots$, the common ratio r is given by

$$r = \frac{\left(\dfrac{1}{6}\right)}{\left(\dfrac{1}{3}\right)} = \frac{1}{2}.$$

Using the formula

$$a_n = a_n r^{n-1} \quad \text{with} \quad a_1 = \frac{1}{3} \quad \text{and} \quad r = \frac{1}{2}$$

we get the general term

$$a_n = \left(\frac{1}{3}\right)\left(\frac{1}{2}\right)^{n-1} = \left(\frac{1}{3}\right)(2^{-1})^{n-1}$$

$$= \left(\frac{1}{3}\right)(2^{-n+1}) = \left(\frac{1}{3}\right)(2^{-n} \cdot 2)$$

which is equivalent to

$$a_n = \left(\frac{2}{3}\right)2^{-n}.$$

(b) Here we are given $a_3 = 8$ and $a_5 = 32$. Using the formula $a_n = a_1 r^{n-1}$ twice, we get

$$\begin{cases} a_3 = a_1 r^{3-1} \\ a_5 = a_1 r^{5-1} \end{cases} \quad \text{or} \quad \begin{cases} 8 = a_1 r^2 \\ 32 = a_1 r^4. \end{cases}$$

By dividing the corresponding terms of the two latter equations, we get

$$\frac{1}{4} = \frac{1}{r^2} \quad \text{or} \quad r^2 = 4.$$

The solution of $r^2 = 4$ includes $r = -2$ or $r = 2$. However, we use only the positive value for r, $r = 2$, because all the terms of the sequence are positive (as stated in the problem).

Substituting $r = 2$ into the first equation in the system, $8 = a_1 r^2$, we obtain

$$8 = a_1 \cdot 2^2, \quad \text{or} \quad a_1 = 2.$$

So the general term of the sequence is

$$a_n = a_1 r^{n-1} = 2 \cdot 2^{n-1} \quad \text{or} \quad a_n = 2^n.$$

EXAMPLE 7 **Graphing a Geometric Sequence**

(a) Graph the given terms for a geometric sequence

$$4, \frac{12}{5}, \frac{36}{25}, \frac{108}{125}, \dots$$

(b) Compare the graph in part (a) to the continuous graph of

$$y = 4\left(\frac{3}{5}\right)^{x-1} \quad \text{where } x \geq 1.$$

Solution   (a) Figure 2a shows the graph of the given sequence.

(b) The graph of $y = 4(3/5)^{x-1}$, where $x \geq 1$ is displayed in Figure 2b. For the given sequence,

$$a_1 = 4 \quad \text{and} \quad r = \frac{\left(\dfrac{12}{5}\right)}{4} = \frac{3}{5}.$$

So the general term is given by

$$a_n = 4\left(\frac{3}{5}\right)^{n-1}.$$

Upon comparing the graph of $a_n = 4(3/5)^{n-1}$ to that of $y = 4(3/5)^{x-1}$, where $x \geq 1$, we conclude that the graph of the geometric sequence consists of discrete points that lie on the graph of the exponential function $y = 4(3/5)^{x-1}$.

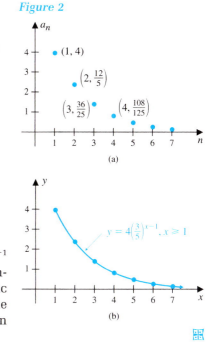

Figure 2

(a)

(b)

Finding the Partial Sum of a Geometric Sequence

Consider the geometric sequence

$$1, 3, 9, 27, \ldots, 3^{n-1}, \ldots$$

By adding successive terms of this sequence, we create another sequence, the sequence of partial sums $\{S_n\}$, as follows:

$$S_1 = 1$$
$$S_2 = 1 + 3 = 4$$
$$S_3 = 1 + 3 + 9 = 13$$
$$S_4 = 1 + 3 + 9 + 27 = 40$$
$$\vdots$$
$$S_n = 1 + 3 + 9 + 27 + \cdots + 3^{n-1}.$$

Using sigma notation, the latter expression can be rewritten as

$$S_n = \sum_{k=1}^{n} 3^{k-1}.$$

To find a formula for the partial sum S_n of a geometric sequence $\{a_n\}$ with first term a_1 and common ratio r we proceed as follows:

$$S_n = a_1 + a_1 r + a_1 r^2 + \cdots + a_1 r^{n-1}$$
$$r S_n = a_1 r + a_1 r^2 + a_1 r^3 + \cdots + a_1 r^n \qquad \text{Multiply each side by } r$$
$$S_n - r S_n = a_1 - a_1 r^n \qquad \text{Substract the two equations}$$

By factoring each side of this latter equation, we have

$$(1 - r)S_n = a_1(1 - r^n) \quad \text{or} \quad S_n = \frac{a_1(1 - r^n)}{1 - r} \quad \text{if } r \neq 1.$$

If r = 1, then the partial sum S$_n$ is given by

$S_n = a_1 + a_1 + ... + a_1$
$\quad = na_1$

Thus the general formula for the partial sum S_n of the geometric sequence $\{a_1 r^{n-1}\}$ is given by

$$S_n = \sum_{k=1}^{n} a_1 r^{k-1} = \frac{a_1(1 - r^n)}{1 - r}, \quad r \neq 1.$$

For example, if $a_1 = 3$ and $r = 3$, then

$$\sum_{k=1}^{n} 3(3^{k-1}) = \frac{3(1 - 3^n)}{1 - 3}$$

$$= \left(-\frac{3}{2}\right)(1 - 3^n).$$

EXAMPLE 8   **Finding a Partial Sum of a Geometric Sequence**

Find the common ratio and the sum of the first ten terms of the geometric sequence

$$\frac{1}{2}, 1, 2, 4, 8, \ldots$$

Solution   Here

$$a_1 = \frac{1}{2} \quad \text{and} \quad \text{the common ratio } r = \frac{1}{\left(\frac{1}{2}\right)} = 2.$$

So if we let $n = 10$ in the formula

$$S_n = \frac{a_1(1 - r^n)}{1 - r}$$

we get the sum of the first ten terms

$$S_{10} = \frac{\frac{1}{2}(1 - 2^{10})}{1 - 2} = \frac{\frac{1}{2}(-1023)}{-1} = \frac{1023}{2} = 511.5.$$

EXAMPLE 9   **Finding Terms of a Geometric Sequence**

The sum of the first five terms of a geometric sequence is 61/27 and the common ratio is $-1/3$. Find the first four terms of the sequence.

Solution   Using the formula for S_n with $r = -1/3$ and $S_5 = 61/27$, we have

$$\frac{61}{27} = \frac{a_1\left[1 - \left(-\frac{1}{3}\right)^5\right]}{1 - \left(-\frac{1}{3}\right)}.$$

Thus

$$\frac{61}{27} = \left[\frac{\left(\dfrac{244}{243}\right)}{\left(\dfrac{4}{3}\right)}\right]a_1 \quad \text{or} \quad \frac{61}{27} = \frac{61}{81}\,a_1.$$

So

$$a_1 = \frac{81}{61} \cdot \frac{61}{27} = 3.$$

Hence the first four terms of the sequence are

$$3,\ 3\left(-\frac{1}{3}\right),\ 3\left(-\frac{1}{3}\right)^2,\ \text{ and }\ 3\left(-\frac{1}{3}\right)^3 \quad \text{or} \quad 3,\ -1,\ \frac{1}{3},\ \text{ and }\ -\frac{1}{9}.$$

Finding the Sum of an Infinite Geometric Series

Suppose we have a sequence $\{a_n\}$ and we use it to form the sequence of partial sums $\{S_n\}$:

$$\begin{aligned}
S_1 &= a_1 \\
S_2 &= a_1 + a_2 \\
S_3 &= a_1 + a_2 + a_3 \\
&\ \ \vdots \qquad \vdots \\
S_n &= a_1 + a_2 + a_3 + \cdots + a_n
\end{aligned}$$

Continuing this process indefinitely, we are led to an expression of the form

$$a_1 + a_2 + a_3 + \cdots + a_n + \cdots.$$

This latter expression is written more compactly using sigma notation as

$$\sum_{k=1}^{\infty} a_k$$

Such an expression is called an **infinite series,** or simply a **series.**

The general study of series is taken up in calculus. For now, we'll only investigate series that are formed from geometric sequences. Such series are called **geometric series.** Examples of geometric series are

$$\sum_{k=1}^{\infty}\left(\frac{1}{2}\right)^{k-1} \quad \text{and} \quad \sum_{k=1}^{\infty}\frac{4}{5}\left(\frac{1}{5}\right)^{k-1}.$$

Let's examine how infinite geometric series are interpreted by considering the series

$$\sum_{k=1}^{\infty}\frac{4}{5}\left(\frac{1}{5}\right)^{k-1} = \frac{4}{5}\left(1 + \frac{1}{5} + \frac{1}{5^2} + \frac{1}{5^3} + \cdots + \frac{1}{5^n} + \cdots\right).$$

This series was formed from the geometric sequence

$$\{a_1 r^{n-1}\}, \text{ where } a_1 = 4/5 \text{ and } r = 1/5.$$

By using the formula for the partial sum,

$$S_n = \sum_{k=1}^{n} a_1 r^{k-1} = \frac{a_1(1 - r^n)}{1 - r}$$

with $a_1 = 4/5$ and $r = 1/5$, we get

$$S_n = \frac{4}{5}\left[\frac{1 - \left(\frac{1}{5}\right)^n}{1 - \left(\frac{1}{5}\right)}\right] = \frac{4}{5} \cdot \frac{5}{4}\left(1 - \frac{1}{5^n}\right) = 1 - \left(\frac{1}{5}\right)^n.$$

Figure 3 shows the graph of the continuous function $f(x) = 1 - (1/5)^x$, where $x \geq 1$. The limit behavior of the graph indicates that as $x \to +\infty, f(x) \to 1$.

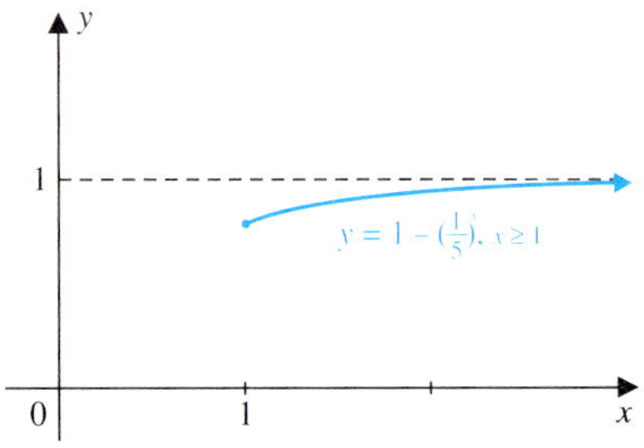

Since the graph of the sequence

$$S_n = 1 - (1/5)^n$$

consists of discrete points on the graph of $f(x) = 1 - (1/5)^x$, where $x \geq 1$, it follows that the limit behavior of S_n is the same as the limit behavior of $f(x)$.

That is, as $n \to +\infty, S_n \to 1$.

But as $n \to +\infty$,

Figure 3

$$S_n \to a_1 + a_2 + a_3 + \cdots a_n + \cdots = \sum_{k=1}^{\infty} a_k.$$

So we assign the value 1 to the given infinite series, and write

$$\sum_{k=1}^{\infty} \frac{4}{5}\left(\frac{1}{5}\right)^{k-1} = 1.$$

We refer to 1 as the *sum of the series.*

In general, given a geometric series $\sum_{k=1}^{\infty} a_1 r^{k-1}$, the partial sum

$$S_n = a_1 \left(\frac{1 - r^n}{1 - r}\right) = \frac{a_1 - a_1 r^n}{1 - r}$$

The use of the term sum and the infinity symbol, ∞, is misleading, because we cannot actually add an infinite number of terms.

can be written as

$$S_n = \frac{a_1}{1 - r} - \frac{a_1 r^n}{1 - r}.$$

Under the condition that $|r| < 1$, it follows that as $n \to +\infty, r^n \to 0$, so

$$S_n \to \frac{a_1}{1 - r} - 0 = \frac{a_1}{1 - r}.$$

This result is summarized as follows:

Sum of an Infinite Geometric Series

If $|r| \geq 1$, the series has no sum.

If $|r| < 1$, then

$$\sum_{k=1}^{\infty} a_1 r^{k-1} = a_1 + a_1 r + a_1 r^2 + \cdots + a_1 r^{n-1} + \cdots = \frac{a_1}{1-r}.$$

EXAMPLE 10   **Finding the Sum of an Infinite Geometric Series**

Find the sum of each infinite geometric series.

(a) $\displaystyle\sum_{k=1}^{\infty} 5\left(-\frac{1}{3}\right)^{k-1}$ (b) $\displaystyle\sum_{k=1}^{\infty} \frac{3}{2^{k-1}}$

Solution   (a) Using the formula $a_1/(1-r)$ for the sum of the series with $a_1 = 5$ and $r = -1/3$, we have

$$\sum_{k=1}^{\infty} 5\left(-\frac{1}{3}\right)^{k-1} = \frac{5}{1-\left(-\dfrac{1}{3}\right)} = \frac{5}{\left(\dfrac{4}{3}\right)} = \frac{15}{4}.$$

(b) The series

$$\sum_{k=1}^{\infty} \frac{3}{2^{k-1}}$$

can be rewritten as

$$\sum_{k=1}^{\infty} \frac{3(1)^{k-1}}{2^{k-1}} \quad \text{or} \quad \sum_{k=1}^{\infty} 3\left(\frac{1}{2}\right)^{k-1}.$$

So in this case, we have a geometric series of the form

$$\sum_{k=1}^{\infty} a_1 r^{k-1}, \quad \text{where} \quad a_1 = 3 \quad \text{and} \quad r = \frac{1}{2}.$$

Thus

$$\sum_{k=1}^{\infty} \frac{3}{2^{k-1}} = \sum_{k=1}^{\infty} 3\left(\frac{1}{2}\right)^{k-1} = \frac{3}{1-\dfrac{1}{2}} = \frac{3}{\left(\dfrac{1}{2}\right)} = 6.$$

Solving Applied Problems

Sequences are used to record and analyze discrete data, as illustrated in the next two examples.

EXAMPLE 11   **Finding the Seating Capacity of a Theater**

A theater has 25 rows of seats. The first row contains 30 seats, the second contains 32 seats, the third contains 34 seats, and so on.

(a) Assuming this pattern continues, find a sequence that models the number of seats in each row of the theater.

(b) Find the number of seats in the twentieth row.

(c) How many total seats are there in the theater?

Solution

(a) If the pattern given by 30, 32, 34, . . . continues, the sequence $\{a_n\}$ that models the situation is arithmetic. Specifically, if we let a_n represent the number of seats in the nth row, then the first term is $a_1 = 30$ and the common difference is $d = 2$, so

$$a_1 = 30, \, a_2 = 32, \, a_3 = 34, \dots, a_n = 30 + (n-1)(2), \dots,$$

where $1 \le n \le 25$.

(b) The number of seats in the twentieth row is given by

$$a_{20} = 30 + (20 - 1)(2)$$
$$= 30 + 38 = 68.$$

(c) Since the sequence $\{a_n\}$ found in part (a) specifies the number of seats in each of the 25 rows, we need to find S_{25}, the sum of the first 25 terms of the sequence. Using the formula

$$S_n = \left(\frac{n}{2}\right)[2a_1 + (n-1)d]$$

with $n = 25$, $a_1 = 30$, and $d = 2$, we get

$$S_{25} = \frac{25}{2}[2(30) + (25 - 1)2] = 1350.$$

Therefore, the theater has a total of 1350 seats.

The following example illustrates how geometric sequences are used in modeling physical phenomena.

EXAMPLE 12   **Modeling the Swing of a Pendulum**

Suppose that the distance traveled by a bob of a pendulum in each swing is 83% of the previous swing (Figure 4).   *Figure 4*

(a) If the bob travels an arc length of 9.5 inches on the first swing and this pattern continues, find a sequence that models the distance traveled by the bob on each swing.

(b) Find the distance traveled on the fifth swing (to two decimal places).

(c) Determine the total distance traveled by the bob after eight swings.

(d) What is the total distance the bob travels as $n \to +\infty$?

9.5 in

Solution

(a) The distances traveled by the bob on each swing follow the pattern

$$1\text{st} = 9.5$$
$$2\text{nd} = 9.5(0.83)$$
$$3\text{rd} = [9.5(0.83)](0.83)$$
$$= 9.5(0.83)^2.$$

If this pattern continues, the sequence that models the situation is geometric; its first term is 9.5 and each subsequent term is 0.83 times the prior term.

Specifically, if we let a_n represent the distance traveled during the nth swing, then $a_1 = 9.5$ and the common ratio is $r = 0.83$

So

$$a_1 = 9.5,$$
$$a_2 = 9.5(0.83),$$
$$a_3 = 9.5(0.83)^2,$$
$$\vdots$$
$$a_n = 9.5(0.83)^{n-1},$$
$$\vdots$$

(b) The distance traveled on the fifth swing is given by

$$a_5 = 9.5(0.83)^4$$
$$= 4.51 \text{ inches (approx.)}.$$

(c) Since each term in the sequence found in part (a) specifies the distance traveled on the nth swing, we need to find S_8, the sum of the first eight terms of the geometric sequence.

Using the formula

$$S_n = \frac{a_1(1 - r^n)}{1 - r}$$

with $n = 8$, $a_1 = 9.5$, and $r = 0.83$, we get

$$S_8 = \frac{9.5(1 - 0.83^8)}{1 - 0.83}$$
$$= 43.30 \text{ (approx.)}.$$

So the bob travels a total of about 43.30 inches by the end of the eighth swing.

(d) The total distance the bob travels as $n \to +\infty$ is represented by the infinite geometric series where $a_1 = 9.5$ and $r = 0.83$ so that

$$\sum_{k=1}^{\infty} 9.5(0.83)^{k-1} = \frac{9.5}{1 - 0.83}$$
$$= 55.88 \text{ (approx.)}.$$

So the bob travels a total distance of about 55.88 inches.

PROBLEM SET 10.2

Mastering the Concepts

In problems 1–6, find the general term of each arithmetic sequence and graph it.

1. $2, 5, 8, 11, \ldots$

2. $-9, -5, -1, 3, \ldots$

3. $12, 9, 6, 3, \ldots$

4. $19, 17, 15, 13, \ldots$

5. The third term is 6 and the eighth term is 16.

6. The fourth term is 6 and the tenth term is -12.

In problems 7–10, which term of the arithmetic sequence is the given term?

  7. $a_n = 85$ for 5, 13, 21, 29, . . .
  8. $a_n = 143$ for $-7, -2, 3, 8, \ldots$
  9. $a_n = -185$ for $3, -1, -5, -9, \ldots$
  10. $a_n = -20$ for $0, -\dfrac{1}{2}, -1, -\dfrac{3}{2} \ldots$

In problems 11–14, find the common difference and the partial sum S_n of each arithmetic sequence for the given value of n.

  11. 1, 4, 7, 10, . . . ; $n = 10$
  12. 50, 45, 40, 35, . . . ; $n = 12$
  13. $-5, -\dfrac{32}{7}, -\dfrac{29}{7}, -\dfrac{26}{7}, \ldots$; $n = 8$
  14. $\dfrac{1}{2}, 1, \dfrac{3}{2}, 2 \ldots$; $n = 10$

In problems 15–18, find the sum of the first 40 terms of the arithmetic sequence that has the given general term.

  15. $a_n = 5n - 4$ **16.** $a_n = -5n - 3$
  17. $a_n = -3n - 2$ **18.** $a_n = 4n - 7$

In problems 19–22, certain information is given about an arithmetic sequence. Find the indicated unknown(s).

  19. n, if $a_1 = 3$, $d = 2$, and $S_n = 143$
  20. S_n, if $a_1 = 38$, $d = -2$, and $n = 25$
  21. d and a_{18}, if $a_1 = 17$ and $S_{18} = 2310$
  22. n and d, if $a_1 = 27$, $a_n = 48$, and $S_n = 1200$

In problems 23–26, find the sum of the first 60 terms of each arithmetic sequence.

  23. 11, 13, 15, 17, . . .   **24.** 2, 7, 12, 17, . . .
  25. $\dfrac{3}{2}, 2, \dfrac{5}{2}, 3, \ldots$ **26.** $-\dfrac{1}{3}, \dfrac{1}{3}, 1, \dfrac{5}{3}, \ldots$

In problems 27–32, find the general term of each geometric sequence.

  27. 2, 6, 18, 54, . . .
  28. $1, \dfrac{1}{5}, \dfrac{1}{25}, \dfrac{1}{125}, \ldots$
  29. $1, 1.03, (1.03)^2, (1.03)^3, \ldots$
  30. $\dfrac{2}{3}, -\dfrac{4}{3}, \dfrac{8}{3}, -\dfrac{16}{3}, \ldots$
  31. The second term is 4 and the twelfth term is 4096.
  32. The third term is 3 and the sixth term is $\dfrac{25}{81}$.

In problems 33–36, find a formula for the general term a_n for each geometric sequence.

  33. 3, 6, 12, 24, . . .   **34.** 1, 5, 25, 125, . . .
  35. $\dfrac{1}{4}, \dfrac{1}{8}, \dfrac{1}{16}, \dfrac{1}{32}, \ldots$ **36.** $6, 2, \dfrac{2}{3}, \dfrac{2}{9}, \ldots$

In problems 37–40, graph the sequence. Find the common ratio and the sum of the first n terms of each geometric sequence for the given value of n.

  37. 5, 20, 80, 320, . . . ; $n = 8$
  38. 54, 36, 24, 16, . . . ; $n = 7$
  39. $-\dfrac{1}{3}, -\dfrac{1}{9}, -\dfrac{1}{27}, -\dfrac{1}{81}, \ldots$; $n = 10$
  $5, -\dfrac{5}{2}, \dfrac{5}{4}, -\dfrac{5}{8}, \ldots$; $n = 8$
  40. $1, 1.04, (1.04)^2, (1.04)^3, \ldots$; $n = 6$

In problems 41–44, find the sum of each infinite geometric series.

  41. $\displaystyle\sum_{k=1}^{\infty} \left(\dfrac{1}{3}\right)^k$ **42.** $\displaystyle\sum_{k=1}^{\infty} 9\left(\dfrac{1}{8}\right)^k$
  43. (a) $\displaystyle\sum_{k=1}^{\infty} \dfrac{4}{2^k}$ **44.** (a) $\displaystyle\sum_{k=1}^{\infty} (0.7)^k$
  (b) $\displaystyle\sum_{k=1}^{\infty} \dfrac{2(-1)^k}{3^{k-1}}$ (b) $\displaystyle\sum_{k=1}^{\infty} (-1)^k \left(\dfrac{2}{3}\right)^{k-1}$

Applying the Concepts

  45. Marching Band: A formation of a marching band has 7 marchers in the front row, 10 in the second row, 13 in the third row, and so on, for 10 rows.
  (a) What kind of sequence does this pattern define? Find the general term of a sequence to model the number of marchers in each row.
  (b) Find the number of marchers in the seventh row.
  (c) How many total marchers are in the band?

  46. Flower Planting: A farmer has a triangular piece of land to grow plants for floral purposes. There are 100 plants in the first row, 105 in the second, 110 in the third, and so on, for 90 rows.
  (a) What kind of sequence does this pattern create? Find the general term of a sequence to model the number of plants in each row.
  (b) Find the number of plants in the twelfth row.
  (c) How many plants are there altogether?

47. **Sales:** A small company had sales of $300,000 during its first year of operation, and the sales increased by $40,000 per year during each successive year.
    (a) List the first five terms of the sequence of increasing sales. What kind of sequence does this pattern create? Find the general term of a sequence to model the sales of the company for each year.
    (b) What were the sales during the sixth year?
    (c) What were the total sales of the company during its first 10 years?

48. **Advertising:** During a sales promotion, a salesperson receives a guaranteed salary plus a bonus of $60 for selling one TV, $90 for selling two TVs, $120 for selling three TVs, and so on, in an arithmetic sequence.
    (a) Find the sequence that represents the potential bonuses.
    (b) How much bonus money will the salesperson earn for selling thirty TVs in 1 week?
    (c) How many TVs must be sold to earn a bonus of $510?
    (d) Use the graph of this sequence to display and discuss the trend of the bonus incentive.

49. **Physics:** An object dropped from a certain height will have fallen vertically 12 feet after the first second, 44 feet after the second second, 76 feet after the third, and so on.
    (a) How many feet will the object fall during the ninth second?
    (b) What is the total number of feet the object has fallen at the end of 9 seconds?

50. **Geometry:** The sum of the interior angles of a triangle is 180°, the sum of the interior angles of a quadrilateral is 360°, the sum is 540° for a pentagon, and so on. Assuming this pattern continues, find the sum of the interior angles for a 10-sided polygon.

51. **Rebounding Ball:** Suppose that a ball dropped from the top of a 100 foot tower rebounds 3/5 of the distance it falls each time.
    (a) Find a geometric sequence that represents the distance traveled on each bounce for this pattern.
    (b) How far (up and down) will the ball have traveled by the time it hits the ground for the eighth time?
    (c) What is the total distance (up and down) this bouncing ball has traveled by the time it hits the ground for the nth time?
    (d) What is the total distance in part (c) as $n \to +\infty$?

52. **Baseball Bonus:** A baseball pitcher agreed to a contract that pays him a guaranteed salary of $750,000 plus a bonus incentive of $10,000 for winning one game, $13,500 for winning two games, $18,225 for winning three games, and so on, up to and including 20 games.
    (a) Represent the bonus incentive by a geometric sequence.
    (b) How much money would he be paid if he won sixteen games?
    (c) How many games must he win to earn a total of about $3 million?
    (d) Use a graph to display and discuss the trend of the bonus incentive.

53. **Football Bonus:** A running back agreed to a contract that pays him a guaranteed salary of $1.8 million plus a bonus incentive of $1000 for scoring one touchdown, $1800 for scoring two touchdowns, $3240 for scoring three touchdowns, and so on, up to and including 12 touchdowns.
    (a) Represent the bonus incentive as a geometric sequence.
    (b) How much money will he be paid if he scores nine touchdowns?
    (c) How many touchdowns must he score to earn a total of at least $2 million?
    (d) Use a graph to display and discuss the trend of the bonus incentive.

54. **Depreciation:** An automobile worth $18,000 depreciates in value each year such that its value at the end of the year is 5/6 of its value at the beginning of the year.
    (a) Represent the depreciation as a sequence.
    (b) How much is the automobile worth after 5 years?
    (c) Assuming the automobile is maintained properly, after how many years will it be worth approximately $6000?
    (d) Use the graph of this sequence to display and discuss the depreciation pattern.

55. **College Enrollment:** In 1999 the enrollment at a college was 25,000. Each successive year the enrollment has dropped to 95% of the preceding year. Suppose this trend continues over a period of 10 years.
    (a) Represent the enrollment as a sequence.
    (b) Given this pattern, predict the enrollment in 2005.
    (c) If this pattern continues, what will the enrollment be in 2009?
    (d) Use the graph of this sequence to discuss the pattern of enrollment.

56. Drug Dosage: A patient receives 10 milligrams of a drug. Each hour 50% of the amount of the drug present in the body is eliminated.

(a) Model the amount of drug eliminated as a sequence.

(b) How much of the drug is present after 6 hours?

(c) If this pattern continues, what will be the amount of drug present after 10 hours?

(d) Use the graph of this sequence to describe the pattern.

(e) Will the drug ever be totally eliminated from the patient's body according to the model found in part (a)? Explain.

Developing and Extending the Concepts

57. The first three terms of an arithmetic sequence have the form $x, 2x + 3, 5x - 2$ for some real number x.

(a) Find x.

(b) Find the numerical value of the tenth term.

58. Find the fifth and tenth terms of the arithmetic sequence

$$t + 1, 3t - 1, 3t + 3, \ldots$$

In problems 59 and 60, find the indicated partial sum for each arithmetic sequence.

59. S_7, for $6, 3b + 1, 6b - 4, \ldots$

60. S_{10}, for $x + 2y, 3y, -x + 4y, \ldots$

In problems 61 and 62, express each sum using sigma notation; then evaluate using the partial sum formula.

61. $\dfrac{1}{2} + \dfrac{1}{4} + \dfrac{1}{8} + \dfrac{1}{16} + \dfrac{1}{32}$

62. $\dfrac{3}{5} + \dfrac{9}{25} + \dfrac{27}{125} + \dfrac{81}{625}$

In problems 63 and 64, find the indicated sum of each geometric sequence.

63. S_6, for $\dfrac{a}{b}, -1, \dfrac{b}{a}, \ldots$

64. S_8, for $\dfrac{k}{b}, \dfrac{k}{b^2}, \dfrac{k}{b^3}$

In problems 65 and 66, rewrite each repeating decimal as a geometric series. Then find its sum. Convert the resulting sum to decimal form to confirm the result.

65. (a) $0.\overline{4}$

(b) $0.\overline{32}$

(c) $0.\overline{561}$

66. (a) $0.\overline{7}$

(b) $0.04\overline{9}$

(c) $0.0\overline{72}$

67. Use the result

$$S_n = \sum_{k=1}^{n} (a_k - a_{k+1})$$

$$= (a_1 - a_2) + (a_2 - a_3) + \cdots + (a_n - a_{n+1})$$

$$= a_1 - a_{n+1}$$

to find a formula for each partial sum.

(a) $S_n = \sum_{k=1}^{n} (2^k - 2^{k+1})$

(b) $S_n = \sum_{k=1}^{n} \left[\left(\dfrac{1}{2}\right)^k - \left(\dfrac{1}{2}\right)^{k+1} \right]$

Objectives

1. Formulate and Apply Mathematical Induction
2. Examine the Pattern in a Binomial Expansion
3. Use Factorial Notation
4. Use the Binomial Theorem

10.3 Mathematical Induction and the Binomial Theorem

In this section we introduce a method of proof called *mathematical induction*. Then we introduce *factorial* notation and present the *binomial theorem*, which provides us with a general formula for the expansion of a binomial.

Formulating and Applying Mathematical Induction

Many results in mathematics and other sciences have been discovered by examining patterns. To illustrate, let's examine the pattern shown in the table for finding the sum of the first n odd positive integers for specific values of n.

n	Sum		Total	n^2
1	1	=	1	1
2	1 + 3	=	4	2^2
3	1 + 3 + 5	=	9	3^2
4	1 + 3 + 5 + 7	=	16	4^2
5	1 + 3 + 5 + 7 + 9 =		25	5^2

We might *conjecture* from this pattern that in general

$$1 + 3 + 5 + \cdots + (2n - 1) = n^2.$$

Obviously, it is impossible to test all positive integer values of n in order to *prove* the validity of this formula. For this kind of situation, we use a method of proof called *mathematical induction.*

The method of **mathematical induction** is based on the following principle:

Principle of Mathematical Induction

Suppose that $S_1, S_2, S_3, \ldots$ is a sequence of statements; that is, suppose that for each positive integer n we have a corresponding statement S_n.
   Assume that the following two conditions hold:

(i) S_1 is true.
(ii) For each fixed positive integer k, the truth of the statement S_k implies the truth of the following statement S_{k+1}.

Then it follows that every statement $S_1, S_2, S_3, \ldots$ is true; that is, S_n is true for all positive integers.

We may understand the logic underlying the principle of mathematical induction more clearly by making the following analogy. Visualize a communications network set up so that each person in a group is put on a numbered list. When a message is to be communicated (for instance, a bad weather report), each person is to contact the next person on the list. We can reach a conclusion about this model in the following way:

(i) Show that the first person is given a message. If we denote this by S_1, then we need to show that S_1 is true.
(ii) (a) Assume that some person on the list, say the kth person, is given the message, denoted by S_k.
     (b) Show that when the kth person gets the message, S_k, then the next $(k + 1)$th person, denoted by S_{k+1}, also gets the message. That is, show that if S_k is true, then S_{k+1} is also true.

Notice that statement (ii) of the definition is given two parts: First assume that S_k is true, and then show that S_{k+1} is true. The fact that the first person gets the message and that the $(k + 1)$th person is informed when the kth person is informed, is enough to support the claim that all persons are informed.
   In other words, if S_1 indicates that the first person receives the message, then S_2 indicates that the first person informs the second. So it follows that when the second person is informed, then the next one, S_3, is also informed, and so on.
   To prove a statement using mathematical induction, we may need to rephrase the statement in a form that is more suitable for proof.

EXAMPLE 1   **Using Mathematical Induction**

Use mathematical induction to prove Property 4 of the summation properties; that is, prove that

$$1 + 2 + 3 + \cdots + n = \frac{n(n + 1)}{2}$$

for any positive integer n.

Solution   Using the principle of mathematical induction, we can prove the statement by verifying conditions (i) and (ii).

Let S_n be the statement

$$1 + 2 + 3 + \cdots + n = \frac{n(n + 1)}{2}.$$

Condition (i), that is, statement S_1 can be verified by direct computation, since

$$S_1 \longrightarrow \boxed{1 = \frac{1 \cdot (1 + 1)}{2}} = \frac{2}{2} = 1$$

is clearly true.

To prove condition (ii), we must show that S_k implies S_{k+1}; that is, we must show that if S_k is *assumed* to be true, then S_{k+1} *must* be true. To this end, assume that S_k is true; that is, assume that the assertion

$$S_k \longrightarrow \boxed{1 + 2 + 3 + \cdots + k = \frac{k(k + 1)}{2}}$$

is true. Since S_k is a true statement, we can add $k + 1$ to both sides of this equation, to get

$$1 + 2 + 3 + \cdots + k + (k + 1) = \frac{k(k + 1)}{2} + (k + 1)$$

$$= (k + 1)\left(\frac{k}{2} + 1\right)$$

$$= (k + 1)\left(\frac{k + 2}{2}\right)$$

$$= \frac{(k + 1)(k + 2)}{2}$$

But the latter statement is precisely S_{k+1}. Hence we have proved condition (ii). So by the principle of mathematical induction, we conclude that S_n is true for any positive integer n. That is,

$$1 + 2 + 3 + \cdots + n = \frac{n(n + 1)}{2}$$

for any positive integer n.

EXAMPLE 2   **Using Mathematical Induction**

Use mathematical induction to prove Property 5 of the summation properties; that is, prove that

$$1^2 + 2^2 + \cdots + n^2 = \frac{n(n + 1)(2n + 1)}{6}$$

for any positive integer n.

Solution   (i) In this situation, the statement S_t is given by

$$S_1 \longrightarrow 1^2 = \frac{1}{6}(1)(1 + 1)(2 \cdot 1 + 1) = \frac{1}{6}(1)(2)(3) = 1$$

which is true.

(ii) Assume that S_k is true; that is, assume that for any fixed positive integer k,

$$S_k \longrightarrow 1^2 + 2^2 + 3^2 + \cdots + k^2 = \frac{k(k + 1)(2k + 1)}{6}.$$

We must now prove that S_{k+1} is true, where S_{k+1} is the statement

$$1^2 + 2^2 + 3^2 + \cdots + k^2 + (k + 1)^2 = \frac{(k + 1)(k + 2)(2k + 3)}{6}.$$

After adding $(k + 1)^2$ to both sides of the equation for S_k, we have

$$1^2 + 2^2 + 3^2 + \cdots + k^2 + (k + 1)^2 = \frac{k(k + 1)(2k + 1)}{6} + (k + 1)^2$$

$$= (k + 1)\left[\frac{k(2k + 1)}{6} + (k + 1)\right]$$

$$= (k + 1)\left(\frac{2k^2 + k + 6k + 6}{6}\right)$$

$$= (k + 1)\left(\frac{2k^2 + 7k + 6}{6}\right)$$

$$= \frac{(k + 1)(k + 2)(2k + 3)}{6}.$$

Hence S_{k+1} is true, and we have proved condition (ii). Thus, by the principle of mathematical induction, we conclude that S_n is true for any positive integer n. That is, for any positive integer n,

$$1^2 + 2^2 + 3^2 + \cdots + n^2 = \frac{n(n + 1)(2n + 1)}{6}.$$

The principle of mathematical induction can also be used to prove certain inequalities.

EXAMPLE 3   **Using Induction to Prove an Inequality**

Use mathematical induction to prove that $2^n > n$ for any positive integer n.

Solution   (i) The statement

$$S_1 \rightarrow \boxed{2^1 > 1}$$

is a true inequality.

(ii) Assume that S_k is true; that is, assume that for any fixed positive integer k,

$$S_k \rightarrow \boxed{2^k > k.}$$

We must prove that S_{k+1} is true, where S_{k+1} is the statement

$$2^{k+1} > k + 1.$$

Multiplying each side of the inequality $2^k > k$ by 2, we have

$$2^k \, 2 > 2k \quad \text{or} \quad 2^{k+1} > k + k.$$

But $k + k > k + 1$ for $k > 1$, so

$$2^{k+1} > k + 1.$$

Thus for any positive integer n, $2^n > n$.

Examining the Pattern in a Binomial Expansion

Our goal here is to construct a formula that generates the expansion of $(a + b)^n$ for a specified positive integer n. The validity of the formula is established by using mathematical induction. Before presenting this formula, we explore the pattern of coefficients that exists in the following binomial expansions obtained by direct calculations:

	$(a + b)^n$
$n = 1$	$(a + b)^1 = a + b$
$n = 2$	$(a + b)^2 = a^2 + 2ab + b^2$
$n = 3$	$(a + b)^3 = a^3 + 3a^2b + 3ab^2 + b^3$
$n = 4$	$(a + b)^4 = a^4 + 4a^3b + 6a^2b^2 + 4ab^3 + b^4$
$n = 5$	$(a + b)^5 = a^5 + 5a^4b + 10a^3b^2 + 10a^2b^3 + 5ab^4 + b^5$

The above expansions lead us to discover the following patterns of the products of the powers of a and b in the expansion of $(a + b)^n$, $n \geq 1$:

1. Each expansion has $n + 1$ terms. The first term is a^n, and the last term is b^n.
2. The power of a decreases by 1 for each term, and the power of b increases by 1 for each term, so that the sum of the exponents of a and b in each term is n.
3. A pattern for the coefficients of the terms of an expanded binomial can be found by writing the coefficients in a triangular array of numbers known as *Pascal's triangle.*

Coefficients of Expanded Form of $(a + b)^n$

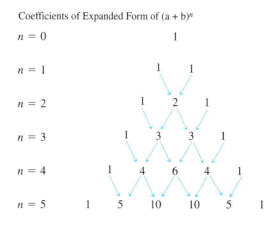

$n = 0$ 1

$n = 1$ 1   1

$n = 2$ 1   2   1

$n = 3$ 1   3   3   1

$n = 4$ 1   4   6   4   1

$n = 5$ 1   5   10   10   5   1

The values in Pascal's triangle are obtained recursively and obey the following patterns:

(i) The first and last coefficient for each expansion is 1.

(ii) Each of the other coefficients can be found by adding the coefficients that are diagonally above it, as indicated by the arrows in the diagram above.

EXAMPLE 4   **Using Pascal's Triangle**

Use the patterns described above to find the expansion of

$$(x + 2y)^3.$$

Solution   We know that the expansion has

$$n + 1 = 3 + 1$$
$$= 4$$

terms.

The first term is $a^3 = x^3$, and the last term is $b^3 = (2y)^3$.

By letting $x = a$ and $2y = b$, we can use the description of the products of the powers of a and b given earlier to get the following pattern:

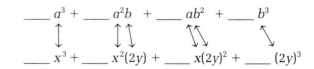

___ a^3 + ___ a^2b + ___ ab^2 + ___ b^3

___ x^3 + ___ $x^2(2y)$ + ___ $x(2y)^2$ + ___ $(2y)^3$

Finally, we use Pascal's triangle to obtain the numerical coefficients for $(a + b)^3$. For $n = 3$:

$$1 \quad 3 \quad 3 \quad 1$$

By combining the patterns of the products of the powers of the variable terms with the patterns of the coefficients, we get the expansion:

$$(x + 2y)^3 = \underline{1}\ x^3 + \underline{3}\ x^2(2y) + \underline{3}\ x(2y)^2 + \underline{1}\ (2y)^3$$
$$= \quad x^3 + \quad 6x^2y + \quad 12xy^2 + \quad 8y^3$$

It is possible to find an explicit formula for the coefficients that Pascal's triangle provides. Before developing such a formula, we need to introduce special notation.

Using Factorial Notation

The symbol $n!$ which is read as *n factorial*, is used to denote the product of the first n consecutive positive integers.

Definition

Factorial Notation

The symbol **$n!$**, or *n* **factorial**, is defined for all nonnegative integers as

$$0! = 1 \quad \text{and} \quad n! = n(n - 1)(n - 2) \cdot \cdots \cdot 2 \cdot 1$$

if n is a positive integer.

For instance,

$$5! = 5 \cdot 4 \cdot 3 \cdot 2 \cdot 1 = 120.$$

Notice that $5!$ can also be written as $5! = 5 \cdot 4!$. In general, $n!$ can be defined recursively as

$$n! = n(n - 1)!$$

for $n \geq 1$.

In calculus it is often necessary to simplify expressions involving factorial notation.

EXAMPLE 5   **Simplifying Expressions Involving Factorials**

Simplify the expression

$$\frac{a_{n+1}}{a_n} \quad \text{if} \quad a_k = \frac{3^k}{k!}.$$

Solution   Since

$$a_n = \frac{3^n}{n!} \quad \text{and} \quad a_{n+1} = \frac{3^{n+1}}{(n + 1)!}$$

it follows that

$$\frac{a_{n+1}}{a_n} = \frac{\dfrac{3^{n+1}}{(n + 1)!}}{\dfrac{3^n}{n!}}$$

$$= \frac{3^{n+1}}{(n + 1)!} \cdot \frac{n!}{3^n}$$

$$= \frac{3^{n+1}}{3^n} \cdot \frac{n!}{(n + 1)n!}$$

$$= \frac{3}{n + 1}.$$

Factorial notation is also used to define the symbol $\binom{n}{k}$.

Definition	Assume k and n are integers such that $0 \le k \le n$. Then
$\binom{n}{k}$	$$\binom{n}{k} = \frac{n!}{k!(n-k)!}.$$

EXAMPLE 6   **Calculating $\binom{n}{k}$**

Calculate each expression.

(a) $\binom{5}{3}$ (b) $\binom{17}{14}$ (c) $\binom{n}{n-1}$ (d) $\binom{n}{0}$ (e) $\binom{n}{n}$

Solution   (a) $\binom{5}{3} = \dfrac{5!}{3!(5-3)!} = \dfrac{5!}{3!2!}$

$$= \frac{5 \cdot 4 \cdot 3!}{3!2!} = 10$$

(b) $\binom{17}{14} = \dfrac{17!}{14!(17-14)!}$

$$= \frac{17!}{14!3!} = \frac{17 \cdot 16 \cdot 15 \cdot 14!}{14!3!} = \frac{17 \cdot 16 \cdot 15}{3} = 680$$

(c) $\binom{n}{n-1} = \dfrac{n!}{(n-1)![n-(n-1)]!}$

$$= \frac{n!}{(n-1)!1!} = \frac{n(n-1)!}{(n-1)!} = n$$

(d) $\binom{n}{0} = \dfrac{n!}{0!(n-0)!} = \dfrac{n!}{1 \cdot n!} = 1$

(e) $\binom{n}{n} = \dfrac{n!}{n!(n-n)!} = \dfrac{n!}{n!0!} = 1$

Using the Binomial Theorem

Now that we have convenient notation, we are ready to write a formula, called the **binomial formula,** that produces all the terms in the expansion of $(a + b)^n$, where n represents a positive integer.

The Binomial Theorem	Let a and b be real numbers, and let n be a positive integer. Then
	$$(a + b)^n = \binom{n}{0}a^n + \binom{n}{1}a^{n-1}b + \cdots + \binom{n}{k}a^{n-k}b^k + \cdots + \binom{n}{n}b^n.$$

Using summation notation, the binomial theorem can be written as

$$(a + b)^n = \sum_{k=0}^{n} \binom{n}{k} a^{n-k} b^k$$

for n a positive integer.

EXAMPLE 7 **Applying the Binomial Theorem**

Use the binomial theorem to determine the expansion of each binomial.

(a) $(x + y)^5$ (b) $\left(3x^2 - \frac{1}{2}\sqrt{y}\right)^4$

Solution (a) By the binomial theorem,

$$(x + y)^5 = \sum_{k=0}^{5} \binom{5}{k} x^{5-k} y^k$$

$$= \binom{5}{0}x^5 + \binom{5}{1}x^4y + \binom{5}{2}x^3y^2 + \binom{5}{3}x^2y^3 + \binom{5}{4}xy^4 + \binom{5}{5}y^5$$

$$= x^5 + \frac{5!}{1!4!}x^4y + \frac{5!}{2!3!}x^3y^2 + \frac{5!}{3!2!}x^2y^3 + \frac{5!}{4!1!}xy^4 + y^5$$

$$= x^5 + 5x^4y + 10x^3y^2 + 10x^2y^3 + 5xy^4 + y^5.$$

(b) By the binomial theorem,

$$\left(3x^2 - \frac{1}{2}\sqrt{y}\right)^4 = \sum_{k=0}^{4} \binom{4}{k} (3x^2)^{4-k}\left(-\frac{1}{2}\sqrt{y}\right)^k$$

$$= \binom{4}{0}(3x^2)^4 + \binom{4}{1}(3x^2)^3\left(-\frac{1}{2}\sqrt{y}\right) + \binom{4}{2}(3x^2)^2\left(-\frac{1}{2}\sqrt{y}\right)^2$$

$$+ \binom{4}{3}(3x^2)\left(-\frac{1}{2}\sqrt{y}\right)^3 + \binom{4}{4}\left(-\frac{1}{2}\sqrt{y}\right)^4$$

$$= (3x^2)^4 - 4(3x^2)^3\left(\frac{1}{2}\sqrt{y}\right) + 6(3x^2)^2\left(\frac{1}{2}\sqrt{y}\right)^2$$

$$- 4(3x^2)\left(\frac{1}{2}\sqrt{y}\right)^3 + \left(\frac{1}{2}\sqrt{y}\right)^4$$

$$= 81x^8 - 54x^6\sqrt{y} + \frac{27}{2}x^4y - \frac{3}{2}x^2y\sqrt{y} + \frac{1}{16}y^2.$$

EXAMPLE 8 **Finding a Term in a Binomial Expansion**

Use the binomial theorem to find the sixth term of the expansion of $(2x - y^2)^8$.

Solution In this situation $n = 8$. The binomial theorem indicates that the sixth term in the expansion is found by setting $k = 5$. So the sixth term is of the form

$$\binom{8}{5}a^{8-5}b^5 = \binom{8}{5}a^3b^5.$$

Thus for the given expression, we have

$$(2x - y^2)^8 = [(2x) + (-y^2)]^8$$

where $a = 2x$ and $b = -y^2$. Hence the sixth term is

$$\binom{8}{5}(2x)^3(-y^2)^5 = -\binom{8}{5}8x^3y^{10}$$

$$= -\frac{8 \cdot 7 \cdot 6}{3 \cdot 2 \cdot 1}(8x^3y^{10}) = -448x^3y^{10}.$$

PROBLEM SET 10.3

Mastering the Concepts

In problems 1–12, use mathematical induction to prove each statement for all positive integers n.

1. $1 + 3 + 5 + \cdots + (2n - 1) = n^2$

2. $1^3 + 2^3 + 3^3 + \cdots + n^3 = \dfrac{n^2(n + 1)^2}{4}$

3. $2 + 4 + 6 + \cdots + 2n = n^2 + n$

4. $\dfrac{1}{1 \cdot 2} + \dfrac{1}{2 \cdot 3} + \dfrac{1}{3 \cdot 4} + \cdots + \dfrac{1}{n(n + 1)} = \dfrac{n}{n + 1}$

5. $4 + 4^2 + 4^3 + \cdots + 4^n = \dfrac{4}{3}(4^n - 1)$

6. $1 + 5 + 5^2 + \cdots + 5^{n-1} = \dfrac{1}{4}(5^n - 1)$

7. $1 + 2 \cdot 2 + 3 \cdot 2^2 + \cdots + n \cdot 2^{n-1} = 1 + (n - 1) \cdot 2^n$

8. $(-1)^1 + (-1)^2 + (-1)^3 + \cdots + (-1)^n = \dfrac{(-1)^n - 1}{2}$

9. $1^2 + 3^2 + 5^2 + \cdots + (2n - 1)^2 = \dfrac{n(2n - 1)(2n + 1)}{3}$

10. $1 \cdot 2 + 2 \cdot 3 + 3 \cdot 4 + \cdots + n(n + 1) = \dfrac{n(n + 1)(n + 2)}{3}$

11. $x^0 + x^1 + x^2 + \cdots + x^n = \dfrac{1 - x^{n+1}}{1 - x}$, for $x \neq 1$

12. $(-1)^{2n-1} = -1$

In problems 13–16, use mathematical induction to prove that each inequality is true.

13. $3^n > n^2$ for all positive integers n

14. $(n + 2)^2 < n^3$ for all positive integers $n \geq 3$

15. $n^2 \leq 2^n$ for integers $n \geq 4$

16. $4^n \geq 4n$ for all positive integers n

In problems 17–22, use the patterns in Pascal's triangle to find the expansion of each binomial.

17. $(x + y)^4$

18. $(x - y)^6$

19. $(2x + 3y)^4$

20. $(2x - 5y)^3$

21. $(5x - 3y)^5$

22. $(3x + 4y)^4$

In problems 23 and 24, the given numbers correspond to a specific row in Pascal's triangle. Write the next row of numbers in the triangle and indicate the exponent n of the expression $(a + b)^n$ corresponding to this new row.

23. 1, 4, 6, 4, 1

24. 1, 5, 10, 10, 5, 1

In problems 25–28, evaluate each expression.

25. (a) $\binom{15}{10}$

(b) $\binom{6}{2}$

26. (a) $\binom{15}{5}$

(b) $\binom{n}{3}$

27. (a) $\binom{52}{5}$

(b) $\binom{52}{50}$

28. (a) $\binom{52}{52}$

(b) $\binom{n}{2}$

In problems 29–32, simplify the expression

$$\left| \frac{a_{n+1}}{a_n} \right|$$

for each given sequence.

29. $a_k = \dfrac{5^k}{(k - 1)!}$

30. $a_k = \dfrac{(-1)^k(k - 1)!}{e^k}$

31. $a_k = \dfrac{3^k}{k!}$

32. $a_k = \dfrac{(-1)^k x^{k+1}}{(k + 1)!}$

In problems 33–36, find the first four terms of the expansion of the given binomial.

33. $(x + y)^{28}$

34. $(x - y)^{49}$

35. $(x - 2y)^{35}$

36. $(x + 3y)^{57}$

In problems 37–42, use the binomial theorem to expand each expression as specified.

37. $(x + 3)^5$; all terms

38. $(2z + x)^4$; all terms

39. $(x - 2)^4$; all terms

40. $\left(\dfrac{1}{a} + \dfrac{x}{2}\right)^3$; all terms

41. $(x + y)^{12}$; the $x^5 y^7$ term

42. $(x - 3y)^7$; the $x^4 y^3$ term

In problems 43–46, use the binomial theorem to find the indicated term.

43. $\left(\dfrac{x^2}{2} + a\right)^{15}$; fourth term

44. $(y^2 - 2z)^{10}$; sixth term

45. $\left(a + \dfrac{x^2}{3}\right)^9$; term containing x^{12}

46. $\left(2\sqrt{y} - \dfrac{x}{2}\right)^{10}$; term containing y^4

Developing and Extending the Concepts

In problems 47–54, use mathematical induction to prove that the given statement is true.

47. 3 is a factor of $4^n - 1$.

48. 4 is a factor of $5^n - 1$.

49. $11^n - 4^n$ is divisible by 7.

50. $n^3 - 4n + 6$ is divisible by 3.

51. $\displaystyle\sum_{k=1}^{n} ar^{k-1} = \dfrac{a(1 - r^n)}{1 - r}$, $r \neq 1$

52. $\displaystyle\sum_{k=1}^{n} kx^{k-1} = \dfrac{1 - x^n}{(1 - x)^2} - \dfrac{nx^n}{1 - x}$, $x \neq 1$

53. $(\cos\theta + i\sin\theta)^n = \cos n\theta + i\sin n\theta$

54. $\ln(x_1 \cdot x_2 \cdot x_3 \cdots x_n) =$ $\ln x_1 + \ln x_2 + \ln x_3 + \cdots + \ln x_n$

55. Verify that:

(a) $\dbinom{n}{0} = \dbinom{n + 1}{0}$

(b) $\dbinom{n}{n} = \dbinom{n + 1}{n + 1}$

(c) $\dbinom{n}{k} = \dbinom{n}{n - k}$

56. Verify that: $\dbinom{n}{k - 1} + \dbinom{n}{k} = \dbinom{n + 1}{k}$

57. Prove or disprove each equation.

(a) $\dfrac{2n!}{(2n)!} = 1$

(b) $\dfrac{(n + 2)!}{(n + 4)!} = \dfrac{1}{n^2 + 7n + 12}$

(c) $(n!)(n!) = (2n)!$

58. Rewrite the binomial expansion of
$$(a + b)^n \quad \text{if} \quad a = b = 1.$$

In problems 59 and 60, use the summation
$$(a + b)^n = \sum_{k=0}^{n} \binom{n}{k} a^{n-k} b^k$$
to write each binomial expansion using summation notation.

59. $(x + y)^5$

60. $(x - 2y)^8$

Objectives

1. Determine the Number of Combinations
2. Determine the Number of Permutations
3. Determine the Probability of an Event

10.4 Combinations, Permutations, and Probability

In applications of mathematics, at times it is necessary to determine the number of different ways to form a collection of certain specific *outcomes* from a given general collection of possible outcomes. Those outcomes that are formed without regard to order are called *combinations,* whereas outcomes that are formed where order is important are called *permutations.* For instance, finding the number of 3-man committees that can be formed from a group of 10 men is an example of finding the number of combinations. However, to find the number of different four-digit license plates that can be made from the set of digits {1,2,3,4,5,6,7,8,9} is an example of permutations, since order is obviously important.

Our purpose in this section is to determine ways of finding the number of different combinations or permutations that can be formed from a given general collection. In addition, we introduce basic notions of *probability.*

Determining the Number of Combinations

Suppose that from a group of 4 people—Amy, Mary, Nancy, and Sam—it is necessary to form a committee of 3 people for a special task. In how many ways can 3 people be selected for the task out of the 4 available? Here, the order of the people selected doesn't matter. What we are actually interested in is how may different ways a group of 3 people can be formed from a given collection of 4 people. The possibilities are listed as follows:

$$\{\text{Amy, Mary, Nancy}\} \ , \ \{\text{Amy, Mary, Sam}\}$$

$$\{\text{Amy, Nancy, Sam}\} \ , \ \{\text{Mary, Nancy, Sam}\}$$

We refer to this list of possibilities as the *combinations* of 4 people taken 3 at a time. The total number of combinations is denoted by the symbol

$$C_3^4$$

so we write

$$C_3^4 = 4.$$

This illustration leads us to the following definition:

Definition
Combination

> Let S be a set consisting of n different elements. A *subcollection* consisting of k of these elements, where $k \leq n$ without regard to the order in which they are chosen or arranged, is called a **combination of n different elements in S taken k at a time.** The symbol
>
> $$C_k^n \quad \text{or} \quad {}_nC_k$$
>
> is used to denote the number of combinations of n elements taken k at a time that can be formed.

For example, the possible combinations of the set of letters

$$\{a, b, c, d\}$$

taken two at a time are as follows:

$$\{a, b\}, \{a, c\}, \{a, d\}, \{b, c\}, \{b, d\}, \{c, d\}.$$

Thus, we write

$$C_2^4 = 6.$$

Most graphers have a built in feature that calculates C_k^n.

Using factorial notation, it can be shown that the number of combinations of n objects taken k at a time, $0 \leq k \leq n$ is given by

$$C_k^n = \frac{n!}{k!(n-k)!}.$$

Note that the value C_k^n is the same as the binomial coefficient $\binom{n}{k}$ used in binomial expansions, that is,

$$C_k^n = \binom{n}{k}.$$

Thus, the number of combinations of four letters a, b, c, and d taken two at a time is given by

$$C_2^4 = \binom{4}{2} = \frac{4!}{2!(4-2)!} = \frac{4!}{2!2!} = 6.$$

Also, the number of 5 person committees that can be formed from a group of 6 people is given by

$$C_5^6 = \binom{6}{5} = \frac{6!}{5!(6-5)!} = 6.$$

EXAMPLE 1   **Finding the Number of Combinations**

(a) Find the number of combinations of 20 things taken 17 at a time.

(b) From a group of 20 employees, 4 are to be selected to work on a special project. In how many different ways can the 4 employees be selected?

Solution   (a) The number of combinations of 20 things taken 17 at time is given by

$$C_{17}^{20} = \binom{20}{17} = \frac{20!}{17!(20-17)!} = \frac{20 \cdot 19 \cdot 18 \cdot 17!}{17! \cdot 3!} = \frac{20 \cdot 19 \cdot 18}{3 \cdot 2 \cdot 1} = 1140.$$

(b) The number of combinations of selecting 4 people to work on a special project from a group of 20 available employees is given by

$$C_4^{20} = \binom{20}{4} = \frac{20!}{4!(20-4)!} = 4845.$$

Determining the Number of Permutations

Consider the set of 4 letters $\{a, b, c, d\}$. Suppose that this set is used to form ordered pairs of two letters without repetition of a letter in the pairings. In how many ways can this be done? The possibilities are listed as follows:

a first	b first	c first	d first
ab	ba	ca	da
ac	bc	cb	db
ad	bd	cd	dc

Thus, there are 12 ways to arrange 2 letters as ordered pairs from the collection $\{a, b, c, d\}$. Each particular arrangement of the letters is called a *permutation* of the letters. The number of permutations of the four letters when two letters are chosen at a time without repetition is denoted by

$$P_2^4.$$

Thus, we write

$$P_2^4 = 12.$$

Note that a permutation is an arrangement of distinct objects in a definite order. The notion of permutations is generalized in the following definition:

Definition

Permutation

Let S be a collection of n distinct objects. An *ordered* arrangement of k, where $0 < k \leq n$, of these objects in a row is called a **permutation of n different objects taken k** at a time. The symbol

$$P_k^n \quad \text{or} \quad {}_nP_k$$

denotes the number of different permutations of n objects taken k at a time.

In the previous example, we could have calculated P_2^4 as follows.

There are 4 ways to select the first letter, followed by 3 ways to select the second letter. So there are $4 \cdot 3 = 12$ possibilities of selecting the 4 letters taken 2 at a time. Thus,

$$P_2^4 = 4 \cdot 3 = 12.$$

In terms of factorials, we have

$$P_2^4 = 4 \cdot 3 = 4 \cdot 3 \cdot 1 = \frac{4 \cdot 3 \cdot 2!}{2!} = \frac{4!}{2!} = 12.$$

These observations are generalized as follows:

The number of permutations of n distinct objects or elements taken k at a time, $0 \leq k \leq n$, is denoted by P_k^n and is given by

$$P_k^n = n(n-1)(n-2)(n-3) \cdots (n-k+1)$$

Most graphers have a feature built in that calculates P_k^n.

or equivalently,

$$P_k^n = \frac{n!}{(n-k)!}.$$

In particular

$$P_n^n = \frac{n!}{(n-n)!} = n!.$$

For instance,

$$P_1^4 = \frac{4!}{(4-1)!} = \frac{4!}{3!} = 4,$$

$$P_3^6 = \frac{6!}{(6-3)!} = \frac{6!}{3!} = 120$$

and

$$P_5^5 = \frac{5!}{(5-5)!} = \frac{5!}{0!} = 120.$$

EXAMPLE 2   **Determining the Number of Permutations**

(a) How many three-digit numbers can be formed from the set of digits {1, 2, 3, 4, 5} if no repetition of digits is allowed?

(b) In how many ways can 5 students be seated in a row of 5 desks?

Solution   (a) Here we have permutations of 5 different digits to be taken 3 at a time. Thus, the number of possibilities is given by

$$P_3^5 = \frac{5!}{(5-3)!} = \frac{5!}{2!} = 5 \cdot 4 \cdot 3 = 60.$$

(b) In this situation, we have 5 different students to be seated in 5 different desks. The number of ways or permutations they can be seated is given by

$$P_5^5 = 5! = 120.$$

The following example relates the concepts we have developed in this section.

EXAMPLE 3   **Writing the Number of Combinations in Terms of the Number of Permutations**

Express C_k^n in terms of P_k^n.

Solution $$C_n^k = \binom{n}{k} = \frac{n!}{k!(n-k)!} = \frac{1}{k!}\left[\frac{n!}{(n-k)!}\right] = \frac{P_k^n}{k!}$$

EXAMPLE 4   **Using Permutations**

It is required to seat 5 men and 4 women in a row of 9 so that the women occupy the even places. How many such arrangements are possible?

Solution   The men may be seated in $P_5^5 = 5! = 120$ ways. The women may be seated in $P_4^4 = 4! = 24$ ways. Each arrangement of men is associated with each arrangement of women. Hence, the number of required arrangements is given by

$$P_5^5 \cdot P_4^4 = 5! \cdot 4! = 2880.$$

Determining the Probability of an Event

Probability is applied in a variety of practical situations in business, economics, and the social and physical sciences. The definition and the rules of probability will be presented here in terms of motivational problems rather than formally. However, before discussing probability, it is necessary to establish some basic terminology.

Activities such as tossing a coin, reading the daily temperature on a thermometer, or drawing a card from a deck of 52 cards are called *experiments*. With any kind of experiment there is associated a set of possible results called *outcomes*.

For example, when a coin is tossed, there are two possible outcomes. They are heads (denoted by H) or tails (denoted by T). If two different coins are tossed simultaneously, the possible outcomes are:

$$(H, H), (H, T), (T, H), \text{ and } (T, T).$$

In each of the previous examples these are the only possible outcomes. The set of all possible outcomes in an experiment is called a **sample space.**

For instance, suppose that an urn contains a white ball (W), a red ball (R), and a green ball (G), and an experiment consists of drawing one ball from the urn and then another one without replacing the first ball that is drawn. To determine the sample space, which is the permutations of 3 objects taken 2 at a time, we note that the first ball removed will be W, R, or G. If the first ball is W, the next ball will be either R or G. Hence, there are two possible outcomes if the first ball is W. Similarly, there are two possible outcomes if the first is either R or G. So, the sample space for this experiment is

$$\{(W, R), (W, G), (R, W), (R, G), (G, W), (G, R)\}.$$

Each element of this set displays the order in which the two balls are drawn. Note that outcome (W, R) is different from outcome (R, W), and that the number of outcomes is given by

$$P_2^3 = 6.$$

In general, we call each subcollection of a sample space an **event.** For example, suppose we roll a single die and observe the number on the face of the die. The sample space S is given by

$$S = \{1, 2, 3, 4, 5, 6\}.$$

If the event is that a number less than 3 shows, then the event is the subcollection $\{1, 2\}$ of the sample space S.

EXAMPLE 5

Describing an Event

A single die is rolled. Write each event.

(a) E_1: The number showing is odd.

(b) E_2: The number showing is even.

(c) E_3: The number showing is greater than 3.

(d) E_4: The number showing is less than 7.

Solution

Each event is written as a subcollection of the sample space $\{1, 2, 3, 4, 5, 6\}$ as follows:

(a) $E_1 = \{1, 3, 5\}$

(b) $E_2 = \{2, 4, 6\}$

(c) $E_3 = \{4, 5, 6\}$

(d) $E_4 = \{1, 2, 3, 4, 5, 6\}$

If a fair die (one that is perfectly balanced and symmetrical) is tossed in an unbiased manner (without trying to force a particular outcome), then any one face of the die has the same chance of showing topside as does any other face. (From here on, in all problems involving dice, we assume that this is the case.) In this situation we are inclined to say, for example, that "the chance that the die will land with 1 up is 1 out of 6." Similarly, we might say, "there are 3 chances out of 6 that the number showing is even," or "the chances that the number showing will be greater than 4 are 2 out of 6." The notion of "chance" leads us to the concept of probability.

Mathematical probability is, for the most part, based on a layperson's interpretation of the word "probability," as illustrated in the preceding paragraph. Accordingly, we define the probability of an event as follows:

Definition

Probability of an Event

> Suppose that an experiment with a finite sample space S is conducted. The probability that an actual outcome belongs to the event E is the ratio of the number of elements in E and the number of elements in S. This ratio is referred to as **the probability of the event E.**
>
> If we denote the number of elements in E by $n(E)$, the number of elements in S by $n(S)$, and let $p(E)$ denote the probability of E, then we have
>
> $$p(E) = \frac{n(E)}{n(S)}.$$

Using this definition, we can deduce the following three properties.

Let E be an event of a finite sample space S, and let $p(E)$ denote the probability of E, then we have:

(i) $p(S) = 1$, since $p(S) = \dfrac{n(S)}{n(S)} = 1$

(ii) If $n(E) = 0$, then $p(E) = \dfrac{n(E)}{n(S)} = \dfrac{0}{n(S)} = 0$,

   so that $p(E) = 0$

(iii) Since $0 \leq n(E) \leq n(S)$, it follows that

$$0 \leq \frac{n(E)}{n(S)} \leq \frac{n(S)}{n(S)} = 1 \text{ or } 0 \leq p(E) \leq 1.$$

EXAMPLE 6   **Finding the Probability of an Event**

Consider the experiment of rolling a die with sample space $S = \{1, 2, 3, 4, 5, 6\}$. What is the probability of the event E?

(a) E is an even number showing.

(b) E is rolling a 4 or a 6.

Solution

(a) The event E is the subset of S consisting of the even numbers, so $E = \{2, 4, 6\}$. Thus,

$$p(E) = \frac{n(E)}{n(S)} = \frac{3}{6} = \frac{1}{2}.$$

(b) The event E of rolling a 4 or 6 is given by $E = \{4, 6\}$. So $n(E) = 2$.

   Thus

$$p(E) = \frac{n(E)}{n(S)} = \frac{2}{6} = \frac{1}{3}.$$

EXAMPLE 7   **Finding the Probability of an Event**

Find the probability of rolling a sum of 8 with a pair of dice with different colors.

Solution   To find the sample space of this experiment, we use ordered pairs of numbers showing the first number, the outcome of the first die, and the second number, the outcome of the second die. Thus, the sample space is:

$$
\left\{
\begin{array}{l}
(1, 1), (1, 2), (1, 3), (1, 4), (1, 5), (1, 6), \\
(2, 1), (2, 2), (2, 3), (2, 4), (2, 5), (2, 6), \\
(3, 1), (3, 2), (3, 3), (3, 4), (3, 5), (3, 6), \\
(4, 1), (4, 2), (4, 3), (4, 4), (4, 5), (4, 6), \\
(5, 1), (5, 2), (5, 3), (5, 4), (5, 5), (5, 6), \\
(6, 1), (6, 2), (6, 3), (6, 4), (6, 5), (6, 6)
\end{array}
\right\}
$$

The sample space S is the set of all ordered pairs, so $n(S) = 36$. The event E of possible outcomes that produces a sum of 8 is given by

$$E = \{(2, 6), (3, 5), (4, 4), (5, 3), (6, 2)\} \text{ with } n(E) = 5.$$

Thus,

$$p(E) = \frac{n(E)}{n(S)} = \frac{5}{36}.$$

EXAMPLE 8   **Probability of Drawing a Ball from an Urn**

Suppose that an urn contains 3 red balls, 4 blue balls, and 2 green balls. Assume that the balls are thoroughly mixed and the experiment consists of selecting a ball out of the urn at random. What is the probability of

(a) drawing a red ball?

(b) drawing a blue ball?

(c) drawing a green ball?

Solution   We set up a sample space by numbering the balls *1, 2, 3, 4, 5, 6, 7, 8, 9,* so that

$$S = \{1, 2, 3, 4, 5, 6, 7, 8, 9\}.$$

Let us designate the red balls by the numbers, *1, 2,* and *3;* the blue balls by the numbers *4, 5, 6,* and *7;* and the green balls by the numbers *8* and *9.*
Suppose the event E_1 consists of drawing a red ball, then $E_1 = \{1, 2, 3\}$, the event E_2 consists of drawing a blue ball, so $E_2 = \{4, 5, 6, 7\}$, the event E_3 consists of drawing a green ball, so $E_3 = \{8, 9\}$.
Thus, we have the following results.

(a) Here $n(S) = 9$ and $n(E_1) = 3$, so that the probability of drawing a red ball is

$$p(E_1) = \frac{n(E_1)}{n(S)} = \frac{3}{9} = \frac{1}{3}.$$

(b) Here $n(S) = 9$ and $n(E_2) = 4$, so that the probability of drawing a blue ball is

$$p(E_2) = \frac{n(E_2)}{n(S)} = \frac{4}{9}.$$

(c) Here $n(S) = 9$ and $n(E_3) = 2$, so that the probability of drawing a green ball is

$$p(E_3) = \frac{n(E_3)}{n(S)} = \frac{2}{9}.$$

Note in Example 8 that

$$p(E_1) + p(E_2) + p(E_3) = \frac{3}{9} + \frac{4}{9} + \frac{2}{9} = 1.$$

That is, the probability of drawing a red, blue, or green ball is 1, which means that the event of drawing a red, green, or blue ball cannot fail to happen. This example illustrates an interesting feature of problems in probability, which is that choosing the sample space correctly is extremely important. If we had chosen the space to be $\{R, B, G\}$, standing for red, blue, and green, respectively, the correct result would have been impossible to find. On the other hand, by realizing that we could give numbers to the balls, we found the remaining work to be straightforward.

Often, listing the outcomes of a sample space or event is difficult or impractical. When this is the case, at times we can use the number of combinations or permutations to determine probabilities.

EXAMPLE 9   **Finding the Probability of a Hand of Cards**

Find the probability of drawing 3 hearts from a standard deck of 52 cards.

Solution   The sample space S consists of all possible 3 card hands. The number of elements in the sample space S is the number of ways of drawing 3 cards from a deck of 52 cards. The order in which the cards are dealt does not distinguish one hand from another. So, the number of ways of drawing 3 cards from a deck of 52 cards is the number of combinations of 52 objects taken 3 at a time, or

$$n(S) = C_3^{52} = \frac{52!}{3!(52 - 3)!} = \frac{50 \cdot 51 \cdot 52}{6} = 22{,}100.$$

Since there are 13 hearts in a deck of 52 cards, we let E represent the event that 3 cards are hearts, so that

$$n(E) = C_3^{13} = \frac{13!}{3!(13 - 3)!} = \frac{13 \cdot 12 \cdot 11}{3!} = 286.$$

Therefore, the probability that all 3 cards are hearts is given by

$$p(E) = \frac{n(E)}{n(S)} = \frac{286}{22{,}100} = \frac{11}{850} \text{ or } 0.0129.$$

Consider the experiment of tossing the same coin 3 times in succession and noting the outcomes. The sample space S is given by

$$S = \{HHH, HHT, HTH, HTT, THH, THT, TTH, TTT\}.$$

Suppose that the event E represents the situation in which exactly 2 of the flips are heads.
Then

$$E = \{HHT, HTH, THH\}.$$

Thus

$$p(E) = \frac{n(E)}{n(S)} = \frac{3}{8}.$$

Let E^C represent the set of all outcomes in S *other than those* in E, that is,

$$E^C = \{HHH, HTT, THT, TTH, TTT\}.$$

Then

$$n(E^C) = 5$$

so that

$$p(E^C) = \frac{n(E^C)}{n(S)} = \frac{5}{8}.$$

The event E^C, that is, all outcomes other than those in E, is called the *complement of E*.

The probabilities of E and E^C are related by

$$p(E^C) = \frac{5}{8} = 1 - \frac{3}{8} = 1 - p(E).$$

In general, we have the following property:

If E denotes the event of an experiment in a sample space S, and E^C is the **complement** of E, then $p(E) = 1 - p(E^C)$ or $p(E^C) = 1 - p(E)$.

EXAMPLE 10

Finding the Probability of a Complement

Suppose that one card is drawn from a standard deck of 52 cards. What is the probability that this card is not a spade?

Solution

A spade can be drawn from a deck of 52 cards in 13 different ways. Let E represent this event, so that $n(E) = 13$.

Thus, the probability of drawing a spade is given by

$$p(E) = \frac{n(E)}{n(S)} = \frac{13}{52} = \frac{1}{4}.$$

So the probability of not drawing a spade is given by

$$p(E^C) = 1 - p(E) = 1 - \frac{1}{4} = \frac{3}{4}.$$

PROBLEM SET 10.4

Mastering the Concepts

In problems 1–6, find the value of each expression.

1. (a) C_1^4
   (b) C_1^{10}
2. (a) C_3^{12}
   (b) C_0^{10}
3. (a) C_{10}^{10}
   (b) C_4^{16}
4. (a) C_5^5
   (b) C_4^{52}
5. (a) C_0^n
   (b) C_n^n
6. (a) C_3^n
   (b) C_5^n

In problems 7–10, find the value of each expression.

7. (a) P_2^7
   (b) P_4^6
8. (a) P_5^8
   (b) P_3^4
9. (a) P_6^{10}
   (b) P_7^7
10. (a) P_3^{12}
   (b) P_3^{14}

In problems 11–14, solve for n in each equation.

11. (a) $C_3^n = C_5^n$
   (b) $C_7^n = C_5^n$
12. (a) $C_1^n = C_4^7$
   (b) $C_{n-1}^n = C_2^6$
13. (a) $P_2^{n+2} = 3\,P_1^{n+1}$
   (b) $P_1^n = 2\,P_3^4$
14. (a) $7\,P_2^n = 84$
   (b) $15\,P_1^n = 1050$

In problems 15–18, evaluate each expression.

15. C_3^5 by listing all combinations of the letters a, b, c, d, e taken 3 at a time.
16. C_2^6 by listing all combinations of the letters a, b, c, d, e, f taken 2 at a time.
17. P_3^4 by listing all permutations of the letters a, b, c, d taken 3 at a time.
18. P_2^6 by listing all permutations of the letters a, b, c, d, e, f taken 2 at a time.
19. (a) Find the number of ways 12 joggers can finish among the first 3 in a race.
   (b) Find the number of ways 12 joggers in a race can finish first, second, and third.
20. In how many ways can 10 basketball players be assigned to lockers 1 and 2 in a row of 10 lockers?

In problems 21 and 22, determine the sample space S and the indicated event E for each experiment.

21. Experiment: Flip a coin 3 times.
   Event: At least 2 heads appear.
22. Experiment: Roll a die.
   Event: An even number appears.

In problems 23–26, determine the probability of each event in a given experiment.

23. Experiment: Roll a die.
   Event: The face showing is an even number that is less than 4.

24. Experiment: Draw 1 card from a standard deck of 52 cards.
   Event: The card is a diamond.
25. Experiment: Roll 2 dice.
   Event: The 2 numbers that turn up are both even.
26. Experiment: Toss a fair coin twice.
   Event: At least one tail appears.

Applying the Concepts

27. Selecting Officers: In how many ways can an organization select a president, vice-president, a secretary, and a treasurer from a group of 30 candidates? Assume that no person can hold more than one position.
28. Forming Three-Digit Numbers: In how many ways can three-digit numbers be formed from the digits *1, 3, 4, 7, 8, 9* if no repetition of digits is allowed?
29. Seating Boys and Girls: In how many ways can 5 boys and 5 girls sit in a row if the boys and girls are assigned to alternate seats beginning with a girl?
30. License Plates: How many different license plates can be made that have 3 different letters followed by 3 different digits?
31. Selecting Lotto Tickets: To win a lottery game, a player must correctly select 6 numbers from numbers 1 through 49. Find the total number of possible selections if the player selects only even numbers.
32. Winning a Basketball Championship Series: The winner of the 2005 NBA Championship series is the team in the finals that wins 4 out of 7 games. In how many different ways can the series be extended to 7 games?
33. Selecting Different Hands of a Poker Game: Find the number of different hands in 5-card poker possible from a standard deck of 52 cards.
34. Selecting Subcommittees: In how many ways can one choose a subcommittee of 3 people from a committee of 10 people?
35. Selecting a Face Card: Find the probability of drawing a face card (jack, queen, or king) from a well-shuffled deck of 52 cards.
36. Selecting a Marble: A jar contains 12 marbles, 6 are red, 4 are black, and 2 are blue.
   (a) If one marble is picked at random, what is the probability that the marble is black?
   (b) If one marble is picked, what is the probability that the marble is blue?

37. **Tossing a Coin:** A fair coin is tossed 5 times. What is the probability of obtaining exactly 4 heads and 1 tail?

38. **Rolling Dice:** Suppose we roll 2 dice. What is the probability of each event?

(a) A sum of 2 turns up.

(b) A sum of 5 turns up.

(c) A sum of 4 turns up.

(d) A sum of 8 turns up.

39. **Gender of Children:** Consider an event for each given situation formed from a list of 8 outcomes that are possible when a couple has 3 children. Find the probability that

(a) Among 3 children, there is exactly 1 boy.

(b) Among 3 children, there are exactly 2 boys.

(c) Among 3 children, all are girls

40. **On Time Flight Arrivals:** A study of 180 randomly selected airline flights showed that 112 arrived on time.

(a) What is the probability of the airline flights arriving late?

(b) What is the probability of the airline flights arriving on time?

41. **Finding Complements:** When a baby is born, the probability that the baby is a boy is 0.612. What is the probability the baby is a girl?

42. **Finding Complements:**

(a) If someone is randomly selected, find the probability that his or her birthday is not in September. Ignore leap years.

(b) In a certain U.S. city, a poll shows that 54% of the city population will vote for a certain candidate. What is the probability of someone voting for a different candidate?

Developing and Extending the Concepts

The following result, called the *additive rule,* may be used to find the probability of the **union** of two events E_1 and E_2, that is, the probability of either E_1 or E_2 occurring:

$$P(E_1 \text{ or } E_2) = P(E_1) + P(E_2) - P(E_1 \text{ and } E_2)$$

For example, if

$$P(E_1) = 0.3, P(E_2) = 0.4, \text{ and } P(E_1 \text{ and } E_2) = 0.1$$

then

$$P(E_1 \text{ or } E_2) = P(E_1) + P(E_2) - P(E_1 \text{ and } E_2)$$

$$= (0.3) + (0.4) - (0.1)$$

$$= 0.6$$

For **mutually exclusive events,** that is, for events E_1 and E_2 that have no outcome in common:

$$P(E_1 \text{ or } E_2) = P(E_1) + P(E_2)$$

When each of two events can occur so that neither one affects the probability of the other, we say the events are **independent.**

For independent events E_1 and E_2, the probability that both E_1 and E_2 occur is given by

$$P(E_1 \text{ and } E_2) = P(E_1) \cdot P(E_2).$$

43. **Drawing a Card:** A card is drawn from a standard deck of 52 cards. What is the probability that the card is a king or a heart?

44. **Selecting a Ball:** A bag contains 4 red balls and 2 blue balls; another bag contains 3 red balls and 5 blue balls. If one ball is drawn from each bag, find the probability that

(a) Both balls are red.

(b) Both balls are blue.

(c) One ball is red and one is blue.

45. **Selecting a Ball:** A bag contains 4 red balls, 5 white balls, and 7 blue balls. If a ball is drawn from the bag and then replaced, and then a second ball is drawn, what is the probability that both balls are white?

46. **Selecting a Marble:** A marble is drawn at random from a box containing 10 red, 30 white, 20 blue, and 15 orange marbles. Find the probability that it is:

(a) orange or red

(b) white

(c) not red or blue

(d) red or blue

(e) not blue

(f) white or blue

(g) not white or blue

47. **Winning a Race:** If the probability that car A wins a race is 1/6 and the probability that car B places second in a race is 1/5, what is the probability that in the same race car A will be first and car B will be second?

48. **Selecting a Ball:** A bag contains 10 red balls, 20 blue balls, and 30 green balls. If 5 balls are drawn from the bag and replaced, find the probability of each situation.

(a) All blue balls are drawn.

(b) At least one green ball is drawn.

49. **Tossing Two Dice:** If two dice are tossed, find the probability of rolling a sum of either 7 or 9.

50. **Selecting a Card:** A card is drawn at random from a standard deck of 52 cards and then returned to the deck. Then the deck is shuffled and again a card is drawn at random. Find the probability of drawing two cards of the same suit.

◈ CHAPTER 10 REVIEW PROBLEM SET

In problems 1 and 2, find the first five terms of each sequence whose general term is a_n.

1. (a) $a_n = n^2 - 1$

(b) $a_n = (-1)^n(3n - 2)$

2. (a) $a_n = 4^{n/2}$

(b) $a_n = \dfrac{1}{n^2 + 3n}$

In problems 3 and 4, find the first five terms of the sequence defined by each recursive formula.

3. (a) $a_1 = 4$, $a_n = a_{n-1} + 4$, $\ n \geq 2$

(b) $a_1 = 8$, $a_n = a_{n-1} + 5$, $\ n \geq 2$

4. (a) $a_1 = 1$, $a_2 = 5$, $a_n = 2a_{n-1} + 3a_{n-2}$, $\ n \geq 3$

(b) $a_0 = 1$, $a_1 = 1$, $a_n = na_{n-1}$, $\ n \geq 2$

In problems 5 and 6, find a formula for the general term of each sequence.

5. (a) $2, 4, 6, 8, 10, \ldots$

(b) $-1, 1, -1, 1, -1, 1, \ldots$

6. (a) $1, 4, 9, 16, 25, \ldots$

(b) $\dfrac{1}{2}, \dfrac{2}{3}, \dfrac{3}{4}, \dfrac{4}{5}, \cdots$

In problems 7 and 8, evaluate each sum.

7. (a) $\displaystyle\sum_{k=1}^{5} k(2k - 1)$ **8.** (a) $\displaystyle\sum_{k=0}^{4} \dfrac{2}{3^k}$

(b) $\displaystyle\sum_{k=1}^{4} 2k^2(k - 3)$ (b) $\displaystyle\sum_{k=2}^{6} (k + 1)(k + 2)$

9. Use the properties of summation to evaluate the expression

$$\sum_{k=1}^{20} (3k^2 - 4k + 7)$$

10. Express the sum $1^2 + 2^2 + 3^2 + 4^2 + 5^2$ in sigma notation.

In problems 11 and 12, find:

(i) The nth term, and graph the sequence.

(ii) The sum of the first n terms of each arithmetic sequence for the given value of n.

11. (a) $2, 6, 10, 14, \ldots$; for $n = 15$

(b) $8, 16, 24, 32, \ldots$; for $n = 10$

12. (a) $5, 13, 21, 29, \ldots$; for $n = 12$

(b) $20, 12, 4, -4, -12, \ldots$, for $n = 20$

In problems 13 and 14, find the nth partial sum S_n of the arithmetic sequence $\{a_n\}$ with common difference d.

13. (a) $a_1 = 2$, $d = 5$, $n = 6$

(b) $a_1 = 15$, $d = -6$, $n = 8$

14. (a) $a_1 = -4$, $d = 8$, $n = 7$

(b) $a_1 = 7$, $d = 5$, $n = 10$

In problems 15 and 16, find the sum of the first n terms S_n of each arithmetic sequence for the given value of n.

15. (a) $\dfrac{3}{4}, \dfrac{1}{4}, -\dfrac{1}{4}, -\dfrac{3}{4}, \cdots$; $n = 12$

(b) $2a + 3b, 3a + 2b, 4a + b, \ldots$; $n = 6$

16. (a) $-8, -3, 2, 7, \ldots$; $n = 30$

(b) $a, a + 2, a + 4, a + 6, \ldots$; $n = 20$

In problems 17 and 18, find the general term a_n of each geometric sequence.

17. (a) $5, 15, 45, 135, \ldots$

(b) $4, 8, 16, 32, \ldots$

18. (a) $\dfrac{1}{4}, \dfrac{1}{8}, \dfrac{1}{16}, \dfrac{1}{32}, \cdots$

(b) $-6, 5, -\dfrac{25}{6}, \cdots$

In problems 19 and 20, graph the sequence. Also, find the common ratio and the sum of the first n terms of each geometric sequence for the given value of n.

19. (a) $48, 96, 192, \ldots$; $n = 8$

(b) $-81, -27, -9, \ldots$; $n = 12$

20. (a) $\dfrac{3}{4}, 3, 12, \ldots$; $n = 10$

(b) $0.2, 0.002, 0.00002, \ldots$; $n = 15$

In problems 21 and 22, find the sum of each infinite geometric series.

21. $\displaystyle\sum_{k=1}^{\infty} 3\left(\dfrac{2}{3}\right)^{k-1}$ **22.** $\displaystyle\sum_{k=0}^{\infty} (-1)^k\left(\dfrac{3}{4}\right)^{k}$

In problems 23 and 24, write each rational number as the quotient of two integers in reduced form.

23. $0.\overline{45}$ **24.** $0.4\overline{22}$

In problems 25 and 26, use mathematical induction to prove that the assertion is true for all positive integers n.

25. $n^2 + 3n$ is an even integer

26. $2 \cdot 4 + 4 \cdot 6 + 6 \cdot 8 + \cdots + 2n(2n + 2) =$

$$\dfrac{n}{3}(2n + 2)(2n + 4)$$

In problems 27 and 28, simplify each expression.

27. (a) $\dbinom{12}{5}$ (b) $\dbinom{34}{30}$

(c) $\dfrac{a_{n+1}}{a_n}$ if $a_k = \dfrac{(k+1)!}{7}$

28. (a) $\dbinom{n+7}{n+5}$ (b) $\dbinom{n+3}{n+1}$

(c) $\dfrac{a_{n+1}}{a_n}$ if $a_k = \dfrac{3^k}{k!}$

In problems 29 and 30, expand each expression.

29. (a) $(2+x)^5$

    (b) $(3x - 4y)^4$

30. (a) $(3x^2 - 2y^2)^7$

    (b) $(2a + 3b)^6$

In problems 31 and 32, find and simplify the specified term in the binomial expansion of the expression.

31. (a) The fourth term of $(2x + y)^9$

    (b) The term containing x^5 in $(3x + y)^{10}$

32. (a) The fifth term of $(x - 2y)^7$

    (b) The term containing x^{10} in $(2x^2 - 3)^{11}$

33. Physics: Assume an object falls vertically 16 feet during the first second, 48 feet during the next second, 80 feet during the third, and so on.

    (a) How far will the object fall during the tenth second?

    (b) What is the total number of feet the object has fallen at the end of 10 seconds?

34. Biology: Suppose that a colony of bacteria reproduces to become 300 bacteria after the first day. After 2 days, they reproduce to become 700 bacteria. After 3 days, they reproduce to become 1100 bacteria, and so on.

    (a) Find a recursive formula for the number of bacteria a_n after n days.

    (b) How many bacteria are there after 7 days?

35. Real Estate: A home purchased for $130,000 appreciates at a rate of 3% per year.

(a) If a_n represents the value of the house after n years, write an explicit formula for a_n.

(b) Find the value of the house after 7 years. Round off to two decimal places.

36. Sports: A runner training for a track meet begins on March 1 by running 1 mile, and increases the length of the run by 0.1 mile a day. Find the length of the run on March 25.

In problems 37 and 38, evaluate each expression.

37. (a) C^{26}_{24}

    (b) P^{26}_4

    (c) $\dfrac{P^{26}_4}{C^{26}_{24}}$

38 (a) C^{n+2}_n

    (b) P^{n+2}_2

    (c) $\dfrac{P^{n+2}_2}{C^{n+2}_n}$

39. A homeowner intends to purchase 3 shrubs, each of a different variety, for her front yard. At the local nursery there are 9 varieties of shrubs available. In how many ways can she make her selections?

40. Assume that 4 different shrubs are to be placed in 4 predug holes. How many arrangements are possible?

41. Find the probability of each event when 2 dice are tossed.

    (a) Both are odd.

    (b) The sum is 7 or 11.

    (c) The sum is not 7 or 11.

    (d) The sum is less than 5.

42. A florist has an inventory of 5 dozen red roses, 4 dozen pink roses, and 7 dozen white roses. If each bouquet consists of 1 dozen roses of the same color and 3 bouquets are purchased, find the probability of each event.

    (a) All 3 bouquets consist of red roses.

    (b) None of the 3 bouquets has red roses.

    (c) All 3 bouquets have the same color roses.

    (d) Two bouquets have white roses and the other bouquet has pink roses.

◈ CHAPTER 10 TEST

1. Write the first five terms of each sequence with the given general term.

   (a) $a_n = 3n - 4$

   (b) $a_n = \dfrac{n + 1}{n^2}$

   (c) $a_1 = 3$; $a_n = -2a_{n-1}$, $n \geq 2$

2. Find the general term a_n of each sequence.

   (a) $6, 10, 14, 18, \ldots$

   (b) $-3, 9, -27, 81, \ldots$

3. Evaluate $\displaystyle\sum_{k=2}^{6} (k^2 + 2k)$.

4. Find the indicated unknown for each arithmetic sequence.

   (a) a_n; if $a_1 = 1$, $d = 2$, $n = 6$

   (b) n; if $a_1 = 13$, $d = -4$, $a_n = -7$

   (c) S_n; if $a_1 = 2$, $a_n = 14$, $n = 7$

5. Find the indicated unknown for each geometric sequence.

   (a) r; if $a_5 = 81$, $S_5 = 61$

   (b) a_1; if $r = -1$, $a_n = -1$, $n = 7$

6. Evaluate the sum of each infinite geometric series.

   (a) $\displaystyle\sum_{k=1}^{\infty} 5\left(\dfrac{4}{5}\right)^{k-1}$ (b) $\displaystyle\sum_{k=1}^{\infty} \dfrac{1}{8 \cdot 3^k}$

7. Use mathematical induction to prove

   $$2 + 2^2 + 2^3 + \cdots + 2^n = 2(2^n - 1).$$

8. Expand $(2x + 1)^5$.

9. Find the seventh term of the expansion of $(x + x^2)^{12}$.

10. The tip of a pendulum travels 15 centimeters on its first swing. On each subsequent swing it travels 86% of the previous distance. How far has it traveled after 8 swings? Round off the answer to two decimal places.

11. Evaluate each expression.

    (a) C_3^7 (b) P_3^7 (c) $\dfrac{1}{3!} P_3^7$

12. In setting a 4-digit code for a newly installed security system, a homeowner can select from the digits 0, 1, 2, 3, 4, 5, 6, 7, 8 and 9. How many combinations are possible for each situation?

    (a) No digit is to be repeated.

    (b) The digits can be repeated.

    (c) The first digit cannot be a zero and no digit is to be repeated.

    (d) The first digit cannot be a zero and the digits can be repeated.

13. Find the probability of each event when 2 dice are tossed.

    (a) The sum is 6.

    (b) The sum is not 6.

    (c) The sum exceeds 10.

    (d) One of the die shows an even number and the other shows a 4.

Appendix I: Preliminary Topics in Algebra

Contents

Objectives

1. Identify Types of Real Numbers
2. Convert Rational Numbers to Decimal Form
3. Use a Calculator to Get Approximate Decimal Values
4. Locate Real Numbers on a Number Line

Appendix I contains reviews of three topics that are necessary for mastery of the material covered in this text. They are:

1. The real number system
2. Dividing polynomials, including synthetic division
3. The complex number system

AI.1 Real Numbers

Identifying Types of Real Numbers

Real numbers are encountered in everyday life and can be thought of as numbers that have decimal representations. Table 1 identifies various types of real numbers. The symbol $\mathbb{R}$ is used to represent the set of real numbers.

TABLE 1   Real Number Sets

Name	Description
Natural numbers, or **positive integers**	Counting numbers
Integers	Includes negative integers, zero, and positive integers
Rational numbers	All numbers that can be written in the form a/b, where a and b are integers, $b \neq 0$
Irrational numbers	Real numbers that are not rational numbers
Real numbers	All rational numbers and irrational numbers

Examples of these types of numbers are given in the following table:

Natural numbers, or positive integers	$1, 2, 3, 10, 15$
Integers	$-3, -2, -1, 0, 1, 2, 3$
Rational numbers	$-3, 0, -\frac{3}{4}, \frac{5}{7}, 5\frac{1}{3}, 0.37$
Irrational numbers	$\sqrt{2}, \sqrt[3]{5}, \pi$
Real numbers	$2, -7, \frac{5}{11}, \sqrt{3}, \pi$

A1

The diagram in Figure 1 indicates how the sets of numbers listed in Table 1 are related to each other. Each set includes all numbers in the set(s) connected to and shown below it.

Figure 1

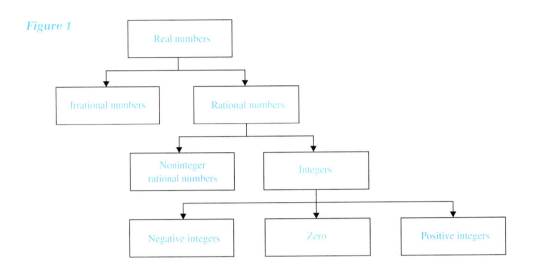

EXAMPLE 1   Use the categories given in Figure 1 to identify the type of each given number.

(a) $-\dfrac{1}{2}$ (b) 0.8   (c) $\sqrt{5}$ (d) 13%   (e) -3

Solution   (a) Since $-\dfrac{1}{2} = \dfrac{-1}{2}$, it follows that the number is rational, and so it is also a real number.

(b) Since $0.8 = \dfrac{8}{10} = \dfrac{4}{5}$, the number is rational and, therefore, it is also real.

(c) The number $\sqrt{5}$ is irrational, therefore it is real.

(d) Since $13\% = \dfrac{13}{100}$, the number is rational and real.

(e) Since -3 is a negative integer, it also is an integer; it is rational because

$$-3 = -\dfrac{3}{1}$$

and, therefore, it is also real.

❖ Practice Exercise 1

Use Figure 1 to indicate the type of each of the given numbers.

(a) $-2\dfrac{1}{5}$ (b) 1.23   (c) $-\sqrt{100}$ (d) $\sqrt{10}$ (e) 0.3%

Converting Rational Numbers to Decimal Form

The set of real numbers consists of rational and irrational numbers. A decimal representation of a rational number a/b, where a and b are integers and $b \neq 0$, is obtained by dividing the numerator a by the denominator b. The result is either a terminating decimal or a nonterminating decimal with a repeating pattern.

EXAMPLE 2   Rewrite each rational number as a decimal number.

(a) $\dfrac{2}{5}$ (b) $\dfrac{2}{3}$ (c) $-\dfrac{15}{11}$

Solution   After dividing the numerator by the denominator in each situation, we get the following results:

(a) $\dfrac{2}{5} = 0.4$, a terminating decimal

The overbar identifies the digits that repeat.   (b) $\dfrac{2}{3} = 0.6666\ldots = 0.\overline{6}$, a repeating decimal

(c) $\dfrac{-15}{11} = -1.363636\ldots = -1.\overline{36}$, a repeating decimal

◈ Practice Exercise 2

Rewrite each number as a decimal number.

(a) $\dfrac{7}{8}$ (b) $-\dfrac{71}{33}$ (c) $4\dfrac{2}{999}$

Using a Calculator to Get Approximate Decimal Values

The decimal representation of an irrational number will neither terminate nor have a repeating decimal pattern. A calculator is commonly used to find decimal approximations of irrational numbers.

Often the symbol $\approx$ is used to indicate an approximation. For instance, rounded off to two decimal places, $\sqrt{127} \approx 11.27$. However, in this text we normally give rounding instructions and then use the equal sign for such approximations. So we write

$$\sqrt{127} = 11.27 \text{ (approx.)}.$$

To round off $\sqrt{127}$ to three decimal places, we get

$$\sqrt{127} = 11.269 \text{ (approx.)}.$$

Rounding $\sqrt{127}$ to four decimal places, we have

$$\sqrt{127} = 11.2694 \text{ (approx.)}.$$

EXAMPLE 3   Use a calculator to write each number in decimal form rounded off to four decimal places.

(a) $\dfrac{23}{17}$

(b) $(3.45)^5$

(c) $\sqrt[3]{75}$

(d) $\dfrac{(5.21)^2\,\pi}{\sqrt{2}+3}$

Solution   Using a calculator, we get the following results:

(a) $\dfrac{23}{17} = 1.3529$ (approx.)

(b) $(3.45)^5 = 488.7598$ (approx.)

(c) $\sqrt[3]{75} = 4.2172$ (approx.)

(d) $\dfrac{(5.21)^2\,\pi}{\sqrt{2}+3} = 19.3184$ (approx.)

❖ Practice Exercise 3

Use a calculator to approximate each expression to four decimal places.

(a) $\left(\dfrac{16}{7}\right)^3$

(b) $\dfrac{3\pi}{\sqrt{2}}$

(c) $\dfrac{\sqrt[5]{73}}{4-\sqrt{3}}$

(d) $\dfrac{7+\sqrt{5}}{7-\sqrt{5}}$

Technically, the number is not the point, nor the point the number. However, it is customary to use the terms real number and point interchangeably.

Locating Real Numbers on a Number Line

A number line establishes a one-to-one correspondence between the real numbers and points on the line. Each real number is associated with a point on the line and, conversely, each point associates with a particular real number called its **coordinate.**

   To establish a number line, we first draw a straight line (usually horizontal) and then select a point to represent 0 and a point to the right of 0 to represent 1. The distance between 0 and 1 establishes the *scale unit* of the number line. Positive integers are located by marking off the scale unit to the right and negative integers are plotted by marking off the scale unit to the left (Figure 2).

Figure 2

Negative direction   Positive direction

   Real numbers can be **located,** or **plotted,** on a number line by using their decimal representations along with integer locations and the proper proportion of the scale unit.

EXAMPLE 4   Locate each of the given numbers on the same number line.

(a) 5.75   (b) $-4\dfrac{1}{4}$ (c) $\sqrt{2}$ (d) π

Solution   The numbers, along with their decimal representations, are located and displayed in Figure 3.

When locating points on a number line, we are primarily interested in displaying the relative positions of various points rather than the precise location of each point.

Figure 3

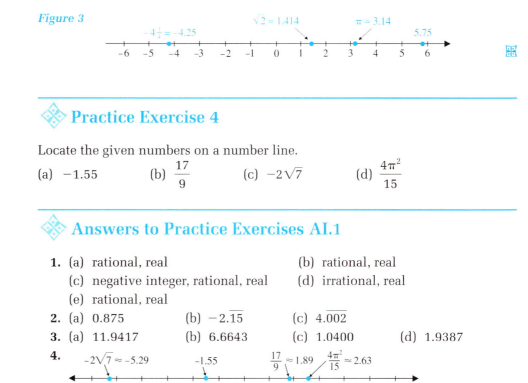

◈ Practice Exercise 4

Locate the given numbers on a number line.

(a) -1.55 　　(b) $\dfrac{17}{9}$ 　　(c) $-2\sqrt{7}$ 　　(d) $\dfrac{4\pi^2}{15}$

◈ Answers to Practice Exercises AI.1

1. (a) rational, real 　　　　　　(b) rational, real
   (c) negative integer, rational, real 　(d) irrational, real
   (e) rational, real
2. (a) 0.875 　　(b) $-2.\overline{15}$ 　　(c) $4.\overline{002}$
3. (a) 11.9417 　　(b) 6.6643 　　(c) 1.0400 　　(d) 1.9387
4.

◈ PROBLEM SET AI.1

In problems 1 and 2, use the categories given in Figure 1 to identify the type of each given number.

1. (a) $-2\dfrac{1}{9}$ 　(b) 0 　(c) 13% 　(d) $\sqrt{20}$

2. (a) $\sqrt{7}$ 　(b) $\dfrac{-5}{1}$ 　(c) $\dfrac{0.3}{11}$ 　(d) $\dfrac{10}{0.2}$

In problems 3 and 4, express each rational number as a terminating or repeating decimal.

3. (a) $\dfrac{3}{4}$ 　(b) $-\dfrac{7}{4}$ 　(c) $5\dfrac{5}{16}$ 　(d) $\dfrac{0.81}{2.7}$

4. (a) $\dfrac{11}{3}$ 　(b) $-\dfrac{16}{5}$ 　(c) $-2\dfrac{5}{33}$ 　(d) $\dfrac{3.2}{0.037}$

In problems 5 and 6, convert each decimal to a fraction in lowest terms.

5. (a) 0.7 　(b) 0.55 　(c) 2.4 　(d) 9.05
6. (a) 0.4 　(b) 0.75 　(c) 3.5 　(d) 1.125

In problems 7 and 8, use a calculator to write each expression in decimal form rounded off to four decimal places.

7. (a) $\dfrac{\sqrt[3]{317}}{11}$ 　　　　(b) $\left(\dfrac{180}{\pi}\right)^2$

   (c) $\dfrac{\sqrt{6}-\sqrt{2}}{2}$ 　　　(d) $\dfrac{4\pi^2+3}{\pi-2}$

8. (a) $\dfrac{-\sqrt{2}}{13}$ 　　　　(b) $\dfrac{\sqrt[3]{5}}{17}$

   (c) $\dfrac{-7+\pi}{\sqrt{3}}$ 　　　(d) $\dfrac{\sqrt{\pi}+3}{\sqrt{\pi}-3}$

In problems 9 and 10, locate the number sets on a number line.

9. (a) The set of positive integers less than 5
   (b) The set of negative integers greater than -4
   (c) The set of rational numbers $-2, -\frac{1}{2}, \frac{2}{3}, 2.3,$ $3\frac{5}{8}, 4.7$
   (d) The set of irrational numbers $\sqrt{3}, \pi, \sqrt[3]{17}$
10. (a) The set of all odd positive integers 1, 3, 5, 7, . . .
    (b) The set of negative integers $-7, -5, -2, -1$
    (c) The set of noninteger rational numbers $-0.3,$ $0.5, \frac{2}{3}, 5\frac{1}{4}$
    (d) The set of irrational numbers $-\sqrt{2}, \sqrt[3]{16}, \sqrt{7}$
11. The formula for the volume V of a sphere with radius r is given by

$$V = \frac{4}{3}\pi r^3.$$

Find the volumes of spheres with the radius 1 inch; 2.7 inches; 11 inches. Round off the answers to two decimal places.

12. The formula for the volume V of a rectangular solid with length l, width w, and height h is given by

$$V = lwh.$$

Find the volume of a concrete slab in the form of a rectangular solid with the length 75.23 feet, width 13.57 feet, and height 1.78 feet. Round off the answer to two decimal places.

13. The formula for the surface area S of a right circular cylinder with the radius r and height h is given by

$$S = 2\pi rh + 2\pi r^2.$$

Find the surface area of a right circular cylinder with the radius of 2.3 centimeters and height of 7.9 centimeters. Round off the answer to two decimal places.

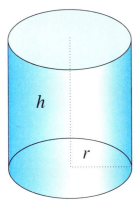

14. The square base of the Great Pyramid, the tomb of Pharaoh Cheops at Al Giza, Egypt, measures 755 feet on each side and has a height of 481 feet (Figure 4). The formula for the volume V is given by

$$V = \frac{1}{3}Bh.$$

where B is the area of the base and h is the height. What is the volume of the pyramid? Round off the answer to two decimal places.

Figure 4

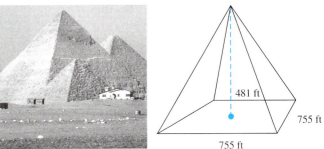

Objectives

1. Perform Long Division of Polynomials
2. Perform Synthetic Division

AI.2 Dividing Polynomials

In this section we perform division of polynomials. Long division of polynomials, like division of real numbers, relies on multiplication and subtraction skills. Synthetic division is introduced as a tool for carrying out division more efficiently for certain situations. We begin with a review of the long division process.

Performing Long Division of Polynomials

An *algorithm* is a step-by-step systematic procedure used to perform a computation. For instance, we use the long division algorithm to divide 372 by 5 to obtain the *quotient* and the *remainder*, as shown below:

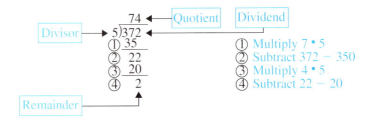

The result of the division can be expressed either in the *multiplicative form* as

$$372 = (5)(74) + 2$$

$$\text{Dividend} = (\text{Divisor})(\text{Quotient}) + \text{Remainder}$$

or in the *fractional form* as

$$\frac{372}{5} = 74 + \frac{2}{5}$$

$$\frac{\text{Dividend}}{\text{Divisor}} = \text{Quotient} + \frac{\text{Remainder}}{\text{Divisor}}$$

The long division of polynomials follows the same pattern, as illustrated in the following example.

EXAMPLE 1   Divide the polynomial

$$F = 2x^3 - 3x^2 + 5x + 2 \text{ by } D = x + 3.$$

Express the result in both multiplicative and fractional forms.

Solution   We note that both the divisor

$$x + 3$$

and the dividend

$$2x^3 - 3x^2 + 5x + 2$$

are arranged in descending powers of *x*. (If they weren't, we would rewrite them so that they were before beginning the process.)

Then we proceed as follows:

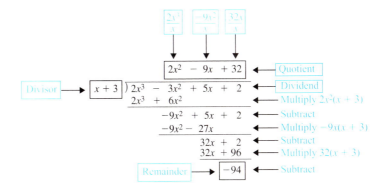

As in the previous numerical example, we can write the result either in multiplicative form:

$$2x^3 - 3x^2 + 5x + 2 = (x + 3)(2x^2 - 9x + 32) + (-94)$$

or in fractional form:

$$\frac{2x^3 - 3x^2 + 5x + 2}{x + 3} = 2x^2 - 9x + 32 + \frac{(-94)}{x + 3}.$$

❖ Practice Exercise 1

Divide the polynomial $4x^3 - 5x^2 + 7x + 3$ by $x + 2$
Write the result in multiplicative form.

At times the division process results in a remainder of zero as shown in the next example.

EXAMPLE 2   Divide the polynomial $x^3 + 6x - 3x^2 - 4$ by $x - 1$.

Express the result in multiplicative form.

Solution   First we note that the divisor and dividend are rearranged in descending powers of x, then we apply the division process as follows:

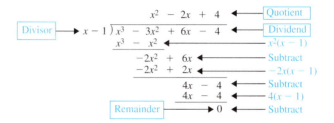

The quotient is $x^2 - 2x + 4$, and the remainder is 0,
so that in multiplication form we have

$$x^3 - 3x^2 + 6x - 4 = (x - 1)(x^2 - 2x + 4).$$

Notice that when we get a remainder of zero, we obtain a factorization and the given polynomial is said to be *divisible*. Thus, as shown in Example 2 above, $x^3 - 3x^2 + 6x - 4$ is divisible by $x - 1$.

❖ Practice Exercise 2

Use long division to show that $7x^3 - 5x^2 + 4x - 156$ is divisible by $x - 3$.
Express the result in multiplicative form.

Example 2 on page A8 and practice exercise 2 illustrate the following property which is stated as follows:

**The Division
Algorithm Result**

> If F and D are polynomials such that D is non zero, then there exist unique polynomials Q and R such that
>
> $$F = D \cdot Q + R$$
>
> where the degree of R is less than the degree of D [R may be 0]. The expression D is called the **divisor**, F is the **dividend**, Q is the **quotient**, and R is the **remainder**.

The result of the division algorithm can also be expressed in the fractional form

$$\frac{F}{D} = Q + \frac{R}{D}.$$

which carries implicit restrictions on the divisor D, since D cannot equal 0.

Performing Synthetic Division

Consider the long division process used to divide the polynomial

$$F = 3x^3 - x^2 + 2x - 18 \text{ by } x - 2.$$

First, notice that the divisor and the dividend are arranged in descending powers of x, then the process is organized as follows:

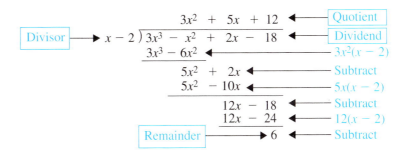

Thus the quotient $Q = 3x^2 + 5x + 12$ and the remainder $R = 6$. Now we can simplify the task of performing such a division by using an algorithm called *synthetic division*, which is an efficient way of dividing a polynomial by a binomial of the form $x - c$ where c is a constant.

Being able to divide a polynomial F by a linear binomial of the form $x - c$ quickly and accurately will be of great help in studying polynomial function in Chapter 3. This kind of division can be carried out more efficiently by a method called **synthetic division.**

In synthetic division, we work with the coefficient(s) of the dividend and divisor to produce a result that gives the coefficients of the quotient and remainder. The process is explained and illustrated in the following table.

The synthetic division method can be used only when the divisor is of the form x − c.

In the illustration on page A9 we used long division to divide the polynomial

$$F = 3x^3 - x^2 + 2x - 18$$

by

$$D = x - 2.$$

Now we perform the same division by using the synthetic division method as follows:

	Explanation of Process	Result
Step 1	Arrange the terms of the polynomial F in descending powers of x, and use the 0 coefficient for any missing powers of x.	$F = 3x^3 - x^2 + 2x - 18$
Step 2	Consider the divisor in the form $x - c$. Write down the value of c (here, $c = 2$), draw a vertical line, and then list the coefficients of the dividend F.	$2 \mid \quad 3 \quad -1 \quad 2 \quad -18$
Step 3	Leave space below the row of coefficients, draw a horizontal line, and copy the leading coefficient of F below the line.	
Step 4	Multiply 3, the leading coefficient, by 2, the value for c, and place the product, 6, above the horizontal line under the second coefficient, -1. Then add -1 to this product and place the result, 5, below the horizontal line.	
	Now multiply 5 by 2 and place the product, 10, above the horizontal line under the third coefficient, 2. Then add 2 to this product and place the result, 12, below the horizontal line.	
	Continue in this way, multiplying 12 by 2 and placing the product, 24, above the horizontal line under the next coefficient, -18. Then add -18 and 24 to obtain 6, placing 6 below the horizontal line. Isolate the very last sum, 6, by drawing a short vertical line.	
Step 5	In the last row, the numbers 3, 5, and 12 are the coefficients of the quotient Q and the last number, 6, is the remainder R.	$Q = 3x^2 + 5x + 12$ $6 = R$

Notice how the long division process used earlier compares with the synthetic division method shown above:

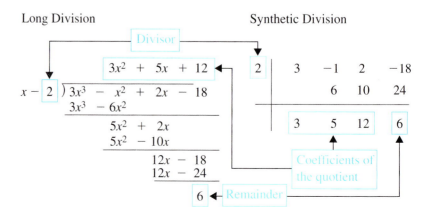

EXAMPLE 3   Use synthetic division to obtain the quotient and the remainder if the polynomial

$$F = 2x^4 - 3x^2 + 5x + 7 \text{ is divided by } D = x + 3.$$

Solution   The terms of F are given in descending powers of x, but we need to use 0 to denote the coefficient of the missing x^3 term. So the coefficients of F are listed as 2, 0, −3, 5, and 7.
The divisor has the form $D = x + 3 = x - (-3) = x - c$, so $c = -3$.
By synthetic division, we get

$$
\begin{array}{r|rrrrr}
-3 & 2 & 0 & -3 & 5 & 7 \\
   &   & -6 & 18 & -45 & 120 \\
\hline
   & 2 & -6 & 15 & -40 & 127 \\
\end{array}
$$

Hence the quotient is $Q = 2x^3 - 6x^2 + 15x - 40$ and the remainder is $R = 127$.

◈ Practice Exercise 3

Use synthetic division to obtain the quotient and remainder if the polynomial F is divided by a binomial D.

(a) $F = 3x^5 - 4x^3 + 7x - 2; D = x - 2$
(b) $F = x^3 + 8; D = x + 2$

◈ Answers to Practice Exercises AI.2

1. $4x^3 - 5x^2 + 7x + 3 = (x + 2)(4x^2 - 13x + 33) + (-63)$
2. $7x^3 - 5x^2 + 4x - 156 = (7x^2 + 16x + 52)(x - 3) + 0$
3. (a) $Q = 3x^4 + 6x^3 + 8x^2 + 16x + 39; R = 76$
    (b) $Q = x^2 - 2x + 4; R = 0$

◈ PROBLEM SET AI.2

In problems 1–8, use long division to divide F by D. Identify the quotient Q and the remainder R. Also express each result in the multiplicative form.

1. $F = 4x^3 - 2x^2 + 7x - 1; D = x + 1$
2. $F = 8x^3 - x^2 - x + 5; D = x - 2$
3. $F = 3x^4 - 2x^3 + 4x^2 + 3x + 7; D = x - 3$
4. $F = 2x^5 + 3x^4 - x^2 + x - 4; D = x + 2$
5. $F = 6x^4 + 10x^2 + 7; D = 2x + 3$
6. $F = 15x^4 + 5x^3 + 6x^2 - 2; D = 5x - 3$
7. $F = -3x^3 - x^2 - 3x - 1; D = 3x + 1$
8. $F = -6x^3 + 25x^2 - 9; D = 3x - 2$

In problems 9–16, use synthetic division to find the quotient Q and the remainder R.

Dividend F is:	Divisor $D(x)$ is:
9. $2x^3 + 17x^2 - 11x - 21$	$x - 3$
10. $x^3 - 2x^2 - 1$	$x + 2$
11. $4x^4 - 3x^3 + 2x^2 + 5$	$x - 1$
12. $10x^4 - 30x^3 + 63$	$x - 6$
13. $-5x^4$	$x + 2$
14. $3x^6 - 7$	$x + 3$
15. $3 + x^2 - 5x^4$	$x + 2$
16. $8 - 2x^3 + 3x^4 - 2x^5$	$x + 3$

In problems 17–20, use synthetic division to write F in the form $F = Q \cdot D + R$.

17. $F = 2x^3 - 5x^2 + 5x + 11$; $D = x - 1$
18. $F = 4x^4 - 2x^3 - 19x^2 + 5$; $D = x + 3$
19. $F = 2x^4 + 3x - 6$; $D = x + 2$
20. $F = 3x^5 - 2x^2 + 7$; $D = x - 3$

In problems 21–24, use synthetic division to find k so that by dividing F by D, the remainder $R = 0$.

21. $F = 3x^3 + 4x^2 + kx - 20$; $D = x - 2$
22. $F = kx^3 - 25x^2 + 47x + 30$; $D = x - 10$
23. $F = 2x^3 - 6x^2 + kx - 20$; $D = x - 3$
24. $F = x^4 - kx^2 + 6$; $D = x + 1$

Objectives

1. Identify Complex Numbers
2. Add, Subtract, and Multiply Complex Numbers
3. Divide Complex Numbers

AI.3 Complex Numbers

Identifying Complex Numbers

Earlier we saw that symbols such as $(-9)^{1/2}$, $\sqrt{-4}$, and $\sqrt{-x}$ (where x is positive) do not represent real numbers because there is no real number whose square is negative.

To define square roots of negative numbers, we have a set of numbers that contains the set of real numbers and also square roots of negative numbers. We call this set of numbers the set of *complex numbers.*

In general, a number of the form

$$a + bi \qquad \text{where } a \text{ and } b \text{ are real numbers}$$

is called a **complex number.**

The symbol i denotes a number whose square is -1, that is, we define i by the equation

$$i^2 = -1$$

and we write

$$\sqrt{-1} = i.$$

For the complex number $z = a + bi$, a is called the **real part** of z and b is called the **imaginary part** of z. Some examples of complex numbers are given in Table 1.

TABLE 1   Examples of Complex Numbers

Number	Real Part	Imaginary Part
$3 + 5i$	3	5
$-1 - 7i$	-1	-7
$8i = 0 + 8i$	0	8
$5 = 5 + 0i$	5	0

Since any real number a can be expressed in the complex number form $a + 0i$, we conclude that the complex number set includes all real numbers.

EXAMPLE 1   Write each expression in the form $a + bi$.

(a) $\sqrt{-4}$ (b) $\sqrt{-25x^2}$, where $x > 0$ (c) i^3 (d) i^4

Solution   (a) $\sqrt{-4} = \sqrt{4(-1)} = \sqrt{4}\sqrt{-1} = 2i = 0 + 2i$

(b) $\sqrt{-25x^2} = \sqrt{25x^2(-1)} = \sqrt{25x^2}\sqrt{-1} = 5xi = 0 + 5xi$

(c) $i^3 = i^2 \cdot i = (-1)i = 0 + (-1)i$

(d) $i^4 = i^2 \cdot i^2 = (-1)(-1) = 1 = 1 + 0i$

◈ Practice Exercise 1

Write each expression in the form $a + bi$.

(a) $2 + \sqrt{-9}$ (b) $\sqrt{-16x^8}$ (c) $8i^2$ (d) $(2i)^3$

Adding, Subtracting, and Multiplying Complex Numbers

Equality, addition, subtraction, and multiplication of two complex numbers $a + bi$ and $c + di$ are defined as follows:

Algebra of Complex Numbers

> (i) **Equality:** $a + bi = c + di$ if and only if $a = c$ and $b = d$
> (ii) **Addition:** $(a + bi) + (c + di) = (a + c) + (b + d)i$
> (iii) **Subtraction:** $(a + bi) - (c + di) = (a - c) + (b - d)i$
> (iv) **Multiplication:** $(a + bi)(c + di) = (ac - bd) + (ad + bc)i$

It is unnecessary to memorize the above definitions. When working with complex numbers, we may treat them the same as polynomials, where i is treated as a variable with the added property that $i^2 = -1$.

EXAMPLE 2   Perform each operation and simplify.

(a) $(3 + 2i) + (5 - 8i)$

(b) $(4 - 7i) - (1 + 2i)$

(c) $(3 - 5i)(2 + 3i)$

Solution   (a) $(3 + 2i) + (5 - 8i) = (3 + 5) + (2 - 8)i = 8 - 6i$

(b) $(4 - 7i) - (1 + 2i) = (4 - 1) + (-7 - 2)i = 3 - 9i$

(c) $(3 - 5i)(2 + 3i) = 6 - 10i + 9i - 15i^2$

$$= 6 - i - 15(-1)$$

$$= 6 + 15 - i = 21 - i$$

◈ Practice Exercise 2

Perform each operation and simplify.

(a) $(6 - 2i) + (-5 + i)$ (b) $(-3 + 2i) - (1 - 2i)$ (c) $(5 + 2i)^2$

Dividing Complex Numbers

We divide complex numbers by applying a process similar to the one we used to rationalize the denominators of radical expressions. In this case, the factor that plays the role of the rationalizing factor is called the *complex conjugate.*

If $z = a + bi$, then the **complex conjugate** is denoted by $\bar{z} = \overline{a + bi}$, and it is defined as

$$\bar{z} = \overline{a + bi} = a - bi$$

For example,

if $z = 3 + 2i$, then its complex conjugate is $\bar{z} = \overline{3 + 2i} = 3 - 2i$;

if $z = -5 - 7i$, then $\bar{z} = -5 + 7i$.

Note that, if $z = a + bi$, then

$$z\bar{z} = (a + bi)(a - bi) = a^2 - b^2 i^2 = a^2 - b^2(-1) = a^2 + b^2$$

which is a real number.

The *division* of two complex numbers is accomplished by multiplying the numerator and the denominator by the conjugate of the denominator. This results in a fraction that contains a real number in the denominator, as the next example shows.

EXAMPLE 3   Perform each division and express the answer in the form $a + bi$.

(a) $\dfrac{1}{5 + 3i}$

(b) $\dfrac{5 - 10i}{3 - 4i}$

Solution   By multiplying the numerator and denominator of each expression by the conjugate of the denominator, we have:

(a) $\dfrac{1}{5 + 3i} = \dfrac{1}{5 + 3i} \cdot \dfrac{5 - 3i}{5 - 3i}$

$= \dfrac{5 - 3i}{25 + 9} = \dfrac{5 - 3i}{34} = \dfrac{5}{34} - \dfrac{3}{34}i = \dfrac{5}{34} + \left(\dfrac{-3}{34}\right)i$

(b) $\dfrac{5 - 10i}{3 - 4i} = \dfrac{5 - 10i}{3 - 4i} \cdot \dfrac{3 + 4i}{3 + 4i} = \dfrac{55 - 10i}{9 + 16} = \dfrac{55 - 10i}{25}$

$= \dfrac{5(11 - 2i)}{25} = \dfrac{11 - 2i}{5} = \dfrac{11}{5} - \dfrac{2}{5}i = \dfrac{11}{5} + \left(\dfrac{-2}{5}\right)i$

❖ Practice Exercise 3

Perform each division and express the answer in the form $a + bi$.

(a) $\dfrac{1}{1 + i}$

(b) $\dfrac{-2 + 3i}{2 - 4i}$

❖ Answers to Practice Exercises AI.3

1. (a) $2 + 3i$ (b) $0 + 4x^4i$ (c) $-8 + 0i$ (d) $0 + (-8)i$
2. (a) $1 - i$ (b) $-4 + 4i$ (c) $21 + 20i$
3. (a) $\dfrac{1}{2} + \left(-\dfrac{1}{2}\right)i$ (b) $\left(-\dfrac{4}{5}\right) + \left(-\dfrac{1}{10}\right)i$

❖ PROBLEM SET AI.3

In problems 1 and 2, write each expression in the form $a + bi$.

1. (a) $\sqrt{-16}$
   (b) $3 + \sqrt{-8}$
   (c) $7 - 3\sqrt{-9}$
   (d) $\dfrac{2 - \sqrt{-4}}{6}$
   (e) $\sqrt{-x^2}$, where $x > 0$
   (f) $i^6 + 5$
2. (a) $2 - \sqrt{-4x^4}$
   (b) $6 - \sqrt{-16y^4}$
   (c) $17 + 5\sqrt{-17}$
   (d) $\dfrac{9 + \sqrt{-27}}{3}$
   (e) $-\sqrt{-x^2}$, where $x < 0$
   (f) $i^7 - 2$

In problems 3–10, perform each operation and simplify.

3. (a) $(3 - 2i) + (7 - 3i)$
   (b) $(4 - i) - (7 - 3i)$
4. (a) $(-3 + 6i) - (3 + 7i)$
   (b) $(7 + 24i) + (-3 - 4i)$
5. (a) $(3 - 6i)(2 + i)$
   (b) $(-5 + 2i)(-3 + 2i)$
6. (a) $(3 + 2i)(1 + i)$
   (b) $(2 + i)(2 - i)$
7. (a) $(2 + i)^2$
   (b) $(3 - 4i)^2$
8. (a) $\dfrac{4 + 3i}{2 - 4i}$ (b) $\dfrac{4 - 3i}{-1 - 2i}$
9. (a) $\dfrac{6 + 4i}{1 + 5i}$ (b) $\dfrac{7 + 6i}{-5 + 2i}$
10. (a) $\dfrac{4 - \sqrt{3}i}{2 + \sqrt{3}i}$ (b) $\dfrac{3 - 5i}{\sqrt{5} + 4i}$

Appendix II: Preliminary Topics in Plane Geometry

Contents

AII.1 Angles

When two lines, ℓ and p, intersect so that the angles formed are right angles, we say that the lines are **perpendicular lines** (Figure 1a). A common notation for this is the symbol $p \perp \ell$. The line ℓ that is perpendicular to $\overline{AB}$ and intersects $\overline{AB}$ at its midpoint M is called the **perpendicular bisector** of $\overline{AB}$ (Figure 1b).

Figure 1

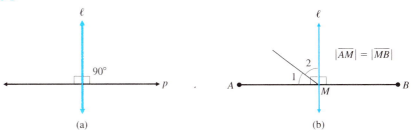

(a)   (b)

Two intersecting lines form four nonstraight angles in the plane. In Figure 2, the four angles are $\angle 1$, $\angle 2$, $\angle 3$, and $\angle 4$. These angles are classified as follows:

1. Angles 1 and 2 (Figure 2) are called **adjacent angles.** Adjacent angles have a common vertex and a common side between them.
2. Nonadjacent angles formed by two intersecting lines, such as angles 1 and 3 or angles 2 and 4, are called **vertical angles** or **opposite angles** (Figure 2). Vertical angles formed by intersecting lines are *equal*.
3. Two angles are called **supplementary angles** if the sum of their measures is 180°. In Figure 2, angles 1 and 2, 2 and 3, 3 and 4, and 4 and 1 are supplementary angles.
4. Two angles are called **complementary angles** if the sum of their measures is 90°. Angles 1 and 2 in Figure 1b are complementary angles.
5. Positive angles less than 90° are **acute angles,** such as angles 1 and 3 in Figure 2. Angles between 90° and 180° are **obtuse angles,** such as angles 2 and 4 in Figure 2.

Figure 2

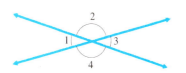

Any line that intersects two or more lines is called a **transversal** of these lines. The angles made by a transversal n that intersects two lines are shown (and numbered) in Figure 3. These angles are named as follows:

6. **Corresponding angles** determined by a transversal are pairs of angles on the "same side" of the transversal and on the "same relative sides" of the given lines. In Figure 3, the pairs of corresponding angles are 6 and 2, 7 and 4, 5 and 1, and 8 and 3.

7. **Interior angles** are those angles between the lines cut by the transversal. Angles 1, 2, 7, and 8 are interior angles in Figure 3.

8. **Alternate interior angles** are certain pairs of angles on opposite sides of the transversal. In Figure 3, the pairs of alternate interior angles are 1 and 7, and 2 and 8.

Figure 3

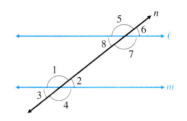

In Figure 4, line a is parallel to line b, and c is a transversal; angles 2 and 7 and angles 1 and 8 are alternate interior angles; and angles 5 and 7, angles 6 and 8, angles 2 and 3, and angles 1 and 4 are corresponding angles.

The following properties are noteworthy:

Figure 4

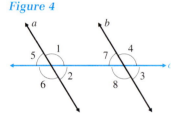

1. **Corresponding angles** (determined by a transversal) of parallel lines are *equal* (Figure 4), and conversely, if two lines are cut by a transversal so that the corresponding angles are equal, then the lines are parallel.

2. **Alternate interior angles** (determined by a transversal) of parallel lines are *equal* (Figure 4), and conversely, if two lines are cut by a transversal so that alternate interior angles are equal, then the lines are parallel.

EXAMPLE 1   **Using Angle Terminology**

In Figure 5, C, E, and F lie on a straight line, and D, E, and A lie on a different straight line. The angle BEF is a right angle.

Figure 5

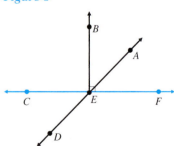

(a) Name a pair of vertical angles.

(b) Name an obtuse angle.

(c) Name an acute angle.

(d) Which angle is the complement of $\angle FEA$?

(e) Which angle is the supplement of $\angle FEA$?

Solution

(a) Angles AEF and CED are a pair of vertical angles.

(b) Angle BED is an obtuse angle.

(c) Angle AEF is an acute angle.

(d) The complement of $\angle FEA$ is $\angle AEB$.

(e) The supplement of $\angle FEA$ is $\angle AEC$.

AII.2 Triangles

Figure 6

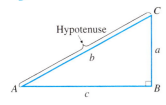

A triangle is a **right triangle** if one of its angles is a right angle—that is, a 90° angle. In Figure 6, *ABC* is a right triangle and ∠*B* is a right angle. In a right triangle, the side opposite the right angle is called the **hypotenuse** (Figure 6). The **Pythagorean theorem** states that *the square of the length of the hypotenuse is equal to the sum of the squares of the lengths of the other two sides.* Thus, in Figure 6, by the Pythagorean theorem,

$$b^2 = a^2 + c^2$$

EXAMPLE 2   **Using the Pythagorean Theorem**

In right triangle *ABC* (Figure 6), suppose that $a = 5$ and $b = 13$. Find *c*.

Solution   By the Pythagorean theorem,

$$c^2 = b^2 - a^2 = 13^2 - 5^2 = 169 - 25 = 144$$

so

$$c = \sqrt{144} = 12$$

Triangles are classified as follows:

1. A triangle is an **isosceles triangle** if two of its sides are equal in length (Figure 7a). It can be shown that the angles opposite the equal sides of an isosceles triangle are equal. In Figure 7a, ∠*B* = ∠*C*.
2. A triangle is an **equilateral triangle** if all three of its sides are equal in length (Figure 7b). All three angles are equal.
3. A triangle is an **oblique triangle** if none of the angles of the triangle is a right triangle (Figure 7c).

Figure 7

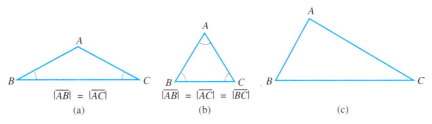

$|\overline{AB}| = |\overline{AC}|$ $|\overline{AB}| = |\overline{AC}| = |\overline{BC}|$
(a)   (b)   (c)

Two triangles, say △*ABC* and △*DEF*, are **congruent,** which is written in symbols as △*ABC* ≅ △*DEF*, if they can be made to coincide. In other words, congruent triangles have the same size and shape (Figure 8). That is, △*ABC* ≅ △*DEF* means that each of the following equations holds:

Figure 8

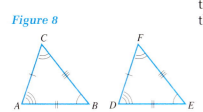

$$\angle A = \angle D \qquad |\overline{CB}| = |\overline{FE}|$$
$$\angle B = \angle E \qquad |\overline{AC}| = |\overline{DF}|$$
$$\angle C = \angle F \qquad |\overline{AB}| = |\overline{DE}|$$

Note that *equal sides are opposite equal angles.* (We use a shorthand notation in Figure 8 to indicate equal sides and equal angles. The single lines drawn through $\overline{AC}$ and $\overline{DF}$, for example, indicate that these sides are equal, as do the double lines through $\overline{AB}$ and $\overline{DE}$, and the triple lines through $\overline{CB}$ and $\overline{FE}$.)

The concept of congruence does not give us any practical way of deciding when two triangles are congruent. The following properties are useful:

1. If three sides of one triangle equal three sides of another triangle (SSS), then the triangles are *congruent.*
2. If two sides and the included angle of one triangle equal two sides and the included angle of another triangle (SAS), then the triangles are *congruent.*
3. If two angles and the included side of one triangle equal two angles and the included side of another triangle (ASA), then the triangles are *congruent.*

EXAMPLE 3   **Determining Congruent Triangles**

In Figure 9, let $\angle 1 = \angle 2$, and $\angle EAB = \angle CBA$. Show that $\triangle AEB \cong \triangle BCA$. What can be concluded about the remaining corresponding parts?

Figure 9

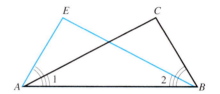

Solution   In the figure, we have

$$|\overline{AB}| = |\overline{BA}|$$

$$\angle 1 = \angle 2$$

$$\angle EAB = \angle CBA$$

Therefore

$$\triangle AEB \cong \triangle BCA \quad \text{(ASA)}$$

Thus,

$$|\overline{AE}| = |\overline{BC}|, \quad |\overline{AC}| = |\overline{BE}| \quad \text{and} \quad \angle E = \angle C$$

Two triangles are **similar** if the triangles have equal angles. Here, $\triangle ABC$ is similar to $\triangle DEF$, which is denoted by $\triangle ABC \sim \triangle DEF$. The corresponding sides are proportional (Figure 10). Since $\triangle ABC \sim \triangle DEF$, we have:

Figure 10

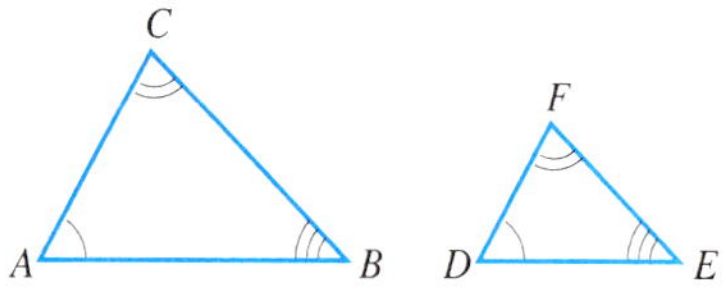

$$\angle A = \angle D, \quad \angle B = \angle E, \quad \text{and} \quad \angle C = \angle F.$$

Also,

$$\frac{|\overline{AB}|}{|\overline{DE}|} = \frac{|\overline{AC}|}{|\overline{DF}|} = \frac{|\overline{BC}|}{|\overline{EF}|}$$

Note that corresponding sides are opposite equal angles.

In order to show that *two triangles are similar, it is enough to show that two angles of one triangle equal two angles of the other triangle.*

EXAMPLE 4   **Determining Similar Triangles**

Suppose that one triangle is inside another triangle and the sides of the triangles are parallel. The smaller triangle has sides 5, 8, and 10 inches in length, and the larger triangle has a side of 15 inches parallel to the 10 inch side (Figure 11). Show that the triangles are similar, and find the lengths of the other two sides of the larger triangle.

Figure 11

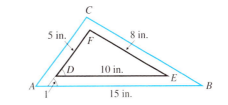

Solution   In Figure 11, $\angle A = \angle 1$ and $\angle D = \angle 1$, because corresponding angles of parallel lines are equal. Hence, $\angle D = \angle A$. Similarly, $\angle E = \angle B$, so that $\triangle DEF \sim \triangle ABC$. Therefore,

$$\frac{|\overline{DE}|}{|\overline{AB}|} = \frac{|\overline{FE}|}{|\overline{CB}|} = \frac{|\overline{DF}|}{|\overline{AC}|}$$

or

$$\frac{10}{15} = \frac{8}{|\overline{CB}|} = \frac{5}{|\overline{AC}|}$$

so that $|\overline{AC}| = 7.5$ inches and $|\overline{CB}| = 12$ inches.

Appendix III: Answers to Selected Problems

Chapter 1: Problem Set 1.1 (on page 9)

1. (a) 1 **(b)** -1 **3. (a)** 1 **(b)** 1 **5.** 2 **7. (a)** 2 **(b)** no solution **9.** -12

11. (a) $5 - \pi$ **(b)** $\sqrt{7} - 2$ **13. (a)** $\begin{cases} x - 5 & \text{if } x \geq 5 \\ 5 - x & \text{if } x < 5 \end{cases}$ **(b)** $\begin{cases} x + 3 & \text{if } x \geq -3 \\ -x - 3 & \text{if } x < -3 \end{cases}$

15. $\begin{cases} x + 5 & \text{if } x \geq -5 \\ -x - 5 & \text{if } x < -5 \end{cases}$ **17.** 8 **19.** 8 **21.** $10|t|$ **23. (a)** $-2, 6$ **(b)** $-12, 6$

25. (a) $-6, 9$ **(b)** $-10, 25$ **27. (a)** $1, 7$ **(b)** $-5, 2$ **29. (a)** $b = a + c$ **(b)** $w = \dfrac{A}{\ell}$

31. (a) $b = \dfrac{2A}{h}$ **(b)** $t = \dfrac{V - 4}{32}$ **33. (a)** $h = \dfrac{2A}{a + b}$ **(b)** $y = \dfrac{12z - 20}{3z}$

35. \$4200 at 7%; \$5800 at 6.25% **37.** \$4000 at 5.5%; \$2200 at 5.8% **39.** 30 cm^3 **41.** 9 mph

43. 3 min. **45. (a)** no solution **(b)** infinite number of solutions **47. (a)** no solution **(b)** $-\dfrac{1}{2}$

49. (a) $-\dfrac{1}{2}$ **(b)** $x \geq 0$

Problem Set 1.2 (on page 20)

1. (a) $-3, 4$ **(b)** $-4, 5$ **3. (a)** $-4, 2$ **(b)** $-3, 2$

5. (a) $-6, 6$ **(b)** $-2\sqrt{2}, 2\sqrt{2}$ **(c)** $-\dfrac{4}{5}, \dfrac{4}{5}$ **(d)** $-3i, 3i$

7. (a) $0, 14$ **(b)** $2\sqrt{3} - \dfrac{1}{2}, -2\sqrt{3} - \dfrac{1}{2}$ **9. (a)** $36; (x + 6)^2$ **(b)** $\dfrac{25}{4}; \left(x - \dfrac{5}{2}\right)^2$

11. (a) $\dfrac{1}{36}; \left(x - \dfrac{1}{6}\right)^2$ **(b)** $\dfrac{9}{64}; \left(x + \dfrac{3}{8}\right)^2$ **13. (a)** $1 \pm \sqrt{3}$ **(b)** $-5 \pm \sqrt{22}$

15. (a) $\dfrac{7}{3}, 1$ **(b)** $\dfrac{4}{5} \pm \dfrac{\sqrt{101}}{5}$ **17. (a)** $0.5, -1$ **(b)** $\dfrac{5 \pm \sqrt{13}}{6}$ **19.** $\dfrac{7.5 \pm \sqrt{82.05}}{3}$

21. (a) $-1, 0, 5$ **(b)** $-1, 0, 3$ **23. (a)** $\pm 2i, \pm 2$ **(b)** $-5, \pm 4$ **25.** $4, -1$ **27.** $-10, 12$

29. (a) 9 **(b)** 5 **31. (a)** 14 **(b)** 2 **33.** $2, 18$ **35.** 6 **37.** 1 **39.** no solution **41.** $-9, 3$

43. 2 cm **45. (a)** 0.11 second **(b)** 6 seconds **47. (a)** $x^2 + x$ **(b)** 5 in. by 6 in.

49. (a) Two distinct real roots **(b)** One real root **(c)** No real roots

51. (a) If $ac = 0$, then $c = 0$. Therefore, $x^2 = 0$ or $x = 0$ **(b)** Two real solutions **53. (a)** $-\dfrac{1}{2}, 1$ **(b)** $1, 16$

55. $-2, 3, \dfrac{1 \pm \sqrt{7}i}{2}$ **57. (a)** $\pm \dfrac{3}{2}\sqrt{4 - x^2}$ **(b)** $-3 \pm \sqrt{9 - (x + 1)^2}$

59. (a) $w^2 + 2$ **(b)** $\dfrac{b^3 - a}{2}$

Problem Set 1.3 (on page 33)

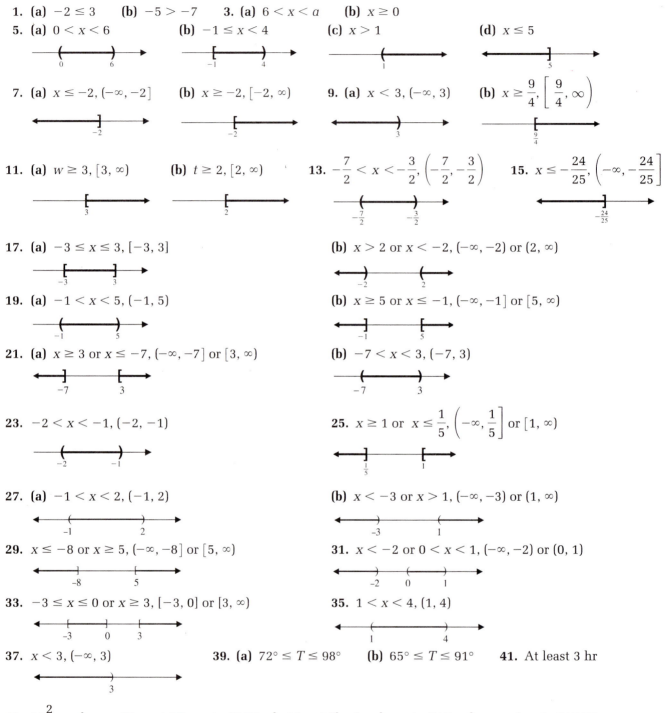

1. **(a)** $-2 \le 3$ **(b)** $-5 > -7$ 3. **(a)** $6 < x < a$ **(b)** $x \ge 0$

5. **(a)** $0 < x < 6$ **(b)** $-1 \le x < 4$ **(c)** $x > 1$ **(d)** $x \le 5$

7. **(a)** $x \le -2, (-\infty, -2]$ **(b)** $x \ge -2, [-2, \infty)$ 9. **(a)** $x < 3, (-\infty, 3)$ **(b)** $x \ge \dfrac{9}{4}, \left[\dfrac{9}{4}, \infty\right)$

11. **(a)** $w \ge 3, [3, \infty)$ **(b)** $t \ge 2, [2, \infty)$ 13. $-\dfrac{7}{2} < x < -\dfrac{3}{2}, \left(-\dfrac{7}{2}, -\dfrac{3}{2}\right)$ 15. $x \le -\dfrac{24}{25}, \left(-\infty, -\dfrac{24}{25}\right]$

17. **(a)** $-3 \le x \le 3, [-3, 3]$ **(b)** $x > 2$ or $x < -2, (-\infty, -2)$ or $(2, \infty)$

19. **(a)** $-1 < x < 5, (-1, 5)$ **(b)** $x \ge 5$ or $x \le -1, (-\infty, -1]$ or $[5, \infty)$

21. **(a)** $x \ge 3$ or $x \le -7, (-\infty, -7]$ or $[3, \infty)$ **(b)** $-7 < x < 3, (-7, 3)$

23. $-2 < x < -1, (-2, -1)$ 25. $x \ge 1$ or $x \le \dfrac{1}{5}, \left(-\infty, \dfrac{1}{5}\right]$ or $[1, \infty)$

27. **(a)** $-1 < x < 2, (-1, 2)$ **(b)** $x < -3$ or $x > 1, (-\infty, -3)$ or $(1, \infty)$

29. $x \le -8$ or $x \ge 5, (-\infty, -8]$ or $[5, \infty)$ 31. $x < -2$ or $0 < x < 1, (-\infty, -2)$ or $(0, 1)$

33. $-3 \le x \le 0$ or $x \ge 3, [-3, 0]$ or $[3, \infty)$ 35. $1 < x < 4, (1, 4)$

37. $x < 3, (-\infty, 3)$ 39. **(a)** $72° \le T \le 98°$ **(b)** $65° \le T \le 91°$ 41. At least 3 hr

43. $16\dfrac{2}{3}$ weeks   45. $x \le 17$ or $x \ge 19.50$; alert is set if price drops to \$17 or less or rises to \$19.50 or more.

47. At least 266 cellular phones must be sold

49. **(a)** $[0, 2)$ **(b)** $(-2, 8]$

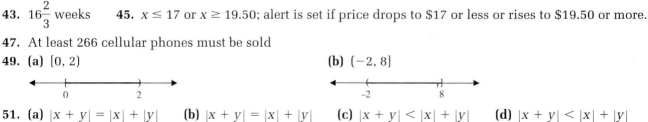

51. **(a)** $|x + y| = |x| + |y|$ **(b)** $|x + y| = |x| + |y|$ **(c)** $|x + y| < |x| + |y|$ **(d)** $|x + y| < |x| + |y|$

Problem Set 1.4 (on page 47)

1. $(3, 4)$: quadrant I; $(-2, 4)$: quadrant II; $(\sqrt{5}, -1)$: quadrant IV: $(0, \sqrt{7})$: y axis

3. **(a)** (i) $d = \sqrt{17} = 4.12$;   (ii) $(-1, 1.5)$ **(b)** (i) $d = \sqrt{61} = 7.81$;   (ii) $(0.5, 2)$

5. **(a)** (i) $d = \sqrt{73} = 8.54$;   (ii) $(-1, -5.5)$ **(b)** (i) $d = \sqrt{0.52} = 0.72$;   (ii) $\left(-\dfrac{1}{10}, \dfrac{2}{5}\right)$

7. **(a)** (i) $d = \sqrt{2} = 1.41$;   (ii) $\left(\dfrac{2t + 1}{2}, \dfrac{2u + 1}{2}\right)$ **(b)** (i) $d = \sqrt{78.6344} = 8.87$;   (ii) $(0.54, 6.77)$

9.   **11.**   **13.**   **15.**

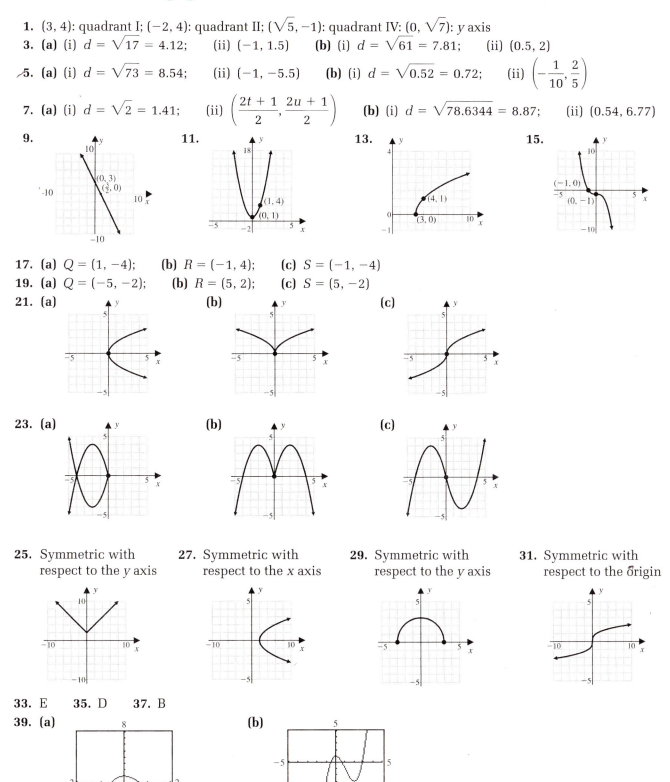

17. **(a)** $Q = (1, -4)$;   **(b)** $R = (-1, 4)$;   **(c)** $S = (-1, -4)$

19. **(a)** $Q = (-5, -2)$;   **(b)** $R = (5, 2)$;   **(c)** $S = (5, -2)$

21. **(a)**   **(b)**   **(c)**

23. **(a)**   **(b)**   **(c)**

25. Symmetric with respect to the y axis   **27.** Symmetric with respect to the x axis   **29.** Symmetric with respect to the y axis   **31.** Symmetric with respect to the origin

33. E   **35.** D   **37.** B

39. **(a)**   **(b)**

41. (a) x intercepts: $-3, 1$ **(b)** x intercepts: $-5, -2, 3$

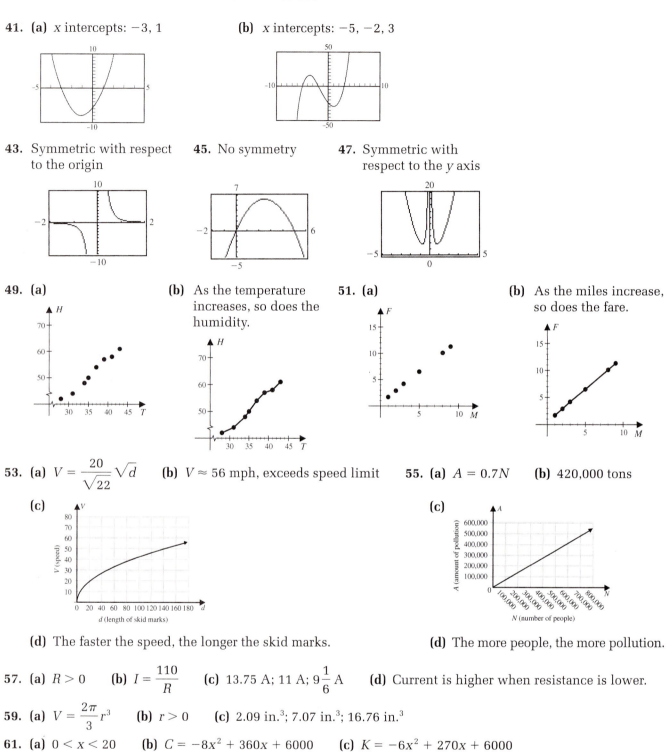

43. Symmetric with respect to the origin

45. No symmetry

47. Symmetric with respect to the y axis

49. (a)   **(b)** As the temperature increases, so does the humidity.

51. (a)   **(b)** As the miles increase, so does the fare.

53. (a) $V = \dfrac{20}{\sqrt{22}}\sqrt{d}$ **(b)** $V \approx 56$ mph, exceeds speed limit

55. (a) $A = 0.7N$ **(b)** 420,000 tons

(c)   **(c)**

(d) The faster the speed, the longer the skid marks.

(d) The more people, the more pollution.

57. (a) $R > 0$ **(b)** $I = \dfrac{110}{R}$ **(c)** 13.75 A; 11 A; $9\dfrac{1}{6}$ A   **(d)** Current is higher when resistance is lower.

59. (a) $V = \dfrac{2\pi}{3}r^3$ **(b)** $r > 0$ **(c)** 2.09 in.³; 7.07 in.³; 16.76 in.³

61. (a) $0 < x < 20$ **(b)** $C = -8x^2 + 360x + 6000$ **(c)** $K = -6x^2 + 270x + 6000$

63. (a) IV   **(b)** I   **(c)** III   **(d)** II

65. (a) $P(1, 5)$ **(b)** yes

67. (a)   **(b)** Since $|\overline{P_1P_2}| + |\overline{P_2P_3}| = |\overline{P_1P_3}|$, points are collinear.

69. y_1 and y_2 are symmetric with respect to the y axis; y_3 and y_4 are symmetric with respect to the y axis; y_2 and y_4 are symmetric with respect to the x axis; y_1 and y_3 are symmetric with respect to the x axis; y_1 and y_4 are symmetric with respect to the origin; y_2 and y_3 are symmetric with respect to the origin.

71. (a) All symmetries   **(b)**

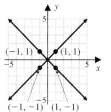

Graph $y = x$ and $y = -x$ in the same coordinate system.

Problem Set 1.5 (on page 63)

1.

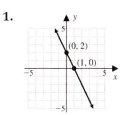

3.

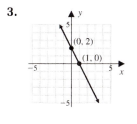

5.

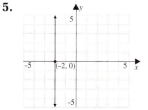

7.

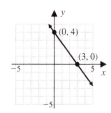

9. $y = -\dfrac{5}{8}x + \dfrac{7}{8}$

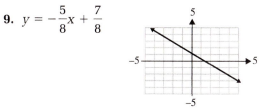

11. $m = \dfrac{5}{3}$; rises to the right   **13.** $m = 0$; horizontal   **15.** $m = -\dfrac{4}{3}$; falls to the right

17. (a) $y + 3 = 3(x - 4)$ **(b)** $y - 3 = 0(x + 2)$ **19. (a)** $y = \dfrac{-2}{3}x$ **(b)** $y = -2x + 2$

21. $y = -2x + 1$ **23.** $y = \dfrac{1}{3}x + \dfrac{1}{3}$ **25.** $y = -\dfrac{1}{2}x + \dfrac{1}{2}$

27. (a) $(x - 3)^2 + (y + 2)^2 = 25$ **(b)** $(x + 1)^2 + (y - 3)^2 = 4$

29. (a) $(x - 2)^2 + (y - 4)^2 = 34$ **(b)** $(x + 1)^2 + (y - 2)^2 = 34$

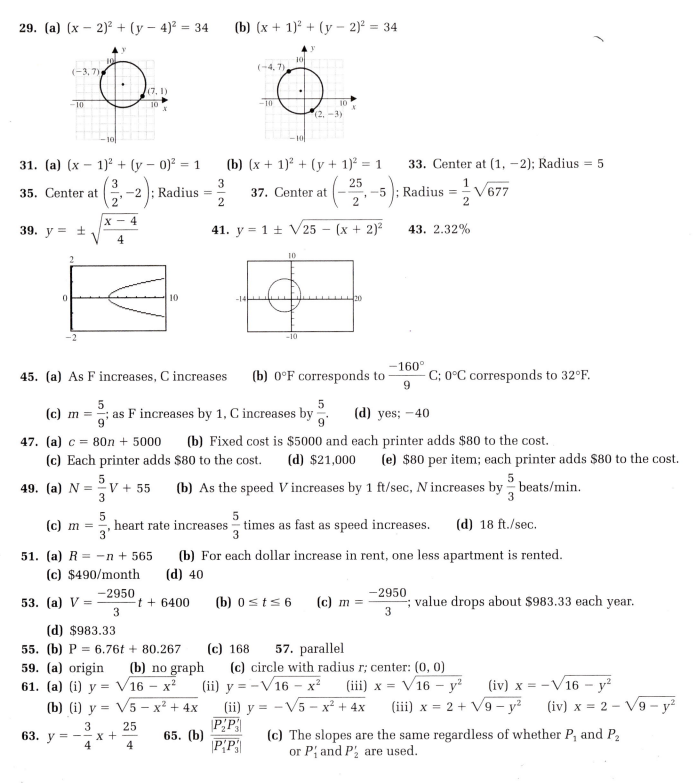

31. (a) $(x - 1)^2 + (y - 0)^2 = 1$ **(b)** $(x + 1)^2 + (y + 1)^2 = 1$ **33.** Center at $(1, -2)$; Radius $= 5$

35. Center at $\left(\dfrac{3}{2}, -2\right)$; Radius $= \dfrac{3}{2}$ **37.** Center at $\left(-\dfrac{25}{2}, -5\right)$; Radius $= \dfrac{1}{2}\sqrt{677}$

39. $y = \pm\sqrt{\dfrac{x - 4}{4}}$ **41.** $y = 1 \pm \sqrt{25 - (x + 2)^2}$ **43.** 2.32%

45. (a) As F increases, C increases   **(b)** 0°F corresponds to $\dfrac{-160°}{9}$ C; 0°C corresponds to 32°F.

(c) $m = \dfrac{5}{9}$; as F increases by 1, C increases by $\dfrac{5}{9}$.   **(d)** yes; -40

47. (a) $c = 80n + 5000$ **(b)** Fixed cost is \$5000 and each printer adds \$80 to the cost.
(c) Each printer adds \$80 to the cost.   **(d)** \$21,000   **(e)** \$80 per item; each printer adds \$80 to the cost.

49. (a) $N = \dfrac{5}{3}V + 55$ **(b)** As the speed V increases by 1 ft/sec, N increases by $\dfrac{5}{3}$ beats/min.

(c) $m = \dfrac{5}{3}$, heart rate increases $\dfrac{5}{3}$ times as fast as speed increases.   **(d)** 18 ft./sec.

51. (a) $R = -n + 565$ **(b)** For each dollar increase in rent, one less apartment is rented.
(c) \$490/month   **(d)** 40

53. (a) $V = \dfrac{-2950}{3}t + 6400$ **(b)** $0 \le t \le 6$ **(c)** $m = \dfrac{-2950}{3}$; value drops about \$983.33 each year.

(d) \$983.33

55. (b) $P = 6.76t + 80.267$ **(c)** 168   **57.** parallel

59. (a) origin   **(b)** no graph   **(c)** circle with radius r; center: $(0, 0)$

61. (a) (i) $y = \sqrt{16 - x^2}$ **(ii)** $y = -\sqrt{16 - x^2}$ **(iii)** $x = \sqrt{16 - y^2}$ **(iv)** $x = -\sqrt{16 - y^2}$
(b) (i) $y = \sqrt{5 - x^2 + 4x}$ **(ii)** $y = -\sqrt{5 - x^2 + 4x}$ **(iii)** $x = 2 + \sqrt{9 - y^2}$ **(iv)** $x = 2 - \sqrt{9 - y^2}$

63. $y = -\dfrac{3}{4}x + \dfrac{25}{4}$ **65. (b)** $\dfrac{|\overline{P_2'P_3'}|}{|\overline{P_1'P_3'}|}$ **(c)** The slopes are the same regardless of whether P_1 and P_2 or P_1' and P_2' are used.

Problem Set 1.6 (on page 79)

1. $y = 0$; $V = (0, 0)$; $F = (2, 0)$; $d: x = -2$ **3.** $x = 0$; $V = (0, 0)$; $F = (0, -1)$; $d: y = 1$

5. $y = 0$; $V = (0, 0)$; $F = \left(-\dfrac{3}{4}, 0\right)$; $d: x = \dfrac{3}{4}$ **7.** $x^2 = 16y$; $x = 0$ **9.** $x^2 = -8y$; $x = 0$

11. As c gets larger, the graphs get closer to the x axis.   **13.** Vertices: $(-4, 0)$, $(4, 0)$; Foci: $(-\sqrt{7}, 0)$, $(\sqrt{7}, 0)$

15. Vertices: $(0, -4)$, $(0, 4)$; Foci: $(0, -2\sqrt{3})$, $(0, 2\sqrt{3})$ **17.** Vertices: $(-\sqrt{10}, 0)$, $(\sqrt{10}, 0)$ Foci: $(-3, 0)$, $(3, 0)$

19. $\dfrac{x^2}{25} + \dfrac{y^2}{16} = 1$ **21.** $\dfrac{x^2}{25} + \dfrac{y^2}{194} = 1$ **23.** $\dfrac{x^2}{\frac{81}{8}} + \dfrac{y^2}{36} = 1$

25.

27. Vertices: $(-3, 0)$, $(3, 0)$; Foci: $(-\sqrt{10}, 0)$, $(\sqrt{10}, 0)$

Asymptotes: $y = -\dfrac{1}{3}x$ and $y = \dfrac{1}{3}x$

29. Vertices: $(0, -3)$, $(0, 3)$; Foci: $(0, -3\sqrt{2})$, $(0, 3\sqrt{2})$; Asymptotes: $y = -x$ and $y = x$

31. Vertices: $\left(-\dfrac{1}{6}, 0\right)$, $\left(\dfrac{1}{6}, 0\right)$; Foci: $\left(-\dfrac{\sqrt{5}}{6}, 0\right)$, $\left(\dfrac{\sqrt{5}}{6}, 0\right)$; Asymptotes: $y = -2x$ and $y = 2x$

33. Vertices: $(-2, 0)$, $(2, 0)$; Foci: $(-3, 0)$, $(3, 0)$; Asymptotes: $y = -\left(\dfrac{\sqrt{5}}{2}\right)x$ and $y = \left(\dfrac{\sqrt{5}}{2}\right)x$

35. $\dfrac{x^2}{256} - \dfrac{y^2}{400} = 1$ **37.** $\dfrac{y^2}{16} - \dfrac{x^2}{9} = 1$ **39.** $\dfrac{x^2}{\frac{7}{3}} - \dfrac{y^2}{35} = 1$

41.

43. (b) $y^2 = 12x$ **(c)** 3 ft   **45. (b)** $x^2 = 2664y$ **(c)** Approx. 3.75 in.

47. 0.75 ft   **49. (b)** $\dfrac{x^2}{169} + \dfrac{y^2}{25} = 1$ **(c)** $\dfrac{25}{13}$ m   **(d)** Approx. 4.87 m

51. (b) $\dfrac{x^2}{8,629,388,130,250,000} + \dfrac{y^2}{8,626,974,768,000,000} = 1$ **(c)** 3,107,000 mi

53. (b) $\dfrac{x^2}{15,625} + \dfrac{y^2}{5625} = 1$ **(c)** 200 ft   **55. (b)** $\dfrac{x^2}{1225} - \dfrac{y^2}{6875} = 1$ **(c)** $(41, 50)$

57. (a) $\dfrac{x^2}{1730.56} - \dfrac{y^2}{6922.24} = 1$ **(b)** 93 million mi

59. (b) (i) 8   **(ii)** 4   **(iii)** 5   **(iv)** $\dfrac{2}{7}$ **61.** $\dfrac{x^2}{25} + \dfrac{y^2}{169} = 1$

63. (a) Circle of radius a and center at the origin   **(b)** Yes; Foci = Center

65. (a) $A = 2xy$ **(b)** $A = \dfrac{3}{2}x\sqrt{16 - x^2}$ **67.** $\dfrac{x^2}{16} - \dfrac{y^2}{9} = 1$ **69. (a)** $y = \pm x$

71. (c) corners of the rectangle: (a, b), $(-a, b)$, $(a, -b)$, $(-a, -b)$
(d) The radius of the circle has the same length as c.

Chapter 1 Review Problem Set (on page 85)

1. (a) 5   **(b)** $\dfrac{3}{2}$ **(c)** $\dfrac{-2}{3}$, 2   **(d)** $\dfrac{-1}{2}$, 1   **3. (a)** -1, $\dfrac{7}{3}$ **(b)** 0   **(c)** $-3, 5$ **(d)** $\dfrac{-3 \pm \sqrt{30}}{3}$

5. (a) $y = \dfrac{7}{a - 5}$ **(b)** $y = \dfrac{-x \pm \sqrt{x^2 - 20}}{2}$

7. (a) $3 - \sqrt{7}$ **(b)** $\begin{cases} 5x - 3 & \text{if } x \geq \dfrac{3}{5} \\ 3 - 5x & \text{if } x < \dfrac{3}{5} \end{cases}$ **(c)** $\begin{cases} 1 - 7x & \text{if } x \leq \dfrac{1}{7} \\ 7x - 1 & \text{if } x > \dfrac{1}{7} \end{cases}$

9. (a) $>$ **(b)** $<$ **(c)** $>$ **(d)** $=$ **11. (a)** $4 \leq x \leq 5$ **(b)** $x \leq -2$ **(c)** $1 < x < 3$

13. (a) $x > 4;\ (4, \infty)$ **(b)** $x < -2;\ (-\infty, -2)$ **(c)** $x < 0;\ (-\infty, 0)$

15. (a) $2 < t < \dfrac{8}{3};\ \left(2, \dfrac{8}{3}\right)$ **(b)** $x \geq 2 \text{ or } x \leq -\dfrac{10}{3};\ \left(-\infty, -\dfrac{10}{3}\right] \text{ or } [2, \infty)$ **(c)** $x \leq -\dfrac{3}{2};\ \left(-\infty, -\dfrac{3}{2}\right]$

17. (a) $(-\infty, -1) \text{ or } (9, \infty)$ **(b)** $(-\infty, 3)$

19. (a) (i) $d = 2\sqrt{10}$ (ii) $(3, -2)$ **(b)** (i) $d = 3\sqrt{10}$ (ii) $\left(\dfrac{3}{2}, \dfrac{1}{2}\right)$

21. (a) Center at $(-1, -3)$; Radius $= 4$ **(b)** Center at $(3, -1)$; Radius $= 6$

23. (a) x intercept $= -5$;   **(b)** x intercepts $= -1, 0, 1$;   **25. (a)** $\dfrac{3}{2}$ **(b)** $-\dfrac{2}{5}$
y intercept $= 5$; y intercept $= 0$;
no symmetry   symmetric with respect
to the y axis

27. (a) x intercept $= -3$; y intercept $= \dfrac{3}{2}$; $m = \dfrac{1}{2}$ **(b)** x intercept $= \dfrac{5}{2}$; y intercept $= \dfrac{5}{3}$; $m = -\dfrac{2}{3}$

29. (a) 21 **(b)** -2 **31. (a)** $y = 3x - 7$ **(b)** $y = 4x - 14$

33. (a) $F = \dfrac{9}{5}C + 32$ **(b)** $125.6°F$ **(d)** For each increase of $1°C$, the Fahrenheit temperature increases $1.8°F$.

35. $\$14,000$ **37.** 75 boys   **39. (a)** $0 \leq t \leq 7$ **(c)** 192 ft   **(d)** At 7 sec   **(e)** 1.70 sec

41. (b) $P = 0.165d + 28.569$ **(c)** $\$400$ **43. (a)** $F = (0, -2)$; $D: y = 2$ **(b)** $F = (9, 0)$; $D: x = -9$

45. (a) Vertices: $(0, -6), (0, 6)$; Foci: $\left(0, -3\sqrt{3}\right), \left(0, 3\sqrt{3}\right)$
   (b) Vertices: $\left(-\sqrt{6}, 0\right), \left(\sqrt{6}, 0\right)$; Foci: $\left(-\sqrt{2}, 0\right), \left(\sqrt{2}, 0\right)$

47. (a) Vertices $(-2, 0), (2, 0)$; Foci: $\left(-\sqrt{13}, 0\right), \left(\sqrt{13}, 0\right)$; Asymptotes: $y = -\dfrac{3}{2}x$ and $y = \dfrac{3}{2}x$
   (b) Vertices $(-2, 0), (2, 0)$; Foci: $\left(-2\sqrt{5}, 0\right), \left(2\sqrt{5}, 0\right)$; Asymptotes: $y = -2x$ and $y = 2x$

49. (a) $\dfrac{x^2}{9} + \dfrac{y^2}{5} = 1$ **(b)** $\dfrac{x^2}{9} - \dfrac{y^2}{16} = 1$

Chapter 1 Test (on page 88)

1. (a) $\dfrac{5}{2}$ **(b)** $-5, 1$ **(c)** $\dfrac{-7}{3}, 1$ **(d)** 6 **(e)** $\dfrac{(5 + w)x}{3}$

2. (a) (i) $-2 < x \leq 5$ (ii) $x \leq 4$ **(b)** (i) $[-2, 4)$ (ii) $[2, \infty)$

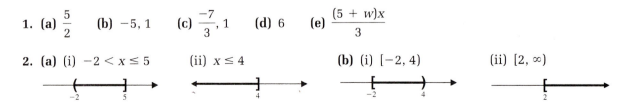

3. (a) $x < -\dfrac{6}{5}$; $\left(-\infty, -\dfrac{6}{5}\right)$ **(b)** $-12 \le x \le 2$; $[-12, 2]$ **(c)** $x > 2$ or $x < -\dfrac{2}{3}$;   **(d)** $-5 \le x \le 2$; $[-5, 2]$

$\left(-\infty, -\dfrac{2}{3}\right)$ or $(2, \infty)$

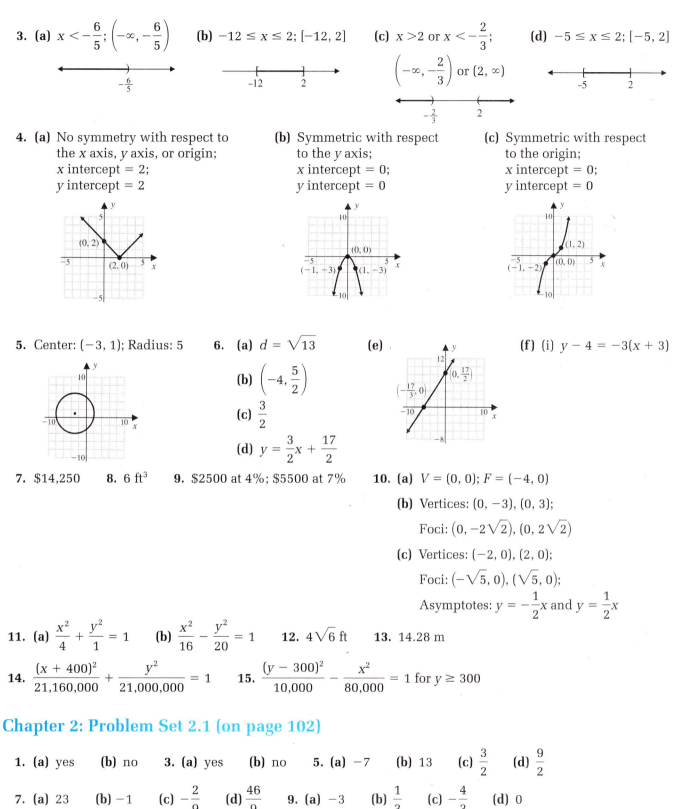

4. (a) No symmetry with respect to the x axis, y axis, or origin; x intercept $= 2$; y intercept $= 2$

(b) Symmetric with respect to the y axis; x intercept $= 0$; y intercept $= 0$

(c) Symmetric with respect to the origin; x intercept $= 0$; y intercept $= 0$

5. Center: $(-3, 1)$; Radius: 5

6. (a) $d = \sqrt{13}$

(b) $\left(-4, \dfrac{5}{2}\right)$

(c) $\dfrac{3}{2}$

(d) $y = \dfrac{3}{2}x + \dfrac{17}{2}$

(e)

(f) (i) $y - 4 = -3(x + 3)$

7. $14,250   **8.** 6 ft^3 **9.** $2500 at 4%; $5500 at 7%

10. (a) $V = (0, 0)$; $F = (-4, 0)$

(b) Vertices: $(0, -3)$, $(0, 3)$;

Foci: $\left(0, -2\sqrt{2}\right)$, $\left(0, 2\sqrt{2}\right)$

(c) Vertices: $(-2, 0)$, $(2, 0)$;

Foci: $\left(-\sqrt{5}, 0\right)$, $\left(\sqrt{5}, 0\right)$;

Asymptotes: $y = -\dfrac{1}{2}x$ and $y = \dfrac{1}{2}x$

11. (a) $\dfrac{x^2}{4} + \dfrac{y^2}{1} = 1$ **(b)** $\dfrac{x^2}{16} - \dfrac{y^2}{20} = 1$ **12.** $4\sqrt{6}$ ft   **13.** 14.28 m

14. $\dfrac{(x + 400)^2}{21,160,000} + \dfrac{y^2}{21,000,000} = 1$ **15.** $\dfrac{(y - 300)^2}{10,000} - \dfrac{x^2}{80,000} = 1$ for $y \ge 300$

Chapter 2: Problem Set 2.1 (on page 102)

1. (a) yes   **(b)** no   **3. (a)** yes   **(b)** no   **5. (a)** -7 **(b)** 13   **(c)** $\dfrac{3}{2}$ **(d)** $\dfrac{9}{2}$

7. (a) 23   **(b)** -1 **(c)** $-\dfrac{2}{9}$ **(d)** $\dfrac{46}{9}$ **9. (a)** -3 **(b)** $\dfrac{1}{3}$ **(c)** $-\dfrac{4}{3}$ **(d)** 0

11. (a) 4   **(b)** 5   **(c)** 4.502   **(d)** 4.133   **13. (a)** 1   **(b)** 5   **(c)** 2   **(d)** 8

15. (a) $-2a - 11$ **(b)** $-2a - 16$ **(c)** $-\dfrac{5a - 23}{a + 3}$ **(d)** $\dfrac{13}{2a + 11}$

17. (a) $2a^2 - a - 2$ **(b)** $2a^2 - 5a - 1$ **18. (c)** $-4a^2 + 1$ **(d)** $2a^4 - 5a^2 + 1$

19. (a) $3 - 14|a|$ **(b)** $21 - 14|a|$ **(c)** $3 - 2|a^2|$ **(d)** $3 - 2a^2$

21. (a) $\dfrac{y + 3}{3y + 10}$ **(b)** $\dfrac{y + 6}{4}$ **(c)** $\dfrac{6y + 18}{y + 9}$ **(d)** $\dfrac{3y + 9}{2y + 18}$

23. (a) 36 **(b)** 1 **(c)** $\dfrac{5}{2}$ **(d)** 25 **25. (a)** 0 **(b)** -3 **(c)** 0 **(d)** -14

27. (a) -7 **(b)** -2 **(c)** -2 **(d)** -119 **29.** $\mathbb{R}$ **31.** $x \neq -4$ **33.** $x \neq \pm 3$ **35.** $\left[\dfrac{2}{3}, \infty\right)$

37. $x \neq 0, 1$ **39. (a)** 4 **(b)** $2t + h$ **41. (a)** $\dfrac{-1}{t(t + h)}$ **(b)** $\dfrac{-3}{(t - 1)(t + h - 1)}$

43. (a) $-32t - 16h$ **(b)** (i) -40 (ii) -36.8 (iii) -33.6
   (c) As the elapsed time increases, the average speed increases.

45. (a) $6t + 3h - 12$ **(b)** (i) 24 (ii) 21 (iii) 20.1
   (c) As the elapsed time increases, the average rate of change of the population increases.

47. (a) function   **(b)** not a function   **49.** For each $x \neq 0$ there are two y values.

51. (a) yes   **(b)** no   **53. (a)** -12 **(b)** $327\dfrac{11}{32}$

Problem Set 2.2 (on page 111)

1. (a) yes   **(b)** no   **3.** Not a function of x.   **5.** Function   **7.** Function   **9.** Function
11. (a) Domain: $\mathbb{R}$; Range: $(-\infty, 5]$ **(b)** Domain: $(-2, 1)$ or $(1, 4)$; Range: $(-\infty, -3]$ or $[3, \infty)$
13. Domain and   **15.** Domain: $[-7, 7]$;   **17.** Domain: $\mathbb{R}$;   **19.** Domain: $\mathbb{R}$;
   range: $\mathbb{R}$ Range $[-7, 0]$ Range: $[2, \infty)$ Range: $[4, \infty)$

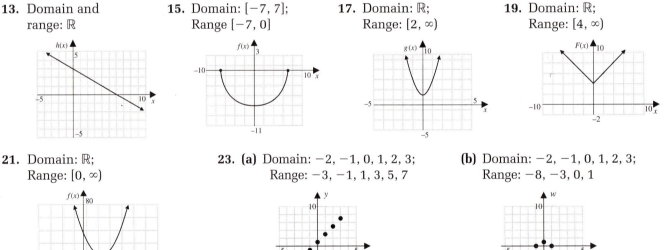

21. Domain: $\mathbb{R}$;   **23. (a)** Domain: $-2, -1, 0, 1, 2, 3$;   **(b)** Domain: $-2, -1, 0, 1, 2, 3$;
   Range: $[0, \infty)$ Range: $-3, -1, 1, 3, 5, 7$ Range: $-8, -3, 0, 1$

25. (a) For every value of V there is one and only one value of F.   **(b)** Domain: $[40, 80]$; Range: $[27.93, 46.33]$
   (c) 46.33 mpg; 36.70 mpg; 29.93 mpg   **(d)** 50 mph; 63.56 mph; 74.83 mph
27. (a) For every value of t there is one and only one value of T.
   (b)   **(c)** Slowly increasing as time goes on.

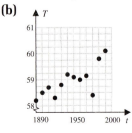

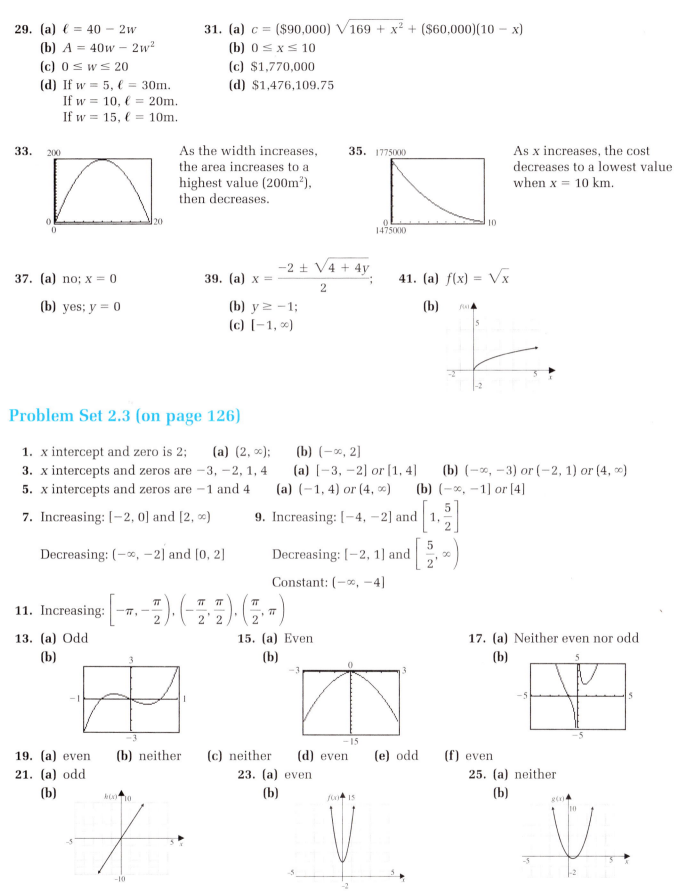

29. (a) $\ell = 40 - 2w$
(b) $A = 40w - 2w^2$
(c) $0 \le w \le 20$
(d) If $w = 5$, $\ell = 30$m.
If $w = 10$, $\ell = 20$m.
If $w = 15$, $\ell = 10$m.

31. (a) $c = (\$90,000) \sqrt{169 + x^2} + (\$60,000)(10 - x)$
(b) $0 \le x \le 10$
(c) $\$1,770,000$
(d) $\$1,476,109.75$

33. 200 · 0 · 20 · 0

As the width increases, the area increases to a highest value (200m²), then decreases.

35. 1775000 · 0 · 1475000 · 10

As x increases, the cost decreases to a lowest value when $x = 10$ km.

37. (a) no; $x = 0$
(b) yes; $y = 0$

39. (a) $x = \dfrac{-2 \pm \sqrt{4 + 4y}}{2}$;
(b) $y \ge -1$;
(c) $[-1, \infty)$

41. (a) $f(x) = \sqrt{x}$
(b)

Problem Set 2.3 (on page 126)

1. x intercept and zero is 2;   (a) $(2, \infty)$;   (b) $(-\infty, 2]$
3. x intercepts and zeros are $-3, -2, 1, 4$ (a) $[-3, -2]$ or $[1, 4]$ (b) $(-\infty, -3)$ or $(-2, 1)$ or $(4, \infty)$
5. x intercepts and zeros are -1 and 4   (a) $(-1, 4)$ or $(4, \infty)$ (b) $(-\infty, -1]$ or $[4]$

7. Increasing: $[-2, 0]$ and $[2, \infty)$
Decreasing: $(-\infty, -2]$ and $[0, 2]$

9. Increasing: $[-4, -2]$ and $\left[1, \dfrac{5}{2}\right]$
Decreasing: $[-2, 1]$ and $\left[\dfrac{5}{2}, \infty\right)$
Constant: $(-\infty, -4]$

11. Increasing: $\left[-\pi, -\dfrac{\pi}{2}\right), \left(-\dfrac{\pi}{2}, \dfrac{\pi}{2}\right), \left(\dfrac{\pi}{2}, \pi\right)$

13. (a) Odd
(b)

15. (a) Even
(b)

17. (a) Neither even nor odd
(b)

19. (a) even   (b) neither   (c) neither   (d) even   (e) odd   (f) even
21. (a) odd
(b)

23. (a) even
(b)

25. (a) neither
(b)

27. Odd;
increasing on ℝ

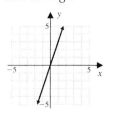

29. Neither;
decreasing on ℝ

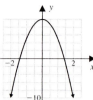

31. Even; decreasing: $[0, \infty)$;
increasing $(-\infty, 0]$

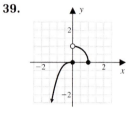

33. Even; decreasing: $[0, 2]$;
increasing $[-2, 0]$

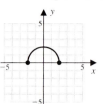

35. Even; decreasing: $[0, \infty)$;
increasing $(-\infty, 0]$

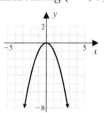

37.

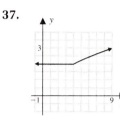

39.

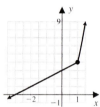

41. Domain: ℝ; Range: ℝ;
no discontinuity

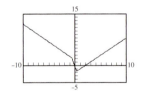

43. Domain: ℝ; Range: $(-\infty, -1)$ or 0;
discontinuities at $-2, 2$

45. Range: all integers

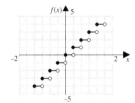

47. 0 and 2; $[0, 2]$

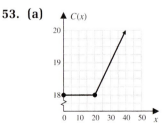

49. -3 and 2; $(-3, 2)$

51. (a) $180,000; $200,000;
$170,000; $150,000; $100,000
(b) 850 items
(c) Increasing: $[0, 4]$
Decreasing: $[4, 10]$
(d) Highest: $200,000 when
$x = 4$ (400 items)
Lowest: $100,000 when
$x = 10$ (1000 items)

53. (a)

(b) $18

(c) $K(x) = \begin{cases} 12 & \text{if } 0 \le x \le 15 \\ 12 + 0.25(x - 15) & \text{if } x > 15 \end{cases}$

(e) $K < C$ for $x < 51\frac{2}{3}$ hundred gallons;

$C < K$ for $x > 51\frac{2}{3}$ hundred gallons.

(d)

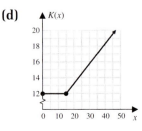

55. (a) $C(N) = \begin{cases} 20 & \text{if } 0 < N \le 1 \\ 38 & \text{if } 1 < N \le 2 \\ 56 & \text{if } 2 < N \le 3 \\ 74 & \text{if } 3 < N \le 4 \end{cases}$ **(b)** $0 < N \le 4$ **(c)**

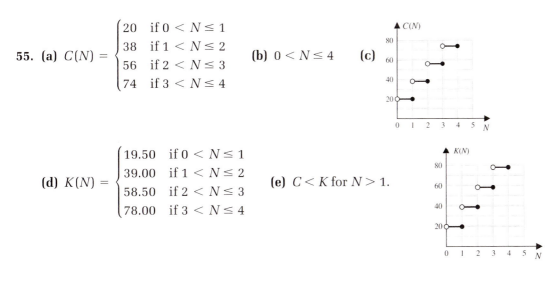

(d) $K(N) = \begin{cases} 19.50 & \text{if } 0 < N \le 1 \\ 39.00 & \text{if } 1 < N \le 2 \\ 58.50 & \text{if } 2 < N \le 3 \\ 78.00 & \text{if } 3 < N \le 4 \end{cases}$ **(e)** $C < K$ for $N > 1$.

57. (a) $R(w) = 0.33 - 0.22 \; [\![1 - w]\!]$ **(b)** $0 < w \le 6$ **(c)**   **(d)** $1.65; \$2.31; \$2.31$

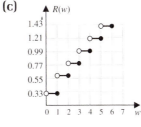

59. (a) no; only if $x = 0$ is in the domain   **(b)**   **(c)** yes

61. (a) no, $x = 2$ would have 2 values for y.   **(b)** yes   **(c)** $k > 5$
63. (a) no, because there would be 2 values for some x's.   **(b)** Yes, consider $y = 0$.
65. $(-3, 82), (-2, 17), (-1, 2), (0, 0), (1, 2), (2, 17), (3, 82)$, yes, $f(-x) = f(x)$

Problem Set 2.4 (on page 142)

1.   **3.**   **5.**

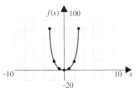

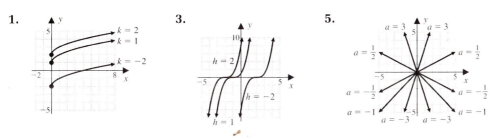

7. Vertically stretch y by 2, then shift up 3 units.
9. Vertically stretch y by 3, shift to right 1 unit, and shift down 2 units.
11. Vertically stretch y by 2, reflect about the x axis, shift to right 1 unit, and shift up 3 units.

13. (a) **(b)** **(c)** **(d)**

(e) **15.** $G(x) = |x - 2|$; $T(x) = -|x + 1|$; $S(x) = |x - 2| + 2$

17. $G(x) = x + 3$ **19.** $G(x) = (x - 2)^2$ **21.** $G(x) = -\sqrt{x - 5}$ **23.** $G(x) = \dfrac{1}{3}[(x + 2)^2 + 1]$

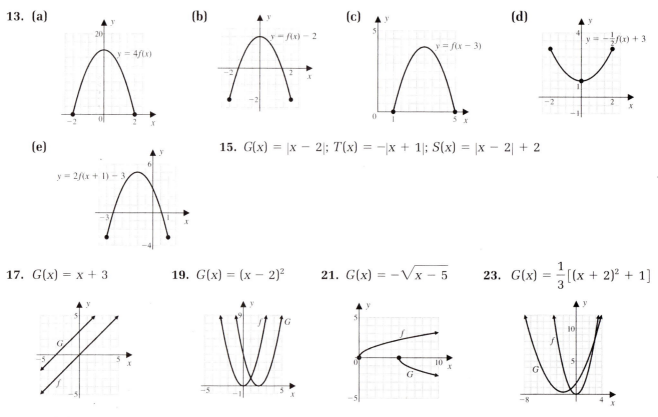

25. Shift the graph of $y = x$ up 2 units; Domain and Range: $\mathbb{R}$

27. Reflect $y = x^3$ about the x axis, and shift up 1 unit; Domain and Range: $\mathbb{R}$

29. Shift the graph of $y = \sqrt{x}$ to left 4 units; Domain: $[-4, \infty)$; Range: $[0, \infty)$

31. Vertically stretch y by 2, and shift to right 1 unit; Domain: $\mathbb{R}$; Range: $[0, \infty)$

33. Vertically stretch y by 5, reflect about the x axis, shift to left 1 unit, and shift up 1 unit; Domain: $[-1, \infty)$; Range: $(-\infty, 1]$

35. Vertically compress y by $\dfrac{1}{3}$, reflect about the x axis, shift to left 1 unit; Domain and Range: $\mathbb{R}$

37. Vertically stretch y by 4, and shift down 1 unit; Domain: $\mathbb{R}$; Range: $[-1, \infty)$

39. Vertically stretch y by 2, shift to right 3 units, and shift up 4 units; Domain: $\mathbb{R}$; Range: $[4, \infty)$

41. Vertically stretch y by 3, shift to right $\dfrac{1}{2}$ unit, and shift down 1 unit; Domain and Range: $\mathbb{R}$

43. Vertically compress y by $\dfrac{1}{2}$; Domain: $[-1, 1]$; Range: $\left[0, \dfrac{1}{2}\right]$

45.   **47. (a)** $(a + 2, b)$ **(b)** $(a - 4, b)$ **(c)** $(-a, -b + 1)$ **49.** a, b, and c if $a > 0$

51. a and c

53. (a) $f(x) = \sqrt{-x - 2}$

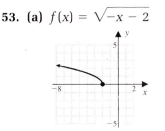

(b) The order of operations is different.

(c) Do transformation in the correct order.

55. (a) $G(x) = \sqrt[3]{x - 1}(x + 1)$

(b)

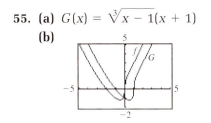

(c) Shift the graph of f to right 1 unit.

57. (a) $G(x) = \dfrac{x^2 + 1}{x}$

(b)

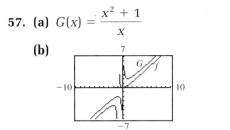

(c) Shift the graph of f to right 1 unit and up 2 units.

59. (a)

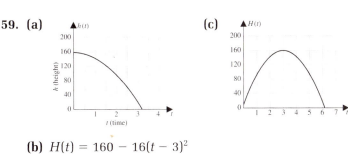

(b) $H(t) = 160 - 16(t - 3)^2$

(c)

Problem Set 2.5 (on page 152)

1. (a) $x - 6$ **(b)** $5x + 8$ **(c)** $-6x^2 - 23x - 7$

3. (a) $5x^2 - 3x + 3$ **(b)** $-3x^2 - 3x + 1$ **(c)** $4x^4 - 12x^3 + 9x^2 - 3x + 2$

5. (a) $3x^3 + 2x + 4$ **(b)** $-x^3 + 8x - 4$ **(c)** $2x^6 + 7x^4 + 4x^3 - 15x^2 + 20x$

7. (a) $\dfrac{x^2 - 13x + 6}{-4x^2 + 23x - 15}$ **(b)** $\dfrac{-x^2 - 3x + 6}{-4x^2 + 23x - 15}$ **(c)** $\dfrac{2x}{-4x^2 + 23x - 15}$

9. (a) $\dfrac{3x + 1}{-2x - 7}, x \neq -\dfrac{7}{2}$ **(b)** $\dfrac{-2x - 7}{3x + 1}, x \neq -\dfrac{1}{3}$ **11. (a)** $\dfrac{x + 3}{x^2 - x + 1}, x \neq -1$ **(b)** $\dfrac{x^2 - x + 1}{x + 3}, x \neq -1, -3$

13. (a) $\dfrac{(x - 1)\sqrt{2 - x}}{x}, x \leq 2, x \neq 0, 1$ **(b)** $\dfrac{x}{(x - 1)\sqrt{2 - x}}, x < 2, x \neq 1$

15. (a) 518   **(b)** 100   **(c)** 398   **(d)** 114   **(e)** 268   **(f)** 1064   **(g)** 2744   **(h)** 394   **(i)** 134   **(j)** -33

17. (a) 2   **(b)** -5 **(c)** 3   **(d)** 2   **(e)** 9   **(f)** -3

19. (a) 15   **(b)** 3   **(c)** -1 **(d)** -15 **21. (a)** $10x - 6$; Domain: $\mathbb{R}$ **(b)** $10x - 3$; Domain: $\mathbb{R}$

23. (a) $98x + 5$; Domain: $[0, \infty)$ **(b)** $7\sqrt{2x^2 + 5}$; Domain: $\mathbb{R}$ **25. (a)** x; Domain: $\mathbb{R}$ **(b)** x; Domain: $\mathbb{R}$

27. (a) $\sqrt{x + 4}$; Domain: $[-4, \infty)$ **(b)** $\sqrt{x - 1} + 5$; Domain: $[1, \infty)$

29. (a) $\dfrac{2x - 5}{4x - 1}$; Domain: $x \neq \dfrac{1}{4}, \dfrac{5}{2}$ **(b)** $\dfrac{9x + 6}{-15x - 8}$; Domain: $x \neq -\dfrac{2}{3}, -\dfrac{8}{15}$

31. One possible answer is $g(x) = 5x - 2; f(x) = x^3$ **33.** One possible answer is $g(t) = t^2 - 2; f(t) = t^{-2}$

35. One possible answer is $g(x) = x + \dfrac{1}{x}; f(x) = \sqrt[3]{x}$ **37. (a)** $-15x + 7$ **(b)** $x = \dfrac{1}{3}$

39. (c) $[0, \infty)$ **(d)** Use only the first quadrant and the origin.   **41. (c)** $x \neq 0$ **(d)** Delete $(0, 0)$ from graph.

43. (a) $f[g(x)] = \sqrt{2 + x}$ **(c)** $[-2, 2]$ **(d)** Use only part of graph from $x = -2$ to $x = 2$.

45. (a) $V = f[g(t)] = 26.52t^3$ in.^{3}/min (approx.); volume increases in proportion to $(\text{time})^3$.

(b) $S = h[g(t)] = 43t^2$ in.^{2}/min (approx.); surface area increases in proportion to $(\text{time})^2$.

47. (a) $x = g(t) = 55t; y = f(x) = \sqrt{90^2 + x^2}$

(b) $f[g(t)] = \sqrt{8100 + 3025t^2}$ is the distance from the ball to first base as a function of time $t \geq 0$.

49. (a) 224,430 gals remain after 3 days; 74,810 gals remain after 5 days
   (b) The number of gals of water in the pond after x days

51. (a) $R(x) = 90x$; revenue exceeds cost for $10 < x \leq 45$ units.

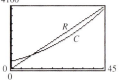

(b) $P(x) = -x^2 + 60x - 500$
A profit occurs for $P > 0$, that is, when x is more than 10 units; a loss occurs when fewer than 10 units are sold.

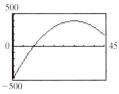

53. (a) Yes; as x increases, f and g increase, so the total $f + g$ increases.
   (b) $f(x) = 3x$; $g(x) = 7x$. Then $(f - g)(x) = -4x$.

55. Yes; all are even.   **57. (a)** yes   **(b)** yes   **59. (a)** $x^{\frac{3}{5}} - 3$ **(b)** $36x^2 - 156x + 173$

61. (a) $g(x) = \frac{1}{3}x$; $h(x) = \frac{-1}{54}x^3 + \frac{1}{3}x^2$ **(b)**

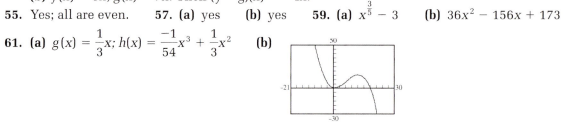

Chapter 2 Review Problem Set (on page 156)

1. (a) function   **(b)** not a function
3. (a) function   **(b)** not a function   **(c)** not a function   **(d)** function
5. (a) -1 **(b)** 1   **(c)** 50   **(d)** 2   **(e)** $2 - 5\sqrt{2}$ **(f)** $2b$ **(g)** 2
7. (a) $-8; 0; 1; 2$ **9. (a)** 5   **11.** x intercepts *or* zeros are $-2, 0,$
   (b) Domain and range: $\mathbb{R}$ **(b)** $6t + 3h$ and 2 $(-\infty, -2]$ *or* $[0, 2]$
   (c)

13. Odd; symmetric about origin
15. (a) left 2 and up 3   **17. (a)** $g(x) = -(x - 3)^2 - 4$
   (b) left 1 and up 3   **(b)** $g(x) = 2\sqrt[3]{x + 5} + 3$
   (c) reflected about the x axis   **(c)** $g(x) = -\sqrt{-x} - 4$
   (d) reflected about the x axis; stretched by 2 and up 2.

19. (a) (i) $2x + 1$ **(ii)** 3   **(iii)** $x^2 + x - 2$ **(iv)** $x + 1$ **(v)** $x + 1$ **(vi)** $\dfrac{(x + 2)}{(x - 1)}$

   (b) (i) $x^3 - x$ **(ii)** $x^3 + x$ **(iii)** $-x^4$ **(iv)** $-x^3$ **(v)** $-x^3$ **(vi)** $\dfrac{x^3}{(-x)} = -x^2$

   (c) (i) $x^2 + 2x + 1$ **(ii)** $x^2 - 2x - 1$ **(iii)** $2x^3 + x^2$ **(iv)** $4x^2 + 4x + 1$ **(v)** $2x^2 + 1$ **(vi)** $\dfrac{x^2}{(2x + 1)}$

   (d) (i) $4x^2$ **(ii)** -2 **(iii)** $4x^4 - 1$ **(iv)** $8x^4 + 8x^2 + 1$ **(v)** $8x^4 - 8x^2 + 3$ **(vi)** $\dfrac{(2x^2 - 1)}{(2x^2 + 1)}$

21. (a) $f(x) = x^5, g(x) = 7x + 2$ **(b)** $f(t) = \sqrt{t}, g(t) = t^2 + 17$

23. (a) $P = \dfrac{-x}{4000} + 6.5$

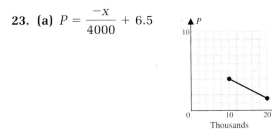

(c) The larger the demand, the lower the price; $2.50 per bag.

(d) 18,000 bags

25. (a) Vertical-line test is applicable.   **(b)** Domain: $[0, 45]$; Range: $[-60, 250]$ **(c)** $-\$50,000$; $0; $250,000; $0
(d) $25,000   **(e)** Increasing profit occurs for $10 \le A \le 25$; decreasing profit occurs for $25 \le A \le 40$; there is a loss if more than $40,000 is spent.

27. (a) $R = 10x$

(b) $P = 10x - 10,000(5 + \sqrt[3]{x + 1})$

(c) Initially there is a loss ($P < 0$), then break even ($P = 0$), and finally a profit ($P > 0$) when $x > 38,877$.

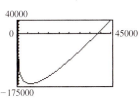

29. (a) $N[T(t)] = -90(3t + 4 + 1)^{-1} + 20$
$= -90(3t + 5)^{-1} + 20$

(b) Number of bacteria at a particular time.

(c)

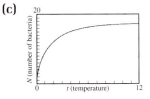

(d) The amount of bacteria increases very slowly.

31. (a) $f(x) = |5x - 3| - |4x + 3| \le 0$
$0, 6; [0, 6]$

(b) $g(x) = |4x + 5| - |3x - 5| > 0$
$-10, 0; (-\infty, -10)$ or $(0, \infty)$

33. (a) $y = 40 - x$
(b) $P = x(40 - x)$
(c) $40 - 2x - 0.01$

Chapter 2 Test (on page 159)

1. (a) function   **(b)** not a function

2. (a)   **(b)**

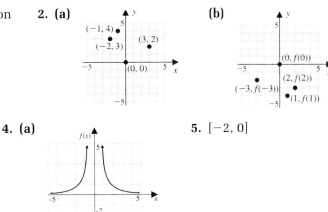

3. (a)

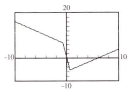

(b) Domain and range: $\mathbb{R}$
(c) increasing: $(-\infty, \infty)$
(d) neither
(e) none
(f) $15t^2 + 15th + h^2$

4. (a)

(b) Domain: $x \ne 0$; Range: $(0, \infty)$
(c) increasing: $(-\infty, 0)$ decreasing: $(0, \infty)$
(d) even
(e) symmetric to y axis
(f) $\dfrac{-3(2 + h)}{t^2(t + h)^2}$

5. $[-2, 0]$

6. Domain: $\mathbb{R}$; Range: $[0, \infty)$

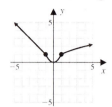

7. **(a)** $-x^2 + 2x - 7$

(b) $-2x^3 + 8x^2 + 2x - 8$

(c) $\dfrac{2x - 8}{1 - x^2}$

(d) $-2x^2 - 6$

(e) -3

(f) 2

8. **(a)** $0; -2; 3; 3; 0$

(b) $[-4, -1]$

(c) $(-\infty, -4]$ and $[2, \infty)$

(d) $[-1, 2]$

(e) $\mathbb{R}$

(f) $\mathbb{R}$

9.

x	1	2	3	4
$f[g(x)]$	1	3	2	4
x	1	2	3	4
$g[f(x)]$	3	2	1	4

10. left 1, stretch by 2, reflected through x axis, and up 3.

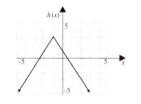

11. right 2, shrink by $\dfrac{1}{3}$, and down 4.

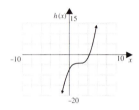

12. **(a)** $c = \sqrt{4 + a^2}$ **(b)** $a = \sqrt{c^2 - 4}$

13. **(a)** $[0, 14]$ **(b)** $t = 1, 12$ **(c)** $t = 6$ **(d)** $A = 50$ mg/dL when $t = 6$.

14. **(a)** $P = 5S - 60{,}000$ **(b)** $\$140{,}000$ **(c)**

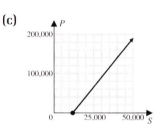

(d) $12{,}000$

15. **(a)** Domain: $(-\infty, 4]$

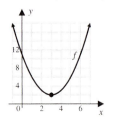

(b) Domain: $[1, \infty)$

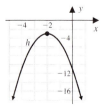

Chapter 3: Problem Set 3.1 (on page 175)

1. **(a)** B   **(b)** C   **(c)** D   **(d)** A

3. Domain: $\mathbb{R}$;
Range: $[2, \infty)$;
Vertex: $(3, 2)$;
Axis of symmetry: $x = 3$

5. Domain: $\mathbb{R}$;
Range: $(-\infty, -3]$;
Vertex: $(-2, -3)$;
Axis of symmetry: $x = -2$

7. Domain: $\mathbb{R}$;
Range: $[1, \infty)$;
Vertex: $(-3, 1)$;
Axis of symmetry: $x = -3$

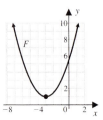

9. $f(x) = (x - 2)^2 - 9$;
Range: $[-9, \infty)$;
Vertex: $(2, -9)$;
Axis of symmetry: $x = 2$

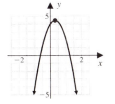

11. $f(x) = -(x + 3)^2 + 1$;
Range: $(-\infty, 1]$;
Vertex: $(-3, 1)$;
Axis of symmetry: $x = -3$

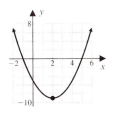

13. $f(x) = 3\left(x - \dfrac{5}{6}\right)^2 - \dfrac{277}{12}$;
Range: $\left[-\dfrac{277}{12}, \infty\right)$;
Vertex: $\left(\dfrac{5}{6}, -\dfrac{277}{12}\right)$;
Axis of symmetry: $x = \dfrac{5}{6}$

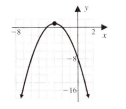

15. $f(x) = -5(x - 0.3)^2 + 4.45$;
Range: $(-\infty, 4.45]$;
Vertex: $(0.3, 4.45)$;
Axis of symmetry: $x = 0.3$

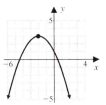

17. $f(x) = -\dfrac{1}{2}(x + 2)^2 + 3$;
Range: $(-\infty, 3]$;
Vertex: $(-2, 3)$;
Axis of symmetry: $x = -2$

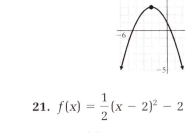

19. **(a)** $y = -x^2 + 4x$
(b) $y = \dfrac{2}{9}x^2 - 2$

21. $f(x) = \dfrac{1}{2}(x - 2)^2 - 2$

23. $f(x) = -x^2 + 1$

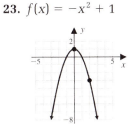

25. $(-\infty, -7)$ or $(2, \infty)$ **27.** $(-\infty, 1]$ or $\left[\dfrac{3}{2}, \infty\right)$ **29.** $\left(-\infty, -\dfrac{5}{2}\right]$ or $\left[\dfrac{4}{3}, \infty\right)$ **31.** Minimum $= -4$

33. Maximum $= -1$ **35.** Minimum $= -\dfrac{9}{8}$ **37.** $V = (-3, 4)$; $y = 4$; right   **39.** $V = (-6, -3)$; $y = -3$; left

41. $V = (0, 3)$; $y = 3$; left   **43.** $V = (-48, 2)$; $y = 2$; right.   **45.** **(a)** $P = 6y + y^2$ **(b)** -3 and 3; -9

47. **(a)** $P(5) = \$750$ thousand; $P(10) = \$1437.50$ thousand; $P(15) = \$2062.50$ thousand
   (b) At 6250 employees, $P = \$4882.81$ (in thousands)
   (c) As x increases, P increases to a maximum at $\$4882.81$ thousand, then decreases.

49. **(a)** $h(50) = 43.50$ ft; $h(100) = 63.50$ ft; $h(200) = 43.50$ ft
   (b) Maximum height 66 ft (approx.); travels 125 ft (approx.) horizontally   **(c)** 253.45 ft (approx.)

51. **(a)** 124.67 ft; 126.68 ft; 116.7 ft   **(b)** Peak at approx. $x = 450.45$, $y = 126.76$ **(c)** 23.24 ft (approx.)

53. **(a)** $A = 500x - x^2$ **(b)** $0 \le x \le 500$ **(c)** 250 ft by 250 ft; 62,500 ft^2

55. 112.5 ft by 225 ft; maximum area is 25,312.5 ft^2 **57.** $y = 2$ ft, $x = 2$ ft

59. **(b)** $y = -7.355x^2 + 29.007x - 0.262$ **61.** **(b)** $N = 3.471t^2 - 12.226t + 24.863$
   (d) $x = 2.5$, $y \approx 26.3$ **(d)** 2010; $t = 9$, $N \approx 196.0$
      $x = 5$, $y \approx -39.1$ 2015; $t = 14$, $N \approx 534.0$
      $x = 10$, $y \approx -445.7$

63. **(a)** Two x intercepts   **(b)** One x intercept   **(c)** No x intercepts   **65.** Two intercepts   **67.** One intercept

Problem Set 3.2 (on page 193)

1. (a) Symmetric with respect to the y axis; resembles $y = x^2$

(b) Symmetric with respect to the origin; resembles $y = x^3$

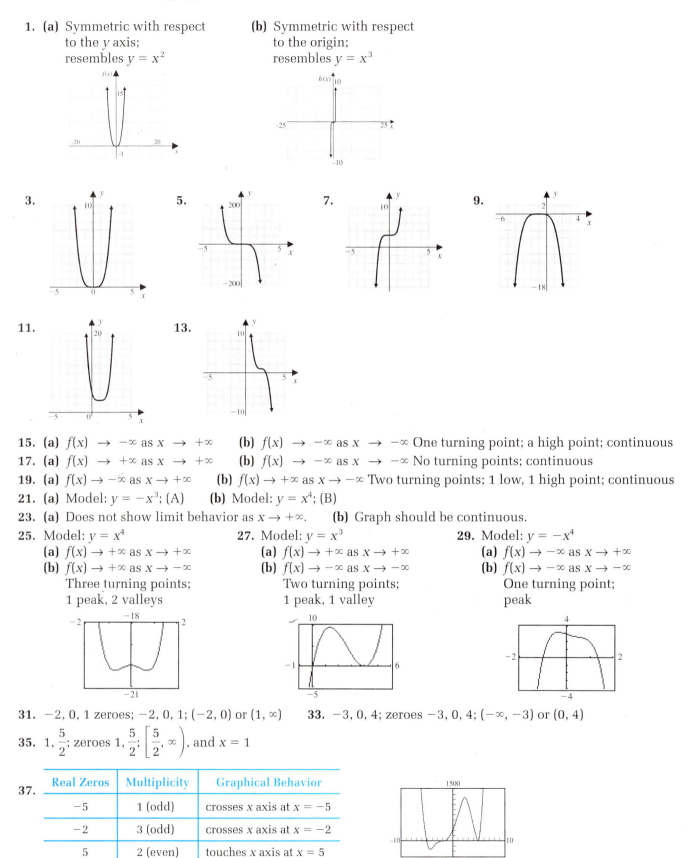

3.

5.

7.

9.

11.

13.

15. (a) $f(x) \to -\infty$ as $x \to +\infty$ **(b)** $f(x) \to -\infty$ as $x \to -\infty$ One turning point; a high point; continuous

17. (a) $f(x) \to +\infty$ as $x \to +\infty$ **(b)** $f(x) \to -\infty$ as $x \to -\infty$ No turning points; continuous

19. (a) $f(x) \to -\infty$ as $x \to +\infty$ **(b)** $f(x) \to +\infty$ as $x \to -\infty$ Two turning points; 1 low, 1 high point; continuous

21. (a) Model: $y = -x^3$; (A) **(b)** Model: $y = x^4$; (B)

23. (a) Does not show limit behavior as $x \to +\infty$. **(b)** Graph should be continuous.

25. Model: $y = x^4$
(a) $f(x) \to +\infty$ as $x \to +\infty$
(b) $f(x) \to +\infty$ as $x \to -\infty$
Three turning points;
1 peak, 2 valleys

27. Model: $y = x^3$
(a) $f(x) \to +\infty$ as $x \to +\infty$
(b) $f(x) \to -\infty$ as $x \to -\infty$
Two turning points;
1 peak, 1 valley

29. Model: $y = -x^4$
(a) $f(x) \to -\infty$ as $x \to +\infty$
(b) $f(x) \to -\infty$ as $x \to -\infty$
One turning point;
peak

31. $-2, 0, 1$ zeroes; $-2, 0, 1$; $(-2, 0)$ or $(1, \infty)$ **33.** $-3, 0, 4$; zeroes $-3, 0, 4$; $(-\infty, -3)$ or $(0, 4)$

35. $1, \dfrac{5}{2}$; zeroes $1, \dfrac{5}{2}$; $\left[\dfrac{5}{2}, \infty\right)$, and $x = 1$

37.

Real Zeros	Multiplicity	Graphical Behavior
-5	1 (odd)	crosses x axis at $x = -5$
-2	3 (odd)	crosses x axis at $x = -2$
5	2 (even)	touches x axis at $x = 5$ and bounces back

39.

Real Zeros	Multiplicity	Graphical Behavior
−3	2 (even)	touches x axis at $x = -3$ and bounces back
0	1 (odd)	crosses x axis at $x = 0$
4	3 (odd)	crosses x axis at $x = 4$

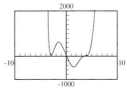

41. (a) 11:00 AM, $f(3) = 7.2°$F
3:00 PM, $f(7) = 26°$F
6:00 PM, $f(10) = 12.8°$F

(b) $t = 2$, 10:00 AM and $t = 11$, 7:00 PM

(d) below zero between 8 AM and 10 AM and between 7 PM and 8 PM; above zero between 10 AM and 7 PM

(c)

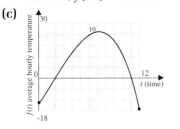

43. (a) $0 \le t \le 48$ **(c)** 3.82% at $t = 40$ **(d)** 6.80% at $t = 9$

45. (a) $V = x(20 - 2x)^2$

(b) $0 < x < 10$

(d) Maximum volume is 592.59 in.^{3} (approx.) when $x = 3.33$ in. (approx.)

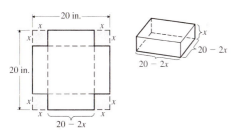

47. (a) For $a > 0$ and $b > 0$: $f(x), g(x) \to +\infty$ as $x \to +\infty$; $f(x), g(x) \to +\infty$ as $x \to -\infty$.
For $a < 0$ and $b < 0$: $f(x), g(x) \to -\infty$ as $x \to +\infty$; $f(x), g(x) \to -\infty$ as $x \to -\infty$.

(b) For $a > 0$ and $b < 0$: $f(x) \to +\infty$ as $x \to +\infty$; $g(x) \to -\infty$ as $x \to +\infty$; $f(x) \to +\infty$ as $x \to -\infty$; $g(x) \to -\infty$ as $x \to -\infty$.
For $a < 0$ and $b > 0$: $f(x) \to -\infty$ as $x \to +\infty$; $g(x) \to +\infty$ as $x \to +\infty$; $f(x) \to -\infty$ as $x \to -\infty$; $g(x) \to +\infty$ as $x \to -\infty$.

49. (a) Concave down on $(-\infty, 0)$; concave up on $(0, \infty)$; point of inflection at $(0, 1)$

(b) Concave down on $(4, \infty)$; concave up on $(-\infty, 4)$; point of inflection at $(4, 1)$

51. $n > 1$

53.

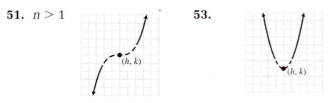

55. They both approach $-\infty$ as $x \to -\infty$ and as $x \to +\infty$.
$R(x) \to 1$ as $x \to -\infty$ and as $x \to +\infty$.

57. (b) $y = 0.0018x^3 - 0.1795x^2 + 6.6212x - 43.8355$ **(d)** 40; 45; 55

Problem Set 3.3 (on page 212)

1. (a) −14   **(b)** 63   **(c)** 241   **3. (a)** Yes; $(x + 2)(x + 4)(x - 6)$ **(b)** Yes; $(x + 4)(x - 6)(x + 2)$
(c) No   **(d)** Yes; $(x - 6)(x + 4)(x + 2)$ **5. (a)** $f(x) = (x - 1)(x - 2)(x - 3)$

7. (a) $f(x) = (x + 2)(2x^3 - 4x^2 + 8x - 13)$ **9.** $f(3) = 0$ **11.** $f(x) = -(x + 3)(x - 1)(x - 3)$

13. $-3, -2, 0, 2$; All are x intercepts.   **15.** $-1, 2, -i, i$; x intercepts: $-1, 2$

17. $-2, 1, 2$; $f(x) = (x - 1)(x - 2)(x + 2)$ **19.** 1; $h(x) = (x - 1)(5x^2 - 7x + 10)$

21. $-1, 1, 2$; $f(x) = (x + 1)(x - 1)^2(x - 2)^2$ **23.** -2 **25.** -5 and 5

27. $-2, 3, -\dfrac{1}{2}, \dfrac{1}{2}$; $f(x) = \dfrac{1}{4}(x + 2)(x - 3)(2x + 1)(2x - 1)$ **29.** $-1, \dfrac{2}{5}, \dfrac{1}{2}$; $f(x) = \dfrac{2}{3}(x + 1)\left(x - \dfrac{2}{5}\right)\left(x - \dfrac{1}{2}\right)$

31. $\dfrac{-1}{2}, -2\sqrt{3}, 2\sqrt{3}$ **33.** $-1, -\dfrac{1}{2}, 1, \dfrac{3}{2}$; $h(x) = 4(x + 1)\left(x + \dfrac{1}{2}\right)(x - 1)\left(x - \dfrac{3}{2}\right)$

35. $-1, 4$; $f(x) = (x + 1)(x - 4)(x^2 + 4)$ **37.** $-\dfrac{1}{2}$; $f(x) = \left(x + \dfrac{1}{2}\right)(x^2 - 4x + 5)$

39. $f(1) = -3, f(3) = 7$ **41.** $f(-3) = -42, f(-2) = 5$ **43.** $f(0.9) = 7.7401, f(2.1) = -7.5959$

45. $c = 0.3477$ **47.** $c = -1.1547$ **49.** $c = 4.7958$ **51.** $c = -1, 2; (2, \infty)$

53. $c = -1.4142; [-1.41, -1]$ or $[1, 4]$ **55. (a)** $V = 21\pi x^2 - \dfrac{\pi}{3}x^3$;   **(b)** $0 < x \le 40$;   **(c)** $x = 36$ ft.

57. (a) $V = 30\pi r^2 - \dfrac{\pi}{3}r^3$ **(b)** $0 < r \le 30$ **(c)** 6 ft.   **59.** 1.83%; 4.13%; 10%

61. (a) 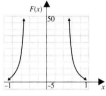   **(c)** $0 < x \le 525$ **(d)** Approx. 8.88 ft

63. False   **65.** True   **67.** False

Problem Set 3.4 (on page 227)

1. (a) Vertical asymptote:
$x = 0$;
Horizontal asymptote:
$y = 0$

(b) Vertical asymptote:
$x = 0$;
Horizontal asymptote:
$y = 0$

3. Vertical asymptote:
$x = 0$;
Horizontal asymptote:
$y = -4$.
As $x \to 0, y \to +\infty$.
As $x \to +\infty$ and as
$x \to -\infty, y \to -4$.

5. Vertical asymptotes:
$x = -2$ and $x = 2$;
Horizontal asymptote:
$y = 0$.
As $x \to -2^-, y \to +\infty$;
as $x \to -2^+, y \to -\infty$.
As $x \to 2^-, y \to +\infty$;
as $x \to 2^+, y \to -\infty$.
As $x \to \infty, y \to 0$;
as $x \to -\infty, y \to 0$.

7. (a) Vertically stretch $1/x$ by
4, shift to the left 3 units.
Vertical asymptote:
$x = -3$;
Horizontal asymptote:
$y = 0$

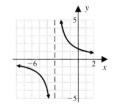

(b) Shift $1/x$ up 3 units.
Vertical asymptote:
$x = 0$;
Horizontal asymptote:
$y = 3$

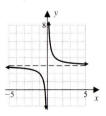

9. (a) Vertically stretch $1/x^2$ by 3, reflect about the x axis, and shift to left 1 unit. Vertical asymptote: $x = -1$; Horizontal asymptote: $y = 0$

(b) Vertically stretch $1/x$ by 2, shift to the right 1 unit and up 3 units. Vertical asymptote: $x = 1$; Horizontal asymptote: $y = 3$

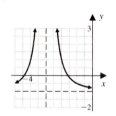

11. (a) Vertically stretch $1/x^2$ by 4, shift to the left 3 units, then down 1 unit. Vertical asymptote: $x = -3$; Horizontal asymptote: $y = -1$

(b) Shift $1/x^2$ to the left 4 units, then up 2 units. Vertical asymptote: $x = -4$; Horizontal asymptote: $y = 2$

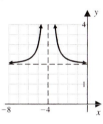

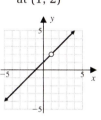

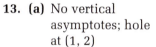

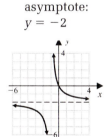

13. (a) No vertical asymptotes; hole at $(1, 2)$

(b) No vertical asymptotes; hole at $(3, 4)$

15. (a) Horizontal asymptote: $y = -2$

(b) Horizontal asymptote: $y = \dfrac{3}{2}$

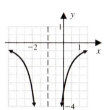

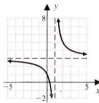

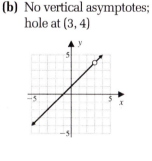

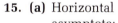

17. (a) C   **(b)** Vertical asymptotes: $x = -1$ and $x = 0$; horizontal asymptote: $y = 0$
19. (a) D   **(b)** Vertical asymptote: $x = -1$; horizontal asymptote: $y = 2$

21.

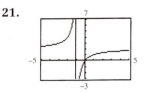

Vertical asymptote: $x = -1$
horizontal asymptote: $y = 2$

23.

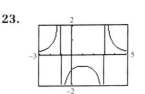

Vertical asymptotes: $x = -1$, $x = 3$
horizontal asymptote: $y = 0$

25.

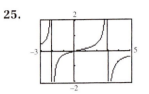

Vertical asymptote: $x = -2$, $x = 3$
horizontal asymptote: $y = 0$

27.

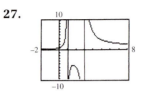

Vertical asymptotes: $x = 1$, $x = 3$
horizontal asymptote: $y = 1$

29. $(-1, 0)$ **31.** $(-\infty, -3)$ or $(1, \infty)$ **33.** $(-2, 0)$ or $(3, \infty)$ **35.** $[1, 3]$

37. (a) $x \geq 0$ **(b)** Horizontal asymptote: $S(x) = 5000$

(c) As expenditures increase, sales approach 5000 units.   **39. (b)** As $x \to 60\%$, the cost approaches ∞.

41. (a) $A = 2x^2 + \dfrac{2000}{x}$, $x > 0$ **(b)** $x \approx 7.94$ cm
$h \approx 7.94$ cm
minimum $A \approx 377.98$ cm^2

43. $-\dfrac{1}{2x}$, $x \neq 2$ **45.** $f(x) = \dfrac{2(x - 1)(x^2 - 9)}{x^3 + 4x}$

47. If k is in the domain of f, then $f(k)$ would have one defined value and $f(x)$ would not approach $+\infty$ or $-\infty$.

49. Oblique asymptote: $y = \dfrac{1}{2}x - 1$ **51. (a)** Oblique asymptote
$y = -4x + 8$

(b) Oblique asymptote
$y = x - 3$

Problem Set 3.5 (on page 236)

1. **(a)** 3 positive real zeros; 1 negative real zero; 0 imaginary zeros
   (b) 3 positive real zeros; 1 negative real zero; 2 imaginary zeros
3. 1 positive real zero; 1 negative real zero; 2 imaginary zeros
5. 1 negative real zero; 2 positive real zeros; 0 imaginary zeros
7. 1 positive real zero; 0 or 2 negative real zeros; 0 or 2 imaginary zeros
9. 0 or 2 positive real zeros; 1 negative real zero; 0 or 2 imaginary zeros
11. 3 or 1 positive real zeros; 1 negative real zero; 0 or 2 imaginary zeros
13. 1 positive real zero; 0 or 2 negative real zeros; 2 or 4 imaginary zeros
15. Upper bound: 2, Lower bound: -2 17. Upper bound: 3, Lower bound: -2
19. Upper bound: 3, Lower bound: -3 21. Upper bound: 2, Lower bound: -1
23. Upper bound: 7, Lower bound: -4 25. zeros of multiplicity 1: $-3i, -2i, 2i, 3i$
27. zeros of multiplicity 2: $-i, i$ 29. zeros of multiplicity 1: $-1 - i, -1 + i, -\sqrt{2}, \sqrt{2}$
31. $f(x) = x^3 - 5x^2 + 5x + 3$ 33. $f(x) = x^2 - 2x + 2$ 35. $f(x) = x^4 - 4x^3 + 24x^2 - 40x + 100$
37. **(a)** no real factors   **(b)** $f(x) = \left[x - \left(\dfrac{1}{2} - \dfrac{\sqrt{3}}{2}i\right)\right]\left[x - \left(\dfrac{1}{2} + \dfrac{\sqrt{3}}{2}i\right)\right]$
39. **(a)** $f(x) = 2x(x^2 + 2)$ **(b)** $f(x) = 2x(x + i\sqrt{2})(x - i\sqrt{2})$
41. **(a)** $f(x) = (x + 3)(x - 3)(x^2 + 9)$ **(b)** $f(x) = (x + 3)(x - 3)(x + 3i)(x - 3i)$
43. **(a)** $f(x) = (x^2 + 1)(3x - 1)$ **(b)** $f(x) = (x + i)(x - i)(3x - 1)$
45. **(a)** i is a zero, but $-i$ is not a zero   **(b)** No, coefficients must be real numbers.   47. $4 - 2i$
49. **(a)** 3   **(b)** 1 real and 2 imaginary   51. **(a)** 4   **(b)** $-2, 2, -2i, 2i$
53. $f(x) = (x - 2)(x + 2)[(x + 1) + \sqrt{3}\,i][(x + 1) - \sqrt{3}\,i][(x - 1) + \sqrt{3}\,i][(x - 1) - \sqrt{3}\,i]$ 55. False

Chapter 3 Review Problem Set (on page 239)

1. Vertex: $(-5, 3)$; axis of symmetry: $x = -5$; Domain: $\mathbb{R}$; Range: $(-\infty, 3]$
3. $f(x) = (x + 2)^2 - 11$; vertex: $(-2, -11)$; axis of symmetry: $x = -2$ 5. $f(x) = (x + 1)^2 - 9$ 7. $\left(-\dfrac{1}{2}, 1\right)$

9.
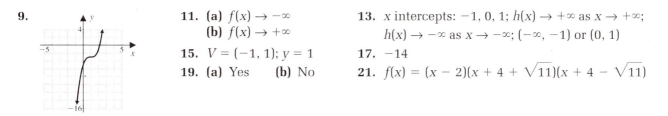

11. **(a)** $f(x) \to -\infty$
    (b) $f(x) \to +\infty$
13. x intercepts: $-1, 0, 1$; $h(x) \to +\infty$ as $x \to +\infty$;
    $h(x) \to -\infty$ as $x \to -\infty$; $(-\infty, -1)$ or $(0, 1)$
15. $V = (-1, 1)$; $y = 1$ 17. -14
19. **(a)** Yes   **(b)** No   21. $f(x) = (x - 2)(x + 4 + \sqrt{11})(x + 4 - \sqrt{11})$

23. 3 positive zeros; 1 negative zero; upper bound: 3; lower bound: -1; $-1 \leq c < 0, 0 \leq c < 1, 1 \leq c < 2$, $2 \leq c < 3$
25. Rational zeros: $-1, 2, 3$ 27. **(a)** $f(0) < 0 < f(2)$ **(b)** 1.9044
29. Zeros: $-2.6458, -1.4142, 1.4142, 2.6458$ (approx.)
31. -1 and 1, each with multiplicity 1; -4 with multiplicity 2; $i, -i$, each with multiplicity 1
33. $5 \pm i\sqrt{11}, \pm i\sqrt{11}$ 35. $f(x) = x^3 - 6x^2 + 13x - 10$
37. 1 positive zero; 2 or 0 negative zeros; 0 or 2 imaginary zeros.
39. **(a)** $f(x) = (x - 1)(x + 1)(x^2 - 8x + 17)$ **(b)** $f(x) = (x - 1)(x + 1)(x - 4 - i)(x - 4 + i)$
41. All real numbers except $0, -3, 3$

43. Vertical asymptote: $x = 1$; horizontal asymptote: $y = -5$; x intercepts: $1 + \sqrt{\dfrac{3}{5}}, 1 - \sqrt{\dfrac{3}{5}}$; y intercept: -2

45. Vertical asymptote: $x = 7$; horizontal asymptote: $y = 4$; x intercept: $-\dfrac{1}{2}$; y intercept: $-\dfrac{2}{7}$

47. (a) $h(0) = 5$, $h(1) = 53$ **(b)** $0 \le t \le 4.08$ (approx.)
    (c) The height increases during the first 2 sec, then decreases until the ball hits the ground.
    (d) 2 sec after being hit   **(e)** 69 ft

49. (a) $N(0) = 20$, $N(10) = 620$, $N(25) = 957$ **(b)** 14.8°C and 35.2°C (approx.)
    (c) Maximum population at 25°C   **(d)** Maximum population is 957 insects (approx.)

51. (a) $0 < x < 100$ **(b)** $p = 2x + \dfrac{200}{x}$ **(c)** $C = 13.2x + \dfrac{1320}{x}$

    (d) Cost is minimum when $x = 10$ ft; minimum cost is \$264.

Chapter 3 Test (on page 242)

1. (a) C   **(b)** A   **(c)** B

2. (a) $f(x) = -2\left(x + \dfrac{5}{4}\right)^2 + \dfrac{49}{8}$ **(b)** Axis of symmetry: $x = -\dfrac{5}{4}$; vertex: $\left(-\dfrac{5}{4}, \dfrac{49}{8}\right)$

    (c) Opens downward; maximum   **(d)** Domain: $\mathbb{R}$; Range: $\left(-\infty, \dfrac{49}{8}\right]$ **(e)** $(-\infty, -3]$ or $\left[\dfrac{1}{2}, \infty\right)$

3. Vertically stretch x^4 by 2, reflect about the x axis, shift to right 1 unit, and shift up 1 unit.
    (a)   **(b)** As $x \to +\infty$, $f(x) \to -\infty$; as $x \to -\infty$, $f(x) \to -\infty$.

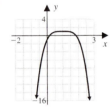

4. 0 (multiplicity 4), 1 (multiplicity 1), -1 (multiplicity 2)

5. (a) $Q(x) = x^4 - 8x^3 + 27x^2 - 56x + 100$; $R = -192$ **(b)** -192 **(c)** $x + 1$ is a factor.

6. (a)   **(b)** 2 negative real zeros; 1 positive real zero

    (c) Rational zeros: $-1, -\dfrac{1}{3}, \dfrac{2}{3}$

    (d) $f(x) = 9(x + 1)\left(x - \dfrac{2}{3}\right)\left(x + \dfrac{1}{3}\right)$

    (e) $(-\infty, -1)$ or $\left(-\dfrac{1}{3}, \dfrac{2}{3}\right)$

7. $f(x)$ **8. (a)** T   **(b)** F   **(c)** F

9. (a) Vertical asymptote: $x = 2$;   **10. (a)** $h(0) = 206$, $h(2) = 1078$,   **11. (a)** $t \ge 0$
    horizontal asymptote: $h(4) = 1822$ **(b)**
    $y = 4$ **(b)**
    (b) x intercept: $\dfrac{5}{4}$
    y intercept: $\dfrac{5}{2}$
    (c)

    (c)   **(c)** As time goes on, the
    (d) 3628.25 ft   amount of information
    that is remembered
    decreases.

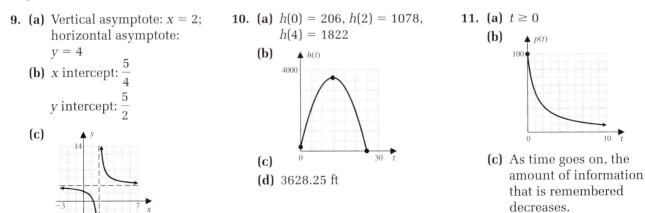

(d) All real numbers except
2; Range: All real
numbers except 4

Chapter 4: Problem Set 4.1 (on page 252)

1. 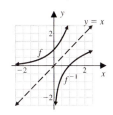   **3.** 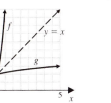   **5.**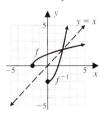

7. (a) Domain of f and Range of f^{-1}: $\mathbb{R}$;
Domain of f^{-1} and Range of f: $(0, \infty)$

(b) Domain of f and Range of f^{-1}: $[-2, \infty)$;
Domain of f^{-1} and Range of f: $[0, \infty)$

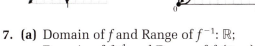

9. (a) Yes   **(b)** No   **11.** One-to-one   **13.** Not one-to-one   **15.** Not one-to-one

17.

x	2	-3	3	5	7	100	-4	1
$f^{-1}(x)$	1	2	3	4	5	6	7	8

Domain of f and Range of f^{-1}: 1, 2, 3, 4, 5, 6, 7, 8;
Domain of f^{-1} and Range of f: -4, -3, 1, 2, 3, 5, 7, 100

19. $\dfrac{x - 5}{7}$ **21.** $\dfrac{3 + x}{x}$ **23.** $x^2 + 3, \ x \geq 0$ **25.** $\sqrt[3]{x + 8}$

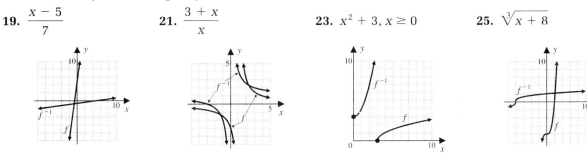

27. f^{-1} does not exist—f fails the horizontal-line test.

29. $(2 - x)^3$ **31.** $\sqrt{-x}, \ x \leq 0$ **33. (a)** f does not have an inverse: g has an inverse   **(b)** $g^{-1}(x) = -x$, $x \geq 0$

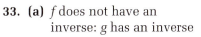

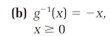

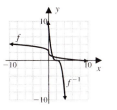   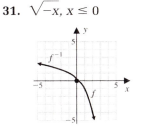   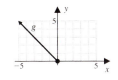

35. (a) f does not have an inverse; g has an inverse   **(b)** $g^{-1}(x) = \sqrt{1 - x^2}$, $0 \leq x \leq 1$

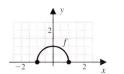   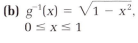

37. (a)

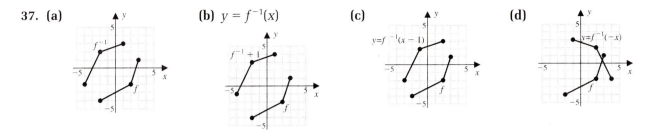

(b) $y = f^{-1}(x)$

(c)

(d)

39. No; fails the horizontal-line test.   **41.** No; fails the horizontal-line test.

43. (a) $W = f(P) = 0.85P$

(b) $P = \dfrac{20}{17}W$

45. (a) $C = f^{-1}(F) = \dfrac{5}{9}(F - 32)$

(b)

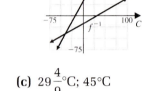

(c) $29\dfrac{4}{9}°C$; $45°C$

47. (a) $y = \$25{,}000 + 0.40x$, $x \geq 0$

(b) $f^{-1}(x) = 2.5x - 62.500$

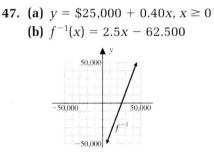

(c) f^{-1} represents sales in terms of the salesperson's income

49. (a) $f^{-1}(x) = \begin{cases} 2x + 8 & \text{if } x < -4 \\ x + 4 & \text{if } x \geq -4 \end{cases}$
Domain and Range of f and f^{-1}: $\mathbb{R}$

(b) f^{-1} does not exist (fails the horizontal line test).

51. Reflect $y = f(x)$ about $y = x$, then reflect the result about $y = x$ to get $f(x)$.

53. (a) $f^{-1}(x) = \dfrac{1}{2}x - 2$

$g^{-1}(x) = \dfrac{x + 1}{3}$

$(f \circ g)^{-1}(x) = \dfrac{x - 2}{6}$

(d) $(g \circ f)^{-1} = f^{-1} \circ g^{-1}$; this is the same result as for $f^{-1} \circ g^{-1}$ in part (b).

55. (a) B   **(b)** A   **(c)** C   **(d)** D

Problem Set 4.2 (on page 265)

1. (a) 8.8250   **(b)** 0.3752   **(c)** 6.3697   **(d)** 85.3490
3. (a) 29.9641   **(b)** 0.0432   **(c)** 4.1133   **(d)** -1.3594
5. (a) 2   **(b)** 0.30   **(c)** 2.25   **(d)** 0.74
7. For g shift the graph of $f(x) = e^x$ 2 units up. For h shift the graph 2 units down.   **9. (a)** 3   **(b)** $\dfrac{1}{2}$

11. (a) $A = 3, b = 3$ **(b)** $A = 1, b = \dfrac{1}{2}$ **13. (a)** B   **(b)** C   **(c)** D   **(d)** A   **15.** 2

17. -1 and 6   **19.** -1

21. Shift the graph of $y = 5^x$ left 1 unit. Domain: $\mathbb{R}$; Range: $(0, \infty)$; increasing; horizontal asymptote: x axis

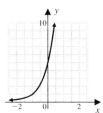

23. Shift the graph of $y = 2^x$ up 3 units. Domain: $\mathbb{R}$; Range: $(3, \infty)$; increasing; horizontal asymptote: $y = 3$

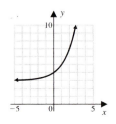

25. Stretch the graph of $y = e^x$ vertically by a factor of 2, reflect about the x axis. Domain: $\mathbb{R}$; Range: $(0, \infty)$; decreasing; horizontal asymptote: x axis

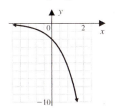

27. Reflect the graph of $y = e^x$ across the y axis and shift down 2 units. Domain: $\mathbb{R}$; Range: $(-2, \infty)$; decreasing; horizontal asymptote: $y = -2$

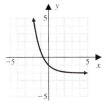

29. (a) $6638.48 **(b)** $6651.82 **(c)** $6660.88 **(d)** $6664.39

31. (a) $S = 5000e^{0.0575t}$ **(b)** $6293 **(c)** Less **33. (a)** $P = 6e^{0.013t}$ **(b)** 6.83 billion

35. (a) 10% per day **(b)** 41,218 **37. (a)** $V = \$19{,}545(0.80)^t$ **(b)** $15,636; $10,007; $6,405

39. 1.63 million; 0.69 million; 0.29 million **41.** 83.86%; 76.80% **43.** 32g; 25.6g; 13.11g

45. (b) $N = 224.474 \cdot 1.021^t$ **(d)** 340

47. (b) When rounded to two decimal places, the regression function values are the same as the table values for each value of t. **(d)** $0.50

49. (a) $f(x) = 1^x = 1$ is a constant function. **(b)** $(-3)^{\frac{1}{2}}$ is not a real number. **51. (a)** $x = 4$ and 3 **(b)** 4

53. (a) $-11, \dfrac{3 \pm \sqrt{7}i}{2}$ **(b)** 1, 2, and 3.

Problem Set 4.3 (on page 279)

1. (a) $\log_5 125 = 3$ **(b)** $\log_{32} 2 = \dfrac{1}{5}$ **(c)** $\log_e 17 = t$ or $\ln 17 = t$ **(d)** $\log_b(13z + 1) = x$

3. (a) $9^2 = 81$ **(b)** $10^{-4} = 0.0001$ **(c)** $c^w = 9$ **(d)** $e^{-1-3x} = \dfrac{1}{2}$

5. (a) -3 **(b)** 4 **(c)** $\dfrac{1}{3}$ **(d)** $\dfrac{2}{3}$ **7. (a)** 0 **(b)** -2 **(c)** 3 **(d)** $\dfrac{1}{3}$

9. (a) 36 **(b)** 16 **11. (a)** 3 **(b)** 8 **13. (a)** 10^{3-y} **(b)** $\dfrac{1}{2}(e^{y-8} - 1)$

15. (a) 1.5933; $10^{1.5933} = 39.2$ **(b)** 6.8680; $e^{6.8680} = 961$ **(c)** -0.1249; $10^{-0.1249} = \dfrac{3}{4}$

(d) -1.4697; $e^{-1.4679} = 0.23$ **(e)** 2.3026; $e^{2.3026} = 10$

17. (a) 10.6974 **(b)** 2.5924 **(c)** 0.0821 **(d)** 0.5112 **19. (a)** 2.0478 **(b)** 2.638

21. (a) $\log_3 x + \log_3(x + 1)$ **(b)** $\log_3 18 - \log_3(x + 2)$ **23. (a)** $4 \log_b(x + 3)$ **(b)** $\log x + \dfrac{1}{2}\log(2x + 1)$

25. (a) $\ln y + \dfrac{2}{3}\ln(3x + 1)$ **(b)** $\ln x + \ln(x + 7)$ **27. (a)** $\log_5 \dfrac{8}{7}$ **(b)** $\log_2\left(\dfrac{55}{2}\right)$

29. (a) $\ln\left(\dfrac{x + 3}{x - 3}\right)$ **(b)** $\ln\left(\dfrac{3x^2 + 7x + 4}{3x^2 - 5x - 12}\right)$ **31. (a)** 5 **(b)** $\dfrac{1}{9}$ **33. (a)** $\dfrac{e^{-7}}{x}$ **(b)** $x^2 - 4$

35. (a) 2.322 **(b)** 0.545 **(c)** -1.469 **(d)** 0.910

37. (a) $A = \dfrac{Pr(1 + r)^n}{(1 + r)^n - 1}$ **(b)** $A = 18{,}000 \dfrac{\dfrac{0.1}{12}\left(1 + \dfrac{0.1}{12}\right)^n}{\left(1 + \dfrac{0.1}{12}\right)^n - 1}$; $580.81; $456.53; $382.45

39. (a) $N = \dfrac{230}{1 + 6.931e^{-0.1702t}}$ **(b)** 29; 102; 158; 221   **41.** 7.7   **43. (a)** 2.5   **(b)** 7.8   **(c)** 6.4

45. (a) 90 db   **(b)** 149.19 db   **49.** Try $a = b = 2$ **53.** Domain: $(2, \infty)$; x intercept: $e + 2$ **55.** 9

Problem Set 4.4 (on page 288)

1. $f^{-1}(x) = \log_5 x$;
Domain: $(0, \infty)$;
Range: $\mathbb{R}$

3. $h^{-1}(x) = \left(\dfrac{1}{10}\right)^x$;
Domain: $\mathbb{R}$;
Range: $(0, \infty)$

5. (a) $b = 4$ **(b)** $b = \dfrac{1}{16}$ **(c)** $b = e$

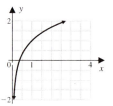

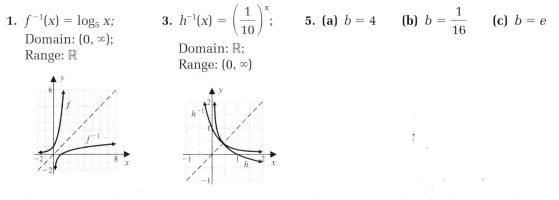

7. $(-\infty, 2)$ **9.** $(-\infty, -1)$ or $(-1, \infty)$ **11.** $(-2, \infty)$ **13.** $(-\infty, 0)$ or $(1, \infty)$ **15.** $(-\infty, -1)$ or $(1, \infty)$

17. $f(x)$ is steeper than $g(x)$; $g(x)$ is steeper than $h(x)$; $f(x) \to +\infty$, $g(x) \to +\infty$ and $h(x) \to +\infty$ as $x \to +\infty$

19. Shift 2 units to the right; Domain: $(2, \infty)$; Vertical asymptote: $x = 2$

21. Reflect the graph g across the x axis; Domain: $(0, \infty)$; Vertical asymptote: $x = 0$

23. $y = \log_3 (x + 1)$

25. Domain: $(0, \infty)$;
vertical asymptote: y axis

27. Domain: $(0, \infty)$;
vertical asymptote: y axis

29. Domain: $(-3, \infty)$;
vertical asymptote: $x = -3$

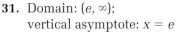

31. Domain: (e, ∞);
vertical asymptote: $x = e$

33. Domain: $(0, \infty)$;
vertical asymptote: y axis

35. Domain: $\left(-\dfrac{1}{5}, \infty\right)$

vertical asymptote: $x = -\dfrac{1}{5}$

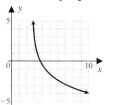

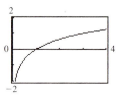

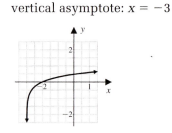

37. Domain: $\left(-\infty, -\dfrac{1}{2}\right)$ or $\left(\dfrac{1}{2}, \infty\right)$ **39.** \$2,352,000; \$3,481,000; \$4,042,000   **41.** $P_1 > P_2$

vertical asymptotes: $x = -\dfrac{1}{2}, x = \dfrac{1}{2}$

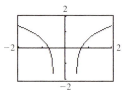

43. Domain: $(2, \infty)$; vertical asymptote: $x = 2$

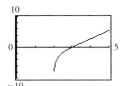

45. Domain: $(-\infty, 1)$ or $(4, \infty)$; vertical asymptotes: $x = 1$, $x = 4$

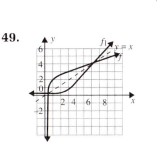

47. Domain: $\left(\dfrac{3}{2}, \infty\right)$, vertical asymptote: $x = \dfrac{3}{2}$

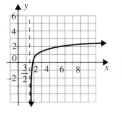

49.

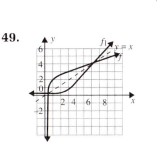

Problem Set 4.5 (on page 295)

1. 1.3869   **3.** -0.3892 **5.** 0.6090   **7.** 1.3026   **9.** 2.5604   **11.** -3.3049 **13.** 0.6129

15. $(-\infty, 2.77)$ **17.** $(-\infty, 1.68)$ **19.** $(-\infty, 0.65)$ **21.** $(-\infty, -0.47)$ **23.** $(11.90, \infty)$ **25.** 3   **27.** 4

29. 7   **31.** $\dfrac{1}{3}$ **33.** 22   **35.** 0; 3.538   **37.** 0   **39.** -0.907; 1.501   **41.** 1.314   **43.** $x < -0.69$

45. $x > 1.70$ **47.** $x \geq 0.12$ **49.** $x > 1$ **51.** (a) 14.77 yr.   (b) 14.68 yr.   (c) 14.62 yr.   (d) 14.59 yr.

53. (a) $t = \dfrac{\log k}{4 \log 1.012125}$ (b) 14.38 yr.; 22.79 yr.; 28.76 yr.   **55.** 39.91 parts   **57.** (a) 300   (b) 6.39 yr.

59. (a) 0.15   (b) 12.64 min.   **61.** no solution   **63.** (a) $x > 0$ (b) $x > 0$ **65.** 0.88

Chapter 4 Review Problem Set (on page 297)

1. (a) $f^{-1}(x) = \dfrac{7 - x}{13}$ (b) $f^{-1}(x) = x^5$ **3.** (a) 2.5   (b) 0.16   (c) 0.760   (d) 4.889   (e) 0.211

5. (a) Shift the graph of $y = 2^x$ up 1 unit. Domain: $\mathbb{R}$; Range: $(1, \infty)$; increasing; horizontal asymptote: $y = 1$
   (b) Reflect the graph of $y = 2^x$ about the x axis. Domain: $\mathbb{R}$; Range: $(-\infty, 0)$; decreasing; horizontal asymptote: x axis
   (c) Vertically stretch the graph of $y = 2^x$ by 3. Domain: $\mathbb{R}$; Range: $(0, \infty)$; increasing; horizontal asymptote: x axis
   (d) Shift the graph of $y = 2^x$ down 3 units. Domain: $\mathbb{R}$; Range: $(-3, \infty)$; increasing; horizontal asymptote: $y = -3$
   (e) Shift the graph of $y = 2^x$ left 1 unit and shift down 3 units. Domain: $\mathbb{R}$; Range: $(-3, \infty)$; increasing; horizontal asymptote: $y = -3$
   (f) Vertically stretch the graph of $y = 2^x$ by 3, reflect about the x axis, shift left 1 unit, and shift up 4 units. Domain: $\mathbb{R}$; Range: $(-\infty, 4)$; decreasing; horizontal asymptote: $y = 4$

7. (a) $\ln a = 3.1$ (b) $\ln(a + b) = x^2$ (c) $\log_5 b = -2.7$ (d) $\ln c = a + b$ (e) $\log t = a - b$

9. (a) $-\dfrac{11}{4}$ (b) $-4, 4$ (c) 2   (d) $\sqrt{\dfrac{3}{2}}$ (e) $\dfrac{8}{(e - 1)}$ (f) 0   (g) $e^{\frac{(b-t)}{3}}$ (h) $\dfrac{1}{2}(u - y)$

11. $\dfrac{1}{2} \ln\left(\dfrac{u + 1}{u - 1}\right)$

13. (a) Shift the graph of $y = \ln x$ horizontally right 3 units. Domain: $x > 3$; Range: $\mathbb{R}$; vertical asymptote: $x = 3$.

   (b) Reflect the graph of $y = \ln x$ across the y axis and shift horizontally left 3 units. Domain: $x < 3$; Range: $\mathbb{R}$; vertical asymptote: $x = 3$.

   (c) Reflect $y = \log x$ across the x axis. Domain: $x > 0$; Range: $\mathbb{R}$; vertical asymptote: $x = 0$ (y axis).

15. (a) -0.535 **(b)** -5.666 **(c)** 0.457

17. (a) $x < 0.46$ **(b)** $0 < x < 0.16$ or $x > 1.14$ **(c)** $x \geq -8.13$

19. (a) At the end of 3, 7, and 10 yr, the accumulated balance will be $11,576.25, $14,071.00, and $16,288.95, respectively.   **(b)** 25.68 yr (approx.)

21. (a) After 2 and 3 yr, the value of the car will be $8437.50 and $6328.13, respectively   **(b)** 1.24 yr (approx.)

23. (a) 95.12 g   **(b)** 1204 yr (approx.)   **25. (a)** 15.5 yr   **(b)** 15.4 yr (approx.)

27. (a) 2.80   **(b)** 10.50   **29. (a)** 298; 333   **(b)** No

31. (a) $c^{-1}(x) = \dfrac{100}{x - 205}$ **(b)** The cost is inversely proportional to the number of people.

Chapter 4 Test (on page 300)

1. (a) $f^{-1}(x) = \dfrac{1}{4}(x + 1)$ **2. (a)** 9   **(b)** 9   **(c)** 0.794   **(d)** 20.832

3. (a)   **(b)** Domain: $\mathbb{R}$; Range: $(0, \infty)$ **(e)** Vertically stretch the graph of f by 2 and shift down 3 units. Horizontal asymptote: $y = -3$.

   (c) Decreasing

   (d) Horizontal asymptote: x axis

4. (a) $\log_6 7 = u$ **(b)** $3^b = x$ **5. (a)** -3 **(b)** $\dfrac{1}{.7}$ **(c)** $\sqrt{x + 3}$ **(d)** x^2

6. (a) $-1; 0; 0.4771; 0.8451$ **(c)** $f^{-1}(x) = 10^x$ **(d)** Reflect the graph of f about the x axis and shift left 7 units. Vertical asymptote: $x = -7$.

   (b) Domain: $(0, \infty)$; Range: $\mathbb{R}$

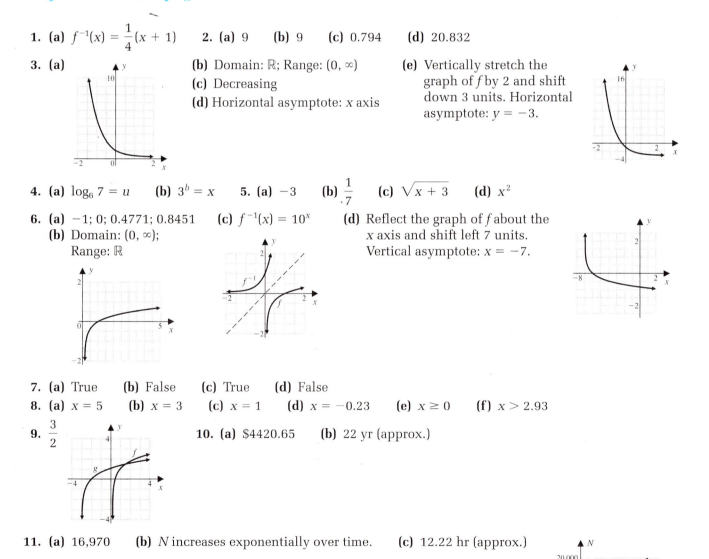

7. (a) True   **(b)** False   **(c)** True   **(d)** False

8. (a) $x = 5$ **(b)** $x = 3$ **(c)** $x = 1$ **(d)** $x = -0.23$ **(e)** $x \geq 0$ **(f)** $x > 2.93$

9. $\dfrac{3}{2}$ **10. (a)** $4420.65   **(b)** 22 yr (approx.)

11. (a) 16,970   **(b)** N increases exponentially over time.   **(c)** 12.22 hr (approx.)

Chapter 5: Problem Set 5.1 (on page 313)

1. (a) $\dfrac{-2\pi}{3}$ (b) $\dfrac{5\pi}{2}$ 3. (a) $\left(\dfrac{1}{2}, \dfrac{-\sqrt{3}}{2}\right)$ (b) $\left(\dfrac{\sqrt{2}}{2}, \dfrac{-\sqrt{2}}{2}\right)$ (c) $\left(\dfrac{\sqrt{3}}{2}, \dfrac{-1}{2}\right)$

5. (a) I   (b) III   (c) IV   (d) I   7. (a) IV   (b) I   (c) III   (d) I

9. (a) Quadrant II; coterminal: $\dfrac{11\pi}{4}, -\dfrac{5\pi}{4}$ (b) Quadrant IV; coterminal: $\dfrac{11\pi}{6}, -\dfrac{13\pi}{6}$

11. (a) Quadrant IV; coterminal: $315°, -405°$ (b) Quadrant II; coterminal: $480°, -240°$

13. (a) Quadrant IV; coterminal: $\dfrac{5\pi}{3}, -\dfrac{\pi}{3}$ (b) Quadrant I; coterminal: $20°, -340°$

15. (a) $16°18'36''$ (b) $-87°48'36''$ (c) $-156°37'48''$ (d) $89°44'24''$ 17. (a) $38.3°$ (b) $65.19°$

(c) $-141.47°$ (d) $244.77°$ 19. (a) $\dfrac{5\pi}{12} = 1.31$ (b) $-\dfrac{3\pi}{4} = -2.36$

21. (a) $-\dfrac{19\pi}{36} = -1.66$ (b) $\dfrac{37\pi}{15} = 7.75$ 23. (a) $\dfrac{3\pi}{8}, 1.18$ (b) $\dfrac{339\pi}{2000}, 0.53$

25. (a) $120°$ (b) $1290°$ 27. (a) $105°$ (b) $-80°$ 29. (a) $286.48°$ (b) $-131.78°$

31. (a) 4.71 in.   (b) 16.49 sq yd   33.   (a) 6.82 m   (b) 6.14 m^2 35. (a) 0.95 yd   (b) 2.39 sq yd

37. (a) 1.15 mi   (b) 6082.12 ft   39. 20.94 ft   41. 10.05 cm/sec

43. (a) 5280 rad/min   (b) 840.34 rotations/min   45. (a) 863.94 in./min   (b) 30.56 rotations/min

47. $\omega = 0.63$ rad/sec; $\nu = 5.03$ ft/sec   49. Radian measure depends on s as well as r.   51. α 53. $\dfrac{11\pi}{12}$

Problem Set 5.2 (on page 323)

1. $\sin t = \dfrac{1}{2}$
   $\csc t = 2$
   $\cos t = -\dfrac{\sqrt{3}}{2}$
   $\sec t = -\dfrac{2\sqrt{3}}{3}$
   $\tan t = -\dfrac{\sqrt{3}}{3}$
   $\cot t = -\sqrt{3}$

3. $\sin t = -\dfrac{12}{13}$
   $\csc t = -\dfrac{13}{12}$
   $\cos t = \dfrac{5}{13}$
   $\sec t = \dfrac{13}{5}$
   $\tan t = -\dfrac{12}{5}$
   $\cot t = -\dfrac{5}{12}$

5. $\sin t = \dfrac{2}{\sqrt{13}}$
   $\csc t = \dfrac{\sqrt{13}}{2}$
   $\cos t = -\dfrac{3}{\sqrt{13}}$
   $\sec t = -\dfrac{\sqrt{13}}{3}$
   $\tan t = -\dfrac{2}{3}$
   $\cot t = -\dfrac{3}{2}$

7. (a) $(0, -1)$; $\sin t = -1$; $\csc t = -1$; $\cos t = 0$; $\sec t$ is undefined; $\tan t$ is undefined; $\cot t = 0$
   (b) Same as part (a)
9. (a) $(1, 0)$; $\sin t = 0$; $\csc t$ is undefined; $\cos t = 1$; $\sec t = 1$; $\tan t = 0$; $\cot t$ is undefined
   (b) $(0, -1)$; $\sin t = -1$; $\csc t = -1$; $\cos t = 0$; $\sec t$ is undefined; $\tan t$ is undefined; $\cot t = 0$

11. $\dfrac{\sqrt{2}}{2}$ 13. $-\dfrac{1}{2}$ 15. $-\dfrac{1}{2}$ 17. $-\dfrac{\sqrt{3}}{2}$ 19. $-\dfrac{\sqrt{2}}{2}$ 21. $-\dfrac{\sqrt{2}}{2}$ 23. 0.1411   25. 0.2837

27. 1.5574   29. -0.6421 31. -2.1943 33. 1.1099   35. (a) $(-0.9041, 0.4274)$

(b) $(-0.5076, 0.8616)$ 37. 6.02 in.; 6.17 in.; 6.43 in.

39. $\sin t = \dfrac{-3\sqrt{10}}{10}$; $\cos t = \dfrac{-\sqrt{10}}{10}$; $\tan t = 3$; $\cot t = \dfrac{1}{3}$; $\sec t = -\sqrt{10}$; $\csc t = \dfrac{-\sqrt{10}}{3}$

41. $(\cos t)^2 \neq \cos t^2 = \cos(t^2)$; for example, $\left[\cos\left(\dfrac{\pi}{3}\right)\right]^2 = \left(\dfrac{1}{2}\right)^2 = \dfrac{1}{4} = 0.25$ whereas $\cos\left[\left(\dfrac{\pi}{3}\right)^2\right] = 0.4566$ (approx.)

43. $\sin 0.1 = 0.0998$, $g(0.1) = 0.0998$; $\sin 0.2 = 0.1987$, $g(0.2) = 0.1987$; $\sin(-0.3) = -0.2955$, $g(-0.3) = -0.2955$; $\sin(-1) = -0.8415$, $g(-1) = -0.8417$

Problem Set 5.3 (on page 331)

1. $\sec t = \dfrac{13}{12}$; $\csc t = \dfrac{13}{5}$; $\tan t = \dfrac{5}{12}$; $\cot t = \dfrac{12}{5}$

3. $\cot t = \dfrac{5}{12}$; $\sec t = \dfrac{-13}{5}$; $\sin t = \dfrac{-12}{13}$; $\csc t = \dfrac{-13}{12}$

5. $\csc t = -\sqrt{2}$; $\sec t = -\sqrt{2}$; $\tan t = 1$; $\cot t = 1$

7. $\sec t = 1$; $\tan t = 0$; $\csc t$ and $\cot t$ are undefined

9. (a) IV   **(b)** IV   **11. (a)** II   **(b)** IV   **13.** $\dfrac{13}{12}$ **15.** $\dfrac{-5}{12}$

17. (a) $\sin(-3.752) = 0.5732 = -\sin 3.752$ **(b)** $\cos(-5.384) = 0.6222 = \cos 5.384$

19. $\cos t = \dfrac{8}{17}$; $\tan t = \dfrac{15}{8}$; $\csc t = \dfrac{17}{15}$; $\sec t = \dfrac{17}{8}$; $\cot t = \dfrac{8}{15}$

21. $\sin t = \dfrac{3}{5}$; $\tan t = -\dfrac{3}{4}$; $\csc t = \dfrac{5}{3}$; $\sec t = -\dfrac{5}{4}$; $\cot t = -\dfrac{4}{3}$

23. $\sin t = \dfrac{8}{17}$; $\cos t = -\dfrac{15}{17}$; $\tan t = -\dfrac{8}{15}$; $\sec t = -\dfrac{17}{15}$; $\cot t = -\dfrac{15}{8}$

25. $\cos t = -\dfrac{3}{5}$; $\sin t = -\dfrac{4}{5}$; $\tan t = \dfrac{4}{3}$; $\cot t = \dfrac{3}{4}$; $\csc t = -\dfrac{5}{4}$

27. $\cos t = -0.82$; $\tan t = -0.69$; $\csc t = 1.75$; $\sec t = -1.22$; $\cot t = -1.44$

29. $\sin t = 0.84$; $\cos t = -0.55$; $\csc t = 1.19$; $\sec t = -1.83$; $\cot t = -0.65$

31. $\dfrac{\sqrt{3}}{2}$ **33.** $\dfrac{\sqrt{3}}{2}$ **35.** 1   **37.** $-\sqrt{2}$ **39.** $\dfrac{\sqrt{3}}{3}$

41. $2\cos t$ **43.** $3\sec t$ **45.** $2\cos t$

Problem Set 5.4 (on page 350)

1. (a) $[-3, 3]$ (b) $\left[\dfrac{11}{3}, \dfrac{13}{3}\right]$ (c) $\left[1, \dfrac{7}{3}\right]$

3. (a) Range: $[-3, 3]$;
   Amplitude = 3;
   Period = 2π

5. (a) Range: $\left[-\dfrac{2}{3}, \dfrac{2}{3}\right]$;
   Amplitude = $\dfrac{2}{3}$;
   Period = 2π

7. (a) Range: $[-1, 1]$;
   Amplitude = 1;
   Period = π

9. (a) Range: $[-1, 1]$;
   Amplitude = 1;
   Period = 6

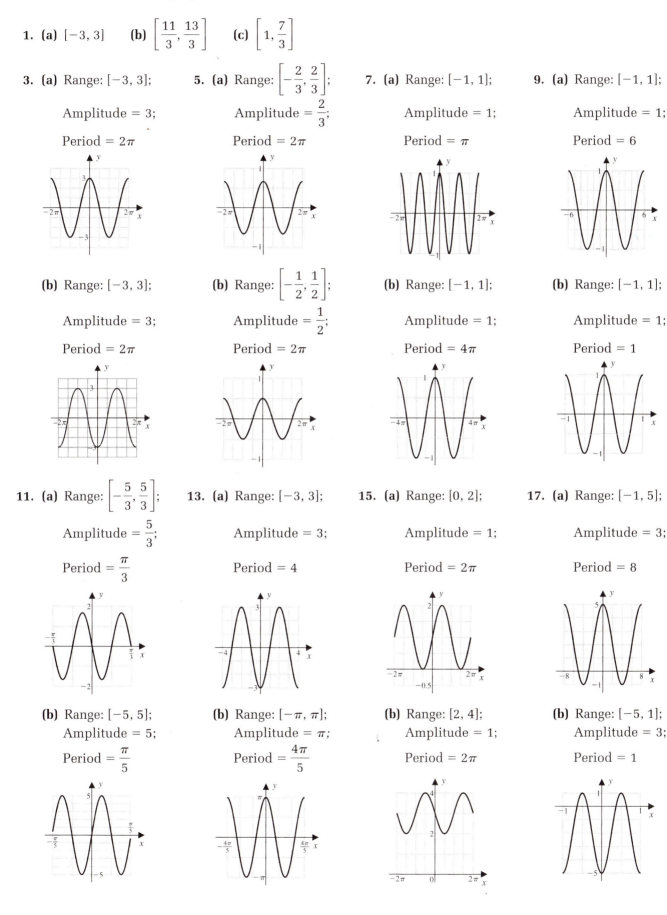

(b) Range: $[-3, 3]$;
   Amplitude = 3;
   Period = 2π

(b) Range: $\left[-\dfrac{1}{2}, \dfrac{1}{2}\right]$;
   Amplitude = $\dfrac{1}{2}$;
   Period = 2π

(b) Range: $[-1, 1]$;
   Amplitude = 1;
   Period = 4π

(b) Range: $[-1, 1]$;
   Amplitude = 1;
   Period = 1

11. (a) Range: $\left[-\dfrac{5}{3}, \dfrac{5}{3}\right]$;
   Amplitude = $\dfrac{5}{3}$;
   Period = $\dfrac{\pi}{3}$

13. (a) Range: $[-3, 3]$;
   Amplitude = 3;
   Period = 4

15. (a) Range: $[0, 2]$;
   Amplitude = 1;
   Period = 2π

17. (a) Range: $[-1, 5]$;
   Amplitude = 3;
   Period = 8

(b) Range: $[-5, 5]$;
   Amplitude = 5;
   Period = $\dfrac{\pi}{5}$

(b) Range: $[-\pi, \pi]$;
   Amplitude = π;
   Period = $\dfrac{4\pi}{5}$

(b) Range: $[2, 4]$;
   Amplitude = 1;
   Period = 2π

(b) Range: $[-5, 1]$;
   Amplitude = 3;
   Period = 1

19. C; B; A

21. Amplitude = 2;
Period = 2π;
Phase shift = $\dfrac{\pi}{4}$

23. Amplitude = 4;
Period = 2π;
Phase shift = $-\dfrac{\pi}{6}$

25. Amplitude = 2;
Period = π;
Phase shift = $\dfrac{\pi}{2}$

27. Amplitude = 2;
Period = 2;
Phase shift = 1

29. Amplitude = 3;
Period = $\dfrac{2\pi}{3}$;
Phase shift = $-\dfrac{2}{3}$

31. Amplitude = 2;
Period = 4;
Phase shift = $\dfrac{1}{2\pi}$

33. Amplitude = 2;
Period = $\dfrac{2\pi}{3}$;
Phase shift = $\dfrac{5}{3}$

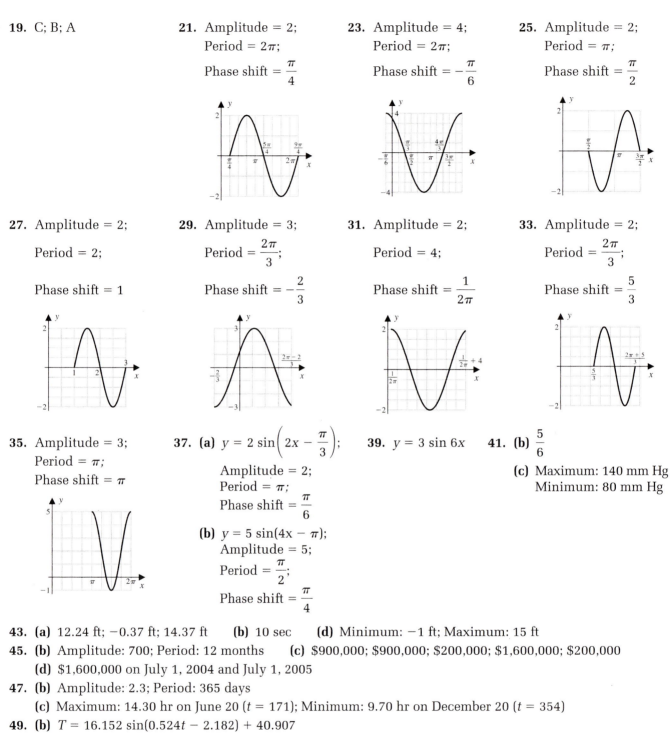

35. Amplitude = 3;
Period = π;
Phase shift = π

37. **(a)** $y = 2 \sin\left(2x - \dfrac{\pi}{3}\right)$;
Amplitude = 2;
Period = π;
Phase shift = $\dfrac{\pi}{6}$
   (b) $y = 5 \sin(4x - \pi)$;
Amplitude = 5;
Period = $\dfrac{\pi}{2}$;
Phase shift = $\dfrac{\pi}{4}$

39. $y = 3 \sin 6x$

41. **(b)** $\dfrac{5}{6}$
   (c) Maximum: 140 mm Hg
Minimum: 80 mm Hg

43. **(a)** 12.24 ft; −0.37 ft; 14.37 ft   **(b)** 10 sec   **(d)** Minimum: −1 ft; Maximum: 15 ft

45. **(b)** Amplitude: 700; Period: 12 months   **(c)** $900,000; $900,000; $200,000; $1,600,000; $200,000
   (d) $1,600,000 on July 1, 2004 and July 1, 2005

47. **(b)** Amplitude: 2.3; Period: 365 days
   (c) Maximum: 14.30 hr on June 20 ($t = 171$); Minimum: 9.70 hr on December 20 ($t = 354$)

49. **(b)** $T = 16.152 \sin(0.524t - 2.182) + 40.907$
   (d) For March ($t = 3$), $T = 31.7$; for June ($t = 6$), $T = 54.2$; for September ($t = 9$), $T = 50.1$

51. **(a)** Amplitude = 1; Period = $\dfrac{1}{100}$ sec; Frequency = 100

   (b) Amplitude = 2; Period = 2000 sec; Frequency = $\dfrac{1}{2000}$

53. **(a)** $y = -50 \cos\sqrt{\dfrac{8}{30}}\,t$ **(c)** Amplitude = 50; Period ≈ 12.17; Frequency ≈ 0.08

55. **(a)** $\dfrac{\pi}{2.2}$ **(b)** $y = 0.025 \sin\left(\dfrac{\pi}{2.2}t\right)$ **57.** They are the same due to the phase shift on $f(x)$.

59. They are the same.

61. (a) No
   (b) $f(x) \to 1$ as $x \to 0$
   (c) $f(x) \to 1$ as $x \to 0^-$, $f(x) \to 1$ as $x \to 0^+$

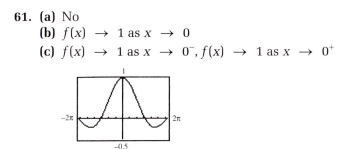

63. Period $= 2\pi$

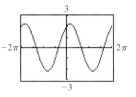

Problem Set 5.5 (on page 365)

1. Period $= \pi$;
   Phase shift $= 0$

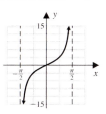

3. Period $= \pi$;
   Phase shift $= 0$

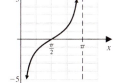

5. Period $= 2\pi$;
   Phase shift $= 0$

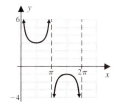

7. Period $= 2\pi$;
   Phase shift $= 0$

9. Period $= \dfrac{\pi}{4}$;
   Phase shift $= 0$

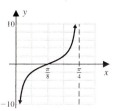

11. Period $= 3\pi$;
   Phase shift $= 0$

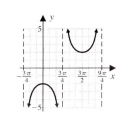

13. Period $= \dfrac{\pi}{2}$;
   Phase shift $= 0$

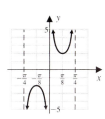

15. Period $= 2\pi$;
   Phase shift $= \dfrac{\pi}{6}$

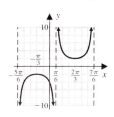

17. Period $= \pi$;
   Phase shift $= \dfrac{\pi}{4}$

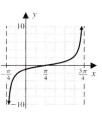

19. Period $= \pi$;
   Phase shift $= \dfrac{\pi}{2}$

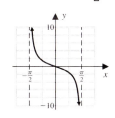

21. Period $= 8\pi$;
   Phase shift $= -\pi$

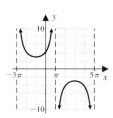

23. Period $= 4$;
   Phase shift $= 1$

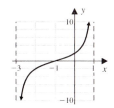

25.

	As x increases from:			
	0 to $\pi/2$	$\pi/2$ to π	π to $(3\pi)/2$	$(3\pi)/2$ to 2π
(a)	Increasing	Increasing	Increasing	Increasing
(b)	Decreasing	Decreasing	Decreasing	Decreasing
(c)	Increasing	Increasing	Decreasing	Decreasing
(d)	Decreasing	Increasing	Increasing	Decreasing

27. (a) 1   **(b)** $-\infty$ **29. (a)** ∞ **(b)** 1   **31.** $y = \tan\left(\dfrac{x}{2}\right)$ **33.** $y = 2\csc(2x + \pi)$

35. π **37.** $\dfrac{\pi}{3}$ **39. (b)** $(0.79, 8.00)$

41.

43. $f(x) \neq g(x)$; $f(x) = -g(x)$ because the cosecant is an odd function.
45. The graphs are the same.
47. (a) $y = \tan(6x)$

  (b) $y = \csc\left(\dfrac{12x}{5}\right)$

51. $f(x) \neq g(x)$ since $f(x)$ is not defined when $\cos x = 0$.

Problem Set 5.6 (on page 372)

1. $\sin\theta = \dfrac{3}{5}$; $\csc\theta = \dfrac{5}{3}$; $\cos\theta = \dfrac{4}{5}$; $\sec\theta = \dfrac{5}{4}$; $\tan\theta = \dfrac{3}{4}$; $\cot\theta = \dfrac{4}{3}$

3. $\sin\theta = \dfrac{12}{13}$; $\csc\theta = \dfrac{13}{12}$; $\cos\theta = -\dfrac{5}{13}$; $\sec\theta = -\dfrac{13}{5}$; $\tan\theta = -\dfrac{12}{5}$; $\cot\theta = -\dfrac{5}{12}$

5. $\sin\theta \approx -0.898$; $\csc\theta \approx -1.113$; $\cos\theta \approx 0.439$; $\sec\theta \approx 2.276$; $\tan\theta \approx -2.044$; $\cot\theta \approx -0.489$

7. $\sin\theta = -\dfrac{1}{2}$; $\csc\theta = -2$; $\cos\theta = -\dfrac{\sqrt{3}}{2}$; $\sec\theta = -\dfrac{2\sqrt{3}}{2}$; $\tan\theta = \dfrac{\sqrt{3}}{3}$; $\cot\theta = \sqrt{3}$

9. $\sin\theta = \cos\theta = \dfrac{\sqrt{2}}{2}$; $\tan\theta = \cot\theta = 1$; $\csc\theta = \sec\theta = \sqrt{2}$

11. $\sin 90° = 1$; $\csc 90° = 1$; $\cos 90° = 0$; $\sec 90°$ is undefined; $\tan 90°$ is undefined; $\cot 90° = 0$

13. $\sin\left(-\dfrac{\pi}{2}\right) = -1$; $\csc\left(-\dfrac{\pi}{2}\right) = -1$; $\cos\left(-\dfrac{\pi}{2}\right) = 0$; $\sec\left(-\dfrac{\pi}{2}\right)$ is undefined; $\tan\left(-\dfrac{\pi}{2}\right)$ is undefined;

$\cot\left(-\dfrac{\pi}{2}\right) = 0$ **15. (a)** $\dfrac{1}{2}$ **(b)** $-\dfrac{\sqrt{2}}{2}$ **17. (a)** $\sqrt{3}$ **(b)** $-\sqrt{3}$ **19. (a)** $\sqrt{2}$ **(b)** $\dfrac{-2\sqrt{3}}{3}$

21. (a) -1 **(b)** $\sqrt{3}$ **23.** $\dfrac{-\sqrt{2}}{2}$ **25.** $-\sqrt{3}$ **27.** -1 **29** $\dfrac{\sqrt{3}}{2}$ **31.** -1

33. (a) $\sin\theta \approx 0.6820$ **(b)** $\sin\theta \approx -0.9949$ **35. (a)** $\sin\theta \approx 0.3420$ **(b)** $\sin\theta \approx -0.0548$
    $\csc\theta \approx 1.4663$ $\csc\theta \approx -1.0051$ $\csc\theta \approx 2.9238$ $\csc\theta \approx -18.2470$
    $\cos\theta \approx -0.7314$ $\cos\theta \approx -0.1011$ $\cos\theta \approx 0.9397$ $\cos\theta \approx 0.9985$
    $\sec\theta \approx -1.3673$ $\sec\theta \approx -9.8955$ $\sec\theta \approx 1.0642$ $\sec\theta \approx 1.0015$
    $\tan\theta \approx -0.9325$ $\tan\theta \approx 9.8448$ $\tan\theta \approx 0.3640$ $\tan\theta \approx -0.0549$
    $\cot\theta \approx -1.0724$ $\cot\theta \approx 0.1016$ $\cot\theta \approx 2.7475$ $\cot\theta \approx -18.2195$

37. (a) $\sin\theta \approx 0.9093$ **(b)** $\sin\theta \approx -0.9636$ **39. (a)** $\sin\theta \approx 0.8739$ **(b)** $\sin\theta = 0$
    $\csc\theta \approx 1.0998$ $\csc\theta \approx -1.0378$ $\csc\theta \approx 1.1443$ $\csc\theta$ is undefined
    $\cos\theta \approx -0.4161$ $\cos\theta \approx 0.2675$ $\cos\theta \approx 0.4861$ $\cos\theta = -1$
    $\sec\theta \approx -2.4030$ $\sec\theta \approx 3.7383$ $\sec\theta \approx 2.0572$ $\sec\theta = -1$
    $\tan\theta \approx -2.1850$ $\tan\theta \approx -3.6021$ $\tan\theta \approx 1.7978$ $\tan\theta = 0$
    $\cot\theta \approx -0.4577$ $\cot\theta \approx -0.2776$ $\cot\theta \approx 0.5562$ $\cot\theta$ is undefined

41. (a)   **(b)** $(184.43, -14.51)$ **(c)** -0.0787

Problem Set 5.7 (on page 382)

1. $\sin \theta = 0.625$; $\cos \theta = 0.781$; $\tan \theta = 0.801$; $\csc \theta = 1.600$; $\sec \theta = 1.281$; $\cot \theta = 1.249$

3. $\sin \theta = 0.324$; $\cos \theta = 0.946$; $\tan \theta = 0.342$; $\csc \theta = 3.086$; $\sec \theta = 1.057$; $\cot \theta = 2.920$

5. $\sin \theta = 0.882$; $\cos \theta = 0.471$; $\tan \theta = 1.875$; $\csc \theta = 1.133$; $\sec \theta = 2.125$; $\cot \theta = 0.533$

7. $\cos \theta = \dfrac{3}{5}$; $\tan \theta = \dfrac{4}{3}$; $\csc \theta = \dfrac{5}{4}$; $\sec \theta = \dfrac{5}{3}$; $\cot \theta = \dfrac{3}{4}$ **9. (a)** $\dfrac{1}{2}$ **(b)** $\dfrac{-\sqrt{2}}{2}$

11. (a) $\sqrt{3}$ **(b)** $-\sqrt{3}$ **13. (a)** $\sqrt{2}$ **(b)** $\dfrac{-2\sqrt{3}}{3}$ **15. (a)** -1 **(b)** $\sqrt{3}$

17. (a) 0.6820   **(b)** 1.1371   **(c)** 0.2225   **(d)** 2.6131

19. (a) 0.7265   **(b)** 0.3127   **(c)** 1.2392   **(d)** 0.4377

21. (a) 33.37°   **(b)** 11.48°   **(c)** 88.10°   **(d)** 45°

23. $\alpha = 40°$, $a = 6.43$, $b = 7.66$ **25.** $\alpha = 47°$, $a = 4.29$, $c = 5.87$ **27.** $\beta = 58.77°$, $a = 909.52$, $c = 1754.20$

29. $\alpha = 72.74°$, $\beta = 17.26°$, $c = 13.82$ **31.** $\alpha = 67.5°$, $\beta = 22.5°$, $b = 0.41$, $c = 1.08$ **33.** 10.68 in.

35. 10.90 ft   **37.** 46.06 ft   **39.** 80.83 mph   **41. (a)** 23.58°   **(b)** 2 ft./sec.

43. $d = 12.5$ in., $\theta = 36.87°$ **45.** 154.51 m   **47.** 84.57°   **49.** 146.71 m

51. Building: 30.57 ft; antenna: 13.95 ft   **53.** 499.00 ft   **55.** For example, let $\alpha = 30°$ and $\beta = 60°$.

57. $T = r \cot\left(\dfrac{\theta}{2}\right)$ **59. (a)–(b)** Side opposite has length 23 sin 35° or approx. 13.19 in.

Problem Set 5.8 (on page 397)

1. (a) $\dfrac{\pi}{4}$ **(b)** $-\dfrac{\pi}{6}$ **3. (a)** $\dfrac{\pi}{6}$ **(b)** $-\dfrac{\pi}{6}$ **5. (a)** $-\dfrac{\pi}{3}$ **(b)** $\dfrac{5\pi}{6}$

7. (a) $\dfrac{\pi}{6}$ **(b)** $\dfrac{\pi}{4}$ **9. (a)** $-\dfrac{\pi}{4}$ **(b)** -1 **11. (a)** 0.22   **(b)** -0.89

13. (a) 0.82   **(b)** -1.55 **15. (a)** 2.36   **(b)** 1.56   **17.** $x = \sin y - 1$; $-\dfrac{\pi}{2} \le y \le \dfrac{\pi}{2}$, $-2 \le x \le 0$

19. $x = \dfrac{1}{2}\cos y + 4$; $0 \le y \le \pi$, $\dfrac{7}{2} \le x \le \dfrac{9}{2}$ **21.** $x = \dfrac{1}{3}\left[2 - \tan\left(\dfrac{y}{3}\right)\right]$; $-\dfrac{3\pi}{2} < y < \dfrac{3\pi}{2}$, x is real

23. $x = \dfrac{1}{2}\cos 3y + 2$; $0 \le y \le \dfrac{\pi}{3}$, $\dfrac{3}{2} \le x \le \dfrac{5}{2}$

25. (a) 0.3   **(b)** 1.3   **(c)** -0.5236 **(d)** 2.6180   **27.** $-1 \le x \le -\dfrac{1}{3}$

29. $3 - \dfrac{3\pi}{2} < x < 3 + \dfrac{3\pi}{2}$ **31. (a)** $\dfrac{3}{5}$ **(b)** $\dfrac{12}{13}$ **33. (a)** 2   **(b)** $\dfrac{-2\sqrt{5}}{5}$

35. (a) $\dfrac{4\sqrt{15}}{15}$ **(b)** $\dfrac{\pi}{3}$ **37. (a)** $\sqrt{1 - x^2}$ **(b)** $\dfrac{\sqrt{1 - x^2}}{x}$ **(c)** $\dfrac{1}{x}$

39. (a) $\dfrac{\pi}{4}$ **(b)** $-\dfrac{\pi}{4}$ **41. (a)** $\dfrac{\pi}{4}$ **(b)** $-\dfrac{\pi}{3}$ **43. (a)** 0.80   **(b)** 1.27

45. (a) 2.00   **(b)** -0.11 **47. (a)** $\theta = \tan^{-1}\left(\dfrac{32}{x}\right) - \tan^{-1}\left(\dfrac{24}{x}\right)$ **(b)** 0.07 rad. (or about 4.25°)

49. (a) $\theta = \tan^{-1}\left(\dfrac{9}{x}\right) - \tan^{-1}\left(\dfrac{5}{x}\right)$ **(b)** 0.18 rad. (or about 10.19°)

51.   **53.**   **55.**

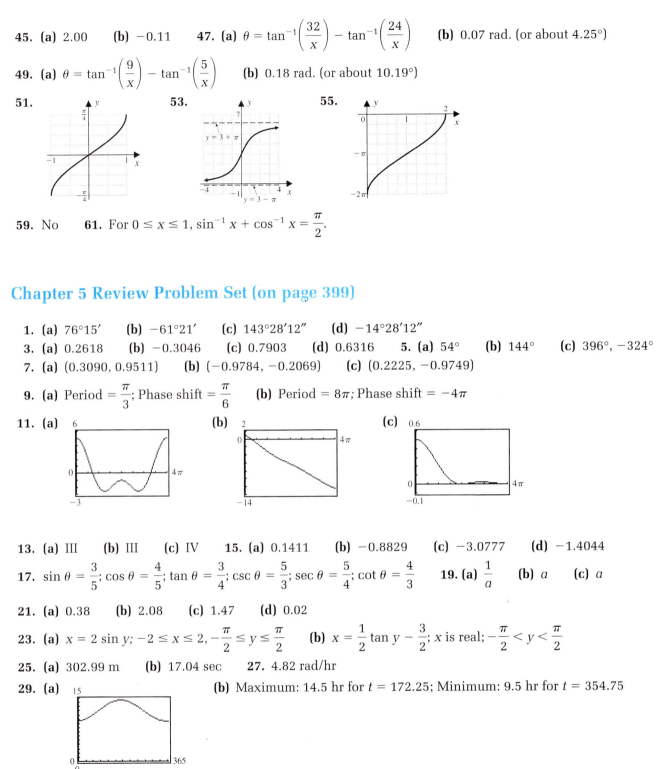

59. No   **61.** For $0 \le x \le 1$, $\sin^{-1} x + \cos^{-1} x = \dfrac{\pi}{2}$.

Chapter 5 Review Problem Set (on page 399)

1. (a) 76°15′   **(b)** $-61°21′$ **(c)** 143°28′12″   **(d)** $-14°28′12″$

3. (a) 0.2618   **(b)** -0.3046 **(c)** 0.7903   **(d)** 0.6316   **5. (a)** 54°   **(b)** 144°   **(c)** 396°, $-324°$

7. (a) (0.3090, 0.9511)   **(b)** $(-0.9784, -0.2069)$ **(c)** $(0.2225, -0.9749)$

9. (a) Period $= \dfrac{\pi}{3}$; Phase shift $= \dfrac{\pi}{6}$ **(b)** Period $= 8\pi$; Phase shift $= -4\pi$

11. (a)   **(b)**   **(c)**

13. (a) III   **(b)** III   **(c)** IV   **15. (a)** 0.1411   **(b)** -0.8829 **(c)** -3.0777 **(d)** -1.4044

17. $\sin\theta = \dfrac{3}{5}$; $\cos\theta = \dfrac{4}{5}$; $\tan\theta = \dfrac{3}{4}$; $\csc\theta = \dfrac{5}{3}$; $\sec\theta = \dfrac{5}{4}$; $\cot\theta = \dfrac{4}{3}$ **19. (a)** $\dfrac{1}{a}$ **(b)** a **(c)** a

21. (a) 0.38   **(b)** 2.08   **(c)** 1.47   **(d)** 0.02

23. (a) $x = 2\sin y$; $-2 \le x \le 2$, $-\dfrac{\pi}{2} \le y \le \dfrac{\pi}{2}$ **(b)** $x = \dfrac{1}{2}\tan y - \dfrac{3}{2}$; x is real; $-\dfrac{\pi}{2} < y < \dfrac{\pi}{2}$

25. (a) 302.99 m   **(b)** 17.04 sec   **27.** 4.82 rad/hr

29. (a)   **(b)** Maximum: 14.5 hr for $t = 172.25$; Minimum: 9.5 hr for $t = 354.75$

31. 73.38 ft

Chapter 5 Test (on page 402)

1. (a) $\dfrac{26\pi}{9}$ or 9.08 (approx.)   **(b)** $\dfrac{540}{7}$ or 77.14° (approx.)   **2. (a)** 1.5; 75 cm^2 **(b)** $\dfrac{270}{\pi}$ or 85.94° (approx.)

3. (a)   **(b)** $\sin\theta = -\dfrac{7\sqrt{58}}{58}$ $\csc\theta = -\dfrac{\sqrt{58}}{7}$

$\cos\theta = \dfrac{3\sqrt{58}}{58}$ $\sec\theta = \dfrac{\sqrt{58}}{3}$

$\tan\theta = -\dfrac{7}{3}$ $\cot\theta = -\dfrac{3}{7}$

4. (a) $\dfrac{\sqrt{5}}{5}$ **(b)** $\sin t = \dfrac{2\sqrt{5}}{5}$; $\cos t = \dfrac{\sqrt{5}}{5}$; $\tan t = 2$

5. (a) $-\dfrac{1}{2}$ **(b)** $-\dfrac{\sqrt{2}}{2}$ **(c)** $-\sqrt{3}$ **(d)** $\dfrac{3\pi}{4}$ **(e)** $\dfrac{4}{5}$ **6. (a)** 0.39   **(b)** 3.48   **(c)** 2.57

7. (a) III   **(b)** IV   **8.** $\sin\theta = \dfrac{2\sqrt{66}}{17}$; $\tan\theta = -\dfrac{2\sqrt{66}}{5}$; $\csc\theta = \dfrac{17\sqrt{66}}{132}$; $\sec\theta = -\dfrac{17}{5}$; $\cot\theta = -\dfrac{5\sqrt{66}}{132}$

9. (a) Amplitude = 3;   **(b)** Amplitude = 2;   **(c)** Amplitude = 1;

Period = $\dfrac{\pi}{2}$;   Period = 4;   Period = 2π;

Phase shift = $\dfrac{\pi}{4}$ Phase shift = $\dfrac{2}{3}$ Phase Shift = $-\pi$

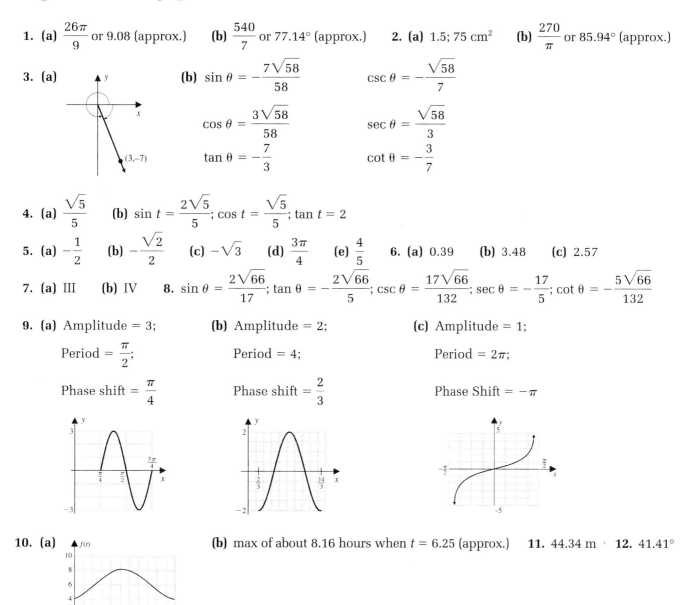

10. (a)   **(b)** max of about 8.16 hours when $t = 6.25$ (approx.)   **11.** 44.34 m   **12.** 41.41°

Chapter 6: Problem Set 6.1 (on page 408)

1. $\cos^2 3\theta$ **3.** 1   **5.** -2 **7.** 4   **9. (a)** $\sin t$ **(b)** $8 - 5\sin^2 t$ **(c)** $\sin^2 t$

11. (a) $2\cos^4\theta$ **(b)** $-\csc\theta$ **13. (a)** $\tan t$ **(b)** $\cos\theta$ **15. (a)** $\csc t$ **(b)** $\sec\theta$

49. False for $t = \dfrac{\pi}{2}$ **51.** False for $t = \dfrac{\pi}{4}$ **53.** QII and III   **55.** Q_I and IV

61. (b) Graphs are the same.   **(c)** Yes.

Problem Set 6.2 (on page 419)

1. (a) $\dfrac{\sqrt{6} - \sqrt{2}}{4} = 0.2588$ **(b)** $\dfrac{\sqrt{6} + \sqrt{2}}{4} = 0.9659$

3. (a) $\dfrac{\sqrt{6} + \sqrt{2}}{4} = 0.9659$ **(b)** $\dfrac{-\sqrt{6} - \sqrt{2}}{4} = -0.9659$ **(c)** $2 + \sqrt{3} = 3.7321$

5. (a) $\dfrac{\sqrt{2} - \sqrt{6}}{4} = -0.2588$ **(b)** $\dfrac{-\sqrt{6} - \sqrt{2}}{4} = -0.9659$ **7. (a)** $\dfrac{\sqrt{3}}{2} = 0.8660$ **(b)** $\tan \dfrac{\pi}{2}$; undefined

9. (a) $-\dfrac{63}{65}$ **(b)** $-\dfrac{16}{65}$ **(c)** $-\dfrac{33}{65}$ **(d)** $\dfrac{56}{65}$ **(e)** $\dfrac{63}{16}$ **(f)** $-\dfrac{33}{56}$

11. (a) -0.204 **(b)** -0.979 **(c)** 0.698 **(d)** 0.716 **(e)** 0.208 **(f)** 0.976 **13.** 1 **15.** 0

17. (a) $\cos t$ **(b)** $\sin t$ **19. (a)** $\dfrac{\tan t + 1}{1 - \tan t}$ **(b)** $\dfrac{\sqrt{2}}{2}(\cos \theta - \sin \theta)$

21. (a) $\dfrac{1}{\cos \theta}$ **(b)** $-\cos t$ **23. (a)** $\cos 8t$ **(b)** $\sin 4t$ **25. (a)** $\cos x$ **(b)** $\tan 7t$

47. $y = \sqrt{2} \cos\left(t - \dfrac{\pi}{4}\right)$ **49.** $y = 2 \cos\left(2\pi t - \dfrac{5\pi}{3}\right)$

51. (a) $h(t) = 0.5 \cos\left(\dfrac{\pi t}{6} - 0.927\right)$ **(b)** Maximum height is 0.5 m when t is about 1.8.

  (c) Minimum height is -0.5 m when t is about 7.8.   **53. (a)** $d(t) = 5 \cos (4t - 0.644)$ **(b)** 5 **(c)** $\dfrac{\pi}{2}$

55. (a) $\sin \gamma$ **(b)** $-\cos \gamma$ **57. (c)** Same; shift the graph of $y = \cos x$ to the right $\dfrac{\pi}{2}$ units.

59. (b) 0 and 1

Problem Set 6.3 (on page 429)

1. $\cos 2t \neq 2 \cos t$ **3. (a)** $\dfrac{24}{25}$ **(b)** $-\dfrac{7}{25}$ **(c)** $-\dfrac{24}{7}$ **5. (a)** $\dfrac{336}{625}$ **(b)** $-\dfrac{527}{625}$ **(c)** $-\dfrac{336}{527}$

7. (a) $-\dfrac{120}{169}$ **(b)** $\dfrac{119}{169}$ **(c)** $-\dfrac{120}{119}$ **9. (a)** $-\dfrac{175}{337}$ **(b)** $-\dfrac{288}{337}$ **(c)** $\dfrac{175}{288}$ **11.** $\sin 6t$ **13.** $\cos 16t$

15. $\cos 10t$ **17.** $\tan 8\theta$ **19. (a)** $\dfrac{\sqrt{3}}{2} = 0.8660$ **(b)** $\dfrac{\sqrt{3}}{2} = 0.8660$ **(c)** $-\dfrac{24}{7} = -3.4286$

21. (a) $\dfrac{\sqrt{2 + \sqrt{2}}}{2} = 0.9239$ **(b)** $\dfrac{\sqrt{2 + \sqrt{3}}}{2} = 0.9659$

23. (a) $\dfrac{\sqrt{2 - \sqrt{3}}}{2} = 0.2588$ **(b)** $\sqrt{3 + 2\sqrt{2}} = 2.4142$

25. (a) $-\dfrac{\sqrt{2 - \sqrt{2}}}{2} = -0.3827$ **(b)** $\dfrac{\sqrt{2 + \sqrt{2}}}{2} = 0.9239$

27. (a) $\dfrac{t}{2}$ in quadrant I: $\dfrac{3}{5}$ **(b)** $\dfrac{t}{2}$ in quadrant I: $\dfrac{4}{5}$ **(c)** $\dfrac{t}{2}$ in quadrant I: $\dfrac{3}{4}$

29. (a) $\dfrac{t}{2}$ in quadrant I: $\dfrac{3}{\sqrt{10}}$ **(b)** $\dfrac{t}{2}$ in quadrant I: $\dfrac{1}{\sqrt{10}}$ **(c)** $\dfrac{t}{2}$ in quadrant I: 3

31. (a) $\dfrac{t}{2}$ in quadrant II: $\dfrac{5}{\sqrt{34}}$ **(b)** $\dfrac{t}{2}$ in quadrant II: $-\dfrac{3}{\sqrt{34}}$ **(c)** $\dfrac{t}{2}$ in quadrant II: $-\dfrac{5}{3}$

33. (a) $\dfrac{t}{2}$ in quadrant II: $\dfrac{1}{\sqrt{17}}$ **(b)** $\dfrac{t}{2}$ in quadrant II: $-\dfrac{4}{\sqrt{17}}$ **(c)** $\dfrac{t}{2}$ in quadrant II: $-\dfrac{1}{4}$

35. (a) $\cos 2t$ **(b)** $\sin 3t$ **49. (b)** 4.70 cm^2 **(c)** $0 < \theta < 180°$ **(d)** Maximum area is 8 cm^2 when $\theta = 90°$

51. (b) 17.09 in² **(c)** $0° < \theta < 90°$

53. (a) $W = 980 - 980\cos(240\pi t)$ **(b)** V, Period $= \dfrac{1}{60}$; I, Period $= \dfrac{1}{60}$; W, Period $= \dfrac{1}{120}$ **(c)** 1960

57. (a) $\left(-\dfrac{7}{25}, -\dfrac{24}{25}\right)$ **(b)** $\left(-\dfrac{2}{\sqrt{5}}, \dfrac{1}{\sqrt{5}}\right)$ if t is positive **59. (a)** $\dfrac{\sec^2 t}{2 - \sec^2 t}$ **(b)** $-\dfrac{5}{4}$

Problem Set 6.4 (on page 441)

3. (a) No solution **(b)** No solution **(c)** $225°, 315°$
(d) $225° + (360°)k$ or $315° + (360°)k$, where k is an integer
5. (a) $\sin 190° = -\sin 10° = -0.1736$ **(b)** $\tan(-113°) = \tan 67° = 2.3559$
(c) $\sec 372° = \sec 12° = 1.0223$ **(d)** $\cos(-1.9) = -0.3233$; $-\cos 1.2416 = -0.3233$
(e) $\cot 7.2 = 0.7665$; $\cot 0.9168 = 0.7665$

7. (a) 2.0944 or $\dfrac{2\pi}{3}$; 4.1888 or $\dfrac{4\pi}{3}$ **(b)** 2.3562 or $\dfrac{3\pi}{4}$; 3.9270 or $\dfrac{5\pi}{4}$

9. (a) $60°$ or $240°$ **(b)** $120°$ or $300°$

11. (a) 1.0472 or $\dfrac{\pi}{3}$; 2.0944 or $\dfrac{2\pi}{3}$ **(b)** $225°$ or $315°$

13. (a) 2.3562 or $\dfrac{3\pi}{4}$; 3.9270 or $\dfrac{5\pi}{4}$ **(b)** 3.1416 or π; 4.7124 or $\dfrac{3\pi}{2}$

15. (a) 0.9500 or 2.1916 **(b)** 2.9916 or 6.1332 **17. (a)** 2.5559 or 3.7273 **(b)** 1.4711 or 4.6127

19. 0, 3.1416 or π **21.** $90°, 240°, 270°, 300°$ **23.** 0, 1.0472 or $\dfrac{\pi}{3}$, 3.1416 or π, 4.1888 or $\dfrac{4\pi}{3}$

25. $180°$ **27.** 1.0472 or $\dfrac{\pi}{3}$, 2.0944 or $\dfrac{2\pi}{3}$, 4.1888 or $\dfrac{4\pi}{3}$; 5.2360 or $\dfrac{5\pi}{3}$

29. $60°, 120°, 240°, 300°$ **31.** 1.0472 or $\dfrac{\pi}{3}$, 1.5708 or $\dfrac{\pi}{2}$, 4.7124 or $\dfrac{3\pi}{2}$, 5.2360 or $\dfrac{5\pi}{3}$

33. (a) $45°, 75°, 165°, 195°, 285°, 315°$ **(b)** $55°, 85°, 175°, 205°, 295°, 325°$

35. (a) 1.5708 or $\dfrac{\pi}{2}$ **(b)** 0.3142 or $\dfrac{\pi}{10}$

37. 0.7854 or $\dfrac{\pi}{4}$, 2.3562 or $\dfrac{3\pi}{4}$, 3.9270 or $\dfrac{5\pi}{4}$, 5.4978 or $\dfrac{7\pi}{4}$ **39.** $45°, 90°, 225°, 270°$

41. 0, 0.5234 or $\dfrac{\pi}{6}$, 2.6180 or $\dfrac{5\pi}{6}$, 3.1416 or π **43.** $0°, 270°$

45. 0.7854 or $\dfrac{\pi}{4}$, 1.5708 or $\dfrac{\pi}{2}$, 2.3562 or $\dfrac{3\pi}{4}$, 4.7124 or $\dfrac{3\pi}{2}$ **47.** 0, 1.0472 or $\dfrac{\pi}{3}$, 5.2360 or $\dfrac{5\pi}{3}$
49. $210°, 270°, 330°$ **51.** 0.6749 or 5.6083 **53. (a)** $32.67°$ **(b)** $62.04°$ **55. (b)** $0° < \theta < 90°$
59. (a) $k < -1$ or $k > 1$ **(b)** $k = -1$ or $k = 1$ **(c)** $-1 < k < 1$ **(d)** No values of k
61. 2.14 radians **63. (a)** 0.201, 0.340, 2.802, 2.940 **(b)** (0, 0.201) or (0.340, 2.802) or (2.940, 6.283)
65. (a) 0.160 **(b)** [0, 0.160] **67. (a)** 0, 1.293, 4.721; $t = 0$ **(b)** [1.293, 4.721]; $t = 0$

Problem Set 6.5 (on page 449)

1. (a) $-\dfrac{1}{4}$ **(b)** -0.2500 **3. (a)** $\dfrac{(-2 - \sqrt{3})}{4}$ **(b)** -0.9330 **5. (a)** $\dfrac{(2 - \sqrt{3})}{4}$ **(b)** 0.0670

7. (a) $\dfrac{\sqrt{2}}{4}$ **(b)** 0.3536 **9.** $\dfrac{1}{2}(\cos 3w - \cos 5w)$ **11.** $\dfrac{1}{2}(\cos 4v + \cos 12v)$ **13.** $\dfrac{1}{2}(\sin 7t + \sin t)$

15. $\frac{1}{2}(\cos t - \cos 9t)$ **17. (a)** $\sqrt{3}\sin\frac{2\pi}{9}$ **(b)** 1.1133   **19. (a)** $\frac{\sqrt{2}}{2}$ **(b)** 0.7071

21. (a) $\frac{\sqrt{2}}{2}$ **(b)** 0.7071   **23. (a)** $\frac{-\sqrt{2}}{2}$ **(b)** -0.7071 **25.** $2\sin 2\theta \cos\theta$ **27.** $-2\cos 6t \sin t$

29. $2\cos\frac{5t}{2}\cos\frac{t}{2}$ **31.** $2\sin\frac{11x}{2}\sin\frac{x}{2}$ **35.** $0, \frac{\pi}{7}, \frac{3\pi}{7}, \frac{2\pi}{3}, \frac{4\pi}{3}, 2\pi$ **37.** $0, \frac{\pi}{2}, \pi, \frac{3\pi}{2}, 2\pi, 3\pi$

39. (a) $P = 0.5\cos(128\pi t) - 0.5\cos(144\pi t)$; frequency of $P = 68$ hertz.   **(b)** $\sin(136\pi t)\sin(8\pi t)$

Chapter 6 Review Problem Set (on page 450)

5. (a) $\sin 5t$ **(b)** $\sin t$ **7. (a)** $\sin t$ **(b)** $-\cos 8\theta$ **9. (a)** $\cos t$ **(b)** $\sqrt{3}\sin t$

11. (a) $\frac{\sqrt{6}-\sqrt{2}}{4}$ **(b)** $\frac{\sqrt{2}+\sqrt{3}}{2}$ **13. (a)** $\frac{\sqrt{6}-\sqrt{2}}{4}$ **(b)** $\frac{-\sqrt{6}-\sqrt{2}}{4}$

15. (a) $\frac{\sqrt{2}-\sqrt{2}}{2}$ **(b)** $\frac{-\sqrt{2}+\sqrt{3}}{2}$

17. (a) $\cos\theta = -\frac{4}{5}$; $\tan\theta = -\frac{3}{4}$; $\csc\theta = \frac{5}{3}$; $\sec\theta = -\frac{5}{4}$; $\cot\theta = -\frac{4}{3}$.

  (b) $\sin\theta = -\frac{\sqrt{55}}{8}$; $\tan\theta = -\frac{\sqrt{55}}{3}$; $\csc\theta = -\frac{8}{\sqrt{55}}$; $\sec\theta = \frac{8}{3}$; $\cot\theta = -\frac{3}{\sqrt{55}}$.

19. (a) $\frac{-5-6\sqrt{110}}{68}$ **(b)** $\frac{2\sqrt{66}-5\sqrt{15}}{68}$ **(c)** $\frac{\sqrt{15}}{8}$ **(d)** $\frac{\sqrt{102}}{17}$ **(e)** $\frac{\sqrt{15}}{7}$

29. (a) $270°$ **(b)** $71.23°$, $288.77°$

31. (a) $\frac{\pi}{2}$, 3.34, 6.08   **(b)** $\frac{\pi}{6}$, $\sin^{-1}\frac{1}{3}$, $\pi - \sin^{-1}\frac{1}{3}$, $\frac{5\pi}{6}$ or $t = 0.34, 0.52, 2.62, 2.80$

33. (a) $-\frac{1}{4}$ **(b)** $\frac{(-2-\sqrt{3})}{4}$ **35. (a)** $\sin 3t - \sin t$ **(b)** $\cos 2t + \cos 8t$

37. (a) $2\cos\frac{5t}{2}\sin\frac{3t}{2}$ **(b)** $2\cos 5t \cos 2t$ **39.** 0.32   **41.** $y = 2\cos\left(x - \frac{\pi}{3}\right)$

43. $y = \sqrt{2}\cos\left(x - \frac{7\pi}{4}\right)$ or $y = \cos\left(x + \frac{\pi}{4}\right)$ **45. (b)** 0.004 sec

47. (a) 25.71 in.   **(b)** $\tan\theta = \frac{7}{6}$; $\theta = 49°$

Chapter 6 Test (on page 452)

1. (a) $\frac{(-\sqrt{6}-\sqrt{2})}{4}$ **(b)** $\frac{(\sqrt{2}-\sqrt{6})}{4}$ **(c)** $2+\sqrt{3}$

2. (a) $-\frac{24}{25}$ **(b)** $\frac{7}{25}$ **(c)** $-\frac{24}{7}$ **(d)** $-\frac{24}{25}$ **(e)** $\frac{7}{25}$ **(f)** $-\frac{336}{625}$ **(g)** $\frac{1}{10}$ **(h)** $-\frac{117}{125}$

3. (a) $\frac{(\sqrt{2}+\sqrt{2})}{2}$ **(b)** $\frac{(\sqrt{2}-\sqrt{2})}{2}$ **(c)** $\sqrt{2}+1$

4. (a) $-\cos 94°$ **(b)** $\frac{1}{2}\sin\frac{10\pi}{7}$ **(c)** $\sin 60°$ **(d)** $\cos\pi$ **(e)** $\cos\frac{t}{2}$

5. (a) 0.52 or $\frac{\pi}{6}$, 2.62 or $\frac{5\pi}{6}$, 3.67 or $\frac{7\pi}{6}$, 5.76 or $\frac{11\pi}{6}$

(b) 15°, 75°, 195°, 255°   **(c)** $\dfrac{\pi}{4}, \dfrac{5\pi}{4}$, 2.16, 5.30; or 0.79, 2.16, 3.93, 5.30

7. −0.32   **8. (a)** $\dfrac{57,600 - x^2}{57,600 + x^2}$ **(b)** 50.96°

Chapter 7: Problem Set 7.1 (on page 464)

1. SAS; $a = 2.05$ **3.** SAS; $b = 5.59$ **5.** SAS; $c = 3.35$ **7.** SSS; $\gamma = 104.48°$ **9.** SSS; $\beta = 50.50°$
11. SSS; $\alpha = 14.15°$ **13.** AAS; $\gamma = 15°$, $b = 28.66$, $c = 10.68$ **15.** ASA; $\beta = 59°$, $a = 34.47$, $c = 22.96$
17. ASA; $\beta = 76.5°$, $a = 13.60$, $c = 17.78$ **19.** AAS; $\alpha = 58.67°$, $a = 4.89$, $b = 4.44$
21. ASA; $\beta = 78°$, $a = 12.49$, $c = 17.15$ **23.** No possible triangle   **25.** No possible triangle
27. Case 1: $\alpha = 58.34°$, $\gamma = 84.99°$, $c = 14.51$; Case 2: $\alpha = 121.66°$, $\gamma = 21.67°$, $c = 5.38$
29. $\beta = 53.65°$, $\gamma = 61.17°$, $c = 20.56$
31. Case 1: $\beta = 54.59°$, $\gamma = 92.08°$, $c = 54.56$; Case 2: $\beta = 125.41°$, $\gamma = 21.26°$, $c = 19.80$
33. SSA; no possible triangle   **35.** SSS; $\alpha = 41.41°$, $\beta = 41.41°$, $\gamma = 97.18°$
37. AAS; $\gamma = 71°$, $b = 66.83$, $c = 63.67$ **39.** AAS; $\gamma = 74.2°$, $a = 37.98$, $c = 37.15$ **41.** 143.96 ft
43. 343.68 m   **45.** 8.23 mi   **47.** 3.52 km   **49.** 1.08 km; 0.81 km   **51.** 1.88 mi   **53.** 1.88 m
55. 42.19 nautical mi   **57. (a)** $d = \sqrt{11943.64t^2 + 3107.75t + 232.44}$

(b) $\dfrac{1}{2}$ hr., 69.08 km; 1 hr, 123.63 km; 75 mins, 150.93 km

(c) $d = 50$ km, $t \approx 0.32$ hr; $d = 100$ km, $t \approx 0.78$ hr; $d = 150$ km, $t \approx 1.24$ hrs.
59. Angle *PQR*: 52.70°; angle *PRQ*: 50.51°; angle *QPR*: 76.79°   **61.** 20.68 in.; 14.91 in.
63. Because of the triangle inequality; $|\cos \theta| > 1$

Problem Set 7.2 (on page 475)

7. Some points are: **(a)** **(i)** (3, −260°), (3, 460°), (3, 820°)   **(ii)** (−3, −80°), (−3, 280°), (−3, 640°)

(b) **(i)** $\left(3, -\dfrac{3\pi}{4} \right)$, $\left(3, \dfrac{5\pi}{4} \right)$, $\left(3, \dfrac{13\pi}{4} \right)$ **(ii)** $\left(-3, -\dfrac{7\pi}{4} \right)$, $\left(-3, \dfrac{9\pi}{4} \right)$, $\left(-3, \dfrac{17\pi}{4} \right)$

9. (a) $(3\sqrt{3}, 3)$ **(b)** $(2\sqrt{2}, -2\sqrt{2})$ **11. (a)** (2.70, −1.30)   **(b)** (−1.37, −3.76)

13. (a) (−1.98, −0.28)   **(b)** (0, −3)   **15. (a)** $\left(2, \dfrac{2\pi}{3} \right)$ **(b)** $\left(12, \dfrac{2\pi}{3} \right)$

17. (a) (5.92, 2.10)   **(b)** (9.22, 4.00)   **19.** $r = 2 \sec \theta$ **21.** $r = -3 \csc \theta$ **23.** $r = \dfrac{2}{3 \cos \theta - \sin \theta}$

25. $r = 5$ **27.** $r = -8 \cos \theta$ **29.** $r = \dfrac{1}{4} \tan \theta \sec \theta$ **31.** $r = 8 \cot \theta \csc \theta$ **33.** $x^2 + y^2 = 9$

35. $y = 2$ **37.** $y = x$ **39.** $(x + 2)^2 + y^2 = 4$ **41.** $y = \dfrac{1}{4} x^2$ **43.** $4x + 3y = 12$

45. $(x - 2)^2 + (y - 1)^2 = 5$
47. Circle   **49.** Line   **51.** Circle   **53.** Cardioid

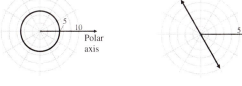

57. $(a, 0)$, $\left(a, \dfrac{\pi}{3} \right)$, $\left(a, \dfrac{2\pi}{3} \right)$, (a, π), $\left(a, \dfrac{4\pi}{3} \right)$, $\left(a, \dfrac{5\pi}{3} \right)$

59. Symmetry: line $\theta = \dfrac{\pi}{2}$ **61.** Symmetry: polar axis   **63.** Symmetry: line $\theta = \dfrac{\pi}{2}$ **65.** Symmetry: pole (and both axes)

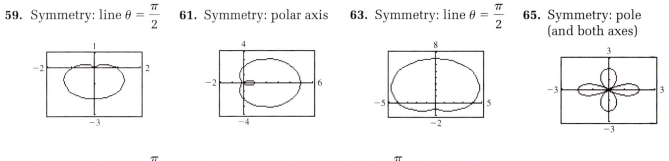

67. Symmetry: line $\theta = \dfrac{\pi}{2}$ **69.** Symmetry: polar axis, line $\theta = \dfrac{\pi}{2}$, and the pole

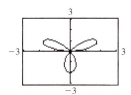

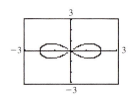

71. (a) $r = 2 \cos 3\theta$ $r = 2 \cos\left[3\left(\theta - \dfrac{\pi}{6}\right)\right]$ $r = 2 \cos\left[3\left(\theta + \dfrac{\pi}{4}\right)\right]$

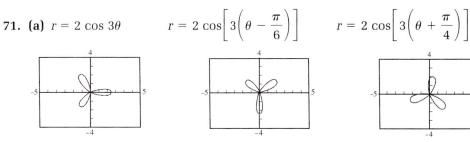

(b) Rotate $r = 2 \cos 3\theta$, $\dfrac{\pi}{6}$ (30°) counterclockwise to get $r = 2 \cos\left[3\left(\theta - \dfrac{\pi}{6}\right)\right]$;

rotate $r = 2 \cos 3\theta$, $\dfrac{\pi}{4}$ (45°) clockwise to get $r = 2 \cos\left[3\left(\theta + \dfrac{\pi}{4}\right)\right]$.

Problem Set 7.3 (on page 486)

3. $\mathbf{u} = \langle -5, -2 \rangle$; $|\mathbf{u}| = \sqrt{29}$ **5.** $\mathbf{u} = \langle 5, -11 \rangle$; $|\mathbf{u}| = \sqrt{146}$ **7.** $\mathbf{u} = \langle -6, -6 \rangle$; $|\mathbf{u}| = 6\sqrt{2}$

9. (a) $\langle 1, 8 \rangle$ **(b)** $\sqrt{65}$ **11. (a)** $\langle -19, -20 \rangle$ **(b)** $\sqrt{761}$ **13. (a)** $\langle 9, -28 \rangle$ **(b)** $\sqrt{865}$

15. (a) $-2\mathbf{i} + 4\mathbf{j}$ **(b)** $3\mathbf{i}$ **17. (a)** $2\mathbf{i} + 5\mathbf{j}$ **(b)** $-4\mathbf{i} + 4\mathbf{j}$ **19.** $|\mathbf{w}| = \sqrt{53}$; approx. 37.18°

21. $-\dfrac{\sqrt{3}}{2}\mathbf{i} + \dfrac{1}{2}\mathbf{j}$; 150°   **23.** $\dfrac{5}{13}\mathbf{i} + \dfrac{12}{13}\mathbf{j}$; 67.38°   **25.** $\dfrac{1}{\sqrt{2}}\mathbf{i} - \dfrac{1}{\sqrt{2}}\mathbf{j}$; −45°   **27.** $\dfrac{1}{\sqrt{26}}\mathbf{i} + \dfrac{5}{\sqrt{26}}\mathbf{j}$; 78.69°

29. $x = 3$, $y = 0$ **31.** $x = 3.54$, $y = 3.54$ **33.** $x = -1.73$, $y = -1$ **35.** $x = -0.09$, $y = 0.49$

37. (a) 55.71 lb   **(b)** 20.49°   **39. (a)** 12.63 N   **(b)** 17.29°

41. (a) 540.03 mph   **(b)** N81.34°E   **(c)** 3.66°

43. (a) 51.22 kph   **(b)** S1.34°E   **(c)** 38.66°   **45.** (13, 1)   **49. (a)** −9   **(b)** 142.1°

51. (a) −12   **(b)** 143.1°   **53. (a)** 14   **(b)** 58.7°   **57.** Yes; vertex angle B is the right angle.

Problem Set 7.4 (on page 496)

1. (a) $|z| = \sqrt{53}$ **(b)** $|z| = \sqrt{29}$ **3. (a)** $|z| = 5$ **(b)** $|z| = \sqrt{34}$

5. (a) 4   **(b)** 2   **(c)** 4   **(d)** $\dfrac{1}{2}$ **(e)** $\dfrac{1}{2}$

7. (a) $5\sqrt{29}$ **(b)** 25   **(c)** $5\sqrt{29}$ **(d)** $\dfrac{5}{\sqrt{29}}$ **(e)** $\dfrac{5}{\sqrt{29}}$

9. $\sqrt{3} + i$ **11.** $2i$ **13.** -8 **15.** $1.97 + 0.35i$

17. (a) $\sqrt{2}\left(\cos\dfrac{5\pi}{4} + i\sin\dfrac{5\pi}{4}\right)$ **(b)** $2\left(\cos\dfrac{7\pi}{6} + i\sin\dfrac{7\pi}{6}\right)$

19. (a) $\sqrt{3}(\cos 0.62 + i\sin 0.62)$ **(b)** $5(\cos\pi + i\sin\pi)$

21. (a) $1\left(\cos\dfrac{2\pi}{3} + i\sin\dfrac{2\pi}{3}\right)$ **(b)** $1\left(\cos\dfrac{\pi}{6} + i\sin\dfrac{\pi}{6}\right)$

23. (a) $8\left(\cos\dfrac{7\pi}{6} + i\sin\dfrac{7\pi}{6}\right) = -4\sqrt{3} - 4i$ **(b)** $2\left(\cos\dfrac{\pi}{2} + i\sin\dfrac{\pi}{2}\right) = 2i$

25. (a) $1(\cos 90° + i\sin 90°) = i$ **(b)** $1[\cos(-30°) + i\sin(-30°)] = \dfrac{\sqrt{3}}{2} - \dfrac{1}{2}i$

27. (a) $2(\cos 180° + i\sin 180°) = -2$ **(b)** $1[\cos(-90°) + i\sin(-90°)] = -i$

29. (a) $6(\cos 90° + i\sin 90°) = 6i$ **(b)** $\dfrac{2}{3}(\cos 10° + i\sin 10°) = 0.66 + 0.12i$

31. $-\dfrac{\sqrt{3}}{2} - \dfrac{1}{2}i$ **33.** $512 - 512\sqrt{3}\,i$ **35.** 256   **37.** $-125{,}000i$ **39.** $-8 - 8\sqrt{3}\,i$

41. $-2^{30} = -1{,}073{,}741{,}824$ **43.** -1 **45.** $0.71 + 0.71i, -0.71 - 0.71i$

47. $2, -1 + 1.73i, -1 - 1.73i$ **49.** $0.91 + 1.78i, -1.41 + 1.41i, -1.78 - 0.91i, 0.31 - 1.98i, 1.98 - 0.31i$

51. $1 + 1.73i, -1.73 + i, -1 - 1.73i, 1.73 - i$

53. $2(\cos 60° + i\sin 60°) = 1 + 1.73i; 2(\cos 180° + i\sin 180°) = -2; 2(\cos 300° + i\sin 300°) = 1 - 1.73i$

55. $3(\cos 45° + i\sin 45°) = 2.12 + 2.12i; 3(\cos 135° + i\sin 135°) = -2.12 + 2.12i;$
  $3(\cos 225° + i\sin 225°) = -2.12 - 2.12i; 3(\cos 315° + i\sin 315°) = 2.12 - 2.12i$

57. $2(\cos 30° + i\sin 30°) = 1.73 + i; 2(\cos 90° + i\sin 90°) = 2i; 2(\cos 150° + i\sin 150°) = -1.73 + i;$
  $2(\cos 210° + i\sin 210°) = -1.73 - i; 2(\cos 270° + i\sin 270°) = -2i; 2(\cos 330° + i\sin 330°) = 1.73 - i$

61. (a) On the circle $x^2 + y^2 = 4$ **(b)** On the circle $x^2 + (y - 1)^2 = 1$

65. (a) $-\dfrac{\sqrt{3}}{2} - \dfrac{1}{2}i$ **(b)** $\dfrac{1}{32} - \dfrac{1}{32}i$

Chapter 7 Review Problem Set (on page 498)

1. (a) $c = 3.66, \alpha = 43.12°, \beta = 106.88°$ **(b)** $\gamma = 60°, a = 8.16, b = 11.15$

3. (a) $\left(\dfrac{5\sqrt{2}}{2}, \dfrac{5\sqrt{2}}{2}\right)$ **(b)** $(\sqrt{3}, 1)$ **(c)** $(-1, -1)$

5. (a) $x^2 + (y + 2)^2 = 4$; circle   **(b)** $\left(x + \dfrac{3}{2}\right)^2 + y^2 = \dfrac{9}{4}$; circle

7. (a)   **(b)**

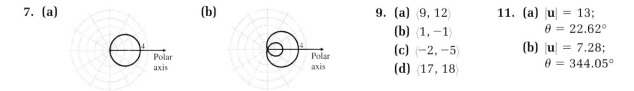

9. (a) $\langle 9, 12 \rangle$
(b) $\langle 1, -1 \rangle$
(c) $\langle -2, -5 \rangle$
(d) $\langle 17, 18 \rangle$

11. (a) $|\mathbf{u}| = 13;$
$\theta = 22.62°$
(b) $|\mathbf{u}| = 7.28;$
$\theta = 344.05°$

13. (a) $\mathbf{u} = \langle 10, -5 \rangle; |\mathbf{u}| = 5\sqrt{5}$ **(b)** $\mathbf{u} = \langle 10, -7 \rangle; |\mathbf{u}| = \sqrt{149}$

15. (a) $2\left(\cos \dfrac{2\pi}{3} + i \sin \dfrac{2\pi}{3} \right)$ **(b)** $\sqrt{2}\left(\cos \dfrac{3\pi}{4} + i \sin \dfrac{3\pi}{4} \right)$

17. (a) $13(\cos 1.18 + i \sin 1.18)$ **(b)** $\sqrt{85}(\cos 319.40° + i \sin 319.40°)$

19. $z_1 z_2 = 6\left(\cos \dfrac{3\pi}{2} + i \sin \dfrac{3\pi}{2} \right) = -6i; \dfrac{z_1}{z_2} = \dfrac{2}{3}\left(\cos \dfrac{\pi}{2} + i \sin \dfrac{\pi}{2} \right) = \dfrac{2}{3}i$

21. $z_1 z_2 = 18(\cos 305° + i \sin 305°) = 10.32 - 14.74i; \dfrac{z_1}{z_2} = 2(\cos 155° + i \sin 155°) = -1.81 + 0.85i$

23. (a) $1(\cos 360° + i \sin 360°) = 1$ **(b)** $36\sqrt{6}(\cos \pi + i \sin \pi) = -36\sqrt{6}$

25. (a) $81(\cos 108° + i \sin 108°) = -25.03 + 77.04i$ **(b)** $25\sqrt{5}\left(\cos \dfrac{5\pi}{8} + i \sin \dfrac{5\pi}{8} \right) = -21.39 + 51.65i$

27. (a) $2\left(\cos \dfrac{\pi}{6} + i \sin \dfrac{\pi}{6} \right); 2\left(\cos \dfrac{5\pi}{6} + i \sin \dfrac{5\pi}{6} \right); 2\left(\cos \dfrac{3\pi}{2} + i \sin \dfrac{3\pi}{2} \right)$

(b) $\sqrt[8]{2}\left(\cos \dfrac{\pi}{16} + i \sin \dfrac{\pi}{16} \right); \sqrt[8]{2}\left(\cos \dfrac{9\pi}{16} + i \sin \dfrac{9\pi}{16} \right);$

$\sqrt[8]{2}\left(\cos \dfrac{17\pi}{16} + i \sin \dfrac{17\pi}{16} \right); \sqrt[8]{2}\left(\cos \dfrac{25\pi}{16} + i \sin \dfrac{15\pi}{16} \right)$

29. 92.54 ft   **31.** 31.72 ft   **33.** 93.73°

Chapter 7 Test (on page 500)

1. $b = 65.01, c = 105.96, \gamma = 103.35°$ **2.** $b = 5.65$ **3. (a)** $\left(2, -2\sqrt{3} \right)$ **(b)** $\left(3\sqrt{2}, \dfrac{3\pi}{4} \right)$

4. (a) $x^2 + y^2 = 25$ **(b)** $x^2 + (y - 2)^2 = 4$ **(c)** $y = x$

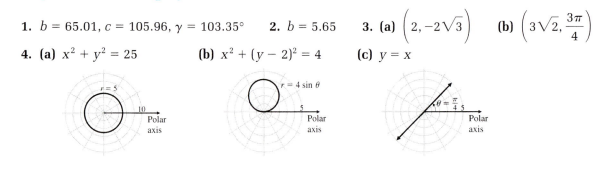

5. (a) $r = \dfrac{5}{3 \cos \theta + 2 \sin \theta}$ **(b)** $r = 4$ **6.** $24i + 20j$ **7.** 56.31°   **8.** $2.42i + 12.27j$

9. $2\sqrt{2} - 2\sqrt{2}\,i$ **10.** $2(\cos 240° + i \sin 240°)$

11. (a) $2(\cos 210° + i \sin 210°) = \sqrt{3} - i \approx -1.732 - i$ **(b)** $\dfrac{1}{2}(\cos 68° + i \sin 68°) \approx 0.1873 + 0.4636i$

12. $4096 + 0i$ **13.** $\dfrac{\sqrt{3}}{2} + \dfrac{1}{2}i; \dfrac{-\sqrt{3}}{2} + \dfrac{1}{2}i; -i$ **14.** 82.7 mi

Chapter 8: Problem Set 8.1 (on page 509)

1. $\left(\dfrac{11}{5}, \dfrac{14}{5}\right)$ **3.** No solution **5.** $\left(\dfrac{11}{7}, \dfrac{5}{7}\right)$ **7.** No solution **9.** $(a, a - 5)$, where a is a real number

11. $\left(\dfrac{7}{3}, -\dfrac{4}{3}\right)$ **13.** $(2, 3, -1)$ **15.** $(4, 3, 2)$ **17.** $(3, 2)$ **19.** $(1, -2)$ **21.** $(y, z) = \left(\dfrac{13}{5}, -\dfrac{7}{5}\right)$

23. $(2, -1)$ **25.** $(1, 1)$ **27.** $\left(-2, \dfrac{10}{3}, \dfrac{2}{3}\right)$ **29.** $\left(-5, \dfrac{5}{3}, \dfrac{28}{3}\right)$ **31.** $(2, 1, 0)$

33. $(-1, -2, 7)$ **35.** $(2, 1)$ **37.** 30 females, 90 males

39. **(a)** $V_0 = 32$ ft./sec, $h_0 = 10$ ft. **(b)** $h = -16t^2 + 32t + 10$

  (c) $h(1) = 26$ ft, $h = 0$ when $t \approx 2.27$ secs. **41.** 2 L 30% solution, 8 L 20% solution

43. Airplane: approx. 617 mph; wind: approx. 49 mph

45. 185 pairs tennis shoes; 75 pairs running shoes

47. \$24,000 at 5% interest; \$11,000 at 7% interest; \$7000 at 9% interest

49. 3 of the 12 passenger vans, 11 of the 10 passenger vans, 1 of the 6 passenger vans

51. A pumps 10,000 gal/hr; B pumps 12,000 gal/hr; C pumps 15,000 gal/hr

53. **(a)** $x = \dfrac{1}{3}, y = -1$ **(b)** $x = \dfrac{1}{4}, y = -\dfrac{1}{2}$

Problem Set 8.2 (on page 522)

1. $\begin{bmatrix} 1 & 6 & \vdots & 7 \\ 3 & 7 & \vdots & 10 \end{bmatrix}$ **3.** $\begin{bmatrix} 2 & 1 & 0 & \vdots & 2 \\ 1 & -1 & 0 & \vdots & 4 \\ 3 & 0 & 1 & \vdots & 6 \end{bmatrix}$ **5.** $(1, 3)$ **7.** $(1, 1, 1)$ **9.** $(5, 3, -2)$

11. **(a)** $\begin{bmatrix} 1 & 0 & 3 \\ 5 & -1 & 5 \\ -3 & 1 & 4 \end{bmatrix}$ **(b)** $\begin{bmatrix} 1 & 0 & 3 \\ -3 & 1 & 4 \\ 15 & -3 & 15 \end{bmatrix}$ **(c)** $\begin{bmatrix} 1 & 0 & 3 \\ -3 & 1 & 4 \\ 8 & -1 & 14 \end{bmatrix}$ **13.** Both (a) and (b)

15. **(b)** **17.** $(1, 1)$ **19.** $(2 + t, 3 + 2t, t)$, where t is a real number **21.** No solution **23.** $(3, 5)$

25. $\left(\dfrac{23}{4}, \dfrac{13}{2}\right)$ **27.** $(1, 2, 3)$ **29.** $(1, 1, 1)$ **31.** $(4, 5, 6)$ **33.** $(1, 1, 0)$ **35.** No solution

37. $\left(\dfrac{7}{5} - \dfrac{7}{5}t, -\dfrac{2}{5} - \dfrac{3}{5}t, t\right)$, where t is a real number **39.** $\left(-\dfrac{23}{13}, -\dfrac{93}{26}, -\dfrac{37}{13}\right)$ **41.** $A = 2, B = 1$

43. $A = -1, B = -6$ **45.** $A = -5, B = 31, C = -26$ **47.** $A = 2, B = -1, C = 3$

49. $A = 6, B = -6, C = 0$ **53.** 10 oz A, 5 oz B, 8 oz C **55.** 500 lb 25%, 2000 lb 35%, 1500 lb 40%

57. $I_1 = \dfrac{33}{167}A, I_2 = \dfrac{87}{167}A, I_3 = \dfrac{54}{167}A$ **59.** $(1, -1, 1, 1)$ **61.** $(3, 2, 1, -1)$

63. **(a)** $2a - b - c \neq 0$ **(b)** Not possible **(c)** $2a - b - c = 0$

65. **(a)** $a = -\sqrt{3}, \sqrt{3}$ **(b)** $a \neq -\sqrt{3}, \sqrt{3}$ **(c)** Not possible **67.** $a = b = c = 1$

Problem Set 8.3 (on page 537)

1. (a) $\begin{bmatrix} 8 & 11 \\ -2 & 7 \end{bmatrix}$ **(b)** $\begin{bmatrix} -2 & -1 \\ 0 & 1 \end{bmatrix}$ **(c)** $\begin{bmatrix} 12 & 20 \\ -4 & 16 \end{bmatrix}$ **(d)** $\begin{bmatrix} -25 & -30 \\ 5 & -15 \end{bmatrix}$ **(e)** $\begin{bmatrix} -13 & -10 \\ 1 & 1 \end{bmatrix}$

3. (a) $\begin{bmatrix} 8 & 12 & 8 \\ 2 & 3 & 0 \end{bmatrix}$ **(b)** $\begin{bmatrix} -2 & -10 & -4 \\ -4 & 3 & 2 \end{bmatrix}$ **(c)** $\begin{bmatrix} 12 & 4 & 8 \\ -4 & 12 & 4 \end{bmatrix}$

(d) $\begin{bmatrix} -25 & -55 & -30 \\ -15 & 0 & 5 \end{bmatrix}$ **(e)** $\begin{bmatrix} -13 & -51 & -22 \\ -19 & 12 & 9 \end{bmatrix}$

5. (a) $\begin{bmatrix} -9 & 8 & 3 \\ -3 & 3 & 7 \\ 6 & -2 & 4 \end{bmatrix}$ **(b)** $\begin{bmatrix} -1 & -4 & -1 \\ -1 & -1 & -3 \\ 0 & 0 & 0 \end{bmatrix}$ **(c)** $\begin{bmatrix} -20 & 8 & 4 \\ -8 & 4 & 8 \\ 12 & -4 & 8 \end{bmatrix}$

(d) $\begin{bmatrix} 20 & -30 & -10 \\ 5 & -10 & -25 \\ -15 & 5 & -10 \end{bmatrix}$ **(e)** $\begin{bmatrix} 0 & -22 & -6 \\ -3 & -6 & -17 \\ -3 & 1 & -2 \end{bmatrix}$

7. (a) $\begin{bmatrix} 19 & -1 \\ 2 & 1 \end{bmatrix}$ **(b)** $\begin{bmatrix} 6 & 9 \\ 7 & 14 \end{bmatrix}$ **9. (a)** $\begin{bmatrix} -9 & -29 \\ 4 & -14 \end{bmatrix}$ **(b)** $\begin{bmatrix} -7 & 10 & -12 \\ 0 & -1 & 11 \\ 5 & -5 & -15 \end{bmatrix}$

11. (a) $\begin{bmatrix} 4 & 3 & -1 \\ 8 & -2 & 3 \\ 6 & 5 & 2 \end{bmatrix}$ **(b)** $\begin{bmatrix} 4 & 3 & -1 \\ 8 & -2 & 3 \\ 6 & 5 & 2 \end{bmatrix}$ **13.** Inverses **15.** Inverses **17.** Inverses

19. $\begin{bmatrix} 2 & -5 \\ -1 & 3 \end{bmatrix}$ **21.** $\begin{bmatrix} 0 & 1 \\ 1 & 1 \end{bmatrix}$ **23.** $\begin{bmatrix} 7 & 2 & -6 \\ -3 & -1 & 3 \\ 2 & 1 & -2 \end{bmatrix}$ **25.** $\begin{bmatrix} 1 & 2 & 0 \\ 0 & 2 & 3 \\ 1 & 3 & 1 \end{bmatrix}$

27. $(2, 1)$ **29.** $(4, -2)$ **31.** $(1, 1)$ **33.** $(-1, 2, 1)$ **35.** $(2, 1, 3)$

37. $\begin{bmatrix} 1 & 3 & 2 & 1 \\ 1 & 2 & 4 & 0 \\ 2 & 6 & 5 & 2 \\ 1 & 3 & 2 & 2 \end{bmatrix} \begin{bmatrix} x \\ y \\ z \\ w \end{bmatrix} = \begin{bmatrix} 4 \\ -3 \\ -1 \\ 2 \end{bmatrix}, A^{-1} = \begin{bmatrix} 12 & 3 & -8 & 2 \\ -2 & -1 & -2 & -1 \\ -2 & 0 & 1 & 0 \\ -1 & 0 & 0 & 1 \end{bmatrix}, X = \begin{bmatrix} 51 \\ -9 \\ -9 \\ -2 \end{bmatrix}.$

39. $\begin{bmatrix} 0.2 & 0.4 & -0.6 & 0 \\ -0.2 & 0.1 & 0.3 & 0.2 \\ 0.2 & 0.9 & -0.9 & 0.1 \\ 0.4 & 0.3 & -0.3 & 0 \end{bmatrix} \begin{bmatrix} x \\ y \\ z \\ w \end{bmatrix} = \begin{bmatrix} 0.2 \\ 0.7 \\ 1.1 \\ 0.7 \end{bmatrix}, A^{-1} = \begin{bmatrix} 2 & 1 & -2 & 3 \\ -11 & 3 & 6 & 1 \\ -8.33 & -1.67 & 3.33 & 1.67 \\ 20 & 10 & -10 & 0 \end{bmatrix}, X = \begin{bmatrix} 1 \\ 3 \\ 2 \\ 0 \end{bmatrix}.$

41. (a) $[2325 \; 1351.5 \; 930]$

 (b) Profit from store 1 is \$2325; profit from store 2 is \$1351.50; profit from store 3 is \$930.

43. (a) $[224{,}000{,}000 \; 302{,}000{,}000]$

 (b) Value of cars shipped to site 1 is \$224,000,000; value of cars shipped to site 2 is \$302,000,000.

47. $x = 1, y = 2, z = 4, w = 10$ **51.** $\begin{bmatrix} -1 & 1 & -1 \\ 1 & -\dfrac{1}{2} & \dfrac{5}{4} \\ 1 & -\dfrac{1}{2} & \dfrac{3}{4} \end{bmatrix}$

Problem Set 8.4 (on page 548)

1. (a) 17 **(b)** -19 **3. (a)** -1 **(b)** 11 **5.** 63 **7.** 0 **9.** 3 **17.** 0 **19.** -20 **21.** -2

23. $\left(\dfrac{1}{3}, \dfrac{2}{3}\right)$ **25.** $(1, 2)$ **27.** $\left(-6, \dfrac{1}{2}\right)$ **29.** $(1, 1, 1)$ **31.** $(3, -1, 2)$ **33.** $\left(\dfrac{17}{2}, -\dfrac{3}{2}, -\dfrac{25}{2}\right)$

35. $\left(-\dfrac{3}{4}, \dfrac{83}{8}, \dfrac{21}{8}\right)$ **37.** $x = -\dfrac{2}{9}, y = -\dfrac{1}{9}, z = \dfrac{24}{9}, w = \dfrac{2}{9}$ **39.** $A = \dfrac{33}{2}$

41. (a) 1   **(b)** 54   **(c)** 2   **43. (a)** $\dfrac{1}{4}$ **(b)** $\dfrac{21 \pm \sqrt{185}}{4}$ **45.** 1, 2   **47.** $a = 38, b = -7, c = 20$

49. (a) $\begin{bmatrix} \dfrac{2}{7} & \dfrac{3}{14} \\ -\dfrac{1}{7} & \dfrac{5}{14} \end{bmatrix}$ **(b)** $\begin{bmatrix} -8 & 7 \\ 7 & -6 \end{bmatrix}$

Problem Set 8.5 (on page 557)

13. $(0, 0), (0, 4), (3, 5), (8, 0)$ **15.** $(0, 0), (0, 5), \left(\dfrac{14}{3}, \dfrac{8}{3}\right), (6, 0)$

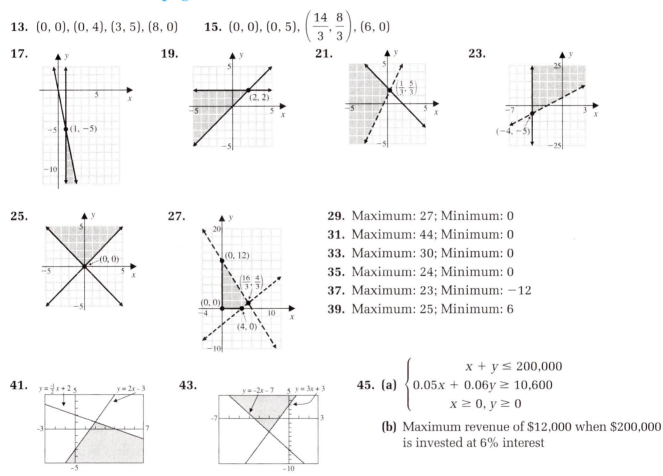

17. **19.** **21.** **23.**

25. **27.**

29. Maximum: 27; Minimum: 0
31. Maximum: 44; Minimum: 0
33. Maximum: 30; Minimum: 0
35. Maximum: 24; Minimum: 0
37. Maximum: 23; Minimum: -12
39. Maximum: 25; Minimum: 6

41. **43.** **45. (a)** $\begin{cases} x + y \le 200{,}000 \\ 0.05x + 0.06y \ge 10{,}600 \\ x \ge 0, y \ge 0 \end{cases}$

(b) Maximum revenue of \$12,000 when \$200,000 is invested at 6% interest

47. (a) Let S and M be the amounts of protein powder made from soybeans and milk, respectively (in grams);
$\begin{cases} 2S + 2M \ge 12 \\ 2S + 4M \ge 8 \\ S \ge 0, M \ge 0 \end{cases}$

(b) Any combination that supplies adequate protein also supplies adequate carbohydrates.

49. 800 tacos and 0 hamburgers   **51.** \$750,000; 5000 barrels of diesel oil and 20,000 barrels of gasoline

53. 100 cord-type and 200 cordless toothbrushes

55.

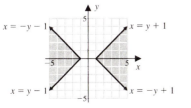

Problem Set 8.6 (on page 565)

1. Two solutions: $(2, 2)$, $\left(-5, \dfrac{25}{2}\right)$ **3.** Two solutions: $(35, 7)$, $(-35, -7)$ **5.** Two solutions: $(2, 6)$, $(3, 4)$

7. Four solutions: $(3, 4)$, $(-3, 4)$, $(4, -3)$, $(-4, -3)$ **9.** Two solutions: $(1, 2)$, $(1, -2)$

11. Two solutions: $(1.09, 2.18)$, $(-1.84, 4.38)$ **13.** Four real solutions: $\left(6, \dfrac{4}{5}\right)$, $\left(6, -\dfrac{4}{5}\right)$, $\left(-6, \dfrac{4}{5}\right)$, $\left(-6, -\dfrac{4}{5}\right)$

15. Four real solutions: $(3, 2)$, $(3, -2)$, $(-3, 2)$, $(-3, -2)$ **17.** Two real solutions: $(1.08, -2.74)$, $(-0.58, 2.24)$

19. Two real solutions: $(-2.72, 4.15)$, $(3.03, 0.32)$ **21.** $(2\sqrt{6}, 1)$, $(2\sqrt{6}, -1)$, $(-2\sqrt{6}, 1)$, $(-2\sqrt{6}, -1)$

23. $(2, 2\sqrt{3})$, $(2, -2\sqrt{3})$, $(-2, 2\sqrt{3})$, $(-2, -2\sqrt{3})$ **25.** $(4, 2)$, $(-4, 2)$ $\left(4\sqrt{\dfrac{5}{3}}, -\dfrac{10}{3}\right)$, $\left(-4\sqrt{\dfrac{5}{3}}, -\dfrac{10}{3}\right)$

27. $(3, 2)$, $(3, -2)$, $(-3, 2)$, $(-3, -2)$ **29.** $(2, 1)$, $(-2, 1)$ **31.** No real solutions

33. $\left(\dfrac{41}{10}, \dfrac{3\sqrt{91}}{10}\right)$, $\left(\dfrac{41}{10}, -\dfrac{3\sqrt{91}}{10}\right)$

35. **37.** **39.** **41.**

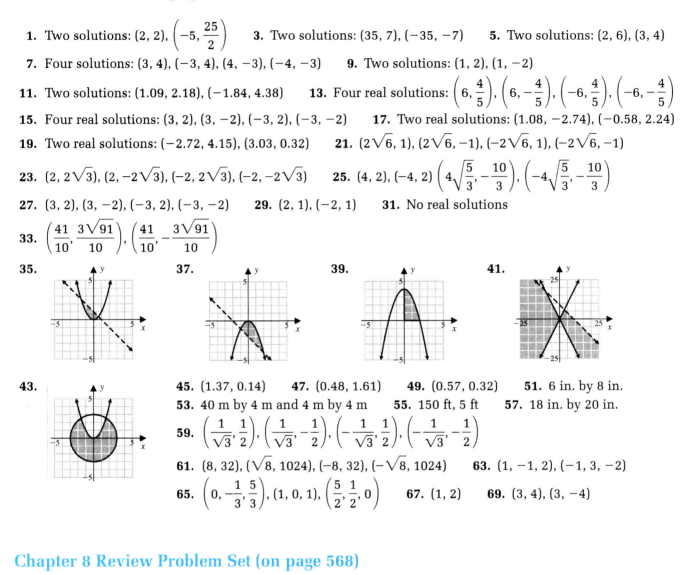

43. **45.** $(1.37, 0.14)$ **47.** $(0.48, 1.61)$ **49.** $(0.57, 0.32)$ **51.** 6 in. by 8 in.
53. 40 m by 4 m and 4 m by 4 m **55.** 150 ft, 5 ft **57.** 18 in. by 20 in.

59. $\left(\dfrac{1}{\sqrt{3}}, \dfrac{1}{2}\right)$, $\left(\dfrac{1}{\sqrt{3}}, -\dfrac{1}{2}\right)$, $\left(-\dfrac{1}{\sqrt{3}}, \dfrac{1}{2}\right)$, $\left(-\dfrac{1}{\sqrt{3}}, -\dfrac{1}{2}\right)$

61. $(8, 32)$, $(\sqrt{8}, 1024)$, $(-8, 32)$, $(-\sqrt{8}, 1024)$ **63.** $(1, -1, 2)$, $(-1, 3, -2)$

65. $\left(0, -\dfrac{1}{3}, \dfrac{5}{3}\right)$, $(1, 0, 1)$, $\left(\dfrac{5}{2}, \dfrac{1}{2}, 0\right)$ **67.** $(1, 2)$ **69.** $(3, 4)$, $(3, -4)$

Chapter 8 Review Problem Set (on page 568)

1. $(-1, -2)$ **3.** $(2, -1)$ **5.** $(-9, 5)$ **7.** $(-1 -3t, 2 - t, t)$, where t is a real number
9. $\left(-\dfrac{2}{9}, \dfrac{28}{9}\right)$ **11.** $(0, 1, 1)$ **13.** $A = \dfrac{9}{8}$, $B = -\dfrac{1}{8}$ **15. (a)** $\begin{bmatrix} 12 & -3 \\ 7 & 7 \end{bmatrix}$ **(b)** $\begin{bmatrix} 14 & -11 \\ 9 & 19 \end{bmatrix}$

17. (a) $\begin{bmatrix} 0 & 5 \\ 11 & -18 \\ 1 & -7 \end{bmatrix}$ **(b)** Not possible **21.** $(-1, 2)$ **23.** 5 **25.** 0 **27.** $(3, 2)$ **29.** $\left(\dfrac{52}{29}, -\dfrac{50}{29}, \dfrac{36}{29}\right)$

31. **33.** Maximum: 21; Minimum: 0 **35.** Maximum: 20; Minimum: 0
37. One solution: $(3, -4)$ **39.** Three solutions: $(-5, \sqrt{11})$, $(6, 0)$, $(-5, -\sqrt{11})$
41. \$2453 in state income tax, \$1387 in city income tax, \$10,410 in federal income tax
43. 7 and 8, 7 and -8, -7 and 8, or -7 and -8 **45.** 40 of A, 60 of B, 90 of C

Chapter 8 Test (on page 570)

1. (a)–(b) $(1, 1)$ **2.** $(4, -1, 3)$ **3. (a)** $\begin{bmatrix} 9 & -5 \\ 8 & 16 \end{bmatrix}$ **(b)** $\begin{bmatrix} 7 & -13 \\ 16 & 6 \end{bmatrix}$ **(c)** $\begin{bmatrix} 3 & 1 \\ 8 & -8 \end{bmatrix}$

4. (b) $(-1, -2)$ **5. (a)** -19 **(b)** $(1, -1)$ **6.** $A = -\dfrac{4}{3}, B = \dfrac{7}{3}$ **7.** $(2, 2), \left(-3, \dfrac{9}{2}\right)$

8. $(0, 4), (2, 2), (3, 0), (0, 0)$ **9.** Maximum: 19; Minimum: 1   **10.** Fixed charge = \$30; hourly rate = \$20

11. A = 40°, B = 60°, C = 80°

Chapter 9: Problem Set 9.1 (on page 580)

1. $V = (3, 2)$; $y = 2$; $F = (5, 2)$; d: $x = 1$

3. $V = (-1, -3)$; $x = -1$; $F = (-1, -4)$; d: $y = -2$

5. $V = (1, -2)$; $y = -2$; $F = \left(-\dfrac{1}{2}, -2\right)$; d: $x = \dfrac{5}{2}$

7. $(y + 1)^2 = -2(x + 3)$; $V = (-3, -1)$; $F = \left(-\dfrac{7}{2}, -1\right)$; $y = -1$; d: $x = -\dfrac{5}{2}$

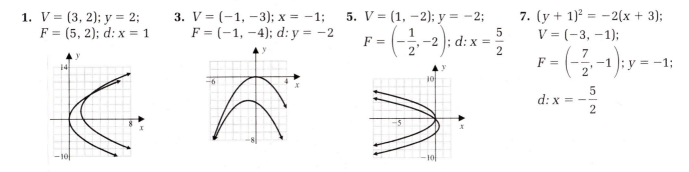

9. $(x + 2)^2 = -2\left(y + \dfrac{1}{6}\right)$; $V = \left(-2, -\dfrac{1}{6}\right)$; $F = \left(-2, -\dfrac{2}{3}\right)$; $x = -2$; d: $y = \dfrac{1}{3}$

11. $(y - 2)^2 = 6\left(x - \dfrac{3}{2}\right)$; $V = \left(\dfrac{3}{2}, 2\right)$; $F = (3, 2)$; $y = 2$; d: $x = 0$ **13.** $(y - 2)^2 = 16(x + 1)$; $y = 2$

15. $(y + 5)^2 = 8(x - 3)$; $y = -5$ **17.** $(x - 1)^2 = 4(y - 3)$; $x = 1$

19. $(x + 3)^2 = 20y$; $x = -3$ **21.** $x^2 = -\dfrac{4}{3}(y - 3)$; $x = 0$

23. (a) $y = -1 - \sqrt{2x - 2}$ or $y = -1 + \sqrt{2x - 2}$

(b) $y = -\dfrac{(x - 3)^2}{6} - 1$

25. $y = -1 - 2\sqrt{6 - 3x}$ or $y = -1 + 2\sqrt{6 - 3x}$

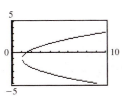

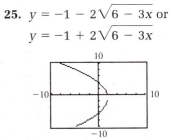

27. Center: $(2, -3)$; Vertices: $(-2, -3), (6, -3)$; Foci: $(2 - \sqrt{7}, -3)$, $\quad (2 + \sqrt{7}, -3)$

29. Center: $(1, -2)$; Vertices: $(1, -10), (1, 6)$; Foci: $(1, -6), (1, 2)$

31. Center: $(-2, 1)$; Vertices: $(-7, 1), (3, 1)$; Foci: $(-5, 1), (1, 1)$

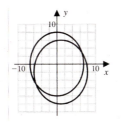

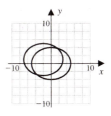

33. Center: $(-1, 1)$; Vertices: $(-1, 1 - \sqrt{15})$, $(-1, 1 + \sqrt{15})$; Foci: $(-1, 1 - 2\sqrt{3})$, $(-1, 1 + 2\sqrt{3})$

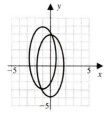

35. $\dfrac{(x + 2)^2}{1} + \dfrac{(y - 3)^2}{4} = 1$; Center: $(2, 3)$; Vertices: $(2, 1)$, $(2, 5)$; Foci: $(2, 3 - \sqrt{3})$, $(2, 3 + \sqrt{3})$

37. $\dfrac{(x + 1)^2}{4} + \dfrac{(y - 1)^2}{1} = 1$; Center: $(-1, 1)$; Vertices: $(-3, 1)$, $(1, 1)$; Foci: $(-1 - \sqrt{3}, 1)$, $(-1 + \sqrt{3}, 1)$

39. $\dfrac{(x + 1)^2}{4} + \dfrac{(y - 2)^2}{9} = 1$; Center: $(-1, 2)$; Vertices: $(-1, -1)$, $(-1, 5)$; Foci: $(-1, 2 - \sqrt{5})$, $(-1, 2 + \sqrt{5})$

41. $\dfrac{(x - 4)^2}{9} + \dfrac{(y - 2)^2}{4} = 1$; Center: $(4, 2)$; Vertices: $(1, 2)$, $(7, 2)$; Foci: $(4 - \sqrt{5}, 2)$, $(4 + \sqrt{5}, 2)$

43. $\dfrac{(x - 3)^2}{4} + \dfrac{(y + 2)^2}{25} = 1$ **45.** $\dfrac{(x + 3)^2}{1} + \dfrac{(y - 1)^2}{25} = 1$ **47.** $\dfrac{(x - 2)^2}{4} + \dfrac{(y - 4)^2}{3} = 1$

49.

51.

53. Center: $(5, 4)$;
Vertices: $(0, 4)$, $(10, 4)$;
Foci: $(5 - \sqrt{29}, 4)$, $(5 + \sqrt{29}, 4)$;
Asymptotes: $y - 4 = -\dfrac{2}{5}(x - 5)$
and $y - 4 = \dfrac{2}{5}(x - 5)$

55. Center: $(-2, -3)$;
Vertices: $(-2, -7)$, $(-2, 1)$;
Foci: $(-2, -3 - 4\sqrt{2})$;
$(-2, -3 + 4\sqrt{2})$;
Asymptotes:
$y + 3 = -(x + 2)$ and
$y + 3 = x + 2$

57. Center: $(1, -2)$;
Vertices: $(-1, -2)$, $(3, -2)$;
Foci: $(1 - \sqrt{13}, -2)$,
$(1 + \sqrt{13}, -2)$;
Asymptotes:
$y + 2 = -\dfrac{3}{2}(x - 1)$ and
$y + 2 = \dfrac{3}{2}(x - 1)$

59. $\dfrac{(y - 1)^2}{4} - \dfrac{(x - 4)^2}{3} = 1$;
Center: $(4, 1)$;
Vertices: $(4, -1)$, $(4, 3)$;
Foci: $(4, 1 - \sqrt{7})$, $(4, 1 + \sqrt{7})$;
Asymptotes: $y - 1 = -\dfrac{2}{\sqrt{3}}(x - 4)$
and $y - 1 = \dfrac{2}{\sqrt{3}}(x - 4)$

61. $\dfrac{(x + 2)^2}{9} - \dfrac{(y + 4)^2}{25} = 1$;
Center: $(-2, -4)$;
Vertices: $(-5, -4)$, $(1, -4)$;
Foci: $(-2 - \sqrt{34}, -4)$,
$(-2 + \sqrt{34}, -4)$;
Asymptotes: $y + 4 = -\dfrac{5}{3}(x + 2)$
and $y + 4 = \dfrac{5}{3}(x + 2)$

63. $\dfrac{(y+1)^2}{3} - \dfrac{(x+1)^2}{\frac{3}{4}} = 1$;

Center: $(-1, -1)$;

Vertices: $(-1, -1 - \sqrt{3})$, $(-1, -1 + \sqrt{3})$;

Foci: $\left(-1, -1 - \dfrac{\sqrt{15}}{2}\right), \left(-1, -1 + \dfrac{\sqrt{15}}{2}\right)$;

Asymptotes: $y + 1 = -2(x + 1)$

and $y + 1 = 2(x + 1)$

65. $\dfrac{(x-1)^2}{9} - \dfrac{(y-2)^2}{4} = 1$;

Center: $(1, 2)$;

Vertices: $(-2, 2)$, $(4, 2)$;

Foci: $(1 - \sqrt{13}, 2)$, $(1 + \sqrt{13}, 2)$;

Asymptotes: $y - 2 = -\dfrac{2}{3}(x - 1)$

and $y - 2 = \dfrac{2}{3}(x - 1)$

67. $\dfrac{(x-3)^2}{9} - \dfrac{(y+4)^2}{4} = 1$ **69.** $\dfrac{(y-5)^2}{1} - \dfrac{(x+1)^2}{3} = 1$ **71.** $\dfrac{(x-4)^2}{4} - \dfrac{(y-3)^2}{5} = 1$

73.

75.

Problem Set 9.2 (on page 588)

1. $\dfrac{\left(y - \dfrac{15}{4}\right)^2}{\dfrac{125}{16}} - \dfrac{(x-3)^2}{\dfrac{25}{4}} = 1$ **3.** $\dfrac{x^2}{25} + \dfrac{y^2}{16} = 1$ **5.** $y^2 = -4(x - 3)$ **7.** $\dfrac{x^2}{36} + \dfrac{y^2}{27} = 1$

9. $\dfrac{x^2}{25} - \dfrac{y^2}{200} = 1$ **11.** $\dfrac{3}{5}$ **13.** $\sqrt{\dfrac{2}{3}}$ **15.** $\dfrac{5}{3}$ **17.** $\dfrac{5}{3}$ **19.** $\sqrt{\dfrac{2}{3}}$ **21.** $\sqrt{2}$ **23.** $r = \dfrac{4}{1 - \sin\theta}$

25. $r = \dfrac{4}{1 - 2\cos\theta}$ **27.** $r = \dfrac{4}{5 + 2\sin\theta}$ **29.** $e = 1$; parabola; $V = \left(0, \dfrac{5}{2}\right)$

31. $e = \dfrac{1}{2}$; ellipse;

Center $\left(\dfrac{2}{3}, 0\right)$;

Vertices: $\left(-\dfrac{2}{3}, 0\right)$, $(2, 0)$

33. $e = 3$; hyperbola;

Center: $\left(\dfrac{5}{8}, 0\right)$;

Vertices: $\left(\dfrac{5}{12}, 0\right)$, $\left(\dfrac{5}{6}, 0\right)$

35. $e = \dfrac{1}{2}$; ellipse;

Center: $\left(-\dfrac{2}{3}, 0\right)$;

Vertices: $(-2, 0)$, $\left(\dfrac{2}{3}, 0\right)$

37. $e = \dfrac{2}{3}$; ellipse;

Center: $\left(0, -\dfrac{12}{5}\right)$;

Vertices: $(0, -6)$, $\left(0, \dfrac{6}{5}\right)$

39. $e = \dfrac{1}{2}$ ellipse;

Center: $\left(-\dfrac{2}{3}, 0\right)$;

Vertices: $\left(-\dfrac{2}{3}, \dfrac{2\sqrt{3}}{3}\right)$, $\left(-\dfrac{2}{3}, -\dfrac{2\sqrt{3}}{3}\right)$

41. $e = \dfrac{1}{2}$; ellipse;

Center: $\left(-\dfrac{1}{3}, 0\right)$;

Vertices: $(-1, 0)$, $\left(\dfrac{1}{3}, 0\right)$

43. (a) $e = \dfrac{1}{59}$ (b) Approx. 94.4 million mi

45. (b) $\dfrac{x^2}{\dfrac{4}{3}} + \dfrac{y^2}{1} = 1$ (c) 2 ft; $\dfrac{4}{\sqrt{3}}$ ft   (d) 3 ft by 3.31 ft (approx.)

47. As e gets close to 0, the conic appears closer to a circle. As e gets larger and larger, the conic appears closer to two parallel vertical lines.

49. The eccentricities are the same. **51. (a)** Two **(b)** Two **(c)** One **53.** $r^2 = \sec 2\theta$

55. In polar coordinates: $(0.74, 0.79)$, $(1.55, 3.93)$; in Cartesian coordinates: $(0.52, 0.52)$, $(-1.09, -1.09)$

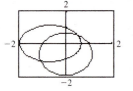

Problem Set 9.3 (on page 598)

1. $\left(-\sqrt{3} - \dfrac{5}{2}, 1 - \dfrac{5\sqrt{3}}{2}\right)$ **3.** $\left(\dfrac{11}{\sqrt{2}}, -\dfrac{9}{\sqrt{2}}\right)$ **5.** $(6.71, 0.40)$

7. $\dfrac{\overline{x}^2}{2} - \dfrac{\overline{y}^2}{2} = 1$ **9.** $\dfrac{\overline{x}^2}{8} + \dfrac{\overline{y}^2}{\frac{8}{5}} = 1$ **11.** $\dfrac{\overline{x}^2}{4} + \dfrac{\overline{y}^2}{1} = 1$ **13.** $\dfrac{\overline{x}^2}{4} + \dfrac{\overline{y}^2}{9} = 1$

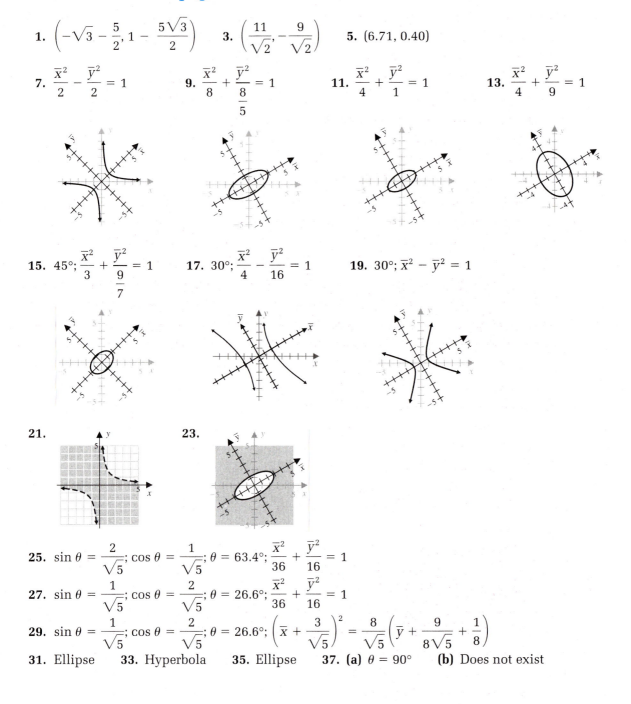

15. $45°$; $\dfrac{\overline{x}^2}{3} + \dfrac{\overline{y}^2}{\frac{9}{7}} = 1$ **17.** $30°$; $\dfrac{\overline{x}^2}{4} - \dfrac{\overline{y}^2}{16} = 1$ **19.** $30°$; $\overline{x}^2 - \overline{y}^2 = 1$

21. **23.**

25. $\sin\theta = \dfrac{2}{\sqrt{5}}$; $\cos\theta = \dfrac{1}{\sqrt{5}}$; $\theta = 63.4°$; $\dfrac{\overline{x}^2}{36} + \dfrac{\overline{y}^2}{16} = 1$

27. $\sin\theta = \dfrac{1}{\sqrt{5}}$; $\cos\theta = \dfrac{2}{\sqrt{5}}$; $\theta = 26.6°$; $\dfrac{\overline{x}^2}{36} + \dfrac{\overline{y}^2}{16} = 1$

29. $\sin\theta = \dfrac{1}{\sqrt{5}}$; $\cos\theta = \dfrac{2}{\sqrt{5}}$; $\theta = 26.6°$; $\left(\overline{x} + \dfrac{3}{\sqrt{5}}\right)^2 = \dfrac{8}{\sqrt{5}}\left(\overline{y} + \dfrac{9}{8\sqrt{5}} + \dfrac{1}{8}\right)$

31. Ellipse **33.** Hyperbola **35.** Ellipse **37. (a)** $\theta = 90°$ **(b)** Does not exist

45. $y = \dfrac{x - \sqrt{16 - 3x^2}}{2}$

or

$y = \dfrac{x + \sqrt{16 - 3x^2}}{2}$

47. $y = 5x - \sqrt{25x^2 - 32}$

or

$y = 5x + \sqrt{25x^2 - 32}$

49. $y = \dfrac{-3x - \sqrt{x^2 + 72}}{2}$

or

$y = \dfrac{-3x + \sqrt{x^2 + 72}}{2}$

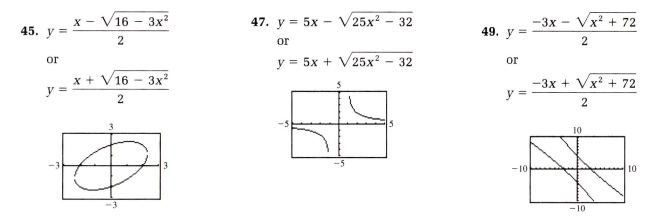

Problem Set 9.4 (on page 605)

1.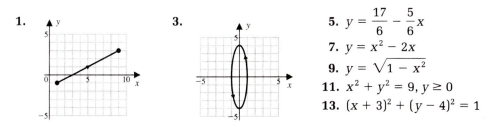

3.

5. $y = \dfrac{17}{6} - \dfrac{5}{6}x$

7. $y = x^2 - 2x$

9. $y = \sqrt{1 - x^2}$

11. $x^2 + y^2 = 9,\ y \ge 0$

13. $(x + 3)^2 + (y - 4)^2 = 1$

15. $x^2 - y^2 = 1,\ x \ge 1$ or $x \le -1$ **17.** $y = \dfrac{1}{x^2},\ x \ge 1$ **19.** $\dfrac{x^2}{4} - \dfrac{y^2}{9} = 1,\ x \ge 2$

21. One possibility: $\begin{cases} x = t \\ y = 3t^2 + 5 \end{cases}$ $0 \le t \le 2$ **23.** One possibility: $\begin{cases} x = t \\ y = t^2 - 4t + 3 \end{cases}$ $-1 \le t \le 1$

25. One possibility: $\begin{cases} x = t \\ y = \dfrac{3}{2}t + \dfrac{1}{2} \end{cases}$ $1 \le t \le 3$ **27.** One possibility: $\begin{cases} x = 3\cos t \\ y = 3\sin t \end{cases}$ $0 \le t \le 2\pi$

29. One possibility: $\begin{cases} x = \cos t \\ y = 1 + \sin t \end{cases}$ $0 \le t \le 2\pi$

31. (a) $\begin{cases} x = (200\cos 50°)t \\ y = -16t^2 + (200\sin 50°)t + 10 \end{cases}$ **(b)** 9.6 sec; 1239 ft   **(d)** 377 ft; 4.8 sec

33. (a) $\begin{cases} x = (125\cos 25°)t \\ y = -16t^2 + (125\sin 25°)t + 4 \end{cases}$ **(b)** Approx. 1.2 ft; no

35. One possibility: $\begin{cases} x = \sqrt{6}\cos t - \sqrt{\dfrac{6}{7}}\sin t \\ y = \sqrt{6}\cos t + \sqrt{\dfrac{6}{7}}\sin t \end{cases}$ $0 \le t \le 2\pi$

37. (b) $y = -\dfrac{2}{5}x + \dfrac{26}{5}$ **(c)** Slope $= \dfrac{\text{Coefficient of } t \text{ values in } y}{\text{Coefficient of } t \text{ values in } x}$ **(d)** The slope is d/b.

39. (a) $y = x$ **(b)** Since t^2 cannot be negative, $x \geq 0$ and $y \geq 0$.

41.   **43.**   **45.**

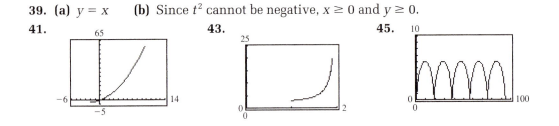

Chapter 9 Review Problem Set (on page 607)

1. $(x + 4)^2 = -6\left(y - \dfrac{3}{2}\right)$ **3.** $(y + 3)^2 = -12(x + 2)$ **5.** $\dfrac{(x - 3)^2}{\sqrt{5} - 1} + \dfrac{(y - 1)^2}{3 + \sqrt{5}} = 1$

7. $\dfrac{(x - 2)^2}{4} - \dfrac{(y - 4)^2}{5} = 1$ **9.** $V = (-1, 2); F = \left(-1, \dfrac{5}{2}\right)$

11. Center: $(1, -2)$; Vertices: $(1, -6)$, $(1, 2)$; Foci: $(1, -2 - \sqrt{7})$, $(1, -2 + \sqrt{7})$

13. Center: $(2, 1)$; Vertices: $(2, -1)$, $(2, 3)$; Foci: $(2, 1 - \sqrt{13})$, $(2, 1 + \sqrt{13})$

15. Parabola; $V = (3, -1); F = \left(\dfrac{7}{2}, -1\right)$ **17.** Ellipse; Vertices: $(5, -1)$ $(5, 5)$; Foci: $(5, 2 - \sqrt{5})$, $(5, 2 + \sqrt{5})$

19. Hyperbola; Vertices: $(2, -10)$, $(2, 0)$; Foci: $\left(2, \dfrac{5\sqrt{13}}{2} - 5\right)$, $\left(2, -\dfrac{5\sqrt{13}}{2} - 5\right)$

21. $\dfrac{x^2}{9} + \dfrac{y^2}{5} = 1$ **23.** $\dfrac{x^2}{64} - \dfrac{y^2}{36} = 1$ **25. (a)** $\dfrac{\sqrt{10}}{4}$ **(b)** $\dfrac{1}{\sqrt{3}}$

27. (a) $r = \dfrac{\dfrac{1}{2}}{1 - \sin\theta}$ **(b)** $r = \dfrac{8}{1 + 2\cos\theta}$

29. (a) $\dfrac{2}{3}$; Ellipse; Center: $\left(\dfrac{6}{5}, 0\right)$; Vertices in Cartesian form: $\left(-\dfrac{3}{5}, 0\right)$, $(3, 0)$

  (b) $\dfrac{5}{2}$; Hyperbola; Center: $\left(0, -\dfrac{25}{21}\right)$; Vertices: $\left(0, -\dfrac{5}{3}\right)$, $\left(0, -\dfrac{5}{7}\right)$

31. Ellipse   **33.** Hyperbola   **35.**   **37.** $y = \dfrac{1}{2}x^2 - \dfrac{3}{2}x$

Chapter 9 Test (on page 609)

1. Vertices: $(-2, -3)$, $(-2, 3)$; Foci: $\left(-2, -\dfrac{(3\sqrt{3})}{2}\right)$, $\left(-2, \dfrac{(3\sqrt{3})}{2}\right)$ **2.** $V = (12, -3); F = \left(\dfrac{49}{4}, -3\right)$

3. $\theta = 45°$; $-\bar{x}^2 + 3\bar{y}^2 = 6$ or $3\bar{x}^2 - \bar{y}^2 = 6$ **4.** $(x - 2)^2 + (y - 1)^2 = 1$

5. $\dfrac{\sqrt{7}}{3}$ **6. (a)** Ellipse   **(b)** $\dfrac{1}{2}$ **(c)** Foci: $(0, 0)$, $\left(0, -\dfrac{8}{3}\right)$ **(d)**

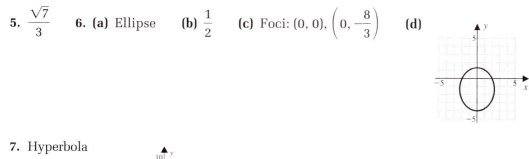

7. Hyperbola

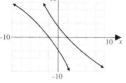

Chapter 10: Problem Set 10.1 (on page 619)

1. Finite; $a_n = 2n - 1$; $f(n) = 2n - 1$; Domain: $\{1, 2, 3, 4, 5\}$; Range: $\{1, 3, 5, 7, 9\}$

3. Infinite; $a_n = n^2$; $f(n) = n^2$; Domain: $\{1, 2, 3, 4, 5 \ldots\}$; Range: $\{1, 4, 9, 16, 25, \ldots\}$

5. Infinite; $a_n = (-1)^{n+1}$; $f(n) = (-1)^{n+1}$; Domain: $\{1, 2, 3, 4, 5, \ldots\}$: Range: $\{-1, 1\}$

7. $5, 7, 9, 11, 13$ **9.** $\dfrac{1}{3}, \dfrac{1}{2}, \dfrac{3}{5}, \dfrac{2}{3}, \dfrac{5}{7}$ **11.** $\dfrac{1}{e}, \dfrac{1}{e^2}, \dfrac{1}{e^3}, \dfrac{1}{e^4}, \dfrac{1}{e^5}$ **13.** $1, \dfrac{1}{2}, \dfrac{1}{4}, \dfrac{1}{8}, \dfrac{1}{16}$ **15.** $-2, -6, -18, -54, -162$

17. $1, 3, 4, 7, 11$ **19.** $a_n = 2n + 1$ **21.** $a_n = \dfrac{1}{(2n + 1)}$ **23.** $a_n = (-1)^{n+1}\, n^2$

25. (a) $5, 8, 11, 14, 17$ **(c)** ∞ **27. (a)** $\dfrac{3}{2}, 2, \dfrac{5}{2}, 3, \dfrac{7}{2}$ **(c)** ∞ **29. (a)** $2, 4, 8, 16, 32$ **(c)** ∞

31. (a) $0, \dfrac{1}{2}, \dfrac{2}{3}, \dfrac{3}{4}, \dfrac{4}{5}$ **(b)**   **(c)** 1

33. (a) $4, \dfrac{13}{4}, \dfrac{28}{9}, \dfrac{49}{16}, \dfrac{76}{25}$ **(b)**   **(c)** 3

35. 275 **37.** 7381 **39.** 83 **41.** $\dfrac{3}{2}$ **43.** 900 **45.** $1{,}691{,}450$

47. (a) $0.84, 0.45, 0.05, -0.19, -0.19$ **(b)**   **(c)** 0

49. (a) 0.46, 0.71, 0.66, 0.41, 0.14   **(b)**   **(c)** 0

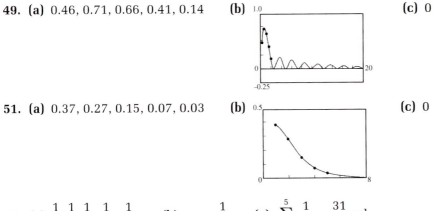

51. (a) 0.37, 0.27, 0.15, 0.07, 0.03   **(b)**   **(c)** 0

53. (a) $\dfrac{1}{2}, \dfrac{1}{4}, \dfrac{1}{8}, \dfrac{1}{16}, \dfrac{1}{32}$ **(b)** $a_n = \dfrac{1}{2^n}$ **(c)** $\displaystyle\sum_{k=1}^{5} \dfrac{1}{2^k} = \dfrac{31}{32}$ gal

(d) The bucket will never be full. Each successive can contains only half as much paint as is needed to fill the bucket.

55. (a) 4, 4.4, 4.84, 5.324, 5.8564   **(b)** $a_n = 4(1.1)^{n-1}$ **(c)** Approx. 8.57 in.

57. $a_n = 8n - 7$ **59. (a)** 2   **(b)** 5   **61.** $a_n = 3, n \ge 2$ **63.** $a_n = 1 + 2n, n \ge 2$

65. $a_n = \dfrac{2}{(n+1)(n+2)}, n \ge 2$ **67.** $a_n = 2^{n-1}, n \ge 2$ **69.** $S_3 = 10, S_6 = 56$ **71.** $S_3 = 46, S_6 = 812$

73. $S_3 = -4, S_6 = 3$ **75. (a)** $S_n = \dfrac{n}{n+1}$ **(b)** $S_n = \dfrac{n}{2n+4}$

Problem Set 10.2 (on page 635)

1. $a_n = 3n - 1$ **3.** $a_n = 15 - 3n$ **5.** $a_n = 2n$ **7.** 11   **9.** 48   **11.** $d = 3; S_{10} = 145$

13. $d = \dfrac{3}{7}; S_8 = -28$ **15.** 3940   **17.** -2540 **19.** 11   **21.** $d = \dfrac{668}{51}; a_{18} = \dfrac{719}{3}$ **23.** 4200

25. 975   **27.** $a_n = 2(3)^{n-1}$ **29.** $a_n = (1.03)^{n-1}$ **31.** $a_n = 2^n$ **33.** Geometric; $a_n = 3(2)^{n-1}$

35. Geometric; $a_n = 2^{-n-1}$ **37.** $r = 4; S_8 = 109{,}225$ **39.** $r = \dfrac{1}{3}; S_{10} = -\dfrac{29{,}524}{59{,}049}$ **41.** $\dfrac{1}{2}$

43. (a) 4   **(b)** $-\dfrac{3}{2}$ **45. (a)** Arithmetic sequence; $a_n = 4 + 3n, 1 \le n \le 10$ **(b)** 25   **(c)** 205

47. (a) 300,000; 340,000; 380,000; 420,000; 460,000; arithmetic sequence; $a_n = 260{,}000 + 40{,}000n$
(b) \$500,000   **(c)** \$4,800,000

49. (a) 268 ft   **(b)** 1260 ft

51. (a) $a_n = 100$ **(b)** 391.6 ft (approx.)   **(c)** $100 + 300(1 - 0.6^{n-1})$ ft   **(d)** 400 ft

53. (a) $a_n = 1000(1.8)^{n-1}$ **(b)** \$1,910,200   **(c)** 10

55. (a) $a_n = 25{,}000(0.95)^{n-1}, 1 \le n \le 10$ **(b)** $a_6 = 18{,}377$ **(c)** $a_{10} = 14{,}968$ **57. (a)** 4   **(b)** 67

59. $63b - 63$ **61.** $\displaystyle\sum_{k=1}^{5} \left(\dfrac{1}{2}\right)^k = \dfrac{31}{32}$ **63.** $\dfrac{a^6 - b^6}{a^5 b + a^4 b^2}$

65. (a) $0.4 + 0.04 + 0.004 + 0.0004 + \ldots = \dfrac{4}{9} = 0.\overline{4}$

(b) $0.32 + 0.0032 + 0.000032 + \ldots = \dfrac{32}{99} = 0.\overline{32}$

(c) $0.561 + 0.000561 + 0.000000561 + \ldots = \dfrac{187}{333} = 0.\overline{561}$

67. (a) $S_n = 2(1 - 2^n)$ **(b)** $S_n = \dfrac{2^n - 1}{2^{n+1}}$

Problem Set 10.3 (on page 647)

1. S_1: $1 = 1^2$; S_k: $1 + 3 + 5 + \cdots + (2k - 1) = k^2$;
S_{k+1}: $1 + 3 + 5 + \cdots + (2k - 1) + (2k + 1) = k^2 + (2k + 1) = (k + 1)^2$
17. $x^4 + 4x^3y + 6x^2y^2 + 4xy^3 + y^4$ **19.** $16x^4 + 96x^3y + 216x^2y^2 + 216xy^3 + 81y^4$
21. $3125x^5 - 9375x^4y + 11{,}250x^3y^2 - 6750x^2y^3 + 2025xy^4 - 243y^5$ **23.** $1, 5, 10, 10, 5, 1$; $n = 5$

25. (a) 3003 **(b)** 15 **27. (a)** $2{,}598{,}960$ **(b)** 1326 **29.** $\dfrac{5}{k}$ **31.** $\dfrac{3}{(k + 1)}$

33. $x^{28} + 28x^{27}y + 378x^{26}y^2 + 3276x^{25}y^3$ **35.** $x^{35} + (-70x^{34}y) + 2380x^{33}y^2 + (-52{,}360x^{32}y^3)$
37. $x^5 + 15x^4 + 90x^3 + 270x^2 + 405x + 243$ **39.** $x^4 - 8x^3 + 24x^2 - 32x + 16$ **41.** $792x^5y^7$

43. $\dfrac{455}{4096}x^{24}a^3$ **45.** $\dfrac{28}{243}a^3x^{12}$

47. S_1: If $n = 1$, then $4^n - 1 = 4^1 - 1 = 3$; so 3 is a factor.
S_k: Assume 3 is a factor of $4^k - 1$; that is, assume that $4^k - 1 = 3p$ for some integer p.
S_{k+1}: Show that 3 is a factor of $4^{k+1} - 1$: $4^{k+1} - 1 = 4^k \cdot 4 - 1 = 4(3p + 1) - 1 = 3 \cdot 4p + 3$, and 3 is a factor of $4^{k+1} - 1$.

57. (a) and **(c)** are false   **59.** $\displaystyle\sum_{k=0}^{5} \binom{5}{k} x^{5-k}y^k$

Problem Set 10.4 (on page 658)

1. (a) 4 **(b)** 10 **3. (a)** 1 **(b)** 1820 **5. (a)** 1 **(b)** 1 **7. (a)** 42 **(b)** 360
9. (a) $151{,}200$ **(b)** 5040 **11. (a)** 8 **(b)** 12 **13. (a)** 1 **(b)** 48
15. 10: {a, b, c}, {a, b, d}, {a, b, e}, {a, c, d}, {a, c, e}, {a, d, e}, {b, c, d}, {b, c, e}, {b, d, e}, {c, d, e}
17. 24: {abc, abd, acd, acb, adb, adc, bcd, bdc, dca, dac, bad, bca, bda, cba, cbd, cad, dab, bac, dcb, cda, dba, cdb, cab, dbc}
19. (a) 220 **(b)** 1320 **21.** $S = \{HHH, HHT, HTH, THH, TTT, TTH, THT, HTT\}$
$E = \{HHH, HHT, HTH, THH\}$

23. $\dfrac{1}{6}$ **25.** $\dfrac{1}{4}$ **27.** $657{,}700$ **29.** $14{,}400$ **31.** $134{,}596$ **33.** $2{,}598{,}960$ **35.** $\dfrac{3}{13}$ **37.** $\dfrac{5}{32}$

39. (a) $\dfrac{3}{8}$ **(b)** $\dfrac{3}{8}$ **(c)** $\dfrac{1}{8}$ **41.** 0.388 **43.** $\dfrac{4}{13}$ **45.** $\dfrac{25}{256}$ **47.** $\dfrac{1}{30}$ **49.** $\dfrac{5}{18}$

Chapter 10 Review Problem Set (on page 660)

1. (a) $0, 3, 8, 15, 24$ **(b)** $-1, 4, -7, 10, -13$ **3. (a)** $4, 8, 12, 16, 20$ **(b)** $8, 13, 18, 23, 28$
5. (a) $a_n = 2n$ **(b)** $a_n = (-1)^n$ **7. (a)** 95 **(b)** 20 **9.** 7910
11. (a) (i) $a_{15} = 58$ (ii) $S_{15} = 450$ **(b)** (i) $a_{10} = 80$ (ii) $S_{10} = 440$ **13. (a)** 87 **(b)** -48
15. (a) -24 **(b)** $27a + 3b$ **17. (a)** $a_n = 5(3)^{n-1}$ **(b)** $a_n = 4(2)^{n-1} = 2^{n+1}$

19. (a) $r = 2$; $S_8 = 12{,}240$ **(b)** $r = \dfrac{1}{3}$; $S_{12} = -\dfrac{265{,}720}{2187}$ **21.** 9 **23.** $\dfrac{45}{99}$

27. (a) 792 **(b)** $46{,}376$ **(c)** $n + 2$
29. (a) $32 + 80x + 80x^2 + 40x^3 + 10x^4 + x^5$ **(b)** $81x^4 - 432x^3y + 864x^2y^2 - 768xy^3 + 256y^4$
31. (a) $5376x^6y^3$ **(b)** $61{,}236x^5y^5$ **33. (a)** 304 ft   **(b)** 1600 ft
35. (a) $a_n = 130{,}000(1.03)^{n-1}$ **(b)** $130{,}000; 133{,}900; 137{,}917; 142{,}054.51; 146{,}316.15; 150{,}705.63; 155{,}226.80$
37. (a) 325 **(b)** $358{,}800$ **(c)** 1104

39. 84 **41. (a)** $\dfrac{1}{4}$ **(b)** $\dfrac{2}{9}$ **(c)** $\dfrac{7}{9}$ **(d)** $\dfrac{1}{6}$

Chapter 10 Test (on page 662)

1. (a) $-1, 2, 5, 8, 11$ **(b)** $2, \dfrac{3}{4}, \dfrac{4}{9}, \dfrac{5}{16}, \dfrac{6}{25}$ **(c)** $3, -6, 12, -24, 48$

2. (a) $a_n = 2 + 4n$ **(b)** $a_n = (-3)^n$ **3.** 130 **4. (a)** $a_6 = 11$ **(b)** $n = 6$ **(c)** $S_7 = 56$

5. (a) $r = -3$ **(b)** $a_1 = -1$ **6. (a)** 25 **(b)** $\dfrac{1}{16}$

7. $S_1: 2^1 = 2(2^1 - 1) = 2 \cdot 1;\ S_k: 2 + 2^2 + 2^3 + \dots + 2^k = 2(2^k - 1);$
$S_{k+1}: 2 + 2^2 + 2^3 + \dots + 2^k + 2^{k+1} = 2(2^k - 1) + 2^{k+1} = 2^{k+1} - 2 + 2^{k+1} = 2(2^{k+1} - 1)$

8. $32x^5 + 80x^4 + 80x^3 + 40x^2 + 10x + 1$ **9** $924x^{18}$ **10.** Approx. 75.08 cm

11. (a) 35 **(b)** 210 **(c)** 35 **12. (a)** 5040 **(b)** $10{,}000$ **(c)** 4536 **(d)** 9000

13. (a) $\dfrac{5}{36}$ **(b)** $\dfrac{31}{36}$ **(c)** $\dfrac{1}{12}$ **(d)** $\dfrac{5}{36}$

Problem Set AI.1 (on page A5)

1. (a) rational, real   **(b)** integer, rational, real   **(c)** rational, real   **(d)** irrational, real

3. (a) 0.75 **(b)** -1.75 **(c)** 5.3125 **(d)** 0.3

5. (a) $\dfrac{7}{10}$ **(b)** $\dfrac{11}{20}$ **(c)** $\dfrac{12}{5}$ **(d)** $\dfrac{181}{20}$

7. (a) 0.6199 **(b)** 3282.8064 **(c)** 0.5176 **(d)** 37.2098

9. (a)   **(b)**

(c)   **(d)**

11. 4.19 in.^{3}; 82.45 in.^{3}; 5575.28 in.^{3} **13.** 147.40 cm^3

Problem Set AI.2 (on page A11)

1. $Q = 4x^2 - 6x + 13,\ R = -14$
$4x^3 - 2x^2 + 7x - 1 = (4x^2 - 6x + 13)(x + 1) - 14$

3. $Q = 3x^3 + 7x^2 + 25x + 78,\ R = 241$
$3x^4 - 2x^3 + 4x^2 + 3x + 7 = (3x^3 + 7x^2 + 25x + 78)(x - 3) + 241$

5. $Q = 6x^3 - \dfrac{9}{2}x^2 + \dfrac{37x}{9} - \dfrac{111}{4};\ R = -\dfrac{305}{4}$

$6x^4 + 10x^2 + 7 = \left(6x^3 - \dfrac{9}{2}x^2 + \dfrac{37x}{4} - \dfrac{111}{4}\right)(2x + 3) - \dfrac{305}{4}$

7. $Q = -x^2 - 1;\ R = 0$
$-3x^3 - x^2 - 3x - 1 = (-x^2 - 1)(3x + 1) + 0$

9. $Q = 2x^2 + 23x + 58;\ R = 153$ **11.** $Q = 4x^3 + x^2 + 3x + 3;\ R = 8$

13. $Q = -5x^3 + 10x^2 - 20x + 40;\ R = -80$ **15.** $Q = -5x^3 + 10x^2 - 19x + 38;\ R = -73$

17. $2x^3 - 5x^2 + 5x + 11 = (2x^2 - 3x + 2)(x - 3) + 13$ **19.** $2x^4 + 3x - 6 = (2x^3 - 4x^2 + 8x - 13)(x + 2) + 20$

21. -10 **23.** $-\dfrac{20}{3}$

Problem Set AI.3 (on page A15)

1. **(a)** $0 + 4i$ **(b)** $3 + 2\sqrt{2}\,i$ **(c)** $7 - 9i$ **(d)** $\dfrac{1}{3} - \dfrac{1}{3}i$ **(e)** $0 + xi$ **(f)** $4 + 0i$

3. **(a)** $10 - 5i$ **(b)** $-3 + 2i$ **5.** **(a)** $12 - 9i$ **(b)** $11 - 16i$ **7.** **(a)** $3 + 4i$ **(b)** $-7 - 24i$

9. **(a)** $1 - i$ **(b)** $-\dfrac{23}{29} - \dfrac{44}{29}i$

Index